Graduate Texts in Physics

Graduate Texts in Physics publishes core learning/teaching material for graduate- and advanced-level undergraduate courses on topics of current and emerging fields within physics, both pure and applied. These textbooks serve students at the MS- or PhD-level and their instructors as comprehensive sources of principles, definitions, derivations, experiments and applications (as relevant) for their mastery and teaching, respectively. International in scope and relevance, the textbooks correspond to course syllabi sufficiently to serve as required reading. Their didactic style, comprehensiveness and coverage of fundamental material also make them suitable as introductions or references for scientists entering, or requiring timely knowledge of, a research field.

More information about this series at http://www.springer.com/series/8431

Hal Tasaki

Physics and Mathematics of Quantum Many-Body Systems

Springer

Hal Tasaki
Gakushuin University
Tokyo, Japan

ISSN 1868-4513 ISSN 1868-4521 (electronic)
Graduate Texts in Physics
ISBN 978-3-030-41264-7 ISBN 978-3-030-41265-4 (eBook)
https://doi.org/10.1007/978-3-030-41265-4

This Springer imprint is published by the registered company Springer Nature Switzerland AG
The registered company address is: Gewerbestrasse 11, 6330 Cham, Switzerland

To my family

Preface

This is a self-contained advanced textbook on quantum many-body systems, which is intended to be accessible to students and researchers in physics, mathematics, quantum information science, and related fields. The prerequisite is undergraduate-level basic knowledge of quantum mechanics, calculus, and linear algebra. We discuss in detail selected topics in quantum spin systems and lattice electron systems, and also describe fundamental concepts and important basic results necessary to understand the advanced topics of the book (and, of course, other related results in the literature).

More specifically, we focus on long-range order and spontaneous symmetry breaking in the antiferromagnetic Heisenberg model in two or higher dimensions (Part I), the Haldane phenomena in antiferromagnetic quantum spin chains and the related notion of symmetry protected topological phase (Part II), and the origin of magnetism in strongly interacting lattice electron systems, namely various versions of the Hubbard model (Part III). Although the selection of the topics is certainly biased by our research interests, we believe that each topic is, by itself, interesting and worth studying. More importantly, each of the topics represents certain non-trivial phenomena or features that we universally encounter in a variety of quantum many-body systems, including quantum field theory, condensed matter systems, cold atoms, and artificial quantum systems designed for future quantum computers. In other words, although most of the systems that we treat in the book are models of magnetism in a broad sense, our interest is not limited to magnetism. We are more interested in universal behaviors of quantum many-body systems.

As the title suggests, we here take the point of view of mathematical physics. Our major goal is to discuss mathematically rigorous results which are of essential importance and interest from the physicists' point of view. We shall also discuss in detal physical intuitions and pictures behind the mathematical results.

We believe it is crucial to insist on mathematically rigorous proofs (when available) since some phenomena in many-body systems are so intricate and subtle that it is not easy for us to reach the right conclusions based only on naive physical intuitions. It is also worth stressing that, in some (but not all) cases, one gets a deeper and clearer understanding of "physics" by appreciating a mathematical proof of a certain physical statement. We hope that the reader will have such experiences by studying some of the theorems and proofs in the present book. We have indeed tried to omit proofs which are too technical, but include those which are enlightening and worth studying. Moreover, most of the proofs discussed in the book have been considerably reorganized and extended so as to make them as elementary and accessible as possible. To give an example, the famous theorem of Lieb's on the half-filled Hubbard model (Theorem 10.4 in p. 350) is among the most significant contributions of modern mathematical physics to the theory of strongly interacting quantum many-body systems. Although the paper containing the theorem has been frequently cited both in theoretical and experimental papers, and the content of the theorem is well-known, it seems that one usually assumes that the proof of the theorem is too difficult to comprehend. We shall, however, present an elementary and detailed exposition of the complete proof which should be understandable to a sufficiently motivated undergraduate physics student with a standard background in mathematics; we do not make use of anything more advanced than diagonalization of Hermitian matrices! (We also should stress that the book is designed in such a manner that one can skip proofs and only appreciate heuristic arguments and rigorous results.)

We do not, on the other hand, go into mathematical formulations which are too advanced, e.g., the operator algebraic formulation of infinitely large quantum many-body systems. Although such sophisticated formalisms have their own merits in deriving stronger results and further extending our physical intuitions, we shall not try to go too much beyond standard formalism of undergraduate quantum mechanics. When treating infinite systems, we try to choose the most elementary formulation, and also carefully introduce and explain necessary notions.

The restriction to non-relativistic lattice quantum systems has a clear advantage that relatively satisfactory rigorous results are available. One may, for example, study phenomena parallel to those treated in Parts I or II in the framework of quantum field theories, or discuss the origin of magnetism, which is the topic of Part III, starting from the many-body Schrödinger equation for all the electrons and the nuclei that form a magnetic material. But our current (theoretical-physical and mathematical) understanding of these frameworks is so poor that we still have to struggle in obtaining very elementary results (or even defining the system itself) if we insist on mathematical rigor; there is no hope of treating interesting physical phenomena. By concentrating on lattice systems, where conceptual issues are considerably simpler, we are able to concentrate on the essence of interesting "physics" and mathematical mechanisms behind it. We shall discuss this point further in Sect. 1.1.

We assume that the reader is familiar with elementary quantum mechanics including the theory of angular momentum. Some experiences in statistical mechanics and condensed matter physics are welcome but by no means necessary. As for mathematics, we only assume basic calculus and linear algebra. Although some mathematical arguments are motivated by functional analysis, we do not require any familiarity with functional analysis (or any other advanced mathematics). We shall frequently refer to $\mathbb{Z}_2$, U(1), or SU(2) symmetry, but we do not require any knowledge in (continuous) group theory. What one should know is explained.

This means that at least a large part of the present book is accessible to sufficiently motivated undergraduate students. The readers with a background in mathematics or quantum information science may notice our heavy use of the theory of quantum mechanical angular momentum. This is nothing but the representation theory of SU(2), but we physicists are so much used to it since undergraduate quantum mechanics classes. For the non-physics-major readers, we have summarized the necessary material about angular momentum in the appendix.

We believe that the material in the present book can be used in several different ways in graduate courses in theoretical or mathematical physics. The author himself has given a half-year course which covers selected topics from Parts I and II, or another course which focuses on topics from Part III. At the time of writing, when many researchers and students are interested in topological phases of matter, a course which covers selected topics from Part II may be attractive.

It is a pleasure to thank Ian Affleck, Takashi Hara, Hosho Katsura, Tom Kennedy, Mahito Kohmoto, Tohru Koma, Elliott Lieb, Andreas Mielke, Yoshiko Ogata, and Akinori Tanaka, who are my collaborators on the subjects related to the topics of the book, for sharing their insights and wisdom with me, and, most of all, for fruitful and enjoyable collaborations. Some of the results from these collaborations are discussed in the present book.

During the preparation of the book, I benefited from useful comments from and discussions with various individuals. I especially wish to thank Hosho Katsura and Akinori Tanaka for their careful readings of the manuscript and for valuable proposals and comments. I also express my gratitude to Ian Affleck, Aron Beekman, Hans Jürgen Briegel, Yuya Dan, Martin Fraas, Yohei Fuji, Keisuke Fujii, Shunsuke Furukawa, Yasuhiro Hatsugai, Takuya Hirano, Chigak Itoi, Tohru Koma, Marius Lemm, Elliott Lieb, Yusuke Masaki, Taku Matsui, Akimasa Miyake, Tadahiro Miyao, Tomonari Mizoguchi, Hisamitsu Mukaida, Bruno Nachtergaele, Fumihiko Nakano, Yoshiko Ogata, Masaki Oshikawa, Glenn Paquette, Louk Rademaker, Robert Raussendorf, Ann Rossilli, Shinsei Ryu, Takahiro Sagawa, Akira Shimizu, Ken Shiozaki, Naoto Shiraishi, Ayumu Sugita, Yuji Tachikawa, Yuhi Tanikawa, Keiji Tasaki, Synge Todo, Masafumi Udagawa, Masahito Ueda, Haruki Watanabe, and Tzu-Chieh Wei for their indispensable contributions.

Last but not least, I wish to thank Mari Okazaki, a renowned Japanese manga artist,[1] for providing the book with her fantastic illustrations, one for the cover, and one for each of the three parts. I asked Mari to visualize (admittedly abstract and intangible) ideas developed in the book freely in her own style. I believe that these imaginative illustrations have given an added charm to the book.

Tokyo, Japan Hal Tasaki
August 28, 2019

[1]Mari Okazaki's works have been published also in China, France, Italy, Korea, Pórtugal, Spain, Taiwan, and the United States.

Contents

1 Introduction 1
1.1 Universality in Macroscopic Physics 1
1.2 Overview of the Book 4
References 10

2 Basics of Quantum Spin Systems 13
2.1 Quantum Mechanics of a Single Spin 13
2.2 Quantum Spin Systems 21
2.3 Time-Reversal and the Kramers Degeneracy 25
2.4 The Ferromagnetic Heisenberg Model 31
2.5 The Antiferromagnetic Heisenberg Model 37
References 44

Part I Long-Range Order and Spontaneous Symmetry Breaking

3 Long-Range Order and Spontaneous Symmetry Breaking in the Classical and Quantum Ising Models 49
3.1 Motivation from the Heisenberg Antiferromagnet 49
3.2 Classical Ising Model 52
3.3 Quantum Ising Model 55
3.4 General Theory of Low-Lying States and SSB 64
References 70

4 Long-Range Order and Spontaneous Symmetry Breaking in the Antiferromagnetic Heisenberg Model 73
4.1 Existence of Long-Range Order 73
4.1.1 Main Results 74
4.1.2 Proof of the Existence of LRO 79

4.2 The "Tower" of Low-Lying States and Spontaneous Symmetry Breaking . . . 93
4.2.1 Main Results . . . 94
4.2.2 Proofs . . . 104
4.3 Ground States of the Infinite System . . . 112
4.3.1 Construction of Ground States . . . 112
4.3.2 Physical Versus Unphysical Ground States . . . 116
4.4 Equilibrium States of the Heisenberg Model . . . 117
4.4.1 Disorder at High-Temperature and in One-Dimension . . . 118
4.4.2 Berezinskii's Harmonic Approximation . . . 120
4.4.3 Absence of Order in Two Dimensions . . . 123
4.4.4 LRO and SSB in Three or Higher Dimensions . . . 130
References . . . 131

5 Long-Range Order and "Spontaneous Symmetry Breaking" in Bose–Einstein Condensates . . . 135
5.1 The Model and the Equivalence to the XY Model . . . 135
5.2 Off-Diagonal Long-Range Order . . . 138
5.3 Symmetry Breaking "Ground States" . . . 140
5.4 Physical Ground States of Bose–Einstein Condensates . . . 142
5.5 "SSB" in Coupled Bose–Einstein Condensates . . . 146
References . . . 149

Part II Haldane Phenomena and Beyond

6 Ground States of the Antiferromagnetic Heisenberg Chains . . . 153
6.1 Haldane "Conjecture" . . . 153
6.2 The Lieb–Schultz–Mattis Theorem . . . 158
6.3 Semi-classical Approach . . . 166
References . . . 172

7 Affleck–Kennedy–Lieb–Tasaki Model . . . 177
7.1 The Model and Main Results . . . 177
7.1.1 The Hamiltonian and the Main Theorem . . . 178
7.1.2 The Exact Ground State . . . 180
7.1.3 The Uniqueness of the Ground State . . . 186
7.1.4 The Proof of the Existence of Gap . . . 188
7.2 Properties of the AKLT Model . . . 190
7.2.1 Hidden Antiferromagnetic Order . . . 191
7.2.2 Matrix Product Representation . . . 194
7.2.3 AKLT Model on Open Chains . . . 205

7.3 Extensions and Related Models 208
7.3.1 Spin Chains with Higher S and the VBS Picture 208
7.3.2 Higher Dimensional Models 210
7.3.3 Briegel–Raussendorf State (Cluster State) and Its Generalizations 214
References 220

8 Haldane Phase 225
8.1 Characterization of the Haldane Phase 225
8.1.1 Topological Phase Transition in Anisotropic Model . . . 226
8.1.2 Hidden Antiferromagnetic Order 229
8.1.3 Emergence of Edge States 234
8.2 Hidden $\mathbb{Z}_2 \times \mathbb{Z}_2$ Symmetry Breaking 238
8.2.1 Phase Diagram of the λ-D Model 239
8.2.2 Nonlocal Unitary Transformation 241
8.2.3 The Picture of Hidden $\mathbb{Z}_2 \times \mathbb{Z}_2$ Symmetry Breaking 244
8.3 Symmetry Protected Topological Phase 251
8.3.1 More on the Picture of the Hidden $\mathbb{Z}_2 \times \mathbb{Z}_2$ Symmetry Breaking 251
8.3.2 Symmetry Protected Topological (SPT) Phase 254
8.3.3 Entanglement and “Topological” Indices for SPT Phases 260
8.3.4 “Topological” Indices for Matrix Product States 264
8.3.5 Lieb–Schultz–Mattis Type Theorem Without Continuous Symmetry 276
8.3.6 Rigorous Index Theorems for SPT Phases 280
8.4 Topological Order in Kitaev’s Toric Code Model 288
References 298

Part III Hubbard Model and the Origin of Magnetism

9 Introduction to the Hubbard Model 305
9.1 What is the Hubbard Model? 305
9.2 Tight-Binding Description of Electrons in a Solid 307
9.2.1 Wave Functions for Electrons 308
9.2.2 Creation and Annihilation Operators in the Wave Function Formalism 310
9.2.3 The Fock Space Representation 314
9.3 Definition of the Hubbard Model 324
9.3.1 Single-Electron Schrödinger Equation 324
9.3.2 Hamiltonian of the Hubbard Model 328
9.3.3 Transformations and Symmetry of the Hubbard Hamiltonian 332

9.4 Bosonic Hubbard Model 337
References 339

10 Half-Filled Models: Lieb's Theorems and the Origin of Antiferromagnetism and Ferrimagnetism 341
10.1 Formal Perturbation Theory for Strong Interaction 341
10.2 Lieb's Theorems 347
10.2.1 Lieb's Theorem for the Attractive Hubbard Model 348
10.2.2 Lieb's Theorem for the Repulsive Hubbard Model 349
10.2.3 Lieb's Ferrimagnetism 354
10.2.4 Proofs of Theorems 10.2 and 10.3 361
10.2.5 Extensions and Other Rigorous Results 367
References 369

11 The Origin of Ferromagnetism 371
11.1 Basic Properties of Ferromagnetism 371
11.1.1 Definition and Basic Theorem 372
11.1.2 Instability of Ferromagnetism 374
11.1.3 Toy Model with Two Electrons 377
11.1.4 Stoner Criterion 379
11.2 Nagaoka's Ferromagnetism 381
11.2.1 Weak Version of Nagaoka's Theorem 382
11.2.2 Nagaoka's Theorem and the Connectivity Condition 385
11.2.3 Instability of Nagaoka's Ferromagnetism 388
11.3 Flat-Band Ferromagnetism 389
11.3.1 Tasaki's Flat-Band Ferromagnetism 390
11.3.2 Mielke's Flat-Band Ferromagnetism 401
11.3.3 Construction of Tight-Binding Models with Flat-Bands 405
11.3.4 General Theory of Flat-Band Ferromagnetism 409
11.4 Ferromagnetism in Non-singular Hubbard Models 414
11.4.1 Wannier State Perturbation Theory 415
11.4.2 Local Stability and the Spin Wave Excitation 421
11.4.3 Ferromagnetism in Non-singular Hubbard Models 425
11.5 Toward Metallic Ferromagnetism 438
11.5.1 Heuristic Arguments 439
11.5.2 Rigorous Results 441
References 452

Appendix A Mathematical Appendices 457
A.1 Dirac Notation 457
A.2 Useful Properties of Operators 462
A.2.1 Commutator and Operator Norm 462
A.2.2 Exponential of an Operator 464
A.2.3 Inequality Between Self-adjoint Operators and Nonnegative Operators 467
A.3 Quantum Mechanical Angular Momentum 470
A.3.1 Definition and Basic Properties 470
A.3.2 SU(2) Invariant Hamiltonian 472
A.3.3 Addition of Angular Momenta 473
A.4 Some Linear Algebra 475
A.4.1 Perron–Frobenius Theorem 475
A.4.2 Decomposition of Matrices 476
A.4.3 Antilinear Operators 478
A.5 Groups and Their Representations 479
A.6 Wigner's Theorem 482
A.6.1 Statement and the Proof 482
A.6.2 Applications 485
A.7 Operator Algebraic Formulation of Infinite Systems 486
References 491

Solutions 493

Index 521

Symbols

▷ $A := B$ or $B =: A$ means that A is defined in terms of B.
▷ $A \simeq B$ means that two quantities A and B are "very close" in a certain sense (usually clear from the context). $A \sim B$ means that A and B are "roughly close" or have the same order of magnitude. We note that this usage may be different from the standard convention in mathematics.
▷ $A \cong B$ means that two mathematical objects (such as linear spaces) A and B are isomorphic, which roughly means that A and B have the same structure.
▷ For a finite set S, we denote the number of its elements as $|S|$.
▷ A condition in a summation (or a product) is specified within brackets as, e.g.,

$$\sum_{\substack{p\in\mathbb{Z} \\ (p \text{ is a prime})}} c_p = c_2 + c_3 + c_5 + c_7 + c_{11} \cdots .$$

▷ c^* denotes the complex conjugate of $c \in \mathbb{C}$.
▷ Λ denotes the set of sites of a general lattice (p. 20).
▷ $\mathscr{B}$ denotes the set of bonds of a general lattice (p. 32). A bond is a pair $\{x,y\}$ of two distinct sites $x, y \in \Lambda$. We identify $\{x,y\}$ with $\{y,x\}$.
▷ $\mathcal{N}(x) = \{y \in \Lambda \,|\, \{x,y\} \in \mathscr{B}\}$ is the set of sites directly connected to a site $x \in \Lambda$ by a bond.
▷ Λ_L is the set of sites of the d-dimensional hyper cubic lattice (p. 51):

$$\Lambda_L = \left\{ (x_1, x_2, \cdots, x_d) \middle| x_i \in \mathbb{Z},\ -\frac{L}{2} < x_i \le \frac{L}{2} \right\} \subset \mathbb{Z}^d \tag{3.1.2}$$

▷ $\mathscr{B}_L$ is the set of bonds of the d-dimensional hyper cubic lattice with periodic boundary conditions (p. 51):

$$\mathscr{B}_L = \{\{x,y\}|x,y \in \Lambda_L, |x-y| = 1\} \tag{3.1.3}$$

▷ For $x = (x_1, x_2, \ldots, x_d) \in \mathbb{Z}^d$, we denote the Euclidean norm as $|x| := \sqrt{(x_1)^2 + \cdots + (x_d)^2}$.

▷ $\hat{A}^\dagger$ denotes the Hermitian conjugate of an operator (i.e., a matrix) $\hat{A}$.

▷ $\hat{\boldsymbol{S}}_x = (\hat{S}_x^{(1)}, \hat{S}_x^{(2)}, \hat{S}_x^{(3)})$ denotes the spin operator acting on site x (p. 21). We have $(\hat{\boldsymbol{S}}_x)^2 = S(S+1)$, where $S = 1/2, 1, 3/2, \ldots$ is the fixed spin quantum number.

▷ $\hat{\boldsymbol{S}}_{\mathrm{tot}} = (\hat{S}_{\mathrm{tot}}^{(1)}, \hat{S}_{\mathrm{tot}}^{(2)}, \hat{S}_{\mathrm{tot}}^{(3)})$ denotes the total spin operator (p. 22). The eigenvalue of $(\hat{\boldsymbol{S}}_{\mathrm{tot}})^2$ and $\hat{S}_{\mathrm{tot}}^{(3)}$ are denoted as $S_{\mathrm{tot}}(S_{\mathrm{tot}}+1)$ and $S_{\mathrm{tot}}^{(3)}$ (or sometimes as M), respectively.

▷ The single-particle Hilbert space (without spin) on lattice Λ is $\mathfrak{h} \cong \mathbb{C}^{|\Lambda|}$. It consists of states $\boldsymbol{\varphi} = (\varphi(x))_{x\in\Lambda}$ with $\varphi(x) \in \mathbb{C}$.

▷ The single-particle Hilbert space (with spin) on lattice Λ is $\tilde{\mathfrak{h}} \cong \mathbb{C}^{2|\Lambda|}$. It consists of states $\boldsymbol{\varphi} = (\varphi(x,\sigma))_{x\in\Lambda,\sigma=\uparrow,\downarrow}$ with $\varphi(x,\sigma) \in \mathbb{C}$.

▷ $\hat{c}_{x,\sigma}$, $\hat{c}_{x,\sigma}^\dagger$, and $\hat{n}_{x,\sigma}$ are the annihilation, creation and number operators, respectively, of an electron at site $x \in \Lambda$ with spin $\sigma = \uparrow, \downarrow$ (p. 314).

▷ $\hat{C}_\sigma(\boldsymbol{\varphi})$ and $\hat{C}_\sigma^\dagger(\boldsymbol{\varphi})$ are the annihilation and creation operators, respectively, of an electron in the state $\boldsymbol{\varphi} \in \mathfrak{h}$ with spin $\sigma = \uparrow, \downarrow$ (p. 321).

▷ $(v_1, \ldots, v_D)^{\mathrm{t}}$ denotes a column vector $\begin{pmatrix} v_1 \\ \vdots \\ v_D \end{pmatrix}$. The symbol t stands for transpose.

▷ A bold symbol like $\boldsymbol{v}$ usually stands for a column vector $(v_1, \ldots, v_D)^{\mathrm{t}}$ with $v_j \in \mathbb{C}$. We denote by $(\boldsymbol{v})_j$ the j-th component of $\boldsymbol{v}$.

▷ The conjugate of a column vector $\boldsymbol{v} = (v_1, \ldots, v_D)^{\mathrm{t}}$ is the row vector $\boldsymbol{v}^\dagger = (v_1^*, \ldots, v_D^*)$. For column vectors $\boldsymbol{v}$ and $\boldsymbol{u} = (u_1, \ldots, u_D)^{\mathrm{t}}$, we denote by $\langle \boldsymbol{v}, \boldsymbol{u} \rangle = \boldsymbol{v}^\dagger \boldsymbol{u} = \sum_{j=1}^D v_j^* u_j$ the inner product, and by $\boldsymbol{u}\boldsymbol{v}\dagger$ (which is the Kronecker product) the matrix such that $(\boldsymbol{u}\boldsymbol{v}\dagger)_{i,j} = u_i v_j^*$.

▷ A sans-serif symbol like A usually stands for a square matrix $\mathsf{A} = (a_{i,j})_{i,j=1,\ldots,D}$. We denote by $(\mathsf{A})_{i,j}$ the i,j entry of A. The identity matrix is denoted as I.

▷ The determinant of A is denoted as $|\mathsf{A}|$.

▷ $\mathsf{A}^\dagger$ and A^{t} denote the Hermitian conjugate and the transpose, respectively, of A. A^* is the matrix obtained by taking the complex conjugate of all the entries of A.

▷ For column vectors $\boldsymbol{v}$ and $\boldsymbol{u}$, we write $\langle \boldsymbol{v}, \mathsf{A}\boldsymbol{u} \rangle = \boldsymbol{v}^\dagger \mathsf{A} \boldsymbol{u} = \sum_{i,j=1}^D v_i^* a_{i,j} u_j$.

See also Appendix A.1 for a summary of the Dirac notation used throughout the book.

Chapter 1
Introduction

1.1 Universality in Macroscopic Physics

It is fair to say that one of the goals of physical science is to understand the world around us on the basis of the fundamental laws of physics. However, if we naively consider the task of understanding the properties of, say, a piece of metal sitting in front of us, we may be led to conclude that such an understanding is simply impossible. Of course we know the basic structure of the atoms composing the metal, and we have some knowledge about the crystalline structure of the metal and the basic properties of the electrons in the metal, including the band structure and the interactions between them. But all of this amounts to nothing more than approximate descriptions. How can we have a precise understanding without knowing, for example, the exact form of the many-electron wave function that spreads over the entire crystal? Moreover, an actual, macroscopically large piece of metal under ordinary conditions will generally not be a perfect crystal. Instead, it will contain many impurities and dislocations. Further, it may have a surface with an irregular form and be interacting with an external world that behaves in a complicated, uncontrolled manner. Focusing on more microscopic levels, we know that the nuclei in the metal are composed of quarks, whose behavior should be described by quantum chromodynamics (while a full QCD calculation of even a small nucleus requires a supercomputer). In addition to such comparatively practical problems, there is the essential limitation that we do not possess an ultimate microscopic theory that could provide an exact description of this piece of metal.

Despite the problems raised above, we would like to argue that a nontrivial understanding should be possible. We believe that the possibility for obtaining such an understanding is due to the universality we repeatedly encounter in macroscopic physics. The purpose of this short section is to (attempt to) convince the reader of this point.

One important aspect of universality is robustness. We find that, in many (but, of course, not all) problems of physics, the behavior of an object as a whole is

H. Tasaki, *Physics and Mathematics of Quantum Many-Body Systems*,
Graduate Texts in Physics, https://doi.org/10.1007/978-3-030-41265-4_1

insensitive to many details regarding both its own composition and the environment within which it exists. For example, the center-of-mass motion of a sufficiently heavy rigid body on Earth is very accurately described by the Newton's second law of motion for a point particle under uniform gravitational field, and in many cases, will be almost completely independent of the internal structure of the object and the presence of other objects in its proximity, other than Earth itself. Indeed it may be that the existence of such robustness is a necessary condition for the possibility of constructing an intelligible body of physical science.[1] Returning to our piece of metal, we wish to be optimistic, and we posit that some of its properties can be understood reasonably well without knowing, say, very precise forms of its electronic orbits, the exact locations of its impurities and dislocations, the shape of its surface, what our next-door neighbor is doing, the position of the moon, or the equations of a "theory of everything."

The other important aspect of universality is that, in many cases, we observe the same phenomenon in a variety of physical systems. A notable example is the phenomenon of wave propagation. While wave propagation is found in many forms, the same wave equation describes wave motion in a variety of media, including air, water, and elastic material, and even in vacuum in the case of electromagnetic field. A conceptually deeper example is provided by thermodynamics. We know that the same set of nontrivial (and mathematically beautiful) laws of thermodynamics apply with great accuracy to essentially any macroscopic system in equilibrium [27, 36]. Due to the general validity of the theoretical descriptions provided by the wave equation and the laws of thermodynamics, in either case, we can make definite predictions concerning the actual behavior of individual systems from this general description, without the need to investigate the particular properties of the individual systems themselves.

With regard to the relationship between a physical system and the behavior it exhibits, the conventional thinking is that, in some sense, the system possesses a primary existence, while the behavior it exhibits possesses a derivative existence, dependent on the system. However, the above discussion hints at another interpretation. Considering the cases of wave propagation, thermodynamic behavior, and the many other types of universal structures observed in the world, we are tempted to imagine that these universal structures themselves possess a primary existence, independent of any individual system, while they become "incarnated" in various concrete forms in actual physical systems.[2] But, irrespective to our philosophical point of view, there is no doubt that it is a task of essential importance to discover and understand universal phenomena and universal structures that are independent of the individual systems in which they are observed. Returning again to our piece of metal, our goal, from the point of view of universality, should be to find

---

[1] We can try to imagine a world in which all minute aspects of all elements are strongly interlinked. In such a world, any prediction of the behavior of a given element would be practically impossible, because this would require detailed knowledge of the behavior of all other elements. But this is not the world we live in. (Indeed it would seem quite likely that no intelligence could evolve in such an unpredictable world.)

[2] See Chap. 1 of [36] for further discussions of the view based on the notion of universality.

characteristic and essential phenomena taking place in it, and then to understand the universal structures behind them.[3]

In our pursuit to advance fundamental science, with the manner of thinking described above, we are led to study classes of systems defined by characteristic types of universal phenomena, rather than actual individual systems. Let us call such a class of systems a "universality class."[4] Within a given universality class, we will have not only actual physical systems and faithful theoretical models describing them (which usually have intractable details), but also some idealized theoretical models that appear to be easier to treat. It should be stressed, however, that such idealized models are not simply "made up" to exhibit the desired properties (for some obvious reasons). Rather, they are nontrivial systems that capture only the essence of the phenomena that we wish to understand.[5] By studying such idealized models, we are able to directly confront the problem of elucidating the essential behavior of interest. Perhaps the best example of such an idealized model is the (classical) Ising model. Although the Ising model is now recognized as a model of a ferromagnet, it is too simple to be a faithful model of any actual magnetic system. Nevertheless, we can learn from the Ising model extremely rich essence of phase transitions and critical phenomena associated with the breakdown of $\mathbb{Z}_2$ symmetry, exhibited by various physical systems, including uniaxial ferromagnets and some quantum field theories. It should be pointed out that, despite its relative simplicity, the Ising model is certainly not easy to solve. However, because with this model, one need not treat some of the very complicated problems involved with more realistic models, such as the overlap of electron orbits that determines the exchange interaction and the ultraviolet divergence that must be removed to realize a well-defined field theory, the core problem that we wish to address—that of describing the collective behavior of infinitely many interacting degrees of freedom—is laid bare. This problem is indeed central to understanding the large-scale behavior of a truly vast range of physical systems.

In this book, we treat selected topics in quantum many-body theory that are directly related to important universal phenomena observed mainly in condensed matter systems. We have chosen topics that are of importance from both physical and mathematical points of view. The Heisenberg model and its variants for spin systems and some versions of the Hubbard model for electron systems, which we study in detail throughout the book, are idealized models representing important universality classes in many-body physics. These models play roles in the study of quantum many-body systems analogous to that played by the Ising model in

[3]Of course we do not argue that this is the only goal. System-dependent properties are important in many applications.

[4]In conventional usage, the term "universality class" refers to a class of statistical mechanical models (or field theories) that exhibit quantitatively identical critical phenomena. Here we are using this terminology in a broader sense.

[5]In most cases, such an idealized model is related to other members of the universality class only through uncontrolled approximations, heuristic arguments, or optimistic hope. It is an extremely challenging problem in theoretical physics to establish firmer connections between complicated and realistic models and simple idealized models. To do so probably requires highly advanced and flexible versions of the renormalization group method.

the study of ferromagnetic systems. We hope that the reader will find the in-depth theoretical studies of these models presented here, which are rooted in our desire to understand a piece of metal sitting in front of us,[6] both fruitful and enjoyable.

1.2 Overview of the Book

We shall give a brief and informal overview of the topics covered in the present book.[7]

We start from a preparatory chapter (Chap. 2), in which we discuss the basics of quantum spin systems. We encourage the reader to first briefly examine this chapter, no matter what his/her main interest is. The reader may skip the details and come back to them later when necessary.

One of the important results discussed in this chapter is the Marshall–Lieb–Mattis theorem (Theorem 2.2 in p. 39) [26, 28]. Consider, for example, the antiferromagnetic Heisenberg model on the d-dimensional $L \times \cdots \times L$ hypercubic lattice with the Hamiltonian

$$\hat{H} = \sum_{x,y} \hat{\boldsymbol{S}}_x \cdot \hat{\boldsymbol{S}}_y, \tag{1.2.1}$$

where x and y are summed over neighboring pairs of sites. (Note that we are using informal notations in this section. We will be more careful in the later sections.) The Marshall–Lieb–Mattis theorem states that, when L is even, the ground state (i.e., the eigenstate with the lowest eigenvalue) of $\hat{H}$ is unique and hence preserves all the symmetry of $\hat{H}$ including, most importantly, the rotational symmetry. We note that in the corresponding ferromagnetic Heisenberg model

$$\hat{H} = -\sum_{x,y} \hat{\boldsymbol{S}}_x \cdot \hat{\boldsymbol{S}}_y, \tag{1.2.2}$$

one gets a large number of ground states in which spins are aligned with each other and pointing in an arbitrarily chosen directions. This comparison already suggests that there is something "more quantum" in antiferromagnets than in ferromagnets, and that the Marshall–Lieb–Mattis theorem touches the essence of this difference.

It may be worth noting that the difference between antiferromagnetic and ferromagnetic systems is apparent even in the simplest possible quantum spin systems, namely, that of two spins with spin quantum number $S = 1/2$. It is easily found that the ground state of the antiferromagnetic Hamiltonian $\hat{H} = \hat{\boldsymbol{S}}_1 \cdot \hat{\boldsymbol{S}}_2$ is the unique singlet state

---

[6] A piece of iron exhibits metallic ferromagnetism, a phenomenon that is still poorly understood theoretically. In the final section of this book, we present a very preliminary attempt at understanding metallic ferromagnetism.

[7] To the reader interested in the connection of our mathematical-physical approach to the standard condensed matter physics, we recommend the book by Fazekas [11].

$$|\Phi_{0,0}\rangle = \frac{1}{\sqrt{2}}\bigl(|\uparrow\rangle_1|\downarrow\rangle_2 - |\downarrow\rangle_1|\uparrow\rangle_2\bigr), \tag{1.2.3}$$

which is rotationally invariant, while the ground states of the ferromagnetic Hamiltonian $\hat{H} = -\hat{\boldsymbol{S}}_1 \cdot \hat{\boldsymbol{S}}_2$ are the triplet states

$$|\Phi_{1,1}\rangle = |\uparrow\rangle_1|\uparrow\rangle_2, \quad |\Phi_{1,0}\rangle = \frac{1}{\sqrt{2}}\bigl(|\uparrow\rangle_1|\downarrow\rangle_2 + |\downarrow\rangle_1|\uparrow\rangle_2\bigr), \quad |\Phi_{1,-1}\rangle = |\downarrow\rangle_1|\downarrow\rangle_2, \tag{1.2.4}$$

and their linear combinations. (The reader who is not familiar with these notions should not be discouraged; they will be explained in the later sections.)

In Part I of this book, we focus on the problem of spontaneous symmetry breaking and long-range order, which are among the most universal phenomena encountered in physical systems with large degrees of freedom.

A prototypical model is the antiferromagnetic Heisenberg model (1.2.1). It is known that the ground state of the model in two or higher dimensions has antiferromagnetic long-range order (LRO), in the sense that spins separated far away on the lattice are still correlated because of the antiferromagnetic interaction. The proof of the existence of LRO, based on the reflection positivity method due to Dyson, Lieb, and Simon [8], is one of the most important achievements of mathematical physics for quantum-many body systems. We shall carefully describe the proof of the theorem by Kennedy, Lieb, and Shastry (Theorem 4.1 in p. 75) [22]. We hope that our detailed account of the method of reflection positivity in quantum spin systems is accessible to a wide range of readers.

Given the existence of LRO, one naively expects that the ground state resembles the Néel state, in which spins are pointing in the alternating directions. See Fig. 3.1 (p. 50). Such a ground state can only appear as a result of spontaneous symmetry breaking (SSB) because the directions of spins should be chosen arbitrarily. However this picture is in conflict with the aforementioned Marshall–Lieb–Mattis theorem. The unique ground state must have complete rotational invariance, and it is impossible for a spin to point in a certain fixed direction. Thus the exact ground state of (1.2.1) does exhibit LRO, but does not exhibit SSB. This conclusion is physically rather mystifying since states with Néel order are observed experimentally at very low temperatures.

This "mystery", which was already known in 1950s, has been almost completely resolved by now. It turns out that the ground state with LRO but without SSB is accompanied by a series of excited states, known as "Anderson's tower", which have very low excitation energies [3, 5]. A physical ground state with both LRO and SSB, which should be observed experimentally at very low temperatures, is not an exact energy eigenstate but is a superposition of the exact ground state and a large number of the low-lying excited states. This in turn means that the exact ground state is a kind of Schrödinger's cat, in the sense that it is a superposition of various physical ground states.

The above picture has been confirmed rigorously by Koma and Tasaki [24], whose works are based on earlier pioneering works on discrete symmetry breaking by

Horsch and von der Linden [16] and Kaplan, Horsch, and von der Linden [19]. We shall discuss these rigorous results carefully, starting from an elementary but important example, namely, the quantum Ising model, and proceed to the difficult case of the Heisenberg model, where one encounters an ever-increasing number of low-lying excited states which form the "tower". We also briefly discuss how LRO naturally leads to SSB in the ground states of infinitely large systems. Although we do not go deeply into operator algebraic formulation of quantum many-body systems, which provides a sophisticated description of infinite systems, we shall discuss some essential points so that the reader can appreciate the flavor of the advanced formulation. In the course of discussion, we also give an important remark about the role of symmetry in the notion of phases. See Figs. 3.5 and 3.6 (p. 63).

One encounters "LRO without SSB" in ground states of various quantum many-body systems in which the order operator and the Hamiltonian do not commute. Examples include not only quantum antiferromagnets, but superconductivity, Bose–Einstein condensation, and a variety of quantum field theories. Since there is a fundamental difference between spin systems and particle systems, we shall discuss in detail the phenomenon of Bose–Einstein condensation in the Bosonic Hubbard model, placing main emphasis on LRO and SSB.

In this part we also give a quick overview of LRO and SSB in quantum Heisenberg model at nonzero temperatures. This includes (probably the first) proof of a stronger version of the famous Hohenberg–Mermin–Wagner theorem about the absence of symmetry breaking in two dimensions.

The topic of Part II, topological phases of matter in quantum spin chains, is no doubt the most fashionable among the subjects treated in this book. As mentioned already, the reader may directly start from this part, after taking a glance at Chap. 2.

The main theme of the study is still the antiferromagnetic Heisenberg model (1.2.1), but the one defined on the one-dimensional lattice (which is often called the chain). It has been known for quite a long time that, unlike in higher dimensions, the ground state of the one-dimensional model does not exhibit any long-range order. The nature of the ground state for the most basic model with $S = 1/2$ had been studied by using the Bethe ansatz method. (Here S denotes the spin quantum number, which takes the values $S = 1/2, 1, 3/2, 2, \ldots$.) It was found that the ground state is critical, i.e., the correlation functions decay with power law, and there are gapless excitations immediately above the ground state.

In 1983, Duncan Haldane, who received the 2016 Nobel prize in physics mainly for this contribution, discovered that there is a qualitative difference between the low energy properties of the models with a half-odd-integral spin and an integral spin [14, 15]. According to Haldane, properties of the spin S Heisenberg antiferromagnetic chain are basically the same as those for $S = 1/2$ when S is a half-odd-integer, such as 3/2 or 5/2. However, when S is an integer, low energy properties are completely different. The correlation functions decay exponentially, and there is a nonvanishing energy gap (now known as the Haldane gap) above the ground state energy. One can say that the ground state of an integer spin chain is disordered.

It seems that Haldane's conclusion was totally against common beliefs of experts of the day. This may be the reason why people referred to his conclusion as the "Haldane conjecture" in the 1980s.[8] The validity of Haldane's conclusion was gradually established through a series of experimental, numerical, theoretical, and mathematical works.

A strong theoretical (and mathematical) support to Haldane's conclusion was provided by Affleck, Kennedy, Lieb, and Tasaki in 1987 [1, 2]. They proposed a model of $S = 1$ antiferromagnetic spin chain whose ground state can be written down explicitly. The Hamiltonian of the model, now called the AKLT model, is given by

$$\hat{H}_{\text{AKLT}} = \sum_{x=1}^{L} \{\hat{\boldsymbol{S}}_x \cdot \hat{\boldsymbol{S}}_{x+1} + \frac{1}{3}(\hat{\boldsymbol{S}}_x \cdot \hat{\boldsymbol{S}}_{x+1})^2\}, \tag{1.2.5}$$

which has extra biquadratic terms when compared to the original Heisenberg Hamiltonian (1.2.1). As we shall discuss in detail in Chap. 7, it was proved that the model possesses the precise properties that Haldane had predicted for integer spin chains, i.e., the ground state is unique, accompanied by a gap, and has exponentially decaying correlations. It is now believed that the AKLT model represents the universality class of models exhibiting Haldane phenomena.

It also turned out that the ground state of the AKLT model is an example of a class of states called the matrix product states (MPS) proposed and formulated by Fannes, Nachtergaele, and Werner in 1989 [9, 10]. The formulation of MPS provides an extremely efficient way of approximately describing a large class of states in one-dimensional quantum many-body systems, and has been playing indispensable roles in condensed matter physics, mathematical physics, and quantum information science. Section 7.2.2 of this book can be read as a tutorial introduction to MPS motivated by the AKLT model.

The ground state of the AKLT model is not only disordered but has two unexpected exotic properties, namely, the hidden antiferromagnetic order and the effective $S = 1/2$ degrees of freedom at the edge of an open chain. It was later pointed out by den Nijs and Rommelses [7] and by Kennedy [20] that these exotic properties are shared by a class of models which includes the Heisenberg model. Kennedy and Tasaki noted that these properties can be interpreted as consequences of spontaneous breakdown of the hidden $\mathbb{Z}_2 \times \mathbb{Z}_2$ symmetry [21].

All these observations suggest that the $S = 1$ antiferromagnetic Heisenberg chain and the AKLT model, which have exotic disordered ground states, belong to a new quantum phase which should be called the "Haldane phase". The true nature of the new phase remained unclear until Gu and Wen pointed out in 2009 that the Haldane phase should be identified as a symmetry protected topological (SPT) phase [13].

---

[8] In mathematics a plausible statement is called a conjecture until it is finally proved and becomes a theorem. It is rather unusual to call a statement in theoretical physics a conjecture, since most of "established facts" in theoretical physics are conjectures from mathematicians' point of view. Mathematically speaking, Haldane's conclusions for the Heisenberg chain (1.2.1) is still a conjecture. (But nobody calls it the "Haldane conjecture" any more.)

Pollmann, Turner, Berg, and Oshikawa soon determined the complete set of symmetry necessary to protect the Haldane phase, and also defined indices (within the MPS formulation) that characterize the topological phases [33]. We present in Sects. 8.3.3 and 8.3.4 a detailed account of their index theory. Finally, in 2018 and 2019, by using sophisticated notions in operator algebraic approach to quantum many-body systems, Ogata developed fully rigorous index theorems, which essentially solves the problem of SPT phases in quantum spin chains [31, 32].

In this part, we also discuss two topics from quantum information science which are closely related to the main theme of the part, namely, the Briegel–Raussendorf state (cluster state) [6] and its generalizations (Sect. 7.3.3) and Kitaev's toric code model (Sect. 8.4) [23]. The latter is of considerable interest as a simple model exhibiting topological order, which is distinct from (and probably more important than) symmetry protected topological order.

In Part III of the book, we turn our attention to the origin of interactions between spins which we see, e.g., in the Hamiltonians (1.2.1) and (1.2.2). We therefore need to take one step further down to the microscopic level, and study the problem of interacting electrons. Such a study was first made back in 1928 by Heisenberg, who argued, based on a simple perturbation theory for a two-electron system, that the Coulomb interaction between electrons and quantum many-body dynamics of fermions lead to ferromagnetic exchange interaction as in (1.2.2). (This is the reason for calling the model the Heisenberg model.)

The Hubbard model was introduced in the 1960s independently by Kanamori [18], Gutzwiller [12], and Hubbard [17] in order to study the origin of ferromagnetism in many-electron systems. It is a tight-binding model with the simple Hamiltonian

$$\hat{H} = \sum_{\substack{x,y \\ \sigma=\uparrow,\downarrow}} t_{x,y}\, \hat{c}^\dagger_{x,\sigma} \hat{c}_{y,\sigma} + U \sum_{x} \hat{n}_{x,\uparrow} \hat{n}_{x,\downarrow} \,, \tag{1.2.6}$$

where the first term represents quantum mechanical hopping of electrons, and the second term describes the on-site Coulomb interaction. We stress that the model is not designed to be a faithful model of realistic systems. It should be regarded as an idealized model that is designed to capture universal phenomena and mechanisms taking place in interacting many-electron systems.

We do not assume that the reader is familiar with the Hubbard model. We shall carefully explain in Sect. 9.2 the description (also known as the "second quantization" formalism) of many-electron systems in terms of creation and annihilation operators, the motivations behind the definition of the Hamiltonian (1.2.6), and the basics about non-interacting fermion systems.

As we see in Chap. 10, the origin of antiferromagnetic interaction (1.2.1) is easily understood, at least heuristically, from Anderson's perturbative argument for half-filled Hubbard models [4]. In 1989, Elliott Lieb proved an important and nontrivial theorem for the Hubbard model at half-filling, which partially confirms the above perturbative picture [25]. Lieb's theorem on the Hubbard model is one of the most important achievements in mathematical physics for quantum many-body systems.

For certain classes of models, the theorem also establishes the emergence of ferrimagnetism and superconductivity. We shall discuss in detail the statement, applications, and the proof of Lieb's theorem. As we have already stressed, a detailed (and hopefully readable) account of the proof of the theorem is one of the main contributions of the present book.

The long final chapter, Chap. 11, is devoted to the emergence of ferromagnetism in the Hubbard model. The first rigorous example of ferromagnetism in the Hubbard model was discovered by Nagaoka in 1966 [30]. Nagaoka's ferromagnetism takes place in a rather singular situation where the number of electrons is one less than the number of lattice sites, and the on-site Coulomb interaction is infinitely large. Then there appears a single "hole" in the configuration, and the motion of the hole generates ferromagnetic coupling of the whole electrons. Although Nagaoka's ferromagnetism is interesting and nontrivial from a theoretical point of view, it is nowadays believed that the mechanism leading to the ferromagnetism does not work in less singular situations with finite interaction and multiple holes.

Almost for a quarter of a century, Nagaoka's ferromagnetism was the only rigorous example of ferromagnetism in the Hubbard model. The situation changed drastically in the early 1990s when first Mielke [29] and then Tasaki [34] proposed essentially different rigorous examples. They considered special classes of tight-binding models in which the corresponding single-electron spectra have a flat lowest band. It was proved that the ground states of the model exhibit ferromagnetism for any nonzero Coulomb interaction when the number of electrons is exactly the same as the degeneracy of the flat band. These examples are now called the flat-band ferromagnetism. The elegant construction Mielke's flat-band models makes use of the notion of line graphs.

It should be noted that all the above rigorous examples of ferromagnetism are singular in some aspects. Nagaoka's ferromagnetism requires infinitely large interaction and exactly one "hole". Flat-band ferromagnetism takes place only when one has a singular band structure with infinitely large density of states. Rigorous examples of ferromagnetism which are free from any such singularities were finally discovered by Tasaki in 1995 [35]. The models were obtained by perturbing Tasaki's flat-band models to make the lowest band dispersive. Then the emergence of ferromagnetism in the ground states was proved for sufficiently large but finite interaction when the width of the lowest band is sufficiently narrow. Tasaki also proved that the model has spin-wave excitations whose dispersion relation precisely recovers the form expected in an insulating ferromagnet. Thus, starting from a well-defined non-singular model of strongly interacting itinerant electron system, it was established that the low energy properties of the model coincide with what are expected in a "healthy" ferromagnetic system. We believe that this is the most satisfactory answer (for the moment) to the problem raised by Heisenberg, i.e., to explain the origin of ferromagnetism in terms of many-body quantum mechanics.

Of course all these are very special examples of ferromagnetic insulators. The problem of the origin of ferromagnetism is still widely open. We end the chapter by discussing our own very preliminary attempts at constructing rigorous examples of metallic ferromagnetism.

References

1. I. Affleck, T. Kennedy, E.H. Lieb, H. Tasaki, Rigorous results on valence-bond ground states in antiferromagnets. Phys. Rev. Lett. **59**, 799 (1987)
2. I. Affleck, T. Kennedy, E.H. Lieb, H. Tasaki, Valence bond ground states in isotropic quantum antiferromagnets. Commun. Math. Phys. **115**, 477–528 (1988), https://projecteuclid.org/euclid.cmp/1104161001
3. P.W. Anderson, An approximate quantum theory of the antiferromagnetic ground state. Phys. Rev. **86**, 694 (1952)
4. P.W. Anderson, New approach to the theory of superexchange interactions. Phys. Rev. **115**, 2–13 (1959)
5. P.W. Anderson, *Basic Notions of Condensed Matter Physics* (Westview Press/Addison-Wesley, Boston, 1997)
6. H.J. Briegel, R. Raussendorf, Persistent entanglement in arrays of interacting particles. Phys. Rev. Lett. **86**, 910 (2001), arXiv:quant-ph/0004051
7. M. den Nijs, K. Rommelse, Preroughening transitions in crystal surfaces and valence-bond phases in quantum spin chains. Phys. Rev. B **40**, 4709 (1989)
8. F.J. Dyson, E.H. Lieb, B. Simon, Phase transitions in quantum spin systems with isotropic and nonisotropic interactions. J. Stat. Phys. **18**, 335–382 (1978)
9. M. Fannes, B. Nachtergaele, R.F. Werner, Exact antiferromagnetic ground states of quantum spin chains. Europhys. Lett. **10**, 633–637 (1989)
10. M. Fannes, B. Nachtergaele, R.F. Werner, Finitely correlated states on quantum spin chains. Commun. Math. Phys. **144**, 443–490 (1992), https://projecteuclid.org/euclid.cmp/1104249404
11. P. Fazekas, *Lecture Notes on Electron Correlation and Magnetism*. Modern Condensed Matter Physics, vol. 5 (World Scientific, Singapore, 1999)
12. M.C. Gutzwiller, Effect of correlation on the ferromagnetism of transition metals. Phys. Rev. Lett. **10**, 159 (1963)
13. Z.-C. Gu, X.-G. Wen, Tensor-entanglement-filtering renormalization approach and symmetry-protected topological order. Phys. Rev. B **80**, 155131 (2009), arXiv:0903.1069
14. F.D.M. Haldane, Continuum dynamics of the 1-D Heisenberg antiferromagnet: identification with the $O(3)$ nonlinear sigma model. Phys. Lett. **93A**, 464–468 (1983), http://www.sciencedirect.com/science/article/pii/037596018390631X
15. F.D.M. Haldane, Nonlinear field theory of large-spin Heisenberg antiferromagnets: semiclassically quantized solitons of the one-dimensional easy-axis Néel state. Phys. Rev. Lett. **50** 1153–1156 (1983). https://doi.org/10.1103/PhysRevLett.50.1153
16. P. Horsch, W. von der Linden, Spin-correlations and low lying excited states of the spin-1/2 Heisenberg antiferromagnet on a square lattice. Z. Phys. B **72**, 181–193 (1988)
17. J. Hubbard, Electron correlations in narrow energy bands. Proc. R. Soc. (Lond.) A **276**, 238 (1963)
18. J. Kanamori, Electron correlation and ferromagnetism of transition metals. Prog. Theor. Phys. **30**, 275 (1963), https://academic.oup.com/ptp/article/30/3/275/1865799
19. T.A. Kaplan, P. Horsch, W. von der Linden, Order parameter in quantum antiferromagnets. J. Phys. Soc. Jpn. **11**, 3894–3898 (1989)
20. T. Kennedy, Exact diagonalization of open spin 1 chains. J. Phys.: Condens. Matter **2**, 5737–5745 (1990)
21. T. Kennedy, H. Tasaki, Hidden $\mathbb{Z}_2 \times \mathbb{Z}_2$ symmetry breaking in Haldane-gap antiferromagnets. Phys. Rev. B **45**, 304–307 (1992)
22. T. Kennedy, E.H. Lieb, B.S. Shastry, Existence of Néel order in some spin-$1/2$ Heisenberg antiferromagnets. J. Stat. Phys. **53**, 1019 (1988)
23. A.Yu. Kitaev, Quantum computations: algorithms and error correction. Russ. Math. Surv. **52** 1191–1249 (1997), http://iopscience.iop.org/article/10.1070/RM1997v052n06ABEH002155/meta

24. T. Koma, H. Tasaki, Symmetry breaking and finite-size effects in quantum many-body systems. J. Stat. Phys. **76**, 745–803 (1994) (The version in the arXiv is more complete than the published version), arXiv:cond-mat/9708132
25. E.H. Lieb, Two theorems on the Hubbard model. Phys. Rev. Lett. **62**, 1201–1204 (1989)
26. E.H. Lieb, D. Mattis, Ordering energy levels in interacting spin chains. J. Math. Phys. **3**, 749–751 (1962)
27. E.H. Lieb, J. Yngvason, The physics and mathematics of the second law of thermodynamics. Phys. Rep. **310**, 1–96 (1999), arXiv:cond-mat/9708200
28. W. Marshall, Antiferromagnetism. Proc. R. Soc. A **232**, 48 (1955)
29. A. Mielke, Ferromagnetism in the Hubbard model on line graphs and further considerations. J. Phys. A **24**, 3311 (1991)
30. Y. Nagaoka, Ferromagnetism in a narrow, almost half-filled s-band. Phys. Rev. **147**, 392 (1966)
31. Y. Ogata, A $\mathbb{Z}_2$-index of symmetry protected topological phases with time reversal symmetry for quantum spin chains, preprint (2018), arXiv:1810.01045
32. Y. Ogata, A $\mathbb{Z}_2$-index of symmetry protected topological phases with reflection symmetry for quantum spin chains, preprint (2019), arXiv:1904.01669.pdf
33. F. Pollmann, A.M. Turner, E. Berg, M. Oshikawa, Entanglement spectrum of a topological phase in one dimension. Phys. Rev. B **81**, 064439 (2010), arXiv:0910.1811
34. H. Tasaki, Ferromagnetism in the Hubbard models with degenerate single-electron ground states. Phys. Rev. Lett. **69**, 1608 (1992)
35. H. Tasaki, Ferromagnetism in Hubbard models. Phys. Rev. Lett. **75**, 4678 (1995), arXiv:cond-mat/9509063
36. H. Tasaki, G. Paquette, *Thermodynamics — A Novel Approach*, to be published from Oxford University Press

Chapter 2
Basics of Quantum Spin Systems

In this preparatory chapter, we define quantum spin systems, and discuss basic features of the ground states of the Heisenberg models. These basic results will be useful throughout the book. After studying this chapter the reader may start with any of the parts I, II, or III. After reviewing basic properties of a single quantum mechanical spin in Sect. 2.1, we introduce general quantum spin systems and fix notations in Sect. 2.2. We review in some detail the notion of time-reversal in Sect. 2.3. Then in Sects. 2.4 and 2.5, we discuss the ferromagnetic and antiferromagnetic Heisenberg models, respectively, placing main emphasis on general properties of their ground states. The Marshall-Lieb-Mattis theorem (Theorem 2.2 in p. 39) for the antiferromagnetic model plays a central role throughout the book.

2.1 Quantum Mechanics of a Single Spin

Let us start by reviewing basic properties of the quantum system consisting of a single spin. We discuss the properties of rotation operators in some detail. We assume that the reader is familiar with elementary theory of quantum mechanical angular momentum. See Appendix A.3 for a brief summary.

Spin operators and states Spin (or, more precisely, spin angular momentum) is a quantum mechanical object described by the spin operator $\hat{\boldsymbol{S}} = (\hat{S}^{(1)}, \hat{S}^{(2)}, \hat{S}^{(3)})$. Throughout the present book, we label the three mutually orthogonal directions as 1, 2, and 3, rather than x, y, and z. The operators $\hat{S}^{(1)}$, $\hat{S}^{(2)}$, and $\hat{S}^{(3)}$ are self-adjoint, and satisfy the commutation relations

$$[\hat{S}^{(\alpha)}, \hat{S}^{(\beta)}] = i \sum_{\gamma=1,2,3} \varepsilon_{\alpha\beta\gamma}\, \hat{S}^{(\gamma)}, \tag{2.1.1}$$

H. Tasaki, *Physics and Mathematics of Quantum Many-Body Systems*, Graduate Texts in Physics, https://doi.org/10.1007/978-3-030-41265-4_2

for any $\alpha, \beta = 1, 2, 3$. See Appendix A.3.1 for the definition of $\varepsilon_{\alpha\beta\gamma}$. We also assume that $\hat{\mathbf{S}}^2 = S(S+1)$ where S is a fixed constant called the spin quantum number.[1] The value of S depends on a specific system in consideration,[2] and is restricted to $1/2, 1, 3/2, 2, \ldots$. It is of fundamental importance that S does not only take positive integers $1, 2, 3 \ldots$, but also positive half-odd-integers $1/2, 3/2, 5/2, \ldots$.

The spin operators act on the $(2S+1)$ dimensional Hilbert space $\mathfrak{h}_0$. A commonly used basis for $\mathfrak{h}_0$ is $\{|\psi^\sigma\rangle\}$ with $\sigma = -S, -S+1, \ldots, S-1, S$. The basis states are fully characterized by the properties

$$\hat{S}^{(3)}|\psi^\sigma\rangle = \sigma|\psi^\sigma\rangle, \tag{2.1.2}$$

$$\hat{S}^{\pm}|\psi^\sigma\rangle = \sqrt{S(S+1) - \sigma(\sigma \pm 1)}\,|\psi^{\sigma\pm1}\rangle, \tag{2.1.3}$$

where $\hat{S}^{\pm} := \hat{S}^{(1)} \pm i\,\hat{S}^{(2)}$, and the normalization condition $\langle\psi^\sigma|\psi^\sigma\rangle = 1$. The interpretation is that $|\psi^S\rangle$ is the state in which the spin is pointing in the positive 3-direction.[3]

For $S = 1/2$ we often write $|\psi^{1/2}\rangle$ and $|\psi^{-1/2}\rangle$ as $|\psi^\uparrow\rangle$ and $|\psi^\downarrow\rangle$, (or sometimes as $|\uparrow\rangle$ and $|\downarrow\rangle$), respectively. Then (2.1.2) and (2.1.3) read

$$\hat{S}^{(3)}|\psi^\uparrow\rangle = \frac{1}{2}|\psi^\uparrow\rangle, \quad \hat{S}^{(3)}|\psi^\downarrow\rangle = -\frac{1}{2}|\psi^\downarrow\rangle \tag{2.1.4}$$

$$\hat{S}^{+}|\psi^\uparrow\rangle = 0, \quad \hat{S}^{-}|\psi^\uparrow\rangle = |\psi^\downarrow\rangle, \quad \hat{S}^{+}|\psi^\downarrow\rangle = |\psi^\uparrow\rangle, \quad \hat{S}^{-}|\psi^\downarrow\rangle = 0. \tag{2.1.5}$$

It is common to identify the basis state $|\psi^\sigma\rangle$ with a column vector

$$|\psi^\sigma\rangle = \begin{pmatrix} 0 \\ \vdots \\ 1 \\ \vdots \\ 0 \end{pmatrix} \tag{2.1.6}$$

whose $(S - \sigma + 1)$-th component is 1 and the rest are 0. Then the operators $\hat{S}^{(1)}$, $\hat{S}^{(2)}$, and $\hat{S}^{(3)}$ are identified with $(2S+1) \times (2S+1)$ Hermitian matrices whose components can be read off from (2.1.2) and (2.1.3). For $S = 1/2$, we find that

[1] To be rigorous this relations should be $\hat{\mathbf{S}}^2 = S(S+1)\hat{1}$ with $\hat{1}$ being the identity operator. We usually omit $\hat{1}$.

[2] In magnetic systems, spin angular momenta come mostly from electron spins, which have spin quantum number 1/2. In some atoms, however, electrons in certain orbits are coupled according to the Hund rule to form a single effective spin with higher S.

[3] If the reader is new to quantum spins, it is a good idea to try examining, for the $S = 1/2$ case, what are the states where the spin is pointing in the positive 1 or 2 directions.

$$\hat{S}^{(1)} = \frac{1}{2}\begin{pmatrix} 0 & 1 \\ 1 & 0 \end{pmatrix}, \quad \hat{S}^{(2)} = \frac{1}{2}\begin{pmatrix} 0 & -i \\ i & 0 \end{pmatrix}, \quad \hat{S}^{(3)} = \frac{1}{2}\begin{pmatrix} 1 & 0 \\ 0 & -1 \end{pmatrix}. \tag{2.1.7}$$

From the matrix representations, one sees that $(\hat{S}^{(\alpha)})^2 = \mathsf{I}/4$ for any α, and $\hat{S}^{(\alpha)}\hat{S}^{(\beta)} + \hat{S}^{(\beta)}\hat{S}^{(\alpha)} = 0$ for any $\alpha \neq \beta$. These (useful) relations hold only for $S = 1/2$. When dealing with $S = 1/2$ systems it is also common to use the Pauli matrices defined by $\hat{\sigma}^{(\alpha)} = 2\hat{S}^{(\alpha)}$, i.e.,

$$\hat{\sigma}^{(1)} = \begin{pmatrix} 0 & 1 \\ 1 & 0 \end{pmatrix}, \quad \hat{\sigma}^{(2)} = \begin{pmatrix} 0 & -i \\ i & 0 \end{pmatrix}, \quad \hat{\sigma}^{(3)} = \begin{pmatrix} 1 & 0 \\ 0 & -1 \end{pmatrix}. \tag{2.1.8}$$

Similarly the operator representations for $S = 1$ are found to be

$$\hat{S}^{(1)} = \frac{1}{\sqrt{2}}\begin{pmatrix} 0 & 1 & 0 \\ 1 & 0 & 1 \\ 0 & 1 & 0 \end{pmatrix}, \quad \hat{S}^{(2)} = \frac{1}{\sqrt{2}}\begin{pmatrix} 0 & -i & 0 \\ i & 0 & -i \\ 0 & i & 0 \end{pmatrix}, \quad \hat{S}^{(3)} = \begin{pmatrix} 1 & 0 & 0 \\ 0 & 0 & 0 \\ 0 & 0 & -1 \end{pmatrix}. \tag{2.1.9}$$

The following problem shows that one can represent any operator by combining the spin operators.

Problem 2.1.a Show that any operator on $\mathfrak{h}_0$ (i.e., any $(2S+1) \times (2S+1)$ matrix) can be written as a polynomial of $\hat{1}$, $\hat{S}^{(1)}$, $\hat{S}^{(2)}$, and $\hat{S}^{(3)}$. [solution→p.493]

Spin rotation operators Let $\alpha = 1, 2, 3$ and $\theta \in \mathbb{R}$. Then $\hat{U}_\theta^{(\alpha)} := \exp[-i\theta\hat{S}^{(\alpha)}]$ is the unitary operator that rotates spin states by angle θ about the α-axis. See Appendix A.2 for the definition and properties of the exponential of an operator. Note that (A.2.20) implies $\hat{U}_\theta^{(\alpha)}\hat{U}_\varphi^{(\alpha)} = \hat{U}_{\theta+\varphi}^{(\alpha)}$. We also have $(\hat{U}_\theta^{(\alpha)})^\dagger = \hat{U}_{-\theta}^{(\alpha)}$ by definition. We use these relations freely in the rest of the book. The rotation operators $\hat{U}_\theta^{(\alpha)} = \exp[-i\theta\hat{S}^{(\alpha)}]$ are elements of a continuous group called SU(2).[4] See Appendix A.5.

Let us now verify the interpretation of $\hat{U}_\theta^{(\alpha)}$ as the spin rotation operator. Take any spin state $|\psi\rangle$ and write its transform as $|\psi'\rangle = \hat{U}_\theta^{(\alpha)}|\psi\rangle$. We see that the (vector valued) expectation value $(\langle\psi|\hat{S}^{(1)}|\psi\rangle, \langle\psi|\hat{S}^{(2)}|\psi\rangle, \langle\psi|\hat{S}^{(3)}|\psi\rangle)$ is transformed into $(\langle\psi'|\hat{S}^{(1)}|\psi'\rangle, \langle\psi'|\hat{S}^{(2)}|\psi'\rangle, \langle\psi'|\hat{S}^{(3)}|\psi'\rangle)$. This transformation should obey the well-known transformation rule for the rotation of three-dimensional vectors. Since this is true for any $|\psi\rangle$, we find that the transformation from $(\hat{S}^{(1)}, \hat{S}^{(2)}, \hat{S}^{(3)})$ into $((\hat{U}_\theta^{(\alpha)})^\dagger\,\hat{S}^{(1)}\hat{U}_\theta^{(\alpha)}, (\hat{U}_\theta^{(\alpha)})^\dagger\,\hat{S}^{(2)}\hat{U}_\theta^{(\alpha)}, (\hat{U}_\theta^{(\alpha)})^\dagger\,\hat{S}^{(3)}\hat{U}_\theta^{(\alpha)})$ should also obey the same rule. To be more precise, we should have

$$\begin{pmatrix} (\hat{U}_\theta^{(\alpha)})^\dagger\,\hat{S}^{(1)}\hat{U}_\theta^{(\alpha)} \\ (\hat{U}_\theta^{(\alpha)})^\dagger\,\hat{S}^{(2)}\hat{U}_\theta^{(\alpha)} \\ (\hat{U}_\theta^{(\alpha)})^\dagger\,\hat{S}^{(3)}\hat{U}_\theta^{(\alpha)} \end{pmatrix} = \mathsf{R}_\theta^{(\alpha)}\begin{pmatrix} \hat{S}^{(1)} \\ \hat{S}^{(2)} \\ \hat{S}^{(3)} \end{pmatrix}, \tag{2.1.10}$$

[4]To be very precise, the rotation operators are representations of elements of SU(2).

for any $\alpha = 1, 2, 3$ and $\theta \in \mathbb{R}$. Here the matrices describing the rotations by θ about the x, y, and z-axes (which we denote here as the 1, 2, and 3-axes) in the Cartesian coordinates are given by

$$\mathsf{R}_\theta^{(1)} = \begin{pmatrix} 1 & 0 & 0 \\ 0 & \cos\theta & -\sin\theta \\ 0 & \sin\theta & \cos\theta \end{pmatrix}, \quad \mathsf{R}_\theta^{(2)} = \begin{pmatrix} \cos\theta & 0 & \sin\theta \\ 0 & 1 & 0 \\ -\sin\theta & 0 & \cos\theta \end{pmatrix},$$
$$\mathsf{R}_\theta^{(3)} = \begin{pmatrix} \cos\theta & -\sin\theta & 0 \\ \sin\theta & \cos\theta & 0 \\ 0 & 0 & 1 \end{pmatrix}, \tag{2.1.11}$$

or by specifying their entries compactly as

$$(\mathsf{R}_\theta^{(\alpha)})_{\beta,\gamma} = \begin{cases} \delta_{\beta,\gamma}\cos\theta - \varepsilon_{\alpha\beta\gamma}\sin\theta & \text{if } \alpha \neq \beta, \\ \delta_{\alpha,\gamma} & \text{if } \alpha = \beta. \end{cases} \tag{2.1.12}$$

Therefore, for the rotation about the 3-axis, for example, we should have

$$(\hat{U}_\theta^{(3)})^\dagger\, \hat{S}^{(1)}\, \hat{U}_\theta^{(3)} = \cos\theta\, \hat{S}^{(1)} - \sin\theta\, \hat{S}^{(2)}, \tag{2.1.13}$$
$$(\hat{U}_\theta^{(3)})^\dagger\, \hat{S}^{(2)}\, \hat{U}_\theta^{(3)} = \sin\theta\, \hat{S}^{(1)} + \cos\theta\, \hat{S}^{(2)}, \tag{2.1.14}$$
$$(\hat{U}_\theta^{(3)})^\dagger\, \hat{S}^{(3)}\, \hat{U}_\theta^{(3)} = \hat{S}^{(3)}, \tag{2.1.15}$$

where the third relation (2.1.15) is readily verified by noting that $\hat{U}_\theta^{(3)} = e^{-i\theta\hat{S}^{(3)}}$ and $\hat{S}^{(3)}$ commute. From (2.1.12) we see that the generalization of the nontrivial relations (2.1.13) and (2.1.14) are

$$(\hat{U}_\theta^{(\alpha)})^\dagger\, \hat{S}^{(\beta)}\, \hat{U}_\theta^{(\alpha)} = \cos\theta\, \hat{S}^{(\beta)} - \sin\theta \sum_{\gamma=1}^{3} \varepsilon_{\alpha\beta\gamma}\, \hat{S}^{(\gamma)}, \tag{2.1.16}$$

for any $\alpha \neq \beta$. We shall verify these transformation rules below.

Proof of (2.1.16) We shall prove the case with $\alpha = 3$, $\beta = 1$, which is (2.1.13). Other cases are essentially the same. Denote the left-hand and the right-hand sides of (2.1.13) as $\hat{L}(\theta)$ and $\hat{R}(\theta)$. Since $\hat{L}(\theta) = e^{i\theta\hat{S}^{(3)}}\hat{S}^{(1)}e^{-i\theta\hat{S}^{(3)}}$, (A.2.19) implies

$$\frac{d}{d\theta}\hat{L}(\theta) = i[\hat{S}^{(3)}, \hat{L}(\theta)]. \tag{2.1.17}$$

We also see that

$$\frac{d}{d\theta}\hat{R}(\theta) = -\sin\theta\, \hat{S}^{(1)} - \cos\theta\, \hat{S}^{(2)} = i[\hat{S}^{(3)}, \hat{R}(\theta)], \tag{2.1.18}$$

where we used the commutation relations (2.1.1). Since $\hat{L}(\theta)$ and $\hat{R}(\theta)$ satisfy the same differential equation, and satisfy $\hat{L}(0) = \hat{R}(0)$, we conclude that $\hat{L}(\theta) = \hat{R}(\theta)$ for any θ. ∎

In the above discussion, we were naturally led to use the convention (A.1.16), i.e., $\hat{A} \to \hat{U}^\dagger \hat{A} \hat{U}$, of the unitary transformation of operators. The other convention (A.1.17), i.e., $\hat{A} \to \hat{U} \hat{A} \hat{U}^\dagger$, also appears in the context of spin rotations.

Let $\boldsymbol{v} = (v_1, v_2, v_3) \in \mathbb{R}^3$ be a (classical) vector, and write $\hat{\boldsymbol{S}} \cdot \boldsymbol{v} = \sum_{\beta=1}^{3} \hat{S}^{(\beta)} v_\beta$. Then, for any $\alpha = 1, 2, 3$ and $\theta \in \mathbb{R}$, we have

$$\begin{aligned} \hat{U}_\theta^{(\alpha)} (\hat{\boldsymbol{S}} \cdot \boldsymbol{v}) (\hat{U}_\theta^{(\alpha)})^\dagger &= \sum_{\beta=1}^{3} (\hat{U}_{-\theta}^{(\alpha)})^\dagger \, \hat{S}^{(\beta)} \hat{U}_{-\theta}^{(\alpha)} \, v_\beta = \sum_{\beta,\gamma=1}^{3} (\mathsf{R}_{-\theta}^{(\alpha)})_{\beta,\gamma} \, \hat{S}^{(\gamma)} \, v_\beta \\ &= \sum_{\beta,\gamma=1}^{3} \hat{S}^{(\gamma)} \, (\mathsf{R}_\theta^{(\alpha)})_{\gamma,\beta} \, v_\beta = \hat{\boldsymbol{S}} \cdot (\mathsf{R}_\theta^{(\alpha)} \boldsymbol{v}), \end{aligned} \tag{2.1.19}$$

where we used (2.1.10), and noted that $\mathsf{R}_\theta^{(\alpha)}$ is an orthogonal matrix. More generally, if $\hat{U}$ is an arbitrary spin rotation[5] and R is the corresponding rotation matrix, then we have

$$\hat{U} (\hat{\boldsymbol{S}} \cdot \boldsymbol{v}) \hat{U}^\dagger = \hat{\boldsymbol{S}} \cdot (\mathsf{R} \boldsymbol{v}), \tag{2.1.20}$$

for any vector $\boldsymbol{v} \in \mathbb{R}^3$. This relation will turn out to be useful in some cases.

Basic properties of spin rotation operators By setting $\theta = \pm\pi$ and $\theta = \pm\pi/2$ in (2.1.16), we get useful relations

$$(\hat{U}_{\pm\pi}^{(\alpha)})^\dagger \, \hat{S}^{(\beta)} \, \hat{U}_{\pm\pi}^{(\alpha)} = -\hat{S}^{(\beta)}, \tag{2.1.21}$$

$$(\hat{U}_{\pm\pi/2}^{(\alpha)})^\dagger \, \hat{S}^{(\beta)} \, \hat{U}_{\pm\pi/2}^{(\alpha)} = \mp \sum_{\gamma=1}^{3} \varepsilon_{\alpha\beta\gamma} \, \hat{S}^{(\gamma)}. \tag{2.1.22}$$

for $\alpha \neq \beta$.

Note that the eigenvalues of $\hat{S}^{(\alpha)}$ are all integers or half-odd-integers when S is an integer or a half-odd-integer, respectively. This implies that

$$\hat{U}_{2\pi}^{(\alpha)} = e^{-2\pi i \hat{S}^{(\alpha)}} = \begin{cases} -\hat{1} & \text{for } S = 1/2, 3/2, \ldots, \\ \hat{1} & \text{for } S = 1, 2, \ldots. \end{cases} \tag{2.1.23}$$

It is remarkable that the rotation by 2π does not give the identity for a half-odd-integer spin.[6] This essential difference between integer and half-odd-integer spins will play important roles in the book, especially in Part II.

The identity (2.1.23) implies an interesting property of the π-rotation operators. Note that (A.2.17) and (2.1.21) imply

---

[5] One can take arbitrary products of $\hat{U}_\theta^{(\alpha)}$ with various $\alpha = 1, 2, 3$ and θ.

[6] This corresponds to the mathematical fact that SU(2) is "twice larger" than the three dimensional rotation group SO(3). See Appendix A.5.

$$
\begin{aligned}
(\hat{U}_\pi^{(\alpha)})^\dagger\,\hat{U}_\pi^{(\beta)}\,\hat{U}_\pi^{(\alpha)} &= (\hat{U}_\pi^{(\alpha)})^\dagger\,\exp[-i\pi\hat{S}^{(\beta)}]\,\hat{U}_\pi^{(\alpha)} = \exp[-i\pi(\hat{U}_\pi^{(\alpha)})^\dagger\,\hat{S}^{(\beta)}\,\hat{U}_\pi^{(\alpha)}] \\
&= \exp[i\pi\hat{S}^{(\beta)}] = (\hat{U}_\pi^{(\beta)})^\dagger,
\end{aligned}
\tag{2.1.24}
$$

if $\alpha \neq \beta$. Operating $\hat{U}_\pi^{(\alpha)}$ from the left, we get $\hat{U}_\pi^{(\beta)}\,\hat{U}_\pi^{(\alpha)} = \hat{U}_\pi^{(\alpha)}\,(\hat{U}_\pi^{(\beta)})^\dagger$, where the right-hand side can be rewritten as $\hat{U}_\pi^{(\alpha)}\,(\hat{U}_\pi^{(\beta)})^\dagger = \hat{U}_\pi^{(\alpha)}\,\hat{U}_\pi^{(\beta)}\,(\hat{U}_{2\pi}^{(\beta)})^\dagger$. Recalling (2.1.23), we find

$$
\hat{U}_\pi^{(\alpha)}\,\hat{U}_\pi^{(\beta)} = \begin{cases} -\hat{U}_\pi^{(\beta)}\,\hat{U}_\pi^{(\alpha)} & \text{for } S = 1/2, 3/2, \ldots, \\ \hat{U}_\pi^{(\beta)}\,\hat{U}_\pi^{(\alpha)} & \text{for } S = 1, 2, \ldots, \end{cases}
\tag{2.1.25}
$$

for any $\alpha \neq \beta$. In other words two π-rotation operators commute when S is an integer and anticommute when S is a half-odd-integer.

For a spin with $S = 1/2$ the rotation operators are written more explicitly as

$$
\hat{U}_\theta^{(\alpha)} = \Bigl(\cos\frac{\theta}{2}\Bigr)\hat{1} - 2i\Bigl(\sin\frac{\theta}{2}\Bigr)\hat{S}^{(\alpha)}.
\tag{2.1.26}
$$

See Problem 2.1.b below for a derivation. We also encourage the reader to (at least) take a look at Problem 2.1.d and its solution; they contain the essence of the rotation of spins. Problem 2.1.e may be interesting by itself.

Problem 2.1.b Derive (2.1.26). (Hint: What is $(2\hat{S}^{(\alpha)})^2$?) [solution→p.493]

Problem 2.1.c Find a relation corresponding to (2.1.26) for $S = 1$. [solution→p.493]

Problem 2.1.d We consider a spin with $S = 1/2$. Express $\hat{U}_\theta^{(\alpha)}$ with $\alpha = 1, 2, 3$ as 2×2 matrices. Take θ and φ such that $0 \leq \theta \leq \pi$ and $0 \leq \varphi < 2\pi$, and let $\boldsymbol{n}(\theta, \varphi) = (\cos\varphi\,\sin\theta, \sin\varphi\,\sin\theta, \cos\theta)$ be the standard unit vector. Then $|\psi_{\theta,\varphi}\rangle = \hat{U}_\varphi^{(3)}\hat{U}_\theta^{(2)}|\psi^\uparrow\rangle$ should be the state where the spin is pointing in the direction $\boldsymbol{n}(\theta, \varphi)$.[7] Express $|\psi_{\theta,\varphi}\rangle$ as a two-component vector, and confirm that it is an eigenstate of $\hat{\boldsymbol{S}} \cdot \boldsymbol{n}(\theta, \varphi)$. [solution→p.494]

Problem 2.1.e By setting $\theta = \varphi = \pi/2$ in Problem 2.1.d, we get the state $|\psi_{(0,1,0)}\rangle$ in which the spin is pointing in the positive 2-direction. But a state in which the spin is pointing in the same direction may be obtained, e.g., by applying the $-\pi/2$ rotation about the 1-axis to $|\psi^\uparrow\rangle$. Compute both $|\psi_{(0,1,0)}\rangle$ and $|\psi'_{(0,1,0)}\rangle = \hat{U}_{-\pi/2}^{(1)}|\psi^\uparrow\rangle$ explicitly and check that they differ by a phase factor. Consider the same setup for a spin with $S = 1$ and confirm that we still get a phase factor.[8] See also Problem 2.2.c (p. 23). [solution→p.494]

The group $\mathbb{Z}_2 \times \mathbb{Z}_2$ and its representations The group $\mathbb{Z}_2 \times \mathbb{Z}_2$, also known as the dihedral group of order 2 (denoted as D_2) or the Klein group, is a finite abelian group which consists of four elements, $\mathbb{Z}_2 \times \mathbb{Z}_2 = \{e, a, b, c\}$. The group is specified by the following multiplication rule:

---

[7] For general S, the corresponding state $|\psi_{\theta,\varphi}\rangle = e^{-i\varphi\hat{S}^{(3)}}e^{-i\theta\hat{S}^{(2)}}|\psi^S\rangle$ is known as the spin coherent state. See, e.g., Chap. 6 of [4].

[8] For a deeper reason for the appearance of the phase factor, see, e.g., Sect. 6.5 (in particular p. 85) of [4].

$$\begin{gathered} ea = ae = a, \quad eb = be = b, \quad ec = ce = c, \\ a^2 = b^2 = c^2 = e, \\ ab = ba = c, \quad bc = cb = a, \quad ca = ac = b \end{gathered} \tag{2.1.27}$$

See Appendix A.5 for more about the group theory.

A representation of the group $\mathbb{Z}_2 \times \mathbb{Z}_2$ is given by the π-rotations about the three axes in the three dimensional space. By setting $\theta = \pi$ in (2.1.11), we have

$$\mathsf{R}_\pi^{(1)} = \begin{pmatrix} 1 & 0 & 0 \\ 0 & -1 & 0 \\ 0 & 0 & -1 \end{pmatrix}, \quad \mathsf{R}_\pi^{(2)} = \begin{pmatrix} -1 & 0 & 0 \\ 0 & 1 & 0 \\ 0 & 0 & -1 \end{pmatrix}, \quad \mathsf{R}_\pi^{(3)} = \begin{pmatrix} -1 & 0 & 0 \\ 0 & -1 & 0 \\ 0 & 0 & 1 \end{pmatrix}. \tag{2.1.28}$$

It is readily verified that the four matrices I, $\mathsf{R}_\pi^{(1)}$, $\mathsf{R}_\pi^{(2)}$, and $\mathsf{R}_\pi^{(3)}$ obey the same multiplication rule as (2.1.27). We see that these matrices form a representation of the group $\mathbb{Z}_2 \times \mathbb{Z}_2$.[9]

Problem 2.1.f What does the relation $\mathsf{R}_\pi^{(1)}\mathsf{R}_\pi^{(2)} = \mathsf{R}_\pi^{(3)}$ mean in the ordinary language? If this is not intuitively obvious to you, try convincing yourself of this fact, probably by performing experiments. [solution→p.495]

We can try similar π-rotations in the spin space. Define $\hat{u}_\alpha := \hat{U}_\pi^{(\alpha)} = e^{-i\pi\hat{S}^{(\alpha)}}$ for $\alpha = 1, 2, 3$ and any S. Then it holds that

$$\hat{u}_1\hat{u}_2 = \hat{u}_3, \quad \hat{u}_2\hat{u}_3 = \hat{u}_1, \quad \hat{u}_3\hat{u}_1 = \hat{u}_2. \tag{2.1.29}$$

These identities may be verfied by using the explicit forms of the matrices (2.1.32), (2.1.33), and (2.1.34), but let us present a general derivation.[10]

Proof By using (A.2.17), (2.1.21) and (2.1.22), we find

$$\begin{aligned} e^{-i\pi\hat{S}^{(1)}}\, e^{-i\pi\hat{S}^{(2)}} &= e^{-i(\pi/2)\hat{S}^{(1)}}\, e^{-i(\pi/2)\hat{S}^{(1)}}\, e^{-i\pi\hat{S}^{(2)}} \\ &= e^{-i(\pi/2)\hat{S}^{(1)}}\,(e^{-i\pi\hat{S}^{(2)}}\, e^{i(\pi/2)\hat{S}^{(1)}}\, e^{i\pi\hat{S}^{(2)}})\, e^{-i\pi\hat{S}^{(2)}} \\ &= e^{-i(\pi/2)\hat{S}^{(1)}}\, e^{-i\pi\hat{S}^{(2)}}\, e^{i(\pi/2)\hat{S}^{(1)}} \\ &= e^{-i\pi\hat{S}^{(3)}}, \end{aligned} \tag{2.1.30}$$

which proves the first relation. The other relation can be proved in the same manner, or by using the first relation and $(\hat{u}_\alpha)^2 = \pm\hat{1}$. ■

The relation (2.1.29) suggests that the three operators $\hat{u}_1, \hat{u}_2, \hat{u}_3$, along with the identity, form a representation of the $\mathbb{Z}_2 \times \mathbb{Z}_2$ group. This is indeed the case when S is an integer, where one has $(\hat{u}_\alpha)^2 = \hat{U}_{2\pi}^{(\alpha)} = \hat{1}$ from (2.1.23). We also have $\hat{u}_\alpha\hat{u}_\beta =$

---

[9]More precisely, one defines a representation ρ by $\rho(e) = \mathsf{I}$, $\rho(a) = \mathsf{R}_\pi^{(1)}$, $\rho(b) = \mathsf{R}_\pi^{(2)}$, and $\rho(c) = \mathsf{R}_\pi^{(3)}$.

[10]We learned this elegant derivation from Akinori Tanaka.

$\hat{u}_\beta\hat{u}_\alpha$ as in (2.1.25). Thus the four operators $\hat{1}$, $\hat{u}_1$, $\hat{u}_2$, and $\hat{u}_3$ form a representation of $\mathbb{Z}_2 \times \mathbb{Z}_2$.[11]

When S is a half-odd-integer, we have $(\hat{u}_\alpha)^2 = -\hat{1}$ and $\hat{u}_\alpha\hat{u}_\beta = -\hat{u}_\beta\hat{u}_\alpha$ if $\alpha \neq \beta$. The multiplication rule is given by

$$\begin{gathered}\hat{1}\hat{u}_1 = \hat{u}_1\hat{1} = \hat{u}_1, \quad \hat{1}\hat{u}_2 = \hat{u}_2\hat{1} = \hat{u}_2, \quad \hat{1}\hat{u}_3 = \hat{u}_3\hat{1} = \hat{u}_3, \\ (\hat{u}_1)^2 = (\hat{u}_2)^2 = (\hat{u}_3)^2 = -\hat{1}, \\ \hat{u}_1\hat{u}_2 = -\hat{u}_2\hat{u}_1 = \hat{u}_3, \quad \hat{u}_2\hat{u}_3 = -\hat{u}_3\hat{u}_2 = \hat{u}_1, \quad \hat{u}_3\hat{u}_1 = -\hat{u}_1\hat{u}_3 = \hat{u}_2,\end{gathered} \tag{2.1.31}$$

which is obviously different from (2.1.27). Note however that if one neglects the multiplicative factors ± 1, then the rule (2.1.31) precisely coincides with the desired (2.1.27) for the $\mathbb{Z}_2 \times \mathbb{Z}_2$ group. Such a "representation" with extra phase factors is called a projective representation of a group.[12] In Sect. 8.3.4 we make use of the notion of projective representations to define "topological" indices for symmetry protected topological (SPT) phases.

Explicit matrix forms of $\hat{u}_1$, $\hat{u}_2$, and $\hat{u}_3$ are obtained, for $S = 1/2$, from $e^{-i\pi\hat{S}^{(\alpha)}} = -2i\hat{S}^{(\alpha)}$ and (2.1.7) as[13]

$$\hat{u}_1 = \begin{pmatrix} 0 & -i \\ -i & 0 \end{pmatrix}, \quad \hat{u}_2 = \begin{pmatrix} 0 & -1 \\ 1 & 0 \end{pmatrix}, \quad \hat{u}_3 = \begin{pmatrix} -i & 0 \\ 0 & i \end{pmatrix}, \tag{2.1.32}$$

and, for $S = 1$, from $e^{-i\pi\hat{S}^{(\alpha)}} = \hat{1} - 2(\hat{S}^{(\alpha)})^2$ and (2.1.9) as

$$\hat{u}_1 = \begin{pmatrix} 0 & 0 & -1 \\ 0 & -1 & 0 \\ -1 & 0 & 0 \end{pmatrix}, \quad \hat{u}_2 = \begin{pmatrix} 0 & 0 & 1 \\ 0 & -1 & 0 \\ 1 & 0 & 0 \end{pmatrix}, \quad \hat{u}_3 = \begin{pmatrix} -1 & 0 & 0 \\ 0 & 1 & 0 \\ 0 & 0 & -1 \end{pmatrix}. \tag{2.1.33}$$

These expressions are generalized to general S as

$$\begin{gathered}\langle\psi^\sigma|\hat{u}_1|\psi^\tau\rangle = (-i)^{2S}\,\delta_{\sigma,-\tau}, \quad \langle\psi^\sigma|\hat{u}_2|\psi^\tau\rangle = (-1)^{S+\sigma}\,\delta_{\sigma,-\tau}, \\ \langle\psi^\sigma|\hat{u}_3|\psi^\tau\rangle = e^{-i\pi\sigma}\,\delta_{\sigma,\tau}.\end{gathered} \tag{2.1.34}$$

Problem 2.1.g Derive the expressions (2.1.34). (Hint: A direct derivation seems to be not easy. It is useful to express a single spin S by using $2S$ spins with spin quantum number 1/2. See Appendix A.3.3 and the expression (2.4.11).) [solution→p.495]

---

[11]More precisely, one defines a representation ρ by $\rho(e) = \hat{1}$, $\rho(a) = \hat{u}_1$, $\rho(b) = \hat{u}_2$, and $\rho(c) = \hat{u}_3$.

[12]See Sect. 8.3.4, in particular (8.3.24), or Footnote 16 (p. 481) in Appendix A.5 for a precise definition.

[13]The identity $e^{-i\pi\hat{S}^{(\alpha)}} = -2i\hat{S}^{(\alpha)}$ (which is valid only for $S = 1/2$) can be derived from (2.1.26). But, by noting that $e^{-i\pi\hat{S}^{(3)}}|\psi^{\pm 1/2}\rangle = e^{\mp i\pi/2}|\psi^{\pm 1/2}\rangle = \mp i|\psi^{\pm 1/2}\rangle = -2i\hat{S}^{(3)}|\psi^{\pm 1/2}\rangle$, the identity for $\alpha = 3$ readily follows. Other cases then follow by symmetry. Similarly the identity $e^{-i\pi\hat{S}^{(\alpha)}} = \hat{1} - 2(\hat{S}^{(\alpha)})^2$ for $S = 1$ can be derived from (S.3) in p. 494, but can also be derived in a similar (simpler) manner.

2.2 Quantum Spin Systems

We are ready to formulate quantum spin systems, idealized models of magnets. Let the lattice Λ be an arbitrary finite set. The elements of Λ are denoted as $x, y, \ldots$, and called sites. A site represents an atom in a magnetic material. We also fix the spin quantum number $S = 1/2, 1, 3/2, \ldots$, and assume that each site (i.e., an atom) carries a spin quantum angular momentum with quantum number S.[14] See Part III, in particular Fig. 9.1, for a further discussion on the relation between quantum spin systems and microscopic pictures of solids.

States We associate with each site x a spin whose Hilbert space $\mathfrak{h}_x$ is identical to $\mathfrak{h}_0$. The basis state of $\mathfrak{h}_x$ corresponding to $|\psi^\sigma\rangle \in \mathfrak{h}_0$ is denoted as[15] $|\psi_x^\sigma\rangle$.

The Hilbert space of the spin system on Λ is $\mathscr{H} := \bigotimes_{x\in\Lambda} \mathfrak{h}_x$, whose dimension is $(2S+1)^{|\Lambda|}$. A natural basis state is

$$|\Psi^{\boldsymbol{\sigma}}\rangle := \bigotimes_{x\in\Lambda} |\psi_x^{\sigma_x}\rangle. \tag{2.2.1}$$

The multi-index $\boldsymbol{\sigma} = (\sigma_x)_{x\in\Lambda}$, with $\sigma_x = -S, \ldots, S-1, S$, determines a spin configuration. For later use we define

$$\overline{\boldsymbol{\sigma}} := \sum_{x\in\Lambda} \sigma_x. \tag{2.2.2}$$

As an example let $\Lambda = \{1, 2\}$ and $S = 1/2$. Then $\boldsymbol{\sigma}$ can be $(\uparrow, \uparrow)$, $(\uparrow, \downarrow)$, $(\downarrow, \uparrow)$, or $(\downarrow, \downarrow)$, and basis states are, for example,

$$|\Psi^{(\uparrow,\uparrow)}\rangle = |\psi_1^\uparrow\rangle \otimes |\psi_2^\uparrow\rangle, \quad |\Psi^{(\uparrow,\downarrow)}\rangle = |\psi_1^\uparrow\rangle \otimes |\psi_2^\downarrow\rangle. \tag{2.2.3}$$

It is common (or even standard) in physics literature to omit the symbol $\otimes$ and write, e.g., $|\Psi^{(\uparrow,\downarrow)}\rangle = |\psi_1^\uparrow\rangle|\psi_2^\downarrow\rangle$ or $|\Psi^{(\uparrow,\downarrow)}\rangle = |\uparrow\rangle_1|\downarrow\rangle_2$.

Spin operators The spin operator $\hat{S}_x^{(\alpha)}$ at site $x \in \Lambda$, where $\alpha = 1, 2, 3$, is a copy of $\hat{S}^{(\alpha)}$ that acts nontrivially only on the space $\mathfrak{h}_x$. It is defined by the action

$$\hat{S}_x^{(\alpha)}\Big(\bigotimes_{y\in\Lambda} |\psi_y^{\sigma_y}\rangle\Big) = \big(\hat{S}^{(\alpha)}|\psi_x^{\sigma_x}\rangle\big) \otimes \Big(\bigotimes_{y\in\Lambda\setminus\{x\}} |\psi_y^{\sigma_y}\rangle\Big), \tag{2.2.4}$$

or, more directly as

$$\hat{S}_x^{(\alpha)} = \hat{1} \otimes \cdots \otimes \hat{1} \otimes \hat{S}^{(\alpha)} \otimes \hat{1} \otimes \cdots \otimes \hat{1}, \tag{2.2.5}$$

---

[14] See Footnote 2 in p. 14.

[15] Probably the notation $|\psi^\sigma\rangle_x$ is more logical (see Appendix A.1), but we use this compact notation.

where $\hat{S}^{(\alpha)}$ acts on $\mathfrak{h}_x$. Note that $\hat{S}_x^{(\alpha)}$ is defined as an operator on the whole Hilbert space $\mathscr{H}$ (rather than $\mathfrak{h}_x$). We write $\hat{\boldsymbol{S}}_x = (\hat{S}_x^{(1)}, \hat{S}_x^{(2)}, \hat{S}_x^{(3)})$. By definition spin operators at different sites commute with each other. From the basic commutation relations (2.1.1), we have

$$[\hat{S}_x^{(\alpha)}, \hat{S}_y^{(\beta)}] = i\,\delta_{x,y} \sum_{\gamma=1,2,3} \varepsilon_{\alpha\beta\gamma}\,\hat{S}_x^{(\gamma)}, \tag{2.2.6}$$

for any $x, y \in \Lambda$ and $\alpha, \beta = 1, 2, 3$.

We define the total spin operator as

$$\hat{\boldsymbol{S}}_{\mathrm{tot}} := \sum_{x\in\Lambda} \hat{\boldsymbol{S}}_x. \tag{2.2.7}$$

We write $\hat{\boldsymbol{S}}_{\mathrm{tot}} = (\hat{S}_{\mathrm{tot}}^{(1)}, \hat{S}_{\mathrm{tot}}^{(2)}, \hat{S}_{\mathrm{tot}}^{(3)})$, and define

$$\hat{S}_{\mathrm{tot}}^{\pm} := \hat{S}_{\mathrm{tot}}^{(1)} \pm i\,\hat{S}_{\mathrm{tot}}^{(2)} = \sum_{x\in\Lambda} \hat{S}_x^{\pm}, \tag{2.2.8}$$

where $\hat{S}_x^{\pm} := \hat{S}_x^{(1)} \pm i\,\hat{S}_x^{(2)}$. An eigenvalue of $(\hat{\boldsymbol{S}}_{\mathrm{tot}})^2$ is denoted as $S_{\mathrm{tot}}(S_{\mathrm{tot}}+1)$, where S_{tot} takes a value in $S_{\mathrm{min}}, S_{\mathrm{min}}+1, \ldots, |\Lambda|S-1, |\Lambda|S$. When $|\Lambda| \neq 1$, we see that $S_{\mathrm{min}} = 0$ if $|\Lambda|$ is even or S is an integer, and $S_{\mathrm{min}} = 1/2$ if S is a half-odd-integer and $|\Lambda|$ is odd. When a state $|\Phi\rangle$ satisfies $(\hat{\boldsymbol{S}}_{\mathrm{tot}})^2|\Phi\rangle = S_{\mathrm{tot}}(S_{\mathrm{tot}}+1)|\Phi\rangle$, we say that the state $|\Phi\rangle$ has total spin S_{tot}.

It is convenient to decompose the whole Hilbert space $\mathscr{H}$ according to the eigenvalue of $\hat{S}_{\mathrm{tot}}^{(3)}$ as

$$\mathscr{H} = \bigoplus_{M=-|\Lambda|S}^{|\Lambda|S} \mathscr{H}_M, \tag{2.2.9}$$

with

$$\mathscr{H}_M := \bigl\{|\Phi\rangle \in \mathscr{H} \bigm| \hat{S}_{\mathrm{tot}}^{(3)}|\Phi\rangle = M|\Phi\rangle\bigr\}. \tag{2.2.10}$$

The sum in (2.2.9) is over the $2|\Lambda|S+1$ eigenvalues of $\hat{S}_{\mathrm{tot}}^{(3)}$, i.e., $M = -|\Lambda|S, -|\Lambda|S+1, \ldots, |\Lambda|S-1, |\Lambda|S$. The eigenvalue M is often denoted as $S_{\mathrm{tot}}^{(3)}$.

As in the case of a single spin (see p. 15), the spin rotation operator is defined as

$$\hat{U}_\theta^{(\alpha)} := \exp[-i\theta\hat{S}_{\mathrm{tot}}^{(\alpha)}] = \prod_{x\in\Lambda} \exp[-i\theta\hat{S}_x^{(\alpha)}], \tag{2.2.11}$$

for $\alpha = 1, 2, 3$ and $\theta \in \mathbb{R}$. These operators describe global rotation of the whole spin system on Λ, and are elements of the group SU(2). See Appendix A.5.

Suppose that an operator $\hat{A}$ satisfies

$$[\hat{A}, \hat{S}_{\mathrm{tot}}^{(\alpha)}] = 0. \tag{2.2.12}$$

Then the definition (A.2.16) of the exponential of an operator implies that $[\hat{A}, \hat{U}_\theta^{(\alpha)}] = 0$, which is equivalent to

$$(\hat{U}_\theta^{(\alpha)})^\dagger \hat{A} \hat{U}_\theta^{(\alpha)} = \hat{A} \quad \text{for any } \theta. \tag{2.2.13}$$

This means that $\hat{A}$ is invariant under any rotation around the α-th axis. See (A.1.16) for the unitary transformation of operators. When (2.2.12) holds for all $\alpha = 1, 2, 3$, we say that the operator $\hat{A}$ is SU(2) invariant. When (2.2.12) holds for one α, say $\alpha = 3$, we say that the operator $\hat{A}$ is U(1) invariant since $\hat{U}_\theta^{(\alpha)}$ with a single α and $\theta \in [0, 2\pi)$ form the group U(1). See Appendix A.5.

Problem 2.2.a For any $\alpha, \beta = 1, 2, 3$ such that $\alpha \neq \beta$, show that $\hat{U}_\pi^{(\alpha)} \hat{U}_\pi^{(\beta)} = \hat{U}_\pi^{(\beta)} \hat{U}_\pi^{(\alpha)}$ if $|\Lambda|S$ is an integer, and $\hat{U}_\pi^{(\alpha)} \hat{U}_\pi^{(\beta)} = -\hat{U}_\pi^{(\beta)} \hat{U}_\pi^{(\alpha)}$ if $|\Lambda|S$ is a half-odd-integer. Let $|\Lambda|S$ be a half-odd-integer, and take any eigenstate $|\Phi\rangle$ of $\hat{U}_\pi^{(\alpha)}$. Show that $|\Phi\rangle$ and $\hat{U}_\pi^{(\beta)}|\Phi\rangle$ are orthogonal for $\beta \neq \alpha$. [solution→p.496]

Problem 2.2.b Consider a system of two spins with $S = 1/2$, and denote the basis states as $|\sigma\rangle_1|\sigma'\rangle_2$ with $\sigma, \sigma' = \uparrow, \downarrow$. (See Appendix A.3.3 for the notation.) Show by explicit calculation (using the results from Problem 2.1.d in p. 18) that

$$\frac{1}{4\pi}\int_0^{2\pi} d\varphi \int_0^{\pi} d\theta \, \sin\theta \, \hat{U}_\varphi^{(3)} \hat{U}_\theta^{(2)} |\uparrow\rangle_1 |\downarrow\rangle_2 = \frac{1}{2}(|\uparrow\rangle_1|\downarrow\rangle_2 - |\downarrow\rangle_1|\uparrow\rangle_2) = \frac{1}{\sqrt{2}}|\Phi_{0,0}\rangle, \tag{2.2.14}$$

where the spin-singlet $|\Phi_{0,0}\rangle$ is defined in (A.3.23). Note that the left-hand side is the average over the solid angle of the states in which the two spins are pointing in the directions $\boldsymbol{n}(\theta, \varphi)$ and $-\boldsymbol{n}(\theta, \varphi)$, respectively. One might then conclude that the resulting state must be completely SU(2) invariant (in the sense that a state $|\Phi\rangle$ satisfies $\hat{U}_\theta^{(\alpha)}|\Phi\rangle = |\Phi\rangle$ for any α and θ). This conclusion is supported by (2.2.14). But this argument is incorrect, and fails in general. Show, again by explicit calculation, that

$$\begin{aligned}\frac{1}{4\pi}\int_0^{2\pi} d\varphi \int_0^{\pi} d\theta \, \sin\theta \, \hat{U}_\varphi^{(3)} \hat{U}_\theta^{(2)} |\uparrow\rangle_1 |\uparrow\rangle_2 &= \frac{\pi}{8}(|\uparrow\rangle_1|\downarrow\rangle_2 + |\downarrow\rangle_1|\uparrow\rangle_2) \\ &= \frac{\pi}{4\sqrt{2}}|\Phi_{1,0}\rangle,\end{aligned} \tag{2.2.15}$$

where $|\Phi_{1,0}\rangle$, one of the spin-triplet, is defined in (A.3.22). It is not SU(2) invariant. [solution→p.496]

Problem 2.2.c The basic reason that we get an SU(2) non-invariant state in (2.2.15) is the presence of the nontrivial phase that we saw in Problem 2.1.e (p. 18). As the SU(2) invariance of (2.2.14) suggests, we do not encounter such a phase when rotating the state $|\uparrow\rangle_1|\downarrow\rangle_2$. To be more precise, fix an arbitrary unit vector $\boldsymbol{n} \in \mathbb{R}^3$, and let $\hat{U}$ be any uniform rotation of two spins[16] such that $\hat{U}\hat{S}_x^{(3)}\hat{U}^\dagger = \hat{\boldsymbol{S}}_x \cdot \boldsymbol{n}$ for

[16]One can take arbitrary products of $\hat{U}_\theta^{(\alpha)}$ with various $\alpha = 1, 2, 3$ and θ.

$x = 1, 2$. Then prove that the rotated state $\hat{U}|\uparrow\rangle_1|\downarrow\rangle_2$ is determined solely by $\boldsymbol{n}$ and does not depend on the choice of $\hat{U}$. We can say that the state $|\uparrow\rangle_1|\downarrow\rangle_2$ transforms exactly as a classical vector under rotation. This is a special case of the observation made in [10]. (This problem may be difficult without hints. See the footnote[17] below for a big hint.) [solution→p.496]

Properties of $\hat{\boldsymbol{S}}_x \cdot \hat{\boldsymbol{S}}_y$ Let us investigate the properties of the operator $\hat{\boldsymbol{S}}_x \cdot \hat{\boldsymbol{S}}_y = \sum_{\alpha=1,2,3} \hat{S}_x^{(\alpha)} \hat{S}_y^{(\alpha)}$ for $x \neq y$, which is the building block of the Heisenberg Hamiltonian to be investigated in the following sections and chapters.

It is useful to note that

$$\hat{\boldsymbol{S}}_x \cdot \hat{\boldsymbol{S}}_y = \frac{1}{2}(\hat{S}_x^+ \hat{S}_y^- + \hat{S}_x^- \hat{S}_y^+) + \hat{S}_x^{(3)} \hat{S}_y^{(3)}. \tag{2.2.16}$$

An important property of the operator is the commutation relation

$$[\hat{\boldsymbol{S}}_x \cdot \hat{\boldsymbol{S}}_y, \hat{S}_{\mathrm{tot}}^{(\alpha)}] = 0, \tag{2.2.17}$$

for any $\alpha = 1, 2, 3$. This means that $\hat{\boldsymbol{S}}_x \cdot \hat{\boldsymbol{S}}_y$ is SU(2) invariant, i.e., invariant under any global rotation of spins. The SU(2) invariance as well as the simplicity of the expression suggests that $\hat{\boldsymbol{S}}_x \cdot \hat{\boldsymbol{S}}_y$ is a natural interaction between two spins at sites x and y. This intuition is supported by the analysis of underlying electron models such as the Hubbard model. See Part III where we focus on the origin of spin-spin interactions.

Proof of (2.2.17) Since all the directions are equivalent, it suffices to check the relation for $\alpha = 3$. By inspection one notes $[\hat{S}_x^\pm, \hat{S}_{\mathrm{tot}}^{(3)}] = [\hat{S}_x^\pm, \hat{S}_x^{(3)}] = \mp \hat{S}_x^\pm$. Then one finds $[\hat{S}_x^+ \hat{S}_y^-, \hat{S}_{\mathrm{tot}}^{(3)}] = [\hat{S}_x^+, \hat{S}_{\mathrm{tot}}^{(3)}]\hat{S}_y^- + \hat{S}_x^+[\hat{S}_y^-, \hat{S}_{\mathrm{tot}}^{(3)}] = 0$. One also sees that $[\hat{S}_x^- \hat{S}_y^+, \hat{S}_{\mathrm{tot}}^{(3)}] = 0$. Since it is obvious that $[\hat{S}_x^{(3)} \hat{S}_y^{(3)}, \hat{S}_{\mathrm{tot}}^{(3)}] = 0$, one gets (2.2.17). ∎

To investigate the eigenvalues of the operator $\hat{\boldsymbol{S}}_x \cdot \hat{\boldsymbol{S}}_y$, we note that

$$\begin{aligned} \hat{\boldsymbol{S}}_x \cdot \hat{\boldsymbol{S}}_y &= \frac{1}{2}\{(\hat{\boldsymbol{S}}_x + \hat{\boldsymbol{S}}_y)^2 - (\hat{\boldsymbol{S}}_x)^2 - (\hat{\boldsymbol{S}}_y)^2\} \\ &= \frac{1}{2}(\hat{\boldsymbol{S}}_x + \hat{\boldsymbol{S}}_y)^2 - S(S+1). \end{aligned} \tag{2.2.18}$$

The eigenvalues of $(\hat{\boldsymbol{S}}_x + \hat{\boldsymbol{S}}_y)^2$ are readily found from the theory of addition of angular momenta. See Appendix A.3.3. The minimum eigenvalue is 0, and the maximum eigenvalue is $2S(2S+1)$. We thus find that the operator $\hat{\boldsymbol{S}}_x \cdot \hat{\boldsymbol{S}}_y$ has

$$\begin{cases} \text{minimum eigenvalue} & -S(S+1) \\ \text{maximum eigenvalue} & S^2. \end{cases} \tag{2.2.19}$$

When the lattice consists only of the two sites x and y, the minimum eigenvalue is non-degenerate and the maximum eigenvalue is $(4S+1)$-fold degenerate.

---

[17]Note that $|\uparrow\rangle_1|\downarrow\rangle_2 = \sqrt{2}(\frac{1}{2} + \hat{S}_1^{(3)})(\frac{1}{2} - \hat{S}_2^{(3)})|\Phi_{0,0}\rangle$.

It is instructive to compare the situation with that of classical spins. A classical spin $\vec{S}_x = (S_x^{(1)}, S_x^{(2)}, S_x^{(3)}) \in \mathbb{R}^3$ is simply a vector with a fixed length $|\vec{S}_x| = S$. For such two classical spins, one obviously has $\vec{S}_x \cdot \vec{S}_y = S^2 \cos\theta$, where θ is the angle between $\vec{S}_x$ and $\vec{S}_y$. The minimum value $-S^2$ of $\vec{S}_x \cdot \vec{S}_y$ is obtained when $\theta = \pi$, i.e., when the two spins are pointing in the opposite directions, and the maximum value S^2 is obtained when $\theta = 0$, i.e., when the two spins are pointing in the same direction. There is a clear symmetry between the minimum and the maximum values. This is obvious since we get one from the other by the transformation $(\vec{S}_x, \vec{S}_y) \to (\vec{S}_x, -\vec{S}_y)$. Note also that there are infinitely many spin configurations (corresponding to the choice of the direction of one spin) associated with either the minimum or the maximum values of $\vec{S}_x \cdot \vec{S}_y$.

The quantum mechanical case (2.2.19) looks very different. One first notes that the minimum and the maximum eigenvalues, i.e., $-S(S+1)$ and S^2, are not symmetric. It is also remarkable that (when there are only two sites x and y) the eigenstate corresponding to the minimum value $-S(S+1)$ is unique, in contrast to the degeneracy one finds for the eigenstates corresponding to S^2 and also for the classical spin configurations corresponding to the minimum and the maximum value of $\vec{S}_x \cdot \vec{S}_y$. The essential difference between the minimum and the maximum eigenvalues reflects an intrinsically quantum mechanical nature of this simple eigenvalue problem. It is also suggested that we should see stronger "quantum effect" in ground states of antiferromagnetic models, whose simplest version is the eigenstate of $\hat{\mathbf{S}}_x \cdot \hat{\mathbf{S}}_y$ corresponding to the minimum eigenvalue $-S(S+1)$.

2.3 Time-Reversal and the Kramers Degeneracy

Here we shall discuss the time-reversal in quantum spin systems. We introduce the time-reversal map, and discuss the Kramers degeneracy in systems with time-reversal symmetry. Although we do not make heavy use of the material in the present book, it is useful to spend some pages since the topic is often confusing to the beginners. The reader may skip this section, and come back when necessary.

Motivation Let us briefly describe the motivation for the abstract definitions presented below. We wish to consider a transformation that reverses the direction of time. More precisely when t is the time variable of a physical system, we want to know how the same system is described if we use $s = -t$ as a new time variable.

Consider, for example, the simplest time-dependent Schrödinger equation for a single particle

$$i\frac{\partial}{\partial t}\psi(t,\boldsymbol{r}) = \left\{-\frac{\Delta}{2m} + V(\boldsymbol{r})\right\}\psi(t,\boldsymbol{r}), \tag{2.3.1}$$

where $\boldsymbol{r} = (x, y, z)$, $\Delta = \partial^2/\partial x^2 + \partial^2/\partial y^2 + \partial^2/\partial z^2$, $V(\boldsymbol{r})$ is a real valued potential, and $\psi(t,\boldsymbol{r})$ is the complex valued wave function. By substituting $t = -s$ and taking the complex conjugate, (2.3.1) turns into the same Schrödinger equation

$$i\frac{\partial}{\partial s}\psi_{\mathrm{R}}(s,\boldsymbol{r}) = \left\{-\frac{\Delta}{2m} + V(\boldsymbol{r})\right\}\psi_{\mathrm{R}}(s,\boldsymbol{r}), \tag{2.3.2}$$

where

$$\psi_{\mathrm{R}}(s,\boldsymbol{r}) := \{\psi(-s,\boldsymbol{r})\}^* \tag{2.3.3}$$

is the desired time reversal transformation in this case. Note that the problem is time reversal invariant since (2.3.1) and (2.3.2) are identical.

In the above case the time reversal transformation (2.3.3) is basically the complex conjugation map. To treat spin systems, we recall that the time reversal switches the sign of the velocity as $\boldsymbol{v} \to -\boldsymbol{v}$ and hence that of the angular momentum as $\boldsymbol{J} \to -\boldsymbol{J}$.[18] Since spin is a form of angular momentum, the desired time reversal map should induce the transformation $\hat{\boldsymbol{S}} \to -\hat{\boldsymbol{S}}$. We thus see that the time reversal map contains more than complex conjugation.

Single spin For clarity we start from the case of a single spin discussed in Sect. 2.1. We introduce the time-reversal map $\hat{\Theta}$ formally and discuss its properties. As we shall see finally in (2.3.14), the map $\hat{\Theta}$ indeed induce the desired transformation $\hat{\boldsymbol{S}} \to -\hat{\boldsymbol{S}}$.

Let $|\varphi\rangle = \sum_{\sigma=-S}^{S} c_\sigma|\psi^\sigma\rangle$ be an arbitrary state of spin S, where $c_\sigma \in \mathbb{C}$ are coefficients. We define the complex conjugation map $\hat{K}$ by its action on $|\varphi\rangle$ as

$$\hat{K}|\varphi\rangle = \sum_{\sigma=-S}^{S} (c_\sigma)^*|\psi^\sigma\rangle. \tag{2.3.4}$$

It should be noted that the map $\hat{K}$ is antilinear, i.e., $\hat{K}(\alpha|\varphi\rangle + \beta|\psi\rangle) = \alpha^*\hat{K}|\varphi\rangle + \beta^*\hat{K}|\psi\rangle$ for any $|\varphi\rangle, |\psi\rangle \in \mathfrak{h}_0$ and $\alpha, \beta \in \mathbb{C}$, and that the definition of $\hat{K}$ depends explicitly on the choice of basis states. See Appendix A.4.3 for basic conventions about antilinear operators (although we shall avoid using the general theory here).

Denoting, as before, the operator for the π rotation about the 2-axis as $\hat{u}_2 = e^{-i\pi\hat{S}^{(2)}}$, we define the time-reversal map by $\hat{\Theta} := \hat{u}_2\hat{K}$. Since the matrix elements of $\hat{u}_2$ are real as in (2.1.32)–(2.1.34), we have $\hat{\Theta} = \hat{K}\hat{u}_2$. From (2.1.34), the action of $\hat{\Theta}$ on the above $|\varphi\rangle$ is found to be

$$\hat{\Theta}|\varphi\rangle = \sum_{\sigma=-S}^{S} (-1)^{S+\sigma}(c_{-\sigma})^*|\psi^\sigma\rangle. \tag{2.3.5}$$

For $S = 1/2$, this means $\hat{\Theta}(a|\psi^\uparrow\rangle + b|\psi^\downarrow\rangle) = -b^*|\psi^\uparrow\rangle + a^*|\psi^\downarrow\rangle$, or, in the vector notation,

$$\hat{\Theta}\begin{pmatrix} a \\ b \end{pmatrix} = \begin{pmatrix} -b^* \\ a^* \end{pmatrix}. \tag{2.3.6}$$

The time-reversal map $\hat{\Theta}$ is also antilinear.[19]

---

[18] In (classical) mechanics the angular momentum $\boldsymbol{J}$ of a particle at $\boldsymbol{r}$ with velocity $\boldsymbol{v}$ is given by $\boldsymbol{J} = \boldsymbol{r} \times \boldsymbol{v}$.

[19] The product of a unitary operator and the complex-conjugation map, such as $\hat{\Theta}$, is called an antiunitary operator. See Appendix A.4.3 for more details.

The antilinear maps $\hat{K}$ and $\hat{\Theta}$ should be used with care since our "intuition" about linear operators almost always fails for these maps. For example, the easily verifiable relation[20]

$$\langle\varphi|\hat{K}|\psi\rangle = \langle\psi|\hat{K}|\varphi\rangle, \tag{2.3.7}$$

for any $|\varphi\rangle$ and $|\psi\rangle$ can never be valid for a linear operator. Also note that, unlike linear operators which may act either on ket or bra states, the antilinear maps like $\hat{K}$ and $\hat{\Theta}$ act only on ket states from the left.[21]

Let S be a half-odd-integer. Then we find from (2.3.5) that

$$\begin{aligned}\langle\varphi|\big(\hat{\Theta}|\varphi\rangle\big) &= \sum_{\sigma=-S}^{S}(-1)^{S+\sigma}(c_\sigma\,c_{-\sigma})^* \\ &= \sum_{\sigma=1/2}^{S}\big\{(-1)^{S+\sigma}+(-1)^{S-\sigma}\big\}(c_\sigma\,c_{-\sigma})^* = 0,\end{aligned} \tag{2.3.8}$$

for any $|\varphi\rangle$, where we noted that 2σ is always odd (and hence $(-1)^{S+\sigma}+(-1)^{S-\sigma}=0$) to get the final equality. Therefore any state $|\varphi\rangle$ and its time-reversal $\hat{\Theta}|\varphi\rangle$ are orthogonal. (See Problem 2.3.a below for a derivation from the general theory.) For an integer S, the states $|\varphi\rangle$ and $\hat{\Theta}|\varphi\rangle$ may or may not be orthogonal.

The inverse of $\hat{\Theta}$ is written as $\hat{\Theta}^{-1} := \hat{K}\hat{u}_2^\dagger$, or, more explicitly, by[22]

$$\hat{\Theta}^{-1}|\varphi\rangle = \sum_{\sigma=-S}^{S}(-1)^{S-\sigma}(c_{-\sigma})^*|\psi^\sigma\rangle. \tag{2.3.9}$$

By definition one has $\hat{\Theta}^{-1}\hat{\Theta}|\varphi\rangle = \hat{\Theta}\,\hat{\Theta}^{-1}|\varphi\rangle = |\varphi\rangle$ for any $|\varphi\rangle$. Also note that $\hat{\Theta}^{-1} = \hat{\Theta}$ if S is an integer, and $\hat{\Theta}^{-1} = -\hat{\Theta}$ if S is a half-odd-integer. Equivalently, we have $\hat{\Theta}^2 = \hat{1}$ if S is an integer, and $\hat{\Theta}^2 = -\hat{1}$ if S is a half-odd-integer.

Let $\hat{A}$ be a linear operator. By operating $\hat{\Theta}$, $\hat{A}$ and then $\hat{\Theta}^{-1}$ on an arbitrary state $|\varphi\rangle = \sum_{\sigma=-S}^{S} c_\sigma|\psi^\sigma\rangle$, we get

$$\begin{aligned}\hat{\Theta}^{-1}\big(\hat{A}(\hat{\Theta}|\varphi\rangle)\big) &= \hat{K}\Big(\hat{u}_2^\dagger\,\hat{A}\,\hat{u}_2\,\hat{K}\big(\sum_\sigma c_\sigma|\psi^\sigma\rangle\big)\Big) = \hat{K}\Big(\hat{u}_2^\dagger\,\hat{A}\,\hat{u}_2\sum_\sigma(c_\sigma)^*|\psi^\sigma\rangle\Big) \\ &= \hat{K}\Big(\sum_{\tau,\tau',\tau'',\sigma}(c_\sigma)^*|\psi^{\tau''}\rangle\langle\psi^{\tau''}|\hat{u}_2^\dagger|\psi^{\tau'}\rangle\langle\psi^{\tau'}|\hat{A}|\psi^\tau\rangle\langle\psi^\tau|\hat{u}_2|\psi^\sigma\rangle\Big) \\ &= \sum_{\tau,\tau',\tau'',\sigma} c_\sigma|\psi^{\tau''}\rangle\langle\psi^{\tau''}|\hat{u}_2^\dagger|\psi^{\tau'}\rangle\big(\langle\psi^{\tau'}|\hat{A}|\psi^\tau\rangle\big)^*\,\langle\psi^\tau|\hat{u}_2|\psi^\sigma\rangle \\ &= \Big\{\sum_{\tau,\tau'}\hat{u}_2^\dagger|\psi^{\tau'}\rangle\big(\langle\psi^{\tau'}|\hat{A}|\psi^\tau\rangle\big)^*\,\langle\psi^\tau|\hat{u}_2\Big\}|\varphi\rangle,\end{aligned} \tag{2.3.10}$$

---

[20] This relation can be interpreted as $\hat{K}^\dagger = \hat{K}$. See Appendix A.4.3.

[21] Let $\langle\varphi|$ and $\hat{A}$ be an arbitrary bra state and a linear operator. Then there exists a bra state $\langle\xi|$ such that $\langle\varphi|(\hat{A}|\psi\rangle) = \langle\xi|\psi\rangle$ for any $|\psi\rangle$. We therefore write $\langle\xi| = \langle\varphi|\hat{A}$. But note, e.g., that there are no $\langle\varphi|$ and $\langle\xi|$ such that $\langle\varphi|(\hat{K}|\psi\rangle) = \langle\xi|\psi\rangle$ for any $|\psi\rangle$.

[22] One can also write $\hat{\Theta}^{-1}$ as $\hat{\Theta}^\dagger$. See Appendix A.4.3.

where we noted that the matrix elements of $\hat{u}_2$ and $\hat{u}_2^\dagger$ are real. Note that, in the rightmost hand side, the operator inside $\{\cdots\}$ is a linear operator. Therefore operating $\hat{\Theta}$, $\hat{A}$ and $\hat{\Theta}^{-1}$ on an arbitrary ket is equivalent to operating the linear operator

$$\hat{\Theta}^{-1}\hat{A}\hat{\Theta} = \sum_{\sigma,\tau=-S}^{S} \hat{u}_2^\dagger|\psi^\sigma\rangle\big(\langle\psi^\sigma|\hat{A}|\psi^\tau\rangle\big)^* \langle\psi^\tau|\hat{u}_2 \tag{2.3.11}$$

on it.[23] We call $\hat{\Theta}^{-1}\hat{A}\hat{\Theta}$ the time-reversal of $\hat{A}$. Since $\hat{\Theta}^{-1}\hat{A}\hat{\Theta}$ is a linear operator, one can use it in the usual manner, e.g., operating it on a bra from the right. We also have the expected relation $\hat{\Theta}^{-1}(\hat{A}\hat{B})\hat{\Theta} = (\hat{\Theta}^{-1}\hat{A}\hat{\Theta})(\hat{\Theta}^{-1}\hat{B}\hat{\Theta})$. Since $\hat{\Theta}^{-1} = \pm\hat{\Theta}$, the operator $\hat{\Theta}\hat{A}\hat{\Theta}^{-1}$ is identical to $\hat{\Theta}^{-1}\hat{A}\hat{\Theta}$.

It is useful to note that (2.3.11) implies $\hat{\Theta}^{-1}\hat{A}\hat{\Theta} = \hat{u}_2^\dagger\,\hat{A}\,\hat{u}_2$ if all the matrix elements $\langle\psi^{\tau'}|\hat{A}|\psi^\tau\rangle$ are real, and $\hat{\Theta}^{-1}\hat{A}\hat{\Theta} = -\hat{u}_2^\dagger\,\hat{A}\,\hat{u}_2$ if all the matrix elements are pure imaginary.

From (2.1.2) and (2.1.3), we see that $\hat{S}^\pm$ and $\hat{S}^{(3)}$ have only real matrix elements. This means that $\hat{S}^{(1)}$ also has real matrix elements, while $\hat{S}^{(2)}$ has purely imaginary matrix elements. We thus find

$$\hat{\Theta}^{-1}\,\hat{S}^{(1)}\hat{\Theta} = \hat{u}_2^\dagger\,\hat{S}^{(1)}\,\hat{u}_2 = -\hat{S}^{(1)}, \quad \hat{\Theta}^{-1}\,\hat{S}^{(3)}\hat{\Theta} = \hat{u}_2^\dagger\,\hat{S}^{(3)}\,\hat{u}_2 = -\hat{S}^{(3)}, \tag{2.3.12}$$

where we used (2.1.21), and

$$\hat{\Theta}^{-1}\,\hat{S}^{(2)}\hat{\Theta} = -\hat{u}_2^\dagger\,\hat{S}^{(2)}\,\hat{u}_2 = -\hat{S}^{(2)}. \tag{2.3.13}$$

To sum, we have

$$\hat{\Theta}^{-1}\,\hat{S}^{(\alpha)}\hat{\Theta} = -\hat{S}^{(\alpha)}, \tag{2.3.14}$$

for any $\alpha = 1, 2, 3$. This is the desired property for the time reversal map that we have discussed in the beginning of the present section.

Multiple spins The above definition and discussion can be generalized to systems of multiple spins in a straightforward manner. Consider a general quantum spin system with spin S on lattice Λ as in Sect. 2.2. Take an arbitrary state $|\Phi\rangle = \sum_\sigma c_\sigma|\Psi^\sigma\rangle$, where $c_\sigma \in \mathbb{C}$ are coefficients, and the basis states $|\Psi^\sigma\rangle$ are defined in (2.2.1). Then the complex conjugation map $\hat{K}$ is again defined by $\hat{K}|\Phi\rangle = \sum_\sigma (c_\sigma)^*|\Psi^\sigma\rangle$, and the time-reversal map $\hat{\Theta}$ and its inverse $\hat{\Theta}^{-1}$ by

$$\hat{\Theta} := \hat{U}_2\,\hat{K} = \hat{K}\,\hat{U}_2, \quad \hat{\Theta}^{-1} = \hat{K}\,\hat{U}_2^\dagger = \hat{U}_2^\dagger\,\hat{K}, \tag{2.3.15}$$

where $\hat{U}_2 := \hat{U}_\pi^{(2)} = \exp[-i\pi\hat{S}_{\mathrm{tot}}^{(2)}]$ is the operation for the global π rotation about the 2-axis. The explicit action on the above $|\Phi\rangle$ is given by

---

[23]The linearity is a consequence of the general fact that the product of two antilinear operators is linear. See Appendix A.4.3.

$$\hat{\Theta}|\Phi\rangle = \sum_{\sigma}\Bigl\{\prod_{x\in\Lambda}(-1)^{S+\sigma_x}\Bigr\}(c_{-\sigma})^*|\Psi^{\sigma}\rangle, \tag{2.3.16}$$

$$\hat{\Theta}^{-1}|\Phi\rangle = \sum_{\sigma}\Bigl\{\prod_{x\in\Lambda}(-1)^{S-\sigma_x}\Bigr\}(c_{-\sigma})^*|\Psi^{\sigma}\rangle, \tag{2.3.17}$$

where $-\boldsymbol{\sigma}$ denotes the spin configuration $(-\sigma_x)_{x\in\Lambda}$. Note that the maps $\hat{\Theta}$ and $\hat{\Theta}^{-1}$ are non-local in the sense that they are defined for the whole system, not as a product of local maps. We again see that $\hat{\Theta}^{-1} = \hat{\Theta}$ (or, equivalently, $\hat{\Theta}^2 = \hat{1}$) if $|\Lambda|\,S$ is an integer, and $\hat{\Theta}^{-1} = -\hat{\Theta}$ (or $\hat{\Theta}^2 = -\hat{1}$) if $|\Lambda|\,S$ is a half-odd-integer.

Let $\hat{A}$ be an arbitrary linear operator. Exactly as (2.3.11) for a single spin, the time-reversal of $\hat{A}$ is given by

$$\hat{\Theta}^{-1}\hat{A}\hat{\Theta} = \hat{\Theta}\hat{A}\hat{\Theta}^{-1} = \sum_{\sigma,\tau}\hat{U}_2^{\dagger}|\Psi^{\sigma}\rangle\bigl(\langle\Psi^{\sigma}|\hat{A}|\Psi^{\tau}\rangle\bigr)^*\,\langle\Psi^{\tau}|\hat{U}_2, \tag{2.3.18}$$

which is a linear operator. It is easy to check that

$$\hat{\Theta}^{-1}\hat{1}\hat{\Theta} = \hat{1}, \tag{2.3.19}$$

$$\hat{\Theta}^{-1}\hat{A}^{\dagger}\hat{\Theta} = (\hat{\Theta}^{-1}\hat{A}\hat{\Theta})^{\dagger}, \tag{2.3.20}$$

$$\hat{\Theta}^{-1}(\alpha\hat{A}+\beta\hat{B})\hat{\Theta} = \alpha^*(\hat{\Theta}^{-1}\hat{A}\hat{\Theta}) + \beta^*(\hat{\Theta}^{-1}\hat{B}\hat{\Theta}), \tag{2.3.21}$$

and

$$\hat{\Theta}^{-1}(\hat{A}\hat{B})\hat{\Theta} = (\hat{\Theta}^{-1}\hat{A}\hat{\Theta})(\hat{\Theta}^{-1}\hat{B}\hat{\Theta}), \tag{2.3.22}$$

for any linear operators $\hat{A}$, $\hat{B}$, and $\alpha, \beta \in \mathbb{C}$.

Although the maps $\hat{\Theta}$ and $\hat{\Theta}^{-1}$ are non-local, the time-reversal (2.3.18) of an operator is local in the sense that it is determined by only considering the support (i.e., the part of the lattice on which the operator acts nontrivially) of the operator. In particular if $\hat{A}$ acts on a single site, then $\hat{\Theta}^{-1}\hat{A}\hat{\Theta}$ defined by (2.3.18) is exactly the same as that defined by (2.3.11). We thus see

$$\hat{\Theta}^{-1}\,\hat{S}_x^{(\alpha)}\hat{\Theta} = -\hat{S}_x^{(\alpha)}, \tag{2.3.23}$$

for any $\alpha = 1, 2, 3$ and $x \in \Lambda$.

In fact the map from a linear operator $\hat{A}$ to its time-reversal $\hat{\Theta}^{-1}\hat{A}\hat{\Theta}$ is fully determined by the general properties (2.3.19), (2.3.20), (2.3.21), and (2.3.22), with the specific rules (2.3.23). Note that this map is independent of the choice of the basis states, while the map $\hat{\Theta}$ is basis-dependent.[24]

---

[24] There is indeed a deep reason behind this observation. Define a map $\Gamma_{\rm tr}$ by $\Gamma_{\rm tr}(\hat{S}_x^{(\alpha)}) = -\hat{S}_x^{(\alpha)}$ for any $x \in \Lambda$ and $\alpha = 1, 2, 3$, $\Gamma_{\rm tr}(\hat{A}^{\dagger}) = \Gamma_{\rm tr}(\hat{A})^{\dagger}$, $\Gamma_{\rm tr}(\alpha\hat{A}+\beta\hat{B}) = \alpha^*\Gamma_{\rm tr}(\hat{A}) + \beta^*\Gamma_{\rm tr}(\hat{B})$, and $\Gamma_{\rm tr}(\hat{A}\hat{B}) = \Gamma_{\rm tr}(\hat{A})\Gamma_{\rm tr}(\hat{B})$ for any operators $\hat{A}$, $\hat{B}$ and $\alpha, \beta \in \mathbb{C}$. Such a map is called an antilinear $*$-automorphism. Then a general result known as Wigner's theorem guarantees that there exists an antiunitary operator $\hat{\Theta}$ such that $\Gamma_{\rm tr}(\hat{A}) = \hat{\Theta}^{-1}\hat{A}\hat{\Theta}$ for any $\hat{A}$. See Appendix A.6.

Time-reversed Schrödinger equation Let us observe that the time-reversal map $\hat{\Theta}$ indeed generates a time-dependent state which evolves backward in time. Consider any spin system with Hamiltonian $\hat{H}$, and let $|\Phi(t)\rangle$ be a solution of the time-dependent Schrödinger equation

$$i\frac{d}{dt}|\Phi(t)\rangle = \hat{H}|\Phi(t)\rangle. \tag{2.3.24}$$

Applying the map $\hat{\Theta}$, we get

$$-i\frac{d}{dt}\hat{\Theta}|\Phi(t)\rangle = \hat{\Theta}\hat{H}|\Phi(t)\rangle = \hat{\Theta}\hat{H}\hat{\Theta}^{-1}\hat{\Theta}|\Phi(t)\rangle = \hat{H}_{\mathrm{R}}\hat{\Theta}|\Phi(t)\rangle, \tag{2.3.25}$$

where we defined the time-reversed Hamiltonian as

$$\hat{H}_{\mathrm{R}} := \hat{\Theta}\hat{H}\hat{\Theta}^{-1} = \hat{\Theta}^{-1}\hat{H}\hat{\Theta}. \tag{2.3.26}$$

Let the time-reversed state be

$$|\Phi_{\mathrm{R}}(s)\rangle = \hat{\Theta}|\Phi(-s)\rangle, \tag{2.3.27}$$

which is nothing but $\hat{\Theta}|\Phi(t)\rangle$ expressed as a function of the reversed time variable $s = -t$. Then (2.3.25) is rewritten as

$$i\frac{d}{ds}|\Phi_{\mathrm{R}}(s)\rangle = \hat{H}_{\mathrm{R}}|\Phi_{\mathrm{R}}(s)\rangle, \tag{2.3.28}$$

which is the Schrödinger equation for the Hamiltonian $\hat{H}_{\mathrm{R}}$ and the reversed time s. Unlike in the simple example where the original Schrödinger equation (2.3.1) and the time-reversed equation (2.3.2) are identiccal, the time-reversed equation (2.3.28) is in general different from the original (2.3.24).

As a very simple example, take the Hamiltonian $\hat{H} = -\boldsymbol{B}\cdot\sum_{x\in\Lambda}\hat{\boldsymbol{S}}_x$, which describes non-interacting spins under a uniform constant magnetic field $\boldsymbol{B} \in \mathbb{R}^3$. From (2.3.23), we find that the time-reversed Hamiltonian is $\hat{H}_{\mathrm{R}} = \boldsymbol{B}\cdot\sum_{x\in\Lambda}\hat{\boldsymbol{S}}_x$. The effect of time-reversal is to reverse the direction of the magnetic field.[25]

Kramers degeneracy Here we only consider a half-odd-integral S, and the lattice Λ with an odd number of sites. This means that $|\Lambda|S$ is a half-odd integer. In this case we can repeat the argument in (2.3.8) to find for any $|\Phi\rangle$ that

[25]This conclusion is intuitively understood if we assume that the magnetic field is generated by an electromagnet. Then the time-reversal changes the direction of the electric current, and hence reverses the magnetic field.

$$\langle\Phi|(\hat{\Theta}|\Phi\rangle) = \sum_{\sigma}\Bigl\{\prod_{x\in\Lambda}(-1)^{S+\sigma_x}\Bigr\}(c_\sigma\, c_{-\sigma})^*$$

$$= \sum_{\substack{\sigma \\ (\sum_{x\in\Lambda}\sigma_x>0)}} \Bigl\{\prod_{x\in\Lambda}(-1)^{S+\sigma_x} + \prod_{x\in\Lambda}(-1)^{S-\sigma_x}\Bigr\}(c_\sigma\, c_{-\sigma})^* = 0, \quad (2.3.29)$$

where we again noted that $2\sum_{x\in\Lambda}\sigma_x$ is always odd. Thus an arbitrary state $|\Phi\rangle$ and its time-reversal $\hat{\Theta}|\Phi\rangle$ are orthogonal. (See Problem 2.3.a below for a derivation from the general theory.)

We now assume that the Hamiltonian $\hat{H}$ of the spin system has time-reversal symmetry in the sense that

$$\hat{H} = \hat{\Theta}^{-1}\hat{H}\hat{\Theta}. \quad (2.3.30)$$

Let $|\Phi\rangle$ be an arbitrary eigenstate of $\hat{H}$, i.e, $\hat{H}|\Phi\rangle = E|\Phi\rangle$. Then we have

$$\hat{H}\hat{\Theta}|\Phi\rangle = \hat{\Theta}\hat{\Theta}^{-1}\hat{H}\hat{\Theta}|\Phi\rangle = \hat{\Theta}\hat{H}|\Phi\rangle = E\hat{\Theta}|\Phi\rangle, \quad (2.3.31)$$

which shows that $\hat{\Theta}|\Phi\rangle$ is also an eigenstate of $\hat{H}$ with the same eigenvalue E. Since $|\Phi\rangle$ and $\hat{\Theta}|\Phi\rangle$ are orthogonal, they form a pair of degenerate energy eigenstates. Such a pair of energy eigenstates is called a Kramers doublet. We thus find that all the energy eigenvalues of $\hat{H}$ are even-fold degenerate. The degeneracy was first noted by Kramers, and later explained by Wigner as a consequence of time-reversal symmetry.

A necessary and sufficient condition for a Hamiltonian $\hat{H}$ to have time-reversal symmetry is that it is a sum of products of even numbers of spin operators. This is easily confirmed by using the basic rules (2.3.22) and (2.3.23) of time-reversal.

The Kramers degeneracy is usually relevant for small quantum spin systems. For example consider an $S = 1/2$ system on $\Lambda = \{1, 2, 3\}$ with Hamiltonian

$$\hat{H} = J_1\hat{S}_1^{(1)}\hat{S}_2^{(1)} + J_2\hat{S}_2^{(2)}\hat{S}_3^{(2)} + J_3\hat{S}_3^{(3)}\hat{S}_1^{(3)}, \quad (2.3.32)$$

where $J_1, J_2, J_3 \in \mathbb{R}$ are arbitrary parameters. Then without performing any calculations, we know, for example, that the ground states are at least two-fold degenerate.

Problem 2.3.a Here we use the notations in Appendix A.4.3. Let V be an antiunitary operator such that $\mathsf{V}^2 = -\mathsf{I}$. Then prove that $\langle \boldsymbol{v}, \mathsf{V}\boldsymbol{v}\rangle = 0$ for any $\boldsymbol{v}$. Note that (2.3.8) and (2.3.29) are special cases of this result. [solution→p.496]

2.4 The Ferromagnetic Heisenberg Model

As a warmup, we shall examine the ground states of the ferromagnetic Heisenberg model.

Definition of the model Let Λ be an arbitrary lattice, i.e., a finite set. A bond $\{x, y\}$ in Λ is a set of two sites $x, y \in \Lambda$ such that $x \neq y$. Note that we identify $\{x, y\}$ with $\{y, x\}$. Let $\mathscr{B}$ be a collection of bonds in Λ. The pair $(\Lambda, \mathscr{B})$ determines the structure of the lattice.[26]

Consider a spin system on Λ with spin S. The Hamiltonian of the ferromagnetic Heisenberg model is

$$\hat{H} = -\sum_{\{x,y\}\in\mathscr{B}} \hat{\boldsymbol{S}}_x \cdot \hat{\boldsymbol{S}}_y. \tag{2.4.1}$$

The minus sign indicates that the Hamiltonian favors configurations where spins (connected by bonds) align with each other. From (2.2.17) one readily finds that

$$[\hat{H}, \hat{S}^{(\alpha)}_{\text{tot}}] = 0, \tag{2.4.2}$$

for any $\alpha = 1, 2, 3$. This means that the Hamiltonian is SU(2) invariant; it satisfies $[\hat{H}, \hat{U}^{(\alpha)}_\theta] = 0$, which is equivalent to

$$(\hat{U}^{(\alpha)}_\theta)^\dagger \hat{H} \hat{U}^{(\alpha)}_\theta = \hat{H}, \tag{2.4.3}$$

for any $\theta \in \mathbb{R}$ and $\alpha = 1, 2, 3$ as in (2.2.13).

Fully polarized ground states Consider the "all-up state"

$$|\Phi^\uparrow\rangle := \bigotimes_{x\in\Lambda} |\psi^S_x\rangle, \tag{2.4.4}$$

in which all the spins in Λ are pointing in the positive 3-direction. We note that this is the unique state which satisfies $(\hat{\boldsymbol{S}}_{\text{tot}})^2|\Phi^\uparrow\rangle = S_{\max}(S_{\max}+1)|\Phi^\uparrow\rangle$ and $\hat{S}^{(3)}_{\text{tot}}|\Phi^\uparrow\rangle = S_{\max}|\Phi^\uparrow\rangle$, with $S_{\max} = |\Lambda|\, S$ being the largest possible magnitude of the total spin. The ferromagnetic nature of the Hamiltonian suggests that $|\Phi^\uparrow\rangle$ is a ground state. This fact can be easily proved as follows by noting that $|\Phi^\uparrow\rangle$ minimizes each term in the Hamiltonian separately.

By using the expression (2.2.16), we find

$$\begin{aligned} -\hat{\boldsymbol{S}}_x \cdot \hat{\boldsymbol{S}}_y|\Phi^\uparrow\rangle &= -\Big\{\frac{1}{2}(\hat{S}^+_x\hat{S}^-_y + \hat{S}^-_x\hat{S}^+_y) + \hat{S}^{(3)}_x\hat{S}^{(3)}_y\Big\}|\Phi^\uparrow\rangle \\ &= -\hat{S}^{(3)}_x\hat{S}^{(3)}_y|\Phi^\uparrow\rangle = -S^2|\Phi^\uparrow\rangle, \end{aligned} \tag{2.4.5}$$

where we used (2.1.2) and (2.1.3), in particular $\hat{S}^+|\psi^S\rangle = 0$, about the action of the spin operators on the basis states. Note that $-S^2$ is the minimum eigenvalue of the operator $-\hat{\boldsymbol{S}}_x \cdot \hat{\boldsymbol{S}}_y$. It then follows immediately from Lemma A.9 (p. 469) about "frustration-free Hamiltonians" that $|\Phi^\uparrow\rangle$ is a ground state and that the ground state energy of $\hat{H}$ is $E_{\text{GS}} = -|\mathscr{B}|S^2$.

---

[26]In the language of graph theory, Λ is the set of vertices and $\mathscr{B}$ is the set of edges.

The ground state $|\Phi^{\uparrow}\rangle$ is a fully polarized ferromagnetic state in which the total spin is pointing in the direction $(0, 0, 1)$. One can construct other ground states by rotating $|\Phi^{\uparrow}\rangle$. Take θ and φ such that $0 \le \theta \le \pi$ and $0 \le \varphi < 2\pi$, and let $\boldsymbol{n}(\theta, \varphi) = (\cos\varphi \sin\theta, \sin\varphi \sin\theta, \cos\theta)$ be the standard unit vector. We then define

$$|\Xi_{\theta,\varphi}\rangle := \hat{U}^{(3)}_{\varphi}\hat{U}^{(2)}_{\theta}|\Phi^{\uparrow}\rangle, \tag{2.4.6}$$

which indeed is the ground state in which the total spin is pointing in the direction $\boldsymbol{n}(\theta, \varphi)$, as we shall see below. We note that $|\Xi_{\theta,\varphi}\rangle$ are physical ground states, which are close to states realizable experimentally in ferromagnets at very low temperatures.[27] The state $|\Xi_{\theta,\varphi}\rangle$ is sometimes called the spin coherent state (within the space of ferromagnetic states). See Footnote 7 in p. 18.

To confirm that $|\Xi_{\theta,\varphi}\rangle$ is a ground state, it suffices to note that (2.4.3) implies

$$\hat{H}|\Xi_{\theta,\varphi}\rangle = \hat{H}\hat{U}^{(3)}_{\varphi}\hat{U}^{(2)}_{\theta}|\Phi^{\uparrow}\rangle = \hat{U}^{(3)}_{\varphi}\hat{U}^{(2)}_{\theta}\hat{H}|\Phi^{\uparrow}\rangle = E_{\mathrm{GS}}|\Xi_{\theta,\varphi}\rangle. \tag{2.4.7}$$

To see that it is pointing in the desired direction, we note that (2.1.20) implies $(\hat{U}^{(3)}_{\varphi}\hat{U}^{(2)}_{\theta})\,\hat{S}^{(3)}_{\mathrm{tot}}\,(\hat{U}^{(3)}_{\varphi}\hat{U}^{(2)}_{\theta})^{\dagger} = \hat{\boldsymbol{S}}_{\mathrm{tot}} \cdot \boldsymbol{n}(\theta, \varphi)$. Then we have

$$\hat{\boldsymbol{S}}_{\mathrm{tot}} \cdot \boldsymbol{n}(\theta, \varphi)|\Xi_{\theta,\varphi}\rangle = \hat{U}^{(3)}_{\varphi}\hat{U}^{(2)}_{\theta}\hat{S}^{(3)}_{\mathrm{tot}}|\Phi^{\uparrow}\rangle = S_{\mathrm{max}}|\Xi_{\theta,\varphi}\rangle. \tag{2.4.8}$$

Recall that we used the same argument in the solution to Problem 2.1.d (p. 18).

The space of ground states The ground states $|\Xi_{\theta,\varphi}\rangle$ with $0 \le \theta \le \pi$ and $0 \le \varphi < 2\pi$ provide a complete description of the whole space of ground states. A more standard way to characterize the space of ground states is to consider states which are eigenstates of $\hat{S}^{(3)}_{\mathrm{tot}}$.

For $M = -|\Lambda|S, -|\Lambda|S + 1, \ldots, |\Lambda|S - 1, |\Lambda|S$, let

$$|\Phi_M\rangle := \frac{(\hat{S}^{-}_{\mathrm{tot}})^{|\Lambda|S-M}|\Phi^{\uparrow}\rangle}{\left\|(\hat{S}^{-}_{\mathrm{tot}})^{|\Lambda|S-M}|\Phi^{\uparrow}\rangle\right\|}, \tag{2.4.9}$$

which obviously satisfies $\hat{S}^{(3)}_{\mathrm{tot}}|\Phi_M\rangle = M|\Phi_M\rangle$ (which is equivalent to $|\Phi_M\rangle \in \mathscr{H}_M$). Since $[\hat{H}, \hat{S}^{-}_{\mathrm{tot}}] = 0$ we also see that $\hat{H}|\Phi_M\rangle = E_{\mathrm{GS}}|\Phi_M\rangle$, and hence $|\Phi_M\rangle$ is also a ground state. (See Theorem A.16 (p. 473) in Appendix A.3 for the corresponding general result.) It can be further shown that these are the only ground states when the lattice is connected.[28]

[27] To be precise this can be true only in three or higher dimensions if the system is macroscopically large. In one or two dimensions, it is known from the Hohenberg-Mermin-Wagner theorem (Theorem 4.24 in p. 124) that the ferromagnetic Heisenberg model does not exhibit ferromagnetic order at nonzero temperatures.

[28] A lattice $(\Lambda, \mathscr{B})$ is connected if for any $x, y \in \Lambda$ such that $x \ne y$, there exists a finite sequence $z_1, z_2, \ldots, z_n \in \Lambda$ with the properties $z_1 = x$, $z_n = y$, and $\{z_j, z_{j+1}\} \in \mathscr{B}$ for $j = 1, \ldots, n-1$.

Theorem 2.1 *Assume that the lattice* $(\Lambda, \mathscr{B})$ *is connected. Then any state* $|\Phi\rangle$ *such that* $\hat{H}|\Phi\rangle = E_{\mathrm{GS}}|\Phi\rangle$ *is written as*

$$|\Phi\rangle = \sum_{M=-|\Lambda|S}^{|\Lambda|S} c_M |\Phi_M\rangle, \tag{2.4.10}$$

with suitable coefficients $c_M \in \mathbb{C}$.

See Problem 2.4.a below for a proof. We also note that (2.4.10) are the only states which have the maximum total spin $S_{\max} = |\Lambda|S$, i.e., states which satisfy $(\hat{S}_{\mathrm{tot}})^2|\Phi\rangle = S_{\max}(S_{\max}+1)|\Phi\rangle$.

In a system with $S = 1/2$, where one has $\sigma_x = \pm 1/2$, the ferromagnetic state (or, equivalently, the state with maximum possible total spin $S_{\max} = |\Lambda|/2$) (2.4.9) has a simple expression (2.4.11). To see this, recall that $|\Phi^{\uparrow}\rangle = \bigotimes_{x\in\Lambda} |\psi_x^{\uparrow}\rangle$, and that $\hat{S}_{\mathrm{tot}}^- = \sum_{x\in\Lambda} \hat{S}_x^-$ acts on the state according to (2.1.5). We then see that $|\Phi_M\rangle$ is a superposition of all basis states $|\Psi^{\sigma}\rangle$ such that $\sum_{x\in\Lambda} \sigma_x = M$ with exactly the same weights. By properly normalizing the state, we find

$$|\Phi_M\rangle = \sqrt{\frac{(S_{\max}+M)!\,(S_{\max}-M)!}{(2S_{\max})!}} \sum_{\substack{\sigma \\ (\overline{\sigma}=M)}} |\Psi^{\sigma}\rangle. \tag{2.4.11}$$

Problem 2.4.a Prove Theorem 2.1. The proof uses an idea presented in the Proof of Theorem 2.2. [solution→p.496]

Problem 2.4.b Express the ground state $|\Phi_M\rangle$ for a given M in terms of the ground states $|\Xi_{\theta,\varphi}\rangle$. [solution→p.496]

Problem 2.4.c Express the ground state $|\Xi_{\theta,\varphi}\rangle$ for given θ and φ in terms of the ground states $|\Phi_M\rangle$. It is enough to solve the case with $S = 1/2$ since a general spin can be represented by using $S = 1/2$ spins. (Hint: Use (2.4.11) and the result of Problem 2.1.d in p. 18.) [solution→p.497]

Spin wave excitations It is also possible to find a class of simple excited states, which are usually called spin wave excitations.

For $x \in \Lambda$ define $\mathcal{N}(x) := \{y \in \Lambda \,|\, \{x, y\} \in \mathscr{B}\}$, which is the set of sites directly connected to x. Let $\boldsymbol{f} = (f_x)_{x\in\Lambda}$ (with $f_x \in \mathbb{C}$) and $\lambda \in \mathbb{R}$ be solutions of the eigenvalue equation (or the "Schrödinger equation")

$$-\sum_{y\in\Lambda} \Delta_{x,y} f_y = \lambda f_x \quad \text{for each } x \in \Lambda, \tag{2.4.12}$$

where the lattice Laplacian $\boldsymbol{\Delta}$, a real symmetric $|\Lambda| \times |\Lambda|$ matrix, is defined by

$$\Delta_{x,y} = \begin{cases} -|\mathcal{N}(x)| & \text{if } x = y, \\ 1 & \text{if } \{x, y\} \in \mathcal{B}, \\ 0 & \text{otherwise.} \end{cases} \tag{2.4.13}$$

We note that there is a simple (but important) relation

$$\sum_{x,y\in\Lambda} (g_x)^* \, \Delta_{x,y} \, g_y = - \sum_{\{x,y\}\in\mathcal{B}} |g_x - g_y|^2, \tag{2.4.14}$$

for arbitrary $g_x \in \mathbb{C}$ with $x \in \Lambda$. See Problem 2.4.d below. Multiplying (2.4.12) with f_x^* and summing over $x \in \Lambda$, one finds

$$\lambda \sum_{x\in\Lambda} |f_x|^2 = \sum_{\{x,y\}\in\mathcal{B}} |f_x - f_y|^2, \tag{2.4.15}$$

where we used (2.4.14). We therefore see that any eigenvalue satifies $\lambda \geq 0$. Noting that the left-hand side of (2.4.12) is written as $\sum_{y\in\mathcal{N}(x)}(f_x - f_y)$, one readily finds that $\boldsymbol{f}$ such that $f_x = \text{constant}$ is an eigenvector satisfying (2.4.12) with $\lambda = 0$. When the lattice $(\Lambda, \mathcal{B})$ is connected, it follows from (2.4.15) that this is the unique eigenvector with $\lambda = 0$, and the eigenvalues of all the remaining $|\Lambda| - 1$ eigenvectors satisfy $\lambda > 0$.[29]

Problem 2.4.d Derive (2.4.14). This is not difficult. The reader unfamiliar with the relation is encouraged to derive it. [solution→p.497]

As an example take the one-dimensional lattice $\Lambda = \{1, 2, \ldots, L\}$ and let $\mathcal{B} = \{\{x, x+1\} \,|\, x = 1, \ldots, L\}$, where we use periodic boundary conditions and identify $L + 1$ with 1. Then the eigenvalue equation (2.4.12) becomes

$$2f_x - f_{x+1} - f_{x-1} = \lambda f_x, \tag{2.4.16}$$

whose solutions are

$$f_x^{(k)} = \frac{1}{\sqrt{L}} e^{ikx}, \quad \lambda^{(k)} = 2(1 - \cos k), \tag{2.4.17}$$

where $k = (2\pi/L)n$ with $n = 0, 1, \ldots, L - 1$ labels each eigenvector and eigenvalue.

Let us see how this eigenvalue problem is related to the spin system. Going back to the general lattice Λ, we define for each eigenvector $\boldsymbol{f}$ satisfying (2.4.12) the corresponding state

$$|\Phi^{\boldsymbol{f}}\rangle := \sum_{x\in\Lambda} f_x \, |\Gamma_x\rangle, \tag{2.4.18}$$

---

[29]This fact follow also from the Perron–Frobenius theorem (Theorem A.18 in p. 475).

where

$$|\Gamma_x\rangle = \Big(\bigotimes_{y\in\Lambda\backslash\{x\}} |\psi_y^S\rangle\Big) \otimes |\psi_x^{S-1}\rangle = \frac{1}{\sqrt{2S}}\,\hat{S}_x^-|\Phi^\uparrow\rangle \tag{2.4.19}$$

is the state obtained by lowering the spin at x from the all-up state (2.4.4). $|\Gamma_x\rangle$ is sometimes called a state with a single magnon at site x. We shall show below that

$$\hat{H}|\Phi^f\rangle = (E_{\rm GS} + \lambda S)|\Phi^f\rangle, \tag{2.4.20}$$

and hence $|\Phi^f\rangle$ is an exact energy eigenstate. Note that $|\Phi^f\rangle$ for constant f_x (for which $\lambda = 0$) is identical to $|\Phi_M\rangle$ of (2.4.9) with $M = |\Lambda|S - 1$, and is a ground state. The eigenstate $|\Phi^f\rangle$ corresponding to $\boldsymbol{f}$ with nonvanishing λ is an excited state, and called a spin wave state or a single magnon excited state. The definition (2.4.18) of $|\Phi^f\rangle$ suggests that f_x can be interpreted as a wave function of a magnon.

Let us show (2.4.20). From (2.1.2) we easily sees that

$$\begin{aligned}\Big(-\sum_{\{u,v\}\in\mathscr{B}} \hat{S}_u^{(3)}\hat{S}_v^{(3)}\Big)|\Gamma_x\rangle &= -\Big\{\big(|\mathscr{B}| - |\mathscr{N}(x)|\big)S^2 + |\mathscr{N}(x)|\,S(S-1)\Big\}|\Gamma_x\rangle \\ &= \big\{-|\mathscr{B}|\,S^2 + |\mathscr{N}(x)|\,S\big\}|\Gamma_x\rangle.\end{aligned} \tag{2.4.21}$$

Noting that (2.1.3) implies $\hat{S}^+|\psi^{S-1}\rangle = \sqrt{2S}|\psi^S\rangle$ and $\hat{S}^-|\psi^S\rangle = \sqrt{2S}|\psi^{S-1}\rangle$, we also see that

$$\Big(-\sum_{\{u,v\}\in\mathscr{B}} \frac{1}{2}\{\hat{S}_u^+\hat{S}_v^- + \hat{S}_u^-\hat{S}_v^+\}\Big)|\Gamma_x\rangle = -S\sum_{y\in\mathscr{N}(x)}|\Gamma_y\rangle. \tag{2.4.22}$$

We then find from (2.4.1) and (2.2.16) that

$$\begin{aligned}(\hat{H} + |\mathscr{B}|\,S^2)\sum_{x\in\Lambda} f_x|\Gamma_x\rangle &= S\sum_{x\in\Lambda}|\mathscr{N}(x)|\,f_x|\Gamma_x\rangle - S\sum_{x\in\Lambda} f_x \sum_{y\in\mathscr{N}(x)}|\Gamma_y\rangle \\ &= S\sum_{x\in\Lambda}\sum_{y\in\mathscr{N}(x)}(f_x - f_y)|\Gamma_x\rangle \\ &= -S\sum_{x,y\in\Lambda}\Delta_{x,y}f_y|\Gamma_x\rangle = \lambda S\sum_{x\in\Lambda} f_x|\Gamma_x\rangle,\end{aligned} \tag{2.4.23}$$

which is the desired (2.4.20). To get the second line we used the symmetry between x and y as

$$\sum_{x\in\Lambda}\sum_{y\in\mathscr{N}(x)} f_x|\Gamma_y\rangle = \sum_{\substack{x,y\in\Lambda\\ \{x,y\}\in\mathscr{B}}} f_x|\Gamma_y\rangle = \sum_{\substack{x,y\in\Lambda\\ \{x,y\}\in\mathscr{B}}} f_y|\Gamma_x\rangle = \sum_{x\in\Lambda}\sum_{y\in\mathscr{N}(x)} f_y|\Gamma_x\rangle. \tag{2.4.24}$$

2.5 The Antiferromagnetic Heisenberg Model

We move onto the more interesting problem of antiferromagnetic spin systems.

Definition of the model Here we make an additional assumption that the lattice is bipartite. This assumption considerably simplifies the problem. A lattice is said to be bipartite if it can be decomposed into two sublattices, A and B, in such way that any bond connects sites in different sublattices. More precisely, the lattice $(\Lambda, \mathscr{B})$ is bipartite if $\Lambda = A \cup B$ with $A \cap B = \emptyset$, and for any $\{x, y\} \in \mathscr{B}$ one has either $x \in A$, $y \in B$ or $x \in B$, $y \in A$. There are many lattices which are not bipartite; the triangle lattice is a typical example. See Fig. 2.1.

Consider a spin system with spin S on a bipartite lattice $(\Lambda, \mathscr{B})$. The Hamiltonian of the antiferromagnetic Heisenberg model is

$$\hat{H} = \sum_{\{x,y\}\in\mathscr{B}} \hat{\boldsymbol{S}}_x \cdot \hat{\boldsymbol{S}}_y. \tag{2.5.1}$$

This Hamiltonian is of course SU(2) invariant, i.e., (2.4.2) and (2.4.3) hold for any $\theta \in \mathbb{R}$ and $\alpha = 1, 2, 3$.

The Néel state Intuitively it is expected that the energy of an antiferromagnetic system is minimized when the neighboring spins are pointing in the opposite directions. Such a spin configuration is possible in a bipartite lattice; one only needs to let all the spins in A sublattice point in the same direction, and all the spins in the B sublattice in the opposite direction.[30] This kind of order is known as the Néel order. See Fig. 2.2.

This motivates us to define the Néel state,

$$|\Phi_{\text{Néel}}\rangle := \Bigl(\bigotimes_{x\in A} |\psi_x^S\rangle\Bigr) \otimes \Bigl(\bigotimes_{y\in B} |\psi_y^{-S}\rangle\Bigr). \tag{2.5.2}$$

Given that the all-up state (2.4.4) is an exact ground state of the ferromagnetic Heisenberg model, the Néel state may seem to be a promising candidate for the ground state of the antiferromagnetic Heisenberg model.

This is not the case, and the situation is more subtle and interesting. Take $\{x, y\} \in \mathscr{B}$ and consider the state $|\psi_x^S\rangle|\psi_y^{-S}\rangle$ which is a part of $|\Phi_{\text{Néel}}\rangle$. By using the expression (2.2.16) for $\hat{\boldsymbol{S}}_x \cdot \hat{\boldsymbol{S}}_y$, we find

$$\begin{aligned}(\hat{\boldsymbol{S}}_x \cdot \hat{\boldsymbol{S}}_y)|\psi_x^S\rangle|\psi_y^{-S}\rangle &= \bigl\{\frac{1}{2}(\hat{S}_x^+\hat{S}_y^- + \hat{S}_x^-\hat{S}_y^+) + \hat{S}_x^{(3)}\hat{S}_y^{(3)}\bigr\}|\psi_x^S\rangle|\psi_y^{-S}\rangle \\ &= \hat{S}_x^{(3)}|\psi_x^S\rangle\,\hat{S}_y^{(3)}|\psi_y^{-S}\rangle + \frac{1}{2}\hat{S}_x^-|\psi_x^S\rangle\,\hat{S}_y^+|\psi_y^{-S}\rangle \\ &= -S^2|\psi_x^S\rangle|\psi_y^{-S}\rangle + S|\psi_x^{S-1}\rangle|\psi_y^{-S+1}\rangle,\end{aligned} \tag{2.5.3}$$

[30] Here we are in fact minimizing the classical Hamiltonian $H = \sum_{\{x,y\}\in\mathscr{B}} \vec{S}_x \cdot \vec{S}_y$. If the lattice is non-bipartite, even the classical problem of minimization becomes nontrivial. This is known as the problem of frustration.

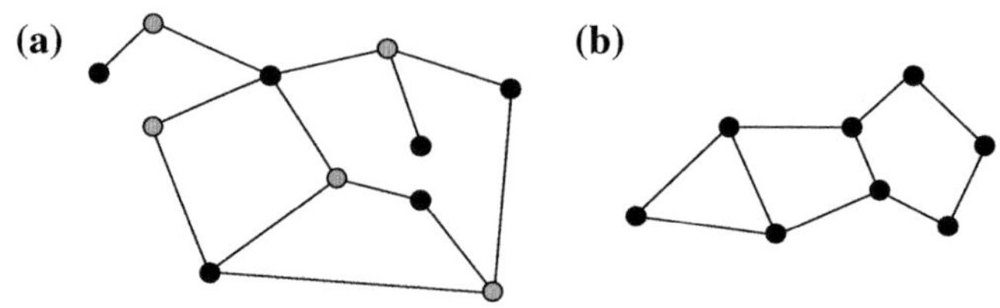

Fig. 2.1 **a** A bipartite lattice. Sites in two sublattices are distinguished by black and gray dots. **b** A non-bipartite lattice (© Hal Tasaki 2020. All Rights Reserved)

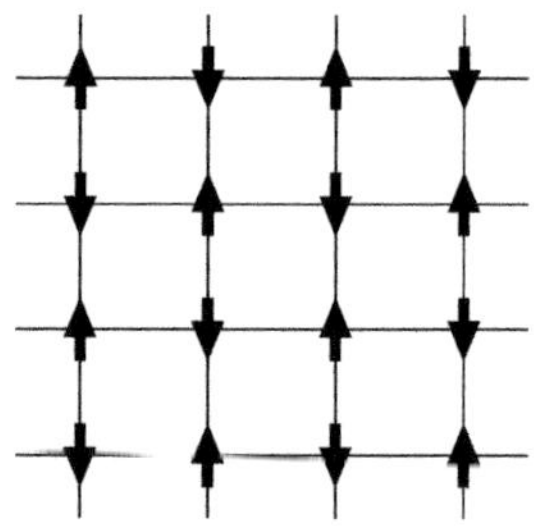

Fig. 2.2 Classical Néel order in the square lattice. Neighboring spins are pointing exactly the opposite directions (© Hal Tasaki 2020. All Rights Reserved)

where we used (2.1.2) and (2.1.3) to get the final expression. This should be compared with (2.4.5) for the ferromagnetic case, where the operator $\hat{S}_x^+\hat{S}_y^- + \hat{S}_x^-\hat{S}_y^+$ simply annihilates the all-up state $|\Phi^{\uparrow}\rangle$. Here the alternating spin configuration in the Néel state resulted in an extra contribution.

$\hat{H}|\Phi_{\text{Néel}}\rangle$ is computed by summing the right-hand side of (2.5.3). It is clear that the contribution $S|\psi_x^{S-1}\rangle|\psi_y^{-S+1}\rangle$ do not cancel, and one finds $\hat{H}|\Phi_{\text{Néel}}\rangle = -|\mathcal{B}|S^2\,|\Phi_{\text{Néel}}\rangle + O(|\mathcal{B}|S)$. The Néel state is not a ground state, not even an eigenstate.[31]

Let E_{GS} be the ground state energy of the Hamiltonian (2.5.1). From the variational principle, we get a simple (but rigorous) upper bound $E_{\text{GS}} \le \langle\Phi_{\text{Néel}}|\hat{H}|\Phi_{\text{Néel}}\rangle = -|\mathcal{B}|S^2$. There is also a simple lower bound of E_{GS} proved by Anderson [2]. See the following problem.

Problem 2.5.a Consider a system of $z+1$ spins $\hat{\boldsymbol{S}}_0, \hat{\boldsymbol{S}}_1, \ldots, \hat{\boldsymbol{S}}_z$ with spin quantum number S. Show that the ground state energy of the Hamiltonian $\hat{h}_0 = \sum_{j=1}^{z} \hat{\boldsymbol{S}}_0 \cdot \hat{\boldsymbol{S}}_j$ is $-S(1+zS)$. [solution→p.497]

Problem 2.5.b By using the result of Problem 2.5.a, derive a lower bound for the ground state energy E_{GS} of the Hamiltonian (2.5.1). (Hint: If the ground state energies of two Hamiltonians $\hat{H}_1$ and $\hat{H}_2$ are E_0, then the ground state energy of $\hat{H}_1 + \hat{H}_2$ is not less than $2E_0$. Note that we do not assume that $\hat{H}_1$ and $\hat{H}_2$ commute. This fact follows from Lemma A.5 in p. 468.) [solution→p.497]

[31] In the classical limit with $S \gg 1$, one may neglect the $O(|\mathcal{B}|S)$ term to find that $\hat{H}|\Phi_{\text{Néel}}\rangle \simeq -|\mathcal{B}|S^2\,|\Phi_{\text{Néel}}\rangle$, and $|\Phi_{\text{Néel}}\rangle$ is a (near) ground state.

Marshall-Lieb-Mattis theorem The foregoing observation shows that to find exact ground states of the antiferromagnetic Heisenberg model is a nontrivial task. The following Marshall-Lieb-Mattis theorem provides a basic characterization of the ground state, which will be the starting point of detailed analysis in the following chapters. Recall that, for a spin configuration $\boldsymbol{\sigma} = (\sigma_x)_{x\in\Lambda}$, we have defined $\overline{\boldsymbol{\sigma}} := \sum_{x\in\Lambda}\sigma_x$ in (2.2.2).

Theorem 2.2 (Marshall-Lieb-Mattis theorem) *Let* $(\Lambda, \mathscr{B})$ *be a connected*[32] *bipartite lattice with* $|A| = |B|$*. Then the ground state,* $|\Phi_{\rm GS}\rangle$*, is unique and has total spin* $S_{\rm tot} = 0$*. It can be expanded in the standard basis* (2.2.1) *as*

$$|\Phi_{\rm GS}\rangle = \sum_{\substack{\boldsymbol{\sigma}\\(\overline{\boldsymbol{\sigma}}=0)}}\Bigl\{\prod_{x\in B}(-1)^{\sigma_x - S}\Bigr\}c_{\boldsymbol{\sigma}}\,|\Psi^{\boldsymbol{\sigma}}\rangle, \tag{2.5.4}$$

with coefficients satisfying $c_{\boldsymbol{\sigma}} > 0$ *for any* $\boldsymbol{\sigma}$ *with* $\overline{\boldsymbol{\sigma}} = 0$.

The theorem was proved by Marshall [7] for the one-dimensional model, and proved in the above complete form by Lieb and Mattis [6]. The proof, which is presented below, is essentially based on the Perron–Frobenius theorem (Theorem A.18 in p. 475). The prefactor $\prod_{x\in B}(-1)^{\sigma_x - S}$ in (2.5.4) could be $\prod_{x\in B}(-1)^{\sigma_x}$, but we included S so as to make it ± 1 for any S.

That the ground state $|\Phi_{\rm GS}\rangle$ has $S_{\rm tot} = 0$ means $(\hat{\boldsymbol{S}}_{\rm tot})^2|\Phi_{\rm GS}\rangle = 0$. This further implies that $\hat{S}^{(\alpha)}_{\rm tot}|\Phi_{\rm GS}\rangle = 0$ for $\alpha = 1, 2, 3$.[33] From the definition (2.2.11) of the rotation operator, we find

$$\hat{U}^{(\alpha)}_\theta|\Phi_{\rm GS}\rangle = |\Phi_{\rm GS}\rangle, \tag{2.5.5}$$

for any $\theta \in \mathbb{R}$ and $\alpha = 1, 2, 3$. We see that the ground state $|\Phi_{\rm GS}\rangle$ is SU(2) invariant (or rotationally invariant).[34]

Problem 2.5.c Show, for the present ground state, that

$$\langle\Phi_{\rm GS}|(\hat{S}^{(\alpha)}_x)^2|\Phi_{\rm GS}\rangle = S(S+1)/3, \tag{2.5.6}$$

for any $\alpha = 1, 2, 3$ and $x \in \Lambda$. (This is easy.) [solution→p.498]

Although the theorem applies to the antiferromagnetic Heisenberg model on any lattice (satisfying the requirements) in any dimensions, the properties of the ground state depend crucially on the dimensionality of the lattice, and sometimes on the spin quantum number S. This will be among the main topics of the following chapters.

The theorem also has an implication on the correlation function as in the following problem. It should be remarked that the inequalities (2.5.7) do not imply the existence

---

[32] See Footnote 28 in p. 33.

[33] This is standard. That $(\hat{\boldsymbol{S}}_{\rm tot})^2|\Phi_{\rm GS}\rangle = 0$ means $0 = \langle\Phi_{\rm GS}|(\hat{\boldsymbol{S}}_{\rm tot})^2|\Phi_{\rm GS}\rangle = \|\hat{S}^{(1)}_{\rm tot}|\Phi_{\rm GS}\rangle\|^2 + \|\hat{S}^{(2)}_{\rm tot}|\Phi_{\rm GS}\rangle\|^2 + \|\hat{S}^{(3)}_{\rm tot}|\Phi_{\rm GS}\rangle\|^2$, which shows $\|\hat{S}^{(\alpha)}_{\rm tot}|\Phi_{\rm GS}\rangle\| = 0$.

[34] In general we say that a state $|\Phi\rangle$ is SU(2) invariant if it holds for any α and θ that $\hat{U}^{(\alpha)}_\theta|\Phi\rangle = c|\Phi\rangle$ with some constant $c \in \mathbb{C}$ (which may depend on α and θ) with $|c| = 1$.

of long-range order. See the beginning of Sect. 3.1 and the remark after Theorem 4.1 in p. 75.

Problem 2.5.d Under the conditions of Theorem 2.2, prove that the two-spin correlation functions in the ground state has the following sign:

$$\langle \Phi_{\rm GS}|\hat{\boldsymbol{S}}_x \cdot \hat{\boldsymbol{S}}_y|\Phi_{\rm GS}\rangle \begin{cases} > 0 & \text{when } x, y \in A \text{ or } x, y \in B \\ < 0 & \text{when } x \in A, y \in B \text{ or } x \in B, y \in A. \end{cases} \tag{2.5.7}$$

(Hint: It is useful to define the unitary operator $\hat{U} := \prod_{x\in B}(-1)^{\hat{S}_x^{(3)}-S}$, with which (2.5.4) is written as $|\Phi_{\rm GS}\rangle = \hat{U}\sum_{\sigma\,(\overline{\sigma}=0)} c_\sigma\,|\Psi^\sigma\rangle$.) [solution→p.498]

We stress that the condition of bipartiteness is essential for the Marshall-Lieb-Mattis theorem. For example, the antiferromagnetic Heisenberg model on the one-dimensional periodic chain with the Hamiltonian $\hat{H}_L = \sum_{x=1}^{L} \hat{\boldsymbol{S}}_x \cdot \hat{\boldsymbol{S}}_{x+1}$, where we set $\hat{\boldsymbol{S}}_{L+1} = \hat{\boldsymbol{S}}_1$, satisfies the condition only when L is even. When L is odd, it is known that the structure of the grounds states (and the excited states) are much more complicated. See, e.g., [3].

Proof of Theorem 2.2 We make a full use of the decomposition (2.2.9). Since $|\Lambda| = |A| + |B| = 2|A|$ is even, the eigenvalue M of $\hat{S}_{\rm tot}^{(3)}$ takes an integer value. Then Theorem A.17 (p. 473) for SU(2) invariant Hamiltonians implies that the whole spectrum of $\hat{H}$ is contained in the subspace $\mathscr{H}_0$. In particular we see that at least one ground state can be found in $\mathscr{H}_0$.

Let us concentrate on $\mathscr{H}_0$ and take the basis states

$$|\tilde{\Psi}^\sigma\rangle := \Big\{\prod_{x\in B}(-1)^{\sigma_x - S}\Big\}|\Psi^\sigma\rangle, \tag{2.5.8}$$

for $\boldsymbol{\sigma}$ such that $\overline{\boldsymbol{\sigma}} = 0$. We then find that (i) $\langle\tilde{\Psi}^\sigma|\hat{H}|\tilde{\Psi}^{\sigma'}\rangle \in \mathbb{R}$ for any $\boldsymbol{\sigma}, \boldsymbol{\sigma}'$, and (ii) $\langle\tilde{\Psi}^\sigma|\hat{H}|\tilde{\Psi}^{\sigma'}\rangle \le 0$ whenever $\boldsymbol{\sigma} \ne \boldsymbol{\sigma}'$. (i) can be easily verified from the definition. To see (ii), note that $\hat{S}_x^+\hat{S}_y^- + \hat{S}_x^-\hat{S}_y^+$ (acting on $|\Psi^\sigma\rangle$) always switches the sign of $\prod_{x\in B}(-1)^{\sigma_x - S}$. (It is useful to first examine the simplest case in (2.5.3) before considering general cases.) Finally we note that the connectivity of the lattice implies that (iii) any $\boldsymbol{\sigma}$ and $\boldsymbol{\sigma}'$ with $\overline{\boldsymbol{\sigma}} = \overline{\boldsymbol{\sigma}'} = 0$ are connected (in the sense of the Perron–Frobenius theorem) by nonvanishing matrix elements of $\langle\tilde{\Psi}^{\sigma''}|\hat{H}|\tilde{\Psi}^{\sigma'''}\rangle$. The essential observation is that one can start from $|\Psi^\sigma\rangle$ and obtain $|\Psi^{\sigma'}\rangle$ (up to a multiplication by a constant) by successively operating $\hat{S}_x^+\hat{S}_y^-$ or $\hat{S}_x^-\hat{S}_y^+$ for properly chosen bonds $\{x, y\} \in \mathscr{B}$. This may be intuitively clear, but we shall present a careful proof below. Then the Perron–Frobenius theorem (Theorem A.18) implies that, in the subspace $\mathscr{H}_0$, the ground state, $|\Phi_{\rm GS}\rangle$, is unique, and is expanded as

$$|\Phi_{\rm GS}\rangle = \sum_{\substack{\sigma \\ (\overline{\sigma}=0)}} c_\sigma\,|\tilde{\Psi}^\sigma\rangle, \tag{2.5.9}$$

with $c_\sigma > 0$. This is the desired (2.5.4).

It remains to determine the total spin of $|\Phi_{\rm GS}\rangle$ and prove the uniqueness in the whole Hilbert space. This may be done in several manners, but we shall follow a clever argument by Lieb and Mattis [6]. Define, on the same lattice, a toy Hamiltonian in which the spins on any pair of sites on different sublattices interact:

$$\hat{H}^{\rm toy} := \frac{1}{|\Lambda|}\sum_{\substack{x\in A\\ y\in B}} \hat{\boldsymbol{S}}_x \cdot \hat{\boldsymbol{S}}_y = \frac{1}{|\Lambda|}\hat{\boldsymbol{S}}_A \cdot \hat{\boldsymbol{S}}_B \tag{2.5.10}$$

Here we defined $\hat{\boldsymbol{S}}_A := \sum_{x\in A}\hat{\boldsymbol{S}}_x$ and $\hat{\boldsymbol{S}}_B := \sum_{y\in B}\hat{\boldsymbol{S}}_y$. The factor $1/|\Lambda|$ is introduced to make the total energy proportional to the system size $|\Lambda|$. It plays no role here, but will be useful when we revisit the toy Hamiltonian in Sect. 4.2. See (4.2.8). Noting that $\hat{\boldsymbol{S}}_{\rm tot} = \hat{\boldsymbol{S}}_A + \hat{\boldsymbol{S}}_B$, we can rewrite $\hat{H}^{\rm toy}$ as

$$\hat{H}^{\rm toy} = \frac{1}{2|\Lambda|}\bigl\{(\hat{\boldsymbol{S}}_{\rm tot})^2 - (\hat{\boldsymbol{S}}_A)^2 - (\hat{\boldsymbol{S}}_B)^2\bigr\}. \tag{2.5.11}$$

The ground state $|\Phi_{\rm GS}^{\rm toy}\rangle$ of $\hat{H}^{\rm toy}$ is readily obtained from elementary facts about addition of angular momenta (see Appendix A.3.3); one only needs to couple all the spins in each sublattice to get the maximum total spin, and then couple these two large spins to realize the zero spin state. In this way, $(\hat{\boldsymbol{S}}_A)^2$ and $(\hat{\boldsymbol{S}}_B)^2$ are maximized, and $(\hat{\boldsymbol{S}}_{\rm tot})^2$ is minimized. We clearly have $(\hat{\boldsymbol{S}}_{\rm tot})^2|\Phi_{\rm GS}^{\rm toy}\rangle = 0$. (See Problem 4.2.1.a in p. 104.)

Note that the toy Hamiltonian (2.5.10) satisfies the conditions for the present therorem. We can make use of the first half of the theorem to see that its ground state $|\Phi_{\rm GS}^{\rm toy}\rangle$ is also expanded as

$$|\Phi_{\rm GS}^{\rm toy}\rangle = \sum_{\substack{\sigma\\ (\overline{\sigma}=0)}} c_\sigma^{\rm toy}\,|\tilde{\Psi}^\sigma\rangle, \tag{2.5.12}$$

with $c_\sigma^{\rm toy} > 0$. This, with (2.5.9), implies $\langle\Phi_{\rm GS}^{\rm toy}|\Phi_{\rm GS}\rangle \neq 0$. Since the uniqueness within $\mathscr{H}_0$ implies that $|\Phi_{\rm GS}\rangle$ is an eigenstate of $(\hat{\boldsymbol{S}}_{\rm tot})^2$, this means $(\hat{\boldsymbol{S}}_{\rm tot})^2|\Phi_{\rm GS}\rangle = 0$. This also implies the uniqueness in the whole Hilbert space since the existence of a degenerate ground state in $\mathscr{H}_M$ with $M \neq 0$ implies that there is a degenerate ground state in $\mathscr{H}_0$ with $S_{\rm tot} \neq 0$. See Theorem A.16 in p. 473. ∎

Proof of Property (iii) The property (iii), i.e., the connectivity of spin configurations, (as well as many properties of similar nature) is usually stated without a proof since they are regarded as "obvious by inspection". Let us give a proof for completeness.[35]

[35] The reader new to this kind of proof is suggested to first consider the case with $S = 1/2$.

Let $x, y \in \Lambda$ and $x \neq y$. For any spin configuration $\boldsymbol{\sigma}$ such that $\sigma_x < S$ and $\sigma_y > -S$, we define a new spin configuration which have $\sigma_x + 1$ at x, $\sigma_y - 1$ at y, and is identical to $\boldsymbol{\sigma}$ on other sites. We call this new configuration $F_{x,y}(\boldsymbol{\sigma})$. We say that two spin configurations $\boldsymbol{\sigma}$ and $\boldsymbol{\sigma}'$ are connected and write $\boldsymbol{\sigma} \sim \boldsymbol{\sigma}'$ if there exists a finite sequence of pairs of sites $(x_1, y_1), \ldots, (x_n, y_n)$ such that $\{x_j, y_j\} \in \mathcal{B}$ for $j = 1, \ldots, n$, and $\boldsymbol{\sigma}' = F_{x_n,y_n} \circ \cdots \circ F_{x_2,y_2} \circ F_{x_1,y_1}(\boldsymbol{\sigma})$.[36] Clearly $\boldsymbol{\sigma}$ and $\boldsymbol{\sigma}'$ are never connected if $\overline{\boldsymbol{\sigma}} \neq \overline{\boldsymbol{\sigma}'}$. We need the following lemma.

Lemma *Let the lattice $(\Lambda, \mathcal{B})$ be connected. Then for any $x, y \in \Lambda$ with $x \neq y$ (which do not necessarily form a bond in $\mathcal{B}$), we have $F_{x,y}(\boldsymbol{\sigma}) \sim \boldsymbol{\sigma}$ for any $\boldsymbol{\sigma}$ such that $\sigma_x < S$ and $\sigma_y > -S$.*

We prove the lemma by induction. Let $d(x, y)$ be the graph theoretic distance, i.e., the minimum number of bonds necessary to connect x and y. For a positive integer r, consider the statement "for any $x \neq y$ such that $d(x, y) \leq r$, we have $F_{x,y}(\boldsymbol{\sigma}) \sim \boldsymbol{\sigma}$ for any $\boldsymbol{\sigma}$ such that $\sigma_x < S$ and $\sigma_y > -S$." The statement is valid by definition for $r = 1$. We thus assume it is valid for r, and prove it for $r + 1$. Take any $x, y \in \Lambda$ such that $d(x, y) = r + 1$. Then there exists a site $z \in \Lambda$ such that $\{x, z\} \in \mathcal{B}$ and $d(z, y) = r$. Take any configuration $\boldsymbol{\sigma}$ such that $\sigma_x < S$ and $\sigma_y > -S$. When $\sigma_z = S$, we observe that $F_{x,y}(\boldsymbol{\sigma}) = F_{z,y}(F_{x,z}(\boldsymbol{\sigma}))$. From the assumption we have $F_{z,y}(F_{x,z}(\boldsymbol{\sigma})) \sim F_{x,z}(\boldsymbol{\sigma})$, and hence $F_{x,y}(\boldsymbol{\sigma}) \sim \boldsymbol{\sigma}$. When $\sigma_z < S$, we observe that $F_{x,y}(\boldsymbol{\sigma}) = F_{x,z}(F_{z,y}(\boldsymbol{\sigma}))$, which again implies $F_{x,y}(\boldsymbol{\sigma}) \sim \boldsymbol{\sigma}$.

We now prove the property (iii), which can be formulated as

Proposition *Let the lattice $(\Lambda, \mathcal{B})$ be connected. Then any spin configurations $\boldsymbol{\sigma}$ and $\boldsymbol{\sigma}'$ with $\overline{\boldsymbol{\sigma}} = \overline{\boldsymbol{\sigma}'}$ are connected.*

We measure the distance between two configurations $\boldsymbol{\sigma}$ and $\boldsymbol{\sigma}'$ by the norm $\|\boldsymbol{\sigma} - \boldsymbol{\sigma}'\|_1 := \sum_{x\in\Lambda} |\sigma_x - \sigma'_x|$. Suppose that $\boldsymbol{\sigma} \neq \boldsymbol{\sigma}'$. Since $\overline{\boldsymbol{\sigma}} = \overline{\boldsymbol{\sigma}'}$ there are $x_1, y_1 \in \Lambda$ such that $\sigma_{x_1} < \sigma'_{x_1}$ and $\sigma_{y_1} > \sigma'_{y_1}$. From the lemma we have $F_{x_1,y_1}(\boldsymbol{\sigma}) \sim \boldsymbol{\sigma}$. Note also that $\|F_{x_1,y_1}(\boldsymbol{\sigma}) - \boldsymbol{\sigma}'\|_1 = \|\boldsymbol{\sigma} - \boldsymbol{\sigma}'\|_1 - 2$, which means that $F_{x_1,y_1}(\boldsymbol{\sigma})$ is closer to $\boldsymbol{\sigma}'$ than $\boldsymbol{\sigma}$. We can repeat this procedure until $\|F_{x_n,y_n} \circ \cdots \circ F_{x_1,y_1}(\boldsymbol{\sigma}) - \boldsymbol{\sigma}'\|_1 = 0$, i.e., $F_{x_n,y_n} \circ \cdots \circ F_{x_1,y_1}(\boldsymbol{\sigma}) = \boldsymbol{\sigma}'$. This shows that $\boldsymbol{\sigma} \sim \boldsymbol{\sigma}'$. ∎

Above we presented the theorem for the most basic case with $|A| = |B|$, but such symmetry was not assumed in the original work of Lieb and Mattis [6]. In general we have the following.

Theorem 2.3 *Let $(\Lambda, \mathcal{B})$ be a connected bipartite lattice with $|A| \geq 1$ and $|B| \geq 1$. Then the ground states have total spin $S_{\mathrm{tot}} = \big||A| - |B|\big|\, S$, and are $2S_{\mathrm{tot}} + 1$ fold degenerate. The ground states are expanded as in* (2.5.4) *with the restriction $\overline{\boldsymbol{\sigma}} = 0$ replaced by $\overline{\boldsymbol{\sigma}} = M$, where $M = -S_{\mathrm{tot}}, \ldots, S_{\mathrm{tot}} - 1, S_{\mathrm{tot}}$.*

Sketch of the Proof We shall be sketchy since the proof is essentially a straightforward modification of that of Theorem 2.2. Again by Theorem A.17 (p. 473), we

[36]This precisely corresponds to the successive operation of $\hat{S}^+_{x_j}\hat{S}^-_{y_j}$ with $j = 1, \ldots, n$ to $|\psi^{\boldsymbol{\sigma}}\rangle$.

can work in the subspace $\mathscr{H}_0$ or $\mathscr{H}_{1/2}$ to find a ground state. We then prove that the ground state is unique (within the subspace) and has the same expansion (2.5.9) with $c_\sigma > 0$. One then uses the same toy Hamiltonian (2.5.10), and shows that the ground state has $S_{\text{tot}} = \big||A| - |B|\big| S$. The degeneracy of the ground states (in the whole Hilbert space) is determined by Theorem A.16 (p. 473). ∎

The above theorem has an interesting implication about antiferromagnetic long-range order. See Theorem 4.4 (p. 78) at the end of Sect. 4.1, and also Theorem 10.8 in p. 359 for the Hubbard model.

Remark We have treated only the simplest Hamiltonian (2.5.1) of the antiferromagnetic Heisenberg model. By examining the proofs, one easily finds that Theorems 2.2 and 2.3, and the inequalities (2.5.7) are valid as they are if we consider the Hamiltonian

$$\hat{H} = \sum_{\{x,y\}\in\mathscr{B}} J_{x,y}\, \hat{\boldsymbol{S}}_x \cdot \hat{\boldsymbol{S}}_y, \tag{2.5.13}$$

where the exchange interaction $J_{x,y}$ satisfies $J_{x,y} = J_{y,x} > 0$ for any $\{x, y\} \in \mathscr{B}$.

Antiferromagnetic model with anisotropy When modeling antiferromagnetic materials it is common to modify the Heisenberg model by introducing uniaxial anisotropy. A typical Hamiltonian is

$$\hat{H} = \sum_{\{x,y\}\in\mathscr{B}} \{\hat{S}_x^{(1)}\hat{S}_y^{(1)} + \hat{S}_x^{(2)}\hat{S}_y^{(2)} + \lambda \hat{S}_x^{(3)}\hat{S}_y^{(3)}\} + D\sum_{x\in\Lambda}(\hat{S}_x^{(3)})^2, \tag{2.5.14}$$

where λ is called the Ising anisotropy parameter and D the crystal field anisotropy parameter.

Note that the Hamiltonian is no longer SU(2) invariant. In fact (2.4.2) and (2.4.3) are valid only for $\alpha = 3$, which means $\hat{H}$ can be simultaneously diagonalizable with $\hat{S}_{\text{tot}}^{(3)}$ but not necessarily with $(\hat{\boldsymbol{S}}_{\text{tot}})^2$. In other words the model only has U(1) invariance. The Marshall-Lieb-Mattis theorem, which relies in an essential manner on the SU(2) invariance, cannot be extended as it is. For the present model, the following analogue of the Marshall-Lieb-Mattis theorem was proved by Mattis [8] and Nishimori [9].[37]

Theorem 2.4 *Let $(\Lambda, \mathscr{B})$ be a connected and bipartite lattice with $|A| = |B|$. Consider a spin system on Λ with Hamiltonian (2.5.14) with (i) $-1 < \lambda \le 1$ and $D \ge 0$ (ii) $\lambda \ge 1$ and $D \le 0$. Then the ground state $|\Phi_{\text{GS}}\rangle$ is unique and satisfies $\hat{S}_{\text{tot}}^{(3)}|\Phi_{\text{GS}}\rangle = 0$.*

Note that for $D = 0$ the theorem applies to models with any $\lambda > -1$, and for $\lambda = 1$ it applies to models with any $D \in \mathbb{R}$.

---

[37]See also [1]. The original works deal with the case $D = 0$. The condition for D was noted in Appendix of [5].

Proof We will show below that, when (i) or (ii) is valid and $(\lambda, D) \neq (1, 0)$, ground states of (2.5.14) are at most doubly degenerate. This implies the desired uniqueness as follows. We know from Theorem 2.2 that the ground state is unique and in $\mathscr{H}_0$ if $(\lambda, D) = (1, 0)$. Suppose that, as one continuously changes (λ, D) from $(1, 0)$, ground states become degenerate. Since the ground state within $\mathscr{H}_0$ is unique (by a straightforward extension of the first half of Theorem 2.2) there must be a degenerate ground state in $\mathscr{H}_M$ with $M \neq 0$. Then the symmetry implies that there is a degenerate ground state in $\mathscr{H}_{-M}$ too, and hence the ground states are (at most) three-fold degenerate. But this is impossible.

We shall prove the claimed bound for the degeneracy. Note that the Hamiltonian

$$\hat{H}' = \sum_{\{x,y\}\in\mathscr{B}} \{\hat{S}_x^{(1)}\hat{S}_y^{(1)} + \lambda\hat{S}_x^{(2)}\hat{S}_y^{(2)} + \hat{S}_x^{(3)}\hat{S}_y^{(3)}\} + D\sum_{x\in\Lambda}(\hat{S}_x^{(2)})^2, \tag{2.5.15}$$

has exactly the same spectrum and degeneracy as $\hat{H}$. Let us rewrite $\hat{H}'$ as

$$\begin{aligned}\hat{H}' = \sum_{\{x,y\}\in\mathscr{B}} &\left\{\frac{1+\lambda}{4}(\hat{S}_x^+\hat{S}_y^- + \hat{S}_x^-\hat{S}_y^+) + \frac{1-\lambda}{4}(\hat{S}_x^+\hat{S}_y^+ + \hat{S}_x^-\hat{S}_y^-) + \hat{S}_x^{(3)}\hat{S}_y^{(3)}\right\}\\ &- \frac{D}{4}\sum_{x\in\Lambda}\{(\hat{S}_x^+)^2 + (\hat{S}_x^-)^2 - \hat{S}_x^+\hat{S}_x^- - \hat{S}_x^-\hat{S}_x^+\}.\end{aligned} \tag{2.5.16}$$

We first consider the case (i), where $(1 \pm \lambda)/4$ and D are nonnegative. Then one can check by inspection that the matrix elements satisfy $\langle\tilde{\Psi}^{\boldsymbol{\sigma}}|\hat{H}'|\tilde{\Psi}^{\boldsymbol{\sigma}'}\rangle \leq 0$ for any $\boldsymbol{\sigma} \neq \boldsymbol{\sigma}'$, where we use the basis states (2.5.8). Note that $(1 + \lambda)/4$ is positive, and also at least one of $(1 - \lambda)/4$ or D is positive because we assume $(\lambda, D) \neq (1, 0)$. One then sees that all $\boldsymbol{\sigma}$ such that $\overline{\boldsymbol{\sigma}} =$ even are connected via nonvanishing matrix elements, and likewise all $\boldsymbol{\sigma}$ such that $\overline{\boldsymbol{\sigma}} =$ odd are connected. The Perron–Frobenius theorem then implies that the ground state in each sector is unique, and hence the whole ground states are at most doubly degenerate.

We next consider the case (ii), where $(1 + \lambda)/4$ is still positive, but $(1 - \lambda)/4$ and D are non-positive. In this case one can verify that $\langle\tilde{\tilde{\Psi}}^{\boldsymbol{\sigma}}|\hat{H}'|\tilde{\tilde{\Psi}}^{\boldsymbol{\sigma}'}\rangle \leq 0$ for any $\boldsymbol{\sigma} \neq \boldsymbol{\sigma}'$, where the new basis states are defined by $|\tilde{\tilde{\Psi}}^{\boldsymbol{\sigma}}\rangle = (-1)^{(\overline{\boldsymbol{\sigma}}-|\Lambda|S)/2}|\tilde{\Psi}^{\boldsymbol{\sigma}}\rangle$. The rest of the proof is the same as above. ∎

References

1. I. Affleck, E.H. Lieb, A proof of part of Haldane's conjecture on spin chains. Lett. Math. Phys. **12**, 57–69 (1986)
2. P.W. Anderson, Limits on the energy of the antiferromagnetic ground state. Phys. Rev. **83**, 1260 (1951)

3. K. Bäerwinkel, P. Hage, H.-J. Schmidt, J. Schnack, Quantum numbers for relative ground states of antiferromagnetic Heisenberg spin rings. Phys. Rev. B **68**, 054422 (2003). arXiv:0303386
4. J.-P. Gazeau, *Coherent States in Quantum Physics* (Wiley-VCH, Weinheim, 2009)
5. Y. Hatsugai, M. Kohmoto, Numerical study of the hidden antiferromagnetic order in the Haldane phase. Phys. Rev. B **44**, 11789–11794 (1991)
6. E.H. Lieb, D. Mattis, Ordering energy levels in interacting spin chains. J. Math. Phys. **3**, 749–751 (1962)
7. W. Marshall, Antiferromagnetism. Proc. R. Soc. A **232**, 48 (1955)
8. D. Mattis, Ground-state symmetry in XY model of magnetism. Phys. Rev. Lett. **42**, 1503 (1979)
9. H. Nishimori, Spin quantum number in the ground state of the Mattis-Heisenberg model. J. Stat. Phys. **26**, 839–845 (1981)
10. L. Rademaker, *Exact Ground State of Lieb-Mattis Hamiltonian as a Superposition of Néel states* (2019), arXiv:1909.09663

Part I
Long-Range Order and Spontaneous Symmetry Breaking

In part I of the book, we focus on the interesting and ubiquitous problem of the relation between long-range order and spontaneous symmetry breaking in the ground states of many-body quantum systems. A central theme is an apparent contradiction between the uniqueness of the ground state and the existence of long-range order; the former seems to inhibit any symmetry breaking while the latter suggests the existence of spontaneous symmetry breaking. The resolution of the "paradox" leads to interesting notions of the "tower" of low-lying excited states and the "physical ground states", which are linear combinations of the low-lying excited states, as we shall discuss in detain in the following chapters.

Although the discussion applies to almost any nontrivial quantum many-body systems with symmetry, we here focus on quantum antiferromagnets and Bose-Einstein condensates, and develop a fully rigorous theory. We stress that these examples represent the essence of the universal phenomena.

Chapter 3
Long-Range Order and Spontaneous Symmetry Breaking in the Classical and Quantum Ising Models

In this chapter, we start by motivating the problem studied throughout Part I, and then discuss simple problems with discrete symmetry. In Sect. 3.1, we briefly discuss what is expected and known in the antiferromagnetic Heisenberg model, and introduce the essential idea of "long-range order (LRO) without spontaneous symmetry breaking (SSB)". In Sect. 3.2, we discuss a similar problem in equilibrium states of the classical Ising model, which is simpler and easier-to-understand. Then, in Sect. 3.3, we study the simplest quantum many-body system that exhibits "LRO without SSB", namely, the quantum Ising model, and see how the above mentioned "paradox" is solved. At the end of this section, we make an important remark about the essential role played by symmetry in the notion of phases. In the final Sect. 3.4, we describe general rigorous theories of LRO and SSB when the relevant symmetry is Ising-like.

3.1 Motivation from the Heisenberg Antiferromagnet

Long-range order (LRO) Consider the antiferromagnetic Heisenberg model with Hamiltonian (2.5.1) on the three dimensional cubic lattice. It is known that, as a result of interaction between a vast number of spins, the ground state of the model exhibits antiferromagnetic long-range order (LRO). More precisely it is proved that the correlation function in the unique ground state $|\Phi_{\text{GS}}\rangle$ behaves as

$$\langle\Phi_{\text{GS}}|\hat{\boldsymbol{S}}_x \cdot \hat{\boldsymbol{S}}_y|\Phi_{\text{GS}}\rangle \simeq \begin{cases} q_0 & \text{if}\, x, y \in A \,\text{or}\, x, y \in B, \\ -q_0 & \text{if}\, x \in A, y \in B \,\text{or}\, x \in B, y \in A, \end{cases} \tag{3.1.1}$$

with the long-range order parameter $q_0 > 0$, provided that the sites x and y are sufficiently far apart. Here A and B are sublattices as in Sect. 2.5. Precise definitions of the lattice and the sublattices will be given in (3.1.2), (4.1.2) and (4.1.3). In short, two spins on the same sublattice tend to point in the same direction, while two spins in different sublattices tend to point in the opposite directions, no matter how

H. Tasaki, *Physics and Mathematics of Quantum Many-Body Systems*,
Graduate Texts in Physics, https://doi.org/10.1007/978-3-030-41265-4_3

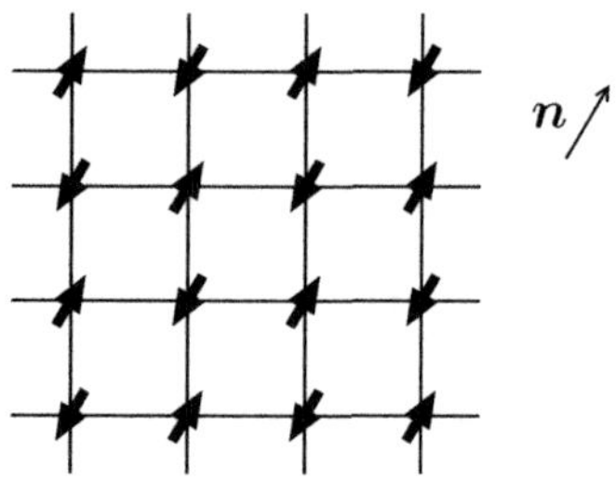

Fig. 3.1 Néel order in the square lattice. An arbitrary direction $\boldsymbol{n}$ is chosen (by the system). Spins in one sublattice are pointing in the direction $\boldsymbol{n}$, and those in the other sublattice are pointing in the direction $-\boldsymbol{n}$. This is an example of spontaneous symmetry breaking (SSB) (© Hal Tasaki 2020. All Rights Reserved)

separated the locations of two spins are. This is an example of nontrivial cooperative phenomenon observed in systems with strongly interacting many degrees of freedom. We shall discuss this statement more precisely in Sect. 4.1.

It should be noted that the inequalities (2.5.7) (which follow from the Marshall–Lieb–Mattis theorem) do not imply (3.1.1). See also the remark after Theorem 4.1 in p. 75.

Spontaneous symmetry breaking (SSB) The existence of long-range order naturally suggests that the ground state also exhibits antiferromagnetic order (or Néel order), i.e., there is a preferred direction, say $\boldsymbol{n}$, and all the spins in the A sublattice almost point in the direction $\boldsymbol{n}$ and all the spins in the B sublattice almost point in the direction $-\boldsymbol{n}$. See Fig. 3.1. In fact this is what is observed, through, e.g., neutron scattering experiments and NMR measurements, in actual quantum antiferromagnets at very low temperatures. In a recent experiment with a system of ultra cold atoms simulating the Heisenberg antiferromagnet, the ordering is observed much more directly [17, 18].

Note that, in this description of a Néel ordered ground state, a direction $\boldsymbol{n}$ should be specified. Since the Hamiltonian (2.5.1) is completely isotropic (or SU(2) invariant), the direction $\boldsymbol{n}$ must be chosen in an arbitrary manner by the system itself. This arbitrary choice is known as spontaneous symmetry breaking (SSB).[1] In this case it is the SU(2) symmetry of the Hamiltonian that is spontaneously broken.

LRO without SSB The uniqueness of the ground state of the antiferromagnetic Heisenberg model, which is guaranteed by the Marshall–Lieb–Mattis theorem (Theorem 2.2 in p. 39), is in a sharp conflict with the idea of SSB. The unique ground state is literally unique; there is no room for the arbitrary choice by the system. In other words, the unique ground state must preserve all the symmetries of the Hamiltonian, and can never exhibit any order in a specific direction. We must therefore conclude that the unique ground state of the three-dimensional antiferromagnetic Heisenberg model exhibits LRO but no SSB. Physically speaking, the situation is rather unnatural, especially because the law of large numbers for a certain macroscopic quantity is inevitably violated in a state with LRO but without SSB.

[1] See [3] for a recent introduction to the concept of SSB from a different perspective.

See (3.2.11), (3.3.12), (3.4.5), and (4.2.1). In fact such states with LRO without SSB are never observed experimentally in magnetic systems; one always observes a state with both LRO and SSB.[2] Nature chooses to break symmetry rather than breaking the law of large numbers.

The purpose of the present and the next chapters is to completely solve this puzzle about the discrepancy between "physically natural ground states" and the exact ground state of a (large but) finite system. It is shown in general that such a ground state with LRO but without SSB is inevitably accompanied by low-lying energy states, which are indeed parts of physical "ground states". We stress that this picture applies universally to almost any quantum many-body systems exhibiting long-range order in which the Hamiltonian and the order operator do not commute. Examples include antiferromagnetism, superconductivity, Bose–Einstein condensate, and any quantum field theory (with LRO), but not ferromagnetism (or ferrimagnetism).

The above mentioned puzzle, i.e., the inconsistency between the uniqueness of the ground state and the expected Néel order was realized since the early days of research on antiferromagnetism. It was mentioned, e.g., in the seminal paper by Anderson [1], who seems to have understood the basic picture. Although this paper published in 1952 is not easy to understand (for us), we find in the introduction of his book [2] published in 1984 a clear discussion about the role of the "tower" of energy eigenstates in the formation of Néel ordered states. Anderson then continues as follows.

> But somehow this is one of those arguments that is, although very simple, and terribly important, not generally available, perhaps because everyone who has ever understood it thinks it too simple to write down. (p. 44, [2])

In Sect. 4.2, we shall "write down" this fascinating and universal picture, emphasizing rigorous results which provide a firm basis for the picture.

In this chapter, we focus on simpler models where the relevant symmetry is discrete, and describe the theory of low-lying states and symmetry breaking. More interesting situation with continuous symmetry is treated in Chap. 4.

d-dimensional hypercubic lattice Before proceeding, we define the d-dimensional hypercubic lattice, which will be the main working ground in the present and the next chapters. See Fig. 3.2.

Let $d = 1, 2, \dots$ be the dimension and L be even. We define the set of sites as

$$\Lambda_L := \Bigl\{(x_1, x_2, \dots, x_d) \Bigm| x_i \in \mathbb{Z},\ -\frac{L}{2} < x_i \le \frac{L}{2}\Bigr\} \subset \mathbb{Z}^d, \tag{3.1.2}$$

which contains $|\Lambda_L| = L^d$ sites. We then define the set of bonds as

$$\mathscr{B}_L := \bigl\{ \{x, y\} \bigm| x, y \in \Lambda_L,\ |x - y| = 1 \bigr\}, \tag{3.1.3}$$

---

[2]But the situation is different in systems exhibiting Bose–Einstein condensation or superconductivity. We will discuss this interesting topic in Chap. 5.

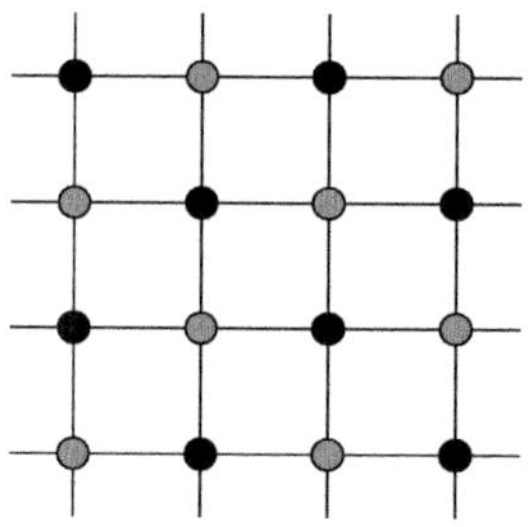

Fig. 3.2 The square lattice, which is the d-dimensional hypercubic lattice with $d = 2$, with $L = 4$. We impose periodic boundary conditions. The lattice is connected and bipartite. Sites in one sublattice are drawn in black and sites in the other in gray (© Hal Tasaki 2020. All Rights Reserved)

where $|\cdots|$ denotes the standard Euclidean norm. We impose periodic boundary conditions and identify $x_i = -L/2$ with $x_i = L/2$.

3.2 Classical Ising Model

In a classical model, a state with LRO but without SSB never appears as a ground state, but is commonly observed (theoretically) as an equilibrium state. It is instructive and useful to recall what is known in the familiar example of the classical ferromagnetic Ising model.

Definition of the model We define the classical ferromagnetic Ising model on the d-dimensional hypercubic lattice Λ_L. With each site $x \in \Lambda_L$ we associate a classical spin $\sigma_x = \pm 1/2$. The collection $\boldsymbol{\sigma} = (\sigma_x)_{x\in\Lambda_L} \in \{-1/2, 1/2\}^{\Lambda_L}$ of the spin variables is called a spin configuration. The Hamiltonian, a function of $\boldsymbol{\sigma}$, is defined as

$$H_0(\boldsymbol{\sigma}) := -\sum_{\{x,y\}\in\mathscr{B}_L} \sigma_x\sigma_y. \tag{3.2.1}$$

The equilibrium state of this model at the inverse temperature $\beta = (k_B T)^{-1} > 0$ is described by the statistical mechanical expectation value[3]

$$\langle A\rangle_{\beta,L} := \frac{1}{Z_L(\beta)}\sum_{\boldsymbol{\sigma}} A(\boldsymbol{\sigma})\, e^{-\beta H_0(\boldsymbol{\sigma})}, \tag{3.2.2}$$

where $A(\boldsymbol{\sigma})$ is an arbitrary function of spin configurations $\boldsymbol{\sigma}$, and the partition function $Z_L(\beta)$ is determined by $\langle 1\rangle_{\beta,L} = 1$.

[3]The same model can be described as a quantum spin system with Hamiltonian $\hat{H}_0 = -\sum_{\{x,y\}\in\mathscr{B}_L} \hat{S}_x^{(3)}\hat{S}_y^{(3)}$. The thermal expectation value of an operator $\hat{A}$ is defined as $\langle\hat{A}\rangle_{\beta,L} := \mathrm{Tr}[\hat{A}e^{-\beta\hat{H}_0}]/\mathrm{Tr}[e^{-\beta\hat{H}_0}]$.

Phase transition The phase transition and the associated critical phenomena in the classical Ising model are among the most intensively studied topics in mathematical physics. By now, we have almost complete understanding of basic features of the phase transition and the structures of the phases, and partial understanding of critical phenomena. See, e.g., [7–9]. For our purpose in the present book only basic knowledge is sufficient.

If $d \geq 2$, there exists a transition point β_c, which depends on d, such that $0 < \beta_c < \infty$. The model is in the disordered phase for $\beta < \beta_c$ and in the ferromagnetic phase for $\beta > \beta_c$. The two phases are characterized by the behavior of the two-point correlation function in the infinite volume defined as[4]

$$\langle \sigma_x \sigma_y \rangle_{\beta,\infty} := \lim_{L \uparrow \infty} \langle \sigma_x \sigma_y \rangle_{\beta,L}, \tag{3.2.3}$$

for $x, y \in \mathbb{Z}^d$. When $\beta < \beta_c$, there is a correlation length $\xi(\beta)$ with $0 < \xi(\beta) < \infty$ such that

$$0 \leq \langle \sigma_x \sigma_y \rangle_{\beta,\infty} \leq e^{-|x-y|/\xi(\beta)}, \tag{3.2.4}$$

for any $x, y \in \mathbb{Z}^d$. This means that the correlation decays quickly with an exponential law, and two spins separated by the distance much larger than $\xi(\beta)$ are essentially not correlated.[5] When $\beta > \beta_c$, on the other hand, it holds that the long-range order parameter is nonvanishing as

$$q(\beta) := \lim_{|x-y| \uparrow \infty} \langle \sigma_x \sigma_y \rangle_{\beta,\infty} > 0, \tag{3.2.5}$$

and one also has

$$\langle \sigma_x \sigma_y \rangle_{\beta,\infty} \geq q(\beta) > 0, \tag{3.2.6}$$

for any $x, y \in \mathbb{Z}^d$. This means that the two spins at sites x and y are, in average, pointing in the same direction, no matter how far the two sites are separated. This is a manifestation of long-range ferromagnetic order.

From the symmetry, on the other hand, one readily finds that $\langle \sigma_x \rangle_{\beta,L} = 0$ for any L and any $\beta < \infty$. This of course means $\langle \sigma_x \rangle_{\beta,\infty} = 0$. The state does not exhibit spontaneous symmetry breaking.

The behavior of order operator It is useful to rewrite the above observation in terms of the order operator

$$\mathcal{O}_L = \sum_{x \in \Lambda_L} \sigma_x, \tag{3.2.7}$$

---

[4]To be rigorous, the existence of the limit in (3.2.3) has been proved only for $\beta \leq \beta_c$ when periodic boundary conditions are used. For $\beta > \beta_c$ one should replace lim by lim sup or lim inf, or understand that a subsequence is taken when necessary. Then the following statements hold rigorously. See, e.g., [9].

[5]More precisely there is power law correction to the exponential decay. The asymptotic decay of the correlation function is (believed to be) given by $\langle \sigma_x \sigma_y \rangle_{\beta,\infty} \simeq (\text{constant})|x - y|^{-(d-1)/2} \times \exp[-|x - y|/\xi(\beta)]$, which is known as the Ornstein-Zernike form. See, e.g., [5].

which is nothing but the total magnetization. By using the translation invariance we see that

$$\langle (\mathcal{O}_L)^2 \rangle_{\beta,L} = \sum_{x,y\in\Lambda_L} \langle \sigma_x\sigma_y \rangle_{\beta,L} = L^d \sum_{x\in\Lambda_L} \langle \sigma_o\sigma_x \rangle_{\beta,L}, \tag{3.2.8}$$

where $o = (0, \ldots, 0)$ is the origin of Λ_L. By recalling the asymptotic behaviors (3.2.4)–(3.2.6), we find that, as L becomes large, the expectation value of $(\mathcal{O}_L/L^d)^2$ behaves as

$$\left\langle \left(\frac{\mathcal{O}_L}{L^d}\right)^2 \right\rangle_{\beta,L} \simeq \begin{cases} L^{-d}\,\chi(\beta) & \text{if } \beta < \beta_c, \\ q(\beta) & \text{if } \beta > \beta_c, \end{cases} \tag{3.2.9}$$

where $\chi(\beta) := \sum_{x\in\Lambda_L} \langle \sigma_o\sigma_x \rangle_{\beta,\infty} < \infty$. We also recall that the symmetry implies

$$\left\langle \frac{\mathcal{O}_L}{L^d} \right\rangle_{\beta,L} = 0, \tag{3.2.10}$$

for any $\beta < \infty$ and L.

This in particular means that, when $\beta > \beta_c$, the fluctuation of the density $\mathcal{O}_L/L^d$ behaves as

$$\sqrt{\left\langle \left(\frac{\mathcal{O}_L}{L^d}\right)^2 \right\rangle_{\beta,L} - \left(\left\langle \frac{\mathcal{O}_L}{L^d} \right\rangle_{\beta,L}\right)^2} \simeq \sqrt{q(\beta)} > 0, \tag{3.2.11}$$

for large L. Recall that in thermodynamics (or in macroscopic physics in general) the density of any bulk quantity (e.g., the energy density, the mass density, and the magnetization density $\mathcal{O}_L/L^d$) should have a definite value. This can be regarded as a consequence of the law of large numbers. The behavior (3.2.11) shows that the present equilibrium state is pathological; we find that the magnetization density $\mathcal{O}_L/L^d$ has nonvanishing fluctuation even in the limit of large L, violating the law of large numbers. The origin of the pathology is of course the existence of LRO not accompanied by SSB.

System with an infinitesimal symmetry breaking field There are several methods to get rid of the pathology and construct a "physical" equilibrium state which exhibits both LRO and SSB. We discuss a physically natural strategy which makes use of an external field which explicitly breaks the symmetry.[6] For the ferromagnetic Ising model the desired symmetry breaking field is nothing but the magnetic field.

Let us define the Hamiltonian of the Ising model under external magnetic field $h \in \mathbb{R}$ as

$$H_h(\boldsymbol{\sigma}) := H_0(\boldsymbol{\sigma}) - h\mathcal{O}_L = -\sum_{\{x,y\}\in\mathcal{B}_L} \sigma_x\sigma_y - h\sum_{x\in\Lambda_L} \sigma_x, \tag{3.2.12}$$

[6] The other strategy is to impose boundary conditions which break the symmetry. The most sophisticated method is to define equilibrium states in the infinite volume, and apply abstract decomposition theories. See Sect. 4.3.

and the corresponding expectation value as

$$\langle A\rangle_{\beta,h,L} := \frac{1}{Z_L(\beta,h)}\sum_{\sigma} A(\sigma)\, e^{-\beta H_h(\sigma)}, \quad \langle 1\rangle_{\beta,h,L} = 1. \tag{3.2.13}$$

It has been proved that this equilibrium state exhibits SSB in the infinite volume limit in the sense that

$$\lim_{h\downarrow 0}\lim_{L\uparrow\infty}\Big\langle\frac{\mathscr{O}_L}{L^d}\Big\rangle_{\beta,h,L} = \begin{cases} 0 & \text{if } \beta \le \beta_{\rm c}, \\ m^*(\beta) > 0 & \text{if } \beta > \beta_{\rm c}, \end{cases} \tag{3.2.14}$$

where the spontaneous magnetization (or the symmetry breaking order parameter) $m^*(\beta)$ satisfies

$$m^*(\beta) = \sqrt{q(\beta)}. \tag{3.2.15}$$

Here we apply a positive magnetic field h to the system, let the system size L tend to infinity, and then let h approach zero. When $\beta > \beta_{\rm c}$, the expectation value of $\mathscr{O}_L/L^d$ remains nonzero even after turning off the external field h. Intuitively one may say that an "infinitesimally small" symmetry breaking field is sufficient to break the symmetry. This is one of the most basic examples of SSB. It should be noted that the order of the two limits in (3.2.14) is crucial; by continuity in h (for a finite L) one finds $\lim_{L\uparrow\infty}\lim_{h\downarrow 0}\langle\mathscr{O}_L/L^d\rangle_{\beta,h,L} = 0$ for any $\beta < \infty$.

For the expectation value of $(\mathscr{O}_L/L^d)^2$, it has been proved that

$$\lim_{h\downarrow 0}\lim_{L\uparrow\infty}\Big\langle\Big(\frac{\mathscr{O}_L}{L^d}\Big)^2\Big\rangle_{\beta,h,L} = \begin{cases} 0 & \text{if } \beta \le \beta_{\rm c}, \\ q(\beta) & \text{if } \beta > \beta_{\rm c}. \end{cases} \tag{3.2.16}$$

This means that the fluctuation of $\mathscr{O}_L/L^d$ is vanishing:

$$\lim_{h\downarrow 0}\lim_{L\uparrow\infty}\sqrt{\Big\langle\Big(\frac{\mathscr{O}_L}{L^d}\Big)^2\Big\rangle_{\beta,h,L} - \Big(\Big\langle\frac{\mathscr{O}_L}{L^d}\Big\rangle_{\beta,h,L}\Big)^2} = 0 \tag{3.2.17}$$

This suggests, and it has been justified rigorously, that the equilibrium state obtained by taking the double limit $\lim_{h\downarrow 0}\lim_{L\uparrow\infty}$ in $\langle\cdots\rangle_{\beta,h,L}$ is a physically natural state in which the density of any bulk quantity has a definite value. We should keep this fact in mind when examining quantum systems in the following sections.

3.3 Quantum Ising Model

Let us come back to the problem of ground states of quantum spin systems. As a warm up we study possibly the simplest system that has a ground state with LRO but without SSB, namely, the quantum Ising model, which is also known as the Ising model under transverse magnetic field. The model also exhibits a prototypical ground state phase transition associated with spontaneous symmetry breaking.

Definition of the model Consider a quantum spin system with $S = 1/2$ on the one-dimensional lattice $\Lambda_L = \{1, 2, \ldots, L\}$. The Hamiltonian of the quantum Ising model is

$$\hat{H} = -\sum_{x=1}^{L-1} \hat{S}_x^{(3)} \hat{S}_{x+1}^{(3)} - \lambda \sum_{x=1}^{L} \hat{S}_x^{(1)}. \tag{3.3.1}$$

The first term precisely corresponds to the classical ferromagnetic Ising model. See Footnote 3 in p. 52. The second term, which describes the external magnetic field with magnitude $\lambda \geq 0$ in the 1-direction, introduces quantum nature to the problem. The field is often referred to as the transverse magnetic field. Note that this field is not a symmetry breaking field. We here took open boundary conditions for technical simplicity.

The present model is highly anisotropic, and has poor rotational symmetry. In fact it is invariant only under the π rotation about the 1-axis, described by the rotation operator $\hat{U}_\pi^{(1)}$ defined in (2.2.11). The model is $\mathbb{Z}_2$ invariant.[7] We call $\hat{U}_\pi^{(1)}$ the symmetry operator of the model. To see the invariance note that (2.1.16) and (2.2.11) imply $(\hat{U}_\pi^{(1)})^\dagger \hat{S}_x^{(3)} \hat{U}_\pi^{(1)} = -\hat{S}_x^{(3)}$ and $(\hat{U}_\pi^{(1)})^\dagger \hat{S}_x^{(1)} \hat{U}_\pi^{(1)} = \hat{S}_x^{(1)}$, which immediately implies $(\hat{U}_\pi^{(1)})^\dagger \hat{H} \hat{U}_\pi^{(1)} = \hat{H}$. We shall study below spontaneous breakdown of this symmetry.

To detect possible symmetry breaking we introduce the order operator

$$\hat{\mathcal{O}}_L = \sum_{x=1}^{L} \hat{S}_x^{(3)}, \tag{3.3.2}$$

which is nothing but $\hat{S}_{\mathrm{tot}}^{(3)}$, and is transformed by the symmetry operator as $(\hat{U}_\pi^{(1)})^\dagger \hat{\mathcal{O}}_L \hat{U}_\pi^{(1)} = -\hat{\mathcal{O}}_L$. To see that this can detect breakdown of the symmetry, let $|\Phi\rangle$ be an arbitrary normalized state which is invariant under the symmetry transformation, i.e., $\hat{U}_\pi^{(1)}|\Phi\rangle = \alpha|\Phi\rangle$ for some $\alpha \in \mathbb{C}$ with $|\alpha| = 1$. Then one has $-\langle\Phi|\hat{\mathcal{O}}_L|\Phi\rangle = \langle\Phi|(\hat{U}_\pi^{(1)})^\dagger \hat{\mathcal{O}}_L \hat{U}_\pi^{(1)}|\Phi\rangle = |\alpha|^2\langle\Phi|\hat{\mathcal{O}}_L|\Phi\rangle = \langle\Phi|\hat{\mathcal{O}}_L|\Phi\rangle$, which implies $\langle\Phi|\hat{\mathcal{O}}_L|\Phi\rangle = 0$. Thus nonvanishing $\langle\Phi|\hat{\mathcal{O}}_L|\Phi\rangle$ indicates that $|\Phi\rangle$ breaks the symmetry.[8]

The case with $\lambda = 0$ Let us start from the trivial case with $\lambda = 0$, which is the classical Ising model. The problem of energy eigenstates is easy. One readily sees that any basis state $|\Psi^\sigma\rangle$ defined in (2.2.1) is an energy eigenstate with eigenvalue $-\sum_{x=1}^{L-1} \sigma_x \sigma_{x+1}$.

In particular there are two ground states

---

[7] $\{\hat{1}, \hat{U}_\pi^{(1)}\}$ can be regarded as (a representation of) the discrete group $\mathbb{Z}_2$. See Appendix A.5.

[8] More generally any nonzero operator $\hat{\mathcal{O}}'_L$ such that $(\hat{U}_\pi^{(1)})^\dagger \hat{\mathcal{O}}'_L \hat{U}_\pi^{(1)} = -\hat{\mathcal{O}}'_L$ has a chance to play the role of the order operator.

$$|\Phi^{\uparrow}\rangle := \bigotimes_{x=1}^{L} |\psi_x^{\uparrow}\rangle, \quad |\Phi^{\downarrow}\rangle := \bigotimes_{x=1}^{L} |\psi_x^{\downarrow}\rangle, \tag{3.3.3}$$

where all the spins are pointing in the positive or the negative 3-direction. Here we denoted the basis states $|\psi_x^{1/2}\rangle$ and $|\psi_x^{-1/2}\rangle$ as $|\psi_x^{\uparrow}\rangle$ and $|\psi_x^{\downarrow}\rangle$, respectively. The ground state energy is $E_{\mathrm{GS}}^{(0)} = -(L-1)/4$. The two ground states (3.3.3) obviously exhibit long-range order (LRO) and also break the symmetry. This is seen from the expectation value of the order operator as follows:

$$\langle \Phi^{\uparrow}|\Big(\frac{\hat{\mathcal{O}}_L}{L}\Big)^2|\Phi^{\uparrow}\rangle = \langle \Phi^{\downarrow}|\Big(\frac{\hat{\mathcal{O}}_L}{L}\Big)^2|\Phi^{\downarrow}\rangle = \frac{1}{4} \quad \text{(LRO)} \tag{3.3.4}$$

$$\langle \Phi^{\uparrow}|\Big(\frac{\hat{\mathcal{O}}_L}{L}\Big)|\Phi^{\uparrow}\rangle = \frac{1}{2}, \quad \langle \Phi^{\downarrow}|\Big(\frac{\hat{\mathcal{O}}_L}{L}\Big)|\Phi^{\downarrow}\rangle = -\frac{1}{2} \quad \text{(symmetry breaking)} \tag{3.3.5}$$

The reason for identifying (3.3.4) as the condition of long-range order is exactly the same as that for the classical Ising model. See (3.2.8) and (3.2.9), and also (3.4.6) below.

The symmetry breaking is also manifest from the action of the symmetry transformation. From (2.1.26) or (S.4) (the latter appears in p. 494 in the solution to Problem 2.1.d) one finds that $\hat{U}_{\pi}^{(1)}|\Phi^{\uparrow}\rangle = (-i)^L|\Phi^{\downarrow}\rangle$ and $\hat{U}_{\pi}^{(1)}|\Phi^{\downarrow}\rangle = (-i)^L|\Phi^{\uparrow}\rangle$, i.e., the two states are not symmetric but related by the symmetry with each other.

The first excited states of the model with $\lambda = 0$ are those with a single "kink", a bond which connects spins pointing in the opposite directions, namely,

$$|\Phi_j^{\uparrow\downarrow}\rangle := \Big(\bigotimes_{x=1}^{j} |\psi_x^{\uparrow}\rangle\Big) \otimes \Big(\bigotimes_{x=j+1}^{L} |\psi_x^{\downarrow}\rangle\Big), \quad |\Phi_j^{\downarrow\uparrow}\rangle := \Big(\bigotimes_{x=1}^{j} |\psi_x^{\downarrow}\rangle\Big) \otimes \Big(\bigotimes_{x=j+1}^{L} |\psi_x^{\uparrow}\rangle\Big), \tag{3.3.6}$$

where $j = 1, \ldots, L-1$ denotes the position of the kink. There are $2(L-1)$ degenerate energy eigenstates with eigenvalue $E_{\mathrm{kink}} = E_{\mathrm{GS}}^{(0)} + (1/2)$. There is an energy gap $1/2$ above the ground state energy.

The case with $0 < \lambda \ll 1$ Let us turn on the positive transverse field λ, but assume that it is extremely small. We expect that low energy properties of the model do not change essentially from the classical model with $\lambda = 0$; there should be two ground states which are obtained by slightly perturbing $|\Phi^{\uparrow}\rangle$ and $|\Phi^{\downarrow}\rangle$.

Let us first note that, by using the Perron–Frobenius theorem (Theorem A.18 in p. 475) as in the proof of the Marshall–Lieb–Mattis theorem (Theorem 2.2 in p. 39), one can easily prove that, for any $\lambda > 0$, the ground state $|\Phi_{\mathrm{GS}}\rangle$ of the Hamiltonian (3.3.1) is unique, and is expanded as

$$|\Phi_{\mathrm{GS}}\rangle = \sum_{\sigma} c_{\sigma} |\Psi^{\sigma}\rangle, \tag{3.3.7}$$

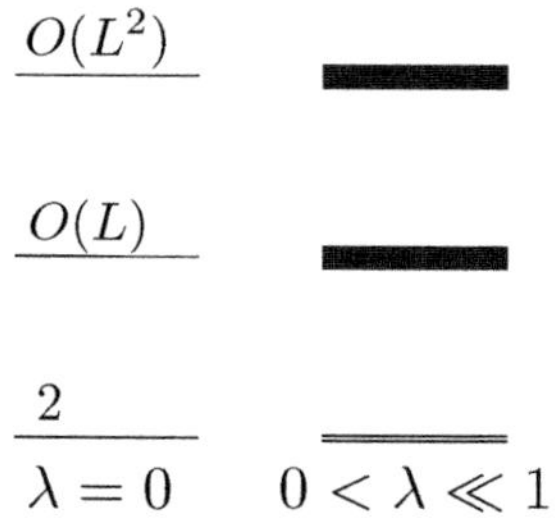

Fig. 3.3 A schematic picture of the low-energy spectra of the Hamiltonian (3.3.1) for $\lambda = 0$ and for $0 < \lambda \ll 1$. In the classical Ising model with $\lambda = 0$, the energy levels are separated by $1/2$. The ground states are doubly degenerate, and the degeneracies of the second and the third levels are of $O(L)$ and $O(L^2)$, respectively. When the perturbation $0 < \lambda \ll 1$ is turned on, the degeneracies are lifted while the overall structure of the spectrum is unchanged (© Hal Tasaki 2020. All Rights Reserved)

with coefficients $c_{\boldsymbol{\sigma}} > 0$ for any $\boldsymbol{\sigma}$. Here $|\Psi^{\boldsymbol{\sigma}}\rangle$ are the standard basis states defined in (2.2.1). It can also be proved that the ground state satisfies $\hat{U}_{\pi}^{(1)}|\Phi_{\mathrm{GS}}\rangle = (-i)^L|\Phi_{\mathrm{GS}}\rangle$, i.e., it is invariant under the symmetry operator.[9]

The uniqueness of the ground state is in conflict with the above guess that there should be two ground states. But this may not be too surprising if one recalls that the two ground states are degenerate when $\lambda = 0$. As the reader should be familiar from elementary quantum mechanics, such a degeneracy is generally lifted by an arbitrarily small perturbation. A perturbative analysis shows that, for $0 < \lambda \ll 1$, the unique ground state (with energy E_{GS}) is given by

$$|\Phi_{\mathrm{GS}}\rangle \simeq \frac{1}{\sqrt{2}}\big(|\Phi^{\uparrow}\rangle + |\Phi^{\downarrow}\rangle\big), \tag{3.3.8}$$

and the first excited state (with energy E_{1st}) by

$$|\Phi_{\mathrm{1st}}\rangle \simeq \frac{1}{\sqrt{2}}\big(|\Phi^{\uparrow}\rangle - |\Phi^{\downarrow}\rangle\big). \tag{3.3.9}$$

See Problem 3.3.a below. It is worth pointing out that both $|\Phi_{\mathrm{GS}}\rangle$ and $|\Phi_{\mathrm{1st}}\rangle$ are Schrödinger's cat-like states, in which macroscopically distinct states are superposed. This fact already indicates that these states might be pathological.

It is also found that the energy difference between the ground state and the first excited state is exponentially small in the system size, i.e., $E_{\mathrm{1st}} - E_{\mathrm{GS}} \simeq 2\lambda^L$. The

---

[9] ***Proof*** We abbreviate $\hat{U}_{\pi}^{(1)}$ as $\hat{U}$. Since the invariance of the Hamiltonian reads $\hat{H}\hat{U} = \hat{U}\hat{H}$, we see $\hat{H}(\hat{U}|\Phi_{\mathrm{GS}}\rangle) = \hat{U}\hat{H}|\Phi_{\mathrm{GS}}\rangle = E_{\mathrm{GS}}\hat{U}|\Phi_{\mathrm{GS}}\rangle$, which means that $\hat{U}|\Phi_{\mathrm{GS}}\rangle$ is also a ground state. The uniqueness then implies $\hat{U}|\Phi_{\mathrm{GS}}\rangle = \alpha|\Phi_{\mathrm{GS}}\rangle$ with some $\alpha \in \mathbb{C}$ with $|\alpha| = 1$. (Note that this conclusion is valid for any nondegenerate energy eigenstate.) For the unique ground state $|\Phi_{\mathrm{GS}}\rangle$, (3.3.7) with $c_{\boldsymbol{\sigma}} > 0$, along with (2.1.26) or (S.4), implies $\alpha = (-i)^L$.

degeneracy for $\lambda = 0$ is lifted, but only slightly.[10] See Fig. 3.3. The state $|\Phi_{1st}\rangle$ is an example of low-lying energy eigenstate, i.e., an energy eigenstate with extremely small excitation energy. See Sect. 2 of [14] for an interesting relation between these near degenerate ground states and boundary modes in the Majorana chain.[11]

The following problem, which justifies the above observations about the near degeneracy, is worth looking at.

Problem 3.3.a When $0 < \lambda \ll 1$, the properties of $|\Phi_{GS}\rangle$ and $|\Phi_{1st}\rangle$ can be well described within the $2L$ dimensional low-energy Hilbert space spanned by $|\Phi^{\uparrow}\rangle$, $|\Phi^{\downarrow}\rangle$, $|\Phi_j^{\uparrow\downarrow}\rangle$, and $|\Phi_j^{\downarrow\uparrow}\rangle$ (with $j = 1, \ldots, L-1$). Carry out the analysis and confirm the above conclusions for $L \gg 1$. [solution→p.498]

The elementary perturbative analysis in the problem is of course not mathematically rigorous, but we stress that it is indeed quite reliable. We expect that it can be made into a rigorous estimate with some extra effort.[12]

Exact ground state versus physical "ground states" The unique ground state (3.3.8) for $0 < \lambda \ll 1$ clearly exhibits LRO as

$$\langle\Phi_{GS}|\Big(\frac{\hat{O}_L}{L}\Big)^2|\Phi_{GS}\rangle \simeq \frac{1}{2}\Big\{\langle\Phi^{\uparrow}|\Big(\frac{\hat{O}_L}{L}\Big)^2|\Phi^{\uparrow}\rangle + \langle\Phi^{\downarrow}|\Big(\frac{\hat{O}_L}{L}\Big)^2|\Phi^{\downarrow}\rangle\Big\} = \frac{1}{4}, \quad (3.3.10)$$

but no SSB as

$$\langle\Phi_{GS}|\Big(\frac{\hat{O}_L}{L}\Big)|\Phi_{GS}\rangle \simeq \frac{1}{2}\Big\{\langle\Phi^{\uparrow}|\Big(\frac{\hat{O}_L}{L}\Big)|\Phi^{\uparrow}\rangle + \langle\Phi^{\downarrow}|\Big(\frac{\hat{O}_L}{L}\Big)|\Phi^{\downarrow}\rangle\Big\} = 0, \quad (3.3.11)$$

where we used (3.3.4) and (3.3.5). This means that, exactly as in (3.2.11), the density $\hat{O}_L/L$ shows nonvanishing fluctuation in this ground state:

$$\sqrt{\langle\Phi_{GS}|\Big(\frac{\hat{O}_L}{L}\Big)^2|\Phi_{GS}\rangle - \Big\{\langle\Phi_{GS}|\Big(\frac{\hat{O}_L}{L}\Big)|\Phi_{GS}\rangle\Big\}^2} \simeq \frac{1}{2} \quad (3.3.12)$$

[10] The same model with periodic boundary conditions can be exactly solved by mapping it to a free fermion system. See [16] or Appendix A of [6]. (Note that the version of [6] on the arXiv is more recent and accurate.) The above conclusions about low-lying energy levels for $0 < \lambda \ll 1$ can be confirmed by the exact solution. The calculation becomes very hard if one uses open boundary conditions as in (3.3.1).

[11] We will encounter a similar correspondence between bulk properties and edge modes in $S = 1$ antiferromagnetic chains in Sect. 8.2.

[12] We note however that a naive convergence estimate (as we learn in elementary quantum mechanics or linear algebra) is not enough for a macroscopic system as the present one. The estimate works for sufficiently small λ, but how small it should be depends on the system size L. In order to get a convergence estimate which is uniform in L, one needs to use an argument which takes into account the locality of the interactions, such as the machinery called the cluster expansion. See, e.g., [13] and references therein.

Exactly as in (3.2.11), we see that the law of large numbers is violated. We expect the law of large numbers (for the density of a physically observable quantity) to hold in any physical description of a macroscopic system, whether it is classical or quantum. Although being the exact unique ground state, $|\Phi_{\rm GS}\rangle$, in which the law of large numbers is violated, cannot be regarded as a physical state of a macroscopic system.

Of course we already know the solution to this puzzle. The all-up state $|\Phi^{\uparrow}\rangle$ and the all-down state $|\Phi^{\downarrow}\rangle$ (with inevitable small modifications) are the physically meaningful "ground states" that are relevant to experimental observations in macroscopic systems. These states are not eigenstates of the Hamiltonian for any finite L, but the physical picture strongly suggests that they represent realistic "ground states" that we expect to observe. We call them physical "ground states".[13] In the next section and the next chapter, we will develop general theories which relate low-energy eigenstates, such as $|\Phi_{\rm GS}\rangle$ and $|\Phi_{\rm 1st}\rangle$, to physical "ground states", such as $|\Phi^{\uparrow}\rangle$ and $|\Phi^{\downarrow}\rangle$. We will also explain in Sect. 4.3 (see Conjecture 4.21 and the discussion following it) how such physical "ground states" are understood within the theory of quantum many-body systems in the infinite volume. See [11, 19] for an interesting application of the idea of physical "ground states" to the problem of chiral molecules.

Before closing the section, we point out some hints for general theories. First, from (3.3.8) and (3.3.9), one finds

$$|\Phi^{\uparrow}\rangle \simeq \frac{1}{\sqrt{2}}\big(|\Phi_{\rm GS}\rangle + |\Phi_{\rm 1st}\rangle\big), \quad |\Phi^{\downarrow}\rangle \simeq \frac{1}{\sqrt{2}}\big(|\Phi_{\rm GS}\rangle - |\Phi_{\rm 1st}\rangle\big), \tag{3.3.13}$$

which suggests that physical "ground states" may in general be obtained as liner combinations of the exact ground state and a low-lying energy eigenstate. Secondly, by noting that $\hat{\mathcal{O}}_L|\Phi^{\uparrow}\rangle = (L/2)|\Phi^{\uparrow}\rangle$ and $\hat{\mathcal{O}}_L|\Phi^{\downarrow}\rangle = -(L/2)|\Phi^{\downarrow}\rangle$, one sees from (3.3.8) and (3.3.9) that

$$\hat{\mathcal{O}}_L|\Phi_{\rm GS}\rangle \simeq \hat{\mathcal{O}}_L\frac{1}{\sqrt{2}}\big(|\Phi^{\uparrow}\rangle + |\Phi^{\downarrow}\rangle\big) \simeq \frac{L}{2\sqrt{2}}\big(|\Phi^{\uparrow}\rangle - |\Phi^{\downarrow}\rangle\big) \simeq \frac{L}{2}|\Phi_{\rm 1st}\rangle. \tag{3.3.14}$$

This suggests that a low-lying energy eigenstate may be constructed (at least approximately) by operating the order operator onto the exact ground state. Finally, by combining (3.3.13) and (3.3.14), we find for the present example that

$$|\Phi^{\uparrow}\rangle \simeq \frac{1}{\sqrt{2}}\Big(|\Phi_{\rm GS}\rangle + \frac{2}{L}\hat{\mathcal{O}}_L|\Phi_{\rm GS}\rangle\Big), \quad |\Phi^{\downarrow}\rangle \simeq \frac{1}{\sqrt{2}}\Big(|\Phi_{\rm GS}\rangle - \frac{2}{L}\hat{\mathcal{O}}_L|\Phi_{\rm GS}\rangle\Big). \tag{3.3.15}$$

These relations are quite interesting since the physical "ground states" are constructed entirely out of the exact ground state $|\Phi_{\rm GS}\rangle$. We will show that analogous construction works in a much more general settings. See (3.4.14) and (4.2.10).

---

[13]The quotation marks indicate that they are not ground states in the standard definition in quantum mechanics.

Phase transition and symmetry Although not directly related to the main topic of the chapter, it is useful to briefly discuss the phase transition which takes place in the ground states of the quantum Ising model (3.3.1). We shall also see an important role played by symmetry in phase transitions and in the notion of phases. The discussion becomes important when we examine more exotic phase transitions in Sect. 8.3. It is useful to first recall that the Hamiltonian is invariant under the symmetry transformation $\hat{U}_\pi^{(1)}$, i.e., it has $\mathbb{Z}_2$ symmetry, and that, for any $\lambda > 0$, the ground state $|\Phi_{\mathrm{GS}}\rangle$ (for a finite lattice) is unique and invariant under $\hat{U}_\pi^{(1)}$.

Let us define $|\psi_x^{\rightarrow}\rangle := \{|\psi_x^{\uparrow}\rangle + |\psi_x^{\downarrow}\rangle\}/\sqrt{2}$ and $|\psi_x^{\leftarrow}\rangle := \{|\psi_x^{\uparrow}\rangle - |\psi_x^{\downarrow}\rangle\}/\sqrt{2}$. It is easily found from (2.1.3) that $\hat{S}_x^{(1)}|\psi_x^{\rightarrow}\rangle = (1/2)|\psi_x^{\rightarrow}\rangle$ and $\hat{S}_x^{(1)}|\psi_x^{\leftarrow}\rangle = -(1/2)|\psi_x^{\leftarrow}\rangle$. Thus these are the states of a single spin pointing in the positive and the negative directions, respectively, in the 1-direction.

Let us consider the situation where the transverse magnetic field $\lambda > 0$ is so large that the Hamiltonian (3.3.1) is approximated as $\hat{H} \simeq -\lambda \sum_{x=1}^{L} \hat{S}_x^{(1)}$. Then the unique ground state is given by $|\Phi_{\mathrm{GS}}\rangle \simeq \bigotimes_{x=1}^{L} |\psi_x^{\rightarrow}\rangle$ whose energy eigenvalue is $E_{\mathrm{GS}} \simeq -\lambda L/2$. More generally any energy eigenstate is written as $\bigotimes_{x=1}^{L} |\psi_x^{\tau_x}\rangle$ with $\tau_x = \rightarrow$ or $\leftarrow$. This in particular means that there is an energy gap $\Delta E \simeq \lambda$ above the ground state energy. Although the above ground state is a simple ordered state with all the spins pointing in the positive 1-direction, it is expressed in the standard basis (2.2.1) as $|\Phi_{\mathrm{GS}}\rangle \simeq \bigotimes_{x=1}^{L}\{|\psi_x^{\uparrow}\rangle + |\psi_x^{\downarrow}\rangle\}/\sqrt{2}$, which (when expanded) appears highly disordered.

Recalling the previous discussion for $0 < \lambda \ll 1$, we find that low energy properties of the model are completely different in the case with $0 < \lambda \ll 1$ and in the case with $\lambda \gg 1$. This indicates that there is a phase transition (or, more precisely, a ground state phase transition) at an intermediate value of λ. Let us briefly summarize what is known about this phase transition. To be precise we say that there is a (ground state) phase transition if the nature of the ground state and low energy excitations changes qualitatively at a certain parameter value when the parameter is varied.[14]

The phase transition is most conveniently characterized by the presence or absence of long-range order as[15]

$$\lim_{L\uparrow\infty} \langle\Phi_{\mathrm{GS}}|\Bigl(\frac{\hat{\mathcal{O}}_L}{L}\Bigr)^2|\Phi_{\mathrm{GS}}\rangle \begin{cases} > 0 & \text{if}\, 0 < \lambda < \lambda_{\mathrm{c}}; \\ = 0 & \text{if}\, \lambda > \lambda_{\mathrm{c}}, \end{cases} \tag{3.3.16}$$

where λ_{c} is the critical value of the transverse magnetic field. From the exact solution it is known that $\lambda_{\mathrm{c}} = 1/2$. See Footnote 10 in p. 59. See also [4] where the "sharpness" of the phase transition is proved in general dimensions both for the ground states and the equilibrium states.

---

[14]This definition does not assume that the two ranges of parameters belong to distinct phases. In fact we do observe a "phase transition" within a single phase. A notable example is the transition between gas and liquid, which are both parts of the same fluid phase. See Fig. 3.5. See also Footnote 18 for the notion of phase. We should say that the term "phase transition" is confusing.

[15]To be rigorous lim should be replaced by lim inf for $0 < \lambda < \lambda_{\mathrm{c}}$.

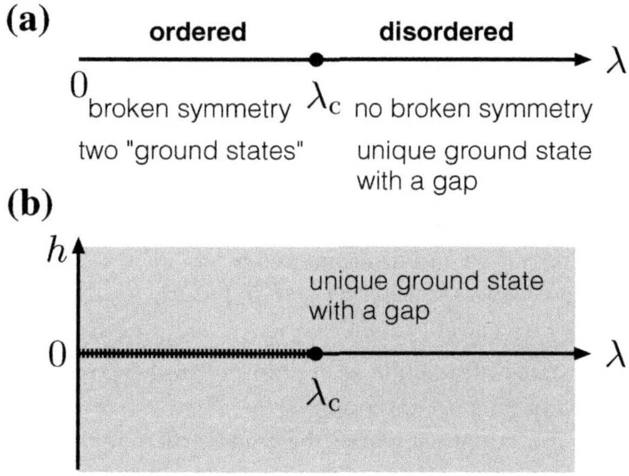

Fig. 3.4 **a** The phase diagram of the ground states of the quantum Ising model (3.3.1) which has $\mathbb{Z}_2$ symmetry. The two phases with and without symmetry breaking is separated by the critical point at λ_c. **b** The phase diagram of the ground states of the quantum Ising model with additional magnetic field (3.3.17) which explicitly breaks the $\mathbb{Z}_2$ symmetry. The model has a unique ground state with a gap except on the line determined by $h = 0$ and $0 \le \lambda \le \lambda_c$. There are no well defined phases in this case

When $\lambda > \lambda_c$, the model is in the disordered phase, where low energy properties are similar to those for $\lambda \gg 1$. The unique ground state $|\Phi_{GS}\rangle$ has no long-range order as in (3.3.16), and the energy gap above the ground state energy does not vanish as L tends to infinity.[16] This means that $|\Phi_{GS}\rangle$ is also a physical ground state for large L. We see that the $\mathbb{Z}_2$ symmetry described by $\hat{U}_\pi^{(1)}$ is not broken in the ground state.

When $0 < \lambda < \lambda_c$, the model is in the ordered phase, where the nature of the ground state and low-energy excited states are qualitatively the same as that for $0 < \lambda \ll 1$. The unique ground state $|\Phi_{GS}\rangle$ (for finite L) exhibits long-range order as in (3.3.16). Then from the general theory described in the next section, we find that there is a low-lying excited state $|\Phi_{1st}\rangle$, and there are two physical "ground states" like (3.3.13) which explicitly break the $\mathbb{Z}_2$ symmetry described by $\hat{U}_\pi^{(1)}$. Thus the ordered phase is also called the symmetry breaking phase.

To sum the phase transition in the quantum Ising model (3.3.1) is a sharp transition between the phases with and without symmetry breaking. The boundary of the two phases is a critical point where the model becomes gapless.[17] See Fig. 3.4a.

This observation suggests that the transition may disappear when the Hamiltonian loses the $\mathbb{Z}_2$ symmetry. For example the Hamiltonian with an additional magnetic field h in the 3-direction

---

[16]To be more precise, there exists a constant $\Delta E > 0$ independent of L, and the first excited energy E_{1st} satisfies $E_{1st} - E_{GS} \ge \Delta E$ for any L.

[17]There appear many low-energy excited states but the unique ground state does not break the symmetry.

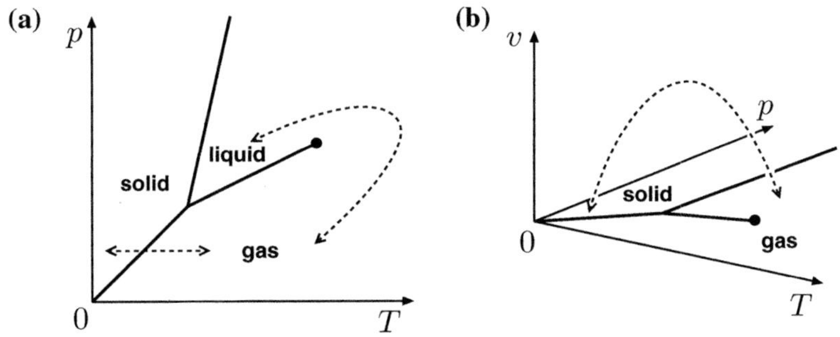

Fig. 3.5 **a** The standard phase diagram of states of matter. Within the T-p parameter space, the liquid and the gas phases are connected without passing through a phase transition, but the solid phase is a distinct phase separated from other two phases by a phase boundary. **b** By adding a new parameter v, the strength of the external (highly fictitious) potential, the solid phase and other two phases can be connected without going through any phase transitions. See Fig. 3.6 for a way to implement this process (© Hal Tasaki 2020. All Rights Reserved)

$$\hat{H} = -\sum_{x=1}^{L-1} \hat{S}_x^{(3)} \hat{S}_{x+1}^{(3)} - \lambda \sum_{x=1}^{L} \hat{S}_x^{(1)} - h \sum_{x=1}^{L} \hat{S}_x^{(3)} \tag{3.3.17}$$

is no longer invariant under the symmetry operator $\hat{U}_\pi^{(1)}$ whenever $h \neq 0$. Figure 3.4b shows the ground state phase diagram of the Hamiltonian (3.3.17). Except for the parameter range with $h = 0$ and $0 \leq \lambda \leq \lambda_c$, the model has a unique ground state accompanied by a gap (which does not vanish as L tends to infinity). This means that there are no distinct phases in the (λ, h) parameter space; there is only a line of singularity at $h = 0$ and $0 \leq \lambda \leq \lambda_c$. It is only in the restricted parameter region with $h = 0$, where the Hamiltonian is invariant under $\hat{U}_\pi^{(1)}$, that the notion of the ordered phase is meaningful.[18] One may say that the ordered phase in the quantum Ising model is "protected" by the $\mathbb{Z}_2$ symmetry, which is present only when $h = 0$.

It is likely that, in traditional examples of phase transitions, protection by certain symmetry is always necessary for the notion of phases to be meaningful.[19] As we shall see below, this remark applies to the standard equilibrium phase diagram of the three states—gas, liquid, and solid—of matter as well.

It is well known that the gas and liquid are not distinct phases since, by going over the triple point, one can go from one to the other without passing through a phase transition point. The solid, in which translation symmetry is spontaneously broken, is usually said to be a distinct phase; one must go through a phase transition point in order to go from the solid phase to the liquid or gas phase. See Fig. 3.5a. This is of course true, provided that we work within the conventional T-p plane, which only covers systems with translation symmetry.

[18]To be precise a phase is a region in the parameter space that is separated from the rest by phase transition points. This definition clearly depends on the choice of the parameter space.

[19]This is true for any phase transitions in classical physics. In quantum many-body systems there can be phases with topological order, which need not be protected by any symmetry. See Sect. 8.4.

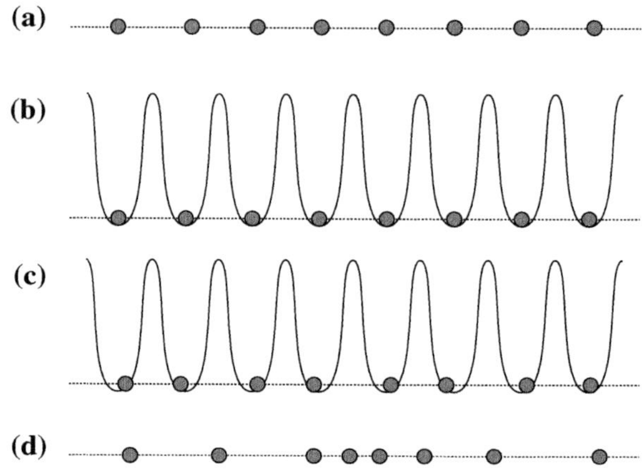

Fig. 3.6 A process, which should be realized using some unknown technology, that connects a solid state to a gas or liquid state without passing through any phase transitions. The actual system must be three dimensional. **a** One starts from the equilibrium state at sufficiently low temperature, where a normal crystalline solid is realized. **b** With temperature fixed, one applies a periodic potential which is designed to reinforce the crystalline structure. **c** With the potential fixed, one increases the temperature. **d** With the temperature fixed, one reduces the potential. The system ends up in the gas or the liquid phase (© Hal Tasaki 2020. All Rights Reserved)

By extending the parameter space as in Fig. 3.5b, we can, at least in principle, go from a solid to a liquid or gas without passing through any phase transition points. Let us describe a gedanken-experiment which realizes such a process. See Fig. 3.6. One starts from the solid phase, where a crystal is formed, and gradually applies, by using some unknown technology, an external single body potential that is periodic and commensurate to the crystalline structure. The potential is designed to reinforce the crystalline order. When the potential is sufficiently strong, one fixes it, and then gradually increases the temperature until it is sufficiently high. Finally, with temperature fixed, one slowly turns off the potential. The system will certainly end up in the liquid or gas phase, but there is no chance for any phase transitions to take place during this hypothetical process.

The essential point of this (highly fictitious) example is that the translation symmetry of the system is explicitly broken by the applied potential. With explicit symmetry breaking one can bypass the phase transition from a symmetry broken phase, i.e., the solid phase, to a symmetric phase, i.e., the gas or the liquid phases. To sum, the solid phase is a distinct phase of matter, as long as it is "protected" by the translation symmetry.

3.4 General Theory of Low-Lying States and SSB

We shall describe in detail a general theory, developed by Horsch and von der Linden [10] and Kaplan, Horsch, and von der Linden [12], which shows that the picture suggested by the simple example in the previous section applies universally to a many-body ground state exhibiting LRO without SSB. When the relevant symmetry is Ising-like (or $\mathbb{Z}_2$), the theory captures essential properties of the ground state, low-lying energy eigenstates, and physical "ground states".

Setting and assumptions Let us consider a quantum spin system on the d-dimensional hyper cubic lattice with spin S and Hamiltonian[20]

$$\hat{H} = \sum_{\{x,y\}\in\mathscr{B}_L} \hat{h}_{x,y}, \tag{3.4.1}$$

where $\hat{h}_{x,y}$ is self-adjoint, acts nontrivially only on spins at x and y (or, more precisely, on $\mathfrak{h}_x \otimes \mathfrak{h}_y$), and satisfies $\|\hat{h}_{x,y}\| = h_0$ with a constant h_0. Suppose that the relevant symmetry breaking is detected by the order operator

$$\hat{\mathcal{O}}_L = \sum_{x\in\Lambda_L} \hat{o}_x, \tag{3.4.2}$$

where $\hat{o}_x$ is self-adjoint, acts nontrivially only on spin at x, and satisfies $\|\hat{o}_x\| = o_0$ with a constant o_0. The self-adjointness of the order operator $\hat{\mathcal{O}}_L$ plays an important role here.

Examples include the quantum Ising model with Hamiltonian (3.3.1) and the order operator $\hat{\mathcal{O}}_L = \sum_{x=1}^{L} \hat{S}_x^{(3)}$, and the antiferromagnetic Heisenberg model on the d-dimensional hypercubic lattice with $d \geq 2$ with the order operator $\hat{\mathcal{O}}_L = \sum_{x\in\Lambda_L} (-1)^x \hat{S}_x^{(3)}$ (see (4.1.5) for the definition of $(-1)^x$).

We now assume that $|\Phi_{\mathrm{GS}}\rangle$ is a normalized ground state of the Hamiltonian (3.4.1), and it exhibits LRO but no SSB in the sense that

$$\langle\Phi_{\mathrm{GS}}|\Big(\frac{\hat{\mathcal{O}}_L}{L^d}\Big)^2|\Phi_{\mathrm{GS}}\rangle \geq q_0 > 0 \quad \text{(LRO)}, \tag{3.4.3}$$

$$\langle\Phi_{\mathrm{GS}}|\Big(\frac{\hat{\mathcal{O}}_L}{L^d}\Big)^n|\Phi_{\mathrm{GS}}\rangle = 0 \quad \text{for}\, n = 1, 3 \quad \text{(no SSB)}, \tag{3.4.4}$$

for sufficiently large L, where the long-range order parameter q_0 is a constant independent of L. We do not assume the uniqueness of the ground state. The condition (3.4.4) for $n = 3$ is new, but is satisfied quite generally.[21] From (3.4.3) and (3.4.4), we find

---

[20]The theory can readily be extended to more general Hamiltonian with short-range interactions. See Problem 3.4.a below.

[21]Suppose that the relevant symmetry is described by a unitary operator $\hat{U}$ such that $\hat{U}^\dagger\hat{H}\hat{U} = \hat{H}$ and $\hat{U}^\dagger\hat{\mathcal{O}}_L\hat{U} = -\hat{\mathcal{O}}_L$. This implies $\hat{U}^\dagger(\hat{\mathcal{O}}_L)^n\hat{U} = -(\hat{\mathcal{O}}_L)^n$ for any odd n. Suppose that the ground state $|\Phi_{\mathrm{GS}}\rangle$ is invariant under $\hat{U}$, i.e., $\hat{U}|\Phi_{\mathrm{GS}}\rangle = c|\Phi_{\mathrm{GS}}\rangle$ with $|c| = 1$. (This is always the case when the ground state is unique. When the ground states are degenerate, one can always find a ground state which is invariant under $\hat{U}$.) Then, for odd n, we see that $\langle\Phi_{\mathrm{GS}}|(\hat{\mathcal{O}}_L)^n|\Phi_{\mathrm{GS}}\rangle = \langle\Phi_{\mathrm{GS}}|\hat{U}^\dagger(\hat{\mathcal{O}}_L)^n\hat{U}|\Phi_{\mathrm{GS}}\rangle = -\langle\Phi_{\mathrm{GS}}|(\hat{\mathcal{O}}_L)^n|\Phi_{\mathrm{GS}}\rangle$, and hence $\langle\Phi_{\mathrm{GS}}|(\hat{\mathcal{O}}_L)^n|\Phi_{\mathrm{GS}}\rangle = 0$.

$$\sqrt{\langle\Phi_{\rm GS}|\Big(\frac{\hat{\mathcal{O}}_L}{L^d}\Big)^2|\Phi_{\rm GS}\rangle - \Big\{\langle\Phi_{\rm GS}|\Big(\frac{\hat{\mathcal{O}}_L}{L^d}\Big)|\Phi_{\rm GS}\rangle\Big\}^2} \geq \sqrt{q_0}, \tag{3.4.5}$$

which is a generalization of (3.3.12). It shows that the order operator $\hat{\mathcal{O}}_L$ violates the law of large numbers, thus suggesting that the exact ground state $|\Phi_{\rm GS}\rangle$ is unphysical. Let us denote the ground state energy as $E_{\rm GS}$.

By writing the left-hand side of (3.4.3) as $L^{-2d}\sum_{x,y,\in\Lambda_L}\langle\Phi_{\rm GS}|\hat{o}_x\hat{o}_y|\Phi_{\rm GS}\rangle$, we find that the condition (3.4.3) for LRO is almost equivalent to

$$\lim_{|x-y|\uparrow\infty}\lim_{L\uparrow\infty}\langle\Phi_{\rm GS}|\hat{o}_x\hat{o}_y|\Phi_{\rm GS}\rangle \geq q_0. \tag{3.4.6}$$

Although the condition (3.4.6) is often used as a criterion for LRO, we shall use (3.4.3) in the present part, mainly for theoretical convenience. We believe that the two criteria are equivalent, but there is no general proof.[22]

Basic variational estimate Suggested by the observation (3.3.14), we introduce the normalized trial state of Horsch and von der Linden [10]:

$$|\Gamma\rangle = \frac{\hat{\mathcal{O}}_L|\Phi_{\rm GS}\rangle}{\|\hat{\mathcal{O}}_L|\Phi_{\rm GS}\rangle\|} \tag{3.4.7}$$

Note that the condition (3.4.4) implies the orthogonality $\langle\Phi_{\rm GS}|\Gamma\rangle = 0$. By recalling $\|\hat{\mathcal{O}}_L|\Phi_{\rm GS}\rangle\|^2 = \langle\Phi_{\rm GS}|(\hat{\mathcal{O}}_L)^2|\Phi_{\rm GS}\rangle$, we estimate the energy expectation value of the trial state $|\Gamma\rangle$ as

$$\begin{aligned}
\langle\Gamma|\hat{H}|\Gamma\rangle - E_{\rm GS} &= \frac{\langle\Phi_{\rm GS}|\hat{\mathcal{O}}_L\hat{H}\hat{\mathcal{O}}_L|\Phi_{\rm GS}\rangle - E_{\rm GS}\langle\Phi_{\rm GS}|(\hat{\mathcal{O}}_L)^2|\Phi_{\rm GS}\rangle}{\langle\Phi_{\rm GS}|(\hat{\mathcal{O}}_L)^2|\Phi_{\rm GS}\rangle} \\
&= \frac{\langle\Phi_{\rm GS}|\hat{\mathcal{O}}_L\hat{H}\hat{\mathcal{O}}_L|\Phi_{\rm GS}\rangle - \frac{1}{2}\langle\Phi_{\rm GS}|(\hat{\mathcal{O}}_L)^2\hat{H}|\Phi_{\rm GS}\rangle - \frac{1}{2}\langle\Phi_{\rm GS}|\hat{H}(\hat{\mathcal{O}}_L)^2|\Phi_{\rm GS}\rangle}{\langle\Phi_{\rm GS}|(\hat{\mathcal{O}}_L)^2|\Phi_{\rm GS}\rangle} \\
&= \frac{\langle\Phi_{\rm GS}|[\hat{\mathcal{O}}_L,[\hat{H},\hat{\mathcal{O}}_L]]|\Phi_{\rm GS}\rangle}{2\langle\Phi_{\rm GS}|(\hat{\mathcal{O}}_L)^2|\Phi_{\rm GS}\rangle},
\end{aligned} \tag{3.4.8}$$

where the final expression in terms of the double commutator can be easily verified by inspection. By using (3.4.1) and (3.4.2), we see

$$[\hat{H},\hat{\mathcal{O}}_L] = \sum_{\{x,y\}\in\mathcal{B}_L}\sum_{z\in\Lambda_L}[\hat{h}_{x,y},\hat{o}_z] = \sum_{\{x,y\}\in\mathcal{B}_L}\big\{[\hat{h}_{x,y},\hat{o}_x] + [\hat{h}_{x,y},\hat{o}_y]\big\}, \tag{3.4.9}$$

where we used the locality of $\hat{h}_{x,y}$ and $\hat{o}_x$. Write $[\hat{h}_{x,y},\hat{o}_x] + [\hat{h}_{x,y},\hat{o}_y]$ as $\hat{a}_{x,y}$. Then we similarly have

---

[22] In the classical Ising model the equivalence is rigorously known. See Sect. 3.2.

$$[\hat{\mathcal{O}}_L, [\hat{H}, \hat{\mathcal{O}}_L]] = \sum_{z\in\Lambda_L}\sum_{\{x,y\}\in\mathcal{B}_L} [\hat{o}_z, \hat{a}_{x,y}] = \sum_{\{x,y\}\in\mathcal{B}_L} \big\{[\hat{o}_x, \hat{a}_{x,y}] + [\hat{o}_y, \hat{a}_{x,y}]\big\}. \tag{3.4.10}$$

By using the properties (A.2.5) and (A.2.6) for the operator norm, one finds $\|[\hat{h}_{x,y}, \hat{o}_x]\| \le 2\|\hat{h}_{x,y}\|\,\|\hat{o}_x\| \le 2h_0 o_0$. Repeating similar estimates, one gets $\Big\|[\hat{\mathcal{O}}_L, [\hat{H}, \hat{\mathcal{O}}_L]]\Big\| \le 16h_0(o_0)^2|\mathcal{B}_L|$, which, with $|\mathcal{B}_L| = dL^d$, implies[23]

$$\langle\Phi_{\rm GS}|[\hat{\mathcal{O}}_L, [\hat{H}, \hat{\mathcal{O}}_L]]|\Phi_{\rm GS}\rangle \le \Big\|[\hat{\mathcal{O}}_L, [\hat{H}, \hat{\mathcal{O}}_L]]\Big\| \le \{16dh_0(o_0)^2\}\,L^d. \tag{3.4.11}$$

Substituting this bound and the assumption (3.4.3) about LRO into (3.4.8), we get

$$0 \le \langle\Gamma|\hat{H}|\Gamma\rangle - E_{\rm GS} \le C\,L^{-d}, \tag{3.4.12}$$

with the constant $C = 8dh_0(o_0)^2/q_0$. The bound (3.4.12) is the most important variational estimate in the present section. It shows that $|\Gamma\rangle$ is a low-lying state, i.e., a state (not necessarily an energy eigenstate) whose energy expectation value approaches the ground state energy as L becomes large. See the discussion after Theorem 4.6 in p. 94 for a more careful characterization of low-lying states.

From an elementary variational argument, one sees that the existence of a low-lying state orthogonal to the ground state implies the existence of a low-lying energy eigenstate, i.e., a low-lying state which is also an energy eigenstate.

Theorem 3.1 (Horsch-von der Linden theorem) *Assume* (3.4.3) *and* (3.4.4). *For sufficiently large L, there exists an energy eigenstate* $|\Psi\rangle$ *such that* $\langle\Phi_{\rm GS}|\Psi\rangle = 0$ *whose energy eigenvalue E satisfies* $E_{\rm GS} \le E \le E_{\rm GS} + C\,L^{-d}$.

The theorem shows that the existence of a ground state with LRO but without SSB inevitably implies the existence of a low-lying energy eigenstate. (The state $|\Psi\rangle$ may be another ground state when the ground states are degenerate.) We should note here that the upper bound $C\,L^{-d}$ of the energy gap is much larger than the exponentially small energy gap $E_{\rm 1st} - E_{\rm GS} \sim 2\lambda^L$ in the quantum Ising model treated in Sect. 3.3. This is a limit of the general theory, which in fact applies to models (such as the antiferromagnetic Heisenberg model) where the energy gap indeed scales as L^{-d}.

Proof of Theorem 3.1 This is elementary. Denote the normalized energy eigenstates as $|\Psi_j\rangle$ $(j = 0, 1, \dots)$ with $|\Psi_0\rangle = |\Phi_{\rm GS}\rangle$ and $\hat{H}|\Psi_j\rangle = E_j|\Psi_j\rangle$. Noting that the expansion $|\Gamma\rangle = \sum_{j\ne0} c_j|\Psi_j\rangle$ implies $\langle\Gamma|\hat{H}|\Gamma\rangle = \sum_{j\ne0}|c_j|^2E_j$ with $\sum_{j\ne0}|c_j|^2 = 1$, one finds from (3.4.12) that there must be $j \ne 0$ such that $E_j - E_{\rm GS} \le C\,L^{-d}$. ∎

Problem 3.4.a The key estimate (3.4.11) can be considerably generalized. Write the Hamiltonian and the order operator as $\hat{H} = \sum_{x\in\Lambda_L}\hat{h}_x$ and $\hat{\mathcal{O}}_L = \sum_{x\in\Lambda_L}\hat{o}_x$, and

[23] In [20] the bound (3.4.11) was used to rule out the possibility of "time crystal", i.e., spontaneous breaking of time translation symmetry, in the ground states of a general class of quantum many-body systems.

assume that both $\hat{h}_x$ and $\hat{o}_x$ act only on sites y such that $|x-y| \le r$, where the range r is a fixed constant. We also assume $\|\hat{h}_x\| \le h_0$ and $\|\hat{o}_x\| \le o_0$ for any $x \in \Lambda_L$ with constants h_0 and o_0. Then show that

$$\langle \Phi_{\mathrm{GS}}|[\hat{\mathcal{O}}_L, [\hat{H}, \hat{\mathcal{O}}_L]]|\Phi_{\mathrm{GS}}\rangle \le 4(2r+1)^d(4r+1)^d h_0(o_0)^2 L^d. \tag{3.4.13}$$

This means that Theorem 3.1 extends to the present general class of models provided that (3.4.3) and $\langle \Phi_{\mathrm{GS}}|\hat{\mathcal{O}}_L|\Phi_{\mathrm{GS}}\rangle = 0$ hold. [solution→p.501]

Low-lying states with LRO and SSB The next step of the general theory is to construct a low-lying state which exhibits both LRO and SSB.

Again, suggested by the observation (3.3.15) in the quantum Ising model, we define another trial state

$$|\Xi_+\rangle = \frac{1}{\sqrt{2}}\big(|\Phi_{\mathrm{GS}}\rangle + |\Gamma\rangle\big), \tag{3.4.14}$$

which obviously satisfies $\langle \Xi_+|\Xi_+\rangle = 1$ and $\langle \Xi_+|\hat{H}|\Xi_+\rangle \le E_{\mathrm{GS}} + (C/2)L^{-d}$. The state $|\Xi_+\rangle$ is a low-lying state.

Recalling the definition (3.4.7) of $|\Gamma\rangle$, we evaluate the expectation value of the order parameter as

$$\begin{aligned}\langle \Xi_+|\hat{\mathcal{O}}_L|\Xi_+\rangle &= \frac{1}{2}\Big\{\Big(\langle \Phi_{\mathrm{GS}}| + \frac{\langle \Phi_{\mathrm{GS}}|\hat{\mathcal{O}}_L}{\|\hat{\mathcal{O}}_L|\Phi_{\mathrm{GS}}\rangle\|}\Big)\hat{\mathcal{O}}_L\Big(|\Phi_{\mathrm{GS}}\rangle + \frac{\hat{\mathcal{O}}_L|\Phi_{\mathrm{GS}}\rangle}{\|\hat{\mathcal{O}}_L|\Phi_{\mathrm{GS}}\rangle\|}\Big)\Big\}\\ &= \frac{\langle \Phi_{\mathrm{GS}}|(\hat{\mathcal{O}}_L)^2|\Phi_{\mathrm{GS}}\rangle}{\|\hat{\mathcal{O}}_L|\Phi_{\mathrm{GS}}\rangle\|} = \sqrt{\langle \Phi_{\mathrm{GS}}|(\hat{\mathcal{O}}_L)^2|\Phi_{\mathrm{GS}}\rangle},\end{aligned} \tag{3.4.15}$$

where we used the assumption (3.4.4) about the absence of SSB in $|\Phi_{\mathrm{GS}}\rangle$. With the assumption (3.4.3) about the existence of LRO, this implies the crucial bound

$$\langle \Xi_+|\frac{\hat{\mathcal{O}}_L}{L^d}|\Xi_+\rangle \ge \sqrt{q_0}, \tag{3.4.16}$$

which shows that the state $|\Xi_+\rangle$ exhibits symmetry breaking. We have thus found that $|\Xi_+\rangle$ is a low-lying state which exhibits both LRO and SSB (see the remark below). Of course the state $|\Xi_-\rangle = (|\Phi_{\mathrm{GS}}\rangle - |\Gamma\rangle)/\sqrt{2}$, which is orthogonal to $|\Xi_L^+\rangle$, is a low-lying state that satisfies $\langle \Xi_-|(\hat{\mathcal{O}}_L/L^d)|\Xi_-\rangle \le -\sqrt{q_0}$.

Given the experience in the quantum Ising model in Sect. 3.3, one might well expect that the low-lying states $|\Xi_+\rangle$ and $|\Xi_-\rangle$, which can be almost identified with $|\Phi^{\uparrow}\rangle$ and $|\Phi^{\downarrow}\rangle$, are physical "ground states". We shall give a precise mathematical formulation of the notion of physical "ground states" in Sect. 4.3 where we treat the system in the infinite volume (see Conjecture 4.21 and the discussion following it). For the moment, it is sufficient to understand that a physical "ground state" is a low-lying state in which the density of any bulk quantity does not fluctuate (when L becomes large).

For the quantum Ising model and similar models with discrete symmetry breaking, the low-lying states $|\Xi_+\rangle$ and $|\Xi_-\rangle$ are indeed physical "ground states". (This fact has been proved rigorously in Appendix B of [15].) See the following Problem 3.4.b. We shall see, however, in Sect. 4.2 that, in the antiferromagnetic Heisenberg model (and other models with continuous symmetry), the states $|\Xi_+\rangle$ and $|\Xi_-\rangle$ are still pathological states in the sense that the density of the order operator shows huge fluctuations. This means that there is a richer story regarding low-lying states with SSB, which will be revealed in Chap. 4.

Remark Note that the Schwarz inequality implies, for any normalized state $|\Phi\rangle$, that

$$\left|\langle\Phi|\frac{\hat{\mathcal{O}}_L}{L^d}|\Phi\rangle\right| \le \sqrt{\langle\Phi|\Big(\frac{\hat{\mathcal{O}}_L}{L^d}\Big)^2|\Phi\rangle}. \tag{3.4.17}$$

This means that the existence of symmetry breaking, which means that the left-hand side is strictly positive, implies the existence of long-range order. Thus "SSB without LRO" is impossible, although we do encounter "LRO without SSB".

Problem 3.4.b Show that the condition

$$\lim_{L\uparrow\infty}\left[\langle\Phi_{\rm GS}|\Big(\frac{\hat{\mathcal{O}}_L}{L^d}\Big)^4|\Phi_{\rm GS}\rangle - \Big\{\langle\Phi_{\rm GS}|\Big(\frac{\hat{\mathcal{O}}_L}{L^d}\Big)^2|\Phi_{\rm GS}\rangle\Big\}^2\right] = 0 \tag{3.4.18}$$

implies that the fluctuation of $\hat{\mathcal{O}}_L/L^d$ in the state $|\Xi_+\rangle$ vanishes as $L\uparrow\infty$, and hence $|\Xi_+\rangle$ can be regarded as a physical "ground state". Note that the condition (3.4.18) essentially says that $(\hat{\mathcal{O}}_L/L^d)^2$ does not fluctuate much. This is a nontrivial condition, which is valid for the quantum Ising model (and other models where $\mathbb{Z}_2$ symmetry is broken) but not for the antiferromagnetic Heisenberg model (and other models where a continuous symmetry is broken). [solution→p.501]

SSB under infinitesimal symmetry breaking field We shall finally show that SSB can be triggered by an infinitesimal symmetry breaking field, exactly as in the equilibrium state of the classical Ising model. The low-lying state $|\Xi_+\rangle$ plays an essential role here.

The following simple but clever variational argument is due to Kaplan, Horsch, and von der Linden [12]. Define the Hamiltonian with symmetry breaking field $h\ge 0$ as

$$\hat{H}_h = \hat{H} - h\hat{\mathcal{O}}_L, \tag{3.4.19}$$

where $\hat{H}$ is the original Hamiltonian in (3.4.1). Note that in the quantum Ising model, where the order operator is taken as $\hat{\mathcal{O}}_L = \sum_{x=1}^{L}\hat{S}_x^{(3)}$, the symmetry breaking field h is nothing but the standard external magnetic field. But in the antiferromagnetic Heisenberg model, where the order operator is $\hat{\mathcal{O}}_L = \sum_{x\in\Lambda_L}(-1)^x\,\hat{S}_x^{(3)}$ (see (4.1.5) for the definition of $(-1)^x$), the field h should have opposite signs for neighboring spins. Such a rapidly oscillating magnetic field, usually called the staggered magnetic field, can never be generated experimentally. It is a fictitious field introduced only to test for possible existence of symmetry breaking.

Let $|\Phi_{\mathrm{GS},h}\rangle$ be a ground state of the new Hamiltonian $\hat{H}_h$. Although we don't know almost anything about $|\Phi_{\mathrm{GS},h}\rangle$, we at least know from the definition of ground state that

$$\langle \Xi_+|\hat{H}_h|\Xi_+\rangle \geq \langle \Phi_{\mathrm{GS},h}|\hat{H}_h|\Phi_{\mathrm{GS},h}\rangle. \tag{3.4.20}$$

By substituting (3.4.19), and arranging the inequality, we find

$$\begin{aligned}\langle \Phi_{\mathrm{GS},h}|\frac{\hat{\mathcal{O}}_L}{L^d}|\Phi_{\mathrm{GS},h}\rangle &\geq \langle \Xi_+|\frac{\hat{\mathcal{O}}_L}{L^d}|\Xi_+\rangle + \frac{1}{hL^d}\{\langle \Phi_{\mathrm{GS},h}|\hat{H}|\Phi_{\mathrm{GS},h}\rangle - \langle \Xi_+|\hat{H}|\Xi_+\rangle\} \\ &\geq \sqrt{q_0} + \frac{1}{hL^d}\{E_{\mathrm{GS}} - \langle \Xi_+|\hat{H}|\Xi_+\rangle\},\end{aligned} \tag{3.4.21}$$

where we used the essential bound (3.4.16) and a trivial inequality $\langle \Phi_{\mathrm{GS},h}|\hat{H}|\Phi_{\mathrm{GS},h}\rangle \geq E_{\mathrm{GS}}$ to get the second inequality. We know that $0 \geq E_{\mathrm{GS}} - \langle \Xi_+|\hat{H}|\Xi_+\rangle \geq -(C/2)L^{-d}$, and hence the second term in the right-most-hand vanishes as $L \uparrow \infty$. Thus the following theorem was proved [12].

Theorem 3.2 (Kaplan–Horsch–von der Linden theorem) *Assume* (3.4.3) *and* (3.4.4). *Then we have*[24]

$$\lim_{h\downarrow 0}\lim_{L\uparrow\infty} \langle \Phi_{\mathrm{GS},h}|\frac{\hat{\mathcal{O}}_L}{L^d}|\Phi_{\mathrm{GS},h}\rangle \geq \sqrt{q_0}. \tag{3.4.22}$$

Note that the order of the limits is again essential, since one obviously has $\lim_{h\downarrow 0}\langle \Phi_{\mathrm{GS},h}|(\hat{\mathcal{O}}_L/L^d)|\Phi_{\mathrm{GS},h}\rangle = 0$ for any finite L by continuity.

The theorem is important since it establishes that an infinitesimal symmetry breaking field can trigger SSB in quantum many-body systems. The state (in the infinite volume limit) obtained by taking the double limit (i.e., first $L \uparrow \infty$ and then $h \downarrow 0$) in $|\Phi_{\mathrm{GS},h}\rangle$ should be one of the desired physical "ground states". Unfortunately we are not able to investigate detailed properties of the limiting state based on the present variational argument.

References

1. P.W. Anderson, An approximate quantum theory of the antiferromagnetic ground state. Phys. Rev. **86**, 694 (1952)
2. P.W. Anderson, *Basic Notions of Condensed Matter Physics*, (Westview Press/Addison-Wesley, Boston, 1997)

---

[24] To be rigorous lim should be replaced by lim inf.

3. A.J. Beekman, L. Rademaker, J. van Wezel, *An Introduction to Spontaneous Symmetry Breaking* (2019), arXiv:1909.01820
4. J.E. Björnberg, G.R. Grimmett, The phase transition of the quantum Ising model is sharp. J. Stat. Phys. **136**, 231–273 (2009), arXiv:0901.0328
5. J. Bricmont, J. Fröhlich, Statistical mechanical methods in particle structure analysis of lattice field theories II: Scalar and surface models. Commun. Math. Phys. **98**, 553–578 (1985), https://projecteuclid.org/euclid.cmp/1103942542
6. P. Calabrese, F.H.L. Essler, M. Fagotti, Quantum quench in the transverse field Ising chain: I. Time evolution of order parameter correlators. J. Stat. Mech. P07016 (2012), arXiv:1204.3911
7. R. Fernández, J. Fröhlich, A.D. Sokal, *Random Walks, Critical Phenomena, and Triviality in Quantum Field Theory* (Springer, Berlin, 1992)
8. S. Friedli, Y. Velenik, *Statistical Mechanics of Lattice Systems: a Concrete Mathematical Introduction* (Cambridge University Press, Cambridge, 2017)
9. T. Hara, H. Tasaki, *Mathematics and Physics of Phase Transitions and Critical Phenomena* (in Japanese) (Kyoritsu, 2015)
10. P. Horsch, W. von der Linden, Spin-correlations and low lying excited states of the spin-1/2 Heisenberg antiferromagnet on a square lattice. Z. Phys. B **72**, 181–193 (1988)
11. G. Jona-Lasinio, C. Presilla, C. Toninelli, Interaction induced localization in a gas of pyramidal molecules. Phys. Rev. Lett. **88**, 123001 (2002). arXiv:cond-mat/0107342
12. T.A. Kaplan, P. Horsch, W. von der Linden, Order Parameter in Quantum Antiferromagnets. J. Phys. Soc. Jpn. **11**, 3894–3898 (1989)
13. T. Kennedy and H. Tasaki, *Hidden symmetry breaking and the Haldane phase in $S = 1$ quantum spin chains*, Comm. Math. Phys. **147**, 431–484 (1992). https://projecteuclid.org/euclid.cmp/1104250747
14. A. Kitaev and C. Laumann, *Topological phases and quantum computation*, Lectures given by Kitaev at the 2008 Les Houches Summer School "Exact methods in low-dimensional physics and quantum computing". arXiv:0904.2771
15. T. Koma and H. Tasaki, *Symmetry breaking and finite-size effects in quantum many-body systems*, J. Stat. Phys. **76**, 745–803 (1994). (The version in the arXiv is more complete than the published version.) arXiv:9708132
16. E.H. Lieb, T. Schultz, D. Mattis, Two soluble models of an antiferromagnetic chain. Ann. Phys. **16**, 407–466 (1961)
17. A. Mazurenko, C.S. Chiu, G. Ji, M.F. Parsons, M. Kanász-Nagy, R. Schmidt, F. Grusdt, E. Demler, D. Greif, and M. Greiner, *A cold-atom Fermi-Hubbard antiferromagnet*, Nature **545**, 462–466 (2017). arXiv:1612.08436
18. M.F. Parsons, A. Mazurenko, C.S. Chiu, G. Ji, D. Greif, and M. Greiner, *Site-resolved measurement of the spin-correlation function in the Fermi-Hubbard model*, Science **353**, 1253–1256 (2016). http://science.sciencemag.org/content/353/6305/1253
19. C. Presilla and G. Jona-Lasinio, *Spontaneous symmetry breaking and inversion-line spectroscopy in gas mixtures*, Phys. Rev. A **91**, 022709 (2015). arXiv:1501.02099
20. H. Watanabe and M. Oshikawa, *Absence of Quantum Time Crystals*, Phys. Rev. Lett. **114**, 251603 (2015). arXiv:1410.2143v3

Chapter 4
Long-Range Order and Spontaneous Symmetry Breaking in the Antiferromagnetic Heisenberg Model

In the present chapter, which is the main chapter of part I, we develop rigorous and almost complete theories about LRO and SSB in the ground states of the antiferromagnetic Heisenberg model in dimensions two or higher. The problem is subtle and difficult compared with that of the quantum Ising model, because the relevant symmetry (which will be eventually broken) is continuous. But the continuity of the symmetry leads to an interesting object known as the "tower of states". Section 4.1 is devoted to the discussion about the existence or absence of LRO in the ground state of the antiferromagnetic Heisenberg model. We give a complete proof of the important theorem on the existence of LRO. Then, in Sect. 4.2, we describe in detail the rigorous theory about the "tower of states" and its relation to SSB. In Sect. 4.3, we briefly discuss the mathematical formulation of quantum spin systems in the infinite volume, and see the implications of the results in Sect. 4.2. In the final Sect. 4.4, which is in a sense independent from the rest of the book, we discuss the existence or absence of order in the equilibrium states of the Heisenberg model. We prove two important theorems about the absence of order in two dimensions. One of them, the improved Hohenberg–Mermin–Wagner theorem (Theorem 4.24 in p. 124), is proved here for the first time.

4.1 Existence of Long-Range Order

The topic of the present section is the existence of long-range order in the ground states of the antiferromagnetic Heisenberg model in two or higher dimensions. In Sect. 4.1.1, we discuss the main theorem and its implications. We describe the proof of the theorem carefully in Sect. 4.1.2. We hope that this provides an accessible account of the powerful method of reflection positivity in quantum many-body systems.

H. Tasaki, *Physics and Mathematics of Quantum Many-Body Systems*,
Graduate Texts in Physics, https://doi.org/10.1007/978-3-030-41265-4_4

4.1.1 Main Results

Let us consider the antiferromagnetic Heisenberg model with spin S on the d-dimensional hyper cubic lattice $(\Lambda_L, \mathcal{B}_L)$ with even L. See (3.1.2) and (3.1.3) for the definition of the lattice. The Hamiltonian is

$$\hat{H} = \sum_{\{x,y\}\in\mathcal{B}_L} \hat{\boldsymbol{S}}_x \cdot \hat{\boldsymbol{S}}_y. \tag{4.1.1}$$

Note that the lattice is bipartite, i.e., we can decompose it as $\Lambda_L = A \cup B$ with

$$A := \Bigl\{(x_1, x_2, \ldots, x_d) \in \Lambda_L \Bigm| \sum_{i=1}^{d} x_i \text{ is even}\Bigr\}, \tag{4.1.2}$$

$$B := \Bigl\{(x_1, x_2, \ldots, x_d) \in \Lambda_L \Bigm| \sum_{i=1}^{d} x_i \text{ is odd}\Bigr\}, \tag{4.1.3}$$

so that there are no bonds in $\mathcal{B}_L$ connecting sites in the same sublattice. We also have $|A| = |B| = L^d/2$. The Marshall–Lieb–Mattis theorem (Theorem 2.2 in p. 39) thus applies to the present model, and guarantees that there is a unique ground state with $S_{\mathrm{tot}} = 0$. We denote by $|\Phi_{\mathrm{GS}}\rangle$ the ground state for each system size L.

We recall that the Hamiltonian (4.1.1) has complete SU(2) invariance as in (2.4.3). We shall argue that a possible breakdown of the SU(2) symmetry associated with antiferromagnetic long-range order (see Sect. 3.1) may be detected by the order operators

$$\hat{\mathcal{O}}_L^{(\alpha)} := \sum_{x\in\Lambda_L} (-1)^x \hat{S}_x^{(\alpha)}, \tag{4.1.4}$$

with $\alpha = 1, 2, 3$. Here the sign factor is defined as

$$(-1)^x := \begin{cases} 1 & x \in A, \\ -1 & x \in B. \end{cases} \tag{4.1.5}$$

To justify this choice, we first let $\hat{A}$ be an arbitrary operator, and set $\Delta\hat{A} = (\hat{U}_\theta^{(\alpha)})^\dagger \hat{A} \hat{U}_\theta^{(\alpha)} - \hat{A}$ for some $\theta \neq 0$. If a state $|\Phi\rangle$ is invariant under $\hat{U}_\theta^{(\alpha)}$, i.e., $\hat{U}_\theta^{(\alpha)}|\Phi\rangle = c|\Phi\rangle$ with some $c \in \mathbb{C}$ such that $|c| = 1$, then we have $\langle\Phi|\Delta\hat{A}|\Phi\rangle = \langle\Phi|(\hat{U}_\theta^{(\alpha)})^\dagger \hat{A}\hat{U}_\theta^{(\alpha)}|\Phi\rangle - \langle\Phi|\hat{A}|\Phi\rangle = |c|^2\langle\Phi|\hat{A}|\Phi\rangle - \langle\Phi|\hat{A}|\Phi\rangle = 0$. Thus nonvanishing $\langle\Phi|\Delta\hat{A}|\Phi\rangle$ is a sign of breakdown of the symmetry. The operator $\Delta\hat{A}$ however itself is not convenient to be used as an order parameter. Observe from the definitions (2.2.11) and (A.2.16) that

$$\Delta\hat{A} = (\hat{U}_\theta^{(\alpha)})^\dagger \hat{A}\hat{U}_\theta^{(\alpha)} - \hat{A} = i\theta\,[\hat{S}_{\mathrm{tot}}^{(\alpha)}, \hat{A}] + O(\theta^2). \tag{4.1.6}$$

This means that the commutator $[\hat{S}^{(\alpha)}_{\text{tot}}, \hat{A}]$ can also be used as an order parameter. Anticipating antiferromagnetic ordering, we now chose $\hat{A} = \hat{\mathcal{O}}^{(\beta)}_L$ with $\beta \neq \alpha$. Then the basic commutation relation (2.2.6) implies that $[\hat{S}^{(\alpha)}_{\text{tot}}, \hat{\mathcal{O}}^{(\beta)}_L] = i\,\varepsilon_{\alpha,\beta,\gamma}\hat{\mathcal{O}}^{(\gamma)}_L$ for the unique γ such that $\gamma \neq \alpha$ and $\gamma \neq \beta$. We have thus confirmed that $\hat{\mathcal{O}}^{(\gamma)}_L$ with any γ which is not equal to α may detect breakdown of the symmetry with respect to $\hat{U}^{(\alpha)}_\theta$.[1]

Our starting point is the following essential theorem, which establishes that the unique ground state $|\Phi_{\text{GS}}\rangle$ exhibits long-range antiferromagnetic order. The theorem is (partly) proved in Sect. 4.1.2.

Theorem 4.1 (Dyson–Lieb–Simon, Neves–Perez, Kennedy–Lieb–Shastry theorem) *For $d \geq 3$ and any S, or $d = 2$ and $S \geq 1$, there exists a positive constant q_0 (which depends only on d and S) such that*

$$\langle \Phi_{\text{GS}}|\Bigl(\frac{\hat{\mathcal{O}}^{(\alpha)}_L}{L^d}\Bigr)^2|\Phi_{\text{GS}}\rangle \geq q_0, \tag{4.1.7}$$

for sufficiently large L and $\alpha = 1, 2, 3$.

Note that the left-hand side of (4.1.7) is independent of α because $|\Phi_{\text{GS}}\rangle$ is SU(2) invariant.[2] The theorem was proved by Neves and Perez [47], except for the difficult case with $d = 3$ and $S = 1/2$, by extending the earlier proof for finite temperatures by Dyson, Lieb, and Simon [12]. The case $d = 3$ and $S = 1/2$ was proved later by Kennedy, Lieb, and Shastry [29], who also presented a new proof which directly applies to the ground state. We review this proof in Sect. 4.1.2.

The positive constant q_0 in the theorem is the long-range order parameter. Note that the inequality (2.5.7) (which is a corollary of the Marshall–Lieb–Mattis theorem) implies $\langle \Phi_{\text{GS}}|(\hat{\mathcal{O}}^{(\alpha)}_L/L^d)^2|\Phi_{\text{GS}}\rangle > 0$ for any L. But this does note rule out the possibility that $\langle \Phi_{\text{GS}}|(\hat{\mathcal{O}}^{(\alpha)}_L/L^d)^2|\Phi_{\text{GS}}\rangle$ decreases as L grows and converges to zero in the limit $L \uparrow \infty$. In fact this is indeed the case for $d = 1$ as we shall see below in (4.1.11). The bound (4.1.7) is stronger in an essential manner, and shows that the expectation value is strictly positive no matter how large L becomes.

Note that Theorem 4.1 does not cover the case with $d = 2$ and $S = 1/2$. Whether the ground state of this model exhibits long-range order or not was once controversial. But by now there have been sufficiently convincing numerical evidences of ordering, and researchers believe that (4.1.7) is still valid with $q_0 > 0$. See, e.g., [56, 64].

Theorem 4.1 establishes that the unique ground state $|\Phi_{\text{GS}}\rangle$ of the Hamiltonian (4.1.1) exhibits LRO. On the other hand the uniqueness of $|\Phi_{\text{GS}}\rangle$ implies that it is SU(2) invariant, and hence

---

[1] Note that the choice of the operator $\hat{A}$ (and hence that of the order operator) is, in a sense, ad hoc. If we set $\hat{A} = \hat{S}^{(\beta)}_{\text{tot}}$, we then get $\hat{S}^{(\gamma)}_{\text{tot}}$ as a candidate of the order operator. This is certainly not useful if there is antiferromagnetic ordering, but is a precise choice if one anticipates ferromagnetic ordering.

[2] To prove this we note that (2.1.22) implies $(\hat{U}^{(\alpha)}_{\pi/2})^\dagger\,\hat{S}^{(\beta)}_x\,\hat{U}^{(\alpha)}_{\pi/2} = -\sum_{\gamma=1}^3 \varepsilon_{\alpha\beta\gamma}\,\hat{S}^{(\gamma)}_x$, and make use of the invariance (2.5.5).

$$\langle\Phi_{\mathrm{GS}}|\frac{\hat{\mathcal{O}}_L^{(\alpha)}}{L^d}|\Phi_{\mathrm{GS}}\rangle = 0, \tag{4.1.8}$$

for any L and $\alpha = 1, 2, 3$. (Equation (4.1.8) readily follows from the above discussion about the construction of the order operators.) The ground state does not exhibit symmetry breaking.

The existence of long-range order in the antiferromagnetic Heisenberg model was first proved by Dyson, Lieb, and Simon, for equilibrium states at sufficiently low temperatures in $d \geq 3$, in their seminal paper in 1978 [12]. See Theorem 4.26 in p. 130. This work was an essential breakthrough in the field of quantum many-body systems. The proof, which is highly nontrivial, is based on a mathematical property called reflection positivity. The notion of reflection positivity was first introduced in the context of quantum field theory (in order to guarantee the unitarity of the theory) [48] and then applied to problems in statistical physics with great success. The interested reader is invited to study the paper by Fröhlich, Simon, and Spencer [19], in which the long-range order in classical spin systems with continuous symmetry was proved for the first time. See also [15, 16, 25], where classical models are treated by using the reflection positivity method. There have been numerous works which extend the reflection positivity method of Dyson, Lieb, and Simon to various quantum spin systems. See [29, 30, 39, 40, 47, 49] and references therein. In the next subsection, we review the proof of the existence of long-range order based on the reflection positivity method directly applied to the ground state [29]. In Sect. 10.2.4, we will discuss the closely related method of spin-space reflection positivity to prove Lieb's theorem for the Hubbard model.

Absence of order in one dimension To sum the above discussion, we know that the ground state of the antiferromagnetic Heisenberg model in two or higher dimensions exhibits long-range order, although a rigorous proof is missing in the single case with $d = 2$ and $S = 1/2$. In one dimension, it is known that strong quantum effect inhibits the model from developing long-range order even for ground states.

Consider the one-dimensional spin S antiferromagnetic Heisenberg model under staggered magnetic field, whose Hamiltonian is

$$\hat{H}_h = \sum_{x=1}^{L}\{\hat{\boldsymbol{S}}_x \cdot \hat{\boldsymbol{S}}_{x+1} - (-1)^x h\, \hat{S}_x^{(3)}\}, \tag{4.1.9}$$

where we labeled the lattice as $\Lambda_L = \{1, 2, \ldots, L\}$. We use periodic boundary conditions to identify $\hat{S}_{L+1}^{(\alpha)}$ with $\hat{S}_1^{(\alpha)}$. Note that the field h is designed to trigger possible symmetry breaking, exactly as in (3.4.19). Let $|\Phi_{\mathrm{GS},h}\rangle$ be a ground state of the Hamiltonian (4.1.9). Then the following was proved by Shastry [58].[3]

Theorem 4.2 (Shastry's theorem) *For any S, one has*

---

[3] We do not prove Theorem 4.2 in the present book. Although the statement is not presented as a mathematical theorem in [58], it can be made rigorous with some effort. See [63] for a more mathematical formulation of Shastry's argument.

$$\lim_{h\downarrow 0}\lim_{L\uparrow\infty}\langle\Phi_{\mathrm{GS},h}|\frac{\hat{\mathcal{O}}_L^{(3)}}{L}|\Phi_{\mathrm{GS},h}\rangle = 0. \tag{4.1.10}$$

Recall that the staggered magnetic field $(-1)^x h\, \hat{S}_x^{(3)}$ is designed precisely to enhance the expectation value of the order operator $\hat{\mathcal{O}}_L^{(3)} = \sum_{x=1}^{L}(-1)^x \hat{S}_x^{(3)}$. The result (4.1.10) clearly indicates that the model never exhibits spontaneous symmetry breaking. See [52] for an earlier (not yet rigorous) work which contains essential ideas, and [46] for a closely related (but different) rigorous result.

We also see immediately from the above theorem and the Kaplan–Horsch–von der Linden theorem (Theorem 3.2 in p. 70) that the model with $h = 0$ does not exhibit long-range order.

Corollary 4.3 *Let $|\Phi_{\mathrm{GS}}\rangle$ be the unique ground state of the Hamiltonian (4.1.9) with $h = 0$. Then one has*

$$\lim_{L\uparrow\infty}\langle\Phi_{\mathrm{GS}}|\Big(\frac{\hat{\mathcal{O}}_L^{(\alpha)}}{L}\Big)^2|\Phi_{\mathrm{GS}}\rangle = 0. \tag{4.1.11}$$

Proof The condition (3.4.4) for Theorem 3.2 (p. 70) is satisfied because of the Marshall–Lieb–Mattis theorem (Theorem 2.2 in p. 39). Since the conclusion (3.4.22) of Theorem 3.2 does not hold, the other condition (3.4.3) must be violated. We thus get (4.1.11) with $\alpha = 3$. The same bound holds for $\alpha = 1$ or 2 because the unique ground state $|\Phi_{\mathrm{GS}}\rangle$ is SU(2) invariant. ∎

Ferrimagnetic long-range order in an asymmetric bipartite lattice Let us finally discuss a simple but interesting implication of the Lieb–Mattis theorem (Theorem 2.3 in p. 42) for a special class of lattices.

As in Sect. 2.5, we consider the general spin S antiferromagnetic Heisenberg model on a connected bipartite lattice $(\Lambda, \mathcal{B})$, which is decomposed into two sublattices as $\Lambda = A \cup B$. There are some lattice structures in which $\big||A| - |B|\big|$, the difference in the number of sites in the two sublattices, is macroscopically large. An example is the Lieb lattice in Fig. 10.2 (p. 355). Then the total spin of the ground states of the antiferromagnetic Heisenberg model (2.5.1), which is equal to $S_{\mathrm{tot}} = \big||A| - |B|\big|\, S$ by Theorem 2.3, is also macroscopically large. We see that the ground states possess a macroscopic magnetic moment. This is a kind of ferromagnetism, but is more precisely called ferrimagnetism, i.e., antiferromagnetism in a system where the total magnetic moments in the two sublattices do not balance.

Furthermore, by only using the fact that S_{tot} of the ground states is macroscopically large, one can prove that the ground states exhibit ferrimagnetic long-range order. Note that, unlike in the case with $|A| = |B|$, the ground states in this case are highly degenerate and explicitly break the SU(2) symmetry. In such a situation, one should better use an SU(2) invariant operator to define the long-range order parameter so that the expectation value is independent of the choice of a ground state. We here make use of the operator

$$(\hat{\boldsymbol{O}}_\Lambda)^2 := \sum_{\alpha=1,2,3}(\hat{\mathcal{O}}_\Lambda^{(\alpha)})^2 = \sum_{x,y\in\Lambda}(-1)^x(-1)^y\,\hat{\boldsymbol{S}}_x\cdot\hat{\boldsymbol{S}}_y, \tag{4.1.12}$$

where $\hat{\mathcal{O}}_{\Lambda}^{(\alpha)}$ is defined as in (4.1.4), with Λ_L replaced by the general bipartite lattice Λ. The operator $(\hat{\boldsymbol{O}}_{\Lambda})^2$ is clearly SU(2) invariant. Then the following theorem was proved by Shen, Qiu, and Tian [59].[4]

Theorem 4.4 *Under the same condition as Theorem 2.3 (p. 42), we have*

$$\langle\Phi_{\rm GS}|(\hat{\boldsymbol{O}}_{\Lambda})^2|\Phi_{\rm GS}\rangle \geq S^2\,(|A|-|B|)^2, \tag{4.1.13}$$

for any ground state $|\Phi_{\rm GS}\rangle$. The left-hand side is independent of the choice of the ground state.

Suppose that one has $\big||A|-|B|\big| = a|\Lambda|$ with a constant $a>0$. For the Lieb lattice (Fig. 10.2) we have $a=1/3$. Then from (4.1.13) we have

$$\langle\Phi_{\rm GS}|\Big(\frac{\hat{\boldsymbol{O}}_{\Lambda}}{|\Lambda|}\Big)^2|\Phi_{\rm GS}\rangle \geq (aS)^2, \tag{4.1.14}$$

which shows the existence of long-range order.

Proof Let $|\Phi_{\rm GS}^{(0)}\rangle$ be the unique ground state with $S_{\rm tot}^{(3)}=0$ or $1/2$. An important ingredient of the proof is the inequality

$$(-1)^x(-1)^y\langle\Phi_{\rm GS}^{(0)}|(\hat{S}_x^{(1)}\hat{S}_y^{(1)}+\hat{S}_x^{(2)}\hat{S}_y^{(2)})|\Phi_{\rm GS}^{(0)}\rangle > 0, \tag{4.1.15}$$

which is proved in the solution of Problem 2.5.d (p. 40). In fact (4.1.15) can be shown in the general setting of Theorem 2.3, and (2.5.7) is then obtained by using the SU(2) invariance of the ground state (which holds only when $|A|=|B|$).

We now observe that

$$\begin{aligned}\langle\Phi_{\rm GS}^{(0)}|(\hat{\boldsymbol{O}}_{\Lambda})^2|\Phi_{\rm GS}^{(0)}\rangle &\geq \langle\Phi_{\rm GS}^{(0)}|\{(\hat{\mathcal{O}}_{\Lambda}^{(1)})^2+(\hat{\mathcal{O}}_{\Lambda}^{(2)})^2\}|\Phi_{\rm GS}^{(0)}\rangle\\ &= \sum_{x,y\in\Lambda}(-1)^x(-1)^y\langle\Phi_{\rm GS}^{(0)}|(\hat{S}_x^{(1)}\hat{S}_y^{(1)}+\hat{S}_x^{(2)}\hat{S}_y^{(2)})|\Phi_{\rm GS}^{(0)}\rangle\\ &\geq \sum_{x,y\in\Lambda}\langle\Phi_{\rm GS}^{(0)}|(\hat{S}_x^{(1)}\hat{S}_y^{(1)}+\hat{S}_x^{(2)}\hat{S}_y^{(2)})|\Phi_{\rm GS}^{(0)}\rangle\\ &= \langle\Phi_{\rm GS}^{(0)}|\{(\hat{S}_{\rm tot}^{(1)})^2+(\hat{S}_{\rm tot}^{(2)})^2\}|\Phi_{\rm GS}^{(0)}\rangle = \langle\Phi_{\rm GS}^{(0)}|\{(\hat{\boldsymbol{S}}_{\rm tot})^2-(\hat{S}_{\rm tot}^{(3)})^2\}|\Phi_{\rm GS}^{(0)}\rangle\\ &\geq S_{\rm tot}(S_{\rm tot}+1)-\frac{1}{4}\geq (S_{\rm tot})^2.\end{aligned} \tag{4.1.16}$$

By using $S_{\rm tot}=\big||A|-|B|\big|\,S$, which is the second ingredient from the Lieb–Mattis theorem, we get (4.1.13). The ground states with other values of $S_{\rm tot}^{(3)}$ is obtained by operating $\hat{S}_{\rm tot}^{\pm}$ to $|\Phi_{\rm GS}^{(0)}\rangle$. Since $(\hat{\boldsymbol{O}}_{\Lambda})^2$ commutes with $\hat{S}_{\rm tot}^{\pm}$, the expectation value $\langle\Phi_{\rm GS}|(\hat{\boldsymbol{O}}_{\Lambda})^2|\Phi_{\rm GS}\rangle$ is independent of the choice of a ground state. ∎

---

[4] Shen, Qiu, and Tian mainly studied the Hubbard model. See Sect. 10.2.3.

4.1.2 Proof of the Existence of LRO

In this section, we prove a part of Theorem 4.1, along the line of the proof given by Kennedy, Lieb, and Shastry [29]. More precisely we give a complete proof of the statement of the theorem for sufficiently large S for each $d \geq 2$, and show how it can be proved for any $S \geq 1$ for $d = 2, 3$. The difficult case with $S = 1/2$ and $d = 3$, which was solved in [29], is not covered.

Although the proof only makes use of elementary linear algebra, the reader may find it not easy to fully digest. The difficulty comes partly from the character of the proof; this is a kind of proof where physically meaningful results are derived from arguments which hardly allow physical interpretations. Nevertheless we are convinced that the proof, being one of the most outstanding achievements in mathematical physics of many-body systems, is worth studying. We have tried to make the proof accessible by starting from physical results, and proceeding, step by step, to more abstract parts. We encourage the motivated reader to go through the proof.[5]

The mathematical core of the whole proof is the expression (4.1.74) for the ground state energy of a Hamiltonian with external field. The expression is nothing more than the standard Schmidt decomposition, but is designed in such a manner that one can make full use of the reflection symmetry of the model. When combined with the elementary inequality $ab \leq (a^2 + b^2)/2$, (4.1.74) leads to the reflection bound (4.1.51), which, through a nontrivial route, leads to the final conclusion about the existence of long-range order. This is an example of a proof based on the reflection positivity method [12, 15, 16, 19, 25, 29, 30, 40, 43, 47–49].

When reflection positivity was introduced in the context of quantum field theory [48], there was a clear physical motivation to construct a proper quantum theory starting from an euclidean description of a field theory. This motivation became rather vague when reflection positivity was applied to classical statistical mechanics [15, 16, 19, 25]. In the context of quantum spin systems [12, 29, 30, 40, 47, 49], the original motivation is meaningless since we already have a quantum theory to begin with. We should regard the reflection positivity method as a mere mathematical technique, which is extremely powerful.

It should be noted that the present proof based on the (spatial) reflection positivity requires the lattice to be completely regular and all the interactions to be identical. Physically speaking, small irregularity in the magnitude of exchange interactions should be irrelevant to macroscopic properties of a spin system. But the mathematical argument used here does not allow such irregular modifications, no matter how small they are.

Long-range order and the infrared bound We study the antiferromagnetic Heisenberg model with the Hamiltonian $\hat{H}$ of (4.1.1) on the d-dimensional $L \times \cdots \times L$ hypercubic lattice Λ_L defined in (3.1.2). It is essential that L is even. In the present

[5] We do not, however, assume any knowledge about the proof in the rest of the book. The reader may skip the whole proof and jump to Sect. 4.2, where we discuss the important notion of the tower of low-lying states.

subsection we abbreviate the ground state expectation value $\langle\Phi_{\rm GS}|(\cdots)|\Phi_{\rm GS}\rangle$ as $\langle\cdots\rangle_L$. We define the k-space corresponding to Λ_L as

$$\mathscr{K}_L := \left\{(k_1,\ldots,k_d)\,\middle|\,k_j = \frac{2\pi n_j}{L},\ n_j\in\mathbb{Z},\ -\frac{L}{2}<n_j\le\frac{L}{2}\right\}\subset(-\pi,\pi]^d. \tag{4.1.17}$$

We let $k^* = (\pi,\cdots,\pi)\in\mathscr{K}_L$. This particular wave number vector plays an essential role in antiferromagnetism. A function $f(x)$ and its Fourier transform $\tilde{f}(k)$ are related by[6]

$$\tilde{f}(k) = \sum_{x\in\Lambda_L} e^{-ik\cdot x} f(x),\quad f(x) = L^{-d}\sum_{k\in\mathscr{K}_L} e^{ik\cdot x}\tilde{f}(k), \tag{4.1.18}$$

where $k\cdot x = \sum_{j=1}^d k_j x_j$.

For $k\in\mathscr{K}_L$, we define the Fourier transform of $\hat{S}_x^{(3)}$ as

$$\tilde{S}_k := L^{-d/2}\sum_{x\in\Lambda_L} e^{-ik\cdot x}\hat{S}_x^{(3)}. \tag{4.1.19}$$

It is then standard that

$$g_L(k) := \langle\tilde{S}_{-k}\tilde{S}_k\rangle_L = L^{-d}\sum_{x,y\in\Lambda_L} e^{-ik\cdot(x-y)}\langle\hat{S}_y^{(3)}\hat{S}_x^{(3)}\rangle_L = \sum_{x\in\Lambda_L} e^{-ik\cdot x}\langle\hat{S}_o^{(3)}\hat{S}_x^{(3)}\rangle_L, \tag{4.1.20}$$

where $o = (0,\ldots,0)$, and we used translation invariance. Note that the right-hand side defines the Fourier transform of the two-point function $\langle\hat{S}_o^{(3)}\hat{S}_x^{(3)}\rangle_L$. The inverse Fourier transformation (4.1.18) gives

$$\langle\hat{S}_o^{(3)}\hat{S}_x^{(3)}\rangle_L = L^{-d}\sum_{k\in\mathscr{K}_L} e^{ik\cdot x} g_L(k). \tag{4.1.21}$$

By evaluating the Fourier transform $g_L(k)$ at $k^* = (\pi,\ldots,\pi)$, we find

$$g_L(k^*) = L^{-d}\sum_{x,y\in\Lambda_L}(-1)^x(-1)^y\langle\hat{S}_x^{(3)}\hat{S}_y^{(3)}\rangle_L = L^d\left\langle\left(\frac{\hat{\mathscr{O}}_L^{(3)}}{L^d}\right)^2\right\rangle_L =: L^d\,q_L. \tag{4.1.22}$$

where the quantity $q_L := \langle(\hat{\mathscr{O}}_L^{(3)}/L^d)^2\rangle_L$ is the measure of the antiferromagnetic long-range order. Our goal is to bound q_L from below. We make use of the inverse Fourier transformation (4.1.21) with $x = o$. By recalling $\langle(\hat{S}_o^{(3)})^2\rangle_L = \langle(\hat{\boldsymbol{S}}_o)^2\rangle_L/3 = S(S+1)/3$ (see Problem 2.5.c in p. 39), we see that

[6] We here use easily verifiable relations $L^{-d}\sum_{x\in\Lambda_L} e^{-ik\cdot x} = \delta_{k,o}$ and $L^{-d}\sum_{k\in\mathscr{K}_L} e^{ik\cdot x} = \delta_{x,o}$.

$$\frac{S(S+1)}{3} = L^{-d} \sum_{k \in \mathcal{K}_L} g_L(k) = q_L + L^{-d} \sum_{k \in \mathcal{K}_L \setminus \{k^*\}} g_L(k). \tag{4.1.23}$$

We here treated the term with $k = k^*$ and terms with $k \neq k^*$ separately. Note that this is reminiscent of the standard derivation of the Bose–Einstein condensation in the free boson gas at thermal equilibrium. Since the left-hand side of (4.1.23) is a constant $S(S+1)/3$, we get a lower bound for the long-range order parameter q_L if we can bound $g_L(k)$ with $k \neq k^*$ from above.

In fact we shall prove in the latter parts of the present subsection (by using the reflection positivity method) an essential inequality called the infrared bound,

$$g_L(k) \leq \sqrt{\frac{e_0}{6d} \frac{\varepsilon(k)}{\varepsilon(k-k^*)}}, \tag{4.1.24}$$

which is valid for any $k \in \mathcal{K}_L \setminus \{k^*\}$. Here we defined

$$\varepsilon(k) := \sum_{j=1}^{d} (1 - \cos k_j), \tag{4.1.25}$$

and e_0 is a constant such that $-E_{\mathrm{GS}} \leq L^d\, e_0$. By using the lower bound of E_{GS} discussed in Problem 2.5.b (p. 38), we can choose

$$e_0 = \frac{S(2dS+1)}{2}. \tag{4.1.26}$$

We point out that the right-hand side of (4.1.24) is very similar to the formula one gets from the standard (and approximate) antiferromagnetic spin wave theory. See, e.g., Sect. 4.1 of [6]. Note, in particular, that, when $|k-k^*| \ll 1$, one has $\varepsilon(k-k^*) \simeq |k-k^*|^2/2$, and hence the right-hand side of (4.1.24) is roughly $(\text{const.})/|k-k^*|$. This is an important conclusion of the spin wave theory. The infrared bound (4.1.24) has a significant physical importance in that it provides a rigorous (partial) justification of the spin wave theory.

By using the infrared bound (4.1.24) in the right-hand side of (4.1.23), we find

$$\frac{S(S+1)}{3} \leq q_L + L^{-d} \sum_{k \in \mathcal{K}_L \setminus \{k^*\}} \sqrt{\frac{e_0}{6d} \frac{\varepsilon(k)}{\varepsilon(k-k^*)}}, \tag{4.1.27}$$

which can be viewed as a lower bound for the long-range order parameter q_L. If we let, for simplicity, $L \uparrow \infty$ and substitute (4.1.26) for e_0, we get

$$\liminf_{L \uparrow \infty} q_L \geq \frac{S(S+1)}{3} - \sqrt{\frac{S(2dS+1)}{12d}}\, I_d, \tag{4.1.28}$$

where we noted here that the sum converges to an integral as

$$L^{-d} \sum_{k \in \mathcal{K}_L \setminus \{k^*\}} \sqrt{\frac{\varepsilon(k)}{\varepsilon(k-k^*)}} \rightarrow \int_{[-\pi,\pi]^d} \frac{d^d k}{(2\pi)^d} \sqrt{\frac{\varepsilon(k)}{\varepsilon(k-k^*)}} =: I_d. \quad (4.1.29)$$

The integrand $\sqrt{\varepsilon(k)/\varepsilon(k-k^*)}$ is singular only at $k = k^*$. As we have seen before, the integral near k^* is approximated as (const.) $\int d^d k \, |k - k^*|^{-1}$, which is convergent unless $d = 1$.[7] We see that I_d is a finite constant for any $d \geq 2$. Then, for each $d \geq 2$, we find that the right-hand side of (4.1.28) becomes strictly positive for sufficiently large S, since the first term $S(S+1)/3$ grows more rapidly than the second term. This proves the statement of Theorem 4.1 for sufficiently large S for each $d \geq 2$. To determine how large S should be, one needs to evaluate the integral (4.1.29). By numerically evaluating I_2 and I_3, one finds that the right-hand side of (4.1.28) is strictly positive if $S \geq 1$ for both $d = 2$ and 3. This is basically the result of Neves and Perez [47].[8]

Kennedy, Lieb, and Shastry [29] made an essential improvement to the argument to cover the important and difficult case with $S = 1/2$ and $d = 3$. The basic idea is to make use of the inverse Fourier transformation (4.1.21) with $|x| = 1$ to extract further information from the infrared bound (4.1.24). See the original paper [29], especially p. 1022.

Derivation of the infrared bound In this part we prove the infrared bound (4.1.24) by using the inequality (4.1.40), which we call the uniform (staggered) field bound.

We start from a preliminary estimate. Let $|\Phi_n\rangle$ and E_n be eigenstates and the corresponding eigenvalues, respectively, of the Hamiltonian (4.1.1), other than the ground state. We have $E_n - E_{\mathrm{GS}} > 0$ for any n since the ground state is unique. Noting that $\langle \Phi_{\mathrm{GS}} | \tilde{S}_k | \Phi_{\mathrm{GS}} \rangle = 0$, we can rewrite $g_L(k)$ as

$$g_L(k) = \langle \Phi_{\mathrm{GS}} | \tilde{S}_{-k} \tilde{S}_k | \Phi_{\mathrm{GS}} \rangle = \sum_n \langle \Phi_{\mathrm{GS}} | \tilde{S}_{-k} | \Phi_n \rangle \langle \Phi_n | \tilde{S}_k | \Phi_{\mathrm{GS}} \rangle = \sum_n \left| \langle \Phi_n | \tilde{S}_k | \Phi_{\mathrm{GS}} \rangle \right|^2. \quad (4.1.30)$$

By using the Schwarz inequality, we find

$$\begin{aligned} \{g_L(k)\}^2 &= \Big\{ \sum_n \left| \langle \Phi_n | \tilde{S}_k | \Phi_{\mathrm{GS}} \rangle \right|^2 \sqrt{E_n - E_{\mathrm{GS}}} \frac{1}{\sqrt{E_n - E_{\mathrm{GS}}}} \Big\}^2 \\ &\leq \Big\{ \sum_n \left| \langle \Phi_n | \tilde{S}_k | \Phi_{\mathrm{GS}} \rangle \right|^2 (E_n - E_{\mathrm{GS}}) \Big\} \Big\{ \sum_n \left| \langle \Phi_n | \tilde{S}_k | \Phi_{\mathrm{GS}} \rangle \right|^2 \frac{1}{E_n - E_{\mathrm{GS}}} \Big\} \\ &= f_L^{(1)}(k) \, f_L^{(-1)}(k), \end{aligned} \quad (4.1.31)$$

[7] That $I_1 = \infty$ suggests that the model with $d = 1$ cannot have long-range order. This is indeed the idea behind Shastry's theorem, Theorem 4.2.

[8] There is a mistake in the calculation in [47], and the authors concluded the existence of order in $d = 2$ only for $S \geq 3/2$. This mistake was pointed out and corrected in Appendix A of [1]. See also endnote 6 of [39].

where we defined

$$f_L^{(\pm 1)}(k) = \sum_n \left|\langle \Phi_n|\tilde{S}_k|\Phi_{\mathrm{GS}}\rangle\right|^2 (E_n - E_{\mathrm{GS}})^{\pm 1}. \tag{4.1.32}$$

$f_L^{(1)}(k)$ is easily evaluated. Observe that

$$\begin{aligned} f_L^{(1)}(k) &= \sum_n \langle \Phi_{\mathrm{GS}}|\tilde{S}_{-k}|\Phi_n\rangle\langle \Phi_n|\tilde{S}_k|\Phi_{\mathrm{GS}}\rangle (E_n - E_{\mathrm{GS}}) \\ &= \langle \tilde{S}_{-k}\hat{H}\tilde{S}_k\rangle_L - \frac{1}{2}\big\{\langle \tilde{S}_{-k}\tilde{S}_k\hat{H}\rangle_L + \langle \hat{H}\tilde{S}_{-k}\tilde{S}_k\rangle_L\big\}. \end{aligned} \tag{4.1.33}$$

Note that we can make an overall replacement $\tilde{S}_k \to \tilde{S}_{-k}$ in $\langle \cdots \rangle_L$ since the ground state is invariant under the reflection $x \to -x$. We can then rewrite (4.1.33) in terms of a double commutator as

$$\begin{aligned} f_L^{(1)}(k) &= \frac{1}{2}\big\{\langle \tilde{S}_{-k}\hat{H}\tilde{S}_k\rangle_L + \langle \tilde{S}_k\hat{H}\tilde{S}_{-k}\rangle_L - \langle \tilde{S}_{-k}\tilde{S}_k\hat{H}\rangle_L - \langle \hat{H}\tilde{S}_k\tilde{S}_{-k}\rangle_L\big\} \\ &= \frac{1}{2}\big\langle [[\tilde{S}_k, \hat{H}], \tilde{S}_{-k}]\big\rangle_L. \end{aligned} \tag{4.1.34}$$

Let $\mathscr{N}(x) = \{y \in \Lambda_L \,|\, \{x, y\} \in \mathscr{B}_L\}$ be the set of sites neighboring x. By noting that $[[\hat{S}_x^{(3)}, \hat{H}], \hat{S}_x^{(3)}] = -\sum_{y\in\mathscr{N}(x)}(\hat{S}_x^{(1)}\hat{S}_y^{(1)} + \hat{S}_x^{(2)}\hat{S}_y^{(2)})$ and $[[\hat{S}_x^{(3)}, \hat{H}], \hat{S}_y^{(3)}] = \hat{S}_x^{(1)}\hat{S}_y^{(1)} + \hat{S}_x^{(2)}\hat{S}_y^{(2)}$ for $y \in \mathscr{N}(x)$, we can evaluate the double commutator in (4.1.34) as

$$[[\tilde{S}_k, \hat{H}], \tilde{S}_{-k}] = L^{-d} \sum_{x\in\Lambda_L} \sum_{y\in\mathscr{N}(x)} (e^{ik\cdot(y-x)} - 1)(\hat{S}_x^{(1)}\hat{S}_y^{(1)} + \hat{S}_x^{(2)}\hat{S}_y^{(2)}). \tag{4.1.35}$$

Note that $E_{\mathrm{GS}} = \langle \hat{H}\rangle_L = |\mathscr{B}_L|\langle \hat{\mathbf{S}}_x \cdot \hat{\mathbf{S}}_y\rangle_L$ for any $\{x, y\} \in \mathscr{B}_L$, since the translation invariance of the ground state (which follows from the uniqueness) implies that $\langle \hat{\mathbf{S}}_x \cdot \hat{\mathbf{S}}_y\rangle_L$ is identical for all $\{x, y\} \in \mathscr{B}_L$. We thus have $\langle \hat{S}_x^{(\alpha)}\hat{S}_y^{(\alpha)}\rangle_L = \langle \hat{\mathbf{S}}_x \cdot \hat{\mathbf{S}}_y\rangle_L/3 = E_{\mathrm{GS}}/(3d\, L^d)$ for $\{x, y\} \in \mathscr{B}_L$, and hence

$$f_L^{(1)}(k) = -\frac{2\,\varepsilon(k)}{3d}\frac{E_{\mathrm{GS}}}{L^d} \leq \frac{2e_0\varepsilon(k)}{3d}, \tag{4.1.36}$$

where the inequality follows from the definition of e_0.

Our main task is then to bound $f_L^{(-1)}(k)$.[9] This is nontrivial, and requires a hard analysis. Fix arbitrary $b_x \in \mathbb{R}$ for each $x \in \Lambda_L$. For an arbitrary $\lambda \in \mathbb{R}$, we define a new Hamiltonian

[9] $f_L^{(-1)}(k)$ is sometimes called the susceptibility since it represents the second order correction to the ground state energy when the external field $h\tilde{S}_k$ is applied. We note however that it is not directly related to the physically relevant magnetic susceptibility of an antiferromagnet, since we are here considering the symmetric ground state in a finite volume.

$$\hat{H}_\lambda = \sum_{\{x,y\}\in\mathscr{B}_L} \Big[\hat{S}^{(1)}_x\hat{S}^{(1)}_y+\hat{S}^{(2)}_x\hat{S}^{(2)}_y + \frac{1}{2}\big\{\hat{S}^{(3)}_x + \hat{S}^{(3)}_y - \lambda(b_x+b_y)\big\}^2\Big] - d\sum_{x\in\Lambda_L}(\hat{S}^{(3)}_x)^2. \tag{4.1.37}$$

This coincides with the original Hamiltonian (4.1.1) when $\lambda = 0$. Let us denote by $E_{\text{GS}}(\lambda)$ the ground state energy of $\hat{H}_\lambda$. To evaluate $E_{\text{GS}}(\lambda)$ by using the standard Rayleigh–Shrödinger perturbation theory, we write (4.1.37) as

$$\begin{aligned}\hat{H}_\lambda &= \hat{H} - \lambda\sum_{\{x,y\}\in\mathscr{B}_L}(b_x+b_y)(\hat{S}^{(3)}_x+\hat{S}^{(3)}_y) + \frac{\lambda^2}{2}\sum_{\{x,y\}\in\mathscr{B}_L}(b_x+b_y)^2\\ &= \hat{H} - \lambda\sum_{x\in\Lambda_L}\tilde{b}_x\hat{S}^{(3)}_x + \frac{\lambda^2}{2}\sum_{\{x,y\}\in\mathscr{B}_L}(b_x+b_y)^2,\end{aligned} \tag{4.1.38}$$

where we wrote $\tilde{b}_x = \sum_{y\in\mathscr{N}(x)}(b_x+b_y)$. Since $\langle\Phi_{\text{GS}}|\hat{S}^{(3)}_x|\Phi_{\text{GS}}\rangle = 0$ by symmetry, there is no first order correction to $E_{\text{GS}}(\lambda)$. By working out the second order perturbation explicitly, we get

$$\begin{aligned}E_{\text{GS}}(\lambda) =& E_{\text{GS}} + \frac{\lambda^2}{2}\sum_{\{x,y\}\in\mathscr{B}_L}(b_x+b_y)^2\\ &-\lambda^2\sum_n\frac{\langle\Phi_{\text{GS}}|\sum_x\tilde{b}_x\hat{S}^{(3)}_x|\Phi_n\rangle\langle\Phi_n|\sum_x\tilde{b}_x\hat{S}^{(3)}_x|\Phi_{\text{GS}}\rangle}{E_n-E_{\text{GS}}} + O(\lambda^3).\end{aligned} \tag{4.1.39}$$

Since we are dealing with a finite system, the perturbation theory always gives reliable results provided that λ is sufficiently small (and how small λ should be depends on the system size L).

A notable and nontrivial result from the reflection positivity method is the bound

$$E_{\text{GS}}(\lambda) \geq E_{\text{GS}}(0) = E_{\text{GS}}, \tag{4.1.40}$$

which is valid for any choice of $b_x \in \mathbb{R}$ and $\lambda \in \mathbb{R}$. The inequality (4.1.40), which we call the uniform (staggered) field bound, may be of interest by itself.[10] By combining the perturbative estimate (4.1.39) with the bound (4.1.40), which is valid for any λ, we obtain

$$\begin{aligned}&\sum_n\frac{\langle\Phi_{\text{GS}}|\sum_x\{\sum_{y\in\mathscr{N}(x)}(b_x+b_y)\}\hat{S}^{(3)}_x|\Phi_n\rangle\langle\Phi_n|\sum_x\{\sum_{y\in\mathscr{N}(x)}(b_x+b_y)\}\hat{S}^{(3)}_x|\Phi_{\text{GS}}\rangle}{E_n-E_{\text{GS}}}\\ &\quad\leq\frac{1}{2}\sum_{\{x,y\}\in\mathscr{B}_L}(b_x+b_y)^2.\end{aligned} \tag{4.1.41}$$

[10] Note that the choice $b_x = (-1)^x b$ is equivalent to $\lambda = 0$, and hence realizes the minimum energy.

for any $b_x \in \mathbb{R}$ (with $x \in \Lambda_L$). Although the inequality was derived only for real b_x, we also find that

$$\sum_n \frac{\langle\Phi_{\rm GS}|\sum_x\{\sum_{y\in\mathcal{N}(x)}(\xi_x^*+\xi_y^*)\}\hat{S}_x^{(3)}|\Phi_n\rangle\langle\Phi_n|\sum_x\{\sum_{y\in\mathcal{N}(x)}(\xi_x+\xi_y)\}\hat{S}_x^{(3)}|\Phi_{\rm GS}\rangle}{E_n - E_{\rm GS}}$$
$$\leq \frac{1}{2}\sum_{\{x,y\}\in\mathcal{B}_L} |\xi_x + \xi_y|^2, \tag{4.1.42}$$

for any $\xi_x \in \mathbb{C}$ (with $x \in \Lambda_L$). To see this we simply write $\xi_x = b_x + ib'_x$ and note that (4.1.42) follows from inequalities (4.1.41) for b_x and b'_x.

We now set $\xi_x = e^{-ik\cdot x}$ for $k \in \mathcal{K}_L\backslash\{k^*\}$. It is straightforward to check that

$$\sum_{y\in\mathcal{N}(x)} (\xi_x + \xi_y) = e^{-ik\cdot x}\, 2\sum_{j=1}^{d}(1+\cos k_j) = 2e^{-ik\cdot x}\,\varepsilon(k-k^*), \tag{4.1.43}$$

$$\sum_{\{x,y\}\in\mathcal{B}_L} |\xi_x + \xi_y|^2 = 2L^d\sum_{j=1}^{d}(1+\cos k_j) = 2L^d\,\varepsilon(k-k^*). \tag{4.1.44}$$

By substituting these into (4.1.42), and recalling the definition (4.1.19) of $\tilde{S}_k$, we find

$$\sum_n \frac{\langle\Phi_{\rm GS}|\tilde{S}_{-k}|\Phi_n\rangle\langle\Phi_n|\tilde{S}_k|\Phi_{\rm GS}\rangle}{E_n - E_{\rm GS}} \leq \frac{1}{4\,\varepsilon(k-k^*)}, \tag{4.1.45}$$

where the left-hand side is precisely $f_L^{(-1)}(k)$. See (4.1.32). Substituting (4.1.36) and (4.1.45) into (4.1.31), we get

$$\{g_L(k)\}^2 \leq \frac{2e_0\varepsilon(k)}{3d}\,\frac{1}{4\,\varepsilon(k-k^*)} = \frac{e_0}{6d}\,\frac{\varepsilon(k)}{\varepsilon(k-k^*)}, \tag{4.1.46}$$

which is the desired infrared bound (4.1.24).

From the reflection bound to the uniform field bound We now introduce the spatial reflection, and show that the uniform field bound (4.1.40) can be derived from another bound (4.1.51), which we call the reflection bound, that follows from reflection positivity. From now on (until the end of the present subsection) we abbreviate Λ_L and $\mathcal{B}_L$ as Λ and $\mathcal{B}$, respectively.

We write the site $x \in \Lambda$ as $x = (x_1, \ldots, x_d)$ with $x_j = -(L/2)+1, \ldots, L/2$. We then define the refection $R : \Lambda \to \Lambda$ about the plane at $x_1 = 1/2$ by

$$R(x_1, x_2, \ldots, x_d) = (1 - x_1, x_2, \ldots, x_d). \tag{4.1.47}$$

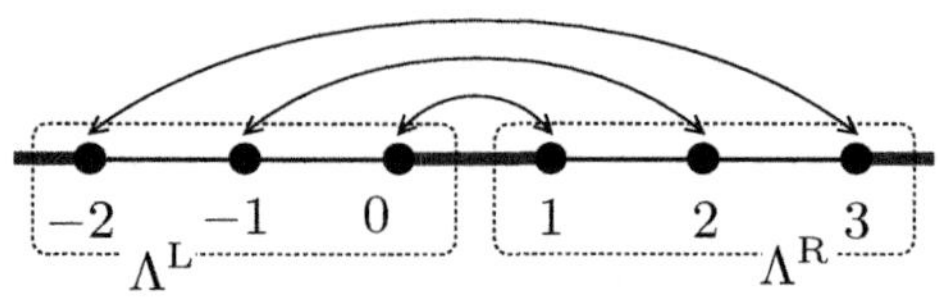

Fig. 4.1 The reflection map R and the sublattices Λ^{L} and Λ^{R} for the lattice with $d = 1$ and $L = 6$. Two thick bonds (0, 1) and (−2, 3) are the elements of $\partial\mathscr{B}$ defined later in (4.1.62)

We decompose the lattice into the left and the right halves as $\Lambda = \Lambda^{\mathrm{L}} \cup \Lambda^{\mathrm{R}}$, where Λ^{L} (which should not be confused with Λ_L) is the set of sites with $x_1 \le 0$ and Λ^{R} is the set of sites with $x_1 > 0$. Note that $R\Lambda^{\mathrm{L}} = \Lambda^{\mathrm{R}}$ and $R\Lambda^{\mathrm{R}} = \Lambda^{\mathrm{L}}$. See Fig. 4.1.

Consider again the modified Hamiltonian (4.1.37), but regard it as parametrized by staggered field $h_x = (-1)^x \lambda b_x$ as

$$\hat{H}_{\boldsymbol{h}} = \sum_{\{x,y\}\in\mathscr{B}} \Big[\hat{S}_x^{(1)}\hat{S}_y^{(1)} + \hat{S}_x^{(2)}\hat{S}_y^{(2)} + \frac{1}{2}\big\{\hat{S}_x^{(3)} + \hat{S}_y^{(3)} - (-1)^x h_x - (-1)^y h_y\big\}^2\Big] - d\sum_{x\in\Lambda_L}(\hat{S}_x^{(3)})^2. \tag{4.1.48}$$

We write the field configuration as $\boldsymbol{h} = (h_x)_{x\in\Lambda}$. We denote the ground state energy of $\hat{H}_{\boldsymbol{h}}$ as $E_{\mathrm{GS}}(\boldsymbol{h})$. Note that if $\boldsymbol{h}^{\mathrm{const}}$ is a constant magnetic field with $h_x^{\mathrm{const}} = h$ for all $x \in \Lambda$, then $E_{\mathrm{GS}}(\boldsymbol{h}^{\mathrm{const}}) = E_{\mathrm{GS}}(0, \ldots, 0) = E_{\mathrm{GS}}$. Thus the uniform field bound (4.1.40) now reads

$$E_{\mathrm{GS}}(\boldsymbol{h}) \ge E_{\mathrm{GS}}(\boldsymbol{h}^{\mathrm{const}}) = E_{\mathrm{GS}}(0, \ldots, 0). \tag{4.1.49}$$

For a given field configuration $\boldsymbol{h} = (h_x)_{x\in\Lambda}$, let us define

$$h_x^{\mathrm{L}} = \begin{cases} h_x & \text{if } x \in \Lambda^{\mathrm{L}}, \\ h_{R(x)} & \text{if } x \in \Lambda^{\mathrm{R}}, \end{cases} \qquad h_x^{\mathrm{R}} = \begin{cases} h_{R(x)} & \text{if } x \in \Lambda^{\mathrm{L}}, \\ h_x & \text{if } x \in \Lambda^{\mathrm{R}}, \end{cases} \tag{4.1.50}$$

and write $\boldsymbol{h}^{\mathrm{L}} = (h_x^{\mathrm{L}})_{x\in\Lambda}$ and $\boldsymbol{h}^{\mathrm{R}} = (h_x^{\mathrm{R}})_{x\in\Lambda}$. Note that the new field configurations are obtained by throwing away the half of the original configuration and replacing it with the reflected copy of the remaining half. See Fig. 4.2. In the next (and the final) part, we prove that

$$E_{\mathrm{GS}}(\boldsymbol{h}) \ge \frac{1}{2}\big\{E_{\mathrm{GS}}(\boldsymbol{h}^{\mathrm{L}}) + E_{\mathrm{GS}}(\boldsymbol{h}^{\mathrm{R}})\big\}, \tag{4.1.51}$$

for any $\boldsymbol{h}$. We call (4.1.51) the reflection bound.

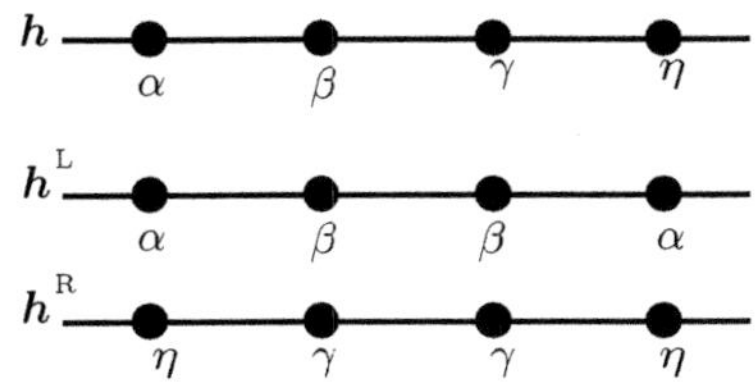

Fig. 4.2 The field configuration $\boldsymbol{h}$ and its reflected copies $\boldsymbol{h}^{\mathrm{L}}$ and $\boldsymbol{h}^{\mathrm{R}}$ for the lattice with $d = 1$ and $L = 4$ (© Hal Tasaki 2020. All Rights Reserved)

Let us see how the uniform field bound (4.1.49) is derived from the reflection bound (4.1.51).[11] Consider first the simplest case with $d = 1$ and $L = 2$, where $\boldsymbol{h}$ is simply (h_0, h_1). By using (4.1.51) for the field $\boldsymbol{h} = (\alpha, \beta)$, we find

$$E_{\mathrm{GS}}(\alpha, \beta) \geq \frac{1}{2}\big\{E_{\mathrm{GS}}(\alpha, \alpha) + E_{\mathrm{GS}}(\beta, \beta)\big\} = E_{\mathrm{GS}}(0, 0), \tag{4.1.52}$$

which is nothing but the desired uniform field bound (4.1.49). For the case with $d = 1$ and $L = 4$, however, what we get directly from (4.1.51) is

$$E_{\mathrm{GS}}(\alpha, \beta, \gamma, \eta) \geq \frac{1}{2}\big\{E_{\mathrm{GS}}(\alpha, \beta, \beta, \alpha) + E_{\mathrm{GS}}(\eta, \gamma, \gamma, \eta)\big\}, \tag{4.1.53}$$

which is not yet what we want. In this case we can also use the translation invariance of the model to see

$$\begin{aligned} E_{\mathrm{GS}}(\alpha, \beta, \beta, \alpha) &= E_{\mathrm{GS}}(\alpha, \alpha, \beta, \beta) \\ &\geq \frac{1}{2}\big\{E_{\mathrm{GS}}(\alpha, \alpha, \alpha, \alpha) + E_{\mathrm{GS}}(\beta, \beta, \beta, \beta)\big\} \\ &= E_{\mathrm{GS}}(0, 0, 0, 0). \end{aligned} \tag{4.1.54}$$

In this way (4.1.53) leads to the uniform field bound $E_{\mathrm{GS}}(\alpha, \beta, \gamma, \eta) \geq E_{\mathrm{GS}}(0, 0, 0, 0)$.

It may be rather clear that the reflection bound (4.1.51) along with the translation invariance leads to the uniform field bound (4.1.49) in general. The reader who is sufficiently convinced may jump to the final part, where the bound (4.1.51) is proved through a very interesting argument.

To prove the uniform field bound (4.1.49) generally it is convenient to state the following "chessboard estimate" [15, 17].

Lemma 4.5 *Let $F : (\mathbb{R}^D)^{2n} \to \mathbb{R}$ be a function of $2n$ vectors in $\mathbb{R}^D$, where D is arbitrary. We assume that F satisfies cyclicity*

$$F(\boldsymbol{f}_1, \ldots, \boldsymbol{f}_{2n-1}, \boldsymbol{f}_{2n}) = F(\boldsymbol{f}_{2n}, \boldsymbol{f}_1, \ldots, \boldsymbol{f}_{2n-1}), \tag{4.1.55}$$

---

[11] There indeed is a clever (but difficult) shortcut which almost directly proves (4.1.49) from (4.1.51). See [29]. We present the following longer proof, from which one can learn an idea useful in different applications of the reflection positivity method.

and the reflection bound

$$F(\boldsymbol{f}_1,\ldots,\boldsymbol{f}_n,\boldsymbol{f}_{n+1},\ldots,\boldsymbol{f}_{2n})$$
$$\geq \frac{1}{2}\big\{F(\boldsymbol{f}_1,\ldots,\boldsymbol{f}_n,\boldsymbol{f}_n,\ldots,\boldsymbol{f}_1)+F(\boldsymbol{f}_{2n},\ldots,\boldsymbol{f}_{n+1},\boldsymbol{f}_{n+1},\ldots,\boldsymbol{f}_{2n})\big\}, \tag{4.1.56}$$

for any $\boldsymbol{f}_1,\ldots,\boldsymbol{f}_{2n}\in\mathbb{R}^D$. *Then we have for any* $\boldsymbol{f}_1,\ldots,\boldsymbol{f}_{2n}$ *that*

$$F(\boldsymbol{f}_1,\ldots,\boldsymbol{f}_{2n})\geq\frac{1}{2n}\sum_{j=1}^{2n}F(\boldsymbol{f}_j,\ldots,\boldsymbol{f}_j). \tag{4.1.57}$$

Proof Define $G(\boldsymbol{f}_1,\ldots,\boldsymbol{f}_{2n})=F(\boldsymbol{f}_1,\ldots,\boldsymbol{f}_{2n})-(2n)^{-1}\sum_{j=1}^{2n}F(\boldsymbol{f}_j,\ldots,\boldsymbol{f}_j)$. Clearly G also satisfies the cyclicity (4.1.55) and the reflection bound (4.1.56).

Fix arbitrary $\boldsymbol{f}_1,\ldots,\boldsymbol{f}_{2n}$. For each $j=1,\ldots,2n$, we take $\boldsymbol{g}_j$ to be one of $\boldsymbol{f}_1,\ldots,\boldsymbol{f}_{2n}$. We thus get $(2n)^{2n}$ strings $(\boldsymbol{g}_1,\ldots,\boldsymbol{g}_{2n})$. Let $G_{\min}=\min_{(\boldsymbol{g}_1,\ldots,\boldsymbol{g}_{2n})}G(\boldsymbol{g}_1,\ldots,\boldsymbol{g}_{2n})$. We shall show $G_{\min}=0$, which implies the desired (4.1.57).

Take a string $(\boldsymbol{g}_1,\ldots,\boldsymbol{g}_{2n})$ such that $G(\boldsymbol{g}_1,\ldots,\boldsymbol{g}_{2n})=G_{\min}$. If we substitute this $G(\boldsymbol{g}_1,\ldots,\boldsymbol{g}_{2n})$ to the left-hand side of (4.1.56) the two $G(\cdots)$ on the right-hand side must also be equal to $G_{\min}$. We thus find

$$G_{\min}=G(\boldsymbol{g}_1,\boldsymbol{g}_2,\ldots,\boldsymbol{g}_n,\boldsymbol{g}_n,\ldots,\boldsymbol{g}_2,\boldsymbol{g}_1)=G(\boldsymbol{g}_1,\boldsymbol{g}_1,\boldsymbol{g}_2,\ldots,\boldsymbol{g}_n,\boldsymbol{g}_n,\ldots,\boldsymbol{g}_3,\boldsymbol{g}_2), \tag{4.1.58}$$

where we used cyclicity (4.1.55). Note that, in the final expression, the first two entries of G are both $\boldsymbol{g}_1$. By substituting this into the left-hand side of (4.1.56), and using cyclicity twice, we get $G_{\min}=G(\boldsymbol{g}_1,\boldsymbol{g}_1,\boldsymbol{g}_1,\boldsymbol{g}_1,\boldsymbol{g}_2,\ldots)$. This procedure can be repeated until we get $G_{\min}=G(\boldsymbol{g}_1,\ldots,\boldsymbol{g}_1)$, which is zero by definition. ∎

To get the uniform field bound (4.1.49) in a general case, we write an arbitrary field configuration $\boldsymbol{h}=(h_x)_{x\in\Lambda}$ as $\boldsymbol{h}=(\tilde{\boldsymbol{h}}_1,\ldots,\tilde{\boldsymbol{h}}_L)$, where $\tilde{\boldsymbol{h}}_j$ is the collection of h_x for all sites $x=(j-(L/2),x_2,\ldots,x_d)$ with arbitrary $x_2,\ldots,x_d$. Then the reflection bound (4.1.51) is written as

$$E_{\mathrm{GS}}(\tilde{\boldsymbol{h}}_1,\ldots,\tilde{\boldsymbol{h}}_L)\geq\frac{1}{2}\big\{E_{\mathrm{GS}}(\tilde{\boldsymbol{h}}_1,\ldots,\tilde{\boldsymbol{h}}_{L/2},\tilde{\boldsymbol{h}}_{L/2},\ldots,\tilde{\boldsymbol{h}}_1)$$
$$+E_{\mathrm{GS}}(\tilde{\boldsymbol{h}}_L,\ldots,\tilde{\boldsymbol{h}}_{(L/2)+1},\tilde{\boldsymbol{h}}_{(L/2)+1},\ldots,\tilde{\boldsymbol{h}}_L)\big\}, \tag{4.1.59}$$

which is precisely the condition (4.1.56). Since the cyclicity is obvious from the translation invariance, we find from (4.1.57) that

$$E_{\mathrm{GS}}(\boldsymbol{h}) \geq \frac{1}{L} \sum_{j=1}^{L} E_{\mathrm{GS}}(\tilde{\boldsymbol{h}}_j, \ldots, \tilde{\boldsymbol{h}}_j). \tag{4.1.60}$$

Note that the field is uniform in the x_1 direction in the configuration $(\tilde{\boldsymbol{h}}_j, \ldots, \tilde{\boldsymbol{h}}_j)$.

We then make use of the rotational invariance of the lattice, and consider the reflection and the corresponding reflection bound in the x_2 direction. The same argument applied to $E_{\mathrm{GS}}(\tilde{\boldsymbol{h}}_j, \ldots, \tilde{\boldsymbol{h}}_j)$ generates a lower bound in terms of field configurations which are uniform in both the x_1 and x_2 directions. Repeating this for all the d directions, we finally get

$$E_{\mathrm{GS}}(\boldsymbol{h}) \geq \frac{1}{L^d} \sum_{x \in \Lambda} E_{\mathrm{GS}}(h_x, \ldots, h_x) = E_{\mathrm{GS}}(0, \ldots, 0), \tag{4.1.61}$$

which is the desired uniform field bound (4.1.49).

Proof of the reflection bound Let us finally prove the reflection bound (4.1.51) by using the method of reflection positivity. We follow Appendix 2 of [43]. The reader may have noticed that the proof is getting less and less intuitive as it proceeds. This final part is most abstract, and, as far as we understand, does not allow natural physical interpretations. Nevertheless the argument is very powerful, and leads to physically significant results, including Theorem 4.1.

Recall the decomposition $\Lambda = \Lambda^{\mathrm{L}} \cup \Lambda^{\mathrm{R}}$. We define the set of bonds connecting Λ^{L} and Λ^{R} as

$$\partial\mathscr{B} = \{(x, y) \,|\, x \in \Lambda^{\mathrm{L}},\ y \in \Lambda^{\mathrm{R}},\ \{x, y\} \in \mathscr{B}\}. \tag{4.1.62}$$

For the later convenience, we defined a bond (x, y) as an ordered pair. See Fig. 4.1.

Let $\boldsymbol{\tau} = (\tau_x)_{x \in \Lambda^{\mathrm{L}}}$ with $\tau_x = -S, \ldots, S$ be a spin configuration on the left half Λ^{L}. We write the basis states for Λ^{L} as

$$|\Psi^{\mathrm{L},\boldsymbol{\tau}}\rangle = \bigotimes_{x \in \Lambda^{\mathrm{L}}} |\psi_x^{\tau_x}\rangle. \tag{4.1.63}$$

For the basis states for the right half, we again use a spin configuration $\boldsymbol{\tau} = (\tau_x)_{x \in \Lambda^{\mathrm{L}}}$ on the left as a label, and write

$$|\Psi^{\mathrm{R},\boldsymbol{\tau}}\rangle = \hat{R}|\Psi^{\mathrm{L},\boldsymbol{\tau}}\rangle, \tag{4.1.64}$$

where $\hat{R}$ is the reflection operator defined by $\hat{R} \bigotimes_{x \in \Lambda^{\mathrm{L}}} |\psi_x^{\tau_x}\rangle = \bigotimes_{x \in \Lambda^{\mathrm{L}}} |\psi_{R(x)}^{\tau_x}\rangle$. This little twist (of using the configuration on the left to label the state on the right) is crucial for the present proof. The basis states for the whole system are $|\Psi^{\mathrm{L},\boldsymbol{\tau}}\rangle\otimes|\Psi^{\mathrm{R},\boldsymbol{\eta}}\rangle$, where both $\boldsymbol{\tau}$ and $\boldsymbol{\eta}$ are spin configurations on Λ^{L}.

For all $x \in \Lambda$ and $\alpha = 1, 2, 3$, we introduce new operators $\hat{T}_x^{(\alpha)}$ by

$$\hat{T}_x^{(1)} = (-1)^x\, \hat{S}_x^{(1)}, \quad \hat{T}_x^{(2)} = i\, \hat{S}_x^{(2)}, \quad \hat{T}_x^{(3)} = (-1)^x\, \hat{S}_x^{(3)}, \tag{4.1.65}$$

which are no longer spin operators. Note that these operators (matrices) are real in the above basis (in the sense that all entries are real). With the new operators, the $\boldsymbol{h}$-modified Hamiltonian (4.1.48) is written as

$$\begin{aligned}\hat{H}_{\boldsymbol{h}} = &\sum_{\{x,y\}\in\mathcal{B}} \Big\{ - \hat{T}_x^{(1)}\hat{T}_y^{(1)} - \hat{T}_x^{(2)}\hat{T}_y^{(2)} + \frac{1}{2}(\hat{T}_x^{(3)} - \hat{T}_y^{(3)} - h_x + h_y)^2\Big\} \\ &- d\sum_{x\in\Lambda}(\hat{T}_x^{(3)})^2.\end{aligned} \tag{4.1.66}$$

We define the Hamiltonians for the left and the right halves as

$$\begin{aligned}\hat{H}_{\boldsymbol{h}}^{\mathrm{L}} = &\sum_{\substack{\{x,y\}\in\mathcal{B}\\(x,y\in\Lambda^{\mathrm{L}})}} \Big\{-\hat{T}_x^{(1)}\hat{T}_y^{(1)} - \hat{T}_x^{(2)}\hat{T}_y^{(2)} + \frac{1}{2}(\hat{T}_x^{(3)} - \hat{T}_y^{(3)} - h_x + h_y)^2\Big\} \\ &- d\sum_{x\in\Lambda^{\mathrm{L}}}(\hat{T}_x^{(3)})^2 + \frac{1}{2}\sum_{(x,y)\in\partial\mathcal{B}}(\hat{T}_x^{(3)} - h_x)^2,\end{aligned} \tag{4.1.67}$$

$$\begin{aligned}\hat{H}_{\boldsymbol{h}}^{\mathrm{R}} = &\sum_{\substack{\{x,y\}\in\mathcal{B}\\(x,y\in\Lambda^{\mathrm{R}})}} \Big\{-\hat{T}_x^{(1)}\hat{T}_y^{(1)} - \hat{T}_x^{(2)}\hat{T}_y^{(2)} + \frac{1}{2}(\hat{T}_x^{(3)} - \hat{T}_y^{(3)} - h_x + h_y)^2\Big\} \\ &- d\sum_{x\in\Lambda^{\mathrm{R}}}(\hat{T}_x^{(3)})^2 + \frac{1}{2}\sum_{(x,y)\in\partial\mathcal{B}}(\hat{T}_y^{(3)} - h_y)^2.\end{aligned} \tag{4.1.68}$$

Then the total Hamiltonian is written as

$$\hat{H}_{\boldsymbol{h}} = \hat{H}_{\boldsymbol{h}}^{\mathrm{L}} + \hat{H}_{\boldsymbol{h}}^{\mathrm{R}} - \sum_{(x,y)\in\partial\mathcal{B}}\Big\{\sum_{\alpha=1,2}\hat{T}_x^{(\alpha)}\hat{T}_y^{(\alpha)} + (\hat{T}_x^{(3)} - h_x)(\hat{T}_y^{(3)} - h_y)\Big\}. \tag{4.1.69}$$

It is essential for the proof that the interaction between the left and the right halves is written in terms of real matrices, and the interactions are of the ferromagnetic form $-\hat{T}_x^{(\alpha)}\hat{T}_y^{(\alpha)}$. In fact these are the conditions that the model is reflection positive.[12]

Since $\hat{H}_{\boldsymbol{h}}$ has real matrix entries in our basis, one can find a normalized ground state

$$|\Phi_{\mathrm{GS},\boldsymbol{h}}\rangle = \sum_{\boldsymbol{\tau},\boldsymbol{\eta}} C_{\boldsymbol{\tau},\boldsymbol{\eta}}|\Psi^{\mathrm{L},\boldsymbol{\tau}}\rangle \otimes |\Psi^{\mathrm{R},\boldsymbol{\eta}}\rangle, \tag{4.1.70}$$

[12] It is worth noting that the ferromagnetic Heisenberg model is not reflection positive [12].

in which the expansion coefficients $C_{\tau,\eta}$ are real.[13] Let us regard $C_{\tau,\eta}$ as entries of a $(2S+1)^{L^d/2} \times (2S+1)^{L^d/2}$ matrix. Then from the singular value decomposition theorem (Theorem A.20 in p. 477) we see that there are two orthogonal matrices, whose entries are denoted as $O^{\mathrm{L}}_{\tau,j}$ and $O^{\mathrm{R}}_{\tau,j}$ with $j = 1, \ldots, (2S+1)^{L^d/2}$, such that

$$C_{\tau,\eta} = \sum_j O^{\mathrm{L}}_{\tau,j}\, \rho_j\, O^{\mathrm{R}}_{\eta,j}, \tag{4.1.71}$$

with $\rho_j \geq 0$. Substituting this expression into (4.1.70), we get

$$|\Phi_{\mathrm{GS},\boldsymbol{h}}\rangle = \sum_{\tau,\eta,j} O^{\mathrm{L}}_{\tau,j}\, \rho_j\, O^{\mathrm{R}}_{\eta,j} |\Psi^{\mathrm{L},\tau}\rangle \otimes |\Psi^{\mathrm{R},\eta}\rangle = \sum_j \rho_j\, |\Xi^{\mathrm{L}}_j\rangle \otimes |\Xi^{\mathrm{R}}_j\rangle, \tag{4.1.72}$$

where we defined

$$|\Xi^{\mathrm{L}}_j\rangle = \sum_\tau O^{\mathrm{L}}_{\tau,j} |\Psi^{\mathrm{L},\tau}\rangle, \quad |\Xi^{\mathrm{R}}_j\rangle = \sum_\tau O^{\mathrm{R}}_{\tau,j} |\Psi^{\mathrm{R},\tau}\rangle. \tag{4.1.73}$$

Note that the orthogonality implies $\langle \Xi^{\mathrm{L}}_j | \Xi^{\mathrm{L}}_k \rangle = \delta_{j,k}$ and $\langle \Xi^{\mathrm{R}}_j | \Xi^{\mathrm{R}}_k \rangle = \delta_{j,k}$. Since $|\Phi_{\mathrm{GS},\boldsymbol{h}}\rangle$ is normalized, we have $\sum_j (\rho_j)^2 = 1$. The expression in the right-hand side of (4.1.72) is known as the Schmidt decomposition of $|\Phi_{\mathrm{GS},\boldsymbol{h}}\rangle$. See also Appendix A.1.

By using the decomposition (4.1.69) of the Hamiltonian and the Schmidt decomposition (4.1.72), we can express the ground state energy as

$$\begin{aligned}
E_{\mathrm{GS}}(\boldsymbol{h}) =& \langle \Phi_{\mathrm{GS},\boldsymbol{h}} | \hat{H}_{\boldsymbol{h}} | \Phi_{\mathrm{GS},\boldsymbol{h}} \rangle \\
=& \sum_j (\rho_j)^2 \bigl\{ \langle \Xi^{\mathrm{L}}_j | \hat{H}^{\mathrm{L}}_{\boldsymbol{h}} | \Xi^{\mathrm{L}}_j \rangle + \langle \Xi^{\mathrm{R}}_j | \hat{H}^{\mathrm{R}}_{\boldsymbol{h}} | \Xi^{\mathrm{R}}_j \rangle \bigr\} \\
& - \sum_{j,k} \rho_j \rho_k \sum_{(x,y)\in\partial\mathscr{B}} \Bigl\{ \sum_{\alpha=1,2} \langle \Xi^{\mathrm{L}}_j | \hat{T}^{(\alpha)}_x | \Xi^{\mathrm{L}}_k \rangle \langle \Xi^{\mathrm{R}}_j | \hat{T}^{(\alpha)}_y | \Xi^{\mathrm{R}}_k \rangle \\
& \qquad + \langle \Xi^{\mathrm{L}}_j | (\hat{T}^{(3)}_x - h_x) | \Xi^{\mathrm{L}}_k \rangle \langle \Xi^{\mathrm{R}}_j | (\hat{T}^{(3)}_y - h_y) | \Xi^{\mathrm{R}}_k \rangle \Bigr\}
\end{aligned} \tag{4.1.74}$$

Note that all the matrix elements are real. Then, by using the elementary inequality $ab \leq (a^2 + b^2)/2$ (for $a, b \in \mathbb{R}$) in the right-hand side, we get

[13] With some extra effort, one can treat models in which $\hat{H}^{\mathrm{L}}_{\boldsymbol{h}}$ or $\hat{H}^{\mathrm{R}}_{\boldsymbol{h}}$ are not real. See Appendix 2 of [43].

$$
\begin{aligned}
E_{\mathrm{GS}}(\boldsymbol{h}) \geq \frac{1}{2}\Biggl(\Bigl[2\sum_j (\rho_j)^2 \langle \Xi_j^{\mathrm{L}} | \hat{H}_{\boldsymbol{h}}^{\mathrm{L}} | \Xi_j^{\mathrm{L}} \rangle \\
&- \sum_{j,k} \rho_j \rho_k \sum_{(x,y)\in\partial\mathscr{B}} \Bigl\{ \sum_{\alpha=1,2} \langle \Xi_j^{\mathrm{L}} | \hat{T}_x^{(\alpha)} | \Xi_k^{\mathrm{L}} \rangle^2 + \langle \Xi_j^{\mathrm{L}} | (\hat{T}_x^{(3)} - h_x) | \Xi_k^{\mathrm{L}} \rangle^2 \Bigr\} \Bigr] \\
&+ \Bigl[2\sum_j (\rho_j)^2 \langle \Xi_j^{\mathrm{R}} | \hat{H}_{\boldsymbol{h}}^{\mathrm{R}} | \Xi_j^{\mathrm{R}} \rangle \\
&- \sum_{j,k} \rho_j \rho_k \sum_{(x,y)\in\partial\mathscr{B}} \Bigl\{ \sum_{\alpha=1,2} \langle \Xi_j^{\mathrm{R}} | \hat{T}_y^{(\alpha)} | \Xi_k^{\mathrm{R}} \rangle^2 + \langle \Xi_j^{\mathrm{R}} | (\hat{T}_y^{(3)} - h_y) | \Xi_k^{\mathrm{R}} \rangle^2 \Bigr\} \Bigr] \Biggr).
\end{aligned}
\tag{4.1.75}
$$

We now define

$$|\tilde{\Xi}_j^{\mathrm{L}}\rangle = \Bigl(\prod_{x\in\Lambda^{\mathrm{R}}} e^{-i\pi \hat{S}_x^{(2)}}\Bigr) \hat{R} |\Xi_j^{\mathrm{L}}\rangle, \tag{4.1.76}$$

which is the copy of $|\Xi_j^{\mathrm{L}}\rangle$ defined for the right half. Note that a site x and its reflection $R(x)$ always belong to different sublattices. Then from the definition (4.1.65) of the $\hat{T}$ operators, we find

$$\hat{R}^\dagger e^{i\pi \hat{S}_x^{(2)}} \hat{T}_x^{(\alpha)} e^{-i\pi \hat{S}_x^{(2)}} \hat{R} = \hat{T}_{R(x)}^{(\alpha)}, \tag{4.1.77}$$

for $x \in \Lambda^{\mathrm{R}}$ and $\alpha = 1, 2, 3$. Here the π rotation about the 2-axis precisely transforms the sing factor $(-1)^x$. Recalling the definition (4.1.50) of reflected copies $\boldsymbol{h}^{\mathrm{L}}$ and $\boldsymbol{h}^{\mathrm{R}}$ of $\boldsymbol{h}$, we thus see that

$$\langle \Xi_j^{\mathrm{L}} | \hat{T}_x^{(\alpha)} | \Xi_k^{\mathrm{L}} \rangle = \langle \tilde{\Xi}_j^{\mathrm{L}} | \hat{T}_y^{(\alpha)} | \tilde{\Xi}_k^{\mathrm{L}} \rangle, \tag{4.1.78}$$

for $(x, y) \in \partial\mathscr{B}$, and

$$\langle \Xi_j^{\mathrm{L}} | \hat{H}_{\boldsymbol{h}}^{\mathrm{L}} | \Xi_j^{\mathrm{L}} \rangle = \langle \tilde{\Xi}_j^{\mathrm{L}} | \hat{H}_{\boldsymbol{h}^{\mathrm{L}}}^{\mathrm{R}} | \tilde{\Xi}_j^{\mathrm{L}} \rangle. \tag{4.1.79}$$

Note also that $\langle \Xi_j^{\mathrm{L}} | \hat{H}_{\boldsymbol{h}}^{\mathrm{L}} | \Xi_j^{\mathrm{L}} \rangle = \langle \Xi_j^{\mathrm{L}} | \hat{H}_{\boldsymbol{h}^{\mathrm{L}}}^{\mathrm{L}} | \Xi_j^{\mathrm{L}} \rangle$ simply by definition.

We similarly define

$$|\tilde{\Xi}_j^{\mathrm{R}}\rangle = \Bigl(\prod_{x\in\Lambda^{\mathrm{L}}} e^{-i\pi \hat{S}_x^{(2)}}\Bigr) \hat{R}^\dagger |\Xi_j^{\mathrm{R}}\rangle, \tag{4.1.80}$$

which is the copy of $|\Xi_j^{\mathrm{R}}\rangle$ for the left half. We have $\langle \Xi_j^{\mathrm{R}} | \hat{T}_y^{(\alpha)} | \Xi_k^{\mathrm{R}} \rangle = \langle \tilde{\Xi}_j^{\mathrm{R}} | \hat{T}_x^{(\alpha)} | \tilde{\Xi}_k^{\mathrm{R}} \rangle$ for $(x, y) \in \partial\mathscr{B}$, $\langle \Xi_j^{\mathrm{R}} | \hat{H}_{\boldsymbol{h}}^{\mathrm{R}} | \Xi_j^{\mathrm{R}} \rangle = \langle \tilde{\Xi}_j^{\mathrm{R}} | \hat{H}_{\boldsymbol{h}^{\mathrm{R}}}^{\mathrm{L}} | \tilde{\Xi}_j^{\mathrm{R}} \rangle$, and $\langle \Xi_j^{\mathrm{R}} | \hat{H}_{\boldsymbol{h}}^{\mathrm{R}} | \Xi_j^{\mathrm{R}} \rangle = \langle \Xi_j^{\mathrm{R}} | \hat{H}_{\boldsymbol{h}^{\mathrm{R}}}^{\mathrm{R}} | \Xi_j^{\mathrm{R}} \rangle$.

We can then rewrite the right-hand side of (4.1.75) as

$$\frac{1}{2}\Biggl(\Bigl[\sum_j(\rho_j)^2\{\langle\Xi_j^{\rm L}|\hat{H}^{\rm L}_{\boldsymbol{h}^{\rm L}}|\Xi_j^{\rm L}\rangle+\langle\tilde{\Xi}_j^{\rm L}|\hat{H}^{\rm R}_{\boldsymbol{h}^{\rm L}}|\tilde{\Xi}_j^{\rm L}\rangle\}$$
$$-\sum_{j,k}\rho_j\rho_k\sum_{(x,y)\in\partial\mathscr{B}}\{\sum_{\alpha=1,2}\langle\Xi_j^{\rm L}|\hat{T}_x^{(\alpha)}|\Xi_k^{\rm L}\rangle\langle\tilde{\Xi}_j^{\rm L}|\hat{T}_y^{(\alpha)}|\tilde{\Xi}_k^{\rm L}\rangle$$
$$+\langle\Xi_j^{\rm L}|(\hat{T}_x^{(3)}-h_x)|\Xi_k^{\rm L}\rangle\langle\tilde{\Xi}_j^{\rm L}|(\hat{T}_y^{(3)}-h_x)|\tilde{\Xi}_k^{\rm L}\rangle\}\Bigr]$$
$$+\Bigl[\sum_j(\rho_j)^2\{\langle\tilde{\Xi}_j^{\rm R}|\hat{H}^{\rm L}_{\boldsymbol{h}^{\rm R}}|\tilde{\Xi}_j^{\rm R}\rangle+\langle\Xi_j^{\rm R}|\hat{H}^{\rm R}_{\boldsymbol{h}^{\rm R}}|\Xi_j^{\rm R}\rangle\}$$
$$-\sum_{j,k}\rho_j\rho_k\sum_{(x,y)\in\partial\mathscr{B}}\{\sum_{\alpha=1,2}\langle\tilde{\Xi}_j^{\rm R}|\hat{T}_x^{(\alpha)}|\tilde{\Xi}_k^{\rm R}\rangle\langle\Xi_j^{\rm R}|\hat{T}_y^{(\alpha)}|\Xi_k^{\rm R}\rangle$$
$$+\langle\tilde{\Xi}_j^{\rm R}|(\hat{T}_x^{(3)}-h_y)|\tilde{\Xi}_k^{\rm R}\rangle\langle\Xi_j^{\rm R}|(\hat{T}_y^{(3)}-h_y)|\Xi_k^{\rm R}\rangle\}\Bigr]\Biggr)$$
$$=\frac{1}{2}\Bigl\{\langle\Phi^{\rm L}|\hat{H}_{\boldsymbol{h}^{\rm L}}|\Phi^{\rm L}\rangle+\langle\Phi^{\rm R}|\hat{H}_{\boldsymbol{h}^{\rm R}}|\Phi^{\rm R}\rangle\Bigr\},\tag{4.1.81}$$

where we defined

$$|\Phi^{\rm L}\rangle=\sum_j\rho_j\,|\Xi_j^{\rm L}\rangle\otimes|\tilde{\Xi}_j^{\rm L}\rangle,\quad |\Phi^{\rm R}\rangle=\sum_j\rho_j\,|\tilde{\Xi}_j^{\rm R}\rangle\otimes|\Xi_j^{\rm R}\rangle.\tag{4.1.82}$$

Note that these states are normalized because $\sum_j(\rho_j)^2=1$. The right hand-side of (4.1.81) is obtained by comparing the expression (4.1.74) with the left-hand side of (4.1.81).

By definition the right-hand side of (4.1.81) cannot be smaller than $\{E_{\rm GS}(\boldsymbol{h}^{\rm L})+E_{\rm GS}(\boldsymbol{h}^{\rm R})\}/2$. We have thus obtained the desired reflection bound (4.1.51), and completed the proof of (a part of) Theorem 4.1.

4.2 The "Tower" of Low-Lying States and Spontaneous Symmetry Breaking

We have confirmed that the ground state $|\Phi_{\rm GS}\rangle$ of the antiferromagnetic Heisenberg model (4.1.1) on the d-dimensional $L\times L$ hypercubic lattice $(\Lambda_L,\mathscr{B}_L)$ (see (3.1.2) and (3.1.3)) exhibits antiferromagnetic LRO as in Theorem 4.1 (p. 75), but does not exhibit SSB because of the uniqueness proved in Theorem 2.2 (p. 39). By combining the lower bound (4.1.7) and the relation $\langle\Phi_{\rm GS}|\hat{\mathscr{O}}_L^{(\alpha)}|\Phi_{\rm GS}\rangle=0$, we see that

$$\sqrt{\langle\Phi_{\rm GS}|\Bigl(\frac{\hat{\mathscr{O}}_L^{(\alpha)}}{L^d}\Bigr)^2|\Phi_{\rm GS}\rangle-\Bigl\{\langle\Phi_{\rm GS}|\Bigl(\frac{\hat{\mathscr{O}}_L^{(\alpha)}}{L^d}\Bigr)|\Phi_{\rm GS}\rangle\Bigr\}^2}\ge\sqrt{q_0},\tag{4.2.1}$$

for $\alpha=1,2,3$, which shows that the order operator $\hat{\mathscr{O}}_L^{(\alpha)}$ violates the law of large numbers, exactly as in (3.2.11), (3.3.12), and (3.4.5). It is then natural to follow the observation in Sect. 3.3 and the general theory in Sect. 3.4, and construct physical "ground states" with both LRO and SSB. However it is clear from the outset that the framework in Sect. 3.4 is not enough to solve the problem. Here we expect Néel order in which an arbitrary direction $\boldsymbol{n}$ should be specified (see Sect. 3.1), or, equivalently, a spontaneous breakdown of continuous SU(2) symmetry. Then there should be infinitely many physical "ground states" corresponding to the infinitely many choices of the direction $\boldsymbol{n}$. The construction in Sect. 3.4, which essentially makes use of the ground state $|\Phi_{\mathrm{GS}}\rangle$ and a single low-lying state $\hat{\mathscr{O}}_L|\Phi_{\mathrm{GS}}\rangle$, cannot give rise to infinitely many states with SSB.[14]

In the present section, we describe the general theory about low-lying states and symmetry breaking in a finite system with continuous symmetry, which was developed by Koma and Tasaki [34, 35], and recently improved by Tasaki [66] and also by Tanaka [62]. We mainly focus on the antiferromagnetic Heisenberg model, and discuss main results in Sect. 4.2.1. Section 4.2.2 is devoted to the proofs of the theorems. Theorems 4.6, 4.9 and 4.11, presented below, may be regarded as (rather sophisticated) extensions of the Horsch-von der Linden theorem (Theorem 3.1 in p. 67) and the essential lower bound (3.4.16), respectively, discussed in Sect. 3.4.

4.2.1 Main Results

The "tower" of low-lying states By using the "raising" and "lowering" order operators

$$\hat{\mathscr{O}}_L^{\pm} := \hat{\mathscr{O}}_L^{(1)} \pm i\,\hat{\mathscr{O}}_L^{(2)} = \sum_{x\in\Lambda_L} (-1)^x \hat{S}_x^{\pm}, \tag{4.2.2}$$

we define a series of trial states

$$|\Gamma_M\rangle = \frac{(\hat{\mathscr{O}}_L^{+})^M|\Phi_{\mathrm{GS}}\rangle}{\|(\hat{\mathscr{O}}_L^{+})^M|\Phi_{\mathrm{GS}}\rangle\|}, \quad |\Gamma_{-M}\rangle = \frac{(\hat{\mathscr{O}}_L^{-})^M|\Phi_{\mathrm{GS}}\rangle}{\|(\hat{\mathscr{O}}_L^{-})^M|\Phi_{\mathrm{GS}}\rangle\|}, \tag{4.2.3}$$

where $M = 1, 2, \ldots$, and $|\Phi_{\mathrm{GS}}\rangle$ is the unique ground state of the antiferromagnetic Heisenberg model (4.1.1). These states should be compared with $|\Gamma\rangle$ defined in (3.4.7). Since $|\Phi_{\mathrm{GS}}\rangle \in \mathscr{H}_0$, we see $|\Gamma_M\rangle \in \mathscr{H}_M$. (Recall that $\mathscr{H}_M$ is the subspace with $\hat{S}_{\mathrm{tot}}^{(3)} = M$. See (2.2.10).) This in particular means that all $|\Gamma_M\rangle$ are orthogonal with each other. Then the following was proved in [35, 66].

Theorem 4.6 *There are constants C_1 and C_2 which depend only on d, S, and q_0. For any L and M such that $|M| \le C_1 L^{d/2}$, the state $|\Gamma_M\rangle$ is well-defined,*[15] *and satisfies*

$$\langle\Gamma_M|\hat{H}|\Gamma_M\rangle \le E_{\mathrm{GS}} + C_2\frac{M^2}{L^d}. \tag{4.2.4}$$

[14]To be more precise, one can make use of three independent low-lying states, namely, $\hat{\mathscr{O}}_L^{(1)}|\Phi_{\mathrm{GS}}\rangle$, $\hat{\mathscr{O}}_L^{(2)}|\Phi_{\mathrm{GS}}\rangle$, and $\hat{\mathscr{O}}_L^{(3)}|\Phi_{\mathrm{GS}}\rangle$. But these are far from enough.

[15]More precisely, the states $(\hat{\mathscr{O}}_L^{+})^M|\Phi_{\mathrm{GS}}\rangle$ or $(\hat{\mathscr{O}}_L^{-})^{|M|}|\Phi_{\mathrm{GS}}\rangle$ are nonzero.

We shall sketch the proof of the theorem in Sect. 4.2.2.

We now argue that the states $|\Gamma_M\rangle$ should be called low-lying states. Note that the translation invariance of (4.2.3) implies that $\langle\Gamma_M|\hat{\mathbf{S}}_x\cdot\hat{\mathbf{S}}_y|\Gamma_M\rangle$, the expectation value of the local Hamiltonian in (4.1.1), is identical for any $\{x,y\}\in\mathcal{B}_L$. This in particular means that $\langle\Gamma_M|\hat{\mathbf{S}}_x\cdot\hat{\mathbf{S}}_y|\Gamma_M\rangle=\langle\Gamma_M|\hat{H}|\Gamma_M\rangle/|\mathcal{B}_L|$. We then see from the bound (4.2.4) that

$$\frac{E_{\mathrm{GS}}}{|\mathcal{B}_L|}\le\langle\Gamma_M|\hat{\mathbf{S}}_x\cdot\hat{\mathbf{S}}_y|\Gamma_M\rangle\le\frac{E_{\mathrm{GS}}}{|\mathcal{B}_L|}+\frac{C_2}{d}\frac{M^2}{L^{2d}},\tag{4.2.5}$$

for any $\{x,y\}\in\mathcal{B}_L$, where we noted that $|\mathcal{B}_L|=d\,L^d$. Since $M^2\le(C_1)^2L^d$, the bounds (4.2.5) imply that, as $L\uparrow\infty$, the local energy $\langle\Gamma_M|\hat{\mathbf{S}}_x\cdot\hat{\mathbf{S}}_y|\Gamma_M\rangle$ converges to the (infinite volume) ground state energy density $\varepsilon_{\mathrm{GS}}$ define by[16]

$$\varepsilon_{\mathrm{GS}}:=\lim_{L\uparrow\infty}\frac{E_{\mathrm{GS}}}{|\mathcal{B}_L|}.\tag{4.2.6}$$

We shall regard the convergence of $\langle\Gamma_M|\hat{\mathbf{S}}_x\cdot\hat{\mathbf{S}}_y|\Gamma_M\rangle$ to $\varepsilon_{\mathrm{GS}}$ as a characterization of low-lying states. See also Theorem 4.20 in p. 115.

We stress that the translation invariance is essential in the above discussion. Suppose, for example, that $|\tilde{\Theta}_L\rangle$ is a state obtained from the ground state $|\Phi_{\mathrm{GS}}\rangle$ by exciting the spin at the origin. One clearly has $\langle\tilde{\Theta}_L|\hat{H}|\tilde{\Theta}_L\rangle-E_{\mathrm{GS}}=O(1)$ and hence $\lim_{L\uparrow\infty}\langle\tilde{\Theta}_L|\hat{H}|\tilde{\Theta}_L\rangle/|\mathcal{B}_L|=\varepsilon_{\mathrm{GS}}$. But the state $|\tilde{\Theta}_L\rangle$ is not a low-lying state since $\langle\tilde{\Theta}_L|\hat{\mathbf{S}}_o\cdot\hat{\mathbf{S}}_x|\tilde{\Theta}_L\rangle$ with $|x|=1$ is larger than $\varepsilon_{\mathrm{GS}}+\Delta E$ for any L with a constant $\Delta E>0$.

Exactly as in Theorem 3.1 (p. 67), the existence of low-lying states in each $\mathcal{H}_M$ immediately implies the existence of low-lying energy eigenstates.

Corollary 4.7 *For any L and M such that $|M|\le C_1L^{d/2}$, there exists an energy eigenstate $|\Psi_M\rangle\in\mathcal{H}_M$ whose energy eigenvalue E_M satisfies*

$$E_{\mathrm{GS}}<E_M\le E_{\mathrm{GS}}+C_2\frac{M^2}{L^d}.\tag{4.2.7}$$

Remarkably we have established that there are $O(L^{d/2})$ distinct low-lying energy eigenstates in the system with linear size L. Comparing this with the simple energy spectrum of the quantum Ising model, where the ground state and a single low-lying energy eigenstate are separated from other energy eigenstates by a nonzero gap (see Fig. 3.3), one sees that the existence of ever-increasing number of low-lying energy eigenstates is a clear manifestation of LRO without SSB related to continuous symmetry.

The existence of such a series of low-lying energy eigenstates and its relation to Néel ordered physical "ground states" had been known much before the theorem.

[16]The existence of the limit is not hard to prove. See, e.g., Appendix of [65].

As we mentioned in Sect. 3.1, the picture is explicitly mentioned in Anderson's 1984 book [3], and may be read off from his 1952 paper [2]. The series of low-lying energy eigenstates is sometimes called Anderson's tower of states.

The tower of states is especially important in the context of numerical diagonalization of quantum spin systems. Unlike in experiments, where a state similar to physical "ground states" should be observed, one directly observes exact ground states and low-lying energy eigenstates in such numerical studies.[17] The peculiar tower structure of the spectrum can be used as an indication for the existence of SSB (in the physical "ground state").[18] See [41] for an overview of Anderson's tower and related problems.[19]

In approximate theories for low-lying energy eigenstates [5, 41] it is argued that the low energy spectrum of the antiferromagnetic Heisenberg model is similar to that of the toy model (2.5.10) with long-range interactions introduced by Lieb and Mattis [42] to prove the Marshall–Lieb–Mattis theorem. From the expression (2.5.11) of the toy Hamiltonian, one readily finds that its low-lying energy levels are given by $E_{\mathrm{GS}} + S_{\mathrm{tot}}(S_{\mathrm{tot}} + 1)/(2|\Lambda|)$ with $S_{\mathrm{tot}} = 0, 1, 2, \ldots$. (See Problem 4.2.1.a in p. 104.) Thus the argued similarity implies that the low-lying energy levels in the full model is given by

$$E_M - E_{\mathrm{GS}} \simeq (\text{constant})\,\frac{M(M+1)}{L^d}, \tag{4.2.8}$$

for $M = 1, 2, \ldots$. This formula in fact fits numerical data quite well. It is also interesting that the scaling M^2/L^d in the rigorous upper bound (4.2.7) basically recovers this behavior. This is in contrast to the case of quantum Ising model, where the actual energy gap is $O(\lambda^L)$ while the upper bound is $O(L^{-1})$. See Sects. 3.3 and 3.4.

Remark The low-lying excited states discussed here should not be confused with spin wave excitations (or, equivalently, the Nambu–Goldston modes), whose excitation energies are proportional to L^{-1} rather than L^{-d}. The states with spin waves should be obtained by modifying the "ground state" with explicit antiferromagnetic order. We may say that the spin wave excitations are relevant to actual experimental observations, while the low-lying excitations are relevant to the exact energy spectrum of a large but finite systems.

Recently Koma has constructed a series of low energy excitations above the infinite volume ground state with explicit SSB in the antiferromagnetic Heisenberg model [32]. As far as we know this is the first rigorous demonstration of the existence

[17]It seems that people started observing the tower structure numerically in the early 90s when sufficiently advanced computers became available. We find, for example, partial data of the tower in Table I of [24], and a complete tower structure in Table I of [31], both for the $S = 1/2$ antiferromagnetic Heisenberg model on the square lattice.

[18]See, e.g., [9], where the existence of LRO (without SSB) in the antiferromagnetic Heisenberg model on the triangular lattice is investigated numerically.

[19]It is unfortunate that our rigorous results in 1994 on Anderson's tower is not mentioned in this review. We succeeded in proving theorems which had confirmed the standard picture, but were not very successful in conveying the result to the most relevant community.

of spin wave excitations in quantum antiferromagnets. See references in [32] for background.

Low-lying states with full symmetry breaking We shall construct low-lying states which break the SU(2) symmetry to the full extent, i.e., candidates for physical "ground states" which have both long-range order and symmetry breaking. The construction is similar to that of $|\Xi_\pm\rangle$ in Sect. 3.4, but here we can (and should) make use of the ever-increasing number of low-lying states in the "tower of states". Koma and Tasaki [34] first constructed such symmetry breaking low-lying states, but, very recently, Tanaka [62] proposed a new definition (4.2.10), which turned out to have much better properties than the previous ones. Here we describe the most complete results based on Tanaka's construction.

Let us define the symmetry breaking order parameter for the ground state by

$$m^* := \lim_{k\uparrow\infty}\lim_{L\uparrow\infty}\left\{\langle\Phi_{\mathrm{GS}}|\Big(\frac{\hat{\mathcal{O}}_L^{(\alpha)}}{L^d}\Big)^{2k}|\Phi_{\mathrm{GS}}\rangle\right\}^{1/(2k)}, \tag{4.2.9}$$

where k takes integral values, and $\alpha = 1, 2, 3$. As we see below in (4.2.23), the condition $q_0 > 0$ guarantees $m^* > 0$. Roughly speaking, m^* is the maximum value that $|\hat{\mathcal{O}}_L^{(\alpha)}/L^d|$ can take in the infinite volume limit.[20] If the reader is not familiar with this type of average, it is instructive to examine the following simple example: Let $a_1, \ldots, a_n$ be real numbers, and suppose that $|a_j|$ takes the maximum value $a_{\max}$ only at $j = j_0$. Take any probability distribution $p_1, \ldots, p_n$ such that $p_j > 0$ for any j, and assume that j is drawn with probability p_j. We denote the corresponding average as $\langle\cdots\rangle$. Noting that $\langle(a_j)^{2k}\rangle = \sum_{j=1}^n p_j(a_j)^{2k} \simeq p_{j_0}(a_{\max})^{2k}$ when k is large, one finds that $\lim_{k\uparrow\infty}\langle(a_j)^{2k}\rangle^{1/(2k)} = a_{\max}$.

Let $M(L) > 0$ be an integer-valued increasing function such that $M(L) + 1 \le C_1 L^{d/2}$. Following Tanaka [62], we define

$$|\Xi_{(1,0,0)}\rangle := \frac{1}{\sqrt{2}}\left\{\frac{(\hat{\mathcal{O}}_L^{(1)})^{M(L)}|\Phi_{\mathrm{GS}}\rangle}{\|(\hat{\mathcal{O}}_L^{(1)})^{M(L)}|\Phi_{\mathrm{GS}}\rangle\|} + \frac{(\hat{\mathcal{O}}_L^{(1)})^{M(L)+1}|\Phi_{\mathrm{GS}}\rangle}{\|(\hat{\mathcal{O}}_L^{(1)})^{M(L)+1}|\Phi_{\mathrm{GS}}\rangle\|}\right\}, \tag{4.2.10}$$

where the subscript $(1, 0, 0)$ indicates that the spins in the A sublattice are pointing in the positive 1-direction. To see the motivation of the definition, suppose that $M(L)$ is even. Then the operator $(\hat{\mathcal{O}}_L^{(1)})^{M(L)}$ modifies $|\Phi_{\mathrm{GS}}\rangle$ by a factor which is positive and large when the absolute value of $\hat{\mathcal{O}}_L^{(1)}$ is large. The other operator $(\hat{\mathcal{O}}_L^{(1)})^{M(L)+1}$ modifies $|\Phi_{\mathrm{GS}}\rangle$ by a factor which is positive and large when $\hat{\mathcal{O}}_L^{(1)}$ is positive and large, and is negative and large when $\hat{\mathcal{O}}_L^{(1)}$ is negative and large. By summing the two modified states as in (4.2.10), we get a state in which the part in the ground state $|\Phi_{\mathrm{GS}}\rangle$ with large positive $\hat{\mathcal{O}}_L^{(1)}$ is magnified.

---

[20] The order of the limits is essential here. If one takes the limit $k \uparrow \infty$ for finite L one simply gets S, which is the maximum possible value of $|\hat{\mathcal{O}}_L^{(\alpha)}/L^d|$. It does not reflect any properties of the ground state.

Since $\hat{\mathcal{O}}_L^{(1)} = (\hat{\mathcal{O}}_L^+ + \hat{\mathcal{O}}_L^-)/2$, we see that $|\Xi_{(1,0,0)}\rangle$ is expanded into a sum of states similar to (but not necessarily the same as) the low-lying state (4.2.3) with $|M| \le M(L)+1$. In fact it was shown that $|\Xi_{(1,0,0)}\rangle$ is also a low-lying state [62].

Theorem 4.8 *With the same constants as in Theorem 4.6, one has*

$$\langle \Xi_{(1,0,0)}|\hat{H}|\Xi_{(1,0,0)}\rangle \le E_{\mathrm{GS}} + C_2\frac{\{M(L)+1\}^2}{L^d} \tag{4.2.11}$$

for any L.

We shall sketch the proof of the theorem at the send of Sect. 4.2.2.

A remarkable property of the low-lying state $|\Xi_{(1,0,0)}\rangle$ is that it is a physical "ground state", which exhibits full symmetry breaking. By using the method developed by Tasaki in [66], Tanaka [62] proved the following theorem. We shall describe the complete proof in Sect. 4.2.2.

Theorem 4.9 *If $M(L)$ diverges to infinity not too rapidly as $L \uparrow \infty$, one has*[21]

$$\lim_{L\uparrow\infty} \langle \Xi_{(1,0,0)}|\frac{\hat{\mathcal{O}}_L^{(1)}}{L^d}|\Xi_{(1,0,0)}\rangle = m^*, \tag{4.2.12}$$

$$\lim_{L\uparrow\infty} \langle \Xi_{(1,0,0)}|\Bigl(\frac{\hat{\mathcal{O}}_L^{(1)}}{L^d}\Bigr)^2|\Xi_{(1,0,0)}\rangle = (m^*)^2, \tag{4.2.13}$$

$$\langle \Xi_{(1,0,0)}|\frac{\hat{\mathcal{O}}_L^{(\alpha)}}{L^d}|\Xi_{(1,0,0)}\rangle = 0, \tag{4.2.14}$$

$$\lim_{L\uparrow\infty} \langle \Xi_{(1,0,0)}|\Bigl(\frac{\hat{\mathcal{O}}_L^{(\alpha)}}{L^d}\Bigr)^2|\Xi_{(1,0,0)}\rangle = 0. \tag{4.2.15}$$

for $\alpha = 2$ or 3.

Note first that (4.2.12) and (4.2.14) show that the state $|\Xi_{(1,0,0)}\rangle$ exhibits symmetry breaking in which the order operator (viewed as a three component vector) is pointing in the $(1,0,0)$ direction. It is essential here that, in the limit $L \uparrow \infty$, the expectation value of $\hat{\mathcal{O}}_L^{(1)}/L^d$ is exactly equal to the order parameter m^*, which is designed to pick up the maximum value of $|\hat{\mathcal{O}}_L^{(\alpha)}/L^d|$. This means that the state $|\Xi_{(1,0,0)}\rangle$ breaks the SU(2) symmetry to the full extent.

The full symmetry breaking also manifests in the expectation values of $(\hat{\mathcal{O}}_L^{(\alpha)}/L^d)^2$, especially in that of $(\hat{\mathcal{O}}_L^{(1)}/L^d)^2$. That we got $(m^*)^2$ in the right-hand side of (4.2.13) is an indication that we are here dealing with a macroscopically "healthy" state. In particular (4.2.12), (4.2.13), (4.2.14), and (4.2.15) together imply that

$$\lim_{L\uparrow\infty} \sqrt{\langle \Xi_{(1,0,0)}|\Bigl(\frac{\hat{\mathcal{O}}_L^{(\alpha)}}{L^d}\Bigr)^2|\Xi_{(1,0,0)}\rangle - \Bigl(\langle \Xi_{(1,0,0)}|\frac{\hat{\mathcal{O}}_L^{(\alpha)}}{L^d}|\Xi_{(1,0,0)}\rangle\Bigr)^2} = 0, \tag{4.2.16}$$

[21] To be rigorous, lim in (4.2.12) and (4.2.13) should be replaced by lim inf.

for $\alpha = 1, 2$, or 3, which means that the fluctuation of the order operator density $\hat{\mathscr{O}}_L^{(\alpha)}/L^d$ vanishes in the limit $L \uparrow \infty$.

We conclude that the low-lying state $|\Xi_{(1,0,0)}\rangle$, which exhibits LRO and full SSB, is the desired physical "ground state". See Sect. 4.3, in particular Conjecture 4.21, for further discussion.

The state $|\Xi_{(1,0,0)}\rangle$ is not an energy eigenstate of the Hamiltonian, and hence changes under the unitary time evolution $e^{-i\hat{H}t}$. However since $|\Xi_{(1,0,0)}\rangle$ is a superposition of energy eigenstates with extremely small excitation energies, it is expected that the time-dependent state $e^{-i\hat{H}t}|\Xi_{(1,0,0)}\rangle$ changes very slowly. We may safely conclude that $|\Xi_{(1,0,0)}\rangle$ is essentially stationary when L is macroscopically large. See also [67, 68] for related discussions about the effect of the tower of states on decoherence of a symmetry breaking state.

We can of course consider essentially the same physical "ground states" in which the order operator (as a vector) is pointing in an arbitrary desired direction. For an arbitrary unit vector $\boldsymbol{n} = (n_1, n_2, n_3) \in \mathbb{R}^3$, let $\hat{\mathscr{O}}_L^{\boldsymbol{n}} := \sum_{\alpha=1}^{3} \hat{\mathscr{O}}_L^{(\alpha)} n_\alpha = \sum_{x \in \Lambda_L} (-1)^x \, \hat{\boldsymbol{S}}_x \cdot \boldsymbol{n}$. Then the desired symmetry breaking low-lying state is defined as

$$|\Xi_{\boldsymbol{n}}\rangle := \frac{1}{\sqrt{2}} \left\{ \frac{(\hat{\mathscr{O}}_L^{\boldsymbol{n}})^{M(L)}|\Phi_{\mathrm{GS}}\rangle}{\|(\hat{\mathscr{O}}_L^{\boldsymbol{n}})^{M(L)}|\Phi_{\mathrm{GS}}\rangle\|} + \frac{(\hat{\mathscr{O}}_L^{\boldsymbol{n}})^{M(L)+1}|\Phi_{\mathrm{GS}}\rangle}{\|(\hat{\mathscr{O}}_L^{\boldsymbol{n}})^{M(L)+1}|\Phi_{\mathrm{GS}}\rangle\|} \right\}. \tag{4.2.17}$$

Note that the normalization $\|(\hat{\mathscr{O}}_L^{\boldsymbol{n}})^{M}|\Phi_{\mathrm{GS}}\rangle\|$ is independent of $\boldsymbol{n}$ and the same as the original $\|(\hat{\mathscr{O}}_L^{(1)})^{M}|\Phi_{\mathrm{GS}}\rangle\|$ because of the SU(2) invariance of $|\Phi_{\mathrm{GS}}\rangle$.

By symmetry, we see that $\langle \Xi_{\boldsymbol{n}}|\hat{\mathscr{O}}_L^{\boldsymbol{n}'}|\Xi_{\boldsymbol{n}}\rangle = 0$ whenever the unit vectors $\boldsymbol{n}$ and $\boldsymbol{n}'$ are orthogonal.[22] By using the translation invariance and the equivalence between the A and the B sublattices, we find that

$$\langle \Xi_{\boldsymbol{n}}|\hat{\boldsymbol{S}}_x|\Xi_{\boldsymbol{n}}\rangle = (-1)^x \, m(L)\, \boldsymbol{n}, \tag{4.2.18}$$

for any $x \in \Lambda_L$, where $\lim_{L\uparrow\infty} m(L) = m^*$.

The states $|\Xi_{\boldsymbol{n}}\rangle$ have an interesting property that they transform under SU(2) rotation exactly as classical vectors. To see this let $\hat{U}$ be an arbitrary uniform spin rotation,[23] and R be the corresponding 3×3 rotation matrix in three dimension. See (2.1.11). Then we have shown in (2.1.20) that $\hat{U}(\hat{\boldsymbol{S}}_x \cdot \boldsymbol{n})\hat{U}^\dagger = \hat{\boldsymbol{S}}_x \cdot (\mathsf{R}\boldsymbol{n})$ for all unit vector $\boldsymbol{n}$ and all $x \in \Lambda_L$. Since this implies $\hat{U}\hat{\mathscr{O}}_L^{\boldsymbol{n}}\hat{U}^\dagger = \hat{\mathscr{O}}_L^{\mathsf{R}\boldsymbol{n}}$, we find from $\hat{U}|\Phi_{\mathrm{GS}}\rangle = |\Phi_{\mathrm{GS}}\rangle$ and (4.2.17) that

$$\hat{U}|\Xi_{\boldsymbol{n}}\rangle = |\Xi_{\mathsf{R}\boldsymbol{n}}\rangle. \tag{4.2.19}$$

---

[22] This can be seen by examining the transformation under $\hat{U}_\pi^{\boldsymbol{n}} = \exp[-i\pi\hat{\boldsymbol{S}}_{\mathrm{tot}} \cdot \boldsymbol{n}]$, the π rotation about the $\boldsymbol{n}$ axis.

[23] One can take arbitrary products of $\hat{U}_\theta^{(\alpha)}$ with various $\alpha = 1, 2, 3$ and θ.

It is notable that the right-hand side is determined solely by the vector $\mathsf{R}\boldsymbol{n}$. This "classical" transformation should be compared with the "quasi-classical" transformation with a nontrivial phase factor we saw in Problem 2.1.e (p. 18).

Recently it was pointed out that the (classical) Néel state (2.5.2) also exhibits the same "classical" transformation property under SU(2) rotation [53]. (This is a generalization of the behavior of the two spin state $|\uparrow\rangle_1|\downarrow\rangle_2$ treated in Problem 2.2.c in p. 23.) Recalling that even the ferromagnetic state (2.4.4) shows a "quasi-classical" transformation property with a nontrivial phase factor,[24] it is possible that the purely classical behavior of the Néel state and $|\Xi_{(1,0,0)}\rangle$ indicate an essential property of symmetry breaking in SU(2) invariant antiferromagnetic systems.

As a direct consequence of the transformation rule (4.2.19), we see for any spin rotation $\hat{U}$ that

$$\hat{U}\frac{1}{4\pi}\int_{|\boldsymbol{n}|=1} d\boldsymbol{n}\,|\Xi_{\boldsymbol{n}}\rangle = \frac{1}{4\pi}\int_{|\boldsymbol{n}|=1} d\boldsymbol{n}\,|\Xi_{\mathsf{R}\boldsymbol{n}}\rangle = \frac{1}{4\pi}\int_{|\boldsymbol{n}|=1} d\boldsymbol{n}\,|\Xi_{\boldsymbol{n}}\rangle, \tag{4.2.20}$$

i.e., the average of the state $|\Xi_{\boldsymbol{n}}\rangle$ over the solid angle is completely SU(2) invariant. (It is useful to compare the result with Problem 2.2.b in p. 23.) The averaged state is of course still a low-lying state and is also translation invariant. Since $|\Phi_{\mathrm{GS}}\rangle$ is a low-lying state with SU(2) and translation invariance, we are led to conjecture that

$$\frac{1}{4\pi}\int_{|\boldsymbol{n}|=1} d\boldsymbol{n}\,|\Xi_{\boldsymbol{n}}\rangle \simeq (\text{const.})|\Phi_{\mathrm{GS}}\rangle, \tag{4.2.21}$$

i.e., the average of the symmetry breaking "ground states" over the solid angle gives the unique ground state with LRO but without SSB. Note that (4.2.21) can be regarded as the SU(2) version of (3.3.8) for the quantum Ising model. For models with U(1) symmetry, we can explicitly prove a relation corresponding to (4.2.21). See (5.3.11) and (3.23) of [66]. Moreover it was shown in [53] that the relation (4.2.21), with $|\Xi_{\boldsymbol{n}}\rangle$ replaced by the classical Néel state, can be explicitly verified for the ground state of the toy model (2.5.10) of Lieb and Mattis [42]. This result can be regarded as the mean-field version of the conjecture (4.2.21). Finally, under a very plausible conjecture stated below as Conjecture 4.12, we can verify the validity of (4.2.21).

Proposition 4.10 *Suppose that the conjectured equality (4.2.26) is valid. Then we have*

$$\lim_{L\uparrow\infty}\left\|\frac{\frac{1}{4\pi}\int_{|\boldsymbol{n}|=1} d\boldsymbol{n}\,|\Xi_{\boldsymbol{n}}\rangle}{\left\|\frac{1}{4\pi}\int_{|\boldsymbol{n}|=1} d\boldsymbol{n}\,|\Xi_{\boldsymbol{n}}\rangle\right\|} - |\Phi_{\mathrm{GS}}\rangle\right\| = 0, \tag{4.2.22}$$

provided that $M(L)$ diverges to infinity not too rapidly as $L \uparrow \infty$.

We shall sketch the proof of the proposition in Sect. 4.2.2.

Long-range order parameter and symmetry breaking order parameter So far we have defined two different order parameters, the long-range order parameter

[24] See, e.g., Sect. 6.5 (in particular p. 85) of [22].

q_0 which characterizes the expectation value of $(\hat{\mathcal{O}}_L^{(\alpha)}/L^d)^2$ as in (4.1.7), and the symmetry breaking order parameter m^* which is the maximum possible value of $|\hat{\mathcal{O}}_L^{(\alpha)}/L^d|$ defined as in (4.2.9). These two are related by the following inequality proved by Koma and Tasaki in [34]. We here state the result without proof. (The reader interested in the proof is recommended to study [66].)

Theorem 4.11 *The two order parameters satisfy*

$$m^* \geq \sqrt{3q_0}. \tag{4.2.23}$$

The factor $\sqrt{3}$, which was absent in the corresponding relations (3.2.15) or (3.4.16) for the classical or quantum Ising models, reflects the SU(2) symmetry of the Heisenberg model.

The appearance of the factor $\sqrt{3}$ is not difficult to understand, at least intuitively. Theorem 4.9 suggests that $\hat{\boldsymbol{o}} = (\hat{\mathcal{O}}_L^{(1)}/L^d, \hat{\mathcal{O}}_L^{(2)}/L^d, \hat{\mathcal{O}}_L^{(3)}/L^d)$ basically behaves as a classical vector of magnitude m^* in physical "ground states" when L is large. In the state $|\Xi_{(1,0,0)}\rangle$, for example, one has $\hat{\boldsymbol{o}} \simeq (m^*, 0, 0)$. In the unique ground state, which is rotationally symmetric, the behavior of $\hat{\boldsymbol{o}}$ is far from that of a classical vector, but we assume that the magnitude m^* may be computed from the expectation value of $(\hat{\boldsymbol{o}})^2$. This leads to the estimate

$$(m^*)^2 \simeq \langle \Phi_{\mathrm{GS}}|(\hat{\boldsymbol{o}})^2|\Phi_{\mathrm{GS}}\rangle = \sum_{\alpha=1}^{3} \langle \Phi_{\mathrm{GS}}|\Big(\frac{\hat{\mathcal{O}}_L^{(\alpha)}}{L^d}\Big)^2|\Phi_{\mathrm{GS}}\rangle \geq 3q_0, \tag{4.2.24}$$

where we noted that (4.1.7) is valid for $\alpha = 1, 2$, and 3.

The above estimate does not only give an interpretation of the inequality (4.2.23), but also leads us to the following conjecture.

Conjecture 4.12 *Define the long-range order parameter* q^* *by*

$$q^* := \lim_{L\uparrow\infty} \langle \Phi_{\mathrm{GS}}|\Big(\frac{\hat{\mathcal{O}}_L^{(\alpha)}}{L^d}\Big)^2|\Phi_{\mathrm{GS}}\rangle, \tag{4.2.25}$$

for any $\alpha = 1, 2, 3$. *We then conjecture that*

$$m^* = \sqrt{3q^*}. \tag{4.2.26}$$

Recall that the long-range order parameter q_0 was introduced in the lower bound (4.1.7). Here q^* is defined as the exact value in the infinite volume limit. The conjectured equality (4.2.26) basically says that both SSB and LRO are manifestations of a common "order" developed in the ground states. We believe that the conjecture is very likely to be valid. In fact the corresponding equality (3.2.15) for the classical Ising model is known to hold rigorously (although the proof is far from being easy).

Unfortunately it seems that we are still very far from proving the conjecture for the quantum antiferromagnetic Heisenberg model.[25]

SSB under infinitesimal symmetry breaking field It is obvious that the theorem of Kaplan, Horsch, and von der Linden (Theorem 3.2 in p. 70) applies to the antiferromagnetic Heisenberg model.

We define the Hamiltonian with symmetry breaking field $h \geq 0$ as

$$\hat{H}_h = \hat{H} - h\hat{\mathcal{O}}_L^{(1)} = \sum_{\{x,y\}\in\mathcal{B}_L} \hat{\boldsymbol{S}}_x \cdot \hat{\boldsymbol{S}}_y - h \sum_{x\in\Lambda_L} (-1)^x \, \hat{S}_x^{(1)}, \tag{4.2.27}$$

and let $|\Phi_{\mathrm{GS},h}\rangle$ be its (not necessarily unique) ground state. One should recall that the symmetry breaking field is not a realistic magnetic field since its direction depends on the sublattice. As we have discussed after (3.4.19) in p. 69, this is a purely theoretical (or fictitious) field called staggered magnetic field.

Repeating the variational proof by using $|\Xi_{(1,0,0)}\rangle$ as a trial state, we see that infinitesimal staggered magnetic field can trigger spontaneous symmetry breaking in the ground state of the antiferromagnetic Heisenberg model.

Theorem 4.13 *We have*[26]

$$\lim_{h\downarrow 0} \lim_{L\uparrow\infty} \langle \Phi_{\mathrm{GS},h} | \frac{\hat{\mathcal{O}}_L^{(1)}}{L^d} | \Phi_{\mathrm{GS},h} \rangle \geq m^* \geq \sqrt{3q_0}. \tag{4.2.28}$$

Unfortunately the theorem does not shed light on the physical mechanism of SSB in the antiferromagnetic Heisenberg model, because we do not expect staggered magnetic field h to appear in realistic systems. See the end of Sect. 4.3 for a totally different approach to understand physical mechanism of SSB.

Remark The left-hand side of (4.2.28) defines yet another order parameter (which should also be called symmetry breaking order parameter). We conjecture that it is identical to m^*, but the proof seems to be very difficult. See [32] for a closely related recent result.

The XXZ model The above results apply, with small modifications, to models with U(1) symmetry. The most important example is the XXZ model with general spin S, whose Hamiltonian is

$$\hat{H} = \sum_{\{x,y\}\in\mathcal{B}_L} \{\hat{S}_x^{(1)}\hat{S}_y^{(1)} + \hat{S}_x^{(2)}\hat{S}_y^{(2)} + \lambda \hat{S}_x^{(3)}\hat{S}_y^{(3)}\}, \tag{4.2.29}$$

---

[25] One can construct an artificial model with SU(2) invariance in which (4.2.26) is violated for an obvious reason. See Remark 4 in p. 195 of [34]. This means that general arguments as found in Sect. 4.2.2 are useless; one needs to invoke a strong argument specific to the antiferromagnetic Heisenberg model on the hypercubic lattice.

[26] To be rigorous lim should be replaced by lim inf.

where λ is the Ising anisotropy parameter. See (2.5.14). From Theorem 2.4 (p. 43), we know that the ground state $|\Phi_{\mathrm{GS}}\rangle$ is unique and satisfies $\hat{S}^{(3)}_{\mathrm{tot}}|\Phi_{\mathrm{GS}}\rangle = 0$ provided that $\lambda > -1$. Since we are interested in the breakdown of continuous symmetry, we do not consider the region with $\lambda > 1$, where the model exhibits (or is expected to exhibit) Ising type long range order and symmetry breaking.

By extending the method of Dyson, Lieb, and Simon (see Theorem 4.1 in p. 75), the existence of LRO in the ground state, i.e, $\langle\Phi_{\mathrm{GS}}|(\hat{\mathcal{O}}^{(1)}_L/L^d)^2|\Phi_{\mathrm{GS}}\rangle = \langle\Phi_{\mathrm{GS}}|(\hat{\mathcal{O}}^{(2)}_L/L^d)^2|\Phi_{\mathrm{GS}}\rangle \geq q_0 > 0$ was proved for the model in $d = 3$ with any $\lambda \in [0, 1]$ and $S = 1/2, 1, \ldots$. In $d = 2$, the existence of LRO was proved for $\lambda \in [0, 0.20)$ if $S = 1/2$ and for $\lambda \in [0, 1]$ if $S = 1$ or larger. See [29, 30, 40, 49] and references therein.

For this model, Theorems 4.6 and 4.9 are valid as they are. Reflecting the U(1) symmetry, the inequality (4.2.23) in Theorem 4.11 is replaced by $m^* \geq \sqrt{2q_0}$. See [34, 66].

Remark about ferrimagnetism It is obvious that there is no "tower" of low-lying states in the ferromagnetic Heisenberg model. As we have seen in Sect. 2.4, the ground states of the ferromagnetic Heisenberg model are $2|\Lambda|S+1$ fold degenerate. Physical ground states (also known as the spin coherent states), where all the spins are pointing in a fixed direction, are given by $|\Phi^{\uparrow}\rangle$ in (2.4.4), and, more generally, by $|\Xi_{\theta,\varphi}\rangle$ in (2.4.6). These physical ground states are also exact ground states of the model. Such a simple behavior is expected since the order operators, which are nothing but the total spin operator $\hat{S}^{(\alpha)}_{\mathrm{tot}}$, commute with the Hamiltonian as in (2.4.2).

As we have seen at the end of Sect. 4.1, the spin S antiferromagnetic Heisenberg model (2.5.1) on a bipartite lattice exhibits ferrimagnetism when the numbers of sites in the two sublattices differ macroscopically. Ferrimagnetism is an interesting intermediate between ferromagnetism and antiferromagnetism. Ferrimagnetism is similar to ferromagnetism in that there are a macroscopic number of degenerate ground states. But ferrimagnetism is similar also to antiferromagnetism in that the Hamiltonian and the order operator (4.1.4) do not commute,[27] and hence we generally do not know the exact ground states.[28]

It is therefore not at all obvious whether degenerate ground states of a ferrimagnet are accompanied by low-lying states. Recently Rademaker, Beekman, and van Wezel [54] studied this problem from several viewpoints and concluded that there is no "tower" of states above the ground states in a ferrimagnetic spin system (or, more generally, in a system exhibiting the type-B SSB).[29] A convincing (and ele-

[27]The operator (4.1.4), with Λ_L replaced by Λ, is a natural order operator for detecting ferrimagnetic order. It should be noted however that ferrimagnetic order may also be characterized by the ferromagnetic order operator, namely, the total spin.

[28]One may say that there are quantum fluctuations in the ground states of antiferromagnetic and ferrimagnetic systems, but not in those of ferromagnetic systems.

[29]It is also argued in [54] that the exact ferrimagnetic ground state with the largest eigenvalue of $\hat{S}^{(3)}_{\mathrm{tot}}$ is a physical (or ergodic) state in which macroscopic quantities exhibit small fluctuations. To prove this conjecture is a challenging problem in mathematical physics.

mentary) evidence comes from the toy model (2.5.10) used by Lieb and Mattis. See Problem 4.2.1.a below.

Problem 4.2.1.a Consider the toy model 2.5.10 with $|A| - |B| = a|\Lambda|$ with $a > 0$, and show that there exists a nonzero energy gap (uniform in the system size $|\Lambda|$) above the ground state energy. [solution→p.502]

4.2.2 *Proofs*

We shall now discuss the proofs of some of the theorems, following [62, 66].[30] The order operator (4.2.31), which enables us to make full use of the U(1) symmetry, plays an important role in the proofs. After stating basic properties of the order operator, we give a complete Proof of Theorem 4.9; the proof is important and not too heavy. We hope that the reader may find some arguments interesting. We skip the Proof of Theorem 4.11 since it is rather technical. See [66]. The Proofs of Theorems 4.6 and 4.8 are rather involved. We here illustrate basic strategy, major difficulties, and basic ideas for overcoming them. Again complete proofs can be found in [62, 66].

Preliminaries Throughout the present subsection, we abbreviate the ground state expectation $\langle\Phi_{\mathrm{GS}}|\cdots|\Phi_{\mathrm{GS}}\rangle$ as $\langle\cdots\rangle$. We use operators per unit volume

$$\hat{o}^{(\alpha)} := \frac{\hat{\mathcal{O}}_L^{(\alpha)}}{V}, \quad \hat{o}^{\pm} := \frac{\hat{\mathcal{O}}_L^{\pm}}{V} = \hat{o}^{(1)} \pm i\hat{o}^{(2)}, \tag{4.2.30}$$

where $V = L^d$ is the volume (or the number of sites). We introduce the new order operator

$$\hat{p} := \frac{1}{2}(\hat{o}^+\hat{o}^- + \hat{o}^-\hat{o}^+) = (\hat{o}^{(1)})^2 + (\hat{o}^{(2)})^2, \tag{4.2.31}$$

which turns out to be useful.

Recalling (4.2.2), we find

$$[\hat{\mathcal{O}}_L^+, \hat{\mathcal{O}}_L^-] = \sum_{x\in\Lambda}[\hat{S}_x^+, \hat{S}_x^-] = 2\hat{S}_{\mathrm{tot}}^{(3)}, \tag{4.2.32}$$

which implies that

$$\left\|[\hat{o}^+, \hat{o}^-]\right\| = \frac{1}{V^2}\left\|[\hat{\mathcal{O}}_L^+, \hat{\mathcal{O}}_L^-]\right\| \le \frac{o_0}{V}, \tag{4.2.33}$$

where we set $o_0 = 2S$. Note that it also holds (somewhat accidentally) that $\|\hat{o}^{\pm}\| \le o_0$. We then easily get the following elementary lemma.

Lemma 4.14 *Let* $s_1, \ldots, s_{2n} = \pm$ *be any sequence such that* $\sum_{j=1}^{2n} s_j = 0$. *Then we have*

$$\left\|\hat{o}^{s_1}\ldots\hat{o}^{s_{2n}} - \hat{p}^n\right\| \le \frac{n^2(o_0)^{2n-1}}{V}. \tag{4.2.34}$$

---

[30] The reader not interested in mathematical details may safely skip this subsection.

Proof Note that $\hat{o}^{s_1}\ldots\hat{o}^{s_{2n}}$ can be rearranged to any other $\hat{o}^{s'_1}\ldots\hat{o}^{s'_{2n}}$ with $\sum_{j=1}^{2n}s'_j=0$ by making at most n^2 exchanges of neighboring $\hat{o}^+$ and $\hat{o}^-$. Then (4.2.33) implies (4.2.34). ∎

Order parameters For a positive integer n, the Schwarz inequality implies

$$\langle\hat{p}^n\rangle^2=\langle\hat{p}^{(n-1)/2}\hat{p}^{(n+1)/2}\rangle^2\le\langle\hat{p}^{n-1}\rangle\langle\hat{p}^{n+1}\rangle, \tag{4.2.35}$$

from which we find

$$\frac{\langle\hat{p}^n\rangle}{\langle\hat{p}^{n-1}\rangle}\le\frac{\langle\hat{p}^{n+1}\rangle}{\langle\hat{p}^n\rangle}. \tag{4.2.36}$$

Recall that, from (4.1.7), we have $\langle\hat{p}\rangle=\langle(\hat{o}^{(1)})^2\rangle+\langle(\hat{o}^{(2)})^2\rangle\ge 2q_0>0$. Thus (4.2.36) implies

$$\frac{\langle\hat{p}^n\rangle}{\langle\hat{p}^{n-1}\rangle}\ge 2q_0, \tag{4.2.37}$$

and hence $\langle\hat{p}^n\rangle\ge(2q_0)^n$.

The following lemma shows the essential property of the ratio $\langle\hat{p}^n\rangle/\langle\hat{p}^{n-1}\rangle$.

Lemma 4.15 *We have*

$$m^*=\lim_{n\uparrow\infty}\lim_{L\uparrow\infty}\sqrt{\frac{\langle\hat{p}^n\rangle}{\langle\hat{p}^{n-1}\rangle}}, \tag{4.2.38}$$

where m^* is defined in (4.2.9).

Note that the limit $n\uparrow\infty$ in (4.2.38) exists because of the monotonicity (4.2.36) and the boundedness $\langle\hat{p}^n\rangle/\langle\hat{p}^{n-1}\rangle\le\|\hat{p}\|$.[31] It is worth noting that (4.2.38) along with (4.2.37) implies that

$$m^*\ge\sqrt{2q_0}. \tag{4.2.39}$$

This is the weaker version of Theorem 4.11, which is optimal for U(1) invariant models.

Proof of Lemma 4.15 The first observation is that (4.2.9) is written as

$$m^*=\lim_{n\uparrow\infty}\lim_{L\uparrow\infty}\sqrt{\frac{\langle(\hat{o}^{(1)})^{2n}\rangle}{\langle(\hat{o}^{(1)})^{2(n-1)}\rangle}}. \tag{4.2.40}$$

This is quite plausible since (4.2.9) suggests that $\langle(\hat{o}^{(1)})^{2n}\rangle\simeq(m^*)^{2n}$. The proof, which we omit, is indeed elementary. See [66].

---

[31] The existence of the limit $L\uparrow\infty$, on the other hand, is not guaranteed in general. To be rigorous one should replace lim with lim inf. The same comment applies to $\lim_{L\uparrow\infty}$ that appear in the rest of this section.

A less trivial part is to show that the limits in the right-hand sides of (4.2.38) and (4.2.40) coincide. Let us write $\hat{o}^{(1)} = (\hat{o}^{+} + \hat{o}^{-})/2$, and observe that

$$\left\langle(\hat{o}^{(1)})^{2n}\right\rangle = \frac{1}{2^{2n}} \sum_{\substack{s_1,\dots,s_{2n}=+ \\ (\sum_j s_j=0)}} \langle \hat{o}^{s_1} \cdots \hat{o}^{s_{2n}} \rangle, \tag{4.2.41}$$

where we noted that the ground state satisfies $\hat{S}^{(3)}_{\text{tot}}|\Phi_{\text{GS}}\rangle = 0$. There are $(2n)!/(n!)^2$ terms in the sum in the right-hand side of (4.2.41). From (4.2.34), we see that each term in the sum is equal to $\langle \hat{p}^n \rangle + O(1/V)$. We then find

$$\left\langle(\hat{o}^{(1)})^{2n}\right\rangle = \frac{1}{2^{2n}} \frac{(2n)!}{(n!)^2} \langle \hat{p}^n \rangle + O\Big(\frac{1}{V}\Big) = \frac{(2n-1)!!}{(2n)!!} \langle \hat{p}^n \rangle + O\Big(\frac{1}{V}\Big), \tag{4.2.42}$$

which implies

$$\frac{\left\langle(\hat{o}^{(1)})^{2n}\right\rangle}{\left\langle(\hat{o}^{(1)})^{2(n-1)}\right\rangle} = \frac{2n-1}{2n} \frac{\langle \hat{p}^n \rangle}{\langle \hat{p}^{n-1} \rangle} + O\Big(\frac{1}{V}\Big). \tag{4.2.43}$$

We thus see

$$\lim_{n\uparrow\infty} \lim_{L\uparrow\infty} \sqrt{\frac{\left\langle(\hat{o}^{(1)})^{2n}\right\rangle}{\left\langle(\hat{o}^{(1)})^{2(n-1)}\right\rangle}} = \lim_{n\uparrow\infty} \lim_{L\uparrow\infty} \sqrt{\frac{\langle \hat{p}^n \rangle}{\langle \hat{p}^{n-1} \rangle}}, \tag{4.2.44}$$

which proves the desired (4.2.38). ∎

Symmetry breaking state We are now ready to prove Theorem 4.9, which shows that the low-lying state $|\Xi_{(1,0,0)}\rangle$ defined in (4.2.10) fully breaks the SU(2) symmetry. For $M = 1, 2, \ldots$ and system size $L = 1, 2, \ldots$, we set

$$|\Xi_L^{(M)}\rangle = \frac{1}{\sqrt{2}} \left\{ \frac{(\hat{o}^{(1)})^M |\Phi_{\text{GS}}\rangle}{\|(\hat{o}^{(1)})^M |\Phi_{\text{GS}}\rangle\|} + \frac{(\hat{o}^{(1)})^{M+1} |\Phi_{\text{GS}}\rangle}{\|(\hat{o}^{(1)})^{M+1} |\Phi_{\text{GS}}\rangle\|} \right\}, \tag{4.2.45}$$

which is essentially the state $|\Xi_{(1,0,0)}\rangle$ defined in (4.2.10), but with $M(L)$ replaced by an independent variable M. Note that $|\Xi_{(1,0,0)}\rangle = |\Xi_L^{(M(L))}\rangle$.

Since the invariance $\hat{U}^{(3)}_{\pi}|\Phi_{\text{GS}}\rangle = |\Phi_{\text{GS}}\rangle$ implies $\langle\Phi_{\text{GS}}|(\hat{o}^{(1)})^K|\Phi_{\text{GS}}\rangle = 0$ for any odd K, we have

$$\langle \Xi_L^{(M)} | \hat{o}^{(1)} | \Xi_L^{(M)} \rangle = \frac{\langle\Phi_{\text{GS}}|(\hat{o}^{(1)})^{2M+2}|\Phi_{\text{GS}}\rangle}{\|(\hat{o}^{(1)})^M |\Phi_{\text{GS}}\rangle\| \, \|(\hat{o}^{(1)})^{M+1} |\Phi_{\text{GS}}\rangle\|} = \sqrt{\frac{\langle(\hat{o}^{(1)})^{2M+2}\rangle}{\langle(\hat{o}^{(1)})^{2M}\rangle}}. \tag{4.2.46}$$

This expression, with (4.2.40), immediately implies

$$\lim_{M\uparrow\infty} \lim_{L\uparrow\infty} \langle \Xi_L^{(M)} | \hat{o}^{(1)} | \Xi_L^{(M)} \rangle = m^*. \tag{4.2.47}$$

We also see that the invariance $\hat{U}_\pi^{(1)}|\Xi_L^{(M)}\rangle = |\Xi_L^{(M)}\rangle$ implies

$$\langle \Xi_L^{(M)}|\hat{o}^{(\alpha)}|\Xi_L^{(M)}\rangle = 0, \tag{4.2.48}$$

for $\alpha = 2$ and 3.

Let us evaluate the expectation value of $(\hat{o}^{(1)})^2 + (\hat{o}^{(2)})^2 = \hat{p}$. Noting that the invariance $\hat{U}_\pi^{(3)}|\Phi_{\rm GS}\rangle = |\Phi_{\rm GS}\rangle$ implies $\langle(\hat{o}^{(1)})^M \hat{p}\,(\hat{o}^{(1)})^{M+1}\rangle = \langle(\hat{o}^{(1)})^{M+1}\hat{p}(\hat{o}^{(1)})^M\rangle = 0$, we have

$$\begin{aligned}\langle \Xi_L^{(M)}|\{(\hat{o}^{(1)})^2 + (\hat{o}^{(2)})^2\}|\Xi_L^{(M)}\rangle \\ = \frac{1}{2}\left\{\frac{\langle(\hat{o}^{(1)})^M \hat{p}\,(\hat{o}^{(1)})^M\rangle}{\langle(\hat{o}^{(1)})^{2M}\rangle} + \frac{\langle(\hat{o}^{(1)})^{M+1} \hat{p}\,(\hat{o}^{(1)})^{M+1}\rangle}{\langle(\hat{o}^{(1)})^{2(M+1)}\rangle}\right\}.\end{aligned} \tag{4.2.49}$$

By using $\hat{o}^{(1)} = (\hat{o}^+ + \hat{o}^-)/2$ and recalling that $\hat{S}_{\rm tot}^{(3)}|\Phi_{\rm GS}\rangle = 0$, we have

$$\begin{aligned}\langle(\hat{o}^{(1)})^M \hat{p}\,(\hat{o}^{(1)})^M\rangle &= 2^{-2M} \sum_{\substack{s_1,\ldots,s_{2M}=\pm \\ (\sum_{j=1}^{2M} s_j=0)}} \langle \hat{o}^{s_1}\cdots\hat{o}^{s_M}\,\hat{p}\,\hat{o}^{s_{M+1}}\cdots\hat{o}^{s_{2M}}\rangle \\ &= 2^{-2M}\frac{(2M)!}{(M!)^2}\langle\hat{p}^{M+1}\rangle + O\Big(\frac{1}{V}\Big),\end{aligned} \tag{4.2.50}$$

where we used (4.2.34). Rewriting the other three expectation values in (4.2.49) similarly, we get

$$\langle \Xi_L^{(M)}|\{(\hat{o}^{(1)})^2 + (\hat{o}^{(2)})^2\}|\Xi_L^{(M)}\rangle = \frac{1}{2}\left\{\frac{\langle\hat{p}^{M+1}\rangle}{\langle\hat{p}^M\rangle} + \frac{\langle\hat{p}^{M+2}\rangle}{\langle\hat{p}^{M+1}\rangle}\right\} + O\Big(\frac{1}{V}\Big). \tag{4.2.51}$$

We then find from (4.2.38) that

$$\lim_{M\uparrow\infty}\lim_{L\uparrow\infty}\langle \Xi_L^{(M)}|\{(\hat{o}^{(1)})^2 + (\hat{o}^{(2)})^2\}|\Xi_L^{(M)}\rangle = (m^*)^2. \tag{4.2.52}$$

We see from (4.2.47) and (4.2.52) that

$$\begin{aligned}(m^*)^2 &= \lim_{M\uparrow\infty}\lim_{L\uparrow\infty}\big(\langle \Xi_L^{(M)}|\hat{o}^{(1)}|\Xi_L^{(M)}\rangle\big)^2 \le \lim_{M\uparrow\infty}\lim_{L\uparrow\infty}\langle \Xi_L^{(M)}|(\hat{o}^{(1)})^2|\Xi_L^{(M)}\rangle \\ &\le \lim_{M\uparrow\infty}\lim_{L\uparrow\infty}\langle \Xi_L^{(M)}|\{(\hat{o}^{(1)})^2 + (\hat{o}^{(2)})^2\}|\Xi_L^{(M)}\rangle = (m^*)^2,\end{aligned} \tag{4.2.53}$$

which implies

$$\lim_{M\uparrow\infty}\lim_{L\uparrow\infty}\langle \Xi_L^{(M)}|(\hat{o}^{(1)})^2|\Xi_L^{(M)}\rangle = (m^*)^2, \tag{4.2.54}$$

and

$$\lim_{M\uparrow\infty}\lim_{L\uparrow\infty}\langle \Xi_L^{(M)}|(\hat{o}^{(2)})^2|\Xi_L^{(M)}\rangle = 0. \tag{4.2.55}$$

Since the directions 2 and 3 are equivalent, the latter implies

$$\lim_{M\uparrow\infty}\lim_{L\uparrow\infty}\langle\Xi_L^{(M)}|(\hat{o}^{(\alpha)})^2|\Xi_L^{(M)}\rangle = 0, \tag{4.2.56}$$

for $\alpha = 2$ and 3.

Note that the relations (4.2.47), (4.2.54), (4.2.48), and (4.2.56) precisely correspond to the desired relations (4.2.12), (4.2.13), (4.2.14), and (4.2.15), respectively, except that we have the double limit $\lim_{M\uparrow\infty}\lim_{L\uparrow\infty}$ instead of the single limit $\lim_{L\uparrow\infty}$ in which $M = M(L)$ varies according to L. Intuitively speaking, the double limit $\lim_{M\uparrow\infty}\lim_{L\uparrow\infty}$ corresponds to the single limit $\lim_{L\uparrow\infty}$ with $M(L)$ that diverges indefinitely slowly. It only remains to extend the relations to $M(L)$ that does not diverge too rapidly, but this is a straightforward consequence of the following lemma.[32]

Lemma 4.16 *If* $f_{L,M} \in \mathbb{R}$ *with* $L, M=1, 2, \ldots$ *satisfies* $\lim_{M\uparrow\infty}\lim_{L\uparrow\infty} f_{L,M}=0$, *then there exists a nondecreasing function* $M(L)$ *such that* $\lim_{L\uparrow\infty} M(L) = \infty$ *and* $\lim_{L\uparrow\infty} f_{L,M(L)} = 0$.

Proof Let $g_M = \lim_{L\uparrow\infty} f_{L,M}$. (We of course have $\lim_{M\uparrow\infty} g_M = 0$.) Take any ε_M for $M = 1, 2, \ldots$ such that $\lim_{M\uparrow\infty}\varepsilon_M = 0$. For each M we take L_M such that $|f_{L,M}-g_M| \le \varepsilon_M$ for any $L \ge L_M$. We can take L_M to be strictly increasing in M. We then define $M(L)=M$ for $L\in(L_{M-1}, L_M]$. It then holds that $|f_{L,M(L)}-g_M| \le \varepsilon_{M(L)}$, which implies $\lim_{L\uparrow\infty} f_{L,M(L)} = 0$. ∎

Finally we see how Proposition 4.10 is proved. Let us define

$$|\Xi_{\boldsymbol{n},L}^{(M)}\rangle = \frac{1}{\sqrt{2}}\left\{\frac{(\hat{\boldsymbol{o}}\cdot\boldsymbol{n})^M|\Phi_{\rm GS}\rangle}{\|(\hat{\boldsymbol{o}}\cdot\boldsymbol{n})^M|\Phi_{\rm GS}\rangle\|} + \frac{(\hat{\boldsymbol{o}}\cdot\boldsymbol{n})^{M+1}|\Phi_{\rm GS}\rangle}{\|(\hat{\boldsymbol{o}}\cdot\boldsymbol{n})^{M+1}|\Phi_{\rm GS}\rangle\|}\right\}, \tag{4.2.57}$$

where $\hat{\boldsymbol{o}} = (\hat{o}^{(1)}, \hat{o}^{(2)}, \hat{o}^{(3)})$. This is essentially the state (4.2.17), but with $M(L)$ replaced by an independent variable M. By using an argument similar to those in the present section, one can show that

$$\frac{1}{4\pi}\int_{|\boldsymbol{n}|=1} d\boldsymbol{n}\,(\hat{\boldsymbol{o}}\cdot\boldsymbol{n})^M = \begin{cases}\dfrac{1}{M+1}(\hat{\boldsymbol{o}}^2)^{M/2} + O\Big(\dfrac{1}{V}\Big) & \text{for even } M,\\ 0 & \text{for odd } M.\end{cases} \tag{4.2.58}$$

See (4.40) of [66]. Assuming M is even, we find from (4.2.57) and (4.2.58) that

$$\frac{1}{4\pi}\int_{|\boldsymbol{n}|=1} d\boldsymbol{n}\,|\Xi_{\boldsymbol{n},L}^{(M)}\rangle = C\,(\hat{\boldsymbol{o}}^2)^{M/2}|\Phi_{\rm GS}\rangle + O\Big(\frac{1}{V}\Big), \tag{4.2.59}$$

where C is a positive constant of $O(V^0)$. We next note that

[32] We should note that, although this general lemma proves the existence of $M(L)$, it does not tell us whether a concrete choice of $M(L)$, say $M(L) = L^2$, is suitable.

$$\left\| \frac{(\hat{\boldsymbol{o}}^2)^{M/2}|\Phi_{\text{GS}}\rangle}{\|(\hat{\boldsymbol{o}}^2)^{M/2}|\Phi_{\text{GS}}\rangle\|} - |\Phi_{\text{GS}}\rangle \right\|^2 = 2\left(1 - \frac{\langle(\hat{\boldsymbol{o}}^2)^{M/2}\rangle}{\sqrt{\langle(\hat{\boldsymbol{o}}^2)^M\rangle}}\right). \tag{4.2.60}$$

Now, by repeating the argument used for the operator $\hat{p}$ in the present section, we can show that $\lim_{L\uparrow\infty}\langle(\hat{\boldsymbol{o}}^2)^n\rangle/\langle(\hat{\boldsymbol{o}}^2)^{n-1}\rangle$ is nondecreasing in n, and $\lim_{n\uparrow\infty}\lim_{L\uparrow\infty}\langle(\hat{\boldsymbol{o}}^2)^n\rangle/\langle(\hat{\boldsymbol{o}}^2)^{n-1}\rangle = (m^*)^2$. Then the conjectured equality $\lim_{L\uparrow\infty}\langle\hat{\boldsymbol{o}}^2\rangle = (m^*)^2$ implies $\lim_{L\uparrow\infty}\langle(\hat{\boldsymbol{o}}^2)^n\rangle = (m^*)^{2n}$ for any n. This implies that the right-hand side of (4.2.60) vanishes in the $L \uparrow \infty$ limit. From (4.2.59) and (4.2.60) we thus have

$$\lim_{L\uparrow\infty} \left\| \frac{\frac{1}{4\pi}\int_{|\boldsymbol{n}|=1} d\boldsymbol{n}\, |\varXi_{\boldsymbol{n},L}^{(M)}\rangle}{\left\|\frac{1}{4\pi}\int_{|\boldsymbol{n}|=1} d\boldsymbol{n}\, |\varXi_{\boldsymbol{n},L}^{(M)}\rangle\right\|} - |\Phi_{\text{GS}}\rangle \right\| = 0, \tag{4.2.61}$$

for each M. Then (4.2.22) with a suitable choice of $M(L)$ follows from Lemma 4.16.

Low-lying states Finally we shall discuss the Proof of Theorems 4.6 and 4.8, which state that $|\Gamma_M\rangle$ in (4.2.3) and $|\varXi_{(1,0,0)}\rangle$ in (4.2.10) are low-lying states. Recall that the corresponding theorem in Sect. 3.4 (Theorem 3.1 in p. 67) was proved easily by noting that the energy expectation value can be written compactly using a double commutator as in (3.4.8). Unfortunately, the Proofs of Theorems 4.6 and 4.8, in which we must control ever-increasing number of low-lying states, are much more involved. Here we follow [66] and discuss the main difficulty and some essential ideas of the Proof of Theorem 4.6. We leave a complete proof to [66]. The Proof of Theorem 4.8 is discussed at the end.

For notational simplicity, we write $\hat{H}' = \hat{H} - E_{\text{GS}}$. Our goal is to bound the energy expectation value

$$\langle\Gamma_M|\hat{H}'|\Gamma_M\rangle = \frac{\left\langle(\hat{o}^-)^M\hat{H}'(\hat{o}^+)^M\right\rangle}{\left\langle(\hat{o}^-)^M(\hat{o}^+)^M\right\rangle}. \tag{4.2.62}$$

The first difficulty is that the numerator in this case cannot be written in terms of a double commutator as in (3.4.8). Recall that it was essential in (3.4.8) that $\hat{H}$ is sandwiched by the same self-adjoint operator $\hat{\mathcal{O}}_L$ from the both sides.

To overcome this difficulty, Koma and Tasaki [35] made use of the $\mathbb{Z}_2$ symmetry of the model. But there is a much simpler method.[33] Since $\left\langle(\hat{o}^+)^M\hat{H}'(\hat{o}^-)^M\right\rangle \geq 0$, we see from (4.2.62) that

[33] We learned this method from Hosho Katsura. A similar technique was used by Sannomiya, Katsura, and Nakayama [57]. See (28) and (29) of [57].

$$\langle\Gamma_L^M|\hat{H}'|\Gamma_L^M\rangle \le \frac{\big\langle(\hat{o}^+)^M\hat{H}'(\hat{o}^-)^M\big\rangle + \big\langle(\hat{o}^-)^M\hat{H}'(\hat{o}^+)^M\big\rangle}{\big\langle(\hat{o}^-)^M(\hat{o}^+)^M\big\rangle}$$
$$= \frac{\big\langle[(\hat{o}^+)^M, [\hat{H}', (\hat{o}^-)^M]]\big\rangle}{\big\langle(\hat{o}^-)^M(\hat{o}^+)^M\big\rangle}, \tag{4.2.63}$$

where the expression in terms of the double commutator is obtained by noting that $\hat{H}'|\Phi_{\mathrm{GS}}\rangle = 0$ and $\langle\Phi_{\mathrm{GS}}|\hat{H}' = 0$. The double commutator is explicitly written as

$$\begin{aligned}
[(\hat{o}^+)^M,&[\hat{H}', (\hat{o}^-)^M]] = \sum_{k=0}^{M-1}[(\hat{o}^+)^M, (\hat{o}^-)^{M-1-k}[\hat{H}', \hat{o}^-](\hat{o}^-)^k] \\
&= \sum_{k,\ell=0}^{M-1}(\hat{o}^+)^{M-1-\ell}[\hat{o}^+, (\hat{o}^-)^{M-1-k}[\hat{H}', \hat{o}^-](\hat{o}^-)^k](\hat{o}^+)^\ell \\
&= \sum_{k,\ell=0}^{M-1}(\hat{o}^+)^{M-1-\ell}(\hat{o}^-)^{M-1-k}[\hat{o}^+, [\hat{H}', \hat{o}^-]](\hat{o}^-)^k(\hat{o}^+)^\ell \\
&\quad + \sum_{k,\ell=0}^{M-1}\sum_{m=0}^{M-k-2}(\hat{o}^+)^{M-1-\ell}(\hat{o}^-)^{M-k-2-m}[\hat{o}^+, \hat{o}^-](\hat{o}^-)^m[\hat{H}', \hat{o}^-](\hat{o}^-)^k(\hat{o}^+)^\ell \\
&\quad + \sum_{k,\ell=0}^{M-1}\sum_{n=0}^{k-1}(\hat{o}^+)^{M-1-\ell}(\hat{o}^-)^{M-1-k}[\hat{H}', \hat{o}^-](\hat{o}^-)^{k-1-n}[\hat{o}^+, \hat{o}^-](\hat{o}^-)^n(\hat{o}^+)^\ell.
\end{aligned} \tag{4.2.64}$$

One might expect that, to get the desired bound (4.2.4), it suffices to bound all these terms by using operator norms, as we did in (3.4.9), (3.4.10), and (3.4.11). But this expectation turns out to be too optimistic. Let us take a look at the first line in the right-hand side of (4.2.64). Exactly as in (3.4.11), we have

$$\Big\|[\hat{o}^+, [\hat{H}', \hat{o}^-]]\Big\| \le \frac{4\zeta^2(o_0)^2 h_0}{V}, \tag{4.2.65}$$

which indeed contains the desired factor $V^{-1} = L^{-d}$, where h_0 is a constant such that $\|\hat{H}'\| \le V h_0$. If we also bound other $\hat{o}^\pm$ by their norms, we can bound the first line of the right-hand side of (4.2.64) by a constant times $M^2(o_0)^{2M}/V$. The denominator of (4.2.63), on the other hand, is bounded from below as $\langle(\hat{o}^-)^M(\hat{o}^+)^M\rangle \ge (2q_0)^M$. Thus we find that

$$\left|\frac{(\text{first line of RHS of (4.2.64)})}{\langle(\hat{o}^-)^M(\hat{o}^+)^M\rangle}\right| \le (\text{const.})\left(\frac{(o_0)^2}{2q_0}\right)^M \frac{M^2}{V}. \tag{4.2.66}$$

Since it must be that $2q_0 < o_0$, we find that the factor $\{(o_0)^2/(2q_0)\}^M$ grows exponentially with M. This means that the upper bound for the energy obtained in this manner can be useful only when one fixes M and lets the system size L grow. This is of course meaningful, but not what we really want. Recall that the desired bound (4.2.4) allows M to be as large as $\sqrt{V} = L^{d/2}$ (which we believe to be optimal).

In order to prove an optimal bound, we have to give up using naive estimates in terms of operator norms. Instead we use "renormalized" bounds which do not contain "bare" factors like $(o_0)^{2M}$ but are expressed in terms of the expectation value $\langle \hat{p}^M \rangle$. To be precise, we shall prove, under the condition $M \leq (\text{const.})\sqrt{V}$, that

$$\frac{1}{2}\langle \hat{p}^M \rangle \leq \left\langle \hat{o}^{s_1} \cdots \hat{o}^{s_{2M}} \right\rangle \leq 2\langle \hat{p}^M \rangle, \tag{4.2.67}$$

and

$$\left| \left\langle \hat{o}^{s_1} \cdots \hat{o}^{s_\ell} \hat{A} \hat{o}^{s_{\ell+1}} \cdots \hat{o}^{s_{2M}} \right\rangle \right| \leq 3 \| \hat{A} \| \left\langle \hat{p}^M \right\rangle, \tag{4.2.68}$$

for any $s_1, \ldots, s_{2M} = \pm$ such that $\sum_j s_j = 0$. Here $\hat{A} = \sum_{x \in \Lambda} \hat{a}_x$ is a self-adjoint operator which commutes with $\hat{S}_{\text{tot}}^{(3)}$, with $\hat{a}_x$ acting only on ζ sites including x. See [66] for details.

We can bound the ground state expectation value of the right-hand side of (4.2.64) by using the upper bound (4.2.68) (or its variant). The denominator of (4.2.63) is bounded from below by (4.2.67) as $\langle (\hat{o}^-)^M (\hat{o}^+)^M \rangle \geq \left\langle \hat{p}^M \right\rangle / 2$. The expectation value $\left\langle \hat{p}^M \right\rangle$ precisely cancels out, and we get an upper bound for $\langle \Gamma_M | \hat{H}' | \Gamma_M \rangle$ without a factor which grows exponentially with M. In this way we prove Theorem 4.6.

The bounds similar to (4.2.68) were proved in [35] by using an involved inductive argument. The reader interested in details of the proof is recommended to study [66], which contains an improved (and considerably simpler) proof.

Let us finally discuss the Proof of Theorem 4.8, which is indeed easier. From the definition (4.2.10) of $|\Xi_{(1,0,0)}\rangle$, we find

$$\begin{aligned} &\langle \Xi_{(1,0,0)} | \hat{H} | \Xi_{(1,0,0)} \rangle \\ &= \frac{1}{2} \left\{ \frac{\langle \Phi_{\text{GS}} | (\hat{\mathcal{O}}_L^{(1)})^M \hat{H} (\hat{\mathcal{O}}_L^{(1)})^M | \Phi_{\text{GS}} \rangle}{\| (\hat{\mathcal{O}}_L^{(1)})^M | \Phi_{\text{GS}} \rangle \|^2} + \frac{\langle \Phi_{\text{GS}} | (\hat{\mathcal{O}}_L^{(1)})^{M+1} \hat{H} (\hat{\mathcal{O}}_L^{(1)})^{M+1} | \Phi_{\text{GS}} \rangle}{\| (\hat{\mathcal{O}}_L^{(1)})^{M+1} | \Phi_{\text{GS}} \rangle \|^2} \right\}, \end{aligned} \tag{4.2.69}$$

where we wrote $M = M(L)$. Cross terms vanish because $|\Phi_{\text{GS}}\rangle$ and $\hat{H}$ are invariant under $\hat{U}_\pi^{(3)}$. It is therefore sufficient to evaluate the expectation value

$$\frac{\langle (\hat{o}^{(1)})^M \hat{H}' (\hat{o}^{(1)})^M \rangle}{\langle (\hat{o}^{(1)})^{2M} \rangle} = \frac{1}{2} \frac{\langle [(\hat{o}^{(1)})^M, [\hat{H}', (\hat{o}^{(1)})^M]] \rangle}{\langle (\hat{o}^{(1)})^{2M} \rangle}, \tag{4.2.70}$$

where we noted that the numerator is written as the double commutator exactly as in (3.4.8). The double commutator is then expanded as

$$\langle[(\hat{o}^{(1)})^M, [\hat{H}', (\hat{o}^{(1)})^M]]\rangle = \sum_{k,\ell=0}^{M-1} \langle(\hat{o}^{(1)})^{2(M-1)-k-\ell}[\hat{o}^{(1)}, [\hat{H}', \hat{o}^{(1)}]](\hat{o}^{(1)})^{k+\ell}\rangle. \tag{4.2.71}$$

At this stage we use the relation $\hat{o}^{(1)} = (\hat{o}^+ + \hat{o}^-)/2$ to write (4.2.71) and the denominator of (4.2.70) in terms of correlation functions of $\hat{o}^+$ and $\hat{o}^-$. We can then use the bounds (4.2.67) and (4.2.68) to evaluate the correlation functions to get the desired bound.

4.3 Ground States of the Infinite System

In this section we study the antiferromagnetic Heisenberg model defined on the infinitely large d-dimensional hyper cubic lattice, and discuss the implication of the results in Sect. 4.2. There is a sophisticated mathematical framework based on operator algebra (or C^*-algebra) to treat quantum many-body systems defined on an infinitely large spatial region. The framework requirers somewhat advanced mathematics which goes beyond the standard mathematical background for physicists. Here we shall avoid going into heavy mathematics, and describe only essential ideas necessary to understand central notions. For the interested reader we give a brief review of the operator algebraic formulation of quantum spin systems in Appendix A.7. The motivated reader is invited to study Bratteli and Robinson's definitive textbook [10, 11] on the subject.

4.3.1 *Construction of Ground States*

We start by briefly discussing the formulation of infinite systems, and then constructing concrete ground states. Let $\mathbb{Z}^d$ be the d-dimensional hyper cubic lattice whose elements (sites) are denote as $x, y, \ldots$. We denote the corresponding set of bonds as

$$\mathcal{B}_\infty := \bigl\{ \{x, y\} \bigm| x, y \in \mathbb{Z}^d,\ |x - y| = 1 \bigr\}. \tag{4.3.1}$$

We also denote the set of even sites as

$$\mathbb{Z}^d_{\text{even}} := \Bigl\{(x_1, x_2, \ldots, x_d) \in \mathbb{Z}^d \Bigm| \sum_{i=1}^{d} x_i \text{ is even} \Bigr\}, \tag{4.3.2}$$

which is the infinite volume version of the A sublattice defined in (4.1.2).

The operator algebra Consider a quantum spin system on $\mathbb{Z}^d$ with spin S. We denote by $\hat{S}^{(1)}_x$, $\hat{S}^{(2)}_x$, and $\hat{S}^{(3)}_x$ the spin operators[34] on site x. The central object is the

[34] As in (2.2.5), the operator $\hat{S}^{(\alpha)}_x$ acts as $\hat{S}^{(\alpha)}$ on the spin at x and as the identity on other spins.

algebra of local operators, which we denote as $\mathfrak{A}_{\mathrm{loc}}$. It is the set of all polynomials (with complex coefficients) of the spin operators $\hat{S}_x^{(\alpha)}$ with $x \in \mathbb{Z}^d$ and $\alpha = 1, 2, 3$. Note that, by definition, each $\hat{A} \in \mathfrak{A}_{\mathrm{loc}}$ is an operator that acts nontrivially only on a finite number of spins, but there is no limit for this finite number.

For $x \in \mathbb{Z}^d$, we define the translation map $\tau_x : \mathfrak{A}_{\mathrm{loc}} \to \mathfrak{A}_{\mathrm{loc}}$ by

$$\tau_x(\hat{S}_{x_1}^{(\alpha_1)} \hat{S}_{x_2}^{(\alpha_2)} \cdots \hat{S}_{x_n}^{(\alpha_n)}) = \hat{S}_{x_1+x}^{(\alpha_1)} \hat{S}_{x_2+x}^{(\alpha_2)} \cdots \hat{S}_{x_n+x}^{(\alpha_n)}, \tag{4.3.3}$$

for any $n, x_1, \ldots, x_n \in \mathbb{Z}^d$, and $\alpha_1, \ldots, \alpha_n = 1, 2, 3$, and by linearity $\tau_x(\alpha \hat{A} + \beta \hat{B}) = \alpha \tau_x(\hat{A}) + \beta \tau_x(\hat{B})$ for any $\hat{A}, \hat{B} \in \mathfrak{A}_{\mathrm{loc}}$ and $\alpha, \beta \in \mathbb{C}$.

States and ground states In quantum spin systems on infinite lattices, we characterize a state by specifying its expectation values rather than identifying it with a vector in a Hilbert space. More precisely, a state $\rho(\cdot)$ is a linear map from $\mathfrak{A}_{\mathrm{loc}}$ to $\mathbb{C}$ which satisfies $\rho(\hat{1}) = 1$ and $\rho(\hat{A}^\dagger \hat{A}) \geq 0$ for any $\hat{A} \in \mathfrak{A}_{\mathrm{loc}}$. The idea is that $\rho(\hat{A})$ is the expectation value of the operator $\hat{A}$ in the state $\rho(\cdot)$.

From now on we concentrate on the antiferromagnetic Heisenberg model, whose finite volume Hamiltonian is (4.1.1). A state $\rho(\cdot)$ is said to be translation invariant if for any $\hat{A}$ and any $x \in \mathbb{Z}^d_{\mathrm{even}}$ it holds that $\rho(\tau_x(\hat{A})) = \rho(\hat{A})$. Here we considered the translation only by x in the even sublattice $\mathbb{Z}^d_{\mathrm{even}}$, anticipating antiferromagnetic order.[35] Recall that a Néel ordered state is invariant under translation by x if and only if $x \in \mathbb{Z}^d_{\mathrm{even}}$. See Fig. 3.1.

By denoting the ground state energy of the model on the finite lattice $(\Lambda_L, \mathcal{B}_L)$ as $E_{\mathrm{GS},L}$, we define, as in (4.2.5), the ground state energy density (or, more precisely, the energy per bond) as

$$\varepsilon_{\mathrm{GS}} := \lim_{L \uparrow \infty} \frac{E_{\mathrm{GS},L}}{|\mathcal{B}_L|}, \tag{4.3.4}$$

where the existence of the limit is easily proved (see, e.g., Appendix A of [65]). Then we can define the notion of translation invariant ground states for the infinite system.

Definition 4.17 A translation invariant state $\omega(\cdot)$ is a ground state of the antiferromagnetic Heisenberg model if and only if $\omega(\hat{\mathbf{S}}_x \cdot \hat{\mathbf{S}}_y) = \varepsilon_{\mathrm{GS}}$ for any $\{x, y\} \in \mathcal{B}_\infty$.

The idea behind the definition must be clear. In a finite volume, one can define a ground state variationally as a normalized state $|\Phi\rangle$ satisfying $\langle \Phi | \hat{H} | \Phi \rangle = E_{\mathrm{GS}}$, where E_{GS} is the lowest eigenvalue of $\hat{H}$. The above is the infinite volume version of this characterization (restricted to translation invariant states). More general characterization of ground states without the assumption of translation invariance is given in Definition A.25 (p. 488) and in Theorem A.26. In particular Definition 4.17 is obtained by restricting the variational characterization in Theorem A.26 to translation invariant states. See Theorem 6.2.58 of [11] for a proof.[36]

---

[35] In general a state is said to be translation invariant if $\rho(\tau_x(\hat{A})) = \rho(\hat{A})$ for any x in $\mathbb{Z}^d$ rather than in $\mathbb{Z}^d_{\mathrm{even}}$. We are here making a slight abuse of the terminology.

[36] Appendix A of [35] contains a (hopefully accessible) discussion about different definitions of ground states in infinite systems.

Ergodic states and physical ground states We now introduce the important concept of ergodic states.[37] This is a precise formulation of macroscopically "healthy" states, in which the density of any bulk quantity takes a definite value. For a self-adjoint $\hat{A} \in \mathfrak{A}_{\text{loc}}$ and L, define the corresponding bulk operator by

$$\hat{A}_L = \sum_{x \in \Lambda_L \cap \mathbb{Z}^d_{\text{even}}} \tau_x(\hat{A}), \tag{4.3.5}$$

where the finite lattice Λ_L is regarded as a subset of $\mathbb{Z}^d$.

Definition 4.18 A translation invariant state $\rho(\cdot)$ is said to be ergodic[38] if it holds that

$$\lim_{L \uparrow \infty} \sqrt{\rho\Bigl(\Bigl(\frac{\hat{A}_L}{L^d}\Bigr)^2\Bigr) - \Bigl\{\rho\Bigl(\frac{\hat{A}_L}{L^d}\Bigr)\Bigr\}^2} = 0, \tag{4.3.6}$$

for any self-adjoint $\hat{A} \in \mathfrak{A}_{\text{loc}}$.

The idea should again be clear. The quantity inside the limit is the fluctuation of the density $\hat{A}_L/L^d$ in the state $\rho(\cdot)$. The fluctuation should vanish as L grows (so that the law of large number holds) if the state is physically "healthy".

We are now ready to define our own notion of physical ground states.

Definition 4.19 A translation invariant ground state $\omega(\cdot)$ of the antiferromagnetic Heisenberg model is a physical ground state when it is ergodic.[39]

Concrete ground states We now construct ground states of the infinite system as limits of finite volume ground states.

Let us construct an infinite volume state by using the unique ground state $|\Phi_{\text{GS}}\rangle$. We define a state $\omega_0(\cdot)$ (in the infinite system) by setting, for each $\hat{A} \in \mathfrak{A}_{\text{loc}}$,

$$\omega_0(\hat{A}) = \lim_{L \uparrow \infty} \langle \Phi_{\text{GS}} | \hat{A} | \Phi_{\text{GS}} \rangle. \tag{4.3.7}$$

Although the quantity $\langle \Phi_{\text{GS}} | \hat{A} | \Phi_{\text{GS}} \rangle$ is not defined if Λ_L is smaller than the support of the operator $\hat{A}$, it is well defined for sufficiently large L since $\hat{A}$ depends only on a finite number of spins. It is expected that the $L \uparrow \infty$ limit in (4.3.7) exists, but the existence of the limit has not been proved. Here, and in what follows, we shall

[37] This terminology has little to do with the ergodic hypothesis, which was once believed to be relevant to the foundation of equilibrium statistical mechanics, but comes from the ergodic theory in mathematics. The ergodic theory of course has its root in Boltzmann's ergodic hypothesis.

[38] See Remark 1 at the end of Sect. 2.5 of [35] for the relation to the standard definition of ergodic states.

[39] To be rigorous, this definition depends on the tacit (but essential) assumption that the physical ground states of the model are invariant under the translation by any $x \in \mathbb{Z}^d_{\text{even}}$, i.e., have period two. There is a logical possibility (which is extremely unlikely but has not been rigorously ruled out yet) that the ground states spontaneously break the translation symmetry and have longer period.

assume that such limits exist. From the abstract result in Theorem A.24 in p. 488, it is known that one can always take a subsequence of L so that such limits exist.

We can also construct infinite volume states from the low-lying states with explicit symmetry breaking defined in (4.2.10) and (4.2.17). For an arbitrary unit vector $\boldsymbol{n} \in \mathbb{R}^3$, we define a state $\omega_{\boldsymbol{n}}(\cdot)$ by setting, for each $\hat{A} \in \mathfrak{A}_{\text{loc}}$,

$$\omega_{\boldsymbol{n}}(\hat{A}) = \lim_{L\uparrow\infty} \langle \Xi_{\boldsymbol{n}} | \hat{A} | \Xi_{\boldsymbol{n}} \rangle. \tag{4.3.8}$$

Theorem 4.20 *The states $\omega_0(\cdot)$ and $\omega_{\boldsymbol{n}}(\cdot)$ are translation invariant ground states.*

The theorem is a direct and simple consequence of the facts that $|\Phi_{\text{GS}}\rangle$ is a finite volume ground state and $|\Xi_{\boldsymbol{n}}\rangle$ are translation invariant low-lying states. For the latter, see also the discussion after Theorem 4.6 in p. 94.

From (4.1.8) and (4.2.18), one readily finds that these ground states satisfy

$$\omega_0(\hat{\boldsymbol{S}}_x) = (0, 0, 0), \tag{4.3.9}$$

$$\omega_{\boldsymbol{n}}(\hat{\boldsymbol{S}}_x) = (-1)^x \, m^* \boldsymbol{n}, \tag{4.3.10}$$

for any $x \in \mathbb{Z}^d$. We see that $\omega_0(\cdot)$ is an infinite volume ground state which exhibits LRO but no SSB, while $\omega_{\boldsymbol{n}}(\cdot)$ are infinite volume ground states which exhibit both LRO and full SSB. This observation leads us to conjecture the following.

Conjecture 4.21 *The ground states $\omega_{\boldsymbol{n}}(\cdot)$ are ergodic, i.e., they are physical ground states in the sense of Definition 4.19.*

This conjecture motivated our identification of low-lying states $|\Xi_{\boldsymbol{n}}\rangle$ as physical "ground states". They are not ground states (of the finite volume Hamiltonian) in the literal sense, but approach physical ground states in the infinite volume limit.

The property (4.2.16) of the low-lying state $|\Xi_{(1,0,0)}\rangle$ is a strong support to the conjecture, but is still insufficient to guarantee the ergodicity. First the relation (4.2.16) deals only with the fluctuation of the order operators while (4.3.6) is for any $\hat{A}$. Secondly the finite volume state on Λ_L is used to evaluate $\hat{\mathcal{O}}_L^{(\alpha)}$ in (4.2.16) while the condition (4.3.6) is for a state in the infinite volume. One needs much stronger estimates to prove the conjecture. So far a statement corresponding to Conjecture 4.21 has been proved only for systems that exhibit discrete symmetry breaking as in the quantum Ising model. In Appendix B of [35], it was proved that the infinite volume ground states constructed from the low-lying states like $|\Xi_\pm\rangle$ of (3.4.14) are ergodic.

It is worth pointing out that if one treats the problem using some kind of mean-field approximation, which is standard in almost any field of many-body physics, the ground state like $\omega_{\boldsymbol{n}}(\cdot)$ is obtained from the outset. This is indeed natural since a mean field theory always starts from the assumption that the order operator behaves as a classical vector without fluctuations.

Note finally that we can construct another infinite volume ground state with both LRO and SSB by using $|\Phi_{\text{GS},h}\rangle$, the ground state of the Hamiltonian (4.2.27) with staggered magnetic field, as

$$\widetilde{\omega}_{(1,0,0)}(\hat{A}) = \lim_{h\downarrow 0} \lim_{L\uparrow\infty} \langle \Phi_{\mathrm{GS},h} | \hat{A} | \Phi_{\mathrm{GS},h} \rangle. \tag{4.3.11}$$

Theorem 4.13 (p. 102) guarantees that this state exhibits SSB. It is expected that $\widetilde{\omega}_{(1,0,0)}(\cdot) = \omega_{(1,0,0)}(\cdot)$, but we even do not know how hard the proof can be.

4.3.2 Physical Versus Unphysical Ground States

Let us now be heuristic (i.e., less rigorous) and discuss the relations between different infinite volume ground states we constructed above. We here assume the validity of Conjecture 4.21 and other plausible statements.

We first summarize what we have observed about these ground states. The ground state $\omega_0(\cdot)$ is the limit of the unique finite volume ground state $|\Phi_{\mathrm{GS}}\rangle$. Since $|\Phi_{\mathrm{GS}}\rangle$ is the only ground state in the sense of textbook-quantum-mechanics, one might think that $\omega_0(\cdot)$ is the most legitimate infinite volume ground state. It turns out however that this is a pathological state with LRO but without SSB, in which the law of large number is violated in the sense that the density of the order operator shows nonvanishing fluctuations. The fact that there is a physical quantity with pathologically large fluctuation usually means that the system is extremely sensitive to a perturbation which couples to this quantity. It is likely that the state is in some sense "unstable"; even extremely small disturbance from the outside can trigger the state to "collapses" into a more "stable" one.[40] The ground state $\omega_{\boldsymbol{n}}(\cdot)$, on the other hand, is the limit of the low-lying state $|\Xi_{\boldsymbol{n}}\rangle$, which is not a ground state, not even an energy eigenstate of the finite volume Hamiltonian. Nevertheless we argued that there is evidence to believe that $\omega_{\boldsymbol{n}}(\cdot)$ is a physical ground state (in the sense of Definition 4.19). We believe that it is a "stable" state that describes a physical state close to experimentally observed ones.

We have conjectured in (4.2.21) that the average of the symmetry breaking "ground states" over the solid angle (almost) gives the exact finite volume ground state $|\Phi_{\mathrm{GS}}\rangle$. The conjecture implies for the ground state expectation value of any $\hat{A} \in \mathfrak{A}_{\mathrm{loc}}$ that

$$\langle \Phi_{\mathrm{GS}} | \hat{A} | \Phi_{\mathrm{GS}} \rangle = \text{const.} \int_{|\boldsymbol{n}|=|\boldsymbol{n}'|=1} d\boldsymbol{n}\, d\boldsymbol{n}' \, \langle \Xi_{\boldsymbol{n}} | \hat{A} | \Xi_{\boldsymbol{n}'} \rangle, \tag{4.3.12}$$

where the constant is independent of $\hat{A}$ (but dependent on L). Note that when $\boldsymbol{n} \neq \boldsymbol{n}'$ the two states $|\Xi_{\boldsymbol{n}}\rangle$ and $|\Xi_{\boldsymbol{n}'}\rangle$, which are ordered in different directions, are macroscopically different. Since $\hat{A}$ can modify the state only locally, it is expected that the matrix element $\langle \Xi_{\boldsymbol{n}} | \hat{A} | \Xi_{\boldsymbol{n}'} \rangle$ for $\boldsymbol{n} \neq \boldsymbol{n}'$ vanishes as $L \uparrow \infty$. This suggests that

---

[40] Note that this instability does not come from an energetic reason since $|\Phi_{\mathrm{GS}}\rangle$ definitely has the lowest energy. Since $|\Phi_{\mathrm{GS}}\rangle$ may be regarded as a superposition of macroscopically distinct states as in (4.2.21) (see also (3.3.8)), this instability may be closely related to the collapse of Schrödinger's cat type states. As far as we understand the mechanism of such collapse is still poorly understood. But see [60] for a formulation of the problem and some promising results.

$$\langle \Phi_{\mathrm{GS}}|\hat{A}|\Phi_{\mathrm{GS}}\rangle \simeq \frac{1}{4\pi}\int_{|\boldsymbol{n}|=1} d\boldsymbol{n}\, \langle \Xi_{\boldsymbol{n}}|\hat{A}|\Xi_{\boldsymbol{n}}\rangle, \tag{4.3.13}$$

for large L, and hence

$$\omega_0(\hat{A}) = \frac{1}{4\pi}\int_{|\boldsymbol{n}|=1} d\boldsymbol{n}\, \omega_{\boldsymbol{n}}(\hat{A}) \quad \text{for any } \hat{A} \in \mathfrak{A}_{\mathrm{loc}}. \tag{4.3.14}$$

Thus, starting from the conjecture (4.2.21), which says that $|\Phi_{\mathrm{GS}}\rangle$ is a superposition of $|\Xi_{\boldsymbol{n}}\rangle$, we have arrived at (4.3.14), which says that the symmetric ground state $\omega_0(\cdot)$ is a statistical mixture of the Néel ordered ground states $\omega_{\boldsymbol{n}}(\cdot)$.

The conjectured equality (4.3.14) sheds light on the mechanism of spontaneous symmetry breaking. As we have stressed several times, the symmetric ground state $\omega_0(\cdot)$ is the infinite volume ground state which we naively expect to observe unless there are special reasons. In the present case, however, we know that there is a definite reason that the state should not be observed; it is a non-ergodic state, which we argued to be "unstable". Now the equality (4.3.14) says that this "unstable" state may be regarded as the mixture of the continuous set of "stable" states $\omega_{\boldsymbol{n}}(\cdot)$. When the state $\omega_0(\cdot)$ is destabilized by external disturbance, it is rather natural that one of its "stable" constituents is chosen. In this manner the system settles to a physical ground state $\omega_{\boldsymbol{n}}(\cdot)$ with arbitrarily chosen direction $\boldsymbol{n}$. This is a natural explanation of spontaneous symmetry breaking in the ground states of the antiferromagnetic Heisenberg model (and also in many other quantum-many body systems).

Remark The conjectured relation (4.3.14) is reminiscent of the abstract decomposition theories developed in mathematical theory of many-body systems [10, 11, 55]. Very roughly speaking the decomposition theories guarantees that any "unphysical" ground state of an infinite system can be expressed as a kind of integral of "physical" ground states. The theories also apply to equilibrium states. See Remark 1 at the end of Sect. 2.5 of [35] for the relation of our observation and general theories.

4.4 Equilibrium States of the Heisenberg Model

If one focuses on applications in condensed matter physics, equilibrium states of macroscopic systems are of essential importance. Although the present book is mostly devoted to the study of ground states, we shall discuss the properties of equilibrium states of the Heisenberg model in some detail in the present section.

The main purpose here is to summarize what is known rigorously about the equilibrium state of the Heisenberg model, both ferromagnetic and antiferromagnetic. But we shall also provide complete proofs of two important theorems about the absence of order in two dimensions. Most of the results presented in the following can be proved in much more general classes of systems, as we shall briefly mention.

In Sect. 4.4.1, we briefly review standard rigorous results about disordered equilibrium states. In Sect. 4.4.2, we discuss Berezinskii's approximate theory for the classical XY model, which sheds light on the essence of phase transitions in spin systems with continuous symmetry. Then, in Sect. 4.4.3, we give detailed discussion on rigorous results which shows the absence of order in the equilibrium states of the two dimensional Heisenberg model. A close analogy between Berezinskii's harmonic approximation and the rigorous method of McBryan and Spencer may be interesting by itself. In Sect. 4.4.4 we briefly review rigorous results about the existence of LRO and SSB in higher dimensions.

We focus on the standard Heisenberg model with spin $S = 1/2, 1, 3/2, \dots$ on the finite d-dimensional hyper cubic lattice $(\Lambda_L, \mathcal{B}_L)$. See (3.1.2) and (3.1.3). The standard Hamiltonians are

$$\hat{H}_h = -\sum_{\{x,y\}\in\mathcal{B}_L} \hat{\boldsymbol{S}}_x \cdot \hat{\boldsymbol{S}}_y - h \sum_{x\in\Lambda_L} \hat{S}_x^{(3)} \tag{4.4.1}$$

for the ferromagnetic case, where $h \in \mathbb{R}$ denotes the external magnetic field, and

$$\hat{H}_h = \sum_{\{x,y\}\in\mathcal{B}_L} \hat{\boldsymbol{S}}_x \cdot \hat{\boldsymbol{S}}_y - h \sum_{x\in\Lambda_L} (-1)^x \hat{S}_x^{(3)} \tag{4.4.2}$$

for the antiferromagnetic case, where $h \in \mathbb{R}$ denotes the staggered magnetic field. As we have discussed below (3.4.19) (see p. 69) the staggered magnetic field should be regarded as a purely theoretical object.

We define the canonical average in the equilibrium state at inverse temperature $\beta = (k_{\mathrm{B}}T)^{-1} \in [0, \infty)$ as

$$\langle \hat{A} \rangle_{\beta,h}^L := \frac{1}{Z_L(\beta, h)} \operatorname{Tr}[\hat{A}\, e^{-\beta \hat{H}_h}], \tag{4.4.3}$$

for any operator $\hat{A}$, where the partition function $Z_L(\beta, h)$ is determined by the normalization condition $\langle \hat{1} \rangle_{\beta,h}^L = 1$.

4.4.1 Disorder at High-Temperature and in One-Dimension

We start from (less interesting) cases where one can generally prove that equilibrium states are disordered. For any $x, y \in \mathbb{Z}^d$, we define the infinite-volume two-spin correlation function under vanishing external field as[41]

[41] The existence of the limit has not been proved. To be rigorous one should replace lim by lim inf or lim sup, or take a suitable subsequence of L. See Theorem A.24 in p. 488 and the discussion that follows.

$$\langle \hat{S}_x^{(\alpha)} \hat{S}_y^{(\alpha)} \rangle_{\beta,0}^{\infty} = \lim_{L\uparrow\infty} \langle \hat{S}_x^{(\alpha)} \hat{S}_y^{(\alpha)} \rangle_{\beta,0}^{L}. \tag{4.4.4}$$

The following theorem says that the one-dimensional models are disordered at any nonzero temperature.

Theorem 4.22 *Consider the model with the Hamiltonian (4.4.1) or (4.4.2) for $d = 1$ and any S. Then one has*

$$\lim_{h\downarrow 0} \lim_{L\uparrow\infty} \langle \hat{S}_x^{(3)} \rangle_{\beta,h}^{L} = 0, \tag{4.4.5}$$

for any $\beta \in [0, \infty)$. There also exist functions $\xi(\beta) \in (0, \infty)$ and $C(\beta) \in (0, \infty)$, which depend on S, such that

$$\left| \langle \hat{S}_x^{(\alpha)} \hat{S}_y^{(\alpha)} \rangle_{\beta,0}^{\infty} \right| \le C(\beta) \exp\left[-\frac{|x-y|}{\xi(\beta)} \right], \tag{4.4.6}$$

for any $\beta \in [0, \infty)$, $\alpha = 1, 2, 3$, and $x, y \in \mathbb{Z}$.

Clearly the relation (4.4.5) means that the equilibrium state does not exhibit SSB. Recall that the order of the two limits is essential here. See, e.g., (3.2.14) in p. 55 and the discussion that follows. The inequality (4.4.6) shows that the correlation function decays exponentially fast, which means that the model is in the disordered phase. Note that $\xi(\beta)$ is an upper bound for the correlation length, i.e., the distance beyond which spins essentially do not correlate.

In two or higher dimensions, the same properties can be proved for sufficiently high temperatures, which corresponds to sufficiently small β.

Theorem 4.23 *Consider the model with the Hamiltonian (4.4.1) or (4.4.2) for $d \ge 2$ and any S. There exists a constant $\beta_0 \in (0, \infty)$ which depends only on d and S. Then one has*

$$\lim_{h\downarrow 0} \lim_{L\uparrow\infty} \langle \hat{S}_x^{(3)} \rangle_{\beta,h}^{L} = 0, \tag{4.4.7}$$

for any $\beta \in [0, \beta_0]$. There also exist functions $\xi(\beta) \in (0, \infty)$ and $C(\beta) \in (0, \infty)$, which depend on d and S, such that

$$\left| \langle \hat{S}_x^{(\alpha)} \hat{S}_y^{(\alpha)} \rangle_{\beta,0}^{\infty} \right| \le C(\beta) \exp\left[-\frac{|x-y|}{\xi(\beta)} \right], \tag{4.4.8}$$

for any $\beta \in [0, \beta_0]$, $\alpha = 1, 2, 3$, and $x, y \in \mathbb{Z}^d$.

In fact Theorems 4.22 and 4.23 represent a very small part of much more general results that can be proved by standard methods. For any $\beta \in [0, \infty)$ and $h \in \mathbb{R}$ if $d = 1$, and for any (β, h) in a certain region (which includes $\beta \in [0, \beta_0]$, $h = 0$) if $d \ge 2$, it can be proved that the expectation value of any local operator is an analytic function of β and h. The relations (4.4.5) and (4.4.7) are simple consequences of this general fact. Under the same conditions, it is proved that the truncated correlation function of any local operators $\hat{A}$ and $\hat{B}$ exhibits exponential decay as

$$\left|\langle \hat{A}\,\tau_x(\hat{B})\rangle^{\infty}_{\beta,h} - \langle \hat{A}\rangle^{\infty}_{\beta,h}\langle \hat{B}\rangle^{\infty}_{\beta,h}\right| \le (\text{constant})\,\exp\Bigl[-\frac{|x|}{\xi(\beta,h)}\Bigr], \tag{4.4.9}$$

with $\xi(\beta,h) \in (0,\infty)$, where τ_x is the translation map defined in (4.3.3). Then (4.4.6) and (4.4.8) are special cases.

These results about disordered equilibrium states are proved for a much larger class of models than the Heisenberg model. Basically any model with translationally invariant short ranged interactions can be treated in one dimension. See [4]. At sufficiently high temperature (in any dimensions), these results can be proved for any model with (not necessarily translation invariant) short ranged interactions. The proof makes use of the technique of cluster expansion. See, e.g., [21, 50, 61].

4.4.2 *Berezinskii's Harmonic Approximation*

The nature of LRO and SSB in equilibrium states also depends crucially on the dimensionality of the system. To see the essence of the dependence, we shall review an approximate theory due to Berezinskii [8] about the correlation function of the classical XY model. Although the theory is simple, it sheds light on essential properties of spin systems with continuous symmetry. In fact the theory was a motivation for the rigorous method of McBryan and Spencer [44], which we shall discuss in Sect. 4.4.3. We also note that it is widely believed that the basic nature of equilibrium states is common for classical and quantum spin systems.

We consider the classical XY model, in which each lattice site $x \in \Lambda_L$ is associated with a classical XY spin, i.e., a two-dimensional vector $\vec{S}_x = (S^{(1)}_x, S^{(2)}_x) \in \mathbb{R}^2$ such that $|\vec{S}_x| = 1$. The Hamiltonian of the ferromagnetic model without external magnetic filed is

$$H^{\mathrm{XY}} = \sum_{\{x,y\}\in\mathcal{B}_L} (1 - \vec{S}_x \cdot \vec{S}_y), \tag{4.4.10}$$

where the constant 1 is inserted for later convenience. It is useful to represent the spin by the angle variable $\theta_x \in [-\pi,\pi)$ as $\vec{S}_x = (\cos\theta_x, \sin\theta_x)$. Then we see that $\vec{S}_x \cdot \vec{S}_y = \cos(\theta_x - \theta_y)$. The Hamiltonian H^{XY} is minimized when $\theta_x - \theta_y = 0$ for any $\{x,y\} \in \mathcal{B}_L$. Thus all the spins point in the same direction in any ground state.[42]

The thermal expectation value of an arbitrary function f of spins is defined as

[42]A ground state of a classical spin system is a spin configuration that minimizes the Hamiltonian.

$$\langle f\rangle_\beta^{\rm XY}:=\frac{1}{Z_L^{\rm XY}(\beta)}\Bigl(\prod_{x\in\Lambda_L}\int_{-\pi}^{\pi}d\theta_x\Bigr)f\,e^{-\beta H^{\rm XY}},\tag{4.4.11}$$

where $Z_L^{\rm XY}(\beta)$ is determined by $\langle 1\rangle_\beta^{\rm XY}=1$.

Now suppose that the temperature is sufficiently low, i.e., β is sufficiently large. We may then (non-rigorously) assume that only those configurations with very small energy, where one has $|\theta_x-\theta_y|\ll 1$ for any $\{x,y\}\in\mathscr{B}_L$, contribute to the integral (4.4.11). Then the XY Hamiltonian (4.4.10) may be approximated by the Gaussian Hamiltonian as

$$H^{\rm XY}\simeq H^{\rm Gauss}=\frac{1}{2}\sum_{\{x,y\}\in\mathscr{B}_L}(\theta_x-\theta_y)^2.\tag{4.4.12}$$

By also extending the range of integral from $(-\pi,\pi)$ to $(-\infty,\infty)$, we shall approximate the expectation value (4.4.11) as

$$\langle f\rangle_\beta^{\rm XY}\simeq\langle f\rangle_\beta^{\rm Gauss}:=\frac{1}{Z_L^{\rm Gauss}(\beta)}\Bigl(\prod_{x\in\Lambda_L}\int_{-\infty}^{\infty}d\theta_x\Bigr)f\,e^{-\beta H^{\rm Gauss}},\tag{4.4.13}$$

where $Z_L^{\rm Gauss}(\beta)$ is determined by[43] $\langle 1\rangle_\beta^{\rm Gauss}=1$. By using the standard property of Gaussian integrals,[44] one finds for $z\in\Lambda_L$ that

$$\langle\vec{S}_o\cdot\vec{S}_z\rangle_\beta^{\rm Gauss}=\langle\cos(\theta_o-\theta_z)\rangle_\beta^{\rm Gauss}=\exp\bigl[-\frac{1}{2}\langle(\theta_o-\theta_z)^2\rangle_\beta^{\rm Gauss}\bigr].\tag{4.4.14}$$

We define (for fixed z) a "classical field" $\boldsymbol{\varphi}^{(o,z)}=(\varphi_x^{(o,z)})_{x\in\Lambda_L}$ by $\varphi_x^{(o,z)}=\langle(\theta_o-\theta_z)\theta_x\rangle_\beta^{\rm Gauss}$, and write (4.4.14) as

$$\langle\vec{S}_o\cdot\vec{S}_z\rangle_\beta^{\rm Gauss}=e^{-\psi_z/2},\tag{4.4.15}$$

with $\psi_z=\varphi_o^{(o,z)}-\varphi_z^{(o,z)}$.

We wish to find the asymptotic behavior of ψ_z. As in (2.4.13), we define the lattice Laplacian $\boldsymbol{\Delta}$ which acts on a function $\boldsymbol{f}=(f_x)_{x\in\Lambda_L}$ as $(\boldsymbol{\Delta}\boldsymbol{f})_x=\sum_{y\in\mathscr{N}(x)}(f_y-f_x)$, where $\mathscr{N}(x)=\{y\in\Lambda_L\,|\,\{x,y\}\in\mathscr{B}_L\}$ is the set of

[43]The integral which determines $Z_L^{\rm Gauss}(\beta)$ is in fact ill-defined since the integrand $\exp[-\beta H^{\rm Gauss}]$ is invariant under the shift $\theta_x\to\theta_x+c$ for all $x\in\Lambda_L$. But this is a well-known problem, which can be resolved by standard methods. All the calculations can be justified, for example, by using the regularized Hamiltonian $H^{\rm Gauss}=\sum_{\{x,y\}\in\mathscr{B}_L}(\theta_x-\theta_y)^2/2+\mu\sum_{x\in\Lambda_L}(\theta_x)^2$ with $\mu>0$, and letting $\mu\downarrow 0$ after evaluating the correlations.

[44]Let f be a Gaussian random variable such that $\langle f\rangle=0$ and $\langle f^2\rangle=a$. It is easily verified that $\langle f^{2n}\rangle=(2n-1)!!\,a^n$. Then one confirms that $\langle\cos f\rangle=\sum_{n=0}^{\infty}(-1)^n\langle f^{2n}\rangle/(2n)!$ $=\sum_{n=0}^{\infty}(-1)^na^n(2n-1)!!/(2n)!=\sum_{n=0}^{\infty}(-1)^na^n/(2^nn!)=\exp(-a/2)$.

sites directly connected to x. Then it is not hard to show that $\boldsymbol{\varphi}^{(o,z)}$ is the solution of the Poisson equation[45]

$$-(\Delta\boldsymbol{\varphi}^{(o,z)})_x = \frac{1}{\beta}(\delta_{x,o} - \delta_{x,z}). \tag{4.4.16}$$

If we use the analogy with electromagnetism, $\boldsymbol{\varphi}^{(o,z)} = (\varphi_x)_{x\in\Lambda_L}$ is the electrostatic potential generated by a positive charge at o and a negative charge at z. For $d = 1$, (4.4.16) is easily solved to give

$$\varphi_x^{(o,z)} = \begin{cases} \psi_z\Big(1 - \dfrac{x}{z}\Big) & \text{for } x = 0, \ldots, z \\ \psi_z \dfrac{x - z}{L - z} & \text{for } x = z, \ldots, L, \end{cases} \tag{4.4.17}$$

where we identify 0 and L. The coefficient ψ_z, which is indeed equal to the desired $\varphi_o^{(o,z)} - \varphi_z^{(o,z)}$, is given by

$$\psi_z = \frac{z(L-z)}{L\beta} \simeq \frac{z}{\beta}, \tag{4.4.18}$$

where the asymptotic form is valid for $|z| \ll L$. The behavior of ψ_z for $1 \ll |z| \ll L$ for other dimensions can be read off from the analogy with electrostatics as[46]

$$\psi_z \simeq \begin{cases} \dfrac{|z|}{\beta} & d = 1, \\ \dfrac{2}{\beta}\Big(\dfrac{1}{2\pi}\log|z| + C_2\Big) & d = 2, \\ \dfrac{2}{\beta}\Big(-\dfrac{1}{4\pi|z|} + C_3\Big) & d = 3. \end{cases} \tag{4.4.19}$$

Note that there is an essential difference in the asymptotic behaviors for $d \leq 2$ and $d > 3$, namely, ψ_z continues to grow as $|z|$ increases for $d \leq 2$, while it converges to a finite value fo $d > 2$. This qualitative difference, which is familiar from electromagnetism, is also relevant to the problem of phase transitions in systems with continuous symmetry.

---

[45]Note first that $\partial \exp[-\beta H^{\text{Gauss}}]/\partial\theta_x = \beta \sum_{y\in\mathcal{N}(x)}(\theta_y - \theta_x)\exp[-\beta H^{\text{Gauss}}]$. Denote the integral over all θ_x as $\int d\Theta(\cdots)$. Then by integration by parts we find $\beta\int d\Theta(\theta_o - \theta_z)\sum_{y\in\mathcal{N}(x)}(\theta_y - \theta_x)\exp[-\beta H^{\text{Gauss}}] = \int d\Theta(\theta_o - \theta_z)\partial\exp[-\beta H^{\text{Gauss}}]/\partial\theta_x = (-\delta_{x,o} + \delta_{x,z})\int d\Theta \exp[-\beta H^{\text{Gauss}}]$. Dividing by the normalization factor $Z_L^{\text{Gauss}}(\beta) = \int d\Theta \exp[-\beta H^{\text{Gauss}}]$, we get $\beta\sum_{y\in\mathcal{N}(x)}\langle(\theta_o - \theta_z)(\theta_y - \theta_x)\rangle_\beta^{\text{Gauss}} = -\delta_{x,o} + \delta_{x,z}$, which is the desired (4.4.16). (To be rigorous this derivation should be first carried out for $\mu > 0$ as in footnote 43.)

[46]We have simply used the standard formulas for electric potential generated by a point charge. Note that we do not need to worry about the divergence near the charge since our problem is defined on a lattice. We believe that the estimate can be made rigorous by a suitable asymptotic analysis, but have not worked out the details.

By substituting these estimates into (4.4.15), we get the following approximations, which may not be very well controlled, for the long distance behavior of the correlation function in the $L \uparrow \infty$ limit of the d-dimensional XY model at low temperatures:

$$\langle \vec{S}_o \cdot \vec{S}_z \rangle_\beta^{\text{XY}} \simeq \begin{cases} e^{-|z|/(2\beta)} & d = 1, \\ e^{-C_2/\beta} \, |z|^{-1/(2\pi\beta)} & d = 2, \\ e^{-C_3/\beta} & d = 3. \end{cases} \tag{4.4.20}$$

It is found that the two point correlation does not decay as $|z| \uparrow \infty$ for $d = 3$, suggesting that there is a long-range order. This conclusion is indeed correct. It was proved by Fröhlich, Simon, and Spencer [19] that a wide class of classical ferromagnetic spin systems, including the XY and the Heisenberg models, exhibit long-range order at sufficiently low temperatures provided that $d \geq 3$.

On the other hand the correlation is found to decay in $d = 1$ and 2. Again this prediction is confirmed rigorously. In $d = 1$, any classical spin system is known to be in the disordered phase at any nonzero temperature. See, e.g., [13, 14, 25]. This is the classical counterpart of Theorem 4.22 in p. 119. In $d = 2$, Berezinskii's approximation suggests that the correlation of the XY model exhibits a peculiar power law decay at low temperatures. The phase transition in the two-dimensional XY model leading to such an "exotic" low temperature phase was later studied in detail by Kosterlitz and Thouless [37, 38], and is now known as the Berezinskii–Kosterlitz–Thouless transition. The existence of such a low temperature phase was proved by Fröhlich and Spencer [20]. The Berezinskii–Kosterlitz–Thouless transition is expected to take place only in the the XY model (and similar models with U(1) symmetry) in two dimensions. The classical Heisenberg model, obtained by replacing $\vec{S}_x$ by a three component unit vector $\vec{S}_x = (S_x^{(1)}, S_x^{(2)}, S_x^{(3)})$, is believed to have exponentially decaying correlation for any $\beta < \infty$ in two dimensions. But this is a notoriously difficult problem, which is not yet understood mathematically.

To sum, although being a rather drastic simple approximation, Berezinskii's harmonic approximation provides a satisfactory picture about the behavior of the XY model and related models with continuous symmetry in dimensions 1, 2, and higher. Interestingly it is known to give a qualitatively wrong prediction in lattices which have fractal dimensions between 1 and 2, say, 1.99 [36].

4.4.3 Absence of Order in Two Dimensions

Let us come back to rigorous results about quantum spin systems, and focus on two dimensional models. Berezinskii's approximation suggests that there is no order at any nonzero temperature. We shall prove two theorems which confirm this conclusion.

Hohenber–Mermin–Wagner theorem The Hohenber–Mermin–Wagner theorem states that the Heisenberg model in two dimensions does not exhibit magnetic ordering at any nonzero temperature. The theorem was proved by Mermin and Wagner [45] by making use of the essential idea of Hohenberg [26], who proved the corresponding statement for quantum particle systems.[47]

To state (an improved version of) the theorem, we introduce the Heisenberg model with the generalized Hamiltonian

$$\hat{H}_h = J \sum_{\{x,y\}\in\mathscr{B}_L} \hat{\boldsymbol{S}}_x \cdot \hat{\boldsymbol{S}}_y - \sum_{x\in\Lambda_L} \boldsymbol{h}_x \cdot \hat{\boldsymbol{S}}_x, \tag{4.4.21}$$

where J is either -1 or 1, which corresponds to the ferromagnetic or the antiferromagnetic model, respectively. The local magnetic field $\boldsymbol{h}_x = (h_x^{(1)}, h_x^{(2)}, h_x^{(3)})$ is written as $\boldsymbol{h}_x = h\,\boldsymbol{\xi}_x$, where $\boldsymbol{\xi}_x$ is an arbitrary vector satisfying $|\boldsymbol{\xi}_x| \le 1$. We shall fix the direction $\boldsymbol{\xi}_x$ for each x, and change only the magnitude $h \ge 0$. The standard magnetic field in (4.4.1) and the staggered magnetic filed in (4.4.2) are included as examples, but we here allow external field with much more complicated spatial modulation. Like the staggered magnetic field, the field $(\boldsymbol{h}_x)_{x\in\Lambda_L}$ is introduced to trigger possible magnetic ordering proportional to $(\boldsymbol{\xi}_x)_{x\in\Lambda_L}$. We then prove the following.

Theorem 4.24 (Improved Hohenberg–Mermin–Wagner theorem) *Consider the model (4.4.21) in two dimensions. Then one has*[48]

$$\lim_{h\downarrow 0}\lim_{L\uparrow\infty} \langle \hat{S}_x^{(\alpha)} \rangle_{\beta,h}^{L} = 0, \tag{4.4.22}$$

for any $\beta \in [0, \infty)$ and $x \in \mathbb{Z}^2$.

The theorem was first proved by Mermin and Wagner [45] for the models (4.4.1) and (4.4.2). This excludes the possibility of standard ferromagnetic or antiferromagnetic order, but does not exclude more exotic ordering, which may be triggered by a more complicated external field. Note that the above improved theorem excludes magnetic ordering of any type.[49]

The original proof by Hohenberg [26] and by Mermin and Wagner [45] is based on an operator inequality called the Bogoliubov inequality. Since the proof relies essentially on the Fourier transformation, the model must have translation invariance. The proof of the improved theorem that we present below, which is based on McBryan and Spencer's complex transformation method [33, 44], does not rely on translation invariance. Although we work on the Hamiltonian (4.4.21) which has translation invariant interactions for simplicity, the proof applies to a much more general class of

---

[47] Mermin and Wagner clearly state in [45] that the essential idea of the proof was due to Hohenberg. It is unfortunate that the result is often referred to as the Mermin–Wagner theorem.

[48] To be rigorous $\lim_{L\uparrow\infty}$ should be $\limsup_{L\uparrow\infty}$.

[49] As far as we know this improved theorem appears for the first time in the present book. The strongest result in this direction may be that by Fröhlich and Pfister [18], which (roughly) states that any infinite volume equilibrium state of a rotationally invariant two-dimensional model preserves the symmetry.

models without translation invariance, including models with random interactions[50] or models defined on fractal lattices [36].

Absence of long-range order For models with vanishing magnetic field, i.e. $h = 0$, where the Hamiltonian (4.4.21) reduces to (4.4.1) or (4.4.2), it is proved that the two-point spin correlation functions do not exhibit any long-range order. This fact was proved by Koma and Tasaki [33] for the Hubbard model, by extending the corresponding result of McBryan and Spencer [44] for classical spin systems. The proof in [33] readily extends to quantum spin systems, as was explicitly noted in [7, 21]. See [27] for a different extension of the McBryan–Spencer method to quantum systems.

The main result is as follows.

Theorem 4.25 (McBryan–Spencer, Koma–Tasaki theorem) *Consider the model (4.4.1) or (4.4.2) in two dimensions with $h = 0$. For any $\beta \in [0, \infty)$ and L, one has*

$$\left|\langle \hat{S}_x^{(\alpha)} \hat{S}_y^{(\alpha)} \rangle_{\beta,0}^L\right| \le 2S^2\, |x - y|^{-\eta(\beta)} \tag{4.4.23}$$

for any $\alpha = 1, 2, 3$ and any $x, y \in \Lambda_L$ such that $|x - y| < L/2$. The exponent $\eta(\beta)$ is an L-independent positive decreasing function of β which behaves as

$$\eta(\beta) \simeq (16CS^2\beta)^{-1} \tag{4.4.24}$$

for $\beta \gg 1$, where C is a numerical constant.

Note that the the upper bound (4.4.23) with a power law decay in the distance $|x - y|$ is reminiscent of the asymptotic behavior (4.4.20) for $d = 2$, found (approximately) for the classical XY model. One can see from the following proof that this is not an accident; the proof is designed so as to extract the behavior predicted by Berezinskii's harmonic approximation. In fact it is expected that the two-point correlation function $\langle \hat{S}_x^{(\alpha)} \hat{S}_y^{(\alpha)} \rangle_{\beta,0}^\infty$ of the two-dimensional Heisenberg model at vanishing magnetic field decays exponentially as $\exp[-|x-y|/\xi(\beta)]$ for any $\beta \in [0, \infty)$. In this sense the upper bound (4.4.23) is still very weak.[51] We also note that Theorem 4.25 readily extends to models without translation invariance exactly as Theorem 4.24. See footnote 50.

Anisotropic models All these results about the absence of order in $d = 2$ crucially depends on the fact that we are concerned with possible spontaneous breakdown of continuous rotational symmetry. Even in two dimensions, discrete symmetry

---

[50] Theorems 4.24 and 4.25 below can be almost automatically extended to a model with the Hamiltonian $\hat{H}_h = \sum_{\{x,y\}\in\mathscr{B}_L} J_{x,y}\, \hat{\boldsymbol{S}}_x \cdot \hat{\boldsymbol{S}}_y - \sum_{x\in\Lambda_L} \boldsymbol{h}_x \cdot \hat{\boldsymbol{S}}_x$, where the interactions $J_{x,y}$ are arbitrary except that $|J_{x,y}| \le 1$.

[51] It is expected that the quantum XY model, which is obtained by setting $\lambda = D = 0$ in (2.5.14), exhibits a power law decay at sufficiently low temperature. Theorem 4.25 automatically extends to this case.

may spontaneously break at sufficiently low temperature. For example, in the two-dimensional anisotropic ferromagnetic model with the Hamiltonian

$$\hat{H} = -\sum_{\{x,y\}\in\mathscr{B}_L} (\hat{S}_x^{(1)}\hat{S}_y^{(1)} + \hat{S}_x^{(2)}\hat{S}_y^{(2)} + \lambda\hat{S}_x^{(3)}\hat{S}_y^{(3)}), \tag{4.4.25}$$

it is proved that there are LRO and SSB at sufficiently low temperatures, provided that $\lambda > 1$ [28].

Basic inequality for the proof In what follows, we present a complete unified Proof of Theorems 4.24 and 4.25. Although the proof is not short, we believe it worth explaining. It is quite interesting to see that Berezinskii's harmonic approximation in Sect. 4.4.2 is partially justified rigorously by considering rotations by purely imaginary angles.

In the present proof the complex rotation operator

$$\hat{R}_{\boldsymbol{\varphi}} := \exp\Bigl[\sum_{x\in\Lambda_L} \varphi_x \hat{S}_x^{(3)}\Bigr] \tag{4.4.26}$$

plays a fundamental role. Here $\boldsymbol{\varphi} = (\varphi_x)_{x\in\Lambda_L}$ with $\varphi_x \in \mathbb{R}$ may be called the "classical field configuration". Note that $\hat{R}_{\boldsymbol{\varphi}}$ may be regarded as the rotation about the 3-axis where the rotation angle at site x is $\theta_x = i\varphi_x$. Of course $\hat{R}_{\boldsymbol{\varphi}}$ is not unitary, but is invertible with its inverse given by $\hat{R}_{\boldsymbol{\varphi}}^{-1} = \exp[-\sum_{x\in\Lambda_L}\varphi_x\hat{S}_x^{(3)}]$.

Let us examine how spin operators and Hamiltonian are transformed by the complex rotation. By noting that

$$e^{-\varphi\hat{S}_x^{(3)}}\,\hat{S}_x^{\pm}\,e^{\varphi\hat{S}_x^{(3)}} = e^{\mp\varphi}\,\hat{S}_x^{\pm}, \tag{4.4.27}$$

we find that

$$\begin{aligned}
\hat{R}_{\boldsymbol{\varphi}}^{-1}(\hat{S}_x^{(1)}\hat{S}_y^{(1)} + \hat{S}_x^{(2)}\hat{S}_y^{(2)})\hat{R}_{\boldsymbol{\varphi}} &= \frac{1}{2}\hat{R}_{\boldsymbol{\varphi}}^{-1}(\hat{S}_x^{+}\hat{S}_y^{-} + \hat{S}_x^{-}\hat{S}_y^{+})\hat{R}_{\boldsymbol{\varphi}} \\
&= \frac{1}{2}(e^{-(\varphi_x-\varphi_y)}\,\hat{S}_x^{+}\hat{S}_y^{-} + e^{\varphi_x-\varphi_y}\,\hat{S}_x^{-}\hat{S}_y^{+}) \\
&= \cosh(\varphi_x-\varphi_y)(\hat{S}_x^{(1)}\hat{S}_y^{(1)} + \hat{S}_x^{(2)}\hat{S}_y^{(2)}) + i\sinh(\varphi_x-\varphi_y)(\hat{S}_x^{(1)}\hat{S}_y^{(2)} - \hat{S}_x^{(2)}\hat{S}_y^{(1)}),
\end{aligned} \tag{4.4.28}$$

and similarly that

$$\hat{R}_{\boldsymbol{\varphi}}^{-1}\,\hat{S}_x^{(1)}\,\hat{R}_{\boldsymbol{\varphi}} = \cosh\varphi_x\,\hat{S}_x^{(1)} - i\sinh\varphi_x\,\hat{S}_x^{(2)}, \tag{4.4.29}$$

$$\hat{R}_{\boldsymbol{\varphi}}^{-1}\,\hat{S}_x^{(2)}\,\hat{R}_{\boldsymbol{\varphi}} = \cosh\varphi_x\,\hat{S}_x^{(2)} + i\sinh\varphi_x\,\hat{S}_x^{(1)}. \tag{4.4.30}$$

Of course $\hat{S}_x^{(3)}$ is not modified by the transformation. From these relations we see that the Hamiltonian (4.4.21) is transformed as

$$\hat{R}_{\varphi}^{-1}\hat{H}_h\hat{R}_{\varphi} = \hat{H}_h + \hat{V} + i\hat{W}, \tag{4.4.31}$$

where

$$\begin{aligned}\hat{V} = J \sum_{\{x,y\}\in\mathscr{B}_L} \{\cosh(\varphi_x - \varphi_y) - 1\}(\hat{S}_x^{(1)}\hat{S}_y^{(1)} + \hat{S}_x^{(2)}\hat{S}_y^{(2)}) \\ - \sum_{x\in\Lambda_L}(\cosh\varphi_x - 1)(h_x^{(1)}\hat{S}_x^{(1)} + h_x^{(2)}\hat{S}_x^{(2)}),\end{aligned} \tag{4.4.32}$$

and

$$\hat{W} = J \sum_{\{x,y\}\in\mathscr{B}_L} \sinh(\varphi_x - \varphi_y)(\hat{S}_x^{(1)}\hat{S}_y^{(2)} - \hat{S}_x^{(2)}\hat{S}_y^{(1)}) + \sum_{x\in\Lambda_L} \sinh\varphi_x (h_x^{(1)}\hat{S}_x^{(2)} - h_x^{(2)}\hat{S}_x^{(1)}) \tag{4.4.33}$$

are self-adjoint.

For an arbitrary operator $\hat{A}$, write $\hat{A}' = \hat{R}_{\varphi}^{-1}\hat{A}\hat{R}_{\varphi}$. Then, by using the cyclicity of trace, we get

$$\mathrm{Tr}[\hat{A}\,e^{-\beta\hat{H}_h}] = \mathrm{Tr}[\hat{R}_{\varphi}^{-1}\hat{A}\hat{R}_{\varphi}\,e^{-\beta\hat{R}_{\varphi}^{-1}\hat{H}_h\hat{R}_{\varphi}}] = \mathrm{Tr}[\hat{A}'\,e^{-\beta(\hat{H}_h+\hat{V}+i\hat{W})}]. \tag{4.4.34}$$

Then, by using the inequality (A.2.24) (proved in Appendix A.2.2), we have

$$\left|\mathrm{Tr}[\hat{A}\,e^{-\beta\hat{H}_h}]\right| \le \|\hat{A}'\|\,\mathrm{Tr}[e^{-\beta(\hat{H}_h+\hat{V})}] \le \|\hat{A}'\|\,e^{\beta\|\hat{V}\|}\,\mathrm{Tr}[e^{-\beta\hat{H}_h}], \tag{4.4.35}$$

where we noted that $-\beta(\hat{H}_h + \hat{V}) \le -\beta\hat{H}_h + \beta\|\hat{V}\|$ and used the bound (A.2.31) (proved in Appendix A.2.3) to get the final expression.

We can therefore bound the expectation value of $\hat{A}$ as

$$\begin{aligned}\left|\langle\hat{A}\rangle_{\beta,h}^{L}\right| &= \frac{\left|\mathrm{Tr}[\hat{A}\,e^{-\beta\hat{H}_h}]\right|}{\mathrm{Tr}[e^{-\beta\hat{H}_h}]} \le \|\hat{A}'\|\,e^{\beta\|\hat{V}\|} \\ &\le \|\hat{A}'\|\,\exp\Bigl[2\beta S^2 \sum_{\{x,y\}\in\mathscr{B}_L}\{\cosh(\varphi_x - \varphi_y) - 1\} + 2\beta h S \sum_{x\in\Lambda_L}(\cosh\varphi_x - 1)\Bigr],\end{aligned} \tag{4.4.36}$$

for any classical field configuration $\boldsymbol{\varphi} = (\varphi_x)_{x\in\Lambda_L}$. This inequality is the basis of the proof.

Proof of Theorem 4.25 Let us first prove Theorem 4.25 by setting $h = 0$, and considering the operator $\hat{A} = \hat{S}_o^+\hat{S}_z^-$ where o is the origin of Λ_L. From (4.4.27), we have $\hat{A}' = e^{-\varphi_0+\varphi_z}\hat{S}_o^+\hat{S}_z^-$, and hence $\|\hat{A}'\| \le 4S^2\,e^{-\varphi_0+\varphi_z}$. Then the main inequality (4.4.36) reads

$$\left|\langle\hat{S}_o^+\hat{S}_z^-\rangle_{\beta,0}^{L}\right| \le 4S^2\,\exp\Bigl[-\varphi_0 + \varphi_z + 2\beta S^2 \sum_{\{x,y\}\in\mathscr{B}_L}\{\cosh(\varphi_x - \varphi_y) - 1\}\Bigr], \tag{4.4.37}$$

where the field configuration $\boldsymbol{\varphi} = (\varphi_x)_{x \in \Lambda_L}$ is arbitrary.

Before proceeding we point out a close relation to Berezinskii's harmonic approximation. Let us be heuristic and assume that the classical field configuration φ_x varies slowly so that one can approximate $\cosh(\varphi_x - \varphi_y) - 1 \simeq (\varphi_x - \varphi_y)^2/2$. Then the argument of exponential in the right-hand side of (4.4.37) becomes $\mathscr{S}[\boldsymbol{\varphi}] \simeq -\varphi_0 + \varphi_z + \beta S^2 \sum_{\{x,y\} \in \mathscr{B}_L} (\varphi_x - \varphi_y)^2$. Since the configuration $\boldsymbol{\varphi}$ is arbitrary we shall chose one which minimizes $\mathscr{S}[\boldsymbol{\varphi}]$, i.e., in which the conditions $\partial \mathscr{S}[\boldsymbol{\varphi}]/\partial \varphi_x = 0$ is satisfied for each $x \in \Lambda_L$. Clearly the condition is written as

$$-(\boldsymbol{\Delta \varphi})_x = \frac{1}{2\beta S^2}(\delta_{x,o} - \delta_{x,z}), \tag{4.4.38}$$

which is nothing but the Poisson equation (4.4.16) (with a modified charge).

Let us continue the proof. The above consideration suggests that it is optimal to chose the classical field configuration $\boldsymbol{\varphi}$ as the solution of the Poisson equation (4.4.38). But it is easier to use, following [51], a more tractable explicit configuration which does almost the same job. Here we shall choose

$$\varphi_x = \begin{cases} 2\eta \log \dfrac{\ell}{|x|+1} & \text{if } |x| \le \ell - 1, \\ 0 & \text{if } |x| > \ell - 1, \end{cases} \tag{4.4.39}$$

where ℓ and η are positive constants to be determined later. Let $\{x, y\} \in \mathscr{B}_L$ (i.e., $|x - y| = 1$) and assume that $|x| \le |y|$. Then we see that

$$\begin{aligned} |\varphi_x - \varphi_y| &= 2\eta \log \frac{|y|+1}{|x|+1} \le 2\eta \log \frac{(|x|+1)+1}{|x|+1} \\ &= 2\eta \log\Bigl(1 + \frac{1}{|x|+1}\Bigr) \le \frac{2\eta}{|x|+1} \le 2\eta. \end{aligned} \tag{4.4.40}$$

Note that the bound is valid even when $\varphi_y = 0$. Noting that $\cosh\theta - 1 \le \{\cosh(2\eta) - 1\}\{\theta/(2\eta)\}^2$ if $|\theta| \le 2\eta$, we find

$$\cosh(\varphi_x - \varphi_y) - 1 \le \frac{\cosh(2\eta) - 1}{(|x|+1)^2}, \tag{4.4.41}$$

for the same x and y. We can thus bound the sum in (4.4.37) as

$$\sum_{\{x,y\} \in \mathscr{B}_L} \{\cosh(\varphi_x - \varphi_y) - 1\} \le 4\{\cosh(2\eta) - 1\} \sum_{\substack{x \in \Lambda_L \\ (|x| \le \ell - 1)}} \frac{1}{(|x|+1)^2}, \tag{4.4.42}$$

where we made an over-counting of bonds by noting that each x belongs to at most four bonds[52] $\{x, y\}$. The sum is bounded by using an integral as

$$\sum_{\substack{x\in\Lambda_L\\(|x|\le\ell-1)}} \frac{1}{(|x|+1)^2} \le C\log\ell, \tag{4.4.43}$$

where C is a numerical constant.

Let us set $\ell = |z|$. Then (4.4.39) implies $\varphi_o = 2\eta\log\ell$ and $\varphi_z = 0$. By substituting (4.4.42) into (4.4.37), we have

$$\left|\langle\hat{S}_o^+\hat{S}_z^-\rangle_{\beta,0}^L\right| \le 4S^2\exp\Big[-\big(2\eta - 8CS^2\beta\{\cosh(2\eta)-1\}\big)\log\ell\Big]. \tag{4.4.44}$$

Let $\eta(\beta)$ be the positive solution of

$$\eta = 8CS^2\beta\{\cosh(2\eta)-1\}. \tag{4.4.45}$$

Note that the solution always exists. For $\beta \gg 1$, we have $\eta(\beta) \ll 1$ and get (4.4.24). By setting η to be $\eta(\beta)$, the bound (4.4.44) becomes

$$\left|\langle\hat{S}_o^+\hat{S}_z^-\rangle_{\beta,0}^L\right| \le 4S^2\exp[-\eta(\beta)\log\ell] = 4S^2\,|z|^{-\eta(\beta)}. \tag{4.4.46}$$

Noting that $|\langle\hat{S}_o^{(1)}\hat{S}_z^{(1)} + \hat{S}_o^{(2)}\hat{S}_z^{(2)}\rangle_{\beta,0}^L| \le |\langle\hat{S}_o^+\hat{S}_z^-\rangle_{\beta,0}^L|$, and using the isotropy and the translation invariance we get the desired bound (4.4.23).

Proof of Theorem 4.24 Let us set $\hat{A} = \hat{S}_o^+$ to prove the improved Hohenberg–Mermin–Wagner theorem (Theorem 4.24 in p. 124). We have $\hat{A}' = e^{-\varphi_o}\hat{S}_o^+$, and hence $\|\hat{A}'\| \le 2S\,e^{-\varphi_o}$. The main inequality (4.4.36) reads

$$\left|\langle\hat{S}_o^+\rangle_{\beta,h}^L\right| \le 2S\exp\Big[-\varphi_o+2\beta S^2\sum_{\{x,y\}\in\mathscr{B}_L}\{\cosh(\varphi_x-\varphi_y)-1\}+2\beta hS\sum_{x\in\Lambda_L}(\cosh\varphi_x-1)\Big]. \tag{4.4.47}$$

We again use the classical field configuration (4.4.39) and set η to be $\eta(\beta)$ to get

$$\left|\langle\hat{S}_o^+\rangle_{\beta,h}^L\right| \le 2S\exp\big[-\eta(\beta)\log\ell + h\,G(\beta,\ell)\big], \tag{4.4.48}$$

where

$$G(\beta,\ell) = 2\beta S\sum_{\substack{x\in\Lambda_L\\(|x|\le\ell-1)}}\cosh\Big(2\eta(\beta)\log\frac{\ell}{|x|+1}\Big) \tag{4.4.49}$$

[52] In fact it is enough to count two bonds for general x, but we must count four neighbors of the origin o.

is a complicated function of β and ℓ. For us it is only necessary that $G(\beta, \ell)$ is well-defined and finite.

Then, for any h satisfying

$$0 \le h \le \frac{\eta(\beta) \log \ell}{2G(\beta, \ell)}, \tag{4.4.50}$$

the bound (4.4.48) implies

$$\left|\langle \hat{S}_o^{(1)} \rangle_{\beta,h}^L\right| \le \left|\langle \hat{S}_o^{+} \rangle_{\beta,h}^L\right| \le 2S \exp\left[-\frac{\eta(\beta)}{2} \log \ell\right] = 2S\, \ell^{-\eta(\beta)/2}. \tag{4.4.51}$$

At this stage we can take the limit[53] $L \uparrow \infty$, and note that the bound (4.4.51) is still valid. Then by letting $\ell \uparrow \infty$ (and recalling that there was nothing special about the site o), we get the desired (4.4.22). Here it is essential that the right-hand side of (4.4.50) is strictly positive for any finite ℓ. It converges to zero as $\ell \uparrow \infty$, but it is not necessary to prove this property.

Problem 4.4.3.a Extend the above proofs to the one-dimensional models, and prove two corresponding theorems. [solution→p.502]

4.4.4 LRO and SSB in Three or Higher Dimensions

Now we move on to three or higher dimensions, where Berezinskii's harmonic approximation (Sect. 4.4.2) suggests the existence of LRO and SSB at sufficiently low temperatures. Although to develop rigorous theories is highly nontrivial, there are rather satisfactory results for the antiferromagnetic Heisenberg model. Consider the standard antiferromagnetic Hamiltonian (4.4.2), and define the order operator $\hat{\mathcal{O}}_L^{(\alpha)}$ as in (4.1.4). Then the following theorem establishes the existence of LRO at vanishing external magnetic field.

Theorem 4.26 (Dyson–Lieb–Simon theorem) *For any $d \ge 3$ and any $S = 1/2, 1, 3/2, \ldots$, there exists $\beta_0 \in (0, \infty)$ and a function $q(\beta)$ such that $q(\beta) > 0$ for any $\beta > \beta_0$. It holds that*

$$\left\langle \left(\frac{\hat{\mathcal{O}}_L^{(\alpha)}}{L^d} \right)^2 \right\rangle_{\beta,0}^L \ge q(\beta), \tag{4.4.52}$$

for any $\beta > \beta_0$ if L is sufficiently large.

The theorem, which is regarded to be one of the most important results in mathematical studies of quantum many-body systems, was proved in [12] by Dyson, Lieb,

---

[53] To be rigorous one should consider $\limsup_{L\uparrow\infty} |\langle \hat{S}_o^{(1)} \rangle_{\beta,h}^L|$.

and Simon, except for the case $d = 3$ and $S = 1/2$. The proof makes an essential use of the reflection positivity. See Sect. 4.1.1. The exceptional case with $d = 3$ and $S = 1/2$ can be treated by using the idea of Kennedy, Lieb, and Shastry [29]. See Sect. 4.1.2.

It is also known rigorously that SSB exists whenever one has LRO. We again consider the Hamiltonian (4.4.2) with staggered magnetic field in the 3-direction.

Theorem 4.27 (Griffiths, Koma-Tasaki theorem) *Under the same condition as in Theorem 4.26, it holds that* [54]

$$\lim_{h\downarrow 0}\lim_{L\uparrow\infty}\left\langle\frac{\hat{\mathcal{O}}_L^{(3)}}{L^d}\right\rangle_{\beta,h}^{L} \geq \sqrt{3\,q(\beta)}. \tag{4.4.53}$$

Note that the factor $\sqrt{3}$ in the right-hand side of (4.4.53) has the same origin as the same factor we saw in Theorem 4.11 in p. 101. Theorem 4.27 was proved by Koma and Tasaki [34] by extending the corresponding theorem proved by Griffiths [23] for models in which the Hamiltonian and the order operator commute.

Although the ferromagnetic Heisenberg model (4.4.1) is also believed to exhibit LRO and SSB at sufficiently low temperatures in $d \geq 3$, there has been no rigorous proof. The method for proving Theorem 4.26 does not apply here because the ferromagnetic Heisenberg model fails to satisfy the condition for reflection positivity. See footnote 12 in p. 90. It is interesting that the analysis of equilibrium states of the ferromagnetic Heisenberg model is so difficult[55] while to characterize its ground states was elementary as we saw in Sect. 2.4.

References

1. I. Affleck, T. Kennedy, E.H. Lieb, H. Tasaki, Valence bond ground states in isotropic quantum antiferromagnets. Comm. Math. Phys. **115**, 477–528 (1988), https://projecteuclid.org/euclid.cmp/1104161001
2. P.W. Anderson, An approximate quantum theory of the antiferromagnetic ground state. Phys. Rev. **86**, 694 (1952)
3. P.W. Anderson, *Basic Notions of Condensed Matter Physics* (Westview Press/Addison-Wesley, Boston, 1997)
4. H. Araki, Gibbs states of a one dimensional quantum lattice. Comm. Math. Phys. **14**, 120–157 (1969), http://projecteuclid.org/euclid.cmp/1103841726
5. P. Azaria, B. Delamotte, D. Mouhanna, Spontaneous symmetry breaking in quantum frustrated antiferromagnets. Phys. Rev. Lett. **70**, 2483–2486 (1993)
6. A.J. Beekman, L. Rademaker, J. van Wezel, An introduction to spontaneous symmetry breaking (2019), arxiv:1909.01820

[54] To be rigorous, lim should be replaced by lim inf.

[55] It is unfortunate that the method based on reflection positivity is essentially the only method we know to prove the existence of long-range order associated with a breakdown of continuous symmetry.

7. C. Benassi, J. Fröhlich, D. Ueltschi, Decay of correlations in 2D quantum systems with continuous symmetry. Ann. Henri Poincaré **18**, 2381–2487 (2017), https://link.springer.com/article/10.1007/s00023-017-0571-4
8. V.L. Berezinskii, Destruction of long-range order in one-dimensional and two-dimensional systems having a continuous symmetry group i. Classical systems. Soviet Phys. JETP **32**, 493–500 (1971), http://jetp.ac.ru/cgi-bin/dn/e_032_03_0493.pdf
9. B. Bernu, C. Lhuillier, L. Pierre, Signature of Néel order in exact spectra of quantum antiferromagnets on finite lattices. Phys. Rev. Lett. **69**, 2590–2593 (1992)
10. O. Bratteli, D.W. Robinson, *Operator Algebras and Quantum Statistical Mechanics I: C^*-and W^*-Algebras. Symmetry Groups. Decomposition of States* (Springer, Berlin, 1987)
11. O. Bratteli, D.W. Robinson, *Operator Algebras and Quantum Statistical Mechanics II: Equilibrium States Models in Quantum Statistical Mechanics* (Springer, Berlin, 1981)
12. F.J. Dyson, E.H. Lieb, B. Simon, Phase transitions in quantum spin systems with isotropic and nonisotropic interactions. J. Stat. Phys. **18**, 335–382 (1978)
13. R. Fernández, J. Fröhlich, A.D. Sokal, *Random Walks, Critical Phenomena, and Triviality in Quantum Field Theory* (Springer, Berlin, 1992)
14. S. Friedli, Y. Velenik, *Statistical Mechanics of Lattice Systems: A Concrete Mathematical Introduction* (Cambridge University Press, Cambridge, 2017)
15. J. Fröhlich, R. Israel, E.H. Lieb, B. Simon, Phase transitions and reflection positivity I: general theory and long range lattice models. Comm. Math. Phys. **62**, 1–34 (1978), https://projecteuclid.org/euclid.cmp/1103904299
16. J. Fröhlich, R. Israel, E.H. Lieb, B. Simon, Phase transitions and reflection positivity II: lattice systems with short range and Coulomb interactions. J. Stat. Phys. **22**, 297–334 (1980)
17. J. Fröhlich, E.H. Lieb, Phase transitions in anisotropic lattice spin systems. Commun. Math. Phys **60**, 233–267 (1978), https://projecteuclid.org/euclid.cmp/1103904129
18. J. Fröhlich, C. Pfister, On the absence of spontaneous symmetry breaking and of crystalline ordering in two-dimensional systems. Comm. Math. Phys. **81**, 277–298 (1980), https://projecteuclid.org/euclid.cmp/1103920246
19. J. Fröhlich, B. Simon, T. Spencer, Infrared bounds, phase transitions, and continuous symmetry breaking. Comm. Math. Phys. **50**, 79–95 (1976), https://projecteuclid.org/euclid.cmp/1103900151
20. J. Fröhlich, T. Spencer, The Kosterlitz-Thouless transition in two-dimensional abelian spin systems and the Coulomb gas. Comm. Math. Phys. **81**, 527–602 (1981), https://projecteuclid.org/euclid.cmp/1103920388
21. J. Fröhlich, D. Ueltschi, Some properties of correlations of quantum lattice systems in thermal equilibrium. J. Math. Phys. **56**, 053302 (2015), arXiv:1412.2534
22. J.-P. Gazeau, *Coherent States in Quantum Physics* (Wiley-VCH, Weinheim, 2009)
23. R.B. Griffiths, Spontaneous magnetization in idealized ferromagnets. Phys. Rev. **152**, 240–246 (1966)
24. M. Gross, E. Sánchez-Velasco, E.D. Siggia, Spin-wave velocity and susceptibility for the two-dimensional Heisenberg antiferromagnet. Phys. Rev. B **40**, 11328–11330 (1989)
25. T. Hara, H. Tasaki, *Mathematics and Physics of Phase Transitions and Critical Phenomena* (in Japanese) (Kyoritsu, Tokyo, 2015)
26. P.C. Hohenberg, Existence of long-range order in one and two dimensions. Phys. Rev. **158**, 383 (1967)
27. K.R. Ito, Clustering in low-dimensional $\mathrm{SO}(N)$-invariant statistical models with long-range interactions. J. Stat. Phys. **29**, 747–760 (1982)
28. T. Kennedy, Long range order in the anisotropic quantum ferromagnetic Heisenberg model. Comm. Math. Phys. **100**, 447–462 (1985), https://projecteuclid.org/euclid.cmp/1104113924
29. T. Kennedy, E.H. Lieb, B.S. Shastry, Existence of Néel order in some spin-$1/2$ Heisenberg antiferromagnets. J. Stat. Phys. **53**, 1019 (1988)
30. T. Kennedy, E.H. Lieb, B.S. Shastry, The XY model has long-range order for all spins and all dimensions greater than one. Phys. Rev. Lett. **61**, 2582 (1988)

31. M. Kikuchi, Y. Okabe, S. Miyashita, On the ground-state phase transition of the spin 1/2 XXZ model on the square lattice. J. Phys. Soc. Jpn. **59**, 492 (1990)
32. T. Koma, Maximum spontaneous magnetization and Nambu–Goldstone mode (2018), arXiv:1712.09018
33. T. Koma, H. Tasaki, Decay of superconducting and magnetic correlations in one- and two-dimensional Hubbard models. Phys. Rev. Lett. **68**, 3248–3251 (1992), arXiv:cond-mat/9709068
34. T. Koma, H. Tasaki, Symmetry breaking in Heisenberg antiferromagnets. Comm. Math. Phys. **158**, 191–214 (1993), https://projecteuclid.org/euclid.cmp/1104254136
35. T. Koma, H. Tasaki, Symmetry breaking and finite-size effects in quantum many-body systems. J. Stat. Phys. **76**, 745–803 (1994). (The version in the arXiv is more complete than the published version), arXiv:cond-mat/9708132
36. T. Koma, H. Tasaki, Classical XY model in 1.99 dimensions. Phys. Rev. Lett. **74**, 3916–3919 (1995), arXiv:cond-mat/9708126
37. J.M. Kosterlitz, D.J. Thouless, Long range order and metastability in two dimensional solids and superfluids. J. Phys. C Solid State Phys. **5**, L124 (1972)
38. J.M. Kosterlitz, D.J. Thouless, Ordering, metastability and phase transitions in two-dimensional systems. J. Phys. C Solid State Phys. **6**, 1181 (1973)
39. K. Kubo, Existence of long-range order in the XY model. Phys. Rev. Lett. **61**, 110 (1988)
40. K. Kubo, T. Kishi, Existence of long-range order in the XXZ model. Phys. Rev. Lett. **61**, 2585 (1988)
41. C. Lhuillier, *Frustrated Quantum Magnets*. Lecture notes at "Ecole de troisieme cycle de Suisse Romande (2002)", arXiv:cond-mat/0502464
42. E.H. Lieb, D. Mattis, Ordering energy levels in interacting spin chains. J. Math. Phys. **3**, 749–751 (1962)
43. E.H. Lieb, B. Nachtergaele, The stability of the Peierls instability for ring-shaped molecules. Phys. Rev. B **51**, 4777–4791 (1995), arXiv:cond-mat/9410100v1
44. O.A. McBryan, T. Spencer, On the decay of correlations in SO(n)-symmetric ferromagnets. Comm. Math. Phys. **53**, 299–302 (1977), https://projecteuclid.org/euclid.cmp/1103900708
45. N.D. Mermin, H. Wagner, Absence of ferromagnetism or antiferromagnetism in one or two-dimensional isotropic heisenberg models. Phys. Rev. Lett. **17**, 1133 (1966); Erratum Phys. Rev. Lett. **17**, 1307 (1966)
46. T. Momoi, Quantum fluctuation in quantum lattice systems with continuous symmetry. J. Stat. Phys. **85**, 193–210 (1996)
47. E.J. Neves, J.F. Perez, Long range order in the ground state of two-dimensional antiferromagnets. Phys. Lett. **114A**, 331–333 (1986)
48. K. Osterwalder, R. Schrader, Axioms for Euclidean Green's functions. Commun. Math. Phys. **31**, 83–112 (1973), https://projecteuclid.org/euclid.cmp/1103858969
49. Y. Ozeki, H. Nishimori, Y. Tomita, Long-range order in antiferromagnetic quantum spin systems. J. Phys. Soc. Jpn. **58**, 82–90 (1989), http://journals.jps.jp/doi/abs/10.1143/JPSJ.58.82
50. Y.M. Park, The cluster expansion for classical and quantum lattice systems. J. Stat. Phys. **27**, 553–576 (1982)
51. P. Picco, Upper bound on the decay of correlations in the plane rotator model with long-range random interaction. J. Stat. Phys. **36**, 489–561 (1984)
52. L. Pitaevskii, S. Stringari, Uncertainty principle, quantum fluctuations, and broken symmetries. J. Low Temp. Phys. **85**, 377–388 (1991)
53. L. Rademaker, Exact ground state of Lieb-Mattis hamiltonian as a superposition of Néel states (2019), arXiv:1909.09663
54. L. Rademaker, A. Beekman, J. van Wezel, Stability and absence of a tower of states in ferrimagnets (2019), arXiv:1909.11381
55. D. Ruelle, *Statistical Mechanics: Rigorous Results* (WSPC/ICP, Singapore, 1999)
56. A. W. Sandvik, Finite-size scaling of the ground-state parameters of the two-dimensional Heisenberg model. Phys. Rev. B **56**, 11678–11690 (1997), arXiv:cond-mat/9707123

57. N. Sannomiya, H. Katsura, Y. Nakayama, Supersymmetry breaking and Nambu–Goldstone fermions with cubic dispersion. Phys. Rev. D **95**, 065001 (2017), arXiv:1612.02285
58. B.S. Shastry, Bounds for correlation functions of the Heisenberg antiferromagnet. J. Phys. A Math. Gen. **25**, L249 (1992)
59. S.-Q. Shen, Z.-M. Qiu, G.-S. Tian, Ferrimagnetic long-range order of the Hubbard model. Phys. Rev. Lett. **72**, 1280 (1994)
60. A. Shimizu, T. Miyadera, Stability of quantum states of finite macroscopic systems against classical noises, perturbations from environments, and local measurements. Phys. Rev. Lett. **89**, 270403 (2002), arXiv:quant-ph/0203106
61. B. Simon, *The Statistical Mechanics of Lattice Gases* (Princeton University Press, Princeton, 1993)
62. A. Tanaka, Low-lying states with full symmetry breaking in the antiferromagnetic Heisenberg model, to be published (2020)
63. K. Tanaka, K. Takeda, T. Idogaki, Absence of spontaneous symmetry breaking in the ground state of one-dimensional spin-orbital model. J. Mag. Mag. Mat. **272–276**, 908–909 (2004)
64. S. Tang, J.E. Hirsch, Long-range order without broken symmetry: two-dimensional Heisenberg antiferromagnet at zero temperature. Phys. Rev. B **39**, 4548–4553 (1989)
65. H. Tasaki, On the local equivalence between the canonical and the microcanonical ensembles for quantum spin systems. J. Stat. Phys. **172**, 905–926 (2018), arXiv:1609.06983
66. H. Tasaki, Long-range order, "tower" of states, and symmetry breaking in lattice quantum systems. J. Stat. Phys. **174**, 735–761 (2019). (The version in the arXiv is more complete than the published version), arXiv:abs/1807.05847
67. J. van Wezel, J. van den Brink, J. Zaanen, An intrinsic limit to quantum coherence due to spontaneous symmetry breaking. Phys. Rev. Lett. **94**, 230401 (2005), arXiv:cond-mat/0408357
68. J. van Wezel, J. Zaanen, J. van den Brink, Relation between decoherence and spontaneous symmetry breaking in many-particle qubits. Phys. Rev. B **74**, 094430 (2006), arXiv:cond-mat/0606140

Chapter 5
Long-Range Order and "Spontaneous Symmetry Breaking" in Bose–Einstein Condensates

In this short chapter, which is the end of Part I, we continue focusing on LRO and SSB in ground states, but in a system of many bosonic particles which exhibits Bose–Einstein condensation (BEC), namely, the system of hard-core bosons on the d-dimensional hypercubic lattice with $d \geq 2$. For background, see, e.g., [4, 14, 15, 17]. We will see that, while the mathematical structure of the ground state and low-lying states is exactly parallel to that in quantum antiferromagnets, its physical interpretation is drastically different. Although we do not discuss superconductivity in the present book, this phenomenon is (at least mathematically) very similar to BEC as long as LRO and SSB are concerned.[1] After defining the problem in Sect. 5.1 and discussing the important notion of off-diagonal long-range order in Sect. 5.2, we show in Sect. 5.3 the results about low-lying states and symmetry breaking, parallel to those obtained for quantum spin systems. In Sect. 5.4, we discuss a fundamental difference between quantum spin systems and particle systems, and develop a picture about physically realistic ground state of a large but finite system exhibiting BEC. Finally, in Sect. 5.5, we describe an interesting phenomenon of "SSB" in coupled Bose–Einstein condensates.

5.1 The Model and the Equivalence to the XY Model

We here study the bosonic version of the Hubbard model. The Hubbard model is a standard model for itinerant electrons in certain solids; it will be the main subject of Part III of the present book. In recent years the bosonic Hubbard model is regarded as a semi-realistic model of bosonic atoms confined in a region with periodic potential. Such a setup can be realized in modern ultra cold atom experiments with optical lattices. See, e.g., [3].

Definition of the model We define the system of bosons on the d-dimensional hypercubic lattice $(\Lambda_L, \mathscr{B}_L)$. The reader new to the description of many-particle systems

[1]This statement is correct only in the traditional theoretical treatment of superconductivity where dynamical electromagnetic field is not included. When dynamical electromagnetic field is included, superconductivity can be understood as a phase with topological order. See [23, 24]

H. Tasaki, *Physics and Mathematics of Quantum Many-Body Systems*,
Graduate Texts in Physics, https://doi.org/10.1007/978-3-030-41265-4_5

in terms of the Fock space representation (also known as the "second quantization" formalism) is suggested to first study Sects. 9.2 and 9.3, where we carefully introduce the description of many-electron systems. We also summarize modifications necessary to treat bosonic systems in Sect. 9.4.

For each site $x \in \Lambda_L$ we denote by $\hat{a}_x$ and $\hat{a}_x^\dagger$ the annihilation and the creation operators, respectively, of a bosonic particle at site x. They satisfy the standard commutation relations

$$[\hat{a}_x, \hat{a}_y] = [\hat{a}_x^\dagger, \hat{a}_y^\dagger] = 0, \quad [\hat{a}_x, \hat{a}_y^\dagger] = \delta_{x,y}, \tag{5.1.1}$$

for any $x, y \in \Lambda_L$. The number operator at site x is defined by $\hat{n}_x = \hat{a}_x^\dagger \hat{a}_x$, and the total number operator by $\hat{N} = \sum_{x \in \Lambda_L} \hat{n}_x$. We denote by $|\Phi_{\text{vac}}\rangle$ the unique state in which there are no particles in the system. It satisfies $\hat{a}_x |\Phi_{\text{vac}}\rangle = 0$ for any $x \in \Lambda_L$.

Fix the total boson number N. When we vary the system size, we fix the density $\rho = N/L^d$, and make both L and N large. Consider the state

$$\hat{a}_{x_1}^\dagger \hat{a}_{x_2}^\dagger \cdots \hat{a}_{x_N}^\dagger |\Phi_{\text{vac}}\rangle, \tag{5.1.2}$$

in which sites $x_1, x_2, \ldots, x_N \in \Lambda_L$ are occupied by bosons. The Hilbert space of N bosons, which we denote as $\tilde{\mathscr{H}}_N$, is spanned by all the states of the form (5.1.2). Here we identify apparently different states that can be obtained from each other by using the commutation relations (5.1.1). For example, $\hat{a}_x^\dagger \hat{a}_y^\dagger |\Phi_{\text{vac}}\rangle$ and $\hat{a}_y^\dagger \hat{a}_x^\dagger |\Phi_{\text{vac}}\rangle$ are the same state for any $x, y \in \Lambda_L$. The standard Hamiltonian of bosons on the lattice is

$$\hat{H}_u = -\sum_{\{x,y\} \in \mathscr{B}_L} (\hat{a}_x^\dagger \hat{a}_y + \hat{a}_y^\dagger \hat{a}_x) + u \sum_{x \in \Lambda_L} \hat{n}_x(\hat{n}_x - 1), \tag{5.1.3}$$

where the first term represents hopping of bosons between neighboring sites, and the second term the on-site repulsive interaction whose strength is $u \geq 0$. This is an idealized (and somewhat standard) model of bosonic atoms trapped in an optical lattice.[2]

Unfortunately the general theory of Koma and Tasaki described in Chap. 4 does not apply to this model.[3] To make the theory applicable we shall assume that $N < |\Lambda_L| = L^d$, and take the limit in which the on-site repulsion becomes infinitely strong, i.e., $u \uparrow \infty$. Then the model reduces to that of hard-core bosons with the hopping Hamiltonian $\hat{H}_0$. More precisely let $\mathscr{H}_N$ be the subspace spanned by the states of the form (5.1.2) with extra condition that $x_i \neq x_j$ whenever $i \neq j$. This is the Hilbert space for hard core bosons. Let $\hat{P}_{\text{hc}}$ be the projection operator onto $\mathscr{H}_N$. Then the $u \uparrow \infty$ limit of the Hamiltonian (5.1.3) is equivalent to the model with the

[2]In actual optical lattices there are always global single-body trapping potential $-\sum_{x \in \Lambda} v_x \hat{n}_x$, which breaks the translation invariance.

[3]The only obstacle is the unboundedness of the boson operators. If one introduces an artificial "cutoff" and restricts number of particles at each site to be less than n_0 (where n_0 is an arbitrary constant independent of L) then the general theorems about low-lying states apply.

effective Hamiltonian

$$\hat{H} = \hat{P}_{\rm hc}\hat{H}_0 = \hat{P}_{\rm hc}\Big\{-\sum_{\{x,y\}\in\mathscr{B}_L}(\hat{a}_x^\dagger\hat{a}_y + \hat{a}_y^\dagger\hat{a}_x)\Big\}, \tag{5.1.4}$$

in the Hilbert space $\mathscr{H}_N$. This fact is usually regarded to be trivial in physics literature (and is indeed trivial), but we prove it as Theorem A.12 (p. 470) for completeness.

Equivalence to the XY model It is well-known and easy to understand that the above model of hard core bosons is equivalent to a $S = 1/2$ quantum spin system called the XY model [16]. The Hamiltonian of the XY model is

$$\hat{H}_{\rm XY} = \sum_{\{x,y\}\in\mathscr{B}_L}\{\hat{S}_x^{(1)}\hat{S}_y^{(1)} + \hat{S}_x^{(2)}\hat{S}_y^{(2)}\} = \frac{1}{2}\sum_{\{x,y\}\in\mathscr{B}_L}(\hat{S}_x^+\hat{S}_y^- + \hat{S}_x^-\hat{S}_y^+), \tag{5.1.5}$$

which is nothing but (2.5.14) with $\lambda = D = 0$, or the XXZ model (4.2.29) with $\lambda = 0$. The equivalence is seen by identifying a site with down spin as empty and a site with up spin as being occupied by a particle. More precisely we make the identification

$$\bigotimes_{x\in\Lambda_L}|\psi_x^\downarrow\rangle \longleftrightarrow |\Phi_{\rm vac}\rangle, \quad (-1)^x\hat{S}_x^+ \longleftrightarrow \hat{a}_x^\dagger, \quad (-1)^x\hat{S}_x^- \longleftrightarrow \hat{a}_x, \tag{5.1.6}$$

where the factor $(-1)^x$ is introduced to change the overall sign of the Hamiltonian. Then we find the exact correspondences

$$2\hat{H}_{\rm XY} \longleftrightarrow \hat{H}, \quad \hat{S}_x^{(3)} + \frac{1}{2} \longleftrightarrow \hat{n}_x, \quad \hat{S}_{\rm tot}^{(3)} + \frac{L^d}{2} \longleftrightarrow \hat{N}. \tag{5.1.7}$$

Although there are complete one-to-one correspondences between the operators, there is a crucial difference in the Hilbert spaces. In spin systems, it is realistic to consider the large Hilbert space which includes all the possible eigenvalues of $\hat{S}_{\rm tot}^{(3)}$. But in systems of particles such as atoms, it is physical to consider the Hilbert space where the total particle number N is fixed. This is because the number of such particles must conserve exactly.[4]

[4]One might say that $\hat{S}_{\rm tot}^{(3)}$ is a conserved quantity in the XY model or the antiferromagnetic Heisenberg model, and there is no essential difference between the system of bosons. One must recall, however, that in actual magnetic systems the conservation of the total spin angular momentum is not an exact law but only an idealization; there always exists LS coupling which couples the spin angular momentum to the lattice degrees of freedom. Therefore it is legitimate in spin systems to look for ground states in the whole Hilbert space with various eigenvalues of $\hat{S}_{\rm tot}^{(3)}$. On the contrary the particle number conservation in systems of, say, atoms is an exact law of nature, which goes back to the baryon number conservation law.

5.2 Off-Diagonal Long-Range Order

In systems of non-interacting bosons, Bose–Einstein condensation (BEC) is precisely characterized by the condensation of macroscopic number of particles into the single-particle ground state. In interacting systems, where the single-particle picture generally breaks down, one cannot use such a criterion to detect BEC. It has been established that the essence of BEC is captured by long-range order of the form

$$\langle\Phi_{\rm GS}|\hat{a}_x^\dagger\hat{a}_y\,|\Phi_{\rm GS}\rangle \to (\text{constant}) > 0 \quad \text{for large enough } |x-y|, \tag{5.2.1}$$

where $|\Phi_{\rm GS}\rangle \in \mathscr{H}_N$ is the ground state of the system of N hard core bosons. Note that the left-hand side is the expectation value of an operator which annihilates a particle at y and then creates one at x. This is quite a strange expectation value, which has no counterpart in classical physics. But the condition (5.2.1) implies that there is condensation into the state (with the wave function $\psi_x = L^{-d/2}$) which spreads uniformly over the entire lattice. The above form of LRO is called the off-diagonal long-range order (ODLRO). See Problem 5.2.a below.

If we introduce the order operators for many-boson systems by

$$\hat{\mathscr{O}}_L^+ := \sum_{x\in\Lambda_L}\hat{a}_x^\dagger, \quad \hat{\mathscr{O}}_L^- := \sum_{x\in\Lambda_L}\hat{a}_x, \tag{5.2.2}$$

the ODLRO (5.2.1) is rewritten as

$$\frac{1}{L^{2d}}\langle\Phi_{\rm GS}|\hat{\mathscr{O}}_L^+\hat{\mathscr{O}}_L^-\,|\Phi_{\rm GS}\rangle \simeq \frac{1}{L^{2d}}\langle\Phi_{\rm GS}|\hat{\mathscr{O}}_L^-\hat{\mathscr{O}}_L^+\,|\Phi_{\rm GS}\rangle \simeq (\text{constant}) > 0. \tag{5.2.3}$$

By defining the self-adjoint order operators as

$$\hat{\mathscr{O}}_L^{(1)} := \frac{\hat{\mathscr{O}}_L^+ + \hat{\mathscr{O}}_L^-}{2}, \quad \hat{\mathscr{O}}_L^{(2)} := \frac{\hat{\mathscr{O}}_L^+ - \hat{\mathscr{O}}_L^-}{2i}, \tag{5.2.4}$$

the ODLRO (5.2.1), (5.2.3) is rewritten into the familiar form

$$\langle\Phi_{\rm GS}|\Bigl(\frac{\hat{\mathscr{O}}_L^{(\alpha)}}{L^d}\Bigr)^2\,|\Phi_{\rm GS}\rangle \ge q_0 \quad \text{for sufficiently large } L, \tag{5.2.5}$$

for $\alpha = 1, 2$, with a constant $q_0 > 0$ independent of L. We here noted that $\langle\Phi_{\rm GS}|(\hat{\mathscr{O}}_L^+)^2\,|\Phi_{\rm GS}\rangle = \langle\Phi_{\rm GS}|(\hat{\mathscr{O}}_L^-)^2\,|\Phi_{\rm GS}\rangle = 0$ because there are exactly N bosons in the ground state.

Problem 5.2.a Consider the free boson model whose Hamiltonian is (5.1.3) with $u = 0$. Here we take the unrestricted Hilbert space $\tilde{\mathscr{H}}_N$. Show that the ground state exhibits ODLRO as in (5.2.1). [solution → p. 503]

The expression (5.2.5) of ODLRO suggests that BEC is related to spontaneous breakdown of a certain symmetry. By examining the definitions (5.2.2), (5.2.4) of the order operators, one finds that the relevant symmetry is the U(1) phase symmetry inherent in quantum mechanics. In elementary quantum mechanics, we learn that nothing is changed if we replace a wave function $\varphi(x, y, z)$ by $e^{-i\theta}\varphi(x, y, z)$ with arbitrary $\theta \in \mathbb{R}$. This is the U(1) phase symmetry.

The multi-particle version of the phase transformation is described by the unitary operator $\hat{U}_\theta = e^{-i\theta\hat{N}}$ with $\theta \in \mathbb{R}$. The definition of $\hat{\mathscr{O}}_L^{\pm}$ readily implies that

$$\hat{U}_\theta^\dagger \hat{\mathscr{O}}_L^+ \hat{U}_\theta = e^{i\theta} \hat{\mathscr{O}}_L^+, \quad \hat{U}_\theta^\dagger \hat{\mathscr{O}}_L^- \hat{U}_\theta = e^{-i\theta} \hat{\mathscr{O}}_L^-, \tag{5.2.6}$$

which shows that $\hat{\mathscr{O}}_L^{(1)}$ and $\hat{\mathscr{O}}_L^{(2)}$ are transformed as

$$\hat{U}_\theta^\dagger \hat{\mathscr{O}}_L^{(1)} \hat{U}_\theta = \cos\theta\, \hat{\mathscr{O}}_L^{(1)} - \sin\theta\, \hat{\mathscr{O}}_L^{(2)}, \quad \hat{U}_\theta^\dagger \hat{\mathscr{O}}_L^{(2)} \hat{U}_\theta = \cos\theta\, \hat{\mathscr{O}}_L^{(2)} + \sin\theta\, \hat{\mathscr{O}}_L^{(1)}. \tag{5.2.7}$$

This means that the doublet $(\hat{\mathscr{O}}_L^{(1)}, \hat{\mathscr{O}}_L^{(2)})$ transforms precisely as a vector. Of course this is not at all a surprise if we note from the identification (5.1.6) that the order operators defined in (5.2.2) and (5.2.4) precisely correspond to the order operators (4.1.4) and (4.2.2) for spin systems.

Let $|\Phi_{\rm GS}\rangle \in \mathscr{H}_N$ be the ground state of the Hamiltonian $\hat{H}$ of (5.1.4). For $d \geq 2$ the ground state is expected to exhibit Bose–Einstein condensation for a wide range of density ρ.[5] Unfortunately rigorous result is available only through those for the XY model. From Theorem 2.4 (p. 43), we know that the unique ground state $|\Phi_{\rm XY,GS}\rangle$ of the Hamiltonian $\hat{H}_{\rm XY}$ defined in (5.1.5) satisfies $\hat{S}_{\rm tot}^{(3)}\,|\Phi_{\rm XY,GS}\rangle = 0$. From the correspondence (5.1.7), we see that results for the ground state of the XY model can be interpreted as that for the hard core boson model at the half-filling, namely, $\rho = 1/2$.

Theorem 5.1 *Let $d \geq 2$ and $\rho = N/L^d = 1/2$. Then there exists a constant $q_0 > 0$ which depends only on d, and* (5.2.5) *is valid.*

The theorem was proved by Kennedy, Lieb, and Shastry [9, 10], and by Kubo and Kishi [13] by extending the reflection positivity method due to Dyson, Lieb, and Simon [5], which was discussed in Sect. 4.1.2. See also [1].

Note, on the other hand, that the ground state can never exhibit corresponding SSB. To see this one simply notes that $|\Phi_{\rm GS}\rangle$ has precisely N particles, and hence

$$\langle\Phi_{\rm GS}|\frac{\hat{\mathscr{O}}_L^{(\alpha)}}{L^d}|\Phi_{\rm GS}\rangle = 0. \tag{5.2.8}$$

Thus the ground state of a many-boson system exhibiting BEC provides us with another example of LRO without SSB.

---

[5]Since this is proved to be the case for $\rho = 1/2$, we expect the same for ρ around 1/2. For $0 < \rho \ll 1$, we also expect BEC since the system should be close to that of free bosons. The same is true for ρ such that $0 < 1 - \rho \ll 1$ since the model with ρ can be exactly mapped to that with $1 - \rho$ by the particle-hole transformation. Then we are tempted to expect that BEC takes place for any ρ with $0 < \rho < 1$.

5.3 Symmetry Breaking "Ground States"

Let us extend the construction in Sect. 4.2 to the present problem of hard core bosons. Here we shall assume that the ground state of the hard core bosons with a fixed particle density $\rho = N/L^d$ exhibits ODLRO as in (5.2.5) with some constant $q_0 > 0$. This is known rigorously (only) for $\rho = 1/2$, but is likely to be the case for any $\rho \in (0, 1)$.[6] The constant q_0 is the (off-diagonal) long-range order parameter.

In order to make our discussion parallel to that in spin systems, we introduce the extended Hilbert space by[7]

$$\mathscr{H} := \bigoplus_{N'=0}^{L^d} \mathscr{H}_{N'}. \tag{5.3.1}$$

The space $\mathscr{H}$ consists of all possible linear combinations of the state (5.1.2) (with the condition that $x_i \neq x_j$ whenever $i \neq j$) with all possible particle number $N = 0, 1, \ldots, L^d$. It should be noted that to consider superpositions of states with different particle numbers is different from using the grand canonical formalism, where one considers statistical mixture of states with different particle numbers. (But see the discussion in Sect. 5.4, especially (5.4.2).)

To get the desired density ρ, we take the Hamiltonian

$$\hat{H}_\mu := \hat{H} - \mu\hat{N} = \hat{P}_{\mathrm{hc}}\Big\{-\sum_{\{x,y\}\in\mathscr{B}_L}(\hat{a}_x^\dagger\hat{a}_y + \hat{a}_y^\dagger\hat{a}_x) - \mu\hat{N}\Big\}, \tag{5.3.2}$$

on the extended Hilbert space $\mathscr{H}$. We then adjust the chemical potential μ so that the ground state $|\Phi_{\mathrm{GS}}\rangle$ of $\hat{H}_\mu$ is in the original Hilbert space $\mathscr{H}_N$ (and hence has the density $\rho = N/L^d$). When the target density is $\rho = 1/2$, we know from Theorem 2.4 (p. 43) that the right choice is $\mu = 0$. Of course the ground state $|\Phi_{\mathrm{GS}}\rangle$ itself is identical to the ground state that we obtain in the formulation with fixed particle number in the previous sections.

We are now ready to repeat the construction in Sect. 4.2. Exactly as in (4.2.3), we define trial states

$$|\Gamma_M\rangle = \frac{\hat{P}_{\mathrm{hc}}(\hat{\mathscr{O}}_L^+)^M\,|\Phi_{\mathrm{GS}}\rangle}{\|\hat{P}_{\mathrm{hc}}(\hat{\mathscr{O}}_L^+)^M\,|\Phi_{\mathrm{GS}}\rangle\|}, \quad |\Gamma_{-M}\rangle = \frac{\hat{P}_{\mathrm{hc}}(\hat{\mathscr{O}}_L^-)^M\,|\Phi_{\mathrm{GS}}\rangle}{\|\hat{P}_{\mathrm{hc}}(\hat{\mathscr{O}}_L^-)^M\,|\Phi_{\mathrm{GS}}\rangle\|}, \tag{5.3.3}$$

for $M = 1, 2, \ldots$. Then the following extension of Theorem 4.6 (p. 94), which shows that $|\Gamma_M\rangle$ are low-lying states of the Hamiltonian $\hat{H}_\mu$, was proved in [21]. See the part "Low-lying states" (p. 109) of Sect. 4.2.2 for the basic idea of the proof.

Theorem 5.2 *Suppose that the ground state* $|\Phi_{\mathrm{GS}}\rangle$ *of* (5.3.2) *with some* μ *exhibits ODLRO as in* (5.2.5) *with some constant* $q_0 > 0$. *Then there are constants* C_1' *and*

---

[6] See the previous footnote 4.

[7] This is a subspace of the Fock space $\bigoplus_{N'=0}^{\infty} \tilde{\mathscr{H}}_{N'}$.

C_2' which depend only on d, ρ, and q_0. For any L and M such that $|M| \le C_1' L^{d/2}$, the state $|\Gamma_M\rangle$ is well-defined (i.e., nonvanishing), and satisfies[8]

$$\langle \Gamma_M | \hat{H}_\mu | \Gamma_M \rangle \le \langle \Phi_{\rm GS} | \hat{H}_\mu | \Phi_{\rm GS} \rangle + C_2' \frac{|M|^3}{L^d}. \tag{5.3.4}$$

We then proceed, as in (4.2.10), to construct low-lying states with explicit symmetry breaking. Here we follow the older construction of Koma and Tasaki [11, 12], and simply sum up the low-lying states defined above. It is likely that this construction is most suited for models with U(1) symmetry. For $\theta \in \mathbb{R}$ and an integer-valued increasing function $M_{\max}(L) > 0$ such that $M_{\max}(L) \le C_1' L^{d/2}$, let

$$|\Xi_\theta\rangle = \frac{1}{\sqrt{2M_{\max}(L)+1}} \sum_{M=-M_{\max}(L)}^{M_{\max}(L)} e^{-iM\theta} \, |\Gamma_M\rangle, \tag{5.3.5}$$

where $|\Gamma_0\rangle = |\Phi_{\rm GS}\rangle$. We see that $|\Xi_\theta\rangle$ is also a low-lying state. Then we have the following counterpart of Theorem 4.9 (p. 98). The proof, which can be found in [21], is similar to that of Theorem 4.9.

Theorem 5.3 *If $M_{\max}(L)$ diverges to infinity not too rapidly as $L \uparrow \infty$, one has*[9]

$$\lim_{L\uparrow\infty} \langle \Xi_\theta | \frac{\hat{\mathcal{O}}_L^{\pm}}{L^d} | \Xi_\theta \rangle = m^* \, e^{\pm i\theta}, \tag{5.3.6}$$

$$\lim_{L\uparrow\infty} \langle \Xi_\theta | \frac{\hat{\mathcal{O}}_L^{(\alpha)}}{L^d} | \Xi_\theta \rangle = \begin{cases} m^* \cos\theta & (\alpha = 1) \\ m^* \sin\theta & (\alpha = 2), \end{cases} \tag{5.3.7}$$

$$\lim_{L\uparrow\infty} \langle \Xi_\theta | \Bigl(\frac{\hat{\mathcal{O}}_L^{(\alpha)}}{L^d}\Bigr)^2 | \Xi_\theta \rangle = \begin{cases} (m^* \cos\theta)^2 & (\alpha = 1) \\ (m^* \sin\theta)^2 & (\alpha = 2). \end{cases} \tag{5.3.8}$$

Here the symmetry breaking order parameter m^* is defined by the same formula (4.2.9) with $\alpha = 1, 2$. As in Theorem 4.11 (p. 101), we can prove the bound $m^* \ge \sqrt{2q_0}$. See (4.2.39). The factor $\sqrt{2}$, which reflects the U(1) symmetry, is understood exactly as in the discussion that follows Theorem 4.11.

Therefore the state $|\Xi_\theta\rangle$ is a low-lying state of the Hamiltonian $\hat{H}_\mu$ (defined in the extended Hilbert space $\mathscr{H}$) that exhibits ODLRO and fully breaks the U(1) phase symmetry. The symmetry breaking is manifest in the remarkable relations

$$\langle \Xi_\theta | \hat{a}_x^\dagger | \Xi_\theta \rangle \simeq m^* \, e^{i\theta}, \quad \langle \Xi_\theta | \hat{a}_x | \Xi_\theta \rangle \simeq m^* \, e^{-i\theta}, \tag{5.3.9}$$

which hold for large L because of (5.3.6). We also note that the $|\Phi_{\rm GS}\rangle$ is obtained from $|\Xi_\theta\rangle$ by

---

[8]The bound is weaker than that in Theorem 4.6 because M^2 is replaced by $|M|^3$. We expect that the same estimate as in Theorem 4.6 is possible, but cannot prove it for some technical reasons.

[9]To be rigorous the limits in these relations should be interpreted properly.

$$|\Phi_{\mathrm{GS}}\rangle = \sqrt{2M_{\max}(L)+1}\,\hat{P}_N\,|\Xi_\theta\rangle\,, \tag{5.3.10}$$

where $\hat{P}_N$ is the projection onto $\mathscr{H}_N$, or by

$$|\Phi_{\mathrm{GS}}\rangle = \frac{\sqrt{2M_{\max}(L)+1}}{2\pi}\int_0^{2\pi} d\theta\,|\Xi_\theta\rangle\,. \tag{5.3.11}$$

Note that (5.3.11) is the U(1) version of the conjectured relation (4.2.21) for the antiferromagnetic Heisenberg model or the observation (3.3.8) for the quantum Ising model.

The relation (5.3.9) shows that the U(1) phase is "pointing" in the specific direction θ in the low-lying state $|\Xi_\theta\rangle$. This situation is mathematically analogous to ordered states in antiferromagnetic spin systems in which the staggered magnetization is pointing in a specific direction. But physical interpretations of the two cases seem very different. We can hardly imagine a state in which the broken U(1) phase is pointing in a definite direction. This point is further discussed in the next section.

5.4 Physical Ground States of Bose–Einstein Condensates

Our discussion so far has been completely parallel to that in Sect. 4.2. Our final conclusion, however, is different. In the present problem of many-boson systems exhibiting BEC, we argue that the U(1) symmetric ground state $|\Phi_{\mathrm{GS}}\rangle$ provides a better description of physical reality than the symmetry breaking "ground states" $|\Xi_\theta\rangle$. A more fundamental remark is that the fixed phase θ associated with the state $|\Xi_\theta\rangle$ is an entirely fictitious object, which has no connection to observable physics. See, e.g., Sect. III.D.1 of [14] for a related discussion.

Let us consider an idealized situation where exactly N bosons are trapped and perfectly confined in an ideal container, and then cooled down to the ground state. It is then obvious that the state of the system must be $|\Phi_{\mathrm{GS}}\rangle$, the ground state with a fixed boson number. If the U(1) symmetry breaking state $|\Xi_\theta\rangle$ emerged as a result of cooling, it would imply the change in the total number of bosons in the system, and hence in the whole universe. That is of course impossible. This is a straightforward and inevitable consequence of the particle number conservation law, which is a definite law for atoms.[10] We also remark that the states $|\Phi_{\mathrm{GS}}\rangle$ and $|\Xi_\theta\rangle$ can be distinguished experimentally if many identical copies of the state are available. If one measure the total number of bosons in the system (by releasing them from the trap), one always gets the constant N for $|\Phi_{\mathrm{GS}}\rangle$, but a fluctuating value for $|\Xi_\theta\rangle$.

One might still wish to describe the system in terms of the symmetry breaking state $|\Xi_\theta\rangle$; it is indeed a standard treatment in mean field theories. This is possible but one should keep the following remarks in mind. Above all one should recall and respect the fact that any physical observable of the system must preserve the

[10] See footnote 4 in p. 137.

particle number. This is a definite consequence of the conservation law. Therefore the expectation value of any observable $\hat{A}$ with respect to $|\Xi_\theta\rangle$ is written as

$$\langle\Xi_\theta|\hat{A}\,|\Xi_\theta\rangle = \frac{1}{2M_{\max}(L)+1}\sum_{M=-M_{\max}(L)}^{M_{\max}(L)} \langle\Gamma_M|\hat{A}\,|\Gamma_M\rangle\,, \tag{5.4.1}$$

where we noted that $\langle\Gamma_M|\hat{A}\,|\Gamma_{M'}\rangle = 0$ whenever $M \neq M'$. It is essential that the right-hand side of (5.4.1) is completely independent of the U(1) phase θ. We conclude that θ is a fictitious quantity which is not related to any physically observable effects. This is in a stark contrast with the corresponding state $|\Xi_{\boldsymbol{n}}\rangle$ for antiferromagnets, defined in (4.2.17), in which the direction $\boldsymbol{n}$ can be measured experimentally.

It is also worth noting that the expectation value (5.4.1) is exactly the same as that obtained from the density matrix

$$\hat{\rho} = \frac{1}{2M_{\max}(L)+1}\sum_{M=-M_{\max}(L)}^{M_{\max}(L)} |\Gamma_M\rangle\,\langle\Gamma_M|. \tag{5.4.2}$$

In conclusion even if there is a reason to use the state $|\Xi_\theta\rangle$, it does not mean that the superposition of states with different particle numbers is relevant to physics. All the physical information is contained in the density matrix (5.4.2), which has nothing to do with the phase θ. The U(1) phase becomes physically meaningful when there are more than one Bose–Einstein condensates, and the notion of relative phase is well defined. See Sect. 5.5. As we see in Problem 5.4.a below, it is also possible to "realize" in certain sense the state $|\Xi_\theta\rangle$ by explicitly considering a system external to the Bose–Einstein condensate. See [19, 20] for background.

Note that all the above remarks do not apply to systems of photons, where the particle number is not conserved. In fact coherent states, which are quite realistic object, is very similar to the states $|\Xi_\theta\rangle$.

Finally we conjecture that the expectation value $\langle\Gamma_M|\hat{A}\,|\Gamma_M\rangle$ which appears in (5.4.1) is almost independent of M when $\hat{A}$ is a local operator (or a sum of local operators) and L is large enough.[11] This in particular means that

$$\langle\Phi_{\rm GS}|\hat{A}\,|\Phi_{\rm GS}\rangle \simeq \langle\Xi_\theta|\hat{A}\,|\Xi_\theta\rangle\,. \tag{5.4.3}$$

This (conjectured) near equality gives a further justification of the use of the symmetry breaking "ground states" $|\Xi_\theta\rangle$. Although $|\Xi_\theta\rangle$ may give a wrong prediction for the total particle number, it gives correct expectation value for reasonable observables.

Problem 5.4.a Consider a system which consists of two parts, the first system exhibiting the Bose–Einstein condensation, and the second (arbitrary) system external to it. The total number of particles in the whole system is fixed to $N_{\rm tot}$, which

[11]For a fixed local $\hat{A}$ and fixed M, the clustering property in the ground state (which is expected, but cannot be proved) implies $\langle\Phi_{\rm GS}|(\hat{\mathcal{O}}_L^-)^M\hat{A}(\hat{\mathcal{O}}_L^+)^M\,|\Phi_{\rm GS}\rangle \simeq \langle\Phi_{\rm GS}|\hat{A}\,|\Phi_{\rm GS}\rangle\,\langle\Phi_{\rm GS}|(\hat{\mathcal{O}}_L^-)^M(\hat{\mathcal{O}}_L^+)^M\,|\Phi_{\rm GS}\rangle$ for sufficiently large L, which is basically the desired independence.

we assume to be sufficiently large. Show that there is a pure state $|\Phi_{\rm tot}\rangle$ of the whole system with the property that

$$\langle\Phi_{\rm tot}|(\hat{A}\otimes\hat{1})\,|\Phi_{\rm tot}\rangle=\langle\Xi_\theta|\hat{A}\,|\Xi_\theta\rangle\,, \tag{5.4.4}$$

where $\hat{A}$ is an arbitrary operator on the first system which preserves the particle number. In this sense the pure state $|\Phi_{\rm tot}\rangle$, which may be in principle realized experimentally, is identical to $|\Xi_\theta\rangle$ when restricted onto the first system. [solution → p. 503]

Stability We argued in Sect. 4.3.2 that the symmetric ground state $|\Phi_{\rm GS}\rangle$ or $\omega_0(\cdot)$ of an antiferromagnetic system should be unstable because the order operator exhibits anomalously large fluctuation, and the state can be regarded as a Schrödinger's cat. The obvious question is why the same argument does not apply to the corresponding state in the system of bosons. To give a short answer there is an essential difference in physically allowed operators.

In boson systems too, the order operators $\hat{\mathcal{O}}_L^{(\alpha)}$ exhibit anomalous fluctuation in $|\Phi_{\rm GS}\rangle$. The ground states should thus collapse easily if exposed to disturbance which couples to the order operator. However a perturbation which couples to the order operator must change the number of particles in the system without changing the number of particles in the rest of the universe. Such an operator is strictly inhibited by the particle number conservation law, or, equivalently, the U(1) symmetry of quantum mechanics. In this way the symmetric ground state $|\Phi_{\rm GS}\rangle$ is protected by the U(1) symmetry from collapse.

Mean field theory Exactly as in spin systems, a mean field theory for many-boson systems directly yields symmetry breaking “ground states” similar to our $|\Xi_\theta\rangle$. The reader familiar with the Bogoliubov theory of interacting bosons may recall that the starting point of the theory is the assumption that the operators $\sum_{x\in\Lambda}\hat{a}_x^\dagger/L^d$ and $\sum_{x\in\Lambda}\hat{a}_x/L^d$ may be replaced by a nonvanishing constant (c-number). See, e.g., Sect. 18 of [6] and Appendices A and D of [15]. Note the close similarity of the relations (5.3.9) to this replacement procedure. The success of the Bogoliubov theory for interacting bosons and the BCS theory (which has a very close spirit) for superconductivity suggests that the symmetry breaking states like $|\Xi_\theta\rangle$ are “theoretically natural” ground states which capture the essence of condensation phenomena.

That mean field theories produce symmetry breaking “ground states” is by no means surprising, especially if we compare the situation with mathematically analogous (or sometimes equivalent) problem of quantum antiferromagnets. In both the systems, we find that the order operators exhibit anomalously large fluctuation in the naive unique ground state $|\Phi_{\rm GS}\rangle$. There is a “theoretically natural” way to cure this pathology and get states with normal fluctuation, namely, to sum up low-lying states from the “tower of states”. See Sect. 4.2. In this process we are (theoretically) forced to extend the Hilbert space. In spin systems, it was simply going out from the sector with $\hat{S}_{\rm tot}^{(3)}=0$, where $|\Phi_{\rm GS}\rangle$ lives, to the whole Hilbert space with various eigenvalues of $\hat{S}_{\rm tot}^{(3)}$. This is quite physical. In boson systems, on the other hand, we had to extend the Hilbert space so as to include superpositions of states with various particle numbers.[12]

Mean field theories are useful in general, and produce qualitatively reliable predictions for systems of weakly interacting bosons [4, 14, 15, 17]. As we have discussed above, however, symmetry breaking states like $|\Xi_\theta\rangle$, which mean-field theories produce, should be handled with some care: The U(1) phase factor θ is not a physically meaningful quantity. One should only compute U(1) invariant observables. The fluctuation of particle number, inherent in symmetry breaking states like $|\Xi_\theta\rangle$, should not be taken literally. We believe that most experts, including of course experimentalists,[13] are well aware of these remarks.

Unfortunately it seems that there still remain some conceptual confusion about the role of symmetry breaking "ground states" in BEC and superconductivity. We suspect that such confusion comes from too literal interpretations of mean-field theories.

For example there often is a question if it is possible to have particle condensation in a system with a fixed number of particles, in which the Bogoliubov type state or the BCS state cannot be defined. The answer is, of course, "yes". As is rigorously demonstrated in Theorem 5.1 (p. 139), we know that a ground state with a definite particle number may exhibit particle condensation, namely, ODLRO. We also recall that a ground state (with ODLO) with a definite particle number is obtained from a Bogoliubov type state by a projection as in (5.3.10) or by a superposition as in (5.3.11).

Another related confusion is that the grand canonical formalism provides a justification for the use of Bogoliubov type states. As we have explicitly demonstrated in (5.4.2), it is true that certain grand-canonical-like density matrices reproduce the expectation values of a Bogoliubov type state $|\Xi_\theta\rangle$. But one should always keep in mind that this is only for operators that preserve the particle number. In particular the information about the U(1) phase θ in the original state is totally lost in the density matrix (5.4.2). We stress once again that a coherent superposition as in (5.3.5) has nothing to do with the grand canonical ensemble.

We believe that the best way to resolve such (and other related) confusion and get the right physical picture is to first study and completely understand the conceptually much easier cases of quantum spin systems, which we discussed carefully in Chaps. 3 and 4. Then, by relying on the mathematical analogy (or equivalence), one may fully appreciate what is going on in the theories of particle condensation phenomena.

Heavy nucleus as a quantum many-body system Interestingly, one encounters a situation very similar to that in BEC in nuclear physics. Nucleus, which are the main objects of study in nuclear physics, are quantum systems of considerable numbers of interacting particles, namely, protons and neutrons (or quarks). It is know that mean field theories are useful in studying heavy nuclei, and produce reliable results

---

[12] This is equivalent to ignoring the distinction between physical and unphysical operators (which are U(1) invariant and non-invariant operators, respectively), and treating all operators generated from $\hat{a}_x^\dagger$ and $\hat{a}_x$.

[13] No experimentalist is interested in measuring $\hat{a}_x^\dagger$.

on various physical quantities. But, exactly as in the case of BEC, mean field theories produce quantum states with explicit symmetry breaking, e.g., states without rotational symmetry or states which are superpositions of different particle numbers. These states, especially the latter, can never be realized in observable steady states of actual nuclei, which are, after all, large but finite quantum systems.

It is worth noting that the "tower of states" also appears very naturally in nuclear physics. Suppose that "physical ground states" (predicted by the mean-filed theory) of a heavy nucleus break the rotational symmetry. Such a symmetry breaking state should be written as a superposition of various states with angular momenta J, where the true (unique) ground state has $J = 0$. It is expected that the energy difference between the true ground state and the excited state with angular momentum J is roughly given by $E_J - E_0 \simeq J(J+1)/(2I)$ where I is the moment of inertia of the nucleus. Since I is large for a heavy nucleus, one clearly see that the excited states with various J form a tower of low-lying energy eigenstates.

For more about nuclear physics, see the clear discussion in Chap. 11 of [18].

5.5 "SSB" in Coupled Bose–Einstein Condensates

We have seen that the broken U(1) phase θ is not a physically meaningful quantity in a single Bose–Einstein condensate. Here we consider a system of two Bose–Einstein condensates coupled weakly by tunneling, and see that a kind of spontaneous symmetry breaking in the relative U(1) phase takes place. This corresponds, e.g., to the experimental setup where bosons are trapped in a double-well potential [2, 7]. The clear interference pattern observed experimentally [2] is a manifestation of a fixed relative U(1) phase.[14]

We consider two exact copies of the d-dimensional hyper cubic lattice Λ_L, and call them Λ_a and Λ_b. Lattice sites are denoted as $(x, \nu) \in \Lambda_\nu$ where $\nu = \mathrm{a}, \mathrm{b}$ and $x \in \Lambda_L$. On each lattice we define the same system of hardcore bosons as before. We further assume that there is a tunneling Hamiltonian which weakly couples the two subsystems on Λ_a and Λ_b. The total Hamiltonian is thus

$$\hat{H}^{\mathrm{tot}}_{\varepsilon} = \hat{H}_\mathrm{a} + \hat{H}_\mathrm{b} + \varepsilon \hat{H}_{\mathrm{tunnel}}. \tag{5.5.1}$$

Here we set, for $\nu = \mathrm{a}, \mathrm{b}$,

$$\hat{H}_\nu = -\hat{P}_{\mathrm{hc}} \sum_{\{x,y\}\in\mathscr{B}_L} \hat{a}^\dagger_{(x,\nu)} \hat{a}_{(y,\nu)}, \tag{5.5.2}$$

[14]The interference pattern is observed after switching off the trapping potential and letting the particles evolve almost freely. We should note that there is an essentially different class of interference phenomena between two Bose–Einstein condensates. It is known that two condensates which have no fixed relative phase (and hence well approximated by $|\Phi_{\mathrm{GS}}\rangle \otimes |\Phi_{\mathrm{GS}}\rangle$) also exhibit interference. See, e.g., [8].

which are the exact copies of (5.1.4), and

$$\hat{H}_{\text{tunnel}} = -\hat{P}_{\text{hc}} \sum_{x\in\Lambda_L} \left(e^{i\varphi}\,\hat{a}^{\dagger}_{(x,\text{a})}\hat{a}_{(x,\text{b})} + e^{-i\varphi}\,\hat{a}_{(x,\text{a})}\hat{a}^{\dagger}_{(x,\text{b})}\right). \tag{5.5.3}$$

Here the phase factor $\varphi \in \mathbb{R}$ is introduced to make clear the physical picture; it is most natural to set $\varphi = 0$.[15]

We treat this problem in a physically realistic Hilbert space where the total number of particles in the coupled system is exactly $2N$. If we denote the copy of the N particle Hilbert space $\mathscr{H}_{N'}$ as $\mathscr{H}^{\nu}_{N'}$ for $\nu = \text{a, b}$, the whole Hilbert space is

$$\mathscr{H}^{\text{tot}}_{2N} = \bigoplus_{K=0}^{2N} \mathscr{H}^{\text{a}}_{K} \otimes \mathscr{H}^{\text{b}}_{2N-K}. \tag{5.5.4}$$

In other words, we assume that the two Bose–Einstein condensates can exchange particles in a coherent manner, while completely isolated from the outside world. This may be a reasonable idealization of realistic situations in cold atom experiments.

Let $\left|\Phi^{\text{tot}}_{\text{GS},\varepsilon}\right\rangle \in \mathscr{H}^{\text{tot}}_{2N}$ be the unique ground state of the total Hamiltonian (5.5.1) with $\varepsilon > 0$. The uniqueness is easily proved by following the proof of the Marshall-Lieb-Mattis theorem (Theorem 2.2 (p. 39)). Then by using the variational argument of Kaplan, Horsch, and von der Linden (see Theorem 3.2 (p. 70)) and the variational state (5.5.12) introduced below, the following theorem was proved in [22].

Theorem 5.4 *Assume the existence of ODLRO (in the single uncoupled system) as in (5.2.5). Then for any $x \in \mathbb{Z}^d$, we have*[16]

$$\lim_{\varepsilon\downarrow 0}\lim_{L\uparrow\infty} \langle \Phi^{\text{tot}}_{\text{GS},\varepsilon}|\hat{a}^{\dagger}_{(x,\text{a})}\hat{a}_{(x,\text{b})}|\Phi^{\text{tot}}_{\text{GS},\varepsilon}\rangle = \tilde{m}^2\, e^{-i\varphi}, \tag{5.5.5}$$

$$\lim_{\varepsilon\downarrow 0}\lim_{L\uparrow\infty} \langle \Phi^{\text{tot}}_{\text{GS},\varepsilon}|\hat{a}_{(x,\text{a})}\hat{a}^{\dagger}_{(x,\text{b})}|\Phi^{\text{tot}}_{\text{GS},\varepsilon}\rangle = \tilde{m}^2\, e^{i\varphi}, \tag{5.5.6}$$

with $\tilde{m} \geq m^* \geq \sqrt{2q_0}$.

We here expect that $\tilde{m} = m^*$, but there is no proof.

The relations (5.5.5) and (5.5.6) indicates that the two condensates are coupled in a coherent manner (or entangled) to have a definite relative U(1) phase.[17] To see this, define, for $\nu = \text{a, b}$ and $x \in \Lambda_L$, the local order operators $\hat{o}^{(1)}_{(x,\nu)}$ and $\hat{o}^{(2)}_{(x,\nu)}$ by

$$\hat{o}^{(1)}_{(x,\nu)} := \frac{\hat{a}^{\dagger}_{(x,\nu)} + \hat{a}_{(x,\nu)}}{2}, \quad \hat{o}^{(2)}_{(x,\nu)} := \frac{\hat{a}^{\dagger}_{(x,\nu)} - \hat{a}_{(x,\nu)}}{2i}. \tag{5.5.7}$$

---

[15] The general case reduces to that with $\varphi = 0$ by replacement $e^{i\varphi}\hat{a}_{(x,\text{b})} \to \hat{a}_{(x,\text{b})}$ for all $x \in \Lambda$.

[16] To be rigorous, the limits in (5.5.5), (5.5.6), (5.5.8), and (5.5.9), whose existence is not proved, should be interpreted properly.

[17] See Appendix A.1 about the notion of entanglement.

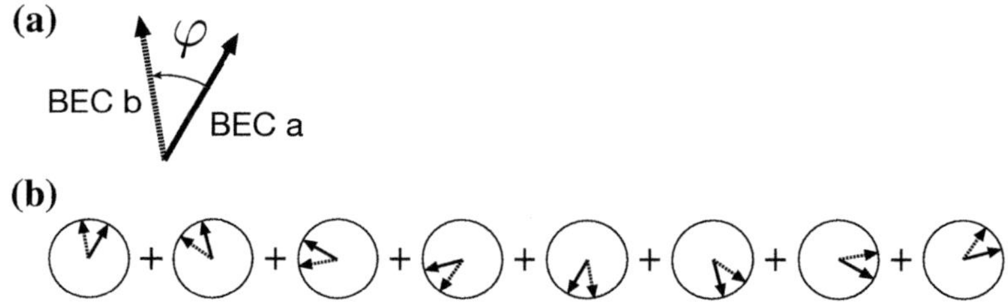

Fig. 5.1 **a** The state $|\Xi_\theta^{\mathrm{a}}\rangle\otimes|\Xi_{\theta+\varphi}^{\mathrm{b}}\rangle$, where each of the two Bose–Einstein condensates has a definite U(1) phase. The relative phase is denoted as φ. The state is not U(1) invariant, and hence does not belong to the physical Hilbert space $\mathscr{H}_{2N}^{\mathrm{tot}}$. **b** By superposing infinitely many states where the two condensates have fixed relative phase φ as in (5.5.11), we get a state in $\mathscr{H}_{2N}^{\mathrm{tot}}$ with a definite relative phase φ (© Hal Tasaki 2020. All Rights Reserved)

Exactly as in (5.2.4) and (5.2.7), the pair $(\hat{o}_{(x,\nu)}^{(1)},\hat{o}_{(x,\nu)}^{(2)})$ transforms as a vector under the operation of the unitary operator $\hat{U}_\theta=e^{-i\theta\hat{N}}$. We of course have $\langle\Phi_{\mathrm{GS},\varepsilon}^{\mathrm{tot}}|(\hat{o}_{(x,\nu)}^{(1)},\hat{o}_{(x,\nu)}^{(2)})|\Phi_{\mathrm{GS},\varepsilon}^{\mathrm{tot}}\rangle=(0,0)$ again by particle number conservation, but (5.5.5) and (5.5.6) imply that

$$\lim_{\varepsilon\downarrow 0}\lim_{L\uparrow\infty}\langle\Phi_{\mathrm{GS},\varepsilon}^{\mathrm{tot}}|\bigl\{\hat{o}_{(x,\mathrm{a})}^{(1)}\hat{o}_{(x,\mathrm{b})}^{(1)}+\hat{o}_{(x,\mathrm{a})}^{(2)}\hat{o}_{(x,\mathrm{b})}^{(2)}\bigr\}|\Phi_{\mathrm{GS},\varepsilon}^{\mathrm{tot}}\rangle=\tilde{m}^2\cos\varphi. \tag{5.5.8}$$

This suggests that the two order operators together behave like two vectors with magnitude $\tilde{m}$ which have a fixed relative angle φ. This is more directly seen by noting that

$$\lim_{\varepsilon\downarrow 0}\lim_{L\uparrow\infty}\langle\Phi_{\mathrm{GS},\varepsilon}^{\mathrm{tot}}|(\hat{o}_{(x,\mathrm{a})}^{(1)},\hat{o}_{(x,\mathrm{a})}^{(2)})\begin{pmatrix}\cos\varphi & \sin\varphi\\ -\sin\varphi & \cos\varphi\end{pmatrix}\begin{pmatrix}\hat{o}_{(x,\mathrm{b})}^{(1)}\\ \hat{o}_{(x,\mathrm{b})}^{(2)}\end{pmatrix}|\Phi_{\mathrm{GS},\varepsilon}^{\mathrm{tot}}\rangle=\tilde{m}^2, \tag{5.5.9}$$

which also follows from (5.5.5) and (5.5.6).

We conclude that, in the ground state obtained in the double limit $\lim_{\varepsilon\downarrow 0}\lim_{L\uparrow\infty}$, the relative U(1) phase between the two condensates has a definite value φ. Since this phase ordering was achieved by "infinitesimal symmetry breaking field" ε in the tunneling Hamiltonian, we can say that this is a kind of spontaneous symmetry breaking. As we noted in the beginning, such ordering of relative phase between two weakly coupled Bose–Einstein condensates can be experimentally observed by means of interference experiments [2].

Let us discuss the construction of the variational state which is used in the Proof of Theorem 5.4. See [22] for details. We need a state in the physical Hilbert space $\mathscr{H}_{2N}^{\mathrm{tot}}$ in which two subsystems have a definite relative phase. Let $|\Xi_\theta^{\mathrm{a}}\rangle$ and $|\Xi_\theta^{\mathrm{b}}\rangle$ be the exact copies of $|\Xi_\theta\rangle$ defined in (5.3.5), which is a "ground state" with explicit U(1) symmetry breaking. A natural candidate for a state with fixed relative phase is the tensor product $|\Xi_\theta^{\mathrm{a}}\rangle\otimes|\Xi_{\theta+\varphi}^{\mathrm{b}}\rangle$, but this state is again a superposition of states with different total particle numbers. We can follow (5.3.10) or (5.3.11) to construct a physical state as

$$|\Xi_{\varphi}^{\text{tot}}\rangle \propto \hat{P}_{2N}^{\text{tot}}\big(|\Xi_{\theta}^{\text{a}}\rangle \otimes |\Xi_{\theta+\varphi}^{\text{b}}\rangle\big), \tag{5.5.10}$$

where $\hat{P}_{2N}^{\text{tot}}$ is the projection onto $\mathscr{H}_{2N}^{\text{tot}}$, or by phase averaging as

$$|\Xi_{\varphi}^{\text{tot}}\rangle \propto \frac{1}{2\pi}\int_0^{2\pi} d\theta\, |\Xi_{\theta}^{\text{a}}\rangle \otimes |\Xi_{\theta+\varphi}^{\text{b}}\rangle. \tag{5.5.11}$$

See Fig. 5.1. The two constructions lead to exactly the same result, and we get

$$|\Xi_{\varphi}^{\text{tot}}\rangle = \frac{1}{\sqrt{2M_{\max}(L)+1}} \sum_{M=-M_{\max}(L)}^{M_{\max}(L)} e^{in\varphi}|\Gamma_M^{\text{a}}\rangle \otimes |\Gamma_{-M}^{\text{b}}\rangle, \tag{5.5.12}$$

where $|\Gamma_M^{\text{a}}\rangle$ and $|\Gamma_M^{\text{b}}\rangle$ are exact copies of low-lying states (5.3.3). $|\Xi_{\varphi}^{\text{tot}}\rangle$ is indeed a "ground state" of $\hat{H}_0^{\text{tot}}$ in which the two condensates have definite relative phase φ.

We see that the definite relative phase φ between the two condensates is realized, as in (5.5.12), by a coherent superposition of states with different divisions of particle numbers between the two subsystems. In other words the two subsystems inevitably entangle if we demand that there is a definite relative phase.

We conjecture that the ground state obtained through the double limit in (5.5.5) or (5.5.6) resembles (or coincides with) the large L limit of $|\Xi_{\varphi}^{\text{tot}}\rangle$, although the proof seems very difficult. We also expect that states realized experimentally in weakly coupled Bose–Einstein condensates resemble $|\Xi_{\varphi}^{\text{tot}}\rangle$.

As an alternative approach to give a meaning of symmetry breaking in coupled Bose–Einstein condensates, a state with a fixed number of particles in a larger system (i.e., the two condensates and the environment) which is "identical" to $|\Xi_{\theta}^{\text{a}}\rangle \otimes |\Xi_{\theta+\varphi}^{\text{b}}\rangle$ is constructed in [20].[18] Here "identical" means that the measurement of any observable which conserves the total number of particles in the two condensates give the same results as $|\Xi_{\theta}^{\text{a}}\rangle \otimes |\Xi_{\theta+\varphi}^{\text{b}}\rangle$. See also [19] for background. It is likely that these states and our $|\Xi_{\varphi}^{\text{tot}}\rangle$ are indistinguishable if we only measure local quantities which preserve the particle number.

References

1. M. Aizenman, E.H. Lieb, R. Seiringer, J.P. Solovej, J. Yngvason, Bose-Einstein quantum phase transition in an optical lattice model. Phys. Rev. A **70**, 023612 (2004), arXiv:cond-mat/0403240
2. M.R. Andrews, C.G. Townsend, H.-J. Miesner, D.S. Durfee, D.M. Kurn, W. Ketterle, Observation of interference between two bose condensates. Science **275**, 637–640 (1997)
3. I. Bloch, J. Dalibard, W. Zwerger, Many-body physics with ultracold gases. Rev. Mod. Phys. **80**, 885–964 (2008), arXiv:0704.3011

[18] The construction is the same as that in Problem 5.4.a (p. 143).

4. F. Dalfovo, S. Giorgini, L.P. Pitaevskii, S. Stringari, Theory of Bose-Einstein condensation in trapped gases. Rev. Mod. Phys. **71**, 463–512 (1999), arXiv:cond-mat/9806038
5. F.J. Dyson, E.H. Lieb, B. Simon, Phase transitions in quantum spin systems with isotropic and nonisotropic interactions. J. Stat. Phys. **18**, 335–382 (1978)
6. A.L. Fetter, J.D. Walecka, *Quantum Theory of Many-Particle Systems* (Dover, Mineola, 2003)
7. M. Gring, M. Kuhnert, T. Langen, T. Kitagawa, B. Rauer, M. Schreitl, J. Schmiedmayer, Relaxation and prethermalization in an isolated quantum system. Science **337**, 1318–1322 (2012), arXiv:1112.0013
8. J. Javanainen, S.M. Yoo, Quantum phase of a Bose-Einstein condensate with an arbitrary number of atoms. Phys. Rev. Lett. **76**, 161 (1996)
9. T. Kennedy, E.H. Lieb, B.S. Shastry, Existence of Néel order in some spin-$1/2$ Heisenberg antiferromagnets. J. Stat. Phys. **53**, 1019 (1988)
10. T. Kennedy, E.H. Lieb, B.S. Shastry, The XY model has long-range order for all spins and all dimensions greater than one. Phys. Rev. Lett. **61**, 2582 (1988)
11. T. Koma, H. Tasaki, Symmetry breaking in Heisenberg antiferromagnets. Comm. Math. Phys. **158**, 191–214 (1993), https://projecteuclid.org/euclid.cmp/1104254136
12. T. Koma and H. Tasaki, Symmetry breaking and finite-size effects in quantum many-body systems. J. Stat. Phys. **76**, 745–803 (1994). (The version in the arXiv is more complete than the published version.), arXiv:cond-mat/9708132
13. K. Kubo, T. Kishi, Existence of long-range order in the XXZ model. Phys. Rev. Lett. **61**, 2585 (1988)
14. A.J. Leggett, Bose-Einstein condensation in the alkali gases: some fundamental concepts. Rev. Mod. Phys. **73**, 307–356 (2001)
15. E.H. Lieb, R. Seiringer, J.P. Solovej, J. Yngvason, *The Mathematics of the Bose Gas and Its Condensation* (Springer, Berlin, 2005), arXiv:cond-mat/0610117
16. T. Matsubara, H. Matsuda, A lattice model of liquid helium I. Prog. Theor. Phys. **16**, 569–582 (1956), https://academic.oup.com/ptp/article/16/6/569/1907565/
17. L. Pitaevskii, S. Stringari, *Bose-Einstein Condensation and Superfluidity* (Oxford University Press, Oxford, 2016)
18. P. Ring, P. Schuck, *The Nuclear Many-Body Problem* (Springer, Berlin, 1980)
19. A. Shimizu, T. Miyadera, Robustness of wave functions of interacting many Bosons in a leaky box. Phys. Rev. Lett. **85**, 688–691 (2000). Errata: Phys. Rev. Lett. **86**, 4422 (2001)
20. A. Shimizu, T. Miyadera, Charge superselection rule does not rule out pure states of subsystems to be coherent superpositions of states with different charges (2001), arXiv:cond-mat/0102429
21. H. Tasaki, Long-range order, "tower" of states, and symmetry breaking in lattice quantum systems. J. Stat. Phys. **174**, 735–761 (2019). (The version in the arXiv is more complete than the published version), arXiv:1807.05847
22. H. Tasaki, Spontaneous symmetry breaking in coupled Bose-Einstein condensates. J. Stat. Phys. **178**, 379–391 (2019), arXiv:1808.09212
23. X.-G. Wen, Colloquium: Zoo of quantum-topological phases of matter. Rev. Mod. Phys. **89**, 41004 (2017), arXiv:1610.03911
24. B. Zeng, X. Chen, D.-L. Zhou, X.-G. Wen, *Quantum Information Meets Quantum Matter: From Quantum Entanglement to Topological Phase in Many-Body Systems*. (Quantum Science and Technology, Springer, Berlin, 2019), arXiv:1508.02595

Part II
Haldane Phenomena and Beyond

In 1983, Haldane made a fascinating and unexpected discovery that there is a qualitative difference in low energy properties of the one-dimensional antiferromagnetic Heisenberg models with half-odd-integer spins and those with integer spins. In particular he concluded that integer spin chains have a unique ground state accompanied by a nonzero energy gap, now known as the Haldane gap. The existence of such a gapped ground state was confirmed experimentally, numerically, and also mathematically through an exact solution of a model proposed by Affleck, Kennedy, Lieb, and Tasaki.

Further studies of integer spin antiferromagnetic chains led to the discovery of various unexpected properties, from which the notions of Haldane phenomena and the Haldane phase were developed. More recently it was found that the Haldane phase provides a prototypical example of a topological phase of matter, more precisely, a nontrivial symmetry protected topological phase.

In part II of the book, we review a variety of topics related to Haldane phenomena in a self-contained manner.

Chapter 6
Ground States of the Antiferromagnetic Heisenberg Chains

In the present chapter, we focus on properties of the ground state and low energy excited states of the one-dimensional antiferromagnetic Heisenberg model, often called the "antiferromagnetic Heisenberg chain". Recall that, according to Shastry's theorem (Theorem 4.2 in p. 76), the ground state does not exhibit long-range order or spontaneous symmetry breaking. In Sect. 6.1, we describe Haldane's conclusion about the qualitative difference between the antiferromagnetic Heisenberg chains with half-odd-integer and integer S, along with necessary background. Then we discuss the origin of the difference from two different points of view. We describe and prove the (generalized) Lieb–Schultz–Mattis theorem in Sect. 6.2, and present a semi-classical picture based on kink dynamics in Sect. 6.3.

6.1 Haldane "Conjecture"

We consider the antiferromagnetic Heisenberg model with spin $S = 1/2, 1, 3/2, \ldots$ on the one-dimensional lattice[1] $\Lambda_L = \{1, 2, \ldots, L\}$, where L is even. The Hamiltonian is

$$\hat{H} = \sum_{x=1}^{L} \hat{\boldsymbol{S}}_x \cdot \hat{\boldsymbol{S}}_{x+1}. \tag{6.1.1}$$

We impose periodic boundary conditions and identify $\hat{\boldsymbol{S}}_{L+1}$ with $\hat{\boldsymbol{S}}_1$. Note that the Marshall–Lieb–Mattis theorem (Theorem 2.2 in p. 39) guarantees that the ground state $|\Phi_{\rm GS}\rangle$ is unique and has $S_{\rm tot} = 0$.

The $S = 1/2$ antiferromagnetic Heisenberg chain The antiferromagnetic Heisenberg chain with $S = 1/2$ is one of the most well-studied models in quantum many-body problems. The model is known to be "exactly solvable" in the sense that some physical quantities can be computed precisely under certain (usually unproved but

[1]Note that the numbering of lattice sites is different from that in (3.1.2).

H. Tasaki, *Physics and Mathematics of Quantum Many-Body Systems*, Graduate Texts in Physics, https://doi.org/10.1007/978-3-030-41265-4_6

plausible) assumptions.[2] The first exact solution was obtained in 1931 by Bethe [11], who introduced the celebrated Bethe ansatz method and computed some energy eigenvalues and eigenstates. Since then many researchers have contributed in enriching the exact solution of this model, and various physical quantities have been obtained. The interested reader is invited to study textbooks such as [37, 55, 57].

Based on these exact computations, the following is believed to be valid for the ground state and low-energy excitations of the model:

(HOI1) The ground state is unique (both for finite and infinite L).
(HOI2) There exists no energy gap above the ground state energy.
(HOI3) The ground state correlation function decays slowly with a power law.

The uniqueness of the ground state in (HOI1) for finite L is known from the Marshall–Lieb–Mattis theorem. The uniqueness for $L \uparrow \infty$ is almost equivalent to the absence of long-range order (LRO) or spontaneous symmetry breaking (SSB) established in Theorem 4.2 in p. 76. It is sometimes said that antiferromagnetic ordering in the ground state is destroyed by strong "quantum fluctuation" in one dimension (although the precise meaning of "quantum fluctuation" is not quite clear in such an assertion).

The absence of gap in (HOI2) roughly means that, in the $L \uparrow \infty$ limit, there is a continuum of excited energies just above the ground state energy.[3] For finite L, there always is a nonvanishing gap above the ground state energy, but this gap scales like $O(1/L)$ as L grows. See Fig. 6.1 and Theorem 6.3 (p. 162).

Finally the two-point correlation function in the ground state is believed (but not yet proved) to have asymptotic behavior

$$\langle \Phi_{\rm GS} | \hat{\mathbf{S}}_x \cdot \hat{\mathbf{S}}_y | \Phi_{\rm GS} \rangle \simeq (\text{constant})\,(-1)^{x-y}\,\frac{\sqrt{\log|x-y|}}{|x-y|}, \tag{6.1.2}$$

for $1 \ll |x-y| \ll L$. The leading behavior comes from $(-1)^{x-y}/|x-y|$, which is the power law decay stated in (HOI3). The logarithmic term $\sqrt{\log|x-y|}$ is a subtle correction [6, 10, 40]. Note that the sign factor $(-1)^{x-y}$ is consistent with the inequalities (2.5.7) which follow from the Marshall–Lieb–Mattis theorem.

The power law decay of correlation function is reminiscent of critical phenomena observed, e.g., at the critical temperature of the classical Ising model in two or higher dimensions [18, 20, 30]. Indeed the ground state of the $S = 1/2$ antiferromagnetic Heisenberg chain is a very well-known and well-studied example of a quantum critical point. See [51] for more about quantum phase transitions and quantum critical points.

From the point of view of quantum field theory, the above (HOI1), (HOI2), and (HOI3) suggest that low energy properties of the model are described by a massless field theory without SSB. Recall that one can create a massless particle from

---

[2] Such results based on unproved assumptions are sometimes referred to as "exact but non-rigorous" results.

[3] There is a set of energy eigenstates, known as the des Cloizeaux–Pearson mode, with excitation energy $\varepsilon_k = (\pi/2)|\sin k|$, where k is the wave number [17].

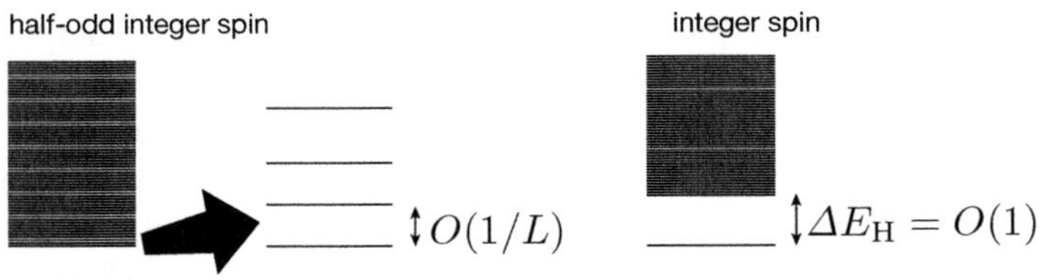

Fig. 6.1 Schematic figures of the energy spectra of the spin S antiferromagnetic Heisenberg chain. When S is a half-odd integer, we have nearly continuous spectrum. The energy gap above the ground state energy is proportional to L^{-1}. When S is an integer, there is an almost L independent nonvanishing energy gap above the ground state energy. This is the Haldane gap ΔE_{H}

the vacuum with infinitesimally small energy cost. This corresponds to the gapless behavior (HOI2). Also the force mediated by a massless particle is long-ranged (like the electromagnetic force) and decays with a power law as in (HOI3).

The antiferromagnetic Heisenberg chain with general spin The antiferromagnetic Heisenberg chain is exactly solvable only for $S = 1/2$. Although we don't know if any researcher before Haldane had seriously investigated the properties of the model for $S \geq 1$, it is very likely that people expected no drastic changes from the simplest $S = 1/2$ case.

In two papers published in 1983 [25, 26], Haldane argued that the low energy properties of the antiferromagnetic Heisenberg chain depend qualitatively on whether the spin quantum number S is a half-odd integer or an integer. The conclusion was so unexpected and drastic that it was called the "Haldane conjecture" in 1980s.

More precisely, Haldane's conclusion was as follows. When the spin quantum number S is a half odd integer, i.e., $S = 1/2, 3/2, 5/2, \ldots$, the properties of the ground state and low-energy excitations are basically the same as (HOI1), (HOI2), and (HOI3) above. Thus the ground state is critical in the language of statistical mechanics, and is massless in the language of field theory.

When S is an integer, i.e., $S = 1, 2, 3, \ldots$, the model has completely different properties:

(I1) The ground state is unique (both for finite and infinite L).
(I2) There exists a nonvanishing energy gap above the ground state energy.
(I3) The ground state correlation function decays exponentially.

The property (I1) is the same as above (HOI1). We will see below that "quantum fluctuation" in this case is much stronger.

The existence of a gap (I2) means that the difference between the ground state energy and the first excitation energy is strictly positive and almost independent of the system size L. See Fig. 6.1. This energy difference ΔE_{H} is called the Haldane gap. According to Haldane's field theoretic analysis [1, 2, 4, 25], the gap behaves as $\Delta E \simeq 2Se^{-\pi S}$ for $S \gg 1$.

The exponential decay in (I3) means that one has the asymptotic behavior

$$\langle \Phi_{\mathrm{GS}} | \hat{\boldsymbol{S}}_x \cdot \hat{\boldsymbol{S}}_y | \Phi_{\mathrm{GS}} \rangle \simeq (\text{constant}) \, \frac{(-1)^{x-y}}{|x-y|^{1/2}} \exp\Bigl[-\frac{|x-y|}{\xi}\Bigr], \tag{6.1.3}$$

for x and y such that $\xi \ll |x - y| \ll L$, where ξ is the correlation length with $0 < \xi < \infty$. Again the main term is $(-1)^{x-y} \exp[-|x - y|/\xi]$. The power law correction $|x - y|^{-1/2}$ is generically found in a disordered ground state of a one-dimensional quantum system (and in equilibrium states of two-dimensional classical systems).[4]

Note that, except for the oscillating factor $(-1)^{x-y}$ intrinsic to an antiferromagnetic system, the asymptotic behavior (6.1.3) is the same as the Ornstein–Zernike form, which we observe in the two-dimensional Ising model at temperatures higher than the transition temperature.[5] Remarkably the ground state is expected to show disordered behavior which is usually observed at high temperatures. Recall that systems at high temperatures show strong thermal fluctuation. It is often said that strong "quantum fluctuation" is the source of this disordered behavior in the ground state of the integer S chains.

From the field theoretic point of view, the above (I1), (I2), (I3) suggest that low energy properties of the model are described by a theory for a massive particle. The energy gap in (I2) precisely corresponds to the minimum energy, mc^2, required to create a particle with mass m. The exponential decay in (I3) also corresponds to the short-range force mediated by a massive particle.

To sum, Haldane's conclusion was that the ground state of the spin S antiferromagnetic Heisenberg chain exhibits critical (or massless) behavior when S is a half-odd integer, and disordered (or massive) behavior when S is an integer. This conclusion came from two somewhat different arguments, both for large S limit of the model, presented in two papers published in 1983. One is based on the analysis of quantization conditions in the effective nonlinear sigma model [25], and the other on semiclassical quantization of solitons [26]. Haldane later announced the reformulation of the first argument in terms of path-integral quantization of nonlinear sigma model with a topological term [27]. Although the argument based on the topological term is famous, it was never written down by Haldane himself, and was worked out by Affleck [1].[6] See also Affleck's review articles [2, 4] and Chap. 7 of [19]. Below in Sect. 6.3, we present a small S version of Haldane's second argument based on soliton (kink) dynamics.

Haldane's conclusion initially came as a surprise to the condensed matter physics community. First of all the qualitative difference between the chains with half-odd-integer and integer spins was probably quite unexpected. From the classical point of view, S is merely the magnitude of a spin. Then why does it matter whether S is an integer or not?[7] Secondly it was probably a common belief, supported by the

---

[4]There is no general proof of this fact. In the quantum Ising model (3.3.1) with sufficiently large λ, the Ornstein–Zernike type behavior of the ground state correlation function was rigorously established by Kennedy [34].

[5]See footnote 5 in p. 53.

[6]We learned that Edward Witten, who heard Haldane's talk in 1983 at the conference in honor of the 60th birthday of Philip Anderson, was one of the first to realize a possible connection between Haldane's argument and the topological angle. Affleck was informed of Haldane's work and its possible topological interpretation by Witten (Ian Affleck, private communication).

[7]It is indeed surprising that Haldane's two published papers [25, 26] deal with large S limit, where the difference between the models with half-odd-integer S and integer S is expected to become

exact solution of the $S = 1/2$ chain and the Lieb–Schultz–Mattis theorem (see next section), that a model with continuous symmetry and a unique ground state should be gapless. This point will be discussed in the next section, where we see that Haldane's conclusion for half-odd-integer S can be (partially) justified. In this sense the more interesting and surprising part of Haldane's work is about integer S spin chains.

The conclusion for the $S = 1$ antiferromagnetic Heisenberg chain was soon confirmed both numerically [12, 43, 47] and experimentally. For example from a numerical diagonalization study in the chain with $L = 22$, the Haldane gap was estimated as $\Delta E_{\mathrm{H}} = 0.41049 \pm 2 \times 10^{-5}$ [22]. The first $S = 1$ quasi-one-dimensional material studied experimentally was $CsNiCl_3$, where neutron scattering experiments showed a clear evidence of the gap [13, 54]. This material however is known to exhibit Néel order below 4.9 K. It was discovered later that another $S = 1$ quasi-one-dimensional material $Ni(C_2H_8N_2)_2NO_2(ClO_4)$ (usually abbreviated as NENP), in which no long-range order has been observed at any temperature,[8] is an ideal material for studying Haldane phenomena. The indication of the gap was first found in susceptibility measurements [49], and then confirmed in neutron scattering experiments [48, 50] and in ESR measurements [16].

The models with higher integral S is much harder to study, especially because the Haldane gap is expected to decrease as S grows. For the chains with $S = 2$, 3, and 4 there are convincing numerical evidences for the existence of a gap [53, 61, 62]. See, in particular, Sect. 5 of [62] for the latest summary of numerical evaluation of the Haldane gap.[9]

Although there is no mathematically rigorous results for integer S models (except for those in related exactly solvable model treated in Chap. 7), we can say that Haldane's conclusion has been well checked and widely accepted. It is by no means a "conjecture".

Remark It is believed that the properties (I2) and (I3), namely, the existence of a gap and the exponential decay of ground state correlation function, generically come together. For a general class of quantum systems it was proved that the existence of a gap implies exponential decay [33].

Problem 6.1.a It is also believed that the properties (HOI2) and (HOI3) (see p. 154) generically come together. Indeed it can be proved rather easily that power law decay

---

infinitesimal, as is suggested by the formula $\Delta E \simeq 2Se^{-\pi S}$ for the Haldane gap for integer S. Now it is known that Haldane had yet another argument which relies on the analogy to the Berezinskii–Kosterlitz–Thouless transition, and does not make use of the large S limit. This work has not been published in a journal but was made public by Haldane himself in 2016 [24]. (According to Haldane one of the referees of the paper stated that it was "in manifest contradiction to fundamental principles of physics" [28].) See also [29].

[8] It is believed that a three dimensional spin system obtained by weakly coupling the $S = 1$ antiferromagnetic Heisenberg chains has a disordered ground state with a gap [3, 38, 52]. It is likely that NENP never develops long-range order at any temperatures.

[9] One finds $\Delta E_{\mathrm{H}} \simeq 0.089$ for $S = 2$, $\Delta E_{\mathrm{H}} \simeq 0.0100$ for $S = 3$, and $\Delta E_{\mathrm{H}} \simeq 7.99 \times 10^{-4}$ for $S = 4$. The result for $S = 4$ was obtained by a Monte Carlo simulation of the chain with 73,728 sites, performed on a supercomputer [62].

of ground state correlation function generally implies that the model is gapless. More precisely, consider a quantum spin system on the d-dimensional hypercubic lattice Λ_L, and assume that the Hamiltonian $\hat{H}$ and the order operator $\hat{\mathcal{O}}_L$ satisfy the conditions as in Problem 3.4.a (p. 67). Further assume that the unique ground state $|\Phi_{\rm GS}\rangle$ of $\hat{H}$ satisfies $\langle\Phi_{\rm GS}|\hat{o}_x|\Phi_{\rm GS}\rangle = 0$ for any x, and

$$\langle\Phi_{\rm GS}|\hat{o}_x\hat{o}_y|\Phi_{\rm GS}\rangle \ge \frac{A}{|x-y|^{d-\kappa}}, \tag{6.1.4}$$

for any $x, y \in \Lambda_L$ such that $x \neq y$ and $|x-y| \le BL$ with an exponent $\kappa \in (0, d)$. Here A, B are (L independent) positive constants. Prove that the first excited energy $E_{\rm 1st}$ satisfies $E_{\rm 1st} - E_{\rm GS} \le C/L^{\kappa}$, where C is a constant independent of L. See the footnote for a hint.[10] [solution→p.504]

6.2 The Lieb–Schultz–Mattis Theorem

We here discuss the Lieb–Schultz–Mattis and related theorems, which suggest that there is a fundamental difference in low energy properties between spin chains with half-odd-integer S and with integer S.

The Lieb–Schultz–Mattis theorem first appeared as Theorem 2 in Appendix B of [39]. The main topic of this paper was spin chains which can be solved exactly by mapping to models of free fermions. The theorem applies to any $S = 1/2$ spin chain whose Hamiltonian is short ranged, translation invariant, and invariant under any rotation about the 3-axis, and whose ground state is unique. It is proved that one can always construct an excited state with an excitation energy $O(L^{-1})$ by applying a gradual "twist" to the ground state. Later, in connection with Haldane's work, the theorem was extended by Affleck and Lieb [5] to a class of spin chains with a general half-odd-integer S.

Main theorem and proof Here we describe and prove the theorem in the context of spin S antiferromagnetic Heisenberg chain, but the extension to more general spin chains is not difficult. See Lemma 6.4.

Let L be finite and even, and denote by $|\Phi_{\rm GS}\rangle$ the unique ground state of the Hamiltonian (6.1.1). Since the Marshall–Lieb–Mattis theorem (Theorem 2.2 in p. 39) shows $\hat{S}^{(3)}_{\rm tot}|\Phi_{\rm GS}\rangle = 0$, we see that the ground state is invariant under any rotation about the 3-axis, i.e.,

$$\hat{U}^{(3)}_\theta|\Phi_{\rm GS}\rangle = |\Phi_{\rm GS}\rangle, \tag{6.2.1}$$

for any $\theta \in \mathbb{R}$, where the rotation operator is $\hat{U}^{(3)}_\theta = \exp[-i\theta\hat{S}^{(3)}_{\rm tot}]$ as in (2.2.11). In other words both the Hamiltonian and the ground state are U(1) invariant. We introduce the "twist" operator, a rotation operator whose rotation angle varies slowly from a site to site:

---

[10]Use the variational estimate described in Sect. 3.4.

$$\hat{U}_{\rm LSM} = \exp[-i\sum_{x=1}^{L}\theta_x \hat{S}_x^{(3)}] \tag{6.2.2}$$

We choose the angle at x as

$$\theta_x = \frac{2\pi}{L}x = \Delta\theta\, x, \tag{6.2.3}$$

with $\Delta\theta = 2\pi/L$. It is essential that the rotation angle at the right end of the chain is $\theta_L = 2\pi$, which is equivalent to 0. Then we define our trial state as

$$|\Phi_{\rm LSM}\rangle = \hat{U}_{\rm LSM}|\Phi_{\rm GS}\rangle. \tag{6.2.4}$$

Because of the invariance (6.2.1) under the rotation with a constant angle, the twisted state $|\Phi_{\rm LSM}\rangle$ should be very close to $|\Phi_{\rm GS}\rangle$. Moreover we see that $|\Phi_{\rm LSM}\rangle$ contains a gradual modification whose wave-length is L. It is then expected that the trial state $|\Phi_{\rm LSM}\rangle$ has a very low excitation energy. This is indeed true.

Lemma 6.1 *For any S and any L, one has*

$$\langle\Phi_{\rm LSM}|\hat{H}|\Phi_{\rm LSM}\rangle - E_{\rm GS} \le \frac{8\pi^2 S^2}{L}. \tag{6.2.5}$$

Proof It is useful to note that

$$e^{i\theta\hat{S}_x^{(3)}}\hat{S}_x^{\pm}e^{-i\theta\hat{S}_x^{(3)}} = e^{\pm i\theta}\hat{S}_x^{\pm}. \tag{6.2.6}$$

This may be derived from (2.1.16), but can be shown directly (and easily) by examining the action onto the basis state $|\psi_x^\sigma\rangle$. Then by using the expression (2.2.16), we find

$$\begin{aligned}\hat{U}_{\rm LSM}^\dagger\,\hat{\boldsymbol{S}}_x\cdot\hat{\boldsymbol{S}}_{x+1}\hat{U}_{\rm LSM} &= \hat{U}_{\rm LSM}^\dagger\Big(\frac{1}{2}\{\hat{S}_x^+\hat{S}_{x+1}^- + \hat{S}_x^-\hat{S}_{x+1}^+\} + \hat{S}_x^{(3)}\hat{S}_{x+1}^{(3)}\Big)\hat{U}_{\rm LSM}\\ &= \frac{1}{2}\{e^{i(\theta_x-\theta_{x+1})}\hat{S}_x^+\hat{S}_{x+1}^- + e^{-i(\theta_x-\theta_{x+1})}\hat{S}_x^-\hat{S}_{x+1}^+\} + \hat{S}_x^{(3)}\hat{S}_{x+1}^{(3)},\end{aligned} \tag{6.2.7}$$

which implies that

$$\hat{U}_{\rm LSM}^\dagger\,\hat{\boldsymbol{S}}_x\cdot\hat{\boldsymbol{S}}_{x+1}\hat{U}_{\rm LSM} - \hat{\boldsymbol{S}}_x\cdot\hat{\boldsymbol{S}}_{x+1} = \frac{1}{2}\big\{(e^{-i\Delta\theta}-1)\,\hat{S}_x^+\hat{S}_{x+1}^- + (e^{i\Delta\theta}-1)\,\hat{S}_x^-\hat{S}_{x+1}^+\big\}. \tag{6.2.8}$$

By summing this over x, we get

$$\hat{U}_{\rm LSM}^\dagger\hat{H}\hat{U}_{\rm LSM} - \hat{H} = \frac{1}{2}\sum_{x=1}^{L}\big\{(e^{-i\Delta\theta}-1)\,\hat{S}_x^+\hat{S}_{x+1}^- + (e^{i\Delta\theta}-1)\,\hat{S}_x^-\hat{S}_{x+1}^+\big\}. \tag{6.2.9}$$

We also find, by the replacement $\theta_x \to -\theta_x$, that

$$\hat{U}_{\rm LSM}\hat{H}\hat{U}_{\rm LSM}^{\dagger}-\hat{H}=\frac{1}{2}\sum_{x=1}^{L}\{(e^{i\Delta\theta}-1)\,\hat{S}_x^{+}\hat{S}_{x+1}^{-}+(e^{-i\Delta\theta}-1)\,\hat{S}_x^{-}\hat{S}_{x+1}^{+}\}. \quad (6.2.10)$$

Now the variational energy is bounded as

$$\begin{aligned}\langle\Phi_{\rm LSM}|\hat{H}|\Phi_{\rm LSM}\rangle-E_{\rm GS}=&\langle\Phi_{\rm GS}|(\hat{U}_{\rm LSM}^{\dagger}\hat{H}\hat{U}_{\rm LSM}-\hat{H})|\Phi_{\rm GS}\rangle\\ \le&\langle\Phi_{\rm GS}|(\hat{U}_{\rm LSM}^{\dagger}\hat{H}\hat{U}_{\rm LSM}-\hat{H})|\Phi_{\rm GS}\rangle\\ &+\langle\Phi_{\rm GS}|(\hat{U}_{\rm LSM}\hat{H}\hat{U}_{\rm LSM}^{\dagger}-\hat{H})|\Phi_{\rm GS}\rangle\\ =&(\cos\Delta\theta-1)\sum_{x=1}^{L}\langle\Phi_{\rm GS}|(\hat{S}_x^{+}\hat{S}_{x+1}^{-}+\hat{S}_x^{-}\hat{S}_{x+1}^{+})|\Phi_{\rm GS}\rangle\\ \le&2(1-\cos\Delta\theta)\sum_{x=1}^{L}\Bigl|\langle\Phi_{\rm GS}|(\hat{S}_x^{(1)}\hat{S}_{x+1}^{(1)}+\hat{S}_x^{(2)}\hat{S}_{x+1}^{(2)})|\Phi_{\rm GS}\rangle\Bigr|\\ \le&(\Delta\theta)^2\,2S^2\,L=\frac{8\pi^2S^2}{L},\end{aligned} \quad (6.2.11)$$

where we used $\cos x\ge 1-x^2/2$, and $|\langle\Phi_{\rm GS}|\hat{S}_x^{(\alpha)}\hat{S}_{x+1}^{(\alpha)}|\Phi_{\rm GS}\rangle|\le\|\hat{S}_x^{(\alpha)}\hat{S}_{x+1}^{(\alpha)}\|\le S^2$ (see Sect. A.2.1 for the definition and properties of operator norm). To get the second line we simply added $\langle\Phi_{\rm GS}|(\hat{U}_{\rm LSM}\hat{H}\hat{U}_{\rm LSM}^{\dagger}-\hat{H})|\Phi_{\rm GS}\rangle$, which is nonnegative from the variational principle. This seemingly unnecessary step made the above estimate simpler.[11] ∎

Having seen that the trial state $|\Phi_{\rm LSM}\rangle$ obtained by giving a gradual twist to $|\Phi_{\rm GS}\rangle$ has extremely low excitation energy, one might question the validity of Haldane's conclusion that the integer S antiferromagnetic Heisenberg chain has a finite gap above the ground state energy. However in order to complete the variational analysis one must also show that the trial state is orthogonal to the ground state. This indeed turns out to be nontrivial, and can be done generally only when S is a half-odd-integer.

Lemma 6.2 *If S is a half-odd integer, one has* $\langle\Phi_{\rm GS}|\Phi_{\rm LSM}\rangle=0$.

Proof Define the translation operator $\hat{T}$ by

[11] We learned this trick from [36]. In fact this step can be avoided, and one can bound the expectation value $\langle\Phi_{\rm GS}|(\hat{U}_{\rm LSM}^{\dagger}\hat{H}\hat{U}_{\rm LSM}-\hat{H})|\Phi_{\rm GS}\rangle$ directly by using a slightly complicated estimate that makes use of an extra symmetry of $|\Phi_{\rm GS}\rangle$ as in [5, 39]. We then get a better bound $\langle\Phi_{\rm LSM}|\hat{H}|\Phi_{\rm LSM}\rangle-E_{\rm GS}\le 4\pi^2S^2/L$ instead of (6.2.11).

In all the early applications of the Lieb–Schultz–Mattis method [5, 39, 45, 69], it was assumed that the system possesses inversion or time-reversal symmetry in addition to the $U(1)$ symmetry and translation invariance. The above estimate shows that the inversion or time-reversal symmetry is in fact not necessary. It is interesting that this (more or less standard) argument had been overlooked until Koma [36] applied the Lieb–Schultz–Mattis method to quantum Hall systems on a quasi one-dimensional strip.

$$\hat{T}\Big(\bigotimes_{x=1}^{L}|\psi_x^{\sigma_x}\rangle\Big)=\bigotimes_{x=1}^{L}|\psi_{x+1}^{\sigma_x}\rangle, \tag{6.2.12}$$

where we identify $L+1$ with 1. We then have $\hat{T}^\dagger\hat{S}_x^{(\alpha)}\hat{T}=\hat{S}_{x-1}^{(\alpha)}$, where we identify $\hat{S}_0^{(\alpha)}$ with $\hat{S}_L^{(\alpha)}$. Since the Hamiltonian is translation invariant, the unique ground state $|\Phi_{\mathrm{GS}}\rangle$ must satisfy $\hat{T}|\Phi_{\mathrm{GS}}\rangle=e^{i\alpha}|\Phi_{\mathrm{GS}}\rangle$ with some $\alpha\in\mathbb{R}$.[12] We then note that

$$\langle\Phi_{\mathrm{GS}}|\Phi_{\mathrm{LSM}}\rangle=\langle\Phi_{\mathrm{GS}}|\hat{U}_{\mathrm{LSM}}|\Phi_{\mathrm{GS}}\rangle=\langle\Phi_{\mathrm{GS}}|\hat{T}^\dagger\hat{U}_{\mathrm{LSM}}\hat{T}|\Phi_{\mathrm{GS}}\rangle. \tag{6.2.13}$$

Recalling the definition (6.2.2) and by using (A.2.17), we find

$$\begin{aligned}\hat{T}^\dagger\hat{U}_{\mathrm{LSM}}\hat{T}&=\hat{T}^\dagger\exp[-i\sum_{x=1}^{L}\theta_x\hat{S}_x^{(3)}]\hat{T}=\exp[-i\sum_{x=1}^{L}\theta_x\hat{S}_{x-1}^{(3)}]=\exp[-i\sum_{x=0}^{L-1}\theta_{x+1}\hat{S}_x^{(3)}]\\&=\exp[-i\sum_{x=0}^{L-1}\theta_x\hat{S}_x^{(3)}]\,\exp[-i\frac{2\pi}{L}\sum_{x=0}^{L-1}\hat{S}_x^{(3)}],\end{aligned} \tag{6.2.14}$$

where we noted that $\theta_{x+1}=\theta_x+2\pi/L$, and extended (6.2.3) to $x=0$. Since $\theta_0=0$, we find

$$\exp[-i\sum_{x=0}^{L-1}\theta_x\hat{S}_x^{(3)}]=\exp[-i\sum_{x=1}^{L-1}\theta_x\hat{S}_x^{(3)}]=e^{i\theta_L\hat{S}_L^{(3)}}\,\hat{U}_{\mathrm{LSM}}. \tag{6.2.15}$$

Noting that $\theta_L=2\pi$, we see

$$e^{i\theta_L\hat{S}_L^{(3)}}=e^{i2\pi\hat{S}_L^{(3)}}=(-1)^{2S}=\begin{cases}-1 & \text{if } S \text{ is a half-odd-integer}\\ 1 & \text{if } S \text{ is an integer,}\end{cases} \tag{6.2.16}$$

which is the remarkable property of the 2π-rotation that we have seen in (2.1.23). Also noting that $\sum_{x=0}^{L-1}\hat{S}_x^{(3)}=\hat{S}_{\mathrm{tot}}^{(3)}$, we have

$$\hat{T}^\dagger\hat{U}_{\mathrm{LSM}}\hat{T}=(-1)^{2S}\hat{U}_{\mathrm{LSM}}\,e^{-i(2\pi/L)\hat{S}_{\mathrm{tot}}^{(3)}}. \tag{6.2.17}$$

Substituting this into (6.2.13), and noting that $\hat{S}_{\mathrm{tot}}^{(3)}|\Phi_{\mathrm{GS}}\rangle=0$, we get

$$\langle\Phi_{\mathrm{GS}}|\Phi_{\mathrm{LSM}}\rangle=\langle\Phi_{\mathrm{GS}}|(-1)^{2S}\hat{U}_{\mathrm{LSM}}|\Phi_{\mathrm{GS}}\rangle=(-1)^{2S}\langle\Phi_{\mathrm{GS}}|\Phi_{\mathrm{LSM}}\rangle, \tag{6.2.18}$$

which implies $\langle\Phi_{\mathrm{GS}}|\Phi_{\mathrm{LSM}}\rangle=0$ if S is a half-odd-integer. ∎

---

[12] The proof is standard (and essentially the same as that described in footnote 9 in p. 58). Translation invariance $\hat{T}\hat{H}\hat{T}^\dagger=\hat{H}$ implies $\hat{T}\hat{H}=\hat{H}\hat{T}$. Thus $\hat{H}(\hat{T}|\Phi_{\mathrm{GS}}\rangle)=\hat{T}\hat{H}|\Phi_{\mathrm{GS}}\rangle=E_{\mathrm{GS}}\hat{T}|\Phi_{\mathrm{GS}}\rangle$, and the uniqueness implies $\hat{T}|\Phi_{\mathrm{GS}}\rangle=e^{i\alpha}|\Phi_{\mathrm{GS}}\rangle$. The Marshall–Lieb–Mattis theorem (Theorem 2.2 in p. 39) implies $e^{i\alpha}=\pm1$, but we do not use this fact.

By combining the above two Lemmas we arrive at the following theorem of Lieb, Schultz, and Mattis [39] and Affleck and Lieb [5].

Theorem 6.3 (Lieb–Schultz–Mattis, Affleck–Lieb theorem) *When S is a half-odd-integer, the first excited energy E_{1st} of the antiferromagnetic Heisenberg chain* (6.1.1) *satisfies*

$$0 < E_{1st} - E_{GS} \leq \frac{8\pi^2 S^2}{L}. \tag{6.2.19}$$

Implication of the theorem Theorem 6.3 states that one can never simultaneously have both the properties (I1) and (I2) (see p. 155) when S is a half-odd-integer. In short it is impossible that the antiferromagnetic Heisenberg model with a half-odd-integer S has a unique ground state with a gap.[13] This is consistent with Haldane's conclusion, and provides its partial justification.

Given the uniqueness of the ground state (for finite L) and Shastry's theorem (Theorem 4.2 in p. 76) which rules out long-range order in the ground state, one is tempted to conclude that Theorem 6.3 rigorously establishes the properties (HOI1) and (HOI2) (see p. 154). Although this is quite likely, we do not know of any rigorous proof.[14]

Theorem 6.3 provides no information at all for models with integer S. In particular note that (6.2.18) in the Proof of Lemma 6.2 gives a trivial relation $\langle \Phi_{GS}|\Phi_{LSM}\rangle = \langle \Phi_{GS}|\Phi_{LSM}\rangle$ if S is an integer; $\langle \Phi_{GS}|\Phi_{LSM}\rangle$ may be nonvanishing or vanishing.

We point out however that if we assume the existence of the Haldane gap, then the overlap $|\langle \Phi_{GS}|\Phi_{LSM}\rangle|$ must approach 1 as L grows. To see this let E_{1st} be the first excitation energy and assume that $E_{1st} \geq E_{GS} + \Delta E_H$ where ΔE_H is independent of L. Then noting that[15]

$$\begin{aligned} \hat{H} &\geq E_{GS}|\Phi_{GS}\rangle\langle\Phi_{GS}| + E_{1st}(1 - |\Phi_{GS}\rangle\langle\Phi_{GS}|) \\ &\geq E_{GS} + \Delta E_H(1 - |\Phi_{GS}\rangle\langle\Phi_{GS}|), \end{aligned} \tag{6.2.20}$$

we see that

$$\langle \Phi_{LSM}|\hat{H}|\Phi_{LSM}\rangle - E_{GS} \geq \Delta E_H\bigl(1 - \bigl|\langle\Phi_{GS}|\Phi_{LSM}\rangle\bigr|^2\bigr). \tag{6.2.21}$$

Since the left-hand side is bounded as in (6.2.5), we find

$$\bigl|\langle\Phi_{GS}|\Phi_{LSM}\rangle\bigr|^2 \geq 1 - \frac{8\pi^2 S^2}{\Delta E_H}\frac{1}{L}, \tag{6.2.22}$$

---

[13]The same statement was proved for the infinite system by Affleck and Lieb [5]. See the end of the present section.

[14]Such a statement should be proved for the infinite system. The proof is difficult since one cannot exclude the possibility that linear combinations of the ground state and low-lying excitations for finite L converge to multiple infinite volume ground states in the limit $L \uparrow \infty$.

[15]See Sect. A.2.3 for the meaning of inequalities between self-adjoint operators.

which implies $|\langle\Phi_{\mathrm{GS}}|\Phi_{\mathrm{LSM}}\rangle| \uparrow 1$ as $L \uparrow \infty$.[16] Although this is an elementary observation based on the nontrivial (and still mathematically unjustified) assumption on the existence of the gap, it is interesting to see that the ground state of an integer S chain is tolerant to the 2π twist. This is consistent with the expectation that the ground state only has short range correlation.

Generalizations of Lemma 6.1 The variational estimate in Lemma 6.1 can be generalized to a much larger class of Hamiltonians with short ranged $U(1)$ invariant interactions. Such a generalization is sometimes useful even without Lemma 6.2. See Problem 7.1.2.c. Let us discuss the generalization although it is not necessary for the present study of the standard antiferromagnetic Heisenberg chain.

Consider a quantum spin system on the chain $\Lambda_L = \{1, 2, \ldots, L\}$ with periodic boundary conditions. We study the general Hamiltonian

$$\hat{H} = \sum_{x=1}^{L} \hat{h}_x, \tag{6.2.23}$$

where the local Hamiltonian $\hat{h}_x$ acts only on sites y such that $|x - y| \le r$, where the range r is a fixed constant. (Note that we take into account periodic boundary conditions when defining the distance $|x - y|$.) We assume $\|\hat{h}_x\| \le h_0$ for any $x \in \Lambda_L$ with a constant h_0. The most crucial assumption is the U(1) invariance, i.e., $(\hat{U}_\theta^{(3)})^\dagger \hat{h}_x \hat{U}_\theta^{(3)} = \hat{h}_x$ for any x and any $\theta \in \mathbb{R}$.

Let $|\Phi_{\mathrm{GS}}\rangle$ be a ground state of $\hat{H}$. We do not assume the uniqueness. We define the LSM state $|\Phi_{\mathrm{LSM}}\rangle$ by (6.2.4) where the twist operator $\hat{U}_{\mathrm{LSM}}$ is define as in (6.2.2). Then we have the following generalization of Lemma 6.1.

Lemma 6.4 *There is a positive constant C which depends only on the constants S, r and h_0 such that*

$$\langle\Phi_{\mathrm{LSM}}|\hat{H}|\Phi_{\mathrm{LSM}}\rangle - E_{\mathrm{GS}} \le \frac{C}{L}, \tag{6.2.24}$$

for any L.

If S is a half-odd-integer and the ground state is translation invariant, i.e., $\hat{T}|\Phi_{\mathrm{GS}}\rangle = e^{i\alpha}|\Phi_{\mathrm{GS}}\rangle$, then Lemma 6.2 is again valid, and we see that $0 \le E_{\mathrm{1st}} - E_{\mathrm{GS}} \le C/L$ as in Theorem 6.3.

Proof of Lemma 6.4 We follow [60] and give a compact proof. Note that we use periodic boundary conditions throughout the proof. As in (6.2.11), we see that

$$\begin{aligned}\langle\Phi_{\mathrm{LSM}}|\hat{H}|\Phi_{\mathrm{LSM}}\rangle - E_{\mathrm{GS}} &= \langle\Phi_{\mathrm{GS}}|(\hat{U}_{\mathrm{LSM}}^\dagger \hat{H}\hat{U}_{\mathrm{LSM}} - \hat{H})|\Phi_{\mathrm{GS}}\rangle \\ &\le \langle\Phi_{\mathrm{GS}}|(\hat{U}_{\mathrm{LSM}}^\dagger \hat{H}\hat{U}_{\mathrm{LSM}} + \hat{U}_{\mathrm{LSM}}\hat{H}\hat{U}_{\mathrm{LSM}}^\dagger - 2\hat{H})|\Phi_{\mathrm{GS}}\rangle\end{aligned}$$

---

[16] It is believed that $\langle\Phi_{\mathrm{GS}}|\Phi_{\mathrm{LSM}}\rangle \simeq -1$ if the ground state exhibits Haldane gap [42].

$$
\begin{aligned}
&= \sum_{x=1}^{L} \langle \Phi_{\mathrm{GS}} | (\hat{U}_{\mathrm{LSM}}^{\dagger} \, \hat{h}_x \, \hat{U}_{\mathrm{LSM}} + \hat{U}_{\mathrm{LSM}} \, \hat{h}_x \, \hat{U}_{\mathrm{LSM}}^{\dagger} - 2\hat{h}_x) | \Phi_{\mathrm{GS}} \rangle \\
&\le \sum_{x=1}^{L} \| \hat{U}_{\mathrm{LSM}}^{\dagger} \, \hat{h}_x \, \hat{U}_{\mathrm{LSM}} + \hat{U}_{\mathrm{LSM}} \, \hat{h}_x \, \hat{U}_{\mathrm{LSM}}^{\dagger} - 2\hat{h}_x \|.
\end{aligned}
\tag{6.2.25}
$$

Note that, from the U(1) invariance and the locality of $\hat{h}_x$, we have

$$
\hat{U}_{\mathrm{LSM}}^{\dagger} \, \hat{h}_x \, \hat{U}_{\mathrm{LSM}} = \hat{U}_{\mathrm{LSM}}^{\dagger} \hat{U}_{\theta_x}^{(3)} \, \hat{h}_x \, (\hat{U}_{\theta_x}^{(3)})^{\dagger} \hat{U}_{\mathrm{LSM}} = \hat{U}_{x,-\Delta\theta} \, \hat{h}_x \, \hat{U}_{x,\Delta\theta}, \tag{6.2.26}
$$

where $\hat{U}_{x,\Delta\theta} = \exp[-i\,\Delta\theta \hat{M}_x]$ is a (local) twist operator which leaves the spin at x unchanged. We wrote $\Delta\theta = 2\pi/L$ as before, and set

$$
\hat{M}_x = \sum_{\substack{y \\ (|y-x|\le r)}} (y - x) \, \hat{S}_y^{(3)}. \tag{6.2.27}
$$

Similarly we have $\hat{U}_{\mathrm{LSM}} \, \hat{h}_x \, \hat{U}_{\mathrm{LSM}}^{\dagger} = \hat{U}_{x,\Delta\theta} \, \hat{h}_x \, \hat{U}_{x,-\Delta\theta}$. We shall show below that

$$
\| \hat{U}_{x,-\Delta\theta} \, \hat{h}_x \, \hat{U}_{x,\Delta\theta} + \hat{U}_{x,\Delta\theta} \, \hat{h}_x \, \hat{U}_{x,-\Delta\theta} - 2\hat{h}_x \| \le B(\Delta\theta)^2, \tag{6.2.28}
$$

with a constant B which depends only on S, r, and h_0. Since the left-hand side is nothing but the summand in the right-hand side of (6.2.25), we get the desired (6.2.24) with $C = (2\pi)^2 B$.

It remains to show (6.2.28). Note that the eigenvalue μ of $\hat{M}_x$ is in the range $-\mu_{\max} \le \mu \le \mu_{\max}$, where $\mu_{\max} = 2(1 + 2 + \cdots + r)S = r(r+1)S$. Let us express $\hat{h}_x$ as a polynomial of $\hat{S}_y^+$, $\hat{S}_y^-$, and $\hat{S}_y^{(3)}$ with y such that $|y - x| \le r$. It is clear that for each monomial $\hat{m}$ one has $[\hat{M}_x, \hat{m}] = \mu \hat{m}$ for some $\mu \in [-2\mu_{\max}, 2\mu_{\max}]$, i.e., $\hat{m}$ changes the eigenvalue of $\hat{M}_x$ by μ. Let $\hat{h}_x^{(\mu)}$ be the sum of all monomials (in the polynomial representation of $\hat{h}_x$) that have a common μ. We then have $\hat{h}_x = \sum_\mu \hat{h}_x^{(\mu)}$ and $[\hat{M}_x, \hat{h}_x^{(\mu)}] = \mu \, \hat{h}_x^{(\mu)}$, which implies[17]

$$
\hat{U}_{x,-\Delta\theta} \, \hat{h}_x^{(\mu)} \, \hat{U}_{x,\Delta\theta} = e^{i\mu\Delta\theta} \, \hat{h}_x^{(\mu)}. \tag{6.2.29}
$$

Then the left-hand side of (6.2.28) is evaluated as

$$
\| \sum_\mu (e^{i\mu\Delta\theta} + e^{-i\mu\Delta\theta} - 2) \hat{h}_x^{(\mu)} \| \le 2 \sum_\mu (1 - \cos \mu\Delta\theta) \| \hat{h}_x^{(\mu)} \| \le h_0 \Big(\sum_\mu \mu^2 \Big) (\Delta\theta)^2, \tag{6.2.30}
$$

where we noted that $\| \hat{h}_x^{(\mu)} \| \le h_0$. ∎

Extensions of the Lieb–Schultz–Mattis argument Let us discuss here in some detail various important extensions of the Lieb–Schultz–Mattis argument.

---

[17] Equation (6.2.29) is easily verified by differentiating both sides by $\Delta\theta$.

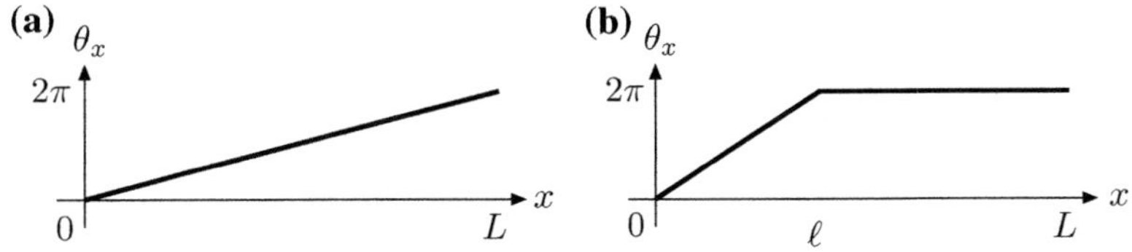

Fig. 6.2 The twist operator $\exp[-i\sum_{x=1}^{L}\theta_x\hat{S}_x^{(3)}]$ is determined by site dependent rotational angle θ_x. **a** In the global twist used here, θ_x gradually varies from 0 to 2π over the whole chain. **b** In the local twist, θ_x varies from 0 to 2π in a finite interval of length ℓ. By fixing ℓ and letting $L \uparrow \infty$, one can discuss the properties of low energy excitations in the infinite chain ()

Affleck and Lieb, in the paper [5] we mentioned above, introduced a very important notion of "local twist", and proved corresponding variational theorem. The original twist operator $\hat{U}_{\mathrm{LSM}}$ defined in (6.2.2), where the rotation angle θ_x varies slowly over the whole lattice, modifies the ground state in a global manner. In order to discuss properties of the system in the infinite volume limit, it is desirable to use an operator which acts locally on the state. This is realized by a local twist operator in which θ_x varies only in a finite (but large) part of the lattice as in Fig. 6.2 (b). In [5], properties of ground states and excitation in infinite chains are discussed using the local twist version of the theorem.

Oshikawa, Yamanaka, and Affleck [45] made an essential observation that the Lieb–Schultz–Mattis argument can be extended to spin chains in which the ground states have nonvanishing magnetization. This extension is relevant to the problem of quantum spin systems under external magnetic field. They found that the trial state constructed in the same manner as (6.2.4) using the global twist is orthogonal to the ground state under the assumption that the "filling factor" $\nu = (M/L) + S$ is not an integer. Here the magnetization M is the eigenvalue of $\hat{S}_{\mathrm{tot}}^{(3)}$ in the ground state. For $M = 0$, one has $\nu = S$, and recovers the result of Affleck and Lieb [5], which is our Lemma 6.2.

Yamanaka, Oshikawa, and Affleck [69] further extended their theory to systems of electrons on the one-dimensional lattice. Now the key quantity ν can be naturally interpreted as the filling factor of electrons. We believe that the two papers [45, 69] made clear the true meaning of the condition that S should be a half-odd-integer in Theorem 6.3. They played essential roles in extending the scope of the Lieb–Schultz–Mattis argument, and making possible many recent applications of the argument to much wider class of systems. Local twist versions of these results were derived in [59].

We also note that Aizenman and Nachtergaele [7] proved a Lieb–Schultz–Mattis type theorem, which, rather interestingly, applies to a different class of quantum spin chains. See also [8] for recent improved results by Aizenman, Duminil-Copin, and Warzel. For an interesting connection between the Lieb–Schultz–Mattis theorem and the Bloch theorem for the current in quantum systems, see, e.g., [64].

To properly extend the Lieb–Schultz–Mattis theorem to systems in two and higher dimensions is quite important, especially because higher dimensional systems may exhibit rich low-energy behavior accompanied by various topological order [66–68, 70] while one dimensional systems do not have intrinsic topological order [14, 15].

See Sect. 8.4. It was known however that a naive extension is problematic since a global twist in a d-dimensional lattice would increase the energy of the ground state by $O(L^{d-2})$ where L is the linear size of the system [39]. The energy increase converges to zero as $L \uparrow \infty$ only for $d = 1$.

A breakthrough was brought by Oshikawa [44], who proposed to make use of a combination of flux insertion and a gauge transformation instead of the twist operation. He showed that if a quantum system has a unique ground state accompanied by a finite energy gap then the filling factor must be an integer.[18] Hastings [31, 32] proposed a similar but different argument (with an additional assumption that the Hamiltonian is written as a real matrix) for extending the Lieb–Schultz–Mattis theorem to higher dimensions. Based on Hastings' idea Nachtergaele and Sims [41] proved a higher dimensional version of the Lieb–Schultz–Mattis theorem which is parallel to the original Theorem 6.3 (p. 162). Recently, Bachmann, Bols, De Roeck and Fraas proved a general index theorem [9], which lead to, among other things, a higher dimensional Lieb–Schultz–Mattis type theorem as was proposed by Oshikawa.

In [46, 63, 65], one finds recent (not yet completely rigorous) refinements of the Lieb–Schultz–Mattis argument (or, more precisely, Oshikawa's argument) which take into account the space group symmetry in higher dimensional systems. It is remarkable that Lieb–Schultz–Mattis type statements for models without U(1) symmetry are presented. See Sect. 8.3.5 for more about such results.

6.3 Semi-classical Approach

Here we will make a simple heuristic observation based on the semi-classical approach and a path integral picture, which sheds light on the difference between the spin chains with $S = 1/2$ and $S = 1$. It can be regarded as a small S version of Haldane's argument based on semiclassical quantization of solitons [26].

Write the Hamiltonian (6.1.1) of the spin S antiferromagnetic Heisenberg chain as

$$\hat{H} = \hat{H}_{\mathrm{c}} + \hat{H}_{\mathrm{q}}, \tag{6.3.1}$$

where $\hat{H}_{\mathrm{c}}$ and $\hat{H}_{\mathrm{q}}$ are the classical part and the quantum part, respectively, defined as

$$\hat{H}_{\mathrm{c}} := \sum_{x=1}^{L} \hat{S}_x^{(3)} \hat{S}_{x+1}^{(3)}, \quad \hat{H}_{\mathrm{q}} := \frac{1}{2} \sum_{x=1}^{L} \{\hat{S}_x^{+} \hat{S}_{x+1}^{-} + \hat{S}_x^{-} \hat{S}_{x+1}^{+}\}. \tag{6.3.2}$$

We here used the expression (2.2.16).

We here develop pictures of low energy properties of the model by treating $\hat{H}_{\mathrm{c}}$ as the main part of the Hamiltonian and treating $\hat{H}_{\mathrm{q}}$ as a "perturbation". Of course this is far from accurate since $\hat{H}_{\mathrm{q}}$ is indeed twice bigger than $\hat{H}_{\mathrm{c}}$. But such a treatment leads us to interesting observations. The reader who wants a firmer justification of the approach is invited to study the path integral method described at the end of the section.

[18] This rephrasing of the Lieb–Schultz–Mattis theorem was also essential for the later development of related theories. See the next paragraph and Sect. 8.3.5.

Since $\hat{H}_c$ is the Hamiltonian for the (classical) antiferromagnetic Ising model, its ground states are readily found to be the Néel state $|\Phi_{\text{Néel}}\rangle$ defined as (2.5.2) and its counterpart with S replaced by $-S$. Here the sublattices are defined as $A = \{2, 4, \ldots, L\}$ and $B = \{1, 3, \ldots, L-1\}$.

The $S = 1/2$ chain Let us examine how the Néel state $|\Phi_{\text{Néel}}\rangle$ of the $S = 1/2$ chain is modified when the "perturbation" $\hat{S}_x^+ \hat{S}_{x+1}^-$ or $\hat{S}_x^- \hat{S}_{x+1}^+$ acts on it. We represent a part of $|\Phi_{\text{Néel}}\rangle$ graphically as

(6.3.3)

where, of course, up and down arrows represent sites with $\sigma_x = 1/2$ and $-1/2$, respectively. When $\hat{S}_x^+ \hat{S}_{x+1}^-$ with some x acts on this state, the configuration changes according to (2.1.5) as

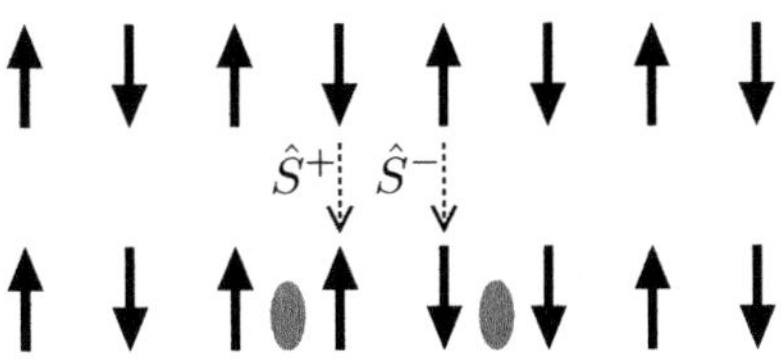

(6.3.4)

and we get an excited state of the classical Hamiltonian $\hat{H}_c$. Note that there are two bonds on which we have to pay extra energy. We have indicated these "unhappy" bonds with gray ovals, and call them kinks. The above process can be interpreted as a pair creation of kinks.

Further action of $\hat{S}_x^+ \hat{S}_{x+1}^-$ modifies the configuration, for example, as

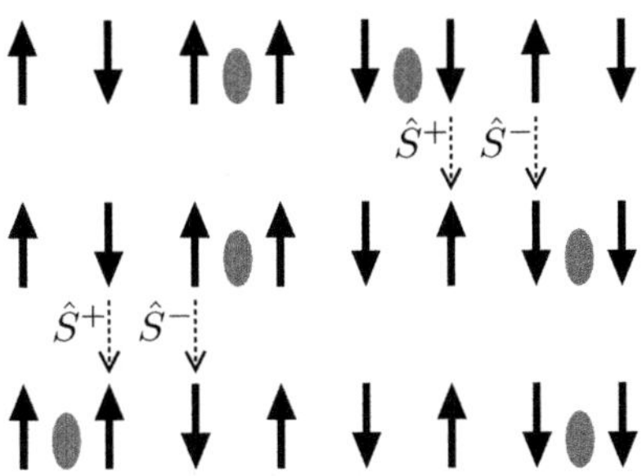

(6.3.5)

We see that the kinks can propagate on the chain. It is also evident that these kinks can be pairwise annihilated if they come close with each other again.

The reader may have noticed that the two kinks are separated by twice the lattice spacing when pair-created, and kinks always hop twice the lattice spacing. This is indeed an essential point. It means that a kink born in an even bond[19] will always live on even bonds, and a kink born on an odd bond will always live on odd bonds. One

[19]One may call a bond right to an even site even.

can say that there are two kinds of kinks, or that kinks have two "colors". As in the following diagram, two kinks with different colors can never be pairwise annihilated even though they come close to each other.

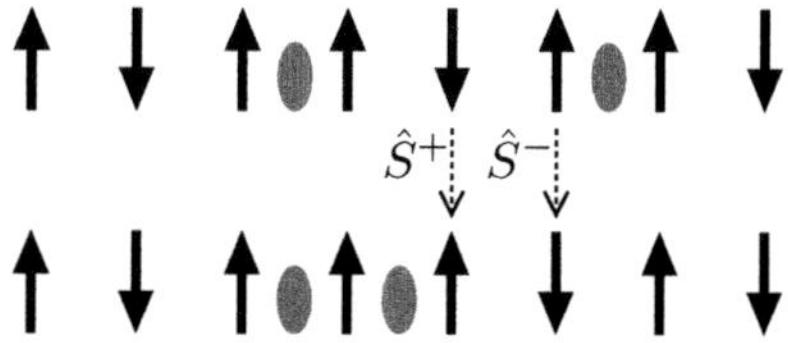

(6.3.6)

In conclusion, the $S = 1/2$ antiferromagnetic Heisenberg model can be viewed as a system of kinks with two colors. Only the kinks with the same color are pairwise created and annihilated. In fact this description is exact. See the path integral representation (6.3.13) derived at the end of the section.

The picture of colored kinks is consistent with the gapless nature of the model. When many kinks are generated, it may happen that neighboring kinks have different colors. Then these kinks cannot be pairwise annihilated, and will remain for a long time. This leads to a slow relaxation, which suggests vanishing energy gap.[20]

The $S = 1$ chain Let us make a similar analysis for the $S = 1$ antiferromagnetic Heisenberg chain. Denoting a site with $\sigma_x = 1, 0$, and -1 as $+$, 0, and $-$, respectively, the Néel state $|\Phi_{\text{Néel}}\rangle$ and the action of the perturbation $\hat{S}_x^+ \hat{S}_{x+1}^-$ on it can be represented as

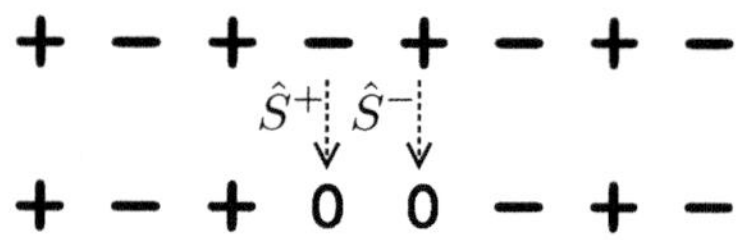

(6.3.7)

where we used (2.1.3). We see that a pair of neighboring 0's is created. Here we can identify 0 with a kink. So this is a pair creation of kinks. Further action of $\hat{S}_x^+ \hat{S}_{x+1}^-$ or $\hat{S}_x^- \hat{S}_{x+1}^+$ generates a "time-evolution" like

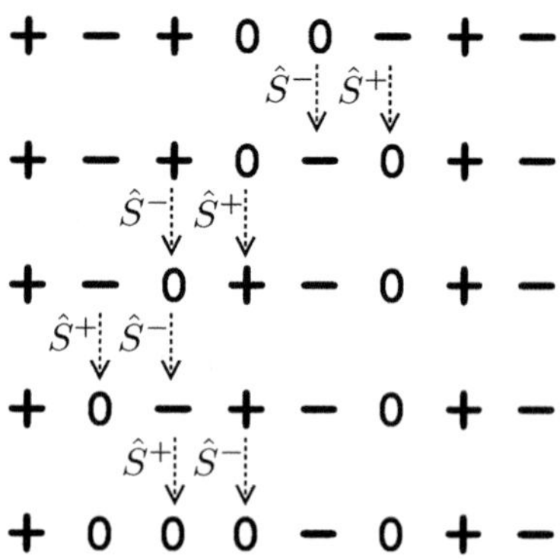

(6.3.8)

[20] We admit this argument is far from sufficient to conclude that the model is gapless.

where we see the propagation of kinks and another pair creation. We finally make an assumption that two kinks are pairwise annihilated through the inverse process of (6.3.7), thus recovering the Néel ordered configuration.[21] Unlike in the $S = 1/2$ case, two kinks are created on neighboring sites, kinks hop the unit lattice spacing, and two kinks which come nearby can be pairwise annihilated. We conclude that kinks in the $S = 1$ chain are colorless.

To sum, we have observed that the $S = 1$ antiferromagnetic Heisenberg chain may be viewed as a system of a single kind of kinks. Kinks are in fact 0's floating in the "sea" of Néel ordered background. They are pairwise created and annihilated, and freely hop around the chain. It is very likely that such (almost free) dynamics of kinks disturbs the system, and destroys the antiferromagnetic long-range order in the Neél state, leading to a disordered ground state with short-range correlation. This picture can be indeed made more precise in Proposition 6.5.

It must be noted however the states described by the above picture of "kinks in the sea of Néel order" does not cover the whole Hilbert space of the model. As can be seen from (6.3.7) and (6.3.8), pair-creation, pair-annihilation (that we allow), and motion of kinks always preserve the alternating ordering of $+$ and $-$. Therefore a state we obtain looks like the following:

$$+ \; - \; + \; 0 \; - \; 0 \; + \; - \; + \; 0 \; 0 \; 0 \; - \; 0 \; + \; - \tag{6.3.9}$$

There is a complete alternating order of $+$ and $-$, but arbitrary numbers of 0's are inserted in arbitrary positions. Such a spin configuration definitely lack long-range Néel order, but possesses a kind of hidden antiferromagnetic order. We shall see that such hidden antiferromagnetic order plays an essential role in the understanding of Haldane phenomena. See Sects. 7.2.1 and 8.1.2.

To state the following theorem, we define the Hilbert space with hidden antiferromagnetic order, which we denote as $\mathscr{H}_{\mathrm{HAF}}$. This is a highly artificial subspace with a nonlocal restriction, and introduced only to develop a physical picture. We define $\mathscr{H}_{\mathrm{HAF}}$ as a subspace (of the Hilbert space of the $S = 1$ chain) spanned by all the basis states $|\Psi^{\boldsymbol{\sigma}}\rangle$ (see (2.2.1) for the definition) which corresponds to spin configurations $\boldsymbol{\sigma} = (\sigma_x)_{x=1,\ldots,L}$ with complete hidden antiferromagnetic order as in (6.3.9).

Then we can state the following preliminary result suggested by the variational analysis of Gómez-Santos [23]. It supports the picture that the dynamics of 0's generates disordered ground state.[22]

Proposition 6.5 *The $S = 1$ antiferromagnetic Heisenberg chain defined on $\mathscr{H}_{\mathrm{HAF}}$ has a unique ground state with a finite energy gap and exponentially decaying correlation function.*

---

[21] We note that this may not be the case in general. In this sense our picture here for the $S = 1$ chain is only approximate.

[22] This proposition was proved by the present author in 1986, but was not published. It was later included as a part of [58].

Thus the Haldane "conjecture" for $S = 1$ is justified rigorously, but within a highly artificial restricted Hilbert space.

Path integral representation As a supplement to the present section, which may be skipped, we discuss the standard procedure to represent the ground state of a spin chain in terms of a "path integral", or, more precisely, a sum of space-time configurations of spins. This consideration provides a basis for the approach based on the "time-evolution" described above. By using the path integral representation, we further write down a representation of ground state expectation values in terms of a two-dimensional classical statistical mechanical system. Proposition 6.5 is proved by using this representation.

Here we focus on the ground state of the antiferromagnetic Heisenberg model, but the method automatically extends to the ground state and the equilibrium state of a general quantum spin system. An early use of such a representation is found in [21], where the existence of a phase transition in some quantum spin systems is proved. The representation is also used as the basis of quantum Monte Carlo method. See, e.g., [56].

Let $|\Phi_{\rm GS}\rangle$ be the ground state of the antiferromagnetic Heisenberg chain with arbitrary S. We fix an arbitrary spin configuration $\boldsymbol{\sigma}^{(0)}$ such that $\overline{\boldsymbol{\sigma}^{(0)}} := \sum_{x=1}^{L} \sigma_x^{(0)}=0$. Then the Marshall–Lieb–Mattis theorem (Theorem 2.2 in p. 39) guarantees that $\langle\Phi_{\rm GS}|\tilde{\Psi}^{\boldsymbol{\sigma}^{(0)}}\rangle > 0$, where $|\tilde{\Psi}^{\boldsymbol{\sigma}}\rangle$ is the basis state with extra sign defined in (2.5.8). Also noting that the ground state is nondegenerate, we find for any finite L that[23]

$$|\Phi_{\rm GS}\rangle = \lim_{\beta\uparrow\infty} \frac{e^{-\beta\hat{H}}|\tilde{\Psi}^{\boldsymbol{\sigma}^{(0)}}\rangle}{\|e^{-\beta\hat{H}}|\tilde{\Psi}^{\boldsymbol{\sigma}^{(0)}}\rangle\|}. \tag{6.3.10}$$

We thus focus on the state $e^{-\beta\hat{H}}|\tilde{\Psi}^{\boldsymbol{\sigma}^{(0)}}\rangle$, and, in particular, look for a representation for the coefficient $A_\beta(\boldsymbol{\sigma}) := \langle\tilde{\Psi}^{\boldsymbol{\sigma}}|e^{-\beta\hat{H}}|\tilde{\Psi}^{\boldsymbol{\sigma}^{(0)}}\rangle$.

Recalling the decomposition (6.3.1) with (6.3.2), and using the Lie product formula (Theorem A.1 in p. 465), we find[24]

$$\begin{aligned} e^{-\beta\hat{H}} &= \lim_{N\uparrow\infty} \left(e^{-\beta\hat{H}_{\rm c}/(2N)} e^{-\beta\hat{H}_{\rm q}/N} e^{-\beta\hat{H}_{\rm c}/(2N)}\right)^N \\ &= \lim_{N\uparrow\infty} \left\{e^{-\beta\hat{H}_{\rm c}/(2N)}\left(1 - \frac{\beta\hat{H}_{\rm q}}{N}\right)e^{-\beta\hat{H}_{\rm c}/(2N)}\right\}^N. \end{aligned} \tag{6.3.11}$$

---

[23] The proof is elementary. Let $|\Psi_j\rangle$ with $j = 1, 2, \ldots, D$ be the normalized energy eigenstate with eigenvalue E_j. We choose $|\Psi_1\rangle = |\Phi_{\rm GS}\rangle$. Then $e^{-\beta\hat{H}}|\tilde{\Psi}^{\boldsymbol{\sigma}^{(0)}}\rangle = \sum_{j=1}^{D} |\Psi_j\rangle e^{-\beta E_j}\langle\Psi_j|\tilde{\Psi}^{\boldsymbol{\sigma}^{(0)}}\rangle = e^{-\beta E_{\rm GS}}\{|\Phi_{\rm GS}\rangle\langle\Phi_{\rm GS}|\tilde{\Psi}^{\boldsymbol{\sigma}^{(0)}}\rangle + \sum_{j=2}^{D} e^{-\beta(E_j-E_{\rm GS})}|\Psi_j\rangle\langle\Psi_j|\tilde{\Psi}^{\boldsymbol{\sigma}^{(0)}}\rangle\}$. Noting that $e^{-\beta(E_j-E_{\rm GS})} \downarrow 0$ as $\beta\uparrow\infty$ if $j \ge 2$, we get (6.3.10).

[24] There are many ways to arrange the product in the right-hand side. One chooses an arrangement suitable for the purpose. See, e.g., [21, 35, 56, 58].

Define

$$A_{\beta,N}(\boldsymbol{\sigma}) := \langle\tilde{\Psi}^{\boldsymbol{\sigma}}|\Big\{e^{-\beta\hat{H}_{\rm c}/(2N)}\Big(1-\frac{\beta\hat{H}_{\rm q}}{N}\Big)e^{-\beta\hat{H}_{\rm c}/(2N)}\Big\}^N|\tilde{\Psi}^{\boldsymbol{\sigma}^{(0)}}\rangle, \tag{6.3.12}$$

which obviously satisfies $A_\beta(\boldsymbol{\sigma}) = \lim_{N\uparrow\infty} A_{\beta,N}(\boldsymbol{\sigma})$. By inserting the completeness relation $\sum_{\boldsymbol{\sigma}} |\tilde{\Psi}^{\boldsymbol{\sigma}}\rangle\langle\tilde{\Psi}^{\boldsymbol{\sigma}}| = \hat{1}$ in between the N products we get the path-integral representation

$$A_{\beta,N}(\boldsymbol{\sigma}) = \sum_{\boldsymbol{\sigma}^{(1)},\boldsymbol{\sigma}^{(2)},\ldots,\boldsymbol{\sigma}^{(N-1)}} \prod_{j=1}^{N} w_{\beta,N}(\boldsymbol{\sigma}^{(j)},\boldsymbol{\sigma}^{(j-1)}), \tag{6.3.13}$$

with

$$\begin{aligned} w_{\beta,N}(\boldsymbol{\sigma},\boldsymbol{\sigma}') &= e^{-\beta\{E_{\rm c}(\boldsymbol{\sigma})+E_{\rm c}(\boldsymbol{\sigma}')\}/(2N)}\langle\tilde{\Psi}^{\boldsymbol{\sigma}}|\Big(1-\frac{\beta\hat{H}_{\rm q}}{N}\Big)|\tilde{\Psi}^{\boldsymbol{\sigma}'}\rangle \\ &= e^{-\beta\{E_{\rm c}(\boldsymbol{\sigma})+E_{\rm c}(\boldsymbol{\sigma}')\}/(2N)}\langle\tilde{\Psi}^{\boldsymbol{\sigma}}|\{1-\frac{\beta}{2N}\sum_{x=1}^{L}(\hat{S}_x^+\hat{S}_{x+1}^- + \hat{S}_x^-\hat{S}_{x+1}^+)\}|\tilde{\Psi}^{\boldsymbol{\sigma}'}\rangle. \end{aligned} \tag{6.3.14}$$

In (6.3.13), $\boldsymbol{\sigma}^{(j)}$ (with $j = 1,\ldots,N-1$) is summed over all the spin configurations,[25] and we set $\boldsymbol{\sigma}^{(N)} = \boldsymbol{\sigma}$. We also introduced the classical energy $E_{\rm c}(\boldsymbol{\sigma}) = \sum_{x=1}^{L}\sigma_x\sigma_{x+1}$. It is clear from the proof of the Marshall–Lieb–Mattis theorem (Theorem 2.2 in p. 39) that the definition of the basis states $|\tilde{\Psi}^{\boldsymbol{\sigma}}\rangle$ implies that $w_{\beta,N}(\boldsymbol{\sigma},\boldsymbol{\sigma}') \geq 0$. We also have $w_{\beta,N}(\boldsymbol{\sigma},\boldsymbol{\sigma}') = w_{\beta,N}(\boldsymbol{\sigma}',\boldsymbol{\sigma})$.

We are now ready to give a precise interpretation to the "time-evolution" of configurations we considered in the previous part of the present section. For $j = 0, 1, \ldots, N$, we regard $\boldsymbol{\sigma}^{(j)}$ as the configuration at time j. The fixed configuration $\boldsymbol{\sigma}^{(0)}$ is then regarded as the initial configuration. Likewise $w_{\beta,N}(\boldsymbol{\sigma}^{(j)},\boldsymbol{\sigma}^{(j-1)})$ is interpreted as the "transition amplitude" from the configuration $\boldsymbol{\sigma}^{(j-1)}$ to $\boldsymbol{\sigma}^{(j)}$. Examining the definition (6.3.14) of $w_{\beta,N}(\boldsymbol{\sigma},\boldsymbol{\sigma}')$, one finds that the possible transitions are realized by operating either $\hat{S}_x^+\hat{S}_{x+1}^-$, $\hat{S}_x^-\hat{S}_{x+1}^+$, or 1 to $|\tilde{\Psi}^{\boldsymbol{\sigma}'}\rangle$. If we choose the initial configuration as $\boldsymbol{\sigma}^{(0)} = (-S, S, \ldots, -S, S)$, which results in $|\tilde{\Psi}^{\boldsymbol{\sigma}^{(0)}}\rangle = |\Phi_{\text{Néel}}\rangle$, we completely recover the picture of "time-evolution" as depicted in (6.3.4)–(6.3.8). Note that the sum over "histories" $(\boldsymbol{\sigma}^{(1)},\boldsymbol{\sigma}^{(2)},\ldots,\boldsymbol{\sigma}^{(N-1)})$ in (6.3.13) precisely corresponds to the sum of over trajectories in the conventional path integral.

From the path integral representation (6.3.13) we can derive the standard representation (6.3.18) of ground state expectation values in terms of expectation values in a classical statistical mechanical system. The representation will be used in Sect. 8.1.2.

Let $\hat{F} = F(\hat{S}_1^{(3)}, \hat{S}_2^{(3)}, \ldots, \hat{S}_L^{(3)})$ be an arbitrary self adjoint operator which depends only on $\hat{S}_x^{(3)}$ with $x = 1, \ldots, L$. Noting that (6.3.10) implies

[25] In fact it is enough to sum over $\boldsymbol{\sigma}^{(j)}$ such that $\overline{\boldsymbol{\sigma}^{(j)}} = 0$.

$$|\Phi_{GS}\rangle = \lim_{\beta\uparrow\infty} \frac{\sum_{\boldsymbol{\sigma}} A_\beta(\boldsymbol{\sigma})|\tilde{\Psi}^{\boldsymbol{\sigma}}\rangle}{\sqrt{\sum_{\boldsymbol{\sigma}}\{A_\beta(\boldsymbol{\sigma})\}^2}}, \tag{6.3.15}$$

we get

$$\langle\Phi_{GS}|\hat{F}|\Phi_{GS}\rangle = \lim_{\beta\uparrow\infty} \frac{\sum_{\boldsymbol{\sigma}} F(\boldsymbol{\sigma})\{A_\beta(\boldsymbol{\sigma})\}^2}{\sum_{\boldsymbol{\sigma}}\{A_\beta(\boldsymbol{\sigma})\}^2}. \tag{6.3.16}$$

By using the symmetry $w_{\beta,N}(\boldsymbol{\sigma},\boldsymbol{\sigma}') = w_{\beta,N}(\boldsymbol{\sigma}',\boldsymbol{\sigma})$, we find from (6.3.13) that the weight $\{A_{\beta,N}(\boldsymbol{\sigma})\}^2$ is written as

$$\{A_{\beta,N}(\boldsymbol{\sigma})\}^2 = \sum_{\boldsymbol{\sigma}^{(1)},\ldots,\boldsymbol{\sigma}^{(N-1)},\boldsymbol{\sigma}^{(N+1)},\ldots,\boldsymbol{\sigma}^{(2N-1)}} \prod_{j=1}^{2N} w_{\beta,N}(\boldsymbol{\sigma}^{(j)},\boldsymbol{\sigma}^{(j-1)}), \tag{6.3.17}$$

where $\boldsymbol{\sigma}^{(0)}$ is fixed as before, and also $\boldsymbol{\sigma}^{(2N)}$ is fixed as $\boldsymbol{\sigma}^{(2N)} = \boldsymbol{\sigma}^{(0)}$. Finally $\boldsymbol{\sigma}^{(N)}$ (in the right-hand side) is set equal to $\boldsymbol{\sigma}$ that appears in the left-hand side. Substituting this into (6.3.16), we obtain the desired representation

$$\langle\Phi_{GS}|\hat{F}|\Phi_{GS}\rangle = \lim_{\beta\uparrow\infty}\lim_{N\uparrow\infty} \frac{\sum_{\boldsymbol{\sigma}^{(1)},\ldots,\boldsymbol{\sigma}^{(2N-1)}} F(\boldsymbol{\sigma}^{(N)}) \prod_{j=1}^{2N} w_{\beta,N}(\boldsymbol{\sigma}^{(j)},\boldsymbol{\sigma}^{(j-1)})}{\sum_{\boldsymbol{\sigma}^{(1)},\ldots,\boldsymbol{\sigma}^{(2N-1)}} \prod_{j=1}^{2N} w_{\beta,N}(\boldsymbol{\sigma}^{(j)},\boldsymbol{\sigma}^{(j-1)})}. \tag{6.3.18}$$

Here the sum is over all $\boldsymbol{\sigma}^{(1)},\ldots,\boldsymbol{\sigma}^{(2N-1)}$ (including $\boldsymbol{\sigma}^{(N)}$), and the "boundary conditions" $\boldsymbol{\sigma}^{(0)} = \boldsymbol{\sigma}^{(2N)}$ are fixed.

The quantity in the limit in (6.3.18) may be interpreted as the expectation value of F in a two-dimensional classical statistical mechanical system, where $(\boldsymbol{\sigma}^{(0)},\boldsymbol{\sigma}^{(1)},\ldots,\boldsymbol{\sigma}^{(2N)})$ (which we interpreted above as a "history") is now regarded as a spin configuration on the (space-time) lattice $\{1,2,\ldots,L\}\times\{0,1,\ldots,2N\}$, and $\prod_{j=1}^{2N} w_{\beta,N}(\boldsymbol{\sigma}^{(j)},\boldsymbol{\sigma}^{(j-1)})$ is the corresponding Boltzmann factor. As we have noted before such a representation is possible in a much more general class of quantum spin systems. A quantum spin system in d-dimension corresponds to a classical statistical mechanical system in $(d+1)$-dimension.

References

1. I. Affleck, The quantum Hall effects, σ-models at $\Theta = \pi$ and quantum spin chains. Nucl. Phys. B **257**, 397–406 (1985)
2. I. Affleck, Quantum spin chains and the Haldane gap. J. Phys. Condens. Matter **1**, 3047 (1989)
3. I. Affleck, Model for quasi-one-dimensional antiferromagnets: application to $CsNiCl_3$. Phys. Rev. Lett. **62** (1989). Errata: Phys. Rev. Lett. **62**, 1927 (1989); Phys. Rev. Lett. **65**, 2477 (1990); Phys. Rev. Lett. **65**, 2835 (1990)
4. I. Affleck, Field theory methods and quantum critical phenomena, in *Fields, Strings and Critical Phenomena, Proceedings of the 1988 Les Houches Summer School*, ed. by E. Brézin, J. Zinn-Justin (North-Holland, Amsterdam, 1990), pp. 563–640

5. I. Affleck, E.H. Lieb, A proof of part of Haldane's conjecture on spin chains. Lett. Math. Phys. **12**, 57–69 (1986)
6. I. Affleck, D. Gepner, H.J. Schulz, T. Ziman, Critical behaviour of spin-s Heisenberg antiferromagnetic chains: analytic and numerical results. J. Phys. A: Math. Gen. **22**, 511 (1989)
7. M. Aizenman, B. Nachtergaele, Geometric aspects of quantum spin states. Commun. Math. Phys. **164**, 17–63 (1994), https://projecteuclid.org/euclid.cmp/1104270709
8. M. Aizenman, H. Duminil-Copin, S. Warzel, Dimerization and Néel order in different quantum spin chains through a shared loop representation, preprint (2020), arXiv:2002.02543
9. S. Bachmann, A. Bols, W. De Roeck, M. Fraas, A many-body index for quantum charge transport. Commun. Math. Phys. (2019), arXiv:1810.07351
10. V. Barzykin, I. Affleck, Finite-size scaling for the spin-Heisenberg antiferromagnetic chain. J. Phys. A: Math. Gen. **32**, 867 (1999), arXiv:cond-mat/9810075
11. H. Bethe, Zur Theorie der Metalle I: Eigenwerte und Eigenfunktionen der linearen Atomkette. Z. Phys. **71**, 205–226 (1931)
12. R. Botet, R. Jullien, M. Kolb, Finite-size-scaling study of the spin-1 Heisenberg-Ising chain with uniaxial anisotropy. Phys. Rev. B **28**, 3914 (1983)
13. W.J.L. Buyers, R.M. Morra, R.L. Armstrong, M.J. Hogan, P. Gelrach, K. Hirakawa, Experimental evidence for the Haldane gap in a spin-1 nearly isotropic, antiferromagnetic chain. Phys. Rev. Lett. **56**, 371 (1986)
14. X. Chen, Z.-C. Gu, X.-G. Wen, Classification of gapped symmetric phases in one-dimensional spin systems. Phys. Rev. B **83**, 035107 (2011), arXiv:1008.3745
15. X. Chen, Z.-C. Gu, X.-G. Wen, Complete classification of 1D gapped quantum phases in interacting spin systems. Phys. Rev. B **84**, 235128 (2011), arXiv:1103.3323
16. M. Date, K. Kindo, Elementary excitation in the Haldane state. Phys. Rev. Lett. **65**, 1659 (1990)
17. J. des Cloizeaux, J.J. Pearson, Spin-wave spectrum of the antiferromagnetic linear chain. Phys. Rev. **128**, 2131 (1962)
18. R. Fernández, J. Fröhlich, A.D. Sokal, *Random Walks, Critical Phenomena, and Triviality in Quantum Field Theory* (Springer, Berlin, 1992)
19. E. Fradkin, *Field Theories of Condensed Matter Physics* (Cambridge University Press, Cambridge, 2013)
20. S. Friedli, Y. Velenik, *Statistical Mechanics of Lattice Systems: a Concrete Mathematical Introduction* (Cambridge University Press, Cambridge, 2017)
21. J. Ginibre, Existence of phase transitions for quantum lattice systems. Commun. Math. Phys. **14**, 205–234 (1969), https://projecteuclid.org/euclid.cmp/1103841776
22. O. Golinelli, Th. Jolicoeur, R. Lacaze, Finite-lattice extrapolations for a Haldane-gap antiferromagnet. Phys. Rev. B **50**, 3037 (1994)
23. G. Gómez-Santos, Variational approach to the XXZ spin-1 linear chain: elementary excitations and Haldane conjecture. Phys. Rev. Lett. **63**, 790 (1989)
24. F.D.M. Haldane, Ground state properties of antiferromagnetic chains with unrestricted spin: integer spin chains as realisations of the $O(3)$ non-linear sigma model, ILL preprint SP-81/95 (1981), arXiv:1612.00076
25. F.D.M. Haldane, Continuum dynamics of the 1-D Heisenberg antiferromagnet: identification with the $O(3)$ nonlinear sigma model. Phys. Lett. A **93**, 464–468 (1983), http://www.sciencedirect.com/science/article/pii/037596018390631X
26. F.D.M. Haldane, Nonlinear field theory of large-spin Heisenberg antiferromagnets: semiclassically quantized solitons of the one-dimensional easy-axis Néel state. Phys. Rev. Lett. **50**, 1153–1156 (1983). https://doi.org/10.1103/PhysRevLett.50.1153
27. F.D.M. Haldane, "Θ physics" and quantum spin chains (abstract). J. Appl. Phys. **57**, 3359 (1985). https://doi.org/10.1063/1.335096
28. F.D.M. Haldane, Topological quantum matter, nobel lecture (2016). See, in particular, from 39m 40s, https://www.youtube.com/watch?v=47Ab1izxH9Y
29. F.D.M. Haldane, Nobel lecture: topological quantum matter. Rev. Mod. Phys. **89**, 040502 (2017)

30. T. Hara, H. Tasaki, *Mathematics and Physics of Phase Transitions and Critical Phenomena* (in Japanese) (Kyoritsu, Tokyo, 2015)
31. M.B. Hastings, Lieb-Schultz-Mattis in higher dimensions. Phys. Rev. B **69**, 104431 (2004), arXiv:0305505
32. M.B. Hastings, Sufficient conditions for topological order in insulators. Eur. Phys. Lett. **70**, 824–830 (2005), arXiv:cond-mat/0411094
33. M.B. Hastings, T. Koma, Spectral gap and exponential decay of correlations. Commun. Math. Phys. **256**, 781–804 (2006), arXiv:math-ph/0507008
34. T. Kennedy, Ornstein-Zernike decay in the ground state of the quantum Ising model in a strong transverse field. Commun. Math. Phys. **137**, 599–615 (1991), https://projecteuclid.org/euclid.cmp/1104202743
35. T. Kennedy, H. Tasaki, Hidden symmetry breaking and the Haldane phase in $S = 1$ quantum spin chains. Commun. Math. Phys. **147**, 431–484 (1992), https://projecteuclid.org/euclid.cmp/1104250747
36. T. Koma, Spectral gaps of quantum Hall systems with interactions. J. Stat. Phys. **99**, 313–381 (2000), arXiv:cond-mat/9809228
37. V.E. Korepin, N.M. Bogoliubov, A.G. Izergin, *Quantum Inverse Scattering Method and Correlation Functions* (Cambridge University Press, Cambridge, 2010)
38. Yu.A. Kosevich, A.V. Chubukov, Fluctuation corrections to the spectra of one- and two-dimensional Heisenberg magnets. Zh. Eksp. Teor. Fiz. **91**, 1105 (1986) [Sov. Phys. JETP **64**, 654 (1986)], http://www.jetp.ac.ru/cgi-bin/e/index/e/64/3/p654?a=list
39. E.H. Lieb, T. Schultz, D. Mattis, Two soluble models of an antiferromagnetic chain. Ann. Phys. **16**, 407–466 (1961)
40. S. Lukyanov, V. Terras, Long-distance asymptotics of spin-spin correlation functions for the XXZ spin chain. Nucl. Phys. B **654**, 323–356 (2003), arXiv:hep-th/0206093
41. B. Nachtergaele, R. Sims, A multi-dimensional Lieb-Schultz-Mattis theorem. Commun. Math. Phys. **276**, 437–472 (2007), arXiv:math-ph/0608046
42. M. Nakamura, S. Todo, Order parameter to characterize valence-bond-solid states in quantum spin chains. Phys. Rev. Lett. **89**, 077204 (2002), arXiv:cond-mat/0112377
43. M.P. Nightingale, H.W.J. Blöte, Gap of the linear spin-1 Heisenberg antiferromagnet: a Monte Carlo calculation. Phys. Rev. B **33**, 659 (1986)
44. M. Oshikawa, Commensurability, excitation gap, and topology in quantum many-particle systems on a periodic lattice. Phys. Rev. Lett. **84**, 1535 (2000), arXiv:cond-mat/9911137
45. M. Oshikawa, M. Yamanaka, I. Affleck, Magnetization plateaus in spin chains: "Haldane gap" for half-integer spins. Phys. Rev. Lett. **78**, 1984 (1997), arXiv:cond-mat/9610168
46. S.A. Parameswaran, A.M. Turner, D.P. Arovas, A. Vishwanath, Topological order and absence of band insulators at integer filling in non-symmorphic crystals. Nat. Phys. **9**, 299–303 (2013), arXiv:1212.0557
47. J.B. Parkinson, J.C. Bonner, Spin chains in a field: crossover from quantum to classical behavior. Phys. Rev. B **32**, 4703 (1985)
48. L.P. Regnault, I. Zaliznyak, J.P. Renard, C. Vettier, Inelastic-neutron-scattering study of the spin dynamics in the Haldane-gap system $Ni(C_2H_8N_2)_2NO_2(ClO_4)$. Phys. Rev. B **50**, 9174 (1994)
49. J.P. Renard, M. Verdaguer, L.P. Regnault, W.A.C. Erkelens, J. Rossat-Mignod, W.G. Stirling, Presumption for a quantum energy gap in the quasi-one-dimensional $S = 1$ Heisenberg antiferromagnet $Ni(C_2\ H_8N_2)_2\ NO_2\ (ClO_4)$. Europhys. Lett. **3**, 945 (1987)
50. J.P. Renard, M. Verdaguer, L.P. Regnault, W.A.C. Erkelens, J. Rossat-Migdot, W.G. Stirling, C. Vettier, Quantum energy gap in two quasi-one-dimensional $S = 1$ Heisenberg antiferromagnets. J. Appl. Phys. **63**, 3538 (1988)
51. S. Sachdev, *Quantum Phase Transitions* (Cambridge University Press, Cambridge, 2011)
52. T. Sakai, M. Takahashi, Effect of the Haldane gap on quasi-one-dimensional systems. Phys. Rev. B **42**, 4537–4543 (1990)
53. U. Schollwöck, Th. Jolicoeur, Haldane gap and hidden order in the $S = 2$ antiferromagnetic quantum spin chain. Europhys. Lett. **30**, 493 (1995), arXiv:cond-mat/9501115

54. M. Steiner, K. Kakurai, J.K. Kjems, D. Petitgrand, R. Pynn, Inelastic neutron scattering studies on 1D near Heisenberg antiferromagnets: a test of the Haldane conjecture. J. Appl. Phys. **61**, 3953 (1987)
55. B. Sutherland, *Beautiful Models — 70 Years of Exactly Solved Quantum Many-Body Problems* (World Scientific, Singapore, 2004)
56. M. Suzuki (ed.), *Quantum Monte Carlo Methods in Condensed Matter Physics* (World scientific, Singapore, 1993)
57. M. Takahashi, *Thermodynamics of One-Dimensional Solvable Models* (Cambridge University Press, Cambridge, 2009)
58. H. Tasaki, Haldane gap in three dimensions: a rigorous example. Phys. Rev. Lett. **64**, 2066–2069 (1990)
59. H. Tasaki, Lieb-Schultz-Mattis theorem with a local twist for general one-dimensional quantum systems. J. Stat. Phys. **170**, 653–671 (2018), arXiv:1708.05186
60. H. Tasaki, Topological phase transition and $\mathbb{Z}_2$ index for $S = 1$ quantum spin chains. Phys. Rev. Lett. *121*, 140604 (2018), arXiv:1804.04337
61. S. Todo, K. Kato, Cluster algorithms for general-S quantum spin systems. Phys. Rev. Lett. **87**, 047203 (2001), arXiv:cond-mat/9911047
62. S. Todo, H. Matsuo, H. Shitara, Parallel loop cluster quantum Monte Carlo simulation of quantum magnets based on global union-find graph algorithm. Comput. Phys. Commun. **239**, 84–93 (2019), arXiv:1810.07485
63. H. Watanabe, The Lieb-Schultz-Mattis-type filling constraints in the 1651 magnetic space groups. Phys. Rev. B **97**, 165117 (2018), arXiv:1802.00587
64. H. Watanabe, A proof of the Bloch theorem for lattice models. J. Stat. Phys. **177**, 717–726 (2019), arXiv:1904.02700
65. H. Watanabe, H.C. Po, A. Vishwanath, M.P. Zaletel, Filling constraints for spin-orbit coupled insulators in symmorphic and nonsymmorphic crystals. Proc. Natl. Acad. Sci. U.S.A. **112**, 14551–14556 (2015), http://www.pnas.org/content/112/47/14551
66. X.-G. Wen, *Quantum Field Theory of Many-Body Systems: From the Origin of Sound to an Origin of Light and Electrons* (Oxford University Press, Oxford, 2007)
67. X.-G. Wen, Colloquium: zoo of quantum-topological phases of matter. Rev. Mod. Phys. **89**, 41004 (2017), arXiv:1610.03911
68. X.-G. Wen, Choreographed entanglement dances: topological states of quantum matter. Science **363**, 834 (2019), http://science.sciencemag.org/content/363/6429/eaal3099
69. M. Yamanaka, M. Oshikawa, I. Affleck, Nonperturbative approach to Luttinger's theorem in one dimension. Phys. Rev. Lett. **79**, 1110 (1997), arXiv:cond-mat/9701141
70. B. Zeng, X. Chen, D.-L. Zhou, X.-G. Wen, *Quantum Information Meets Quantum Matter: From Quantum Entanglement to Topological Phase in Many-Body Systems* (Quantum Science and Technology, Springer, Berlin, 2019), arXiv:1508.02595

Chapter 7
Affleck–Kennedy–Lieb–Tasaki Model

In 1987, Affleck, Kennedy, Lieb, and Tasaki proposed a one-dimensional $S = 1$ antiferromagnetic quantum spin model, now called the AKLT model, whose ground state can be written down explicitly. It was shown that the energy spectrum of the model has a finite gap above the ground state energy, and the ground state correlation function decays exponentially [4, 5]. These results agree with Haldane's conclusion for the integer spin antiferromagnetic Heisenberg chain. But note that this does not prove the Haldane "conjecture" since the model is different from the Heisenberg model. Nevertheless this work proves that there exists an $S = 1$ chain that has a unique disordered ground state with a gap, and hence provides a strong support to Haldane's conclusion. It also provides a starting point of further studies of various phenomena and concepts related to the Haldane gap, as we shall see in Chap. 8. Now it is commonly believed that the AKLT model is (in some sense) at the "center" of the Haldane phase. In Sect. 7.1 we give a detailed discussion of the main results about the AKLT model. In Sect. 7.2 we study some unexpected features of the model, which will turn out to be essential for the study of Haldane phenomena. In Sect. 7.3 we discuss extensions and closely related models.

7.1 The Model and Main Results

We discuss and prove basic properties of the AKLT model in the present section. After giving a precise definition of the model and stating the main theorem in Sect. 7.1.1, we discuss in Sect. 7.1.2 how the exact ground state of the model, called the VBS state, is constructed. We then prove the uniqueness of the ground state and the existence of a gap in Sects. 7.1.3 and 7.1.4, respectively.

H. Tasaki, *Physics and Mathematics of Quantum Many-Body Systems*,
Graduate Texts in Physics, https://doi.org/10.1007/978-3-030-41265-4_7

7.1.1 The Hamiltonian and the Main Theorem

The Affleck–Kennedy–Lieb–Tasaki (AKLT) model [4, 5] is an $S = 1$ quantum spin chain with Hamiltonian

$$\hat{H}_{\text{AKLT}} = \sum_{x=1}^{L} \{\hat{\boldsymbol{S}}_x \cdot \hat{\boldsymbol{S}}_{x+1} + \frac{1}{3}(\hat{\boldsymbol{S}}_x \cdot \hat{\boldsymbol{S}}_{x+1})^2\}, \tag{7.1.1}$$

where we use the periodic boundary condition $\hat{\boldsymbol{S}}_{L+1} = \hat{\boldsymbol{S}}_1$. The model is still antiferromagnetic, and is SU(2) invariant. But the Marshall–Lieb–Mattis theorem does not apply to this model because of the biquadratic term $(\hat{\boldsymbol{S}}_x \cdot \hat{\boldsymbol{S}}_{x+1})^2/3$. Let us note in passing that, in models with $S = 1/2$, adding the biquadratic term does not define a new model; there is an identity $(\hat{\boldsymbol{S}}_x \cdot \hat{\boldsymbol{S}}_{x+1})^2 = (3/16) - \hat{\boldsymbol{S}}_x \cdot \hat{\boldsymbol{S}}_{x+1}/2$.[1]

As we shall see in Sect. 7.1.2, one can explicitly write down the exact ground state of the AKLT Hamiltonian (7.1.1).[2] The ground state is called the valence-bond solid (VBS) state and denoted as $|\Phi_{\text{VBS}}\rangle$. By making full use of the exact ground state, the following was proved in [4, 5].

Theorem 7.1 *For any L (which can be either even or odd), the ground state of the Hamiltonian* (7.1.1) *is unique. The energy gap above the ground state energy is not less than* ΔE_0 *for sufficiently large L, where* ΔE_0 *is a positive constant independent of L. The correlation function in the ground state satisfies*

$$\lim_{L\uparrow\infty} \frac{\langle \Phi_{\text{VBS}} | \hat{\boldsymbol{S}}_x \cdot \hat{\boldsymbol{S}}_y | \Phi_{\text{VBS}} \rangle}{\langle \Phi_{\text{VBS}} | \Phi_{\text{VBS}} \rangle} = 4\,(-3)^{-|x-y|}, \tag{7.1.2}$$

for any x, y *such that* $|x - y| \geq 1$.

We shall give a complete proof of the theorem in the following sections. We construct the exact ground state $|\Phi_{\text{VBS}}\rangle$ in Sect. 7.1.2. Then we prove the uniqueness of the ground state in Sect. 7.1.3, prove the existence of a gap in Sect. 7.1.4, and evaluate the ground state correlation function (by using the matrix product representation) in Sect. 7.2.2. In these sections, we do not reproduce the original proof in [4, 5], but try to present different arguments which are easier to understand, and, more importantly, interesting by themselves.

The constant ΔE_0 is a lower bound for the Haldane gap for the Hamiltonian (7.1.1).[3] Although it was only proved that $\Delta E_0 > 0$ in the original work [5], an explicit lower bound $\Delta E_0 \geq 0.496$ was proved later by Knabe in [44]. We will reproduce this proof in Sect. 7.1.4. The existence of a gap was also proved by Fannes,

---

[1] From (2.2.19), we see that $\hat{\boldsymbol{S}}_x \cdot \hat{\boldsymbol{S}}_{x+1}$ takes two eigenvalues, $-3/4$ and $1/4$. The identity then follows.

[2] Some exact excited states were obtained in [9, 64, 65]. Unlike in integrable models, it is not possible to obtain general excited states.

[3] From numerical works, the energy gap of the AKLT model for large L is estimated to be about 0.7. See, e.g., [32, 34].

Nachtergaele, and Werner within the general theory of matrix product states [27, 28]. See Sect. 7.2.2. A powerful general method for proving the existence of a gap, called the martingale method, was later developed by Nachtergaele [67, 79].

The right-hand side of (7.1.2) can be written as $4(-1)^{x-y}e^{-|x-y|/\xi}$ with $\xi = 1/\log 3 \simeq 0.91$. We thus have exponentially decaying correlation as Haldane concluded. Compared with the general formula (6.1.3), however, we see that the power law correction $|x-y|^{-1/2}$ (which is necessary for the Ornstein–Zernike form) is missing here.[4] This suggests that the model is in some sense special since a disordered ground state of a one-dimensional quantum spin system generically exhibits the behavior (6.1.3). It is believed that the correction $|x-y|^{-1/2}$ is restored by adding a generic small perturbation to the Hamiltonian (7.1.1). See, e.g., [69] and references therein.

The Hamiltonian (7.1.1) is a special case of the general $S = 1$ Hamiltonian with bilinear and biquadratic interaction

$$\hat{H}_\theta^{\mathrm{BLBQ}} = \sum_{x=1}^{L}\big\{\cos\theta\,(\hat{\boldsymbol{S}}_x \cdot \hat{\boldsymbol{S}}_{x+1}) + \sin\theta\,(\hat{\boldsymbol{S}}_x \cdot \hat{\boldsymbol{S}}_{x+1})^2\big\}. \tag{7.1.3}$$

This is the most general translation invariant Hamiltonian for $S = 1$ quantum spin chain with SU(2) invariant nearest neighbor interaction. The model is known to have surprisingly rich ground state phase diagram. It is in the Haldane phase in a finite interval of θ including the Heisenberg antiferromagnet with $\theta = 0$ and the AKLT model with $\tan\theta = 1/3$. It also has a dimerized phase, a ferromagnetic phase, a critical phase, and probably some more exotic phases. See, e.g., [46] and references therein.

Infinite chain It is possible to extend Theorem 7.1, which is valid for an arbitrary finite L, to the corresponding system on the infinite chain. We emphasize that to consider infinite systems is a kind of idealization, which make us possible to directly concentrate on the universal features of macroscopic systems. See Appendix A.7 for a brief review of the operator algebraic formulation of quantum spin systems on infinite lattices.

In the original paper [5], the uniqueness of the infinite volume ground state within translation invariant states was proved. (See Definition 4.17 in p. 113.) Later, Matsui [53] proved the following stronger theorem, which we do not prove here. It fully characterizes the ground state of the AKLT model on the infinite chain.

Theorem 7.2 *The AKLT model on the infinite chain has a unique ground state accompanied by a nonzero gap in the sense of Definitions A.25 and A.27.*

The ground state of the infinite chain of course coincides with the proper infinite volume limit of the finite volume ground state, i.e., the VBS state (7.1.12).

Stability theorems Theorem 7.1 establishes that a disordered unique ground state with a gap is possible, but only for the special Hamiltonian (7.1.1). The gapful nature

---

[4] See footnote 4 in p. 156.

of the model suggests that basic features of the ground state are stable against small perturbations. This is indeed the case but the proof turned out to be highly nontrivial. The difficulty comes partly from the fact that the ground state is not simply disordered but has a certain hidden nontrivial structure as we shall see in Sect. 7.2.1.

To state a stability theorem, let

$$\hat{H}_\varepsilon := \hat{H}_{\text{AKLT}} + \varepsilon \sum_{x=1}^{L} \hat{v}_x, \tag{7.1.4}$$

where $\hat{v}_o$ is an arbitrary self-adjoint operator which acts only on a finite number of spins, and $\hat{v}_x = \hat{T}^x\, \hat{v}_o\, (\hat{T}^\dagger)^x$ is its translation. An example is the Hamiltonian (7.1.3) with $\tan\theta \simeq 1/3$. We state the important theorem by Yarotsky [91] without proof. It was proved by using a sophisticated version of the cluster expansion.

Theorem 7.3 *Suppose that $|\varepsilon|$ is sufficiently small. There exists a positive constant $\Delta E_\varepsilon > 0$ independent of L. For any L, the ground state of the Hamiltonian* (7.1.4) *is unique, and the energy gap above the ground state energy is not less than ΔE_ε. Correlation functions in the ground state decay exponentially.*

Such a statement about the stability is quite important since it establishes that the behavior found in the exactly solvable model (7.1.1) is robust and hence is physically meaningful. The first stability theorem related to the Haldane gap was proved in [42], but the theorem applies only to a modified version of the AKLT model with bond alternation. Recently a more general stability theorem was proved by Michalakis and Zwolak [57], which contains Theorem 7.3 as a special case.

7.1.2 The Exact Ground State

Let us construct the exact ground state of the AKLT model (7.1.1). The most important observation is the following simple identity:

$$\hat{\boldsymbol{S}}_x \cdot \hat{\boldsymbol{S}}_{x+1} + \frac{1}{3}(\hat{\boldsymbol{S}}_x \cdot \hat{\boldsymbol{S}}_{x+1})^2 = 2\, \hat{P}_2[\hat{\boldsymbol{S}}_x + \hat{\boldsymbol{S}}_{x+1}] - \frac{2}{3} \tag{7.1.5}$$

By $\hat{P}_J[\hat{\boldsymbol{J}}]$, where $\hat{\boldsymbol{J}}$ is a general angular momentum, we denote the projection operator onto the space of $|\Psi\rangle$ such that $\hat{\boldsymbol{J}}^2|\Psi\rangle = J(J+1)|\Psi\rangle$. Recall that, in the $S = 1$ chain, the total spin of two sites x and $x+1$ can be either 0, 1, or 2. (See Appendix A.3.3.) Thus $\hat{P}_2[\hat{\boldsymbol{S}}_x + \hat{\boldsymbol{S}}_{x+1}]$ is the projection onto the space where the total spin takes the maximum value, which is 2. The identity (7.1.5) is readily verified by noting that

$$\begin{aligned}\hat{\mathbf{S}}_x \cdot \hat{\mathbf{S}}_{x+1} &= \frac{1}{2}\{(\hat{\mathbf{S}}_x + \hat{\mathbf{S}}_{x+1})^2 - (\hat{\mathbf{S}}_x)^2 - (\hat{\mathbf{S}}_{x+1})^2\} \\ &= \frac{1}{2}\Big\{2(2+1)\,\hat{P}_2[\hat{\mathbf{S}}_x + \hat{\mathbf{S}}_{x+1}] + 1(1+1)\,\hat{P}_1[\hat{\mathbf{S}}_x + \hat{\mathbf{S}}_{x+1}]\Big\} - 2 \\ &= 3\,\hat{P}_2[\hat{\mathbf{S}}_x + \hat{\mathbf{S}}_{x+1}] + \hat{P}_1[\hat{\mathbf{S}}_x + \hat{\mathbf{S}}_{x+1}] - 2. \end{aligned} \tag{7.1.6}$$

The expression (7.1.5) suggests that the AKLT Hamiltonian (7.1.1) describes antiferromagnetic interaction which is in some sense weaker than the standard antiferromagnetic interaction $\hat{\mathbf{S}}_x \cdot \hat{\mathbf{S}}_{x+1}$. This is because $\hat{P}_2[\hat{\mathbf{S}}_x + \hat{\mathbf{S}}_{x+1}]$ gives a penalty only to states of the two spins with total spin 2, and treats states with total spin 0 or 1 equally, while the interaction $\hat{\mathbf{S}}_x \cdot \hat{\mathbf{S}}_{x+1}$ gives a certain penalty to states with total spin 1 and a larger penalty to those with total spin 2.

The identity (7.1.5) implies that the Hamiltonian $\hat{H}_{\mathrm{AKLT}}$ of (7.1.1) is equivalent to the new Hamiltonian

$$\hat{H}'_{\mathrm{AKLT}} = \sum_{x=1}^{L} \hat{P}_2[\hat{\mathbf{S}}_x + \hat{\mathbf{S}}_{x+1}]. \tag{7.1.7}$$

In particular the two Hamiltonians share the same ground state. In the following we explicitly construct a state $|\Phi_{\mathrm{VBS}}\rangle$, called the VBS (valence-bond solid) state, which satisfies $\hat{P}_2[\hat{\mathbf{S}}_x + \hat{\mathbf{S}}_{x+1}]|\Phi_{\mathrm{VBS}}\rangle = 0$ for each x. Since $\hat{P}_2[\hat{\mathbf{S}}_x + \hat{\mathbf{S}}_{x+1}] \geq 0$, Lemma A.9 (p. 469) about frustration-free Hamiltonians implies that $|\Phi_{\mathrm{VBS}}\rangle$ is an exact ground state of $\hat{H}'_{\mathrm{AKLT}}$ and hence of $\hat{H}_{\mathrm{AKLT}}$.

Construction of the VBS state Before constructing the VBS state $|\Phi_{\mathrm{VBS}}\rangle$, we review the standard procedure to represent a spin with $S = 1$ in terms of two spins with $S = 1/2$.[5] Take two spins with $S = 1/2$, and call them L and R (which stand for left and right, respectively). The state space of the spins is spanned by the four basis states $|\psi_{\mathrm{L}}^{\sigma}\rangle|\psi_{\mathrm{R}}^{\sigma'}\rangle$, where $\sigma, \sigma' = \uparrow, \downarrow$. We then define the symmetrization operator $\mathscr{S}$ by

$$\mathscr{S}\big(|\psi_{\mathrm{L}}^{\sigma}\rangle|\psi_{\mathrm{R}}^{\sigma'}\rangle\big) = \frac{1}{2}\big\{|\psi_{\mathrm{L}}^{\sigma}\rangle|\psi_{\mathrm{R}}^{\sigma'}\rangle + |\psi_{\mathrm{L}}^{\sigma'}\rangle|\psi_{\mathrm{R}}^{\sigma}\rangle\big\}. \tag{7.1.8}$$

The symmetrized state always has total spin 1. To see this, one only needs to check that $\mathscr{S}(|\psi_{\mathrm{L}}^{\uparrow}\rangle|\psi_{\mathrm{R}}^{\downarrow}\rangle) = \mathscr{S}(|\psi_{\mathrm{L}}^{\downarrow}\rangle|\psi_{\mathrm{R}}^{\uparrow}\rangle) = (|\psi_{\mathrm{L}}^{\uparrow}\rangle|\psi_{\mathrm{R}}^{\downarrow}\rangle + |\psi_{\mathrm{L}}^{\downarrow}\rangle|\psi_{\mathrm{R}}^{\uparrow}\rangle)/2$ is one of the triplet (A.3.22). If we denote the basis states of a spin $S = 1$ by $|\psi^{+}\rangle$, $|\psi^{0}\rangle$, and $|\psi^{-}\rangle$, we have

$$\begin{gathered}\mathscr{S}\big(|\psi_{\mathrm{L}}^{\uparrow}\rangle|\psi_{\mathrm{R}}^{\uparrow}\rangle\big) = |\psi^{+}\rangle, \quad \mathscr{S}\big(|\psi_{\mathrm{L}}^{\downarrow}\rangle|\psi_{\mathrm{R}}^{\downarrow}\rangle\big) = |\psi^{-}\rangle, \\ \mathscr{S}\big(|\psi_{\mathrm{L}}^{\uparrow}\rangle|\psi_{\mathrm{R}}^{\downarrow}\rangle\big) = \mathscr{S}\big(|\psi_{\mathrm{L}}^{\downarrow}\rangle|\psi_{\mathrm{R}}^{\uparrow}\rangle\big) = \frac{1}{\sqrt{2}}|\psi^{0}\rangle. \end{gathered} \tag{7.1.9}$$

[5] In actual magnetic materials, a spin $S = 1$ comes from two $S = 1/2$'s (of electrons) coupled by Hund's rule. But this mathematical construction is valid irrespective of the origin of the spin $S = 1$.

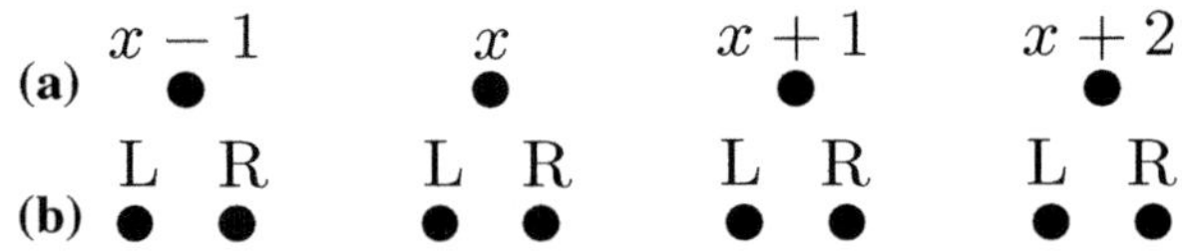

Fig. 7.1 **a** Part of the original chain with L sites. **b** The corresponding part of the duplicated chain with $2L$ sites

Since $\mathscr{S}^2 = \mathscr{S}$, the operator $\mathscr{S}$ is the projection onto the subspace with total spin 1. If we further denote by $\hat{\mathbf{S}}_\mathrm{L}$ and $\hat{\mathbf{S}}_\mathrm{R}$ the $S = 1/2$ spin operators acting on $|\psi_\mathrm{L}^\sigma\rangle$ and $|\psi_\mathrm{R}^\sigma\rangle$, respectively, and by $\hat{\mathbf{S}}$ the $S = 1$ operator acting on $|\psi^\sigma\rangle$, it holds that

$$\hat{\mathbf{S}}\,\mathscr{S} = \mathscr{S}\,(\hat{\mathbf{S}}_\mathrm{L} + \hat{\mathbf{S}}_\mathrm{R}). \tag{7.1.10}$$

This operator identity is trivial if one recalls that the $S = 1$ spin is obtained by adding two $S = 1/2$ spins.[6] See Appendix A.3.3.

In order to construct a state on the chain with L sites $x = 1, \ldots, L$, we consider the duplicated chain with $2L$ sites obtained by doubling each site x. The new sites corresponding to x are denoted as (x, L) and (x, R). See Fig. 7.1. We then consider a quantum spin system of $S = 1/2$ spins on the duplicated chain, and define

$$|\Phi_\text{pre-VBS}\rangle := \bigotimes_{x=1}^{L} \frac{1}{\sqrt{2}}\bigl\{|\psi_{x,\mathrm{R}}^\uparrow\rangle|\psi_{x+1,\mathrm{L}}^\downarrow\rangle - |\psi_{x,\mathrm{R}}^\downarrow\rangle|\psi_{x+1,\mathrm{L}}^\uparrow\rangle\bigr\}, \tag{7.1.11}$$

which is a product of singlet pairs (A.3.23) of two $S = 1/2$'s. (See Appendix A.3.3 if the reader is unfamiliar with the notion of the spin-singlet.) Let us call this state the pre-VBS state. See Fig. 7.2a, where the diagram ●——● represents a singlet pair. Being a linear combination of two different spin configurations ↑↓ and ↓↑, a singlet pair can be the source of "quantum fluctuation". A singlet pair is sometimes called a valence bond since two electrons forming a bonding orbit have vanishing total spin. Note that the pre-VBS state (7.1.11) is a simple tensor product of local states, and spins not connected by valence bonds are uncorrelated. In particular spins on sites x and y are completely independent if $|x - y| \geq 2$.

We can then construct a state on the original $S = 1$ chain with L sites by applying the symmetrization operator (7.1.8) to each site x as

$$|\Phi_\text{VBS}\rangle = \Bigl(\bigotimes_{x=1}^{L} \mathscr{S}_x\Bigr)|\Phi_\text{pre-VBS}\rangle. \tag{7.1.12}$$

[6] ***Proof*** Because of the SU(2) invariance, it suffices to show that $\hat{S}^{(3)}\mathscr{S} = \mathscr{S}(\hat{S}_\mathrm{L}^{(3)} + \hat{S}_\mathrm{R}^{(3)})$. This is easily verified by operating the both sides to $|\psi_\mathrm{L}^\sigma\rangle|\psi_\mathrm{R}^{\sigma'}\rangle$.

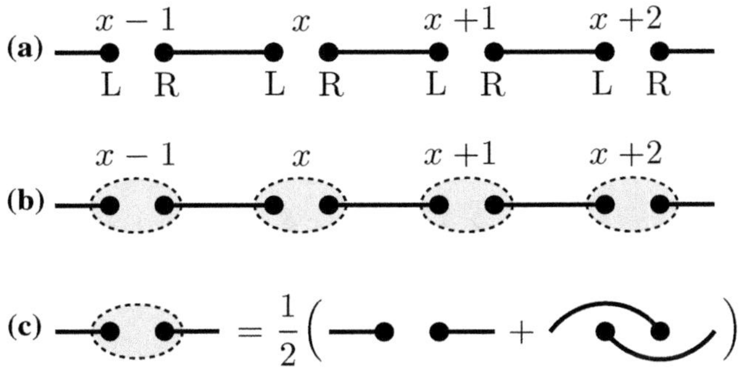

Fig. 7.2 A black dot in the figure represents a spin with $S = 1/2$, and ●——● denotes the singlet pair (also called a valence bond) $\{|\uparrow\rangle|\downarrow\rangle - |\downarrow\rangle|\uparrow\rangle\}/\sqrt{2}$ consisting of two $S = 1/2$'s. **a** The pre-VBS state $|\Phi_{\text{pre-VBS}}\rangle$ is a simple tensor product of valence bonds. **b** The valence-bond solid (VBS) state $|\Phi_{\text{VBS}}\rangle$ is obtained by applying the symmetrization operator, which is depicted in (c), to each pair of $S = 1/2$ spins on the same (original) site. Thus $|\Phi_{\text{VBS}}\rangle$ is a state of the original $S = 1$ chain with L sites (© Hal Tasaki 2020. All Rights Reserved)

See Fig. 7.2b, c. Note that the state $|\Phi_{\text{VBS}}\rangle$ is not normalized. See (7.2.25). This state was named the valence-bond solid (VBS) state, for the obvious reason that here valence bonds (whose ends are symmetrized) compactly cover the whole lattice.[7,8] But we shall argue in Sect. 7.2.1 that the VBS state has properties which resemble that of liquid.

We stress that VBS state (7.1.12) is no longer a simple tensor product state since we applied symmetrization operator to each site. As is suggested in Fig. 7.2c, the symmetrization induces correlation between sites which are not necessarily directly connected by valence bonds. This results in the nonvanishing correlation as in (7.1.2). See Problems 7.1.2.a and 7.1.2.b below.

Let us show that the VBS state (7.1.12) satisfies $\hat{P}_2[\hat{\boldsymbol{S}}_x + \hat{\boldsymbol{S}}_{x+1}]|\Phi_{\text{VBS}}\rangle = 0$. A key observation is that, in our construction of the $S = 1$ chain from the $S = 1/2$ duplicated chain, the spin operator $\hat{\boldsymbol{S}}_x + \hat{\boldsymbol{S}}_{x+1}$ for two $S = 1$'s is identical to $\hat{\boldsymbol{S}}_{x,\text{L}} + \hat{\boldsymbol{S}}_{x,\text{R}} + \hat{\boldsymbol{S}}_{x+1,\text{L}} + \hat{\boldsymbol{S}}_{x+1,\text{R}}$ for the corresponding four $S = 1/2$'s. To be more precise, we note that (7.1.10) leads to the operator identity

$$(\hat{\boldsymbol{S}}_x + \hat{\boldsymbol{S}}_{x+1})\Bigl(\bigotimes_{x=1}^{L} \mathscr{S}_x\Bigr) = \Bigl(\bigotimes_{x=1}^{L} \mathscr{S}_x\Bigr)(\hat{\boldsymbol{S}}_{x,\text{L}} + \hat{\boldsymbol{S}}_{x,\text{R}} + \hat{\boldsymbol{S}}_{x+1,\text{L}} + \hat{\boldsymbol{S}}_{x+1,\text{R}}). \quad (7.1.13)$$

---

[7] We believe that the term "valence-bond solid state" was first used in [4] to mean the specific class of states, which includes (7.1.12), introduced in [4, 5]. But recently it is also common to call a simple tensor product of valence bonds the VBS state. See footnote 10 below. The state (7.1.12) is also known as the AKLT state.

[8] This construction has been implemented in an optical experiment, and the "photonic VBS state" was generated [40]. The experiment is motivated by measurement based quantum computation with the VBS state [17, 60, 63].

From this we find that

$$\begin{aligned}\hat{P}_2[\hat{\boldsymbol{S}}_x + \hat{\boldsymbol{S}}_{x+1}]|\Phi_{\text{VBS}}\rangle &= \hat{P}_2[\hat{\boldsymbol{S}}_x + \hat{\boldsymbol{S}}_{x+1}]\Big(\bigotimes_{x=1}^{L}\mathscr{S}_x\Big)|\Phi_{\text{pre-VBS}}\rangle \\ &= \Big(\bigotimes_{x=1}^{L}\mathscr{S}_x\Big)\hat{P}_2[\hat{\boldsymbol{S}}_{x,\text{L}} + \hat{\boldsymbol{S}}_{x,\text{R}} + \hat{\boldsymbol{S}}_{x+1,\text{L}} + \hat{\boldsymbol{S}}_{x+1,\text{R}}]\,|\Phi_{\text{pre-VBS}}\rangle \\ &= 0, \end{aligned} \tag{7.1.14}$$

where the final equality follows by noting that the spins on (x, R) and $(x+1, \text{L})$ already form a spin singlet in $|\Phi_{\text{pre-VBS}}\rangle$, and the remaining two spins on (x, L) and $(x+1, \text{R})$ can have total spin 0 or 1.

It still remains to show that $|\Phi_{\text{VBS}}\rangle$ is nonvanishing since the projection (symmetrization) in (7.1.12) may annihilate the state. To see that $|\Phi_{\text{VBS}}\rangle \neq 0$ when L is even, it suffices to see that $\langle\Phi_{\text{Néel}}|\Phi_{\text{VBS}}\rangle \neq 0$, where $|\Phi_{\text{Néel}}\rangle$ is the Néel state defined in (2.5.2) with the sublattices $A = \{2, 4, \ldots, L\}$ and $B = \{1, 3, \ldots, L-1\}$.[9] To compute the overlap, think about expanding the state (7.1.11) into the sum of 2^L product states. Clearly the Néel state is obtained when one picks up the term $|\psi^{\uparrow}_{x,\text{R}}\rangle|\psi^{\downarrow}_{x+1,\text{L}}\rangle$ for $x \in A$ and $|\psi^{\downarrow}_{x,\text{R}}\rangle|\psi^{\uparrow}_{x+1,\text{L}}\rangle$ for $x \in B$. Recalling (7.1.9), we see that $\langle\Phi_{\text{Néel}}|\Phi_{\text{VBS}}\rangle = (-1/2)^{L/2}$. We shall compute the normalization $\langle\Phi_{\text{VBS}}|\Phi_{\text{VBS}}\rangle$ explicitly in Sect. 7.2.2. See (7.2.25).

We thus conclude that the VBS state (7.1.12) is an exact ground state of the AKLT Hamiltonian (7.1.1). Theorem 7.1 (p. 178) is proved by making full use of the exact ground state. In the following Sects. 7.1.3 and 7.1.4, we prove the uniqueness of the ground state and the existence of an energy gap. The reader interested in physical properties of the AKLT model and the VBS state may skip the proof and jump to Sect. 7.2 (after taking a look at the following discussion about the Majumdar–Ghosh model).

The following two exercises are elementary but may be illuminating.

Problem 7.1.2.a Let $L = 2$. Write down $|\Phi_{\text{pre-VBS}}\rangle$ and $|\Phi_{\text{VBS}}\rangle$ explicitly. (Be careful about the periodic boundary condition.) Compute $(\hat{\boldsymbol{S}}_1 + \hat{\boldsymbol{S}}_2)^2|\Phi_{\text{VBS}}\rangle$. What does the result mean? [solution→p.504]

Problem 7.1.2.b Let $L = 3$, and write down $|\Phi_{\text{pre-VBS}}\rangle$ and $|\Phi_{\text{VBS}}\rangle$ explicitly. Examine the total spin on sites 1 and 2, and show that there are no components with $S_{\text{tot}} = 2$. [solution→p.505]

The Majumdar–Ghosh model Before proceeding let us discuss an important model of an $S = 1/2$ spin chain, called the Majumdar–Ghosh model, which resembles the AKLT model in certain aspects. Consider a quantum spin system with spin $S = 1/2$ on a chain with L sites, where L is even. The Hamiltonian of the Majumdar–Ghosh model [51, 52] is

[9] When L is odd, one can use the state which almost is the Néel state, with one extra 0. The argument is then essentially the same.

$$\hat{H}_{\mathrm{MG}} = \sum_{x=1}^{L} \{ \hat{\boldsymbol{S}}_x \cdot \hat{\boldsymbol{S}}_{x+1} + \frac{1}{2} \hat{\boldsymbol{S}}_x \cdot \hat{\boldsymbol{S}}_{x+2} \}, \tag{7.1.15}$$

where we use periodic boundary conditions $\hat{\boldsymbol{S}}_x = \hat{\boldsymbol{S}}_{x+L}$. Note that there are nearest neighbor antiferromagnetic interaction and weaker next-nearest-neighbor antiferromagnetic interaction.

Like the AKLT Hamiltonian, the Majumdar–Ghosh Hamiltonian (7.1.15) is exactly rewritten by using projection operators as

$$\hat{H}_{\mathrm{MG}} = \frac{3}{4} \sum_{x=1}^{L} \{ \hat{P}_{3/2}[\hat{\boldsymbol{S}}_x + \hat{\boldsymbol{S}}_{x+1} + \hat{\boldsymbol{S}}_{x+2}] - \frac{1}{2} \}, \tag{7.1.16}$$

where $\hat{P}_{3/2}[\hat{\boldsymbol{S}}_x + \hat{\boldsymbol{S}}_{x+1} + \hat{\boldsymbol{S}}_{x+2}]$ is the projection onto the subspace where the total spin at sites x, $x+1$, and $x+2$ takes the highest value 3/2. This means that if there exists a state $|\Phi\rangle$ with the property that $\hat{P}_{3/2}[\hat{\boldsymbol{S}}_x + \hat{\boldsymbol{S}}_{x+1} + \hat{\boldsymbol{S}}_{x+2}]|\Phi\rangle = 0$ for any $x = 1, \ldots, L$ then $|\Phi\rangle$ is an exact ground state, and the ground state energy is $E_{\mathrm{GS}} = -3L/8$. In fact the two dimer states[10]

$$|\Phi^{\mathrm{odd}}_{\mathrm{dimer}}\rangle = \bigotimes_{y=1}^{L/2} \frac{1}{\sqrt{2}} \Big(|\psi^{\uparrow}_{2y-1}\rangle |\psi^{\downarrow}_{2y}\rangle - |\psi^{\downarrow}_{2y-1}\rangle |\psi^{\uparrow}_{2y}\rangle \Big), \tag{7.1.17}$$

$$|\Phi^{\mathrm{even}}_{\mathrm{dimer}}\rangle = \bigotimes_{y=1}^{L/2} \frac{1}{\sqrt{2}} \Big(|\psi^{\uparrow}_{2y}\rangle |\psi^{\downarrow}_{2y+1}\rangle - |\psi^{\downarrow}_{2y}\rangle |\psi^{\uparrow}_{2y+1}\rangle \Big), \tag{7.1.18}$$

have the desired property. To see this note that, in these states, any three contiguous sites x, $x+1$, and $x+2$ contain exactly one spin singlet, and hence the total spin of the three sites is alway 1/2.

The Hamiltonian (7.1.15) and the states (7.1.17), (7.1.18) are introduced and studied by Majumdar and Ghosh [51, 52]. In [81], it was proved that (7.1.17) and (7.1.18) are ground states. Note that, unlike the VBS state, the dimer states are simple tensor products in which sufficiently separated spins are not correlated at all. We also remark that the two dimer states are not exactly orthogonal but are almost orthogonal as $\langle \Phi^{\mathrm{odd}}_{\mathrm{dimer}} | \Phi^{\mathrm{even}}_{\mathrm{dimer}} \rangle = 2(-1/2)^{L/2}$.

It was proved that the two states (7.1.17) and (7.1.18) (and their liner combinations) are the only ground states of the Hamiltonian (7.1.15) [20], and further that there is a nonvanishing energy gap above the ground state energy [5]. This resembles the Haldane gap, but the crucial difference is that the ground states are doubly degenerate in this model. We also note that the Lieb–Schultz–Mattis argument (with

[10] Here "dimer", as opposed to "monomer", stands for a pair of neighboring sites. The product state of singlet pairs which cover the whole lattice site is called the dimer state (or, more precisely, the dimer covering state). It is sometimes called the VBS (valence-bond solid) state. See footnote 7 above.

proper extension as in Lemma 6.4 in p. 163) applies to this model. The existence of a gap does not lead to contradiction, again because the ground states are degenerate. This situation is worth investigating. See Problem 7.1.2.c.

In [6] Aizenman, Duminil-Copin, and Warzel study a class of quantum spin chains with nearest neighbor interactions that have spontaneously dimerized ground states, and discuss interesting connection to another class of spin chains that have Néel ordered ground states.

Problem 7.1.2.c Here we will examine how the Lieb–Schultz–Mattis argument (Theorem 6.3 in p. 162) applies to the Majumdar–Ghosh model. First try twisting one of the dimer states as $|\Phi'\rangle = \hat{U}_{\mathrm{LSM}}|\Phi^{\mathrm{odd}}_{\mathrm{dimer}}\rangle$. We see from Lemma 6.4 that the state $|\Phi'\rangle$ has energy expectation value close to E_{GS}. Compute the overlap $\langle\Phi^{\mathrm{odd}}_{\mathrm{dimer}}|\Phi'\rangle$, especially its limiting value as $L\uparrow\infty$. Why doesn't Lemma 6.2 apply? Next consider the translation invariant ground state $|\Phi^{+}_{\mathrm{dimer}}\rangle = |\Phi^{\mathrm{odd}}_{\mathrm{dimer}}\rangle + |\Phi^{\mathrm{even}}_{\mathrm{dimer}}\rangle$, and examine the nature of its twist $|\Phi''\rangle = \hat{U}_{\mathrm{LSM}}|\Phi^{+}_{\mathrm{dimer}}\rangle$. In particular how does the state $|\Phi''\rangle$ look like? [solution→p.506]

7.1.3 The Uniqueness of the Ground State

After studying the above construction of the VBS state and the proof that it is an exact ground state of the Hamiltonian (7.1.1), the reader might think it rather obvious that the VBS state is the unique ground state. The "proof" goes as follows. Since a ground state $|\Phi_{\mathrm{GS}}\rangle$ must satisfy $\hat{P}_2[\hat{\boldsymbol{S}}_x + \hat{\boldsymbol{S}}_{x+1}]|\Phi_{\mathrm{GS}}\rangle = 0$, the pair of spins at x and $x+1$ (when expressed as four $S = 1/2$'s) must contain at least one singlet-pair. Because this is true for each x, the only imaginable way to get a ground state is to fill the lattice with singlet-pairs as in Fig. 7.2b. What else can we do?

The actual proof is indeed not very different from this heuristic argument. Let us describe a proof by Kennedy, Lieb, and Tasaki [41]. We start from the following (almost trivial) Lemma which locally characterizes a ground state.

Lemma 7.4 *A state $|\Phi\rangle$ (of the $S = 1$ chain with L sites) satisfies*

$$\hat{P}_2[\hat{\boldsymbol{S}}_x + \hat{\boldsymbol{S}}_{x+1}]|\Phi\rangle = 0 \tag{7.1.19}$$

if and only if it is written as

$$|\Phi\rangle = (\mathscr{S}_x \otimes \mathscr{S}_{x+1})\Big(\big\{|\psi^{\uparrow}_{x,\mathrm{R}}\rangle|\psi^{\downarrow}_{x+1,\mathrm{L}}\rangle - |\psi^{\downarrow}_{x,\mathrm{R}}\rangle|\psi^{\uparrow}_{x+1,\mathrm{L}}\rangle\big\} \otimes |\Xi\rangle\Big) \tag{7.1.20}$$

with some state $|\Xi\rangle$.

Proof It is obvious from the consideration in the previous section that $|\Phi\rangle$ written as (7.1.20) satisfies (7.1.19). We shall count the dimensions to show the converse. In the 9 dimensional state space of two $S = 1$'s, the subspaces with total spin 0, 1, and 2 have dimensions 1, 3, and 5, respectively. Thus the dimension of the space of

$|\Phi\rangle$ satisfying (7.1.19) is $(1+3)\times 3^{L-2}$. Next note that $|\Phi\rangle$ obtained from (7.1.20) with $|\Xi\rangle$ given by

$$|\Xi\rangle = |\psi^{\sigma}_{x,\mathrm{L}}\rangle|\psi^{\sigma'}_{x+1,\mathrm{R}}\rangle \otimes \Bigl(\bigotimes_{y\in\Lambda\backslash\{x,x+1\}} |\psi^{\sigma_y}_y\rangle\Bigr) \tag{7.1.21}$$

with all possible $\sigma, \sigma' = \uparrow, \downarrow$ and $\sigma_y = 0, \pm 1$ (with $y \in \Lambda\backslash\{x, x+1\}$) are linearly independent. Thus the dimension of the space of states written as (7.1.20) is $2^2 3^{L-2}$. Since the dimensions coincide we see that the two spaces are identical. ∎

We shall make the latter half of the argument (i.e., "what else can we do?") rigorous by using classical results in mathematics. The Weyl representation [88] of a single $S = 1$ spin plays an essential role. We associate with each basis state a monomial in the two variables u and v as

$$|\psi^+\rangle \leftrightarrow u^2, \quad |\psi^0\rangle \leftrightarrow \sqrt{2}\,uv, \quad |\psi^-\rangle \leftrightarrow v^2. \tag{7.1.22}$$

Then a general spin state is represented as a homogeneous polynomial of order 2 in u and v. Note a close relation with the representation (7.1.9) in terms of two $S = 1/2$'s. Here symmetrization is automatically implemented since $uv = vu$. It is easy to check that

$$\hat{S}^+ \leftrightarrow u\frac{\partial}{\partial v}, \quad \hat{S}^- \leftrightarrow v\frac{\partial}{\partial u}, \quad \hat{S}^{(3)} \leftrightarrow \frac{1}{2}\Bigl(u\frac{\partial}{\partial u} - v\frac{\partial}{\partial v}\Bigr) \tag{7.1.23}$$

precisely recover the action of spin operators (although we do not use this representation here).

We extend the representation to the states of $S = 1$ spin chain with L sites. Then an arbitrary state $|\Phi\rangle$ is represented as a polynomial $\Phi(\boldsymbol{u}, \boldsymbol{v})$ in $\boldsymbol{u} = (u_1, \ldots, u_L)$ and $\boldsymbol{v} = (v_1, \ldots, v_L)$ with suitable homogeneity. Let $|\Phi\rangle$ be a ground state. Since it must satisfy[11] (7.1.19) for each x, the above lemma implies that the corresponding polynomial can be written as

$$\Phi(\boldsymbol{u}, \boldsymbol{v}) = (u_x v_{x+1} - v_x u_{x+1})\, \Xi(\boldsymbol{u}, \boldsymbol{v}) \tag{7.1.24}$$

with a suitable polynomial $\Xi(\boldsymbol{u}, \boldsymbol{v})$. Recall that $\Phi(\boldsymbol{u}, \boldsymbol{v})$ is a polynomial of degree $2L$ and satisfies (7.1.24) for each $x = 1, 2, \ldots, L$ (with of course different $\Xi(\boldsymbol{u}, \boldsymbol{v})$). Since the factorization of a polynomial is unique,[12] we find

$$\Phi(\boldsymbol{u}, \boldsymbol{v}) = (\text{constant}) \prod_{x=1}^{L} (u_x v_{x+1} - v_x u_{x+1}), \tag{7.1.25}$$

[11] This should be obvious. To be rigorous, we are using Lemma A.10 in p. 469.

[12] This is guaranteed since the ring of polynomials (with complex coefficients) is in a unique factorization domain [89].

which is nothing but the polynomial representation of the VBS state (7.1.12). This proves the uniqueness stated in Theorem 7.1.

The polynomial representation (7.1.25) of the VBS state was first obtained by Arovas, Auerbach, and Haldane [10] by using the Schwinger boson representation of spin states (which is closely related to the Weyl representation). It is remarkable that the representation (7.1.25) resembles the Laughlin wave function which describes the fractional quantum Hall effect. See, e.g., Chap. 13 of [33].

7.1.4 The Proof of the Existence of Gap

We here present an elementary and elegant proof by Knabe [44] of the existence of a nonvanishing gap in the AKLT model. Although Knabe's method applies only to a class of frustration-free Hamiltonians, the argument is very interesting. It enables us to rigorously show the existence of a gap in an indefinitely large system provided that a small system has a sufficiently large energy gap.[13]

Let $\hat{H}$ be any Hamiltonian whose ground state energy is exactly zero. The ground states may be degenerate. Then it is easy to see that $\hat{H}^2 \geq \varepsilon \hat{H}$ for some $\varepsilon > 0$ is equivalent to the lower bound $E_j \geq \varepsilon$ for any energy eigenvalue $E_j \neq 0$. Recall that the Hamiltonian $\hat{H}'_{\mathrm{AKLT}}$ of (7.1.7) has a vanishing ground state energy. We shall show that $(\hat{H}'_{\mathrm{AKLT}})^2 \geq \varepsilon \hat{H}'_{\mathrm{AKLT}}$ with some $\varepsilon > 0$ for a sufficiently large system size L. This proves the existence of a nonzero gap, i.e., the Haldane gap, in the AKLT model.

In the present subsection we abbreviate the projection operator $\hat{P}_2[\hat{\boldsymbol{S}}_x + \hat{\boldsymbol{S}}_{x+1}]$ as $\hat{P}_{x,x+1}$. Thus $\hat{H}'_{\mathrm{AKLT}} = \sum_{x=1}^{L} \hat{P}_{x,x+1}$. Throughout the proof we use periodic boundary conditions and identify x with $x + L$. We here assume that L is even, but the case with odd L is similar (and easier). Noting that $(\hat{P}_{x,x+1})^2 = \hat{P}_{x,x+1}$, we find

$$(\hat{H}'_{\mathrm{AKLT}})^2 = \Bigl(\sum_{x=1}^{L} \hat{P}_{x,x+1}\Bigr)\Bigl(\sum_{x=1}^{L} \hat{P}_{x,x+1}\Bigr) = \hat{H}'_{\mathrm{AKLT}} + \sum_{r=1}^{L/2} \hat{C}_r, \tag{7.1.26}$$

with

$$\hat{C}_1 = \sum_{x=1}^{L} \{\hat{P}_{x,x+1}\hat{P}_{x+1,x+2} + \hat{P}_{x+1,x+2}\hat{P}_{x,x+1}\}, \tag{7.1.27}$$

$$\hat{C}_r = 2\sum_{x=1}^{L} \hat{P}_{x,x+1}\hat{P}_{x+r,x+r+1}, \quad r = 2, \ldots, (L/2) - 1, \tag{7.1.28}$$

$$\hat{C}_{L/2} = \sum_{x=1}^{L} \hat{P}_{x,x+1}\hat{P}_{x+(L/2),x+(L/2)+1}. \tag{7.1.29}$$

[13] See [7, 35, 47–50] for refinements and extensions of Knabe's method, e.g., to systems with open boundaries or in higher dimensions. A related but different method for proving the existence of a finite gap in higher dimensional models is employed in [1, 71, 72]. See Sect. 7.3.2.

Note that, if $r \geq 2$, the two projection operators $\hat{P}_{x,x+1}$ and $\hat{P}_{x+r,x+r+1}$ commute, and hence $\hat{C}_r \geq 0$.

For $1 < \ell < L/2$ and $x = 1, \ldots, L$, let

$$\hat{h}_{x,x+\ell} = \sum_{j=1}^{\ell} \hat{P}_{x+j-1,x+j}, \tag{7.1.30}$$

which is nothing but the AKLT Hamiltonian on the open chain $\{x, x+1, \ldots, x+\ell\}$ with $\ell + 1$ sites. Note that

$$\sum_{x=1}^{L} \hat{h}_{x,x+\ell} = \ell\, \hat{H}'_{\text{AKLT}}. \tag{7.1.31}$$

This simple relation will be useful. As in (7.1.26), we have

$$\begin{aligned}(\hat{h}_{x,x+\ell})^2 = \hat{h}_{x,x+\ell} &+ \sum_{j=1}^{\ell-1} \{\hat{P}_{x+j-1,x+j}\hat{P}_{x+j,x+j+1} + \hat{P}_{x+j,x+j+1}\hat{P}_{x+j-1,x+j}\} \\ &+ \sum_{r=2}^{\ell-1}\sum_{j=1}^{\ell-r} 2\hat{P}_{x+j-1,x+j}\hat{P}_{x+j+r-1,x+j+r}.\end{aligned} \tag{7.1.32}$$

If we sum the summands of the second and the third terms over x, we get

$$\begin{aligned}&\sum_{x=1}^{L} \{\hat{P}_{x+j-1,x+j}\hat{P}_{x+j,x+j+1} + \hat{P}_{x+j,x+j+1}\hat{P}_{x+j-1,x+j}\} \\ &\qquad = \sum_{y=1}^{L} \{\hat{P}_{y,y+1}\hat{P}_{y+1,y+2} + \hat{P}_{y+1,y+2}\hat{P}_{y,y+1}\} = \hat{C}_1,\end{aligned} \tag{7.1.33}$$

$$\sum_{x=1}^{L} 2\hat{P}_{x+j-1,x+j}\hat{P}_{x+j+r-1,x+j+r} = 2\sum_{y=1}^{L} \hat{P}_{y,y+1}\hat{P}_{y+r,y+r+1} = \hat{C}_r. \tag{7.1.34}$$

Thus by summing (7.1.32), we get

$$\sum_{x=1}^{L} (\hat{h}_{x,x+\ell})^2 = \ell\, \hat{H}'_{\text{AKLT}} + \sum_{r=1}^{\ell-1} (\ell - r)\hat{C}_r, \tag{7.1.35}$$

where we also used (7.1.31). Let us rewrite this relation as

$$
\begin{aligned}
\frac{1}{\ell-1}\sum_{x=1}^{L}(\hat{h}_{x,x+\ell})^2 &= \frac{1}{\ell-1}\hat{H}'_{\mathrm{AKLT}} + \hat{H}'_{\mathrm{AKLT}} + \hat{C}_1 + \sum_{r=2}^{\ell-1}\frac{\ell-r}{\ell-1}\hat{C}_r \\
&\le \frac{1}{\ell-1}\hat{H}'_{\mathrm{AKLT}} + \hat{H}'_{\mathrm{AKLT}} + \sum_{r=1}^{L/2}\hat{C}_r \\
&= \frac{1}{\ell-1}\hat{H}'_{\mathrm{AKLT}} + (\hat{H}'_{\mathrm{AKLT}})^2,
\end{aligned}
\tag{7.1.36}
$$

where we used $\hat{C}_r \ge 0$ for $r \ge 2$ to get the second line, and finally used (7.1.26). We thus arrive at the inequality

$$
(\hat{H}'_{\mathrm{AKLT}})^2 \ge \frac{1}{\ell-1}\Bigl\{\sum_{x=1}^{L}(\hat{h}_{x,x+\ell})^2 - \hat{H}'_{\mathrm{AKLT}}\Bigr\}. \tag{7.1.37}
$$

Note that the ground state energy of the Hamiltonian $\hat{h}_{x,x+\ell}$ is also zero.[14] Denoting by $\varepsilon_\ell > 0$ the energy gap of $\hat{h}_{x,x+\ell}$, we have $(\hat{h}_{x,x+\ell})^2 \ge \varepsilon_\ell\,\hat{h}_{x,x+\ell}$. By substituting this bound into (7.1.37), we get

$$
(\hat{H}'_{\mathrm{AKLT}})^2 \ge \frac{1}{\ell-1}\Bigl\{\varepsilon_\ell\sum_{x=1}^{L}\hat{h}_{x,x+\ell} - \hat{H}'_{\mathrm{AKLT}}\Bigr\} = \frac{\ell}{\ell-1}\Bigl(\varepsilon_\ell - \frac{1}{\ell}\Bigr)\hat{H}'_{\mathrm{AKLT}}, \tag{7.1.38}
$$

which means that $\hat{H}'_{\mathrm{AKLT}}$ has an energy gap not less than $\{\ell/(\ell-1)\}\{\varepsilon_\ell - (1/\ell)\}$. This is a remarkable conclusion since ε_ℓ (for not too large ℓ) can be computed numerically, while the lower bound for the gap is valid for any (sufficiently large) L. In fact Knabe [44] found that $\varepsilon_5 \simeq 0.398451$, which shows that the energy gap ΔE of $\hat{H}'_{\mathrm{AKLT}}$ satisfies $\Delta E \ge 0.248$ for any $L > 10$.

7.2 Properties of the AKLT Model

The AKLT model does not only provide a rigorous example supporting Haldane's conclusion, but also exhibits some unexpected interesting properties. We discuss the existence of hidden antiferromagnetic order in Sect. 7.2.1, and the four-fold degeneracy in open chains in Sect. 7.2.3. These properties will turn out to characterize the essence of the Haldane phase, and to be related to deeper notions of hidden symmetry breaking and symmetry protected topological phase, as we shall discuss in Chap. 8.

In Sect. 7.2.2, we give a detailed discussion of the matrix product representation of the VBS state. This long subsection can be read as a tutorial introduction to matrix product states.

---

[14] The ground states are degenerate. See Sect. 7.2.3.

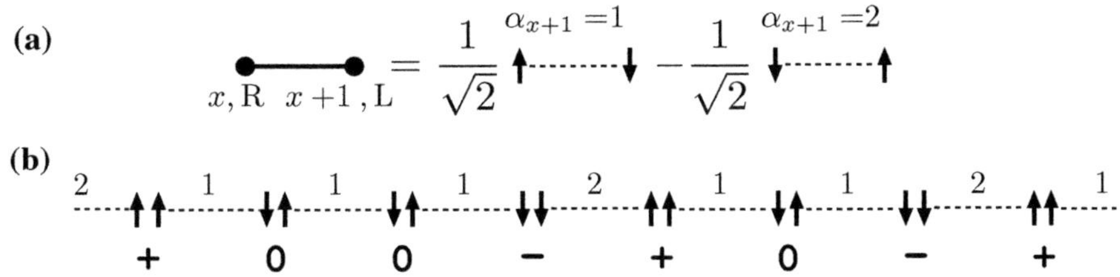

Fig. 7.3 **a** A valence bond is a sum of two product states. We distinguish the two by the index $\alpha_{x+1} = 1, 2$. **b** A typical term in the expansion of the VBS state. The number 1 or 2 on each bond is the index α_{x+1}. After applying the symmetrization $\mathscr{S}_x$ at each site (according the rules (7.1.9)) we obtain a spin configuration $+00-+0-+$, where we see a clear hidden antiferromagnetic order, i.e., $+$ and $-$ alternate with arbitrary number of 0's in between them (© Hal Tasaki 2020. All Rights Reserved)

7.2.1 Hidden Antiferromagnetic Order

Representation in the standard basis The VBS state is compactly represented as (7.1.11) and (7.1.12) in terms of the pre-VBS state and the projection at each site. It turns out quite meaningful to represent the same state in the standard basis state $|\Psi^{\sigma}\rangle$ defined in (2.2.1) (with $\sigma_x = 0, \pm 1$) as

$$|\Phi_{\mathrm{VBS}}\rangle = \sum_{\sigma} c_{\sigma}\,|\Psi^{\sigma}\rangle. \tag{7.2.1}$$

To do this, one only needs to expand the product in the definition (7.1.11) of the pre-VBS state into a sum of simple product states, and then apply the symmetrization operator $\mathscr{S}_x$ at each site by using the rules (7.1.9). See Problems 7.1.2.a and 7.1.2.b, and (S.63) in p. 505. Since one chooses either $|\psi^{\uparrow}_{x,\mathrm{R}}\rangle|\psi^{\downarrow}_{x+1,\mathrm{L}}\rangle/\sqrt{2}$ or $-|\psi^{\downarrow}_{x,\mathrm{R}}\rangle|\psi^{\uparrow}_{x+1,\mathrm{L}}\rangle/\sqrt{2}$ for each bond, there appear 2^L terms in the expansion. Recall that there are 3^L distinct spin configurations $\boldsymbol{\sigma}$ in the $S = 1$ chain. We conclude that there are (at least) $3^L - 2^L$ configurations $\boldsymbol{\sigma}$ with $c_{\sigma} = 0$. This is in a clear contrast with the antiferromagnetic Heisenberg model, where the Marshall–Lieb–Mattis theorem (Theorem 2.2 in p. 39) implies $c_{\sigma} \neq 0$ for any $\boldsymbol{\sigma}$.

To perform the above explained expansion systematically, it is convenient to associate with each bond $\{x, x+1\}$ an index $\alpha_{x+1} = 1, 2$; one sets $\alpha_{x+1} = 1$ when $|\psi^{\uparrow}_{x,\mathrm{R}}\rangle|\psi^{\downarrow}_{x+1,\mathrm{L}}\rangle/\sqrt{2}$ is chosen and $\alpha_{x+1} = 2$ when $-|\psi^{\downarrow}_{x,\mathrm{R}}\rangle|\psi^{\uparrow}_{x+1,\mathrm{L}}\rangle/\sqrt{2}$ is chosen. See Fig. 7.3a. The expansion is obtained by considering all the possible assignments of $\alpha_x = 1$ or 2 for $x = 1, \ldots, L$.

Figure 7.3b shows a typical term in the expansion and the corresponding spin configuration $\boldsymbol{\sigma}$. Clearly the spin configuration exhibits hidden antiferromagnetic order similar to that we discussed in Sect. 6.3. As we shall discuss below, a spin configuration we obtain from the expansion looks, for example, like

$$+ \; - \; + \; 0 \; - \; 0 \; + \; - \; + \; 0 \; 0 \; 0 \; - \; 0 \; + \; - \tag{7.2.2}$$

If we delete all the 0's from the string, we are left with a perfect alternating sequence of $+$ and $-$. It should be noted, however, that this alternating pattern does not result in a long-range antiferromagnetic order since 0's are inserted randomly in between $+$ and $-$.

Although the VBS state appears to be solid-like when expressed in terms of valence-bonds as in Fig. 7.2b, it looks very much different in the present description. The state looks more like liquid, where $+$ and $-$, which have moderate densities[15] (in the background of 0), exhibit large fluctuations. One may say that the VBS state is a "quantum spin liquid with hidden antiferromagnetic order".[16]

The best way to understand the appearance of the hidden order may be to work out simple examples. Suppose that one has $\sigma_1 = 1$, which means that bond indices are chosen as $\alpha_1 = 2$ and $\alpha_2 = 1$ to have the configuration $\overset{2}{\cdots\cdots}\uparrow\uparrow\overset{1}{\cdots\cdots}$. Let us proceed to the right. The configuration around sites 1 and 2 now looks like $\overset{2}{\cdots\cdots}\uparrow\uparrow\overset{1}{\cdots\cdots}\downarrow ?$. Since we already have a $\downarrow$ at site 2, it is impossible to have $\sigma_2 = 1$; either $\sigma_2 = 0$ or $\sigma_2 = -1$ is possible. Suppose that we choose $\alpha_3 = 1$ to have $\sigma_2 = 0$. Then the configuration becomes $\overset{2}{\cdots\cdots}\uparrow\uparrow\overset{1}{\cdots\cdots}\downarrow\uparrow\overset{1}{\cdots\cdots}\downarrow ?$. Note here that the configuration at site 3 (the right-most site) is exactly the same as that for site 2 in the previous step. Then again the possible choice is between $\sigma_3 = 0$ or $\sigma_3 = -1$. One can keep on choosing $\alpha = 1$ to have an arbitrary number of 0's. But the only way to stop this is to choose $\alpha_{x+1} = 2$ at some x to have $\sigma_x = -1$. One ends up with a spin configuration like $+0000000-$ with clear hidden antiferromagnetic order.

This construction can be continued and one finds that only those configurations with complete hidden antiferromagnetic order are generated. We conclude that the coefficient $c_{\boldsymbol{\sigma}}$ in the expansion (7.2.1) is nonvanishing if and only if the configuration $\boldsymbol{\sigma}$ has complete hidden antiferromagnetic order. In the next Sect. 7.2.2 we shall see a useful expression (7.2.11) of $c_{\boldsymbol{\sigma}}$.

In Sect. 6.3, we encountered spin configurations with hidden antiferromagnetic order by examining low energy dynamics of the $S = 1$ antiferromagnetic Heisenberg chain based on the picture of kinks. It is remarkable that seemingly similar feature is observed in the exact ground state of the AKLT Hamiltonian (7.1.1).

String order parameter den Nijs and Rommelse [23] introduced a quantity which measure the presence of hidden antiferromagnetic order.

To motivate the definition, note first that in the configuration

---

[15] Exactly as in Problem 2.5.c (p. 39), one can show that $\langle \Phi_{\text{VBS}} | (\hat{S}_x^{(3)})^2 | \Phi_{\text{VBS}} \rangle = 2/3$. This means that the average densities of $+$, $-$, and 0 are all equal to $1/3$. See also Problem 7.2.2.c (p. 200).

[16] The term "quantum spin liquid" is used to indicate a variety of spin systems which exhibit liquid-like behavior. See, e.g., the review articles [59, 76].

$$+ \; - \; + \; - \; + \; - \; + \; - \; + \; - \; + \; - \; + \; - \; + \; - \tag{7.2.3}$$

which has complete antiferromagnetic order, we always have $(-1)^{y-x}\sigma_x\sigma_y = 1$. Thus possible long-range antiferromagnetic order in a state $|\Phi\rangle$ can be detected from the behavior of the correlation function $(-1)^{y-x}\langle\Phi|\hat{S}_x^{(3)}\hat{S}_y^{(3)}|\Phi\rangle$. In particular the (long-range) Néel order parameter is defined as[17]

$$\mathscr{O}^{(3)}_{\text{Néel}}(\Phi) := \lim_{y-x\uparrow\infty}\lim_{L\uparrow\infty}(-1)^{y-x}\langle\Phi|\hat{S}_x^{(3)}\hat{S}_y^{(3)}|\Phi\rangle. \tag{7.2.4}$$

That $\mathscr{O}^{(3)}_{\text{Néel}}(\Phi) \neq 0$ indicates the presence of Néel order in the state $|\Phi\rangle$.

Likewise, in a configuration like (7.2.2) with complete hidden antiferromagnetic order, we have

$$(-1)^{(\text{the number} \pm \text{between } x \text{ and } y)}\sigma_x\sigma_y = -1. \tag{7.2.5}$$

Note also that the factor $(-1)^{(\text{the number} \pm \text{between } x \text{ and } y)}$ can be written as $\exp[i\pi\sum_{z=x+1}^{y-1}\sigma_z]$ where we assumed $x < y$. We then define the string correlation function for a state $|\Phi\rangle$ (on the chain with L sites) and $\alpha = 1, 2, 3$ by

$$\mathscr{S}^{(\alpha)}_{x,y}(\Phi) := -\langle\Phi|\,\hat{S}_x^{(\alpha)}\,\exp\bigl[i\pi\sum_{z=x+1}^{y-1}\hat{S}_z^{(\alpha)}\bigr]\,\hat{S}_y^{(\alpha)}\,|\Phi\rangle, \tag{7.2.6}$$

where $0 < x < y < L$. The name "string" indicates that there is a string of operators $e^{i\pi\hat{S}_z^{(\alpha)}}$ in between x and y. Then the den Nijs–Rommelse string order parameter is defined as

$$\mathscr{O}^{(\alpha)}_{\text{string}}(\Phi) := -\lim_{y-x\uparrow\infty}\lim_{L\uparrow\infty}\mathscr{S}^{(\alpha)}_{x,y}(\Phi). \tag{7.2.7}$$

From the above consideration, we see that $\mathscr{O}^{(3)}_{\text{string}}(\Phi) > 0$ if configurations like (7.2.2) are dominant in the expansion of the state $|\Phi\rangle$.

For the VBS state, the analysis in the first half of the present section shows the existence of hidden antiferromagnetic order in the 3-direction, i.e., $\mathscr{O}^{(3)}_{\text{string}}(\Phi_{\text{VBS}}) > 0$. But the SU(2) invariance of the state implies that $\mathscr{O}^{(\alpha)}_{\text{string}}(\Phi_{\text{VBS}})$ is independent of $\alpha = 1, 2, 3$. An explicit computation shows that[18]

$$\mathscr{O}^{(1)}_{\text{string}}(\Phi_{\text{VBS}}) = \mathscr{O}^{(2)}_{\text{string}}(\Phi_{\text{VBS}}) = \mathscr{O}^{(3)}_{\text{string}}(\Phi_{\text{VBS}}) = \frac{4}{9}. \tag{7.2.8}$$

See Problem 7.2.2.d. (We of course have $\mathscr{O}^{(\alpha)}_{\text{Néel}}(\Phi_{\text{VBS}}) = 0$.)

---

[17] This corresponds to the definition (3.4.6) of long-range order, which is expected to be equivalent to the definition (3.4.3) used throughout Part I. To be rigorous the existence of the limits in (7.2.4) and (7.2.7) are note proved in general.

[18] Here the existence of the limits in (7.2.7) can be proved.

7.2.2 Matrix Product Representation

Let us examine the above procedure for obtaining the expansion (7.2.1) more quantitatively, and derive the representation (7.2.11) in terms of a product of matrices.[19] This representation turns out to be not only interesting but also quite useful for practical calculations as we see below. But, more importantly, the VBS state represented as (7.2.11) was the first nontrivial example of a very important class of states of quantum spin chains now called matrix product states (MPS), about which we shall briefly discuss below.[20] The formulation of MPS provides an efficient way of describing a large class of states in one-dimensional quantum many-body systems, and has been playing indispensable roles in condensed matter physics, mathematical physics, and quantum information science. The matrix product representation was discovered in 1989 by Fannes, Nachtergaele, and Werner[21] [26–28], who also developed a complete general theory of matrix product states.[22]

The present section serves also as a tutorial introduction to matrix product states. We first continue the preceding discussion and derive the matrix product representation of the VBS state.[23] We then show how the norm and the two-point correlation function are calculated. The techniques explained here apply to general matrix product states. Finally we summarize some important properties of general matrix product states.

Matrix product representation of $c_{\boldsymbol{\sigma}}$ Recall that, in the previous construction, any sequence of bond indices $\alpha_1, \ldots, \alpha_L$ uniquely determines the corresponding spin configuration $\boldsymbol{\sigma} = (\sigma_1, \ldots, \sigma_L)$. More specifically σ_x is determined by the two indices α_x and α_{x+1} as

$$\sigma_x = \begin{cases} 1 & \text{if}\,\alpha_x = 2, \alpha_{x+1} = 1; \\ -1 & \text{if}\,\alpha_x = 1, \alpha_{x+1} = 2; \\ 0 & \text{if}\,\alpha_x = \alpha_{x+1} = 1\,\text{or}\,\alpha_x = \alpha_{x+1} = 2. \end{cases} \tag{7.2.9}$$

---

[19] The reader anxious to study another exotic property of the AKLT model may skip this subsection and jump to Sect. 7.2.3 (and come back here when necessary).

[20] The structure essentially equivalent to matrix product states was already known in the context of classical statistical mechanics. The work by Baxter [13] is an early example.

[21] In the original works [26–28], matrix product states were called finitely correlated states. The term matrix product states was introduced later in [43]. See also [82, 83], where the representation was rediscovered and put into a slightly different context.

[22] The general construction of matrix product states given in [28] has two parents: the AKLT model and Accardi's proposal for the construction of quantum Markov chains [2, 3]. The relationship between the two is clarified in [66]. Note, however, that there are several definitions of quantum Markov chains that continue to be used in parallel. For example, in the quantum information community a common definition is given in terms of the case of equality in the strong subadditivity property of entropy [38].

[23] The derivation is by no means the unique or the most convenient way to derive the matrix product representation of the VBS state. See Sect. 4.1.5 of [77] for a simple derivation which starts from the expression (7.1.11) of the pre-VBS state (which is (84) of [77]), and [80] for a derivation based on the Schwinger boson representation of [10].

To get a precise formula for $c_{\boldsymbol{\sigma}}$ we need to keep track of the factor $\pm 1/\sqrt{2}$ coming from the valence bond in (7.1.11) (i.e., $1/\sqrt{2}$ if $\alpha = 1$ and $-1/\sqrt{2}$ if $\alpha = 2$), and the factor 1 or $1/\sqrt{2}$ from the rules (7.1.9) for symmetrization (i.e., 1 if $\sigma = \pm 1$ and $1/\sqrt{2}$ if $\sigma = 0$). We shall assign the former factor to the site at the right of the valence bond, and the latter factor to the site at which symmetrization is performed. Then the factor assigned to a site x is summarized as the following.

	α_x	α_{x+1}	σ_x	factor for x
2 ↑↑ 1	2	1	1	$-\frac{1}{\sqrt{2}} \times 1 = -\frac{1}{\sqrt{2}}$
1 ↓↓ 2	1	2	-1	$\frac{1}{\sqrt{2}} \times 1 = \frac{1}{\sqrt{2}}$
1 ↓↑ 1	1	1	0	$\frac{1}{\sqrt{2}} \times \frac{1}{\sqrt{2}} = \frac{1}{2}$
2 ↑↓ 2	2	2	0	$-\frac{1}{\sqrt{2}} \times \frac{1}{\sqrt{2}} = -\frac{1}{2}$

Let us introduce a quantity $A^{\sigma}_{\alpha,\alpha'}$ with $\alpha, \alpha' = 1, 2$ and $\sigma = 0, \pm 1$. We "record" the above factors into this quantity as

$$A^{+}_{2,1} = -\frac{1}{\sqrt{2}}, \quad A^{-}_{1,2} = \frac{1}{\sqrt{2}}, \quad A^{0}_{1,1} = \frac{1}{2}, \quad A^{0}_{2,2} = -\frac{1}{2}. \tag{7.2.10}$$

We set $A^{\sigma}_{\alpha,\alpha'} = 0$ for other combinations. Then the coefficient $c_{\boldsymbol{\sigma}}$ in the expansion (7.2.1) can be represented as

$$c_{\boldsymbol{\sigma}} = \sum_{\alpha_1,\alpha_2,\ldots,\alpha_L=1,2} A^{\sigma_1}_{\alpha_1,\alpha_2} A^{\sigma_2}_{\alpha_2,\alpha_3} \cdots A^{\sigma_L}_{\alpha_L,\alpha_1}. \tag{7.2.11}$$

Although the sum runs over the 2^L possible sequences $(\alpha_1, \ldots, \alpha_L)$ of the bond indices, the summand is nonvanishing only for the unique $(\alpha_1, \ldots, \alpha_L)$ which generates the given spin configuration $\boldsymbol{\sigma}$.[24]

Clearly the expression (7.2.11) can be interpreted as the trace of a matrix product. Define three 2×2 matrices A^+, A^0, and A^- by $(\mathsf{A}^{\sigma})_{\alpha,\alpha'} := A^{\sigma}_{\alpha,\alpha'}$, or, more specifically, by

$$\mathsf{A}^+ = \begin{pmatrix} 0 & 0 \\ -\frac{1}{\sqrt{2}} & 0 \end{pmatrix}, \quad \mathsf{A}^0 = \begin{pmatrix} \frac{1}{2} & 0 \\ 0 & -\frac{1}{2} \end{pmatrix}, \quad \mathsf{A}^- = \begin{pmatrix} 0 & \frac{1}{\sqrt{2}} \\ 0 & 0 \end{pmatrix}. \tag{7.2.12}$$

Then (7.2.11) is written as

$$c_{\boldsymbol{\sigma}} = \mathrm{Tr}[\mathsf{A}^{\sigma_1} \mathsf{A}^{\sigma_2} \cdots \mathsf{A}^{\sigma_L}]. \tag{7.2.13}$$

We can also write the VBS state in the matrix product form as

[24] The spin configuration $\boldsymbol{\sigma} = (0, \ldots, 0)$ is the only exception. It is generated both by $(1, \ldots, 1)$ and $(2, \ldots, 2)$.

$$
\begin{aligned}
|\Phi_{\mathrm{VBS}}\rangle &= \sum_{\boldsymbol{\sigma},\boldsymbol{\alpha}} A^{\sigma_1}_{\alpha_1,\alpha_2} A^{\sigma_2}_{\alpha_2,\alpha_3} \cdots A^{\sigma_L}_{\alpha_L,\alpha_1} |\Psi^{\boldsymbol{\sigma}}\rangle \\
&= \sum_{\boldsymbol{\sigma}} \mathrm{Tr}[\mathsf{A}^{\sigma_1}\mathsf{A}^{\sigma_2} \cdots \mathsf{A}^{\sigma_L}]\, |\Psi^{\boldsymbol{\sigma}}\rangle. \qquad (7.2.14)
\end{aligned}
$$

It is common (and indeed useful) to represent the element $A^{\sigma}_{\alpha,\alpha'}$ graphically as

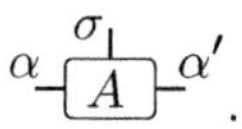

. Then the matrix product representation (7.2.11) of the coefficient can be neatly expressed as

$c_{\boldsymbol{\sigma}} =$

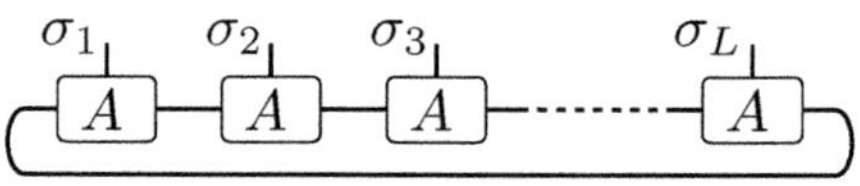

(7.2.15)

where lines connecting two boxes correspond to contracted indices (which are $\alpha_1, \ldots, \alpha_L$ in this case) and open lines correspond to free indices (which are $\sigma_1, \ldots, \sigma_L$ in this case).

Problem 7.2.2.a Starting from the representation (7.2.13) or (7.2.15), confirm that the coefficient $c_{\boldsymbol{\sigma}}$ is nonvanishing only when the spin configuration $\boldsymbol{\sigma}$ has complete hidden antiferromagnetic order. (Hint: Try multiplying the vector $(1, 0)^{\mathrm{t}}$, where t denotes the transpose, by A^+, A^0, or A^- successively.) [solution→p.507]

Problem 7.2.2.b By using a different basis one can rewrite the matrix product representation (7.2.14) of $|\Phi_{\mathrm{VBS}}\rangle$ in a remarkably symmetric form. Define an orthonormal basis $\{|p^{(1)}\rangle, |p^{(2)}\rangle, |p^{(3)}\rangle\}$ of a single $S = 1$ spin by

$$
|p^{(1)}\rangle := \frac{1}{\sqrt{2}}\begin{pmatrix} -1 \\ 0 \\ 1 \end{pmatrix}, \quad |p^{(2)}\rangle := \frac{1}{\sqrt{2}}\begin{pmatrix} i \\ 0 \\ i \end{pmatrix}, \quad |p^{(3)}\rangle := \begin{pmatrix} 0 \\ 1 \\ 0 \end{pmatrix}, \qquad (7.2.16)
$$

which correspond to the p_{x}, p_{y}, and p_{z} orbitals.[25] For any sequence $\boldsymbol{\gamma} = (\gamma_1, \gamma_2, \ldots, \gamma_L)$ with $\gamma_x = 1, 2, 3$, we define a basis state for the whole chain by $|\mathscr{P}^{\boldsymbol{\gamma}}\rangle = \bigotimes_{x=1}^{L} |p_x^{(\gamma_x)}\rangle$, where $|p_x^{(\gamma)}\rangle$ is the copy of $|p^{(\gamma)}\rangle$ at site x. By using (7.2.14), find the matrix product representation of $|\Phi_{\mathrm{VBS}}\rangle$ which uses the basis states $|\mathscr{P}^{\boldsymbol{\gamma}}\rangle$ instead of $|\Psi^{\boldsymbol{\sigma}}\rangle$. [solution→p.507]

Calculation of $\langle\Phi_{\mathrm{VBS}}|\Phi_{\mathrm{VBS}}\rangle$ As an application of the matrix product representation, we compute the normalization factor $\langle\Phi_{\mathrm{VBS}}|\Phi_{\mathrm{VBS}}\rangle$ of the VBS state. By recalling the expansion (7.2.1) and the representation (7.2.11), we readily get

---

[25] These states satisfy $\hat{S}^{(\alpha)}|p^{(\alpha)}\rangle = 0$ and $\exp[-i(\pi/2)\hat{S}^{(\alpha)}]\,|p^{(\beta)}\rangle = \sum_{\gamma=1,2,3} \varepsilon_{\alpha\beta\gamma}\,|p^{(\gamma)}\rangle$ for any $\alpha, \beta, \gamma = 1, 2, 3$, where $\varepsilon_{\alpha\beta\gamma}$ is the Levi-Civita symbol (see (A.3.1)).

$$\langle \Phi_{\rm VBS}|\Phi_{\rm VBS}\rangle = \sum_{\sigma}(c_\sigma)^2 = \sum_{\sigma}\sum_{\substack{\alpha_1,\ldots,\alpha_L=1,2\\ \beta_1,\ldots,\beta_L=1,2}} A^{\sigma_1}_{\alpha_1,\alpha_2}\cdots A^{\sigma_L}_{\alpha_L,\alpha_1}\, A^{\sigma_1}_{\beta_1,\beta_2}\cdots A^{\sigma_L}_{\beta_L,\beta_1}. \tag{7.2.17}$$

This is again expressed graphically as[26]

$\langle \Phi_{\rm VBS}|\Phi_{\rm VBS}\rangle =$ 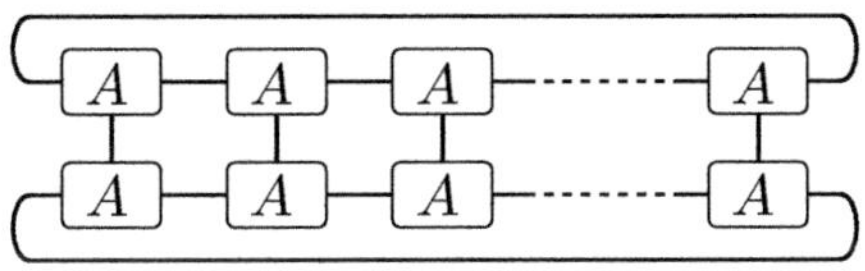

(7.2.18)

Note that there are no open lines since all the indices are contracted. The diagram motivates us to sum over the spin variable σ first and define a larger matrix by

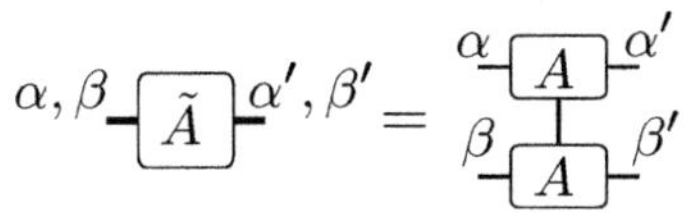

(7.2.19)

or, more precisely, by

$$(\tilde{\mathsf{A}})_{\alpha,\beta;\alpha',\beta'} = \tilde{A}_{\alpha,\beta;\alpha',\beta'} := \sum_{\sigma=-1,0,1} A^{\sigma}_{\alpha,\alpha'}\, A^{\sigma}_{\beta,\beta'}. \tag{7.2.20}$$

where $\tilde{\mathsf{A}}$ is a 4×4 matrix called the transfer matrix. Note that (α,β) labels the row index and (α',β') labels the column index of the matrix element $\tilde{A}_{\alpha,\beta;\alpha',\beta'}$. Then the desired normalization factor is compactly expressed as

$\langle \Phi_{\rm VBS}|\Phi_{\rm VBS}\rangle =$ 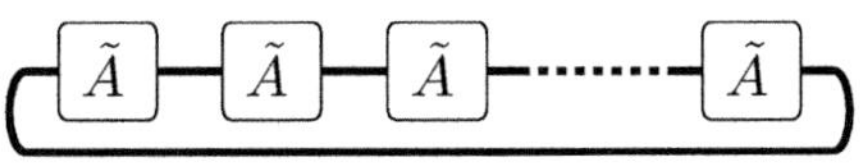

(7.2.21)

or

$$\langle \Phi_{\rm VBS}|\Phi_{\rm VBS}\rangle = \sum_{\substack{\alpha_1,\ldots,\alpha_L=1,2\\ \beta_1,\ldots,\beta_L=1,2}} \tilde{A}_{\alpha_1,\beta_1;\alpha_2,\beta_2}\,\tilde{A}_{\alpha_2,\beta_2;\alpha_3,\beta_3}\cdots\tilde{A}_{\alpha_L,\beta_L;\alpha_1,\beta_1} = {\rm Tr}[\tilde{\mathsf{A}}^L]. \tag{7.2.22}$$

Clearly the final expression ${\rm Tr}[\tilde{\mathsf{A}}^L]$ is advantageous for our calculation.

To see how the transfer matrix $\tilde{\mathsf{A}}$ is determined, examine the case with $(\alpha,\alpha') = (\beta,\beta') = (2,1)$, where the definition (7.2.20) with (7.2.10) gives

$$\tilde{A}_{2,2;1,1} = \sum_{\sigma=-1,0,1} A^{\sigma}_{2,1}A^{\sigma}_{2,1} = A^{+}_{2,1}A^{+}_{2,1} = \frac{1}{2}. \tag{7.2.23}$$

[26] It is natural to represent (in general) $(A^{\sigma}_{\alpha,\alpha'})^*$ as 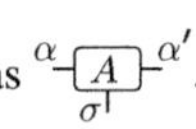.

Note that the only one value of σ, i.e., $\sigma = 1$, contributes to the sum. The evaluation of other components proceeds in similar manner, and is summarized in the following table. The first column represents the unique value of σ which contributes to each component.

σ	α, α'	β, β'	$\tilde{A}_{\alpha,\beta;\alpha',\beta'}$	α, β	α', β'
+	2,1	2,1	1/2	2,2	1,1
−	1,2	1,2	1/2	1,1	2,2
0	1,1	1,1	1/4	1,1	1,1
0	2,2	2,2	1/4	2,2	2,2
0	1,1	2,2	−1/4	1,2	1,2
0	2,2	1,1	−1/4	2,1	2,1

From the table we read off that $\tilde{\mathsf{A}}$ is the following real symmetric matrix.

$$\tilde{\mathsf{A}} = \begin{array}{c} \\ 1,1 \\ 2,2 \\ 1,2 \\ 2,1 \end{array} \begin{array}{c} \begin{array}{cccc} 1,1 & 2,2 & 1,2 & 2,1 \end{array} \\ \begin{pmatrix} \frac{1}{4} & \frac{1}{2} & 0 & 0 \\ \frac{1}{2} & \frac{1}{4} & 0 & 0 \\ 0 & 0 & -\frac{1}{4} & 0 \\ 0 & 0 & 0 & -\frac{1}{4} \end{pmatrix} \end{array} \tag{7.2.24}$$

The eigenvalues of $\tilde{\mathsf{A}}$ is readily obtained to be $3/4$ and $-1/4$ where the latter is three-fold degenerate. Then the desired normalization factor (7.2.22) is calculated as

$$\langle \Phi_{\mathrm{VBS}} | \Phi_{\mathrm{VBS}} \rangle = \mathrm{Tr}[\tilde{\mathsf{A}}^L] = \left(\frac{3}{4}\right)^L + 3\left(-\frac{1}{4}\right)^L . \tag{7.2.25}$$

Calculation of the two-point correlation function We continue to evaluate the correlation function to show the expression (7.1.2) stated in Theorem 7.1 in p. 178. Let $r \geq 1$ and observe that

$$\begin{aligned} \langle \Phi_{\mathrm{VBS}} | \hat{S}_1^{(3)} \hat{S}_{r+1}^{(3)} | \Phi_{\mathrm{VBS}} \rangle &= \sum_{\sigma} \sigma_1 \sigma_{r+1} (c_{\sigma})^2 \\ &= \sum_{\sigma} \sigma_1 \sigma_{r+1} \sum_{\substack{\alpha_1,\ldots,\alpha_L=1,2 \\ \beta_1,\ldots,\beta_L=1,2}} A^{\sigma_1}_{\alpha_1,\alpha_2} \cdots A^{\sigma_L}_{\alpha_L,\alpha_1} A^{\sigma_1}_{\beta_1,\beta_2} \cdots A^{\sigma_L}_{\beta_L,\beta_1} \\ &= \sum_{\substack{\alpha_1,\ldots,\alpha_L=1,2 \\ \beta_1,\ldots,\beta_L=1,2}} \tilde{B}_{\alpha_1,\beta_1;\alpha_2,\beta_2} \tilde{A}_{\alpha_2,\beta_2;\alpha_3,\beta_3} \cdots \tilde{A}_{\alpha_r,\beta_r;\alpha_{r+1},\beta_{r+1}} \tilde{B}_{\alpha_{r+1},\beta_{r+1};\alpha_{r+2},\beta_{r+2}} \times \\ &\qquad \times \tilde{A}_{\alpha_{r+2},\beta_{r+2};\alpha_{r+3},\beta_{r+3}} \cdots \tilde{A}_{\alpha_L,\beta_L;\alpha_1,\beta_1} \\ &= \mathrm{Tr}[\tilde{\mathsf{B}}\tilde{\mathsf{A}}^{r-1}\tilde{\mathsf{B}}\tilde{\mathsf{A}}^{L-r-1}], \end{aligned} \tag{7.2.26}$$

which, for $r = 2$, is expressed graphically as

$$\langle \Phi_{\mathrm{VBS}} | \hat{S}_1^{(3)} \hat{S}_3^{(3)} | \Phi_{\mathrm{VBS}} \rangle =$$

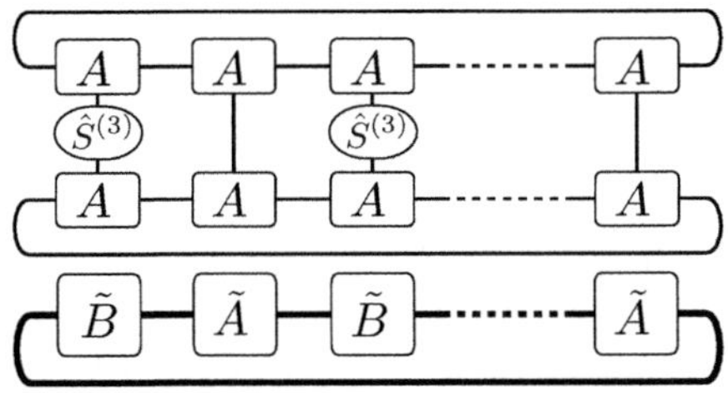

(7.2.27)

Here the matrix $\tilde{\mathsf{B}}$ is defined graphically as

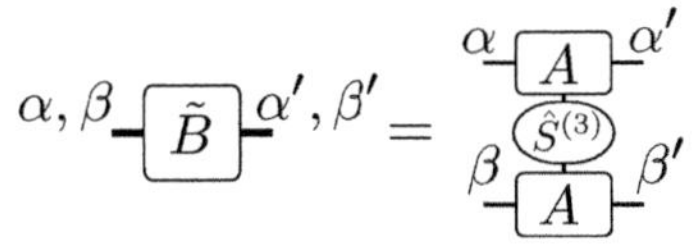

(7.2.28)

or by an explicit formula for the $\tilde{B}_{\alpha,\beta;\alpha',\beta'} = (\tilde{\mathsf{B}})_{\alpha,\beta;\alpha',\beta'}$ as[27]

$$\tilde{B}_{\alpha,\beta;\alpha',\beta'} := \sum_{\sigma,\sigma'=-1,0,1} A^{\sigma}_{\alpha,\alpha'} \langle \psi^{\sigma} | \hat{S}^{(3)} | \psi^{\sigma'} \rangle A^{\sigma'}_{\beta,\beta'} = \sum_{\sigma=-1,0,1} \sigma \, A^{\sigma}_{\alpha,\alpha'} \, A^{\sigma}_{\beta,\beta'}. \quad (7.2.29)$$

The entries of matrix $\tilde{\mathsf{B}}$ is determined, as in $\tilde{\mathsf{A}}$, as

$$\tilde{\mathsf{B}} = \begin{array}{c} \\ 1{,}1 \\ 2{,}2 \\ 1{,}2 \\ 2{,}1 \end{array} \!\! \begin{array}{c} \begin{matrix} 1{,}1 & 2{,}2 & 1{,}2 & 2{,}1 \end{matrix} \\ \begin{pmatrix} 0 & -\frac{1}{2} & 0 & 0 \\ \frac{1}{2} & 0 & 0 & 0 \\ 0 & 0 & 0 & 0 \\ 0 & 0 & 0 & 0 \end{pmatrix} \end{array}, \quad (7.2.30)$$

which means that B is essentially a 2×2 matrix. Thus (7.2.26) becomes

$$\langle \Phi_{\mathrm{VBS}} | \hat{S}_1^{(3)} \hat{S}_{r+1}^{(3)} | \Phi_{\mathrm{VBS}} \rangle = \mathrm{Tr} \left[\begin{pmatrix} 0 & -\frac{1}{2} \\ \frac{1}{2} & 0 \end{pmatrix} \begin{pmatrix} \frac{1}{4} & \frac{1}{2} \\ \frac{1}{2} & \frac{1}{4} \end{pmatrix}^{r-1} \begin{pmatrix} 0 & -\frac{1}{2} \\ \frac{1}{2} & 0 \end{pmatrix} \begin{pmatrix} \frac{1}{4} & \frac{1}{2} \\ \frac{1}{2} & \frac{1}{4} \end{pmatrix}^{L-r-1} \right]. \quad (7.2.31)$$

This is easily evaluated by diagonalizing the 2×2 matrix $\begin{pmatrix} \frac{1}{4} & \frac{1}{2} \\ \frac{1}{2} & \frac{1}{4} \end{pmatrix}$ as follows. Let $\mathsf{O} = \frac{1}{\sqrt{2}} \begin{pmatrix} 1 & 1 \\ 1 & -1 \end{pmatrix}$, which satisfies $\mathsf{O}^2 = \mathsf{I}$, and $\begin{pmatrix} \frac{1}{4} & \frac{1}{2} \\ \frac{1}{2} & \frac{1}{4} \end{pmatrix} = \mathsf{O} \begin{pmatrix} \frac{3}{4} & 0 \\ 0 & -\frac{1}{4} \end{pmatrix} \mathsf{O}$. By noting

[27] In (7.2.27), (7.2.28), and (7.2.29), we used a notation that is valid when the operator $\hat{S}^{(3)}$ is replaced by an arbitrary local operator. The double sum in the second expression of (7.2.29) is not necessary if we are interested only in $\hat{S}^{(3)}$.

that $\begin{pmatrix} \frac{1}{4} & \frac{1}{2} \\ \frac{1}{2} & \frac{1}{4} \end{pmatrix}^s = \mathsf{O}\begin{pmatrix} (\frac{3}{4})^s & 0 \\ 0 & (-\frac{1}{4})^s \end{pmatrix}\mathsf{O}$, and $\mathsf{O}\begin{pmatrix} 0 & -\frac{1}{2} \\ \frac{1}{2} & 0 \end{pmatrix}\mathsf{O} = \begin{pmatrix} 0 & \frac{1}{2} \\ -\frac{1}{2} & 0 \end{pmatrix}$, we can evaluate (7.2.31) as

$$\begin{aligned}\langle \Phi_{\mathrm{VBS}}|\hat{S}_1^{(3)}\hat{S}_{r+1}^{(3)}|\Phi_{\mathrm{VBS}}\rangle \\ &= \mathrm{Tr}\left[\begin{pmatrix} 0 & \frac{1}{2} \\ -\frac{1}{2} & 0 \end{pmatrix}\begin{pmatrix} (\frac{3}{4})^{r-1} & 0 \\ 0 & (-\frac{1}{4})^{r-1} \end{pmatrix}\begin{pmatrix} 0 & \frac{1}{2} \\ -\frac{1}{2} & 0 \end{pmatrix}\begin{pmatrix} (\frac{3}{4})^{L-r-1} & 0 \\ 0 & (-\frac{1}{4})^{L-r-1} \end{pmatrix}\right] \\ &= \left(-\frac{1}{4}\right)^r\left(\frac{3}{4}\right)^{L-r-1} + \left(-\frac{1}{4}\right)^{L-r}\left(\frac{3}{4}\right)^{r-1}\end{aligned} \tag{7.2.32}$$

By combining this estimate with (7.2.25), we find

$$\lim_{L\uparrow\infty}\frac{\langle \Phi_{\mathrm{VBS}}|\hat{S}_1^{(3)}\hat{S}_{r+1}^{(3)}|\Phi_{\mathrm{VBS}}\rangle}{\langle \Phi_{\mathrm{VBS}}|\Phi_{\mathrm{VBS}}\rangle} = \lim_{L\uparrow\infty}\frac{(-\frac{1}{4})^r(\frac{3}{4})^{L-r-1}}{(\frac{3}{4})^L} = \frac{4}{3}\left(-\frac{1}{3}\right)^r. \tag{7.2.33}$$

From the rotational invariance and the translation invariance, we get

$$\lim_{L\uparrow\infty}\frac{\langle \Phi_{\mathrm{VBS}}|\hat{\boldsymbol{S}}_x\cdot\hat{\boldsymbol{S}}_{x+r}|\Phi_{\mathrm{VBS}}\rangle}{\langle \Phi_{\mathrm{VBS}}|\Phi_{\mathrm{VBS}}\rangle} = 4\left(-\frac{1}{3}\right)^r, \tag{7.2.34}$$

which is the desired (7.1.2).

Problem 7.2.2.c Let $\hat{P}_x^0 := |\psi_x^0\rangle\langle\psi_x^0| = 1 - (\hat{S}_x^{(3)})^2$ be the projection on to the 0 state at x. For any $n = 1, 2, \ldots$, take arbitrary distinct n sites $x_1, x_2, \ldots, x_n$, and compute the expectation value

$$\lim_{L\uparrow\infty}\frac{\langle \Phi_{\mathrm{VBS}}|\hat{P}_{x_1}^0\hat{P}_{x_2}^0\cdots\hat{P}_{x_n}^0|\Phi_{\mathrm{VBS}}\rangle}{\langle \Phi_{\mathrm{VBS}}|\Phi_{\mathrm{VBS}}\rangle}. \tag{7.2.35}$$

Note that $x_1, x_2, \ldots, x_n$ should be fixed when letting $L \uparrow \infty$. Discuss the implication of the result. [solution→p.507]

Problem 7.2.2.d Using the implication discussed in the previous problem, confirm (7.2.8). [solution→p.508]

General matrix product states Let us briefly discuss general matrix product states, and their significance in the study of one-dimensional quantum many-body systems. We also introduce the important notion of injective matrix product states, and discuss the corresponding uniqueness theorem (Theorem 7.6). They will play essential roles in Sects. 8.3.4 and 8.3.5.

Consider a spin S quantum spin chain on $\Lambda_L = \{1, 2, \ldots, L\}$. We let A^σ for $\sigma = -S, -S+1, \ldots, S$ be $D \times D$ matrices with complex elements. The matrices do not depend on L. Here the inner dimension (or the bond dimension) D is an arbitrary integer independent of S. Then the matrix product state $|\Phi\rangle$ corresponding to A^σ is defined by generalizing (7.2.14) in a straightforward manner as

$$
\begin{aligned}
|\Phi\rangle &= \sum_{\sigma} \sum_{\alpha_1,\ldots,\alpha_L=1}^{D} A^{\sigma_1}_{\alpha_1,\alpha_2} A^{\sigma_2}_{\alpha_2,\alpha_3} \cdots A^{\sigma_L}_{\alpha_L,\alpha_1} |\Psi^{\sigma}\rangle \\
&= \sum_{\sigma} \mathrm{Tr}[\mathsf{A}^{\sigma_1}\mathsf{A}^{\sigma_2} \cdots \mathsf{A}^{\sigma_L}]\, |\Psi^{\sigma}\rangle. \qquad (7.2.36)
\end{aligned}
$$

It may be obvious that the method of calculating the normalization and the correlations explained above for the VBS state can be applied to the general state $|\Phi\rangle$. Let us also note that a matrix product representation (7.2.36) is not unique, i.e., a single state $|\Phi\rangle$ may be represented by using more than one sets of matrices.

As an instructive example, let $S = 1/2$, $D = 3$, and define

$$
\mathsf{A}^{\uparrow} = \begin{pmatrix} 0 & -1 & 0 \\ 0 & 0 & \frac{1}{\sqrt{2}} \\ 0 & 0 & 0 \end{pmatrix}, \quad \mathsf{A}^{\downarrow} = \begin{pmatrix} 0 & 0 & 0 \\ \frac{1}{\sqrt{2}} & 0 & 0 \\ 0 & 1 & 0 \end{pmatrix}. \qquad (7.2.37)
$$

It is found by inspection that the corresponding matrix product state for even L is[28]

$$
\begin{aligned}
|\Phi\rangle =& \bigotimes_{y=1}^{L/2} \frac{1}{\sqrt{2}}\Big(|\psi^{\uparrow}_{2y}\rangle|\psi^{\downarrow}_{2y+1}\rangle - |\psi^{\downarrow}_{2y}\rangle|\psi^{\uparrow}_{2y+1}\rangle\Big) \\
&+ \bigotimes_{y=1}^{L/2} \frac{1}{\sqrt{2}}\Big(|\psi^{\uparrow}_{2y-1}\rangle|\psi^{\downarrow}_{2y}\rangle - |\psi^{\downarrow}_{2y-1}\rangle|\psi^{\uparrow}_{2y}\rangle\Big), \qquad (7.2.38)
\end{aligned}
$$

which is the linear combination (denoted as $|\Phi^{+}_{\text{dimer}}\rangle = |\Phi^{\text{odd}}_{\text{dimer}}\rangle + |\Phi^{\text{even}}_{\text{dimer}}\rangle$ in Problem 7.1.2.c) of two dimer states (7.1.17) and (7.1.18).

Problem 7.2.2.e Let $S = 1/2$. Find a matrix product representation of the linear combination of two ferromagnetic states

$$
|\Phi\rangle = \bigotimes_{x=1}^{L} |\psi^{\uparrow}_{x}\rangle + \bigotimes_{x=1}^{L} |\psi^{\downarrow}_{x}\rangle, \qquad (7.2.39)
$$

which we encountered in (3.3.8). [solution→p.508]

Problem 7.2.2.f Let $S = 1/2$. Find a matrix product representation of the linear combination of two Néel states

[28] Suppose we start from the vector $(0, 1, 0)^{\mathrm{t}}$. Multiplying it by $\mathsf{A}^{\downarrow}$, we get $(0, 0, 1)^{\mathrm{t}}$. Since further multiplication by $\mathsf{A}^{\downarrow}$ only yields zero, we multiply the result by $\mathsf{A}^{\uparrow}$ to get $(0, 1/\sqrt{2}, 0)^{\mathrm{t}}$. Note that the vector is now proportional to the initial one. Going back to the initial vector $(0, 1, 0)^{\mathrm{t}}$, we can also multiply it by $\mathsf{A}^{\uparrow}$ to get $(-1, 0, 0)^{\mathrm{t}}$, and then by $\mathsf{A}^{\downarrow}$ to get $(0, -1/\sqrt{2}, 0)^{\mathrm{t}}$. In this way we have formed a singlet pair $\{|\uparrow\rangle|\downarrow\rangle - |\downarrow\rangle|\uparrow\rangle\}/\sqrt{2}$.

$$|\Phi\rangle = \bigotimes_{x=1}^{L/2} |\psi^{\uparrow}_{2x-1}\rangle|\psi^{\downarrow}_{2x}\rangle + \bigotimes_{x=1}^{L/2} |\psi^{\downarrow}_{2x-1}\rangle|\psi^{\uparrow}_{2x}\rangle. \tag{7.2.40}$$

[solution→p.508]

The definition (7.2.36) of general matrix product states may first seem as a mere clever way of writing down a class of states of quantum spin chains. But this turns out to be a serious underestimate. The concept of matrix product states brought about a revolution in the way we study one-dimensional quantum many-body systems. It is believed that one can essentially classify all translation invariant states in quantum spin chains by studying matrix product states.

The following theorem, which we state without proof,[29] characterizes a class of matrix product states which plays an essential role in various applications.

Theorem 7.5 (injective matrix product states) *Consider a matrix product state* (7.2.36) *defined by a collection of matrices* $(\mathsf{A}^\sigma)_{\sigma=-S,\ldots,S}$ *that satisfies*

$$\sum_{\sigma=-S}^{S} \mathsf{A}^\sigma(\mathsf{A}^\sigma)^\dagger = \lambda \mathsf{I}, \tag{7.2.41}$$

with a constant $\lambda > 0$. *As in* (7.2.20), *we define the* $D^2 \times D^2$ *transfer matrix by*

$$(\tilde{\mathsf{A}})_{\alpha,\beta;\alpha',\beta'} := \sum_{\sigma=-S}^{S} (A^\sigma_{\alpha,\alpha'})^* \, A^\sigma_{\beta,\beta'}. \tag{7.2.42}$$

Then the following properties are equivalent:

(i) There exists ℓ_0 *such that the set of matrices* $\mathsf{A}^{\sigma_1}\mathsf{A}^{\sigma_2}\ldots\mathsf{A}^{\sigma_{\ell_0}}$ *with all possible* $\sigma_1,\ldots,\sigma_{\ell_0}$ *span the whole space of* $D \times D$ *matrices.*
(ii) There exists ℓ_0 *such that for any* $\ell \geq \ell_0$ *the set of matrices* $\mathsf{A}^{\sigma_1}\mathsf{A}^{\sigma_2}\ldots\mathsf{A}^{\sigma_\ell}$ *with all possible* $\sigma_1,\ldots,\sigma_\ell$ *span the whole space of* $D \times D$ *matrices.*
(iii) λ *is a nondegenerate eigenvalue of the transfer matrix* $\tilde{\mathsf{A}}$, *and any other eigenvalue* λ_j *satisfies* $|\lambda_j| < \lambda$.[30]

A collection of matrices $(\mathsf{A}^\sigma)_{\sigma=-S,\ldots,S}$ is said to be injective (or primitive) if it satisfies (7.2.41) and (i), (ii), (iii). We also call the corresponding matrix product state (or, more precisely, the corresponding matrix product representation) injective (or primitive). Physically speaking, the characterization (iii) implies that an injective matrix product state is not a superposition of macroscopically different states, and has short range correlation (or entanglement) in the sense that all correlation functions decay exponentially. See [93] for introductory discussion, and [28, 70] for details.

---

[29] See [28, 70]. For mathematical details see also Sect. 6.3 of [90].

[30] The eigenvector corresponding to λ is $\delta_{\alpha',\beta'}$ since $\sum_{\alpha',\beta'}(\tilde{\mathsf{A}})_{\alpha,\beta;\alpha',\beta'}\,\delta_{\alpha',\beta'} = \sum_{\alpha',\sigma}(A^\sigma_{\alpha,\alpha'})^*\,A^\sigma_{\beta,\alpha'} = \sum_\sigma(\mathsf{A}^\sigma(\mathsf{A}^\sigma)^\dagger)_{\beta,\alpha} = \lambda\,\delta_{\alpha,\beta}$.

The injectivity of a given set of matrices may be verified (or falsified) by direct calculation. The set of matrices (7.2.12) for the VBS state is found to be injective,[31] while that in (7.2.37) to be not.[32]

In general a matrix product state can be represented in terms of several different sets of matrices. It is known however that the representation as an injective matrix product state is essentially unique, as is stated in the following theorem (which we do not prove) due to Fannes, Nachtergaele, and Werner [31]. See also Theorem 7 of [70]. The uniqueness theorem plays an essential role in the classification of symmetry protected topological phases, which we discuss in Sect. 8.3.4. (See also Sect. 8.3.5.)

Theorem 7.6 *Suppose that two collections of matrices* $(\mathsf{A}^\sigma)_{\sigma=-S,\ldots,S}$ *and* $(\mathsf{B}^\sigma)_{\sigma=-S,\ldots,S}$ *are injective and generate the same matrix product state in the sense that*

$$\sum_{\sigma} \mathrm{Tr}[\mathsf{A}^{\sigma_1} \cdots \mathsf{A}^{\sigma_L}] \, |\Psi^{\boldsymbol{\sigma}}\rangle = \sum_{\sigma} \mathrm{Tr}[\mathsf{B}^{\sigma_1} \cdots \mathsf{B}^{\sigma_L}] \, |\Psi^{\boldsymbol{\sigma}}\rangle, \tag{7.2.43}$$

for any L. Then they have a common dimension D, and there exists a $D \times D$ *unitary matrix* U*, which is unique up to a phase factor, such that*

$$\mathsf{B}^\sigma = \mathsf{U}^\dagger \mathsf{A}^\sigma \mathsf{U}, \tag{7.2.44}$$

for all σ.

Let us give a very brief summary of other essential properties of matrix product states. First the set of matrix product states is known to be weak-$*$ dense in the set of all translation invariant states of a quantum spin chain [30]. This means that the expectation value of any local observable in an arbitrary given translation invariant state can be approximated with a desired precision by a matrix product state with sufficiently large inner dimension D.[33] Moreover it has been shown that a unique ground state accompanied by a finite energy gap can always be efficiently approximated by a matrix product state. See, e.g., [8].

Secondly, for any injective matrix product state, one can construct a short ranged Hamiltonian, called the parent Hamiltonian, whose unique ground state coincides with the given matrix product state.[34] It is also known that the ground state is

---

[31] The condition (7.2.41) is valid with $\lambda = 3/4$. The four matrices $\mathsf{A}^0\mathsf{A}^+$, $\mathsf{A}^0\mathsf{A}^-$, $\mathsf{A}^+\mathsf{A}^-$, and $\mathsf{A}^-\mathsf{A}^+$ are enough to span the whole 2×2 matrices; the condition (i) is verified with $\ell_0 = 2$.

[32] The condition (7.2.41) is valid with $\lambda = 1$. But we find, with the aid of a computer, that the transfer matrix $\tilde{\mathsf{A}}$ has eigenvalues ± 1.

[33] To be precise, let $\rho[\cdot]$ be an arbitrary translation invariant state on the infinite chain. (See Appendix A.7 for the notation.) Then one can take a sequence $\rho_j[\cdot]$ with $j = 1, 2, \ldots$ of matrix product states on the infinite chain such that $\lim_{j\uparrow\infty} \rho_j[\hat{A}] = \rho[\hat{A}]$ for each $\hat{A} \in \mathfrak{A}_{\mathrm{loc}}$.

[34] A general construction (which may not give the simplest parent Hamiltonian) is as follows. Take the open chain $\{x, \ldots, x + \ell_0\}$ with $\ell_0 + 1$ sites, where ℓ_0 is the constant determined in Theorem 7.5, and consider the corresponding matrix product states (7.2.45) with all possible boundary terms. Let $\hat{P}_x$ be the projection onto the space spanned by all these matrix product states. Then $\hat{H} = -\sum_x \hat{P}_x$ gives a desired parent Hamiltonian.

accompanied by a nonvanishing energy gap. This general statement provides another proof of the existence of a gap in the AKLT model (see Theorem 7.1 in p. 178).

It is worth noting that, when a matrix product state has exponentially decaying correlation function, the decay law is purely exponential at large distances. This does not agree with the Ornstein–Zernike form (6.1.3) with the power law correction $|x - y|^{-1/2}$, which is expected in generic disordered ground states of one dimensional quantum spin systems.[35] Although matrix product states can approximate any state with short range correlation, the approximation is not precise enough to recover the power law correction.

Most of these (and other) properties of matrix product states are stated and proved in the original and the definitive paper by Fannes, Nachtergaele, and Werner [28]. Unfortunately the paper is written in a mathematical style and requires advanced background in algebraic formulation of quantum many-body systems. We recommend only mathematically minded reader to study the paper.[36] A standard reference on basic mathematical properties of matrix product states for physicists and quantum information scientists is [70]. As a practical introduction from a physical point of view, see Chap. 8 of [93]. Reference [77] is an interesting review which focuses on the relation between matrix product states and a numerical method called the density matrix renormalization group (DMRG). For an amazing application of the matrix product representation to the measurement based quantum computation, see [25, 36, 37].

We have so far concentrated on translation invariant matrix product states generated by a single set of matrices A^σ. It is obvious that one can generalize the construction to non-translation invariant states by using site dependent matrices.

One can also write down matrix product states on open chains. In addition to the set of matrices A^σ, one prepares a row vector $\boldsymbol{\ell}^\sigma$ (with component ℓ^σ_α) and a column vector $\boldsymbol{r}^\sigma$ (with component r^σ_α) to describe boundary effects, and define

$$\begin{aligned}|\Phi\rangle &= \sum_{\boldsymbol{\sigma}} \sum_{\alpha_1,\ldots,\alpha_L}^{D} \ell^{\sigma_1}_{\alpha_2} A^{\sigma_2}_{\alpha_2,\alpha_3} \cdots A^{\sigma_{L-1}}_{\alpha_{L-1},\alpha_L} r^{\sigma_L}_{\alpha_L} |\Psi^{\boldsymbol{\sigma}}\rangle \\ &= \sum_{\boldsymbol{\sigma}} (\boldsymbol{\ell}^{\sigma_1} \mathsf{A}^{\sigma_2} \cdots \mathsf{A}^{\sigma_{L-1}} \boldsymbol{r}^{\sigma_L}) |\Psi^{\boldsymbol{\sigma}}\rangle. \end{aligned} \tag{7.2.45}$$

The normalization factor and correlation functions can be evaluated (by drawing diagrams if necessary) almost as before. See Problem 7.2.3.c below.

Remark In general one can associate with a unique gapped ground state of a spin chain a mathematical object which is very close to the set of matrices defining a matrix product state [14–16, 54–56]. More precisely one can define operators s_σ with $\sigma = -S, \ldots, S$, which form a representation of the Cuntz algebra, in the von Neumann algebra $\pi(\mathfrak{A}_R)''$ constructed from the ground state of a spin chain restricted

[35] See footnote 4 in p. 156.

[36] It may be useful to know that our $\tilde{\mathsf{A}}$ and $\tilde{\mathsf{B}}$ become $\mathbb{E}_{\mathbb{1}}$ and $\mathbb{E}_{\hat{S}^{(3)}}$, respectively, in [28].

to the right half-infinite chain, provided that the ground state is pure, translation invariant, and satisfies the condition called the split property. A remarkable fact is that the operators s_σ may be regarded as infinite dimensional analogues of matrices defining a matrix product state. See also the remark after Theorem 8.6 in p. 276.

7.2.3 AKLT Model on Open Chains

Ground states In Sect. 7.1, we studied the AKLT model with Hamiltonian (7.1.1) on the periodic chain and found that the model has a unique disordered ground state. Interestingly the nature of the ground states is qualitatively different for the same model on an open chain, i.e., the one-dimensional lattice with open boundary conditions. More precisely we consider the $S = 1$ quantum spin system on $\Lambda_L = \{1, 2, \ldots, L\}$ with the Hamiltonian

$$\hat{H}_{\mathrm{AKLT}}^{\mathrm{open}} = \sum_{x=1}^{L-1} \{\hat{\boldsymbol{S}}_x \cdot \hat{\boldsymbol{S}}_{x+1} + \frac{1}{3}(\hat{\boldsymbol{S}}_x \cdot \hat{\boldsymbol{S}}_{x+1})^2\}. \tag{7.2.46}$$

The ground states of the model are constructed almost in the same manner as we did for the periodic chain in Sect. 7.1.2. We again prepare a system of $S = 1/2$'s on the duplicated chain, and define the pre-VBS states as

$$|\Phi_{\text{pre-VBS}}^{\sigma_\mathrm{L},\sigma_\mathrm{R}}\rangle := |\psi_{1,\mathrm{L}}^{\sigma_\mathrm{L}}\rangle \otimes \Big(\bigotimes_{x=1}^{L-1} \frac{1}{\sqrt{2}}\big\{|\psi_{x,\mathrm{R}}^{\uparrow}\rangle|\psi_{x+1,\mathrm{L}}^{\downarrow}\rangle - |\psi_{x,\mathrm{R}}^{\downarrow}\rangle|\psi_{x+1,\mathrm{L}}^{\uparrow}\rangle\big\}\Big) \otimes |\psi_{L,\mathrm{R}}^{\sigma_\mathrm{R}}\rangle, \tag{7.2.47}$$

where $\sigma_\mathrm{L}, \sigma_\mathrm{R} = \uparrow, \downarrow$. Note that we here have freedom to assign arbitrary spins at the two ends of the chain. See Fig. 7.4a. We then define the corresponding VBS states by

$$|\Phi_{\mathrm{VBS}}^{\sigma_\mathrm{L},\sigma_\mathrm{R}}\rangle = \Big(\bigotimes_{x=1}^{L} \mathscr{S}_x\Big)|\Phi_{\text{pre-VBS}}^{\sigma_\mathrm{L},\sigma_\mathrm{R}}\rangle. \tag{7.2.48}$$

See Fig. 7.4b.

Since each bond $\{x, x+1\}$ with $x = 1, \ldots, L-1$ contains a singlet pair in the VBS state (7.2.48), we see, exactly as in (7.1.14), that $\hat{P}_2[\hat{\boldsymbol{S}}_x + \hat{\boldsymbol{S}}_{x+1}]|\Phi_{\mathrm{VBS}}^{\sigma_\mathrm{L},\sigma_\mathrm{R}}\rangle = 0$ for any $x = 1, \ldots, L-1$. We thus find that the VBS state $|\Phi_{\mathrm{VBS}}^{\sigma_\mathrm{L},\sigma_\mathrm{R}}\rangle$ with any $\sigma_\mathrm{L}, \sigma_\mathrm{R} = \uparrow, \downarrow$ is an exact ground state of the Hamiltonian (7.2.46) of the open AKLT chain. It is not hard to prove that the four ground states $|\Phi_{\mathrm{VBS}}^{\uparrow,\uparrow}\rangle$, $|\Phi_{\mathrm{VBS}}^{\uparrow,\downarrow}\rangle$, $|\Phi_{\mathrm{VBS}}^{\downarrow,\uparrow}\rangle$, and $|\Phi_{\mathrm{VBS}}^{\downarrow,\downarrow}\rangle$ are linearly independent (Problem 7.2.3.a), and also that they are the only ground states (Problem 7.2.3.b). We conclude that the ground states of the open AKLT chain are four-fold degenerate. It is also possible to prove that there is a nonvanishing energy

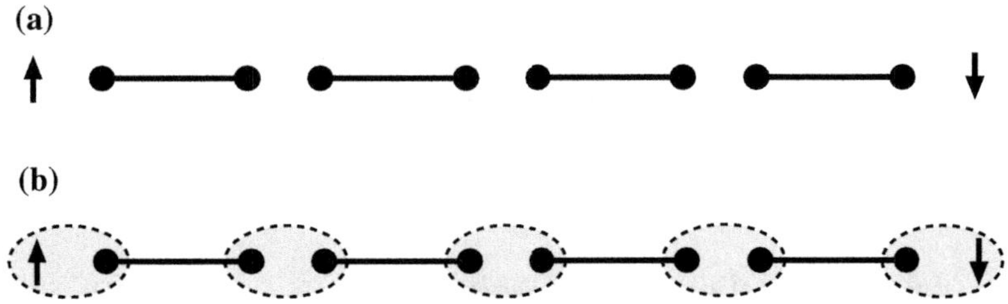

Fig. 7.4 **a** The pre-VBS state $|\Phi^{\uparrow,\downarrow}_{\text{pre-VBS}}\rangle$ and **b** the corresponding VBS state $|\Phi^{\uparrow,\downarrow}_{\text{VBS}}\rangle$ on the open chain with $L = 5$. See Fig. 7.2 for the interpretation of the symbols in the diagram. It should be obvious from the figures that there is freedom to assign ↑ or ↓ at the two ends of the chain (© Hal Tasaki 2020. All Rights Reserved)

gap above the ground state energy. For example, one can extend the method of Knabe discussed in Sect. 7.1.4 to open chains in a rather straightforward manner.[37]

To see the nature of these ground states, it is useful to compute the single spin expectation value in one of the ground states, say, $|\Phi^{\uparrow,\uparrow}_{\text{VBS}}\rangle$. Then we find

$$\langle\Phi^{\uparrow,\uparrow}_{\text{VBS}}|\hat{S}^{(1)}_x|\Phi^{\uparrow,\uparrow}_{\text{VBS}}\rangle = \langle\Phi^{\uparrow,\uparrow}_{\text{VBS}}|\hat{S}^{(2)}_x|\Phi^{\uparrow,\uparrow}_{\text{VBS}}\rangle = 0, \tag{7.2.49}$$

which immediately follows from the symmetry, and

$$\lim_{L\uparrow\infty} \frac{\langle\Phi^{\uparrow,\uparrow}_{\text{VBS}}|\hat{S}^{(3)}_x|\Phi^{\uparrow,\uparrow}_{\text{VBS}}\rangle}{\langle\Phi^{\uparrow,\uparrow}_{\text{VBS}}|\Phi^{\uparrow,\uparrow}_{\text{VBS}}\rangle} = -2(-3)^{-x} \quad \text{for any } x = 1, 2, \ldots. \tag{7.2.50}$$

See Problem 7.2.3.c. The effect of the extra up spin is not strictly localized at the edge, but decays exponentially as one moves away from the edge. Noting that $\sum_{x=1}^{\infty} -2(-3)^{-x} = 1/2$, we see that there indeed is a spin with $S = 1/2$, exponentially localized at the left edge of the chain.

We can say that the ground state space of the open AKLT chain is described by two free spins with $S = 1/2$ at each edge. This is an example of edges states, which are universally observed in (quantum) many-body systems with certain "topological" nature [33, 87].

The appearance of effective spin with $S = 1/2$ might look obvious if one thinks in terms of the pre-VBS state of the duplicated $S = 1/2$ chain. One should note however that the duplication is a mere mathematical technique, and we are here dealing with a system of spins with quantum number 1. The effective $S = 1/2$ should thus be regarded as "fractional spin" generated by intricate interaction between $S = 1$ spins.

[37] We learned this from Hosho Katsura. See [48] for recent refinements of the method.

Problem 7.2.3.a Prove that the four ground states are linearly independent. This may be done in many ways, but see the footnote if the reader needs a hint.[38] [solution→p.508]

Problem 7.2.3.b Prove that the open AKLT chain has exactly four ground states by extending the method of Sect. 7.1.3. [solution→p.508]

Problem 7.2.3.c Find a matrix product representation (7.2.45) of $|\Phi_{\rm VBS}^{\uparrow,\uparrow}\rangle$, and verify (7.2.50). [solution→p.508]

Infinite volume ground states Let us relabel the lattice sites, and write the one-dimensional lattice as

$$\Lambda_L = \{-\tfrac{L}{2}+1, \ldots, \tfrac{L}{2}-1, \tfrac{L}{2}\}, \tag{7.2.51}$$

which is indeed our original definition in (3.1.2). We consider the same AKLT model with open boundary conditions on Λ_L, and study the infinite volume limit as formulated in Sect. 4.3. Exactly as in Sect. 4.3, we can construct an infinite volume ground state through the limit

$$\omega(\hat{A}) = \lim_{L\uparrow\infty} \frac{\langle\Phi_{\rm VBS}^{\sigma_{\rm L},\sigma_{\rm R}}|\hat{A}|\Phi_{\rm VBS}^{\sigma_{\rm L},\sigma_{\rm R}}\rangle}{\langle\Phi_{\rm VBS}^{\sigma_{\rm L},\sigma_{\rm R}}|\Phi_{\rm VBS}^{\sigma_{\rm L},\sigma_{\rm R}}\rangle}, \tag{7.2.52}$$

where $\hat{A} \in \mathfrak{A}_{\rm loc}$ is an arbitrary operator which acts only on spins in a finite interval.

It can be proved rather easily that the limiting ground state $\omega(\cdot)$ is independent of the choice of the boundary spins $\sigma_{\rm L}$ and $\sigma_{\rm R}$. This is natural if one recalls (7.2.50). Since the effect of the boundary spins decay exponentially as one moves away from the boundary, there can be no effect on the expectation value in the limit $L \uparrow \infty$. The infinite volume ground state $\omega(\cdot)$ is also identical to that obtained from the ground state (7.1.12) for a periodic chain. This is consistent with the uniqueness of the infinite volume ground state of the AKLT model stated in Theorem 7.2 in p. 179.

To sum the AKLT model on an open chain has four-fold degenerate ground states for finite L, but has a unique ground state in the limit $L \uparrow \infty$. This stands in instructive contrast to the antiferromagnetic Heisenberg model in two or higher dimensions, which has a unique finite volume ground state, but has infinitely many (symmetry breaking) ground states in the infinite volume limit. See Chap. 4. These two examples clearly shows that the number of ground states is a very subtle notion. We summarize the observation in the following table.

	finite volume	*infinite volume*
open chain AKLT	four GS	unique GS
AF Heisenberg model ($d \geq 2$)	unique GS	infinitely many GS

---

[38]Examine the representation (7.2.1), and see what type of spin configurations one gets for each state.

7.3 Extensions and Related Models

We briefly discuss some extensions of the VBS state and the AKLT model. Extensions to quantum spin chains with higher S discussed in Sect. 7.3.1 provide further support to Haldane's conclusion about the qualitative difference between antiferromagnetic Heisenberg chains with half-odd-integer S and integer S. We discuss some results in higher dimensions in Sect. 7.3.2. Finally, in Sect. 7.3.3, we discuss the Briegel–Raussendorf state (cluster state) and its generalization, which have some similarity to the VBS state.

7.3.1 *Spin Chains with Higher S and the VBS Picture*

Consider a quantum spin chain with general spin S. It is obvious that the construction of the VBS state in Sect. 7.1.2 can be generalized to models with larger S. For example a single spin with $S = 2$ can be represented by symmetrizing four spins with $S = 1/2$. Then, again denoting by  a singlet pair of two $S = 1/2$'s, we define the VBS state for $S = 2$ chain as

$|\Phi^{S=2}_{\mathrm{VBS}}\rangle =$ 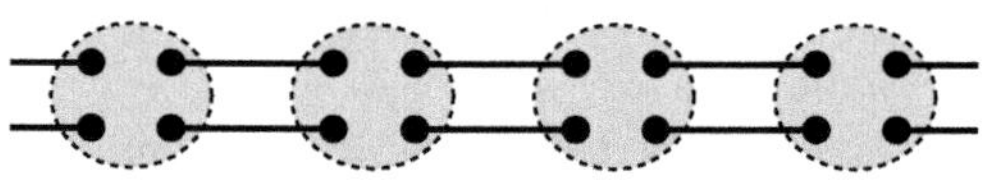 (7.3.1)

where 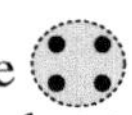 means symmetrized spins.[39] We also easily see that $|\Phi^{S=2}_{\mathrm{VBS}}\rangle$ is an exact ground state of the Hamiltonian

$$\hat{H}_{\mathrm{AKLT},\,S=2} = \sum_{x=1}^{L} \{a\,\hat{P}_4[\hat{\boldsymbol{S}}_x + \hat{\boldsymbol{S}}_{x+1}] + b\,\hat{P}_3[\hat{\boldsymbol{S}}_x + \hat{\boldsymbol{S}}_{x+1}] + c\}, \tag{7.3.2}$$

where $a, b > 0$ and $c \in \mathbb{R}$ are arbitrary constants. As before $\hat{P}_J[\hat{\boldsymbol{S}}_x + \hat{\boldsymbol{S}}_{x+1}]$ is the projection onto the subspace with the total spin $S_{\mathrm{tot}} = J$. Recall that S_{tot} can be 4, 3, 2, 1, or 0. We are giving penalties to the two highest values. As in the case of $S = 1$, we can write down the Hamiltonian as a polynomial of spin operators, for example, as

$$\hat{H}_{\mathrm{AKLT},\,S=2} = \sum_{x=1}^{L} \Big\{\hat{\boldsymbol{S}}_x \cdot \hat{\boldsymbol{S}}_{x+1} + \frac{2}{9}(\hat{\boldsymbol{S}}_x \cdot \hat{\boldsymbol{S}}_{x+1})^2 + \frac{1}{63}(\hat{\boldsymbol{S}}_x \cdot \hat{\boldsymbol{S}}_{x+1})^3\Big\}, \tag{7.3.3}$$

39 One can of course write down a formula corresponding to (7.3.1), exactly as we did in (7.1.12) for the $S = 1$ chain.

which corresponds to the choice $a = 10$, $b = 10/7$, and $c = -10/7$.[40] But this is a very artificial Hamiltonian. The main point here is that the VBS state can be defined.

Problem 7.3.1.a Show that (7.3.3) gives (7.3.2). [solution→p.509]

What about chains with $S = 3/2$? Since an $S = 3/2$ is represented by three $S = 1/2$'s as 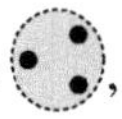, natural states constructed by using valence bonds of unit length are

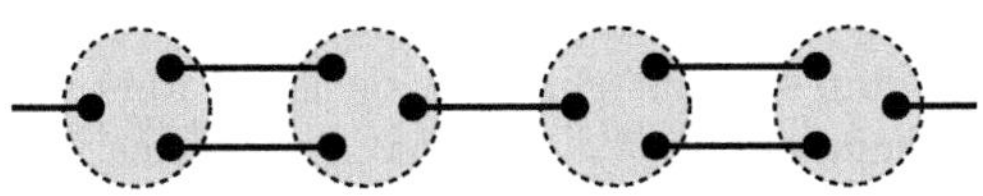

(7.3.4)

or

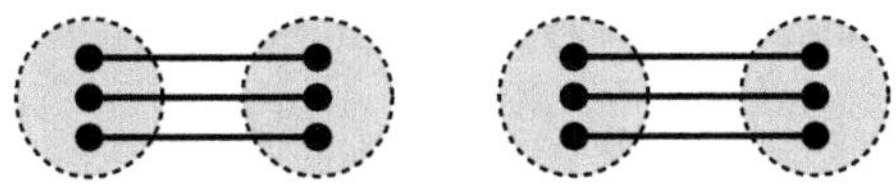

(7.3.5)

which are not translation invariant. Suppose that there is a translation invariant Hamiltonian which has the state (7.3.4) as a ground state. Then the translation invariance implies that the translation of (7.3.4) is also a ground state, and hence ground states are (at least) doubly degenerate. The situation resembles that in the Majumdar–Ghosh model (7.1.15) for the $S = 1/2$ chain. Recall that the AKLT model with $S = 1$ has a unique disordered ground state with a gap.

The above consideration can be generalized. When S is an integer, one can construct a translation invariant VBS state by using valence bonds of unit length. When S is a half-odd integer one must sacrifice translation invariance if only short-ranged valence-bonds are allowed. This is what happens in the Majumdar–Ghosh model with doubly degenerate disordered ground states with a gap. Ground states which are not (even approximately) described by short-ranged valence bonds may be essentially different. An obvious example is the unique ground state of the $S = 1/2$ antiferromagnetic Heisenberg chain, which has power law decaying correlation (6.1.2).

This observation is consistent with Haldane's conclusion that there is a qualitative difference in low energy properties between antiferromagnetic Heisenberg chains with integer S and half-odd-integer S. In particular it is remarkable that the VBS states with short-ranged valence bonds can be constructed if and only if S is an integer. This fact provides a strong support (but not a proof!) to Haldane's main claim that an integer spin quantum antiferromagnetic chain may have a unique disordered ground state with a gap.

[40]More generally (7.3.2) is expressed in terms of a polynomial including $(\hat{\boldsymbol{S}}_x \cdot \hat{\boldsymbol{S}}_{x+1})^4$.

7.3.2 Higher Dimensional Models

VBS states on general lattices It should be obvious by now that the construction of VBS states automatically extends to any lattice in any dimensions, as was discussed in the original AKLT papers [4, 5]. Let us go back to the general notation of Sect. 2.4, and denote a general lattice as $(\Lambda, \mathscr{B})$. Here Λ is the set of sites $x, y, \ldots$, and $\mathscr{B}$ is the set of bonds $\{x, y\} = \{y, x\}$, etc. For each site $x \in \Lambda$, let $\mathscr{N}(x) := \{y \in \Lambda \mid \{x, y\} \in \mathscr{B}\}$ be the set of sites directly connected to x by a bond.

We consider a quantum spin system on Λ by associating with each site x a spin whose spin quantum number is $\bar{S}_x = |\mathscr{N}(x)|/2$. To define the VBS state, we first split a site x into $|\mathscr{N}(x)|$ sites which we denote as $(x; y)$ with $y \in \mathscr{N}(x)$. Note that $(x; y)$ and $(y; x)$ are distinct. We then consider an extended spin system by associating a spin $S = 1/2$ with each site $(x; y)$. This precisely corresponds to the introduction of the duplicated $S = 1/2$ chain in Sect. 7.1.2. We then define the VBS state on $(\Lambda, \mathscr{B})$ by[41]

$$|\Phi_{\rm VBS}^{(\Lambda,\mathscr{B})}\rangle = \Bigl(\bigotimes_{x\in\Lambda} \mathscr{S}_x\Bigr)\Bigl(\bigotimes_{\{x,y\}\in\mathscr{B}} \frac{1}{\sqrt{2}}\bigl\{|\psi^{\uparrow}_{(x;y)}\rangle|\psi^{\downarrow}_{(y;x)}\rangle - |\psi^{\downarrow}_{(x;y)}\rangle|\psi^{\uparrow}_{(y;x)}\rangle\bigr\}\Bigr), \quad (7.3.6)$$

where $\mathscr{S}_x$ is the symmetrization operator for $|\mathscr{N}(x)|$ spins on sites $(x; y)$ with $y \in \mathscr{N}(x)$. This is a straightforward extension of the definition (7.1.12) of the $S = 1$ VBS state. See Fig. 7.5 for an example.

It is found that the VBS state (7.3.6) is the unique ground state of the generalized AKLT Hamiltonian

$$\hat{H}_{\rm AKLT}^{(\Lambda,\mathscr{B})} = \sum_{\{x,y\}\in\mathscr{B}} \hat{P}_{\bar{S}_x+\bar{S}_y}[\hat{\boldsymbol{S}}_x + \hat{\boldsymbol{S}}_y]. \quad (7.3.7)$$

Note that the projection $\hat{P}_{\bar{S}_x+\bar{S}_y}[\hat{\boldsymbol{S}}_x + \hat{\boldsymbol{S}}_y]$ gives a penalty to the largest total spin on two sites x and y. The uniqueness of the ground state was proved by Kennedy, Lieb, and Tasaki [41]. The proof is easily done by extending the method presented in Sect. 7.1.3.

Although the above construction of the VBS state (7.3.6) and the corresponding Hamiltonian (7.3.7) is parallel to that for spin chains, the properties of these general models are far from obvious. Very little has been understood about the properties of the general VBS states. In fact, as we shall briefly discuss below, it is known that even the basic nature of the state crucially depends on the dimension and the lattice structure.

The AKLT model on the hexagonal lattice We have a rather satisfactory understanding of the $S = 3/2$ model on the hexagonal lattice of Fig. 7.5. Let $(\Lambda, \mathscr{B})$ be

---

[41] For simplicity we only discuss VBS states were each bond of the lattice carries one valence bond. Clearly one can define more general VBS states with multiple valence bonds on each lattice bond as in (7.3.1).

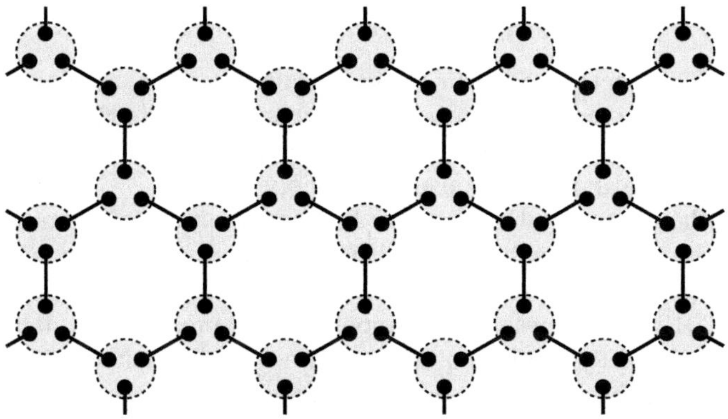

Fig. 7.5 The $S = 3/2$ VBS state (7.3.6) on the hexagonal lattice is the unique ground state of the AKLT model (7.3.8). As in Fig. 7.2, a black dot in the figure represents a spin with $S = 1/2$, and ●——● denotes a singlet pair (or a valence bond). Three spins with $S = 1/2$ are symmetrized to represent $S = 3/2$. It is known that the ground state has exponentially decaying correlations, and is accompanied by a nonzero gap (© Hal Tasaki 2020. All Rights Reserved)

the hexagonal lattice with periodic boundary conditions,[42] and consider a quantum spin system obtained by associating with each site a spin $S = 3/2$. Then the VBS state $|\Phi_{\mathrm{VBS}}^{(\Lambda,\mathscr{B})}\rangle$ defined in (7.3.6) is the unique ground state of the AKLT Hamiltonian

$$\hat{H}_{\mathrm{AKLT}}^{(\Lambda,\mathscr{B})} = \sum_{\{x,y\}\in\mathscr{B}} \hat{P}_3[\hat{\boldsymbol{S}}_x + \hat{\boldsymbol{S}}_y]. \tag{7.3.8}$$

The following was proved by by Affleck, Kennedy, Lieb, and Tasaki [5] and Kennedy, Lieb, and Tasaki [41]. See [41] for the proof.

Theorem 7.7 *For any $x, y \in \Lambda$, we have*

$$0 \le (-1)^{D(x,y)} \frac{\langle\Phi_{\mathrm{VBS}}^{(\Lambda,\mathscr{B})}|\hat{\boldsymbol{S}}_x \cdot \hat{\boldsymbol{S}}_y|\Phi_{\mathrm{VBS}}^{(\Lambda,\mathscr{B})}\rangle}{\langle\Phi_{\mathrm{VBS}}^{(\Lambda,\mathscr{B})}|\Phi_{\mathrm{VBS}}^{(\Lambda,\mathscr{B})}\rangle} \le C\, e^{-D(x,y)/\xi}, \tag{7.3.9}$$

with some positive constants C and ξ independent of the system size. Here $D(x, y)$ denotes the graph theoretic distance, i.e., the minimum number of bonds necessary to connect x and y. Moreover the translation invariant infinite volume ground state (in the sense of Definition 4.17) of the Hamiltonian (7.3.8) *is unique,*[43] *and coincides with the infinite volume state defined by*

$$\omega(\hat{A}) := \lim_{L\uparrow\infty} \frac{\langle\Phi_{\mathrm{VBS}}^{(\Lambda,\mathscr{B})}|\hat{A}|\Phi_{\mathrm{VBS}}^{(\Lambda,\mathscr{B})}\rangle}{\langle\Phi_{\mathrm{VBS}}^{(\Lambda,\mathscr{B})}|\Phi_{\mathrm{VBS}}^{(\Lambda,\mathscr{B})}\rangle}, \tag{7.3.10}$$

---

[42] We assume that the boundary conditions are chosen so that the lattice is bipartite.

[43] Note that our uniqueness theorem is limited to translation invariant states. We still do not have a stronger uniqueness theorem comparable to Theorem 7.2 in p. 179.

for any local operator $\hat{A} \in \mathfrak{A}_{\mathrm{loc}}$. (See Sect. 4.3.1 for details.)

Thus the AKLT model (7.3.8) on the hexagonal lattice has a unique disordered ground state, which may be regarded as a two-dimensional version of Haldane's disordered ground state in spin chains. It was thus conjectured that the Hamiltonian (7.3.8) has a nonvanishing gap above the ground state energy.

To show the existence of a gap in the hexagonal lattice AKLT model, however, turned out to be a very difficult problem; it had remained unsolved for more than thirty years since the model was proposal and conjectured to have a gap in 1987.[44] Very recently in 2019, while I was almost completing the present book, there was rapid progress, which led to solutions of the problem. First, in January, Abdul-Rahman, Lemm, Lucia, Nachtergaele, and Young [1] developed a new method based on an inequality due to Fannes, Nachtergaele, and Werner [28], and proved the existence of a gap in a closely related model, namely, the AKLT model on the "decorated" hexagonal lattice. This method was improved in [71]. Then, in October, Lemm, Sandvik, and Wang [49] demonstrated that the the AKLT model on the hexagonal lattice has a nonzero energy gap. This was done by combining a refinement of Knabe's method (see Sect. 7.1.4) and a numerical calculation with a sophisticated DMRG (density matrix renormalization group) method of a model on a lattice with 36 sites.[45] Finally, in November, Pomata and Wei [72] showed the existence of a nonzero energy gap in the AKLT model on the hexagonal and other lattices with coordination number 3. This was done by combining a refined version of the method of [1] and numerical results obtained by the Lanczos method.[46]

Given that the $S = 3/2$ AKLT model on the hexagonal lattice has a unique disordered ground state, one might be tempted to conjecture that the same is true for the $S = 3/2$ antiferromagnetic Heisenberg model on the hexagonal lattice. It turns out however that one can prove by using the method of reflection positivity by Dyson, Lieb, and Simon [24] and Neves and Perez [68], which was the main topic of Sect. 4.1, that the latter model has long-range Néel order in its ground state. See Appendix of [5]. This means that a phase transition takes place when one continuously changes the Hamiltonian from the AKLT Hamiltonian (7.3.8) to the Heisenberg Hamiltonian (2.5.1).

There is an interesting application (which was totally unexpected when the model was found) of the hexagonal lattice AKLT model to measurement based quantum computation. See [61, 84–86]. See also [45] for a proposal of a realistic system on the hexagonal lattice whose ground state resembles the VBS state.

---

[44]It was demonstrated in [12, 21, 22] that whether a given quantum spin system has a gap or not (in the infinite volume) can be an undecidable problem. Although these results are highly nontrivial and deep, the difficulty related to the gap in the hexagonal lattice AKLT model has nothing to do with the undecidability. (And the existence of a gap was finally (almost) decided. See below.)

[45]First part of the work is fully rigorous, but the second part is numerical. Therefore the lower bound for the energy gap, although being reliable and conclusive, is not yet mathematically rigorous.

[46]Again the first part is rigorous and the second part is numerical. But this work is closer to (or may be regarded as) a computer-aided proof since only the Lanczos method is used in the numerical part.

Other models One can define the VBS state with $S = 2$ on the square lattice. It was proved in [41] that the state has an exponentially decaying correlation function (but the uniqueness of the ground state is not yet proved). This is consistent with the conjecture by Arovas, Auerbach, and Haldane [10] that the VBS state in two dimensions is always disordered.[47]

In dimensions higher than two, the situation is much more complicated. It is believed that the behavior of the VBS state (7.3.6) depends crucially on the lattice structure. It is known that the state is disordered when the coordination number of the lattice is small enough [41]. But it is expected that the VBS state on some lattices exhibits long range Néel order, as is confirmed by a calculation [5] and by a proof [29] for the VBS states on certain tree-like lattices.

Tensor networks Obviously the general VBS state (7.3.6) can be represented in a similar manner as the matrix product representation (7.2.14) for the one-dimensional VBS state. In this case one associates with each site x a suitable tensor[48] $A^{\sigma}_{x;(\alpha_{x,y})_{y\in\mathcal{N}(x)}}$, where $\alpha_{x,y} = \alpha_{y,x} = 1, 2$, and represents the VBS state as

$$|\Phi^{(\Lambda,\mathcal{B})}_{\mathrm{VBS}}\rangle = \sum_{\boldsymbol{\sigma}}\Big(\prod_{\{x,y\}\in\mathcal{B}}\ \sum_{\alpha_{x,y}=1,2}\Big)\Big(\prod_{x\in\Lambda} A^{\sigma_x}_{x;(\alpha_{x,y})_{y\in\mathcal{N}(x)}}\Big)|\Psi^{\boldsymbol{\sigma}}\rangle. \tag{7.3.11}$$

This is an example of the class of states called tensor networks, which is a straightforward generalization of matrix product states. Tensor networks are widely used to represent and study low-energy states in many-body quantum systems. But, unlike matrix product states which are theoretically guaranteed to (at least approximately) represent the ground states of a large class of one-dimensional systems, the role of tensor product states is not yet clear. See. e.g., Chap. 9 of [93], and also [62].

Problem 7.3.2.a Let $(\Lambda_L, \mathcal{B}_L)$ be the square lattice, defined by (3.1.2) and (3.1.3) with $d = 2$, and consider the $S = 2$ VBS state

$$|\Phi^{(\Lambda_L,\mathcal{B}_L)}_{\mathrm{VBS}}\rangle = \Big(\bigotimes_{x\in\Lambda_L}\mathcal{S}_x\Big)\Big[\bigotimes_{\substack{x\in\Lambda_L\\ \alpha=\mathrm{x},\mathrm{y}}}\frac{1}{\sqrt{2}}\{|\psi^{\uparrow}_x\rangle|\psi^{\downarrow}_{x+\boldsymbol{e}_\alpha}\rangle - |\psi^{\downarrow}_x\rangle|\psi^{\uparrow}_{x+\boldsymbol{e}_\alpha}\rangle\}\Big], \tag{7.3.12}$$

where $\boldsymbol{e}_{\mathrm{x}} = (1, 0)$ and $\boldsymbol{e}_{\mathrm{y}} = (0, 1)$. Find a tensor $A^{\sigma}_{\alpha_1,\alpha_2,\alpha_3,\alpha_4}$ and represent the VBS state as a tensor network

---

[47] One might wonder if the two-dimensional VBS state exhibits exotic properties similar to those of its one-dimensional counterpart. It is obvious from the construction that the state defined on a finite lattice with open boundary is accompanied by edge spins that live on boundary sites. On the other hand it is likely that the state does not have a simple hidden order as in the one-dimensional VBS state. This observation is consistent with the modern picture that the two dimensional VBS state has only weak SPT (symmetry protected topological) order. See the end of Sect. 8.3.4.

[48] If all sites are identical the tensor is simply $A^{\sigma}_{(\alpha_{x,y})_{y\in\mathcal{N}(x)}}$.

$$|\Phi_{\mathrm{VBS}}^{(\Lambda_L,\mathscr{B}_L)}\rangle = \sum_{\sigma}\Bigl(\prod_{\{x,y\}\in\mathscr{B}_L}\sum_{\alpha_{x,y}=1,2}\Bigr)\Bigl(\prod_{x\in\Lambda_L} A^{\sigma_x}_{\alpha_{x,x+e_x},\alpha_{x,x+e_y},\alpha_{x,x-e_x},\alpha_{x,x-e_y}}\Bigr)|\Psi^{\sigma}\rangle. \tag{7.3.13}$$

[solution→p.509]

7.3.3 *Briegel–Raussendorf State (Cluster State) and Its Generalizations*

In this section we focus on a class of simple but interesting states studied extensively in the context of measurement based quantum computation [73]. The most basic is the Briegel–Raussendorf state, which is usually called the cluster state [18, 19] or the graph state [39].[49] We also discuss its generalization known as the hyper graph state [75]. See [84] for a recent review on quantum spin models in measurement based quantum computation.

Although the states studied in this section are not extensions of the VBS state, they have close similarities with the AKLT model and the VBS state. Some of the states also provide prototypical examples of states in a nontrivial symmetry protected topological phase in two or higher dimensions. See the end of Sect. 8.3.4.

Briegel–Raussendorf state (cluster state) in one dimension Let us first discuss the most basic state, namely, the Briegel–Raussendorf state (cluster state) in one dimension. We consider an $S = 1/2$ quantum spin system on the chain $\Lambda_L = \{1, 2, \ldots, L\}$. Take the simple tensor product state where all the spins are pointing in the 1-direction, and expand it as

$$|\Phi_{\rightarrow}\rangle := \bigotimes_{x=1}^{L}\frac{|\psi_x^{\uparrow}\rangle + |\psi_x^{\downarrow}\rangle}{\sqrt{2}} = 2^{-L/2}\sum_{\sigma}|\Psi^{\sigma}\rangle, \tag{7.3.14}$$

where $\boldsymbol{\sigma} = (\sigma_1, \ldots, \sigma_L)$ (with $\sigma_x = \uparrow, \downarrow$) is summed over all possible spin configurations. Let us introduce the following sign factor:

$$s(\sigma, \sigma') = \begin{cases} -1 & \text{if}\, \sigma = \sigma' = \downarrow, \\ 1 & \text{otherwise} \end{cases} \tag{7.3.15}$$

Then the Briegel–Raussendorf state (cluster state) is defined by

$$|\Phi_{\mathrm{BR}}\rangle := 2^{-L/2}\sum_{\sigma}\Bigl\{\prod_{x=1}^{L} s(\sigma_x, \sigma_{x+1})\Bigr\}|\Psi^{\sigma}\rangle, \tag{7.3.16}$$

---

[49]See [11] for experimental realizations of the Briegel–Raussendorf state.

where we assumed the periodic boundary condition $\sigma_{x+L} = \sigma_x$. It is clear that the state (7.3.16) is also written as

$$|\Phi_{\rm BR}\rangle = 2^{-L/2}\sum_{\boldsymbol{\sigma}}(-1)^{N_{\downarrow\downarrow}(\boldsymbol{\sigma})}|\Psi^{\boldsymbol{\sigma}}\rangle, \tag{7.3.17}$$

where $N_{\downarrow\downarrow}(\boldsymbol{\sigma})$ denotes the number of bonds (pairs of neighboring sites) with two $\downarrow$-spins in the configuration $\boldsymbol{\sigma}$. Note that only the phase factors have been modified compared with the trivial product state (7.3.14). This implies that the state $|\Phi_{\rm BR}\rangle$ behaves in exactly the same manner as the product state (7.3.14) when functions of $\hat{S}^{(3)}_x$ are measured. Moreover it can be shown that $\langle\Phi_{\rm BR}|\hat{A}\hat{B}|\Phi_{\rm BR}\rangle = \langle\Phi_{\rm BR}|\hat{A}|\Phi_{\rm BR}\rangle\langle\Phi_{\rm BR}|\hat{B}|\Phi_{\rm BR}\rangle$ for any operators $\hat{A}$ and $\hat{B}$ whose supports are separated by more than the distance two. See Problem 7.3.3.b below. The Briegel–Raussendorf state is therefore said to have zero correlation length. Nevertheless the state exhibits nontrivial entanglement properties that originates from the modified phases. See, e.g., (7.3.23) below. The Briegel–Raussendorf state on a regular lattice (see (7.3.29) below) including the chain is usually called the cluster state.[50]

There is a simple Hamiltonian whose unique ground state is the Briegel–Raussendorf state. For notational simplicity, let us use the Pauli matrices $\hat{\sigma}^{(\alpha)}_x = 2\hat{S}^{(\alpha)}_x$. See (2.1.8). We first note that, for any $x = 1, \ldots, L$, it holds that

$$\hat{\sigma}^{(3)}_{x-1}\hat{\sigma}^{(1)}_x\hat{\sigma}^{(3)}_{x+1}|\Phi_{\rm BR}\rangle = |\Phi_{\rm BR}\rangle. \tag{7.3.18}$$

Proof of (7.3.18) Fix x and let $\boldsymbol{\sigma}' = (\sigma_1, \ldots, \sigma_{x-1}, \sigma_{x+1}, \ldots, \sigma_L)$ be an arbitrary spin configuration on $\Lambda_L\backslash\{x\}$. Then the Briegel–Raussendorf state (7.3.16) is rewritten as

$$\begin{aligned}|\Phi_{\rm BR}\rangle = 2^{-L/2}\sum_{\boldsymbol{\sigma}'}\Bigl\{\prod_{y\in\Lambda_L\backslash\{x-1,x\}} s(\sigma_y, \sigma_{y+1})\Bigr\} \bigotimes_{y\in\Lambda_L\backslash\{x\}} |\psi^{\sigma_y}_y\rangle \\ \otimes \Bigl\{s(\sigma_{x-1}, \uparrow)s(\uparrow, \sigma_{x+1})|\psi^{\uparrow}_x\rangle + s(\sigma_{x-1}, \downarrow)s(\downarrow, \sigma_{x+1})|\psi^{\downarrow}_x\rangle\Bigr\}.\end{aligned} \tag{7.3.19}$$

We note that the state in the summand is

$$\begin{aligned}s(\sigma_{x-1}, \uparrow)s(\uparrow, \sigma_{x+1})|\psi^{\uparrow}_x\rangle + s(\sigma_{x-1}, \downarrow)s(\downarrow, \sigma_{x+1})|\psi^{\downarrow}_x\rangle \\ = \begin{cases}|\psi^{\uparrow}_x\rangle + |\psi^{\downarrow}_x\rangle & \text{if } \sigma_{x-1}\sigma_{x+1} = 1, \\ |\psi^{\uparrow}_x\rangle - |\psi^{\downarrow}_x\rangle & \text{if } \sigma_{x-1}\sigma_{x+1} = -1,\end{cases}\end{aligned} \tag{7.3.20}$$

where the states in the right-hand side are the eigenstate of $\hat{\sigma}^{(1)}_x$ with eigenvalues 1 and -1, respectively. This readily implies the desired (7.3.18). ∎

---

[50] In the original paper by Briegel and Raussendorf [19], the term "cluster state" was introduced primarily to indicate the state (7.3.29) defined on a connected percolation cluster. See also [18].

See Problem 7.3.3.a below for a different (quantum information theoretic) derivation of (7.3.18).

Since $\hat{\sigma}^{(3)}_{x-1}\hat{\sigma}^{(1)}_{x}\hat{\sigma}^{(3)}_{x+1} \le 1$, we see that $|\Phi_{\rm BR}\rangle$ is a ground state of the frustration-free Hamiltonian (see Appendix A.2.3)

$$\hat{H}_{\rm BR} = -\sum_{x=1}^{L} \hat{\sigma}^{(3)}_{x-1}\hat{\sigma}^{(1)}_{x}\hat{\sigma}^{(3)}_{x+1}, \tag{7.3.21}$$

with the ground state energy $E_{\rm GS} = -L$. In fact we can further show that $|\Phi_{\rm BR}\rangle$ is the unique ground state of the Hamiltonian (7.3.21), and that the energy gap above $E_{\rm GS}$ is exactly 2. We prove these properties below in a more general context, by showing that the Hamiltonian (7.3.21) is transformed into a trivial Hamiltonian (7.3.37) by a simple unitary transformation. See Theorem 7.8 and its proof. In spite of such simplicity, the Briegel–Raussendorf state exhibits nontrivial behavior similar to that of the AKLT model as we shall see now.

Recalling that $(\hat{\sigma}^{(\alpha)}_x)^2 = \hat{1}$, we see that

$$\prod_{\ell=1}^{n}(\hat{\sigma}^{(3)}_{2\ell-1}\hat{\sigma}^{(1)}_{2\ell}\hat{\sigma}^{(3)}_{2\ell+1}) = \hat{\sigma}^{(3)}_{1}(\prod_{\ell=1}^{n}\hat{\sigma}^{(1)}_{2\ell})\hat{\sigma}^{(3)}_{2n+1}, \tag{7.3.22}$$

provided that $2n+1 \le L$. We then find from (7.3.18) that

$$\langle\Phi_{\rm BR}|\hat{\sigma}^{(3)}_{1}\,(\prod_{\ell=1}^{n}\hat{\sigma}^{(1)}_{2\ell})\,\hat{\sigma}^{(3)}_{2n+1}|\Phi_{\rm BR}\rangle = 1. \tag{7.3.23}$$

Recall, on the other hand, that

$$\langle\Phi_{\rm BR}|\hat{\sigma}^{(3)}_{x}\hat{\sigma}^{(3)}_{y}|\Phi_{\rm BR}\rangle = 0, \tag{7.3.24}$$

for any $x \ne y$. Although the Briegel–Raussendorf state does not have any long-range order (or any correlations measured by $\hat{\sigma}^{(3)}_x$) it has "hidden order" measured by a string correlation function as in (7.3.23). The situation is quite analogous to that for the VBS state.[51]

Another similarity of the Briegel–Raussendorf state to the AKLT model is the emergence of edge degrees of freedom. Consider the Hamiltonian corresponding to (7.3.21) defined on an open chain:

[51] In Sects. 8.2.3 and 8.3.2, in connection with the notion of symmetry protected topological (SPT) phase, we argue that the invariance under the $\mathbb{Z}_2 \times \mathbb{Z}_2$ transformation is an essential property of the VBS state and the AKLT model. The state (7.3.16) and the corresponding Hamiltonian (7.3.21) also have $\mathbb{Z}_2 \times \mathbb{Z}_2$ symmetry. More precisely, when L is even, the Hamiltonian (7.3.21) commutes with both $\hat{U}_{\rm odd} := \prod_{\ell=1}^{L/2}\hat{\sigma}^{(1)}_{2\ell-1}$ and $\hat{U}_{\rm even} := \prod_{\ell=1}^{L/2}\hat{\sigma}^{(1)}_{2\ell}$. For the aspect of the Briegel–Raussendorf as a state in a nontrivial symmetry protected topological phase, see, e.g., [78].

$$\hat{H}_{\mathrm{BR}}^{\mathrm{open}} = -\sum_{x=2}^{L-1} \hat{\sigma}_{x-1}^{(3)} \hat{\sigma}_x^{(1)} \hat{\sigma}_{x+1}^{(3)} \tag{7.3.25}$$

Then one easily finds that the four states

$$|\Phi_{\mathrm{BR}}^{\sigma_1,\sigma_L}\rangle := 2^{-(L-2)/2} \sum_{\sigma_2,\ldots,\sigma_{L-1}=\uparrow,\downarrow} \Bigl\{\prod_{x=1}^{L-1} s(\sigma_x,\sigma_{x+1})\Bigr\} |\Psi^{\boldsymbol{\sigma}}\rangle, \tag{7.3.26}$$

with $\sigma_1, \sigma_L = \uparrow, \downarrow$ are ground states of (7.3.25). There appear free $S = 1/2$ spins strictly localized at the boundaries.

The Briegel–Raussendorf state (7.3.16) also has a matrix product representation (7.2.36) with

$$\mathsf{A}^{\uparrow} = \begin{pmatrix} 1 & 1 \\ 0 & 0 \end{pmatrix}, \quad \mathsf{A}^{\downarrow} = \begin{pmatrix} 0 & 0 \\ 1 & -1 \end{pmatrix}. \tag{7.3.27}$$

It is instructive to examine how these matrices generate the desired phase factors.[52]

Problem 7.3.3.a For $x \neq y$, define the operator $\hat{C}_{x,y}$ by $\hat{C}_{x,y}|\Psi^{\boldsymbol{\sigma}}\rangle = s(\sigma_x,\sigma_y)|\Psi^{\boldsymbol{\sigma}}\rangle$ for any $\boldsymbol{\sigma}$. This is indeed a special case of (7.3.32). (In quantum information $\hat{C}_{x,y}$ is called the controlled-Z operator.) Note that the Briegel–Raussendorf state (7.3.16) is written as $|\Phi_{\mathrm{BR}}\rangle = (\prod_{x=1}^{L} \hat{C}_{x,x+1})|\Phi_{\rightarrow}\rangle$. Show that

$$\hat{C}_{x,y}\, \hat{\sigma}_x^{(1)} = \hat{\sigma}_x^{(1)} \hat{\sigma}_y^{(3)} \hat{C}_{x,y}. \tag{7.3.28}$$

Prove the relation (7.3.18) by using (7.3.28). It is useful to note that $\hat{\sigma}_x^{(1)}|\Phi_{\rightarrow}\rangle = |\Phi_{\rightarrow}\rangle$. [solution→p.510]

Briegel–Raussendorf state on a general lattice It is apparent that the state (7.3.17) and the corresponding Hamiltonian (7.3.21) can readily be extended to the $S = 1/2$ spin system on an arbitrary lattice specified by the set of sites Λ and the set of bonds $\mathcal{B}$. (See, e.g., Sect. 2.4 for the notation.) The Briegel–Raussendorf state on $(\Lambda, \mathcal{B})$ is given by [19, 39]

$$\begin{aligned} |\Phi_{\mathrm{BR}}\rangle &:= 2^{-|\Lambda|/2} \sum_{\boldsymbol{\sigma}} \Bigl\{ \prod_{\{x,y\}\in\mathcal{B}} s(\sigma_x,\sigma_y) \Bigr\} |\Psi^{\boldsymbol{\sigma}}\rangle \\ &= 2^{-|\Lambda|/2} \sum_{\boldsymbol{\sigma}} (-1)^{N_{\downarrow\downarrow}(\boldsymbol{\sigma})} |\Psi^{\boldsymbol{\sigma}}\rangle, \end{aligned} \tag{7.3.29}$$

[52] Let us start from the vector $(1,0)^{\mathrm{t}}$. The vector does not change when we multiply it (repeatedly) by $\mathsf{A}^{\uparrow}$. When we multiply $(1,0)^{\mathrm{t}}$ by $\mathsf{A}^{\downarrow}$, it changes to $(0,1)^{\mathrm{t}}$, passing information to the left that there is a $\downarrow$. A multiplication of $(0,1)^{\mathrm{t}}$ by $\mathsf{A}^{\uparrow}$ simply reverts it to $(1,0)^{\mathrm{t}}$. But when we multiply $(0,1)^{\mathrm{t}}$ by $\mathsf{A}^{\downarrow}$, we get $(0,-1)^{\mathrm{t}}$ which contains the desired phase factor (and still the information that there is a $\downarrow$ is passed to the left).

where $N_{\downarrow\downarrow}(\boldsymbol{\sigma})$ is the number of bonds with two down spins. In the literature, the Briegel–Raussendorf state (7.3.29) on a regular lattice is called the cluster state, while that on a general lattice is called the graph state [39].

It again follows from Theorem 7.8 that $|\Phi_{\rm BR}\rangle$ is the unique ground state of the Hamiltonian

$$\hat{H}_{\rm BR} = -\sum_{x\in\Lambda} \hat{\sigma}_x^{(1)} \prod_{y\in\mathcal{N}(x)} \hat{\sigma}_y^{(3)}, \tag{7.3.30}$$

where $\mathcal{N}(x) = \{y \in \Lambda \,|\, \{x, y\} \in \mathcal{B}\}$ is the set of sites directly connected to x by a bond.

Generalized Briegel–Raussendorf state (hyper graph state) The simplicity of the Briegel–Raussendorf state allows a generalization known as the hyper graph state [75]. Let the lattice Λ be a set of N sites. We do not specify any geometric structures. We again consider a spin system with $S = 1/2$ on Λ. For any subset $A \subset \Lambda$, we define the sign factor

$$s_A(\boldsymbol{\sigma}) = \begin{cases} -1 & \text{if } \sigma_x = \downarrow \text{ for all } x \in A, \\ 1 & \text{otherwise,} \end{cases} \tag{7.3.31}$$

and the corresponding operator $\hat{C}_A$ by the action onto the basis states as

$$\hat{C}_A|\Psi^{\boldsymbol{\sigma}}\rangle = s_A(\boldsymbol{\sigma})|\Psi^{\boldsymbol{\sigma}}\rangle. \tag{7.3.32}$$

By a cluster c we mean any subset of Λ which contains more than one site.[53] Typical examples of a cluster include (the set of sites in) a bond, a triangle, or a plaquette. Let $\mathcal{C}$ be a set of clusters such that for any $x \in \Lambda$ there is at least one $c \in \mathcal{C}$ with $c \ni x$. Then the corresponding generalized Briegel–Raussendorf state or the hyper graph state is defined as

$$|\Phi_{\mathcal{C}}\rangle := 2^{-N/2} \sum_{\boldsymbol{\sigma}} \Bigl(\prod_{c\in\mathcal{C}} s_c(\boldsymbol{\sigma})\Bigr) |\Psi^{\boldsymbol{\sigma}}\rangle. \tag{7.3.33}$$

Note that this reduces to the original Briegel–Raussendorf state (7.3.29) if we take clusters to be bonds.[54] The state (7.3.33) also has zero correlation length. See Problem 7.3.3.b below.

---

[53]Note that our usage of the term cluster is very much different from that in the original paper by Briegel and Raussendorf [19].

[54]We find that the names "graph state" and "hyper graph state" are too general and do not characterize the particular states (7.3.29) or (7.3.33) very well. Since these states are characterized by the quantum mechanical phase associated with each cluster, we feel that the name "cluster-phase state" is somewhat more appropriate. In this terminology one may specify the type of cluster by talking about the bond-phase state, the triangle-phase state, or the plaquette-phase state. The bond-phase state is of course the Briegel–Raussendorf state (7.3.29). (It is unfortunate that the term "cluster phase" (without hyphen) is used, e.g., in [74] to mean a symmetry protected topological phase that includes the cluster state.)

The Hamiltonian for the state (7.3.33) is given by

$$\hat{H}_{\mathscr{C}} := -\sum_{x\in\Lambda}\Bigl(\prod_{\substack{c\in\mathscr{C}\\(c\ni x)}}\hat{C}_{c\backslash\{x\}}\Bigr)\hat{\sigma}_x^{(1)}. \tag{7.3.34}$$

Noting that $\hat{C}_{\{y\}} = \hat{\sigma}_y^{(3)}$, one sees that this is a generalization of the Hamiltonian (7.3.21). We have the following theorem.

Theorem 7.8 *The state* $|\Phi_{\mathscr{C}}\rangle$ *is the unique ground state of the Hamiltonian* (7.3.34) *with energy* $E_{\mathrm{GS}} = -N$. *The energy gap above* E_{GS} *is equal to 2.*

Proof Note first that the state (7.3.33) is written as

$$|\Phi_{\mathscr{C}}\rangle = \hat{U}_{\mathscr{C}}|\Phi_{\rightarrow}\rangle, \tag{7.3.35}$$

where $|\Phi_{\rightarrow}\rangle = 2^{-N/2}\sum_{\sigma}|\Psi^{\sigma}\rangle$ is the trivial product state corresponding to (7.3.14), and the unitary operator $\hat{U}_{\mathscr{C}}$ is defined by

$$\hat{U}_{\mathscr{C}} = \prod_{c\in\mathscr{C}}\hat{C}_c. \tag{7.3.36}$$

One clearly has $(\hat{U}_{\mathscr{C}})^2 = \hat{1}$. Next recall that $|\Phi_{\rightarrow}\rangle$ is the unique ground state of the trivial Hamiltonian

$$\hat{H}_{\rightarrow} = -\sum_{x\in\Lambda}\hat{\sigma}_x^{(1)}. \tag{7.3.37}$$

The ground state energy is $-N$, and the energy gap is 2. We shall show that

$$\hat{U}_{\mathscr{C}}\hat{H}_{\rightarrow}\hat{U}_{\mathscr{C}} = \hat{H}_{\mathscr{C}}, \tag{7.3.38}$$

from which and (7.3.35) the desired theorem readily follow.

The key for (7.3.38) is the identity

$$\hat{C}_A\hat{\sigma}_x^{(1)}\hat{C}_A = \hat{C}_{A\backslash\{x\}}\hat{\sigma}_x^{(1)}, \tag{7.3.39}$$

for any subset A which includes x. To show this consider first the subspace with $\hat{C}_{A\backslash\{x\}} = 1$, where (7.3.39) is obvious since $\hat{C}_A$ always gives 1. In the subspace with $\hat{C}_{A\backslash\{x\}} = -1$, one finds that $\hat{C}_A$ is identical with $\hat{\sigma}_x^{(3)}$. Then (7.3.39) follows from the identity $\hat{\sigma}_x^{(3)}\hat{\sigma}_x^{(1)}\hat{\sigma}_x^{(3)} = -\hat{\sigma}_x^{(1)}$. By noting that the operators $\hat{C}_c$ commute with each other and $(\hat{C}_c)^2 = \hat{1}$, we get (7.3.38). ∎

Problem 7.3.3.b For an operator $\hat{A}$, denote by $\Lambda_{\hat{A}}$ its support (i.e., the set of sites on which $\hat{A}$ acts nontrivially). We define the corresponding set of clusters by $\mathscr{C}_{\hat{A}} = \{c\in\mathscr{C}\,|\,c\cap\Lambda_{\hat{A}}\neq\emptyset\}$, and the enlarged support by $\bar{\Lambda}_{\hat{A}} = \{x\in\Lambda\,|\,x\in c$ for some

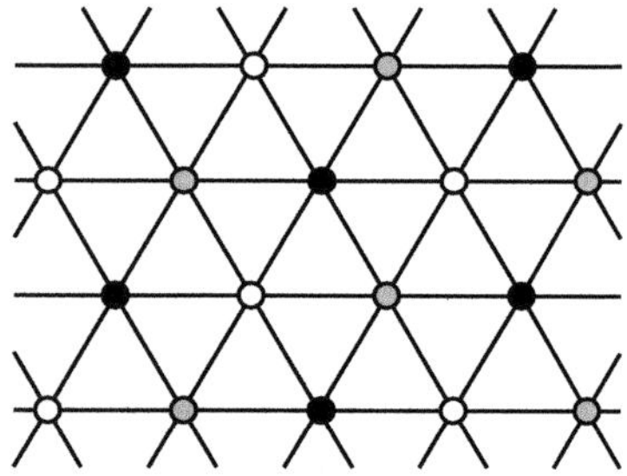

Fig. 7.6 The triangular lattice is decomposed into three sublattices indicated by black, gray, and white circles (© Hal Tasaki 2020. All Rights Reserved)

$c \in \mathscr{C}_{\hat{A}}\}$. Let $\hat{A}$ and $\hat{B}$ be arbitrary operators such that $\bar{\Lambda}_{\hat{A}} \cap \bar{\Lambda}_{\hat{B}} = \emptyset$. (For the one-dimensional Briegel–Raussendorf state, this means that the distance between the supports $\Lambda_{\hat{A}}$ and $\Lambda_{\hat{B}}$ is greater than two.) Prove that

$$\langle \Phi_{\mathscr{C}} | \hat{A}\hat{B} | \Phi_{\mathscr{C}} \rangle = \langle \Phi_{\mathscr{C}} | \hat{A} | \Phi_{\mathscr{C}} \rangle \langle \Phi_{\mathscr{C}} | \hat{B} | \Phi_{\mathscr{C}} \rangle, \tag{7.3.40}$$

which means that there is no correlation between $\hat{A}$ and $\hat{B}$. (Hint: The expression (7.3.35) is useful.) [solution→p.510]

Problem 7.3.3.c Let Λ be the triangular lattice with suitable periodic boundary conditions. We take $\mathscr{C}$ as the set of all unit triangles, and consider the corresp-nding generalized Briegel–Raussendorf state[55] (or the hyper graph state) (7.3.33) and the Hamiltonian (7.3.34). This model provides one of the simplest examples of nontrivial symmetry protected topological phase in two dimensions [58, 92]. (See the end of Sect. 8.3.4). Write the Hamiltonian of the model explicitly. What is the analogue of the string correlation function in this model? As shown in Fig. 7.6 the triangular lattice (with suitable periodic boundary conditions) is decomposed into three sublattices as $\Lambda = S_1 \cup S_2 \cup S_3$. Show that the Hamiltonian is invariant under the unitary transformation $\hat{U}_\nu = \prod_{x \in S_\nu} \hat{\sigma}_x^{(1)}$ for any $\nu = 1, 2, 3$. Thus the model has $\mathbb{Z}_2 \times \mathbb{Z}_2 \times \mathbb{Z}_2$ symmetry [92]. [solution→p.510]

References

1. H. Abdul-Rahman, M. Lemm, A. Lucia, B. Nachtergaele, A. Young, *A class of two-dimensional AKLT models with a gap* (2019), arXiv:1901.09297
2. L. Accardi, Topics in quantum probability. Phys. Rep. **77**, 169–192 (1981)
3. L. Accardi, A. Frigerio, Markovian cocycles. Proc. R. Ir. Acad. **83A**(2), 251–263 (1983)
4. I. Affleck, T. Kennedy, E.H. Lieb, H. Tasaki, Rigorous results on valence-bond ground states in antiferromagnets. Phys. Rev. Lett. **59**, 799 (1987)
5. I. Affleck, T. Kennedy, E.H. Lieb, H. Tasaki, Valence bond ground states in isotropic quantum antiferromagnets. Commun. Math. Phys. **115**, 477–528 (1988), https://projecteuclid.org/euclid.cmp/1104161001

[55]The state is called the triangle-phase state in the terminology proposed in footnote 54.

6. M. Aizenman, H. Duminil-Copin, S. Warzel, *Dimerization and Néel order in different quantum spin chains through a shared loop representation*, preprint (2020), arxiv:2002.02543
7. A. Anshu, *Improved local spectral gap thresholds for lattices of finite dimension* (2019), arXiv:1909.01516v1
8. I. Arad, Z. Landau, U. Vazirani, T. Vidick, Rigorous RG algorithms and area laws for low energy eigenstates in 1D. Commun. Math. Phys. **356**, 65–105 (2017), arXiv:1602.08828
9. D.P. Arovas, Two exact excited states for the $S = 1$ AKLT chain. Phys. Lett. A **137**, 431–433 (1989)
10. D.P. Arovas, A. Auerbach, F.D.M. Haldane, Extended Heisenberg models of antiferromagnetism: analogies to the fractional quantum Hall effect. Phys. Rev. Lett. **60**, 531 (1988)
11. W. Asavanant, Y. Shiozawa, S. Yokoyama, B. Charoensombutamon, H. Emura, R.N. Alexander, S. Takeda, J. Yoshikawa, N.C. Menicucci, H. Yonezawa, A. Furusawa, Generation of time-domain-multiplexed two-dimensional cluster state. Science **366**, 373–376 (2019), arXiv:1903.03918
12. J. Bausch, T. Cubitt, A. Lucia, D. Perez-Garcia, *Undecidability of the spectral gap in one dimension* (2018), arXiv:1810.01858
13. R.J. Baxter, Dimers on a rectangular lattice. J. Math. Phys. **9**, 650 (1968)
14. O. Bratteli, P. Jorgensen, Endomorphisms of $B(H)$ II. Finitely correlated states on O_n. J. Funct. Anal. **145**, 323–373 (1997)
15. O. Bratteli P. Jorgensen, A. Kishimoto, R.F. Werner, Pure states on $\mathcal{O}_d$. J. Oper. Theory **43**, 97–143 (2000)
16. O. Bratteli, P. Jorgensen, G. Price, Endomorphisms of $B(\mathcal{H})$: quantization, nonlinear partial differential equations, and operator algebra, in *Proceedings of Symposia in Pure Mathematics*, vol. 59 (1996), pp. 93–138, https://www.duo.uio.no/handle/10852/43152
17. G.K. Brennen, A. Miyake, Measurement-based quantum computer in the gapped ground state of a two-body hamiltonian. Phys. Rev. Lett. **101**, 010502 (2008), arXiv:0803.1478
18. H.J. Briegel, Cluster states, in *Compendium of Quantum Physics*, ed. by D. Greenberger, K. Hentschel, F. Weinert (Springer, Berlin, 2009), https://link.springer.com/chapter/10.1007/978-3-540-70626-7_30
19. H.J. Briegel, R. Raussendorf, Persistent entanglement in arrays of interacting particles. Phys. Rev. Lett. **86**, 910 (2001), arXiv:quant-ph/0004051
20. W.J. Caspers, K.M. Emmett, W. Magnus, The Majumdar-Ghosh chain. Twofold ground state and elementary excitations. J. Phys. A: Math. Gen. **17**, 2687 (1984)
21. T.B. Cubitt, D. Perez-Garcia, M.M. Wolf, Undecidability of the spectral gap. Nature **528**, 207–211 (2015), arXiv:1502.04135
22. T.B. Cubitt, D. Perez-Garcia, M.M. Wolf, *Undecidability of the spectral gap (full version)* (2015), arXiv:1502.04573
23. M. den Nijs, K. Rommelse, Preroughening transitions in crystal surfaces and valence-bond phases in quantum spin chains. Phys. Rev. B **40**, 4709 (1989)
24. F.J. Dyson, E.H. Lieb, B. Simon, Phase transitions in quantum spin systems with isotropic and nonisotropic interactions. J. Stat. Phys. **18**, 335–382 (1978)
25. D.V. Else, I. Schwarz, S.D. Bartlett, A.C. Doherty, Symmetry-protected phases for measurement-based quantum computation. Phys. Rev. Lett. **108**, 240505 (2012), arXiv:1201.4877
26. M. Fannes, B. Nachtergaele, R.F. Werner, Exact antiferromagnetic ground states of quantum spin chains. Europhys. Lett. **10**, 633–637 (1989)
27. M. Fannes, B. Nachtergaele, R.F. Werner, Valence bond states on quantum spin chains as ground states with spectral gap. J. Phys. A Math. Gen. **24**, L185–190 (1991)
28. M. Fannes, B. Nachtergaele, R.F. Werner, Finitely correlated states on quantum spin chains. Commun. Math. Phys. **144**, 443–490 (1992), https://projecteuclid.org/euclid.cmp/1104249404
29. M. Fannes, B. Nachtergaele, R.F. Werner, Ground states of VBS models on Cayley trees. J. Stat. Phys. **66**, 939–973 (1992)
30. M. Fannes, B. Nachtergaele, R.F. Werner, Abundance of translation invariant pure states on quantum spin chains. Lett. Math. Phys. **25**, 249–258 (1992)

31. M. Fannes, B. Nachtergaele, R.F. Werner, Finitely correlated pure states. J. Funct. Anal. **120**, 511–534 (1994), ftp://ace2.ma.utexas.edu/pub/mp_arc/converted/92/92-132.ps.gz
32. G. Fáth, J. Sólyom, Solitonic excitations in the Haldane phase of an $S = 1$ chain. J. Phys.: Condens. Matter **5**, 8983 (1993), arXiv:cond-mat/9309011
33. E. Fradkin, *Field Theories of Condensed Matter Physics* (Cambridge University Press, Cambridge, 2013)
34. A. Garcia-Saez, A. Murg, T.-C. Wei, Spectral gaps of Affleck-Kennedy-Lieb-Tasaki hamiltonians using tensor network methods. Phys. Rev. B **88**, 245118 (2013), arXiv:1308.3631
35. D. Gosset, E. Mozgunov, Local gap threshold for frustration-free spin systems. J. Math. Phys. **57**, 091901 (2016), arXiv:1512.00088
36. D. Gross, *Computational power of quantum many-body states and some results on discrete phase spaces*, Ph.D. thesis, London University (2008), http://www.thp.uni-koeln.de/gross/files/diss.pdf
37. D. Gross, J. Eisert, Quantum computational webs. Phys. Rev. A **82**, 040303(R) (2010), arXiv:0810.2542
38. P. Hayden, R. Jozsa, D. Petz, A. Winter, Structure of states which satisfy strong subadditivity of quantum entropy with equality. Commun. Math. Phys. **246**, 359–374 (2004), arXiv:quant-ph/0304007
39. M. Hein, W. Dür, J. Eisert, R. Raussendorf, M. Van den Nest, H.-J. Briegel, Entanglement in graph states and its applications, in *Proceedings of the International School of Physics "Enrico Fermi" on "Quantum Computers, Algorithms and Chaos*, Varenna, Italy (2005), arXiv:quant-ph/0602096
40. R. Kaltenbaek, J. Lavoie, B. Zeng, S.D. Bartlett, K.J. Resch, Optical one-way quantum computing with a simulated valence-bond solid. Nat. Phys. **6**, 850–854 (2010), https://www.nature.com/articles/nphys1777
41. T. Kennedy, E.H. Lieb, H. Tasaki, A two-dimensional isotropic quantum antiferromagnet with unique disordered ground state. J. Stat. Phys. **53**, 383–415 (1988)
42. T. Kennedy, H. Tasaki, Hidden symmetry breaking and the Haldane phase in $S = 1$ quantum spin chains. Commun. Math. Phys. **147**, 431–484 (1992), https://projecteuclid.org/euclid.cmp/1104250747
43. A. Klümper, A. Schadschneider, J. Zittartz, Matrix product ground states for one-dimensional spin-1 quantum antiferromagnets. Europhys. Lett. **24**, 293 (1993)
44. S. Knabe, Energy gaps and elementary excitations for certain VBS-quantum antiferromagnets. J. Stat. Phys. **52**, 627–638 (1988)
45. M. Koch-Janusz, D.I. Khomskii, E. Sela, Affleck-Kennedy-Lieb-Tasaki state on a honeycomb lattice from t_{2g} orbitals. Phys. Rev. Lett. **114**, 247204 (2015), arXiv:1501.03512
46. A. Laeuchli, G. Schmid, S. Trebst, Spin nematic correlations in bilinear-biquadratic $S = 1$ *spin chains*, Phys. Rev. B **74**, 144426 (2006), arXiv:cond-mat/0607173
47. M. Lemm, *Finite-size criteria for spectral gaps in D-dimensional quantum spin systems*, (2019), arXiv:1902.07141
48. M. Lemm, E. Mozgunov, *Spectral gaps of frustration-free spin systems with boundary* (2018), arXiv:1801.08915
49. M. Lemm, A. Sandvik, L. Wang, *Existence of a spectral gap in the AKLT model on the hexagonal lattice* (2019), arXiv:1910.11810
50. M. Lemm, A. Sandvik, S. Yang, The AKLT model on a hexagonal chain is gapped. J. Stat. Phys. (2019), arXiv:1904.01043
51. C.K. Majumdar, Antiferromagnetic model with known ground state. J. Phys. C **3**, 911–915 (1970)
52. C.K. Majumdar, D.K. Ghosh, On next nearest-neighbor interaction in linear chain. I, II. J. Math. Phys. **10**, 1388–1402 (1969)
53. T. Matsui, Translational symmetry breaking and soliton sectors for massive quantum spin models in $1 + 1$ dimensions. Commun. Math. Phys. **189**, 127–144 (1997)
54. T. Matsui, The split property and the symmetry breaking of the quantum spin chain. Commun. Math. Phys. **218**, 393–416 (2001)

55. T. Matsui, Boundedness of entanglement entropy and split property of quantum spin chains. Rev. Math. Phys. 1350017 (2013), arXiv:1109.5778
56. T. Matsui, A characterization of finitely correlated pure states. Infin. Dimens. Anal. Q. Probab. **1**, 647–661 (1998)
57. S. Michalakis, J.P. Zwolak, Stability of frustration-free hamiltonians. Commun. Math. Phys. **322**, 277–302 (2013), arXiv:1109.1588
58. J. Miller, A. Miyake, Hierarchy of universal entanglement in 2D measurement-based quantum computation. npj Q. Inf. **2**, 16036 (2016), https://www.nature.com/articles/npjqi201636
59. G. Misguich, Quantum spin liquids, *Lectures Given at the Les Houches Summer School on Exact Methods in Low-dimensional Statistical Physics and Quantum Computing* (2008), arXiv:0809.2257
60. A. Miyake, *Quantum computation on the edge of a symmetry-protected topological order*, Phys. Rev. Lett. **105**, 040501 (2010), arXiv:1003.4662
61. A. Miyake, Quantum computational capability of a 2D valence bond solid phase. Ann. Phys. **326**, 1656–1671 (2011), arXiv:1009.3491
62. A. Molnar, Y. Ge, N. Schuch, J.I. Cirac, *A generalization of the injectivity condition for Projected entangled pair states* (2017), arXiv:1706.07329
63. T. Morimae, V. Dunjko, E. Kashefi, Ground state blind quantum computation on AKLT state. Q. Inf. Comput. **15**, 0200–0234 (2015), arXiv:1009.3486
64. S. Moudgalya, S. Rachel, B.A. Bernevig, N. Regnault, Exact excited states of non-integrable models. Phys. Rev. B **98**, 235155 (2018), arXiv:1708.05021
65. S. Moudgalya, N. Regnault, B.A. Bernevig, Entanglement of exact excited states of AKLT models: exact results, many-body scars and the violation of strong ETH. Phys. Rev. B **98**, 235156 (2018), arXiv:1806.09624
66. B. Nachtergaele, Working with quantum Markov states and their classical analogues, in *Quantum Probability and Applications V*, Lecture Notes in Mathematics, vol. 1442, ed. by L. Accardi, W. von Waldenfelds (Springer, Berlin, 1990), pp. 267–285, https://link.springer.com/chapter/10.1007
67. B. Nachtergaele, The spectral gap for some spin chains with discrete symmetry breaking. Commun. Math. Phys. **175**, 565–606 (1996), https://projecteuclid.org/euclid.cmp/1104276093
68. E.J. Neves, J.F. Perez, Long range order in the ground state of two-dimensional antiferromagnets. Phys. Lett. **114A**, 331–333 (1986)
69. K. Nomura, Onset of incommensurability in quantum spin chains. J. Phys. Soc. Jpn. **72**, 476–478 (2003), arXiv:cond-mat/0210143
70. D. Perez-Garcia, F. Verstraete, M.M. Wolf, J.I. Cirac, Matrix product state representations. Q. Inf. Comput. **7**, 401 (2007), arXiv:quant-ph/0608197
71. N. Pomata, T.-C. Wei, AKLT models on decorated square lattices are gapped. Phys. Rev. B **100**, 094429 (2019), arXiv:1905.01275
72. N. Pomata, T.-C. Wei, *Demonstrating the AKLT spectral gap on 2D degree-3 lattices* (2019), arXiv:1911.01410
73. R. Raussendorf, H.-J. Briegel, A one-way quantum computer. Phys. Rev. Lett. **86**, 5188 (2001), arXiv:quant-ph/0010033
74. R. Raussendorf, C. Okay, D.-S. Wang, D.T. Stephen, H. Poulsen-Nautrup, A computationally universal phase of quantum matter. Phys. Rev. Lett. **122**, 090501 (2019), arXiv:1803.00095
75. M. Rossi, M. Hube, D. Bruß, C. Macchiavello, Quantum hypergraph states. New J. Phys. **15**, 113022 (2013), http://iopscience.iop.org/article/10.1088/1367-2630/15/11/113022/meta
76. L. Savary, L. Balents, Quantum spin liquids. Rep. Prog. Phys. **80**, 016502 (2017), arXiv:1601.03742
77. U. Schollwöck, The density-matrix renormalization group in the age of matrix product states. Ann. Phys. **326**, 96 (2011), arXiv:1008.3477
78. W. Son, L. Amico, V. Vedral, Topological order in 1D Cluster state protected by symmetry. Q. Inf. Process. **11**, 1961–1968 (2012), arXiv:1111.7173
79. W. Spitzer, S. Starr, Improved bounds on the spectral gap above frustration-free ground states of quantum spin chains. Lett. Math. Phys. **63**, 65–177 (2003), arXiv:math-ph/0212029

80. K. Totsuka, M. Suzuki, Matrix formalism for the VBS-type models and hidden order. J. Phys.: Condens. Matter **7**, 1639–1662 (1995)
81. P.M. van den Broeck, Exact value of the ground state energy of the linear antiferromagnetic Heisenberg chain with nearest and next-nearest neighbour interactions. Phys. Lett. A **77**, 261–262 (1980)
82. G. Vidal, Efficient classical simulation of slightly entangled quantum computations. Phys. Rev. Lett. **91**, 147902 (2003), arXiv:quant-ph/0301063
83. G. Vidal, Efficient simulation of one-dimensional quantum many-body systems. Phys. Rev. Lett. **93**, 040502 (2004), arXiv:quant-ph/0310089
84. T.-C. Wei, Quantum spin models for measurement-based quantum computation. Adv. Phys.: X **1**, 1461026 (2018), https://www.tandfonline.com/doi/full/10.1080/23746149.2018.1461026
85. T.-C. Wei, I. Affleck, R. Raussendorf, Affleck-Kennedy-Lieb-Tasaki state on a honeycomb lattice is a universal quantum computational resource. Phys. Rev. Lett. **106**, 070501 (2011), arXiv:1102.5064
86. T.-C. Wei, I. Affleck, R. Raussendorf, The 2D AKLT state on the honeycomb lattice is a universal resource for quantum computation. Phys. Rev. A **86**, 032328 (2012), arXiv:1009.2840
87. X.-G. Wen, *Quantum Field Theory of Many-Body Systems: From the Origin of Sound to an Origin of Light and Electrons* (Oxford University Press, Oxford, 2007)
88. H. Weyl, *The Theory of Groups and Quantum Mechanics* (Dover Publication, Dover, 2014)
89. Wikipedia, *Unique factorization domain*, https://en.wikipedia.org/wiki/Unique_factorization_domain
90. M.M. Wolf, *Quantum Channels & Operations Guided Tour*, Unpublished lecture note (2012), https://www-m5.ma.tum.de/foswiki/pub/M5/Allgemeines/MichaelWolf/QChannelLecture.pdf
91. D.A. Yarotsky, Ground states in relatively bounded quantum perturbations of classical lattice systems. Commun. Math. Phys. **261**, 799–819 (2006), arXiv:math-ph/0412040
92. B. Yoshida, Topological phases with generalized global symmetries. Phys. Rev. B **93**, 155131 (2016), arXiv:1508.03468
93. B. Zeng, X. Chen, D.-L. Zhou, X.-G. Wen, *Quantum Information Meets Quantum Matter: From Quantum Entanglement to Topological Phase in Many-Body Systems*, Quantum Science and Technology (Springer, Berlin, 2019), arXiv:1508.02595

Chapter 8
Haldane Phase

Theoretical studies of the $S = 1$ antiferromagnetic spin chains exhibiting the Haldane gap have led to an understanding that the corresponding ground states are not simply disordered but exhibit rich phenomena, which should be called the Haldane phenomena (Sect. 8.1). It was then pointed out that all of the Haldane phenomena can be naturally understood as consequences of spontaneous breakdown of hidden $\mathbb{Z}_2 \times \mathbb{Z}_2$ symmetry (Sect. 8.2). But it was soon realized that the picture of hidden $\mathbb{Z}_2 \times \mathbb{Z}_2$ symmetry breaking is far from enough to fully characterize antiferromagnetic spin chains with the Haldane gap. The notion of symmetry protected topological (SPT) phase (Sect. 8.3), which was proposed by Gu and Wen in 2009 and refined by Pollmann, Turner, Berg, and Oshikawa in 2010, finally gave satisfactory understanding of the phenomena. Very recently, in 2018 and 2019, fully rigorous index theorems based on operator algebraic techniques that characterize SPT phases were developed by Ogata. The rigorous index theorems essentially complete the study of the Haldane phase. In this chapter we also discuss closely related idea of topological order, and give an introductory review of Kitaev's toric code model (Sect. 8.4).

8.1 Characterization of the Haldane Phase

In order to characterize the Haldane phase, we here study the anisotropic model (8.1.1), and see that the model exhibits a topological phase transition between the large-D phase and the Haldane phase (Sect. 8.1.1). We argue that the Haldane phase can be characterized by two exotic properties, namely, the presence of hidden antiferromagnetic order (Sect. 8.1.2) and the emergence of edge spins (Sect. 8.1.3). These two properties, along with the existence of the gap, should be called the Haldane phenomena.

H. Tasaki, *Physics and Mathematics of Quantum Many-Body Systems*,
Graduate Texts in Physics, https://doi.org/10.1007/978-3-030-41265-4_8

8.1.1 Topological Phase Transition in Anisotropic Model

We start our discussion of the Haldane phase by investigating low energy properties of the anisotropic $S = 1$ antiferromagnetic Hamiltonian

$$\hat{H}_D = \sum_{x=1}^{L} \{\hat{\boldsymbol{S}}_x \cdot \hat{\boldsymbol{S}}_{x+1} + D(\hat{S}_x^{(3)})^2\}, \tag{8.1.1}$$

where $D \geq 0$ is the crystal field anisotropy parameter. We take periodic boundary conditions, and assume that L is even. Recall that Theorem 2.4 (p. 43) guarantees that the ground state $|\Phi_{\text{GS}}\rangle$ of the model is unique and satisfies $\hat{S}_{\text{tot}}^{(3)}|\Phi_{\text{GS}}\rangle = 0$.

As a warmup consider the trivial Hamiltonian

$$\hat{H}_{\text{trivial}} = D\sum_{x=1}^{L}(\hat{S}_x^{(3)})^2, \tag{8.1.2}$$

with $D > 0$, which is obtained by dropping the Heisenberg interaction from (8.1.1). The Hamiltonian is diagonalized in the standard basis (2.2.1) as $\hat{H}_{\text{trivial}}|\Psi^{\sigma}\rangle = D\{\sum_{x=1}^{L}(\sigma_x)^2\}|\Psi^{\sigma}\rangle$. Thus the ground state is unique and given by $|\Phi_0\rangle := \bigotimes_{x=1}^{L} |\psi_x^0\rangle$, where all the spins are in the 0 state. The ground state energy is $E_{\text{GS}} = 0$. The first excited states are those where one spin is in the state $+$ or $-$ and all the rest are in 0. Clearly the energy eigenvalue is $E_{\text{1st}} = D$. We see that this trivial model has a unique disordered ground state[1] accompanied by a nonvanishing energy gap $D > 0$.

Low energy properties of the trivial model (8.1.2) are stable under small perturbation to the Hamiltonian. In particular the anisotropic model (8.1.1) with $D \gg 1$ exhibits essentially the same low energy properties as the model (8.1.2), as stated in the following theorem.

Theorem 8.1 *There is a constant $D_0 > 0$, and the following is valid when $D \geq D_0$. The energy gap $\Delta E(D)$ (above the ground state energy) of the Hamiltonian* (8.1.1) *is not less than $\Delta E_0(D) > 0$, which is a function independent of L. The ground state correlation function $\langle\Phi_{\text{GS}}|\hat{S}_x^{(\alpha)}\hat{S}_y^{(\alpha)}|\Phi_{\text{GS}}\rangle$ decays exponentially in the distance $|x - y|$. The ground state in the $L \uparrow \infty$ limit is unique and accompanied by a gap.*

The theorem is proved by applying standard methods of rigorous perturbation theory (based on a cluster expansion) for quantum spin systems. See, e.g., [49]. We must make D_0 large to prove the theorem rigorously (we once estimated that $D_0 = 28$ is enough), but it is expected that the statements in the theorem are all valid for any $D > D_c \simeq 1$ where the model is in the large-D phase. See below.

---

[1]One may interpret $|\Phi_0\rangle := \bigotimes_{x=1}^{L} |\psi_x^0\rangle$ as a state which possesses nematic order. Here we say that the state is disordered to mean that it preserves the symmetry of the Hamiltonian and has short-range correlation.

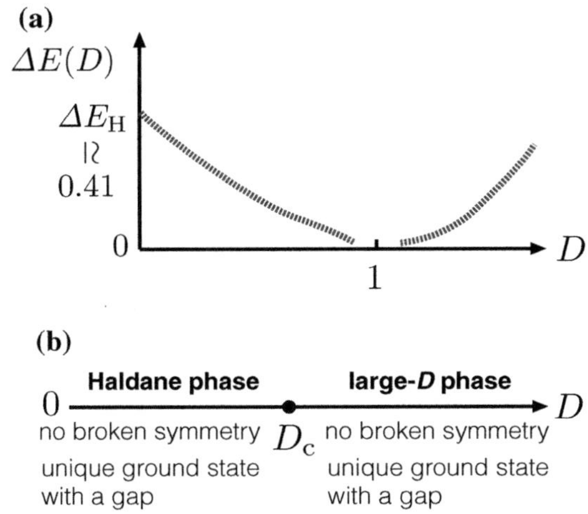

Fig. 8.1 **a** A schematic graph of the energy gap $\Delta E(D)$ of the anisotropic $S = 1$ antiferromagnetic Hamiltonian (8.1.1) obtained numerically [8]. The gap decreases for small $D > 0$, touches zero at $D = D_{\mathrm{c}} \simeq 1$, and increases for $D > D_{\mathrm{c}}$. **b** The corresponding phase diagram. The large-D phase and the Haldane phase are separated by the gapless point at $D = D_{\mathrm{c}}$. In both the phases, the ground states are unique, disordered, and accompanied by a nonzero gap, and no symmetry is spontaneously broken. This is an example of a topological phase transition (© Hal Tasaki 2020. All Rights Reserved)

We have thus seen that, when $D \gg 1$, the model (8.1.1) has a unique disordered ground state with a nonvanishing gap. The situation is apparently similar to what is expected for the $S = 1$ antiferromagnetic Heisenberg chain, except that everything is trivial in the case $D \gg 1$. A natural question is whether the trivial ground state for $D \gg 1$ is "smoothly connected" to the highly nontrivial ground state of the Heisenberg chain. The answer to this question of course depends on what one means by "smoothly connected".[2] The consideration has finally led to the deep notion of symmetry protected topological phase, which we shall discuss in Sect. 8.3.

As a first step for answering the question, we examine low energy properties of the model (8.1.1) when D is varied. Figure 8.1a shows the behavior of the energy gap $\Delta E(D)$ of the Hamiltonian (8.1.1) (for sufficiently large L) obtained numerically [8]. For $D = 0$, where the model reduces to the $S = 1$ antiferromagnetic Heisenberg chain, the gap $\Delta E(0)$ is equal to the Haldane gap $\Delta E_{\mathrm{H}} \simeq 0.41$. When $D \gg 1$, the perturbative analysis shows that $\Delta E(D) \simeq D$, and hence the gap $\Delta E(D)$ is increasing in D. Interestingly, as we increase D, the gap $\Delta E(D)$ is found to decrease when D is positive and small, approach zero around $D_{\mathrm{c}} \simeq 1$, and then start increasing. This remarkable numerical observation suggests that the model (8.1.1) undergoes a (ground state) phase transition at $D = D_{\mathrm{c}}$, where the model becomes gapless, i.e., the gap above the ground state energy vanishes. It is natural to call the phase with

[2] The final answer will turn out to be "no" if the Hamiltonian has proper symmetry, but "yes" if not, as will be discussed in Sect. 8.3.2.

$D > D_c$ the large-D phase, and the phase with $0 \le D \le D_c$ the Haldane phase. See Fig. 8.1b. An essential question then is how one can characterize these two phases.

At the end of Sect. 3.3, we discussed the ground state phase transition in the quantum Ising model, where we also encountered two phases separated by a gapless model. See Fig. 3.4. For the quantum Ising model, we found that the presence/absence of spontaneous $\mathbb{Z}_2$ symmetry breaking (or, equivalently, the corresponding long-range order) precisely characterizes the two phases. In the present model, however, the ground state is expected to be unique, disordered, and accompanied by a nonvanishing gap except at $D = D_c$. We do not see any indication of spontaneous symmetry breaking or long-range order for any D. In particular, we expect that the Néel order parameter[3]

$$\mathscr{O}^{(\alpha)}_{\text{Néel}}(\Phi_{\text{GS}}) := \lim_{y-x\uparrow\infty} \lim_{L\uparrow\infty} (-1)^{y-x} \langle \Phi_{\text{GS}} | \hat{S}^{(\alpha)}_x \hat{S}^{(\alpha)}_y | \Phi_{\text{GS}} \rangle \tag{8.1.3}$$

is vanishing for any $D \ge 0$ and $\alpha = 1, 2, 3$. Nevertheless there exists a gapless model separating the two "phases", which seem to be essentially different. We have a natural intuition that the large-D phase, whose part is completely controlled by easy Theorem 8.1, is rather trivial while the Haldane phase, which includes the antiferromagnetic Heisenberg chain, is highly nontrivial.

A ground state phase transition which is not characterized by symmetry breaking is nowadays refereed to as a topological phase transition. Correspondingly the nontrivial phase for small D, the Haldane phase, is regarded as a topological phase. We remark that here the term "topological" should not be interpreted in the strict mathematical sense, but should be regarded as a synonym for "exotic", "non-standard", or, more specifically, "not characterized by symmetry breaking".

As we discussed at the end of Sect. 3.3, the solid phase of an ordinary matter, for example, is a well-defined phase if and only if the system possesses translation invariance. Then the solid phase is precisely characterized by spontaneous breakdown of the translation symmetry. A final goal of the present study is to have a similar universal characterization of the Haldane phase. In Sect. 8.3 we shall see that such a characterization is provided by the novel notion of symmetry protected topological (SPT) phases.

In the rest of the present section we shall take the first step toward a universal characterization of the Haldane phase. We shall examine in which sense the Haldane phase differs from the large-D phase, and is indeed nontrivial. Remarkably, we will find that the two exotic properties observed in the VBS state, namely, the presence of hidden antiferromagnetic order (Sect. 7.2.1) and the emergence of effective edge spins (Sect. 7.2.3) characterize the Haldane phase.

Remark There indeed is an interesting correspondence between "topology" in the above broad sense and topology in mathematics. Most notably, the celebrated periodic table of topological insulators and topological superconductors for free fermions is exactly the same as the classification of vector bundles known in the topological

---

[3] See footnote 17 in p. 193.

K-theory [54, 85]. In particular, for the integer quantum Hall system, the topological number given in the K-theory reduces to a more elementary concept, namely, the Chern number obtained from the integration of the curvature [57].[4] From the field theoretical point of view, it is understood that indices such as the Chern number for the quantum Hall effect is a special example of much more general "topological" indices characterizing various "topological" phases. See also the beginning of Sect. 8.3.3.

8.1.2 Hidden Antiferromagnetic Order

Hidden order and string order parameter We have already encountered the notion of hidden antiferromagnetic order twice, first in Sect. 6.3, where we developed a heuristic path-integral-like picture for the $S = 1$ antiferromagnetic Heisenberg chain, and next in Sect. 7.2.1, where we investigated the representation of the VBS state in the standard basis with $+$, $-$, and 0.

In 1989, based on the analogy with statistical mechanical models of crystal surfaces, den Nijs and Rommelse argued that the ground state of the $S = 1$ antiferromagnetic Heisenberg chain also possesses hidden antiferromagnetic order [22]. More precisely they discovered a new phase in crystal surfaces called the disordered flat phase, and noted that the phase precisely corresponds to disordered ground states with hidden antiferromagnetic order in $S = 1$ quantum spin chains.

One should recall however that, unlike the VBS state in which only those spin configurations $\boldsymbol{\sigma}$ with complete hidden antiferromagnetic order contribute to the expansion (7.2.1), the unique ground state of the Heisenberg antiferromagnetic chain is expanded as in (2.5.4) with all $\boldsymbol{\sigma}$ such that $\overline{\boldsymbol{\sigma}} = \sum_{x=1}^{L} \sigma_x = 0$ having nonzero contributions. This is a consequence of the Marshall–Lieb–Mattis theorem (Theorem 2.2 in p. 39). Therefore hidden antiferromagnetic order in this case cannot be exact as in the VBS state, but should appear as a result of a statistical average. An expected typical configuration is like the following:

$$+ \quad - \quad + \quad 0 \quad - \quad 0 \quad + \quad + \quad - \quad 0 \quad 0 \quad 0 \quad - \quad 0 \quad + \quad - \tag{8.1.4}$$

By removing all 0's, we get a sequence

[4] It is worthwhile to clarify the history of topological approaches to the integer quantum Hall effect. In 1982, Thouless, Kohmoto, Nightingale, and den Nijs [95] noted that the Hall conductance of a class of two-dimensional electron systems may be written as a certain integral over k-space, and the integral should be quantized. But no connection to topology was made in this seminal work. Possible connection of the quantization condition to topology was pointed out by Avron, Seiler, and Simon [3], who took the point of view of homotopy. Then Simon [90] discussed the aspect of holonomy, and suggested a possible relation to the Chern class. It was Kohmoto [57] who introduced a rigid differential geometric formulation of the problem (which is commonly used these days as a computable scheme) and identified the relevant integer as the Chern number.

$$+ \; - \; + \; - \; + \; (+ \; -) \; - \; + \; - \tag{8.1.5}$$

which do not have a complete alternating pattern of $+$ and $-$, but still has long-range order. A pair of $-$ and $+$ surrounded by a dotted oval is a local defect which does not destroy the long-range order.[5] Note that a defect consisting of a single $+$ or $-$ would destroy the (hidden) antiferromagnetic order.

Such incomplete hidden antiferromagnetic order can also be measured by the den Nijs–Rommelse string order parameter (7.2.7) introduced in Sect. 7.2.1. In what follows we write the string order parameter $\mathscr{O}^{(\alpha)}_{\mathrm{string}}(\Phi_{\mathrm{GS}})$, where $|\Phi_{\mathrm{GS}}\rangle$ is the ground state of (8.1.1) with anisotropy D, simply as $\mathscr{O}^{(\alpha)}_{\mathrm{string}}(D)$. The argument of den Nijs and Rommelse [22] implies that $\mathscr{O}^{(\alpha)}_{\mathrm{string}}(D) > 0$ for $\alpha = 1, 2, 3$ for the Heisenberg point $D = 0$ and its neighbors. Although this claim has not been proved rigorously, it has been confirmed by other theoretical considerations [48, 49, 92] as well as numerical works [38], where it was reported that $\mathscr{O}^{(\alpha)}_{\mathrm{string}}(0) \simeq 0.37$.

It is then quite natural to expect that the hidden antiferromagnetic order distinguishes the Haldane phase from the large-D phase. It is conjectured that one has $\mathscr{O}^{(\alpha)}_{\mathrm{string}}(D) > 0$ for $\alpha = 1, 2, 3$ for any D in the range $0 \le D < D_{\mathrm{c}}$. On the other hand it can be easily proved that $\mathscr{O}^{(\alpha)}_{\mathrm{string}}(D) = 0$ for $\alpha = 1, 2, 3$ if D is large enough that Theorem 8.1 is applicable. It is natural to expect that $\mathscr{O}^{(\alpha)}_{\mathrm{string}}(D) = 0$ for $\alpha = 1, 2, 3$ for any D such that $D > D_{\mathrm{c}}$. This conjecture is consistent with numerical results. See, e.g, [38]. See also [71], where closely related order parameter is investigated. We conclude that the string order parameter $\mathscr{O}^{(\alpha)}_{\mathrm{string}}(D)$ is the precise order parameter which distinguishes the nontrivial Haldane phase from the trivial large-D phase.[6] The picture of "quantum spin liquid with hidden antiferromagnetic order" for the VBS state discussed in Sect. 7.2.1 thus applies to the ground state of the anisotropic model (8.1.1) provided that it is in the Haldane phase.

Stochastic geometric argument Although a mathematical proof is still lacking, it is quite plausible that one has $\mathscr{O}^{(\alpha)}_{\mathrm{string}}(\Phi_{\mathrm{GS}}) > 0$ for the $S = 1$ antiferromagnetic Heisenberg chain and models close to it. Here we use the path integral representation, and give an argument which supports the conjecture [92]. The argument is basically that of den Nijs and Rommelse [22], presented in the language of spin systems. We will see that the phenomenon of percolation of loops formed by $+$ and $-$ is relevant to the hidden antiferromagnetic order.

Recall the path integral representation (6.3.18) of ground state expectation values in terms of the sum over "histories" $(\boldsymbol{\sigma}^{(0)}, \boldsymbol{\sigma}^{(1)}, \ldots, \boldsymbol{\sigma}^{(2N)})$, i.e., spin configurations on the space-time lattice $\{1, 2, \ldots, L\} \times \{0, 1, \ldots, 2N\}$. The representation applies to the present model (8.1.1) if we replace the definition of the classical energy $E_{\mathrm{c}}(\boldsymbol{\sigma})$ by

[5] A local defect need not occupy only two sites. It can be larger but should be confined locally.

[6] The string order parameter (7.2.7) can never be measured experimentally in standard condensed matter systems, but may be measured in systems of ultra cold atoms. See [25].

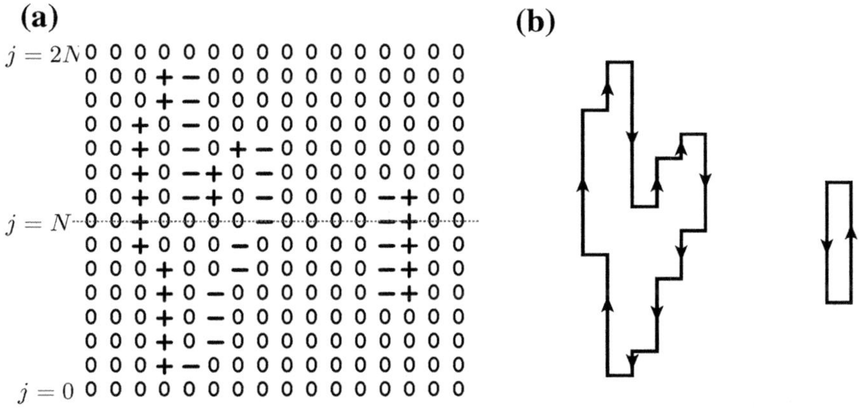

Fig. 8.2 **a** A history (in the space-time lattice) which contributes to the sum in the path integral representation (6.3.18) for ground state correlation functions. It is generated by pair creations and pair annihilations of $+$ and $-$, and hopping of $+$ and $-$ (in the background of 0's) by the unit distance. Because the initial and the final configurations at $j = 0$ and $2N$ are $00\cdots0$, any history consists of closed loops formed by $+$ and $-$. The spin configuration relevant to the correlation function is that on the equal time slice with $j = N$, which is indicated by the dotted line. The configuration is read off as 00+0000−0000−+00. **b** The same history is more conveniently represented by simple oriented loops, where up-going line represents $+$ and down-going $-$ (© Hal Tasaki 2020. All Rights Reserved)

$$E_{\rm c}(\boldsymbol{\sigma}) = \sum_{x=1}^{L} \bigl\{\sigma_x \sigma_{x+1} + D(\sigma_x)^2\bigr\}. \tag{8.1.6}$$

We shall choose boundary configurations as $\boldsymbol{\sigma}^{(0)} = \boldsymbol{\sigma}^{(2N)} = (0, \ldots, 0)$.

Histories $(\boldsymbol{\sigma}^{(0)}, \boldsymbol{\sigma}^{(1)}, \ldots, \boldsymbol{\sigma}^{(2N)})$ which contribute to the sum in (6.3.18) can be read off from the expression (6.3.14) of the weight $w_{\beta,N}(\boldsymbol{\sigma}, \boldsymbol{\sigma}')$ by focusing on the motion of $+$ and $-$ in the background of 0's.[7] When acting on a configuration $\boldsymbol{\sigma}$ containing two adjacent 0's, $\hat{S}_x^+ \hat{S}_{x+1}^-$ or $\hat{S}_x^- \hat{S}_{x+1}^+$ creates a pair of neighboring $+$ and $-$. If a configuration contains the pattern $+0$, then $\hat{S}_x^- \hat{S}_{x+1}^+$ changes it to $0+$, thus moving $+$ to the right by the unit distance. It is clear that $+$ or $-$ can be moved to the left or right in the same manner. Finally a pair of neighboring $+$ and $-$ can be annihilated to yield 00 by the action of $\hat{S}_x^+ \hat{S}_{x+1}^-$ or $\hat{S}_x^- \hat{S}_{x+1}^+$. Recall that any history starts from the configuration $\boldsymbol{\sigma}^{(0)} = (0, \ldots, 0)$ and ends in the same configuration $\boldsymbol{\sigma}^{(2N)} = (0, \ldots, 0)$. This means that $+$ and $-$ appearing in any history should be created pairwise at some moments, and then should be completely annihilated pairwise before the final moment $j = 2N$ is reached. Clearly such a history, no matter how complicated, is a collection of closed loops formed by $+$ and $-$ as in Fig. 8.2a. It is much more convenient to represent these loops of $+$ and $-$ by simple

[7] This picture is complementary to the approximate picture in terms of kinks discussed in Sect. 6.3. See (6.3.7) and (6.3.8).

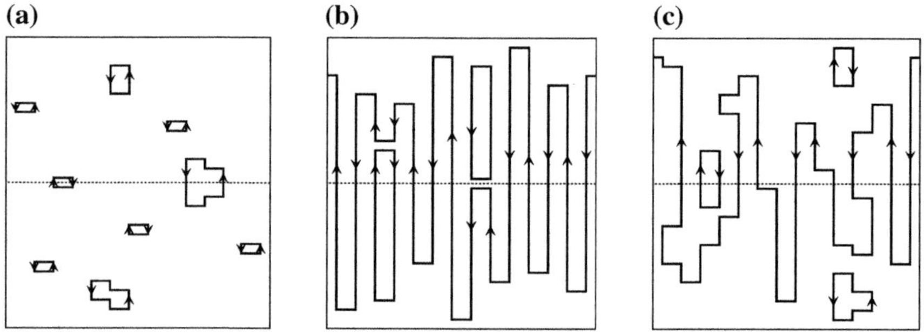

Fig. 8.3 Typical histories (which contribute to the path integral representation (6.3.18)) in the gas, solid, and liquid phases. **a** In the gas phase, which is realized in the anisotropic model (8.1.1) with $D \gg 1$, only small loops appear sparsely. The configuration $\boldsymbol{\sigma}^{(N)}$ obtained at the constant time slice with $j = N$ indicated by the dotted line is found to be 00+−00000−0+00 in this history. Clearly the corresponding ground state is disordered. **b** In the solid phase, which is realized in the model (8.2.1) with $\lambda \gg 1$, the space-time lattice is densely filled by large loops with antiferromagnetic ordering. The configuration $\boldsymbol{\sigma}^{(N)}$ is found to be +−+−+−+00−+−+−. The ground state exhibits Néel order. **c** In the liquid phase, which is conjectured to be realized in the Haldane phase, there is a gigantic loop which spreads over the whole space-time lattice. The configuration $\boldsymbol{\sigma}^{(N)}$ is 0++−−−+0−0+−0+−. The gigantic loop generates hidden antiferromagnetic order, but not Néel order. In the problem of statistical mechanics of crystal surfaces, this phase corresponds to the disordered flat phase discovered by den Nijs and Rommelse [22]

oriented loops as in Fig. 8.2b, where an up-going line represents a continuing series of +'s, and a down-going line represents that of −'s.

According to the representation (6.3.18), the ground state expectation value of a function $\hat{F}$ is determined by the spin configuration $\boldsymbol{\sigma}^{(N)}$ on "the constant time slice" with $j = N$. See Fig. 8.2a, where the slice is shown by the dotted line. We see that the relevant configuration is $\boldsymbol{\sigma}^{(N)} =$ (00+0000−0000−+00) in this figure. In the language of loops, the configuration $\boldsymbol{\sigma}^{(N)}$ is determined from the intersection of the horizontal line at $j = N$ and the loops.

Let us investigate the nature of histories which give dominant contributions to the representation (6.3.18) for a given value of the parameter D. Recall that the weight (6.3.14) contains the classical weight $e^{-\beta\{E_c(\boldsymbol{\sigma})+E_c(\boldsymbol{\sigma}')\}/(2N)}$ as well as the operators $\hat{S}^+_x\hat{S}^-_{x+1}$ and $\hat{S}^-_x\hat{S}^+_{x+1}$. We first suppose that $D \gg 1$. Since the classical energy (8.1.6) increases with the number of + and −, we see that histories with many + and − are suppressed. Consequently dominant contributions come from histories only with small loops distributed with a low density as in Fig. 8.3a. In other words the classical statistical mechanical system of loops described by (6.3.18) is in the gas phase.

In such a gas-like history, the relevant configuration $\boldsymbol{\sigma}^{(N)}$ mainly consists of 0's with some pairs of + and − which belong to a single loop. These + and − typically stay close to each other because loops are typically small. Thus the spin configuration $\boldsymbol{\sigma}^{(N)}$ also looks like that of a low density gas consisting of pairs of + and −. For the history in Fig. 8.3a, we have $\boldsymbol{\sigma}^{(N)} =$ (00+−00000−0+00). It is clear that the ground state is disordered, and has no Néel order or string order. In fact Theorem 8.1 can be proved by making this observation into a rigorous estimate.

We next look at the opposite case where the density of loops is extremely high, and the statistical mechanical system of loops is in its solid phase. Since the classical energy (8.1.6) contains nearest neighbor antiferromagnetic interaction, histories with dominant contributions should have an alternating arrangement of up-going and down-going lines as in Fig. 8.3b. The corresponding spin configuration at $j = N$ is found to be $\boldsymbol{\sigma}^{(N)} = (+-+-+-+00-+-+-)$. It is clear that the corresponding ground state exhibits Néel order, i.e., $\mathscr{O}^{(3)}_{\text{Néel}}(\Phi_{\text{GS}}) > 0$. We do not expect the solid phase to be realized in the present model (8.1.1) for any values of $D \geq 0$ which we are now interested in. This phase is realized in the ground state of (8.1.1) with negative D or that of the model (8.2.1) with Ising anisotropy when $\lambda \gg 1$. See Sect. 8.2.1.

We finally turn to the most interesting case with small $D \geq 0$, where the model is expected be in the Haldane phase. Of course this case is much more delicate and difficult than the previous two cases. Note that, for the Heisenberg chain with $D = 0$, we know from Problem 2.5.c (p. 39) that $\langle \Phi_{\text{GS}} | (\hat{S}^{(\alpha)}_x)^2 | \Phi_{\text{GS}} \rangle = 2/3$ for any x. This implies that, in the relevant spin configuration $\boldsymbol{\sigma}^{(N)}$ in the path-integral representation (6.3.18), the averaged densities of $+$, $-$, and 0 are exactly $1/3$ (in suitable limits of large β and N). In other words, loops occupy $2/3$ of the whole space-time lattice (except for the initial and the final region where the density decreases) in a typical history. This means that a typical history is neither in the gas phase (with low density) or solid phase (with density close to 1), but in an intermediate phase. We expect this phase to be liquid like, where the density of loops is moderately large, but still there remains sufficient vacancy for the shapes and the positions of loops to exhibit large fluctuations. The same picture should hold for sufficiently small $D > 0$ as well.[8]

We further make a plausible but highly nontrivial assumption that the moderately large density of loops causes the phenomenon of percolation.[9] More specifically we assume that, in a typical history, there appears a single gigantic loop which spreads over the whole lattice with a nonzero density as in Fig. 8.3c. We argue that this assumption implies that the corresponding ground state has hidden antiferromagnetic order (in the 3-direction). To see this imagine how the gigantic loop intersects with the constant time slice with $j = N$. Because of the initial and final conditions

---

[8]In fact the same picture is maintained, again for small $D > 0$, if we replace $D(\hat{S}^{(3)}_x)^2$ with $D(\hat{S}^{(1)}_x)^2$ (which, in the path-integral representation, induces the transitions $+ \leftrightarrow -$). This observation shows that $\mathscr{O}^{(\alpha)}_{\text{string}}(D) > 0$ should hold also for $\alpha = 1, 2$.

[9]If the reader is unfamiliar with the percolation theory, here is a quick description of the simplest model, namely, the independent site percolation on the square lattice. Take the infinitely large square lattice $\mathbb{Z}^2$, and fix $p \in [0, 1]$, which is the only parameter of the model. Then each site of the lattice is colored black with probability p or colored white with the remaining probability $1 - p$. This is done independently for each site. After all the sites are colored, one declares that two neighboring black sites are connected with each other, and investigates the nature of clusters formed by connected black sites. It is known that the model exhibits a sharp phase transition at the critical probability $p_c \simeq 0.59$. When $p < p_c$ any cluster has a finite size with probability one. When $p > p_c$ there appears, with probability one, a single infinitely large cluster which spreads over the whole lattice with a nonzero density. (There are also finite clusters in the same configuration.) The appearance of the infinite cluster is the phenomenon of percolation.

$\boldsymbol{\sigma}^{(0)} = \boldsymbol{\sigma}^{(2N)} = (0, \ldots, 0)$, the gigantic loop can move between the upper half $j > N$ and the lower half $j < N$ (of the whole space-time lattice) only by crossing the constant time slice at $j = N$. This means that the direction of the crossings precisely alternates between up and down. See Fig. 8.3c. Since the positions of the crossings fluctuate, this generates a complete hidden antiferromagnetic order as seen in the configuration (6.3.9), but not the Néel order. In a typical history there must be smaller loops other than the gigantic loop. When a small loop intersects with the constant time slice at $j = N$, a pair of $+$ and $-$ is generated, which may or may not be compatible with the hidden antiferromagnetic order maintained by the gigantic loop. An incompatible pair becomes a local defect in the spin configuration, like the one we have seen in (8.1.4) and (8.1.5). In fact from the history in Fig. 8.3c, we get the configuration $\boldsymbol{\sigma}^{(N)} = (0{+}{+}{-}{-}{-}{+}0{-}0{+}{-}0{+}{-})$, where we clearly see hidden antiferromagnetic order with a local defect. We have thus recovered the picture of "quantum spin liquid with hidden antiferromagnetic order" for the ground state of model (8.1.1) with small $D \geq 0$ (which includes the $S = 1$ antiferromagnetic Heisenberg chain). In the original argument by den Nijs and Rommelse [22], which is based on analogy with statistical mechanics of crystal surfaces, this phase is identified as the disordered flat phase.

8.1.3 Emergence of Edge States

The spectrum of open chain and edge spins In 1990, Kennedy studied the $S = 1$ antiferromagnetic Heisenberg model on open chains with Hamiltonian

$$\hat{H} = \sum_{x=1}^{L-1} \hat{\boldsymbol{S}}_x \cdot \hat{\boldsymbol{S}}_{x+1}, \tag{8.1.7}$$

by using numerical diagonalization, and discovered a remarkable difference in the spectra of periodic and open chains [46]. The model on open chains, to which the Marshall–Lieb–Mattis theorem (Theorem 2.2 in p. 39) also applies, has a unique ground state with $S_{\text{tot}} = 0$. Immediately above the ground state, however, there is a triplet of energy eigenstates with $S_{\text{tot}} = 1$, whose excitation energy is quite small. In fact the excitation energy of the triplet is expected to vanish exponentially as the chain length L is increased. Above these near four-fold degenerate "ground states", there is a nonvanishing gap, i.e., the Haldane gap. See Fig. 8.4. The triplet of low energy excitations is called the Kennedy triplet.

The interpretation of this observation about the spectra of Hamiltonians should be clear if one recalls the picture of edge spins in the VBS state on open chains. See Sect. 7.2.3. In low energy eigenstates of the $S = 1$ antiferromagnetic Heisenberg model on open chains, it is expected that effective $S = 1/2$ spin degrees of freedom emerge at the two ends of the chain, while the bulk of the chain is "frozen" in a VBS-like state. See Fig. 7.4b in p. 206. It is then natural that the two $S = 1/2$ edge

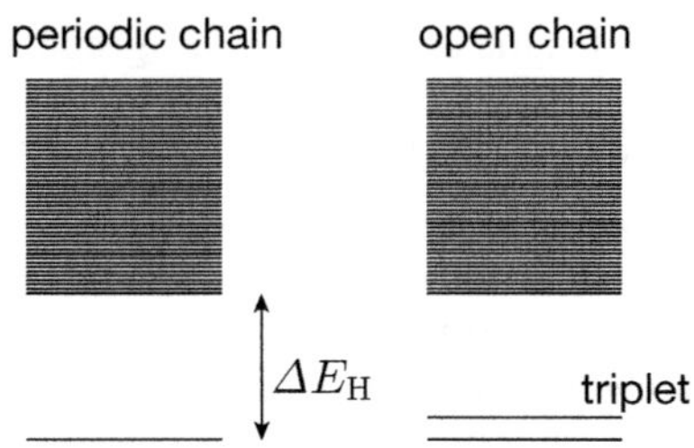

Fig. 8.4 The energy spectra of the $S = 1$ antiferromagnetic Heisenberg model on periodic and open chains. In an open chain, there appears a triplet of low-lying energy excitations (Kennedy triplet) whose excitation energy is exponentially small in L. The near four-fold degeneracy of "ground states" is a consequence of emergent $S = 1/2$ degrees of freedom at the two edges of the chain (© Hal Tasaki 2020. All Rights Reserved)

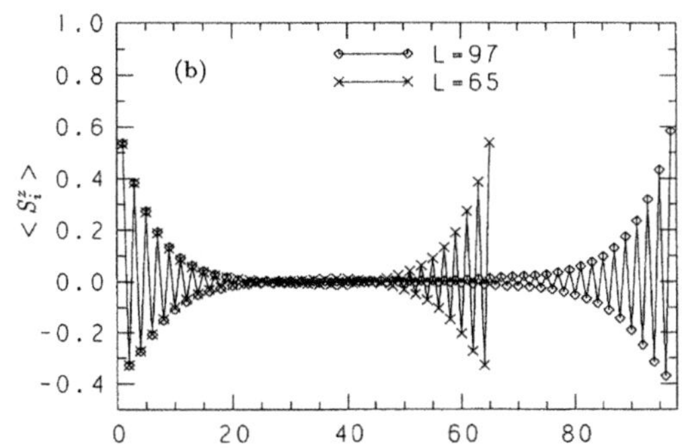

Fig. 8.5 The expectation values $\langle \Phi_{1st}|\hat{S}_x^{(3)}|\Phi_{1st}\rangle$ in the first excited state $|\Phi_{1st}\rangle$ with $S_{tot}^{(3)} = 1$ of the $S = 1$ antiferromagnetic Heisenberg model on open chains with $L = 65$ and $L = 97$ are plotted. The horizontal axis represents the lattice site $x = 1, \ldots, L$. Edge spins are clearly seen. Reprinted with permission from S. Miyashita and A. Yamamoto, Phys. Rev. **B**, 48 (2), 913 (1993) [67]. Copyright (1993) by the American Physical Society

spins, which we describe by operators $\hat{\mathbf{S}}_{\rm L}$ and $\hat{\mathbf{S}}_{\rm R}$, are weakly coupled by an SU(2) invariant effective Hamiltonian $\hat{H}_{\rm coup} = J_{\rm coup}(L)\,\hat{\mathbf{S}}_{\rm L} \cdot \hat{\mathbf{S}}_{\rm R}$ with $J_{\rm coup}(L) > 0$. Then one readily finds (see (2.2.18) and (2.2.19)) that the ground state is a singlet with the coupling energy $-3J_{\rm coup}(L)/4$ and the excited states are triplet with the coupling energy $J_{\rm coup}(L)/4$. The excitation energy of the triplet is thus $J_{\rm coup}(L)$. Since the coupling between $\hat{\mathbf{S}}_{\rm L}$ and $\hat{\mathbf{S}}_{\rm R}$ should be caused by small overlap of the two effective spins, it is expected that the coupling constant $J_{\rm coup}(L)$ approaches zero exponentially as the system size L is increased. This is consistent with the expected behavior of the excitation energy of the Kennedy triplet. Recall that we have exact four-fold degeneracy in the VBS state. This means that the coupling $J_{\rm coup}(L)$ accidentally becomes zero in the AKLT model.

The existence of the effective edge spins can be most directly observed from numerical calculations [67]. Figure 8.5 shows the expectation value $\langle \Phi_{1st}|\hat{S}_x^{(3)}|\Phi_{1st}\rangle$ in the first excited state $|\Phi_{1st}\rangle$ with $S_{tot}^{(3)} = 1$. One clearly sees the indication of spins exponentially localized at the two edges. This graph should be compared with (7.2.50) for the VBS state.

If we consider the anisotropic model (8.1.1) with varying D, it is expected that the near four-fold degeneracy of "ground states" persists throughout the Haldane phase.[10] In the large-D phase, on the other hand, it is expected that the model on an open chain also has a simple spectrum where the ground state energy (of the unique ground state) is accompanied by a finite gap. This can be rigorously proved when D is large enough to satisfy the condition for Theorem 8.1.

Remarkably, Hagiwara, Katsumata, Affleck, Halperin, and Renard [33] made an unambiguous experimental observation of the effective edge spins. They made an electron-spin-resonance (ESR) measurement of NENP where small amount of Ni is replaced by Cu. The measurement result is consistent with the picture that Cu, which has a spin $S = 1/2$, breaks the antiferromagnetic chain and introduces "edges".

Relation to hidden antiferromagnetic order The emergence of edge spins is closely related to the other exotic property, the existence of hidden antiferromagnetic order. In particular we can show that hidden antiferromagnetic order inevitably implies the existence of near four-fold degenerate "ground states".

We start from a simple heuristic argument. Take an $S = 1$ model in the Haldane phase on an open chain, and consider a typical spin configuration $\boldsymbol{\sigma}$ that appears in the expansion (2.5.4) of the ground state $|\Phi_{\mathrm{GS}}\rangle$. We should find non-complete hidden antiferromagnetic order in the configuration. As we did before we first omit all 0's from the configuration to get a string of $+$ and $-$. We next locate all local defects in the string, as we did in (8.1.5), and remove all of them. We end up with a complete alternating sequence of $+$ and $-$. Such a sequence is clearly classified as one of the four types, $+-\cdots+-$, $+-\cdots-+$, $-+\cdots+-$, and $-+\cdots-+$. This, although only roughly, suggests that there are four independent states with low energy.

The connection between hidden antiferromagnetic order and near four-fold degeneracy can be made more precise by using, quite interestingly, the ideas discussed in Part I of the present book. We again take the $S = 1$ model (8.1.1) on the chain $\Lambda_L = \{1, \ldots, L\}$ with open boundary conditions, and define the string order operator for $\alpha = 1, 2, 3$ as

$$\hat{\mathcal{O}}^{(\alpha)}_{\mathrm{string}} := \sum_{x=1}^{L} \hat{S}^{(\alpha)}_x \exp\Bigl[i\pi \sum_{y=1}^{x-1} \hat{S}^{(\alpha)}_y\Bigr]. \tag{8.1.8}$$

Note that $\hat{\mathcal{O}}^{(\alpha)}_{\mathrm{string}}$ is self-adjoint because $(e^{i\pi \hat{S}^{(\alpha)}_y})^\dagger = e^{-i\pi \hat{S}^{(\alpha)}_y} = e^{i\pi \hat{S}^{(\alpha)}_y}$ for $S = 1$. See (2.1.23). We then observe that

[10]Since the Hamiltonian is no longer SU(2) invariant for $D \neq 0$, the Kennedy triplet splits into a doublet and a singlet.

$$
\begin{aligned}
(\hat{\mathscr{O}}^{(\alpha)}_{\text{string}})^2 &= \sum_{x,y=1}^{L} \Big(\hat{S}^{(\alpha)}_x e^{i\pi \sum_{z=1}^{x-1} \hat{S}^{(\alpha)}_z}\Big)\Big(\hat{S}^{(\alpha)}_y e^{i\pi \sum_{w=1}^{y-1} \hat{S}^{(\alpha)}_w}\Big) \\
&= \sum_{x=1}^{L} (\hat{S}^{(\alpha)}_x)^2 - 2 \sum_{\substack{x,y=1 \\ (x<y)}}^{L} \hat{S}^{(\alpha)}_x e^{i\pi \sum_{z=x+1}^{y-1} \hat{S}^{(\alpha)}_z} \hat{S}^{(\alpha)}_y,
\end{aligned}
\tag{8.1.9}
$$

where we noted that $\hat{S}^{(\alpha)}_x \, e^{i\pi \hat{S}^{(\alpha)}_x} = -\hat{S}^{(\alpha)}_x$ for $S = 1$.[11] Recall that the string correlation function $\mathscr{S}^{(\alpha)}_{x,y}(\Phi)$ in (7.2.6) is the expectation value of $-\hat{S}^{(\alpha)}_x e^{i\pi \sum_{z=x+1}^{y-1} \hat{S}^{(\alpha)}_z} \hat{S}^{(\alpha)}_y$.

We now assume that the unique ground state $|\Phi_{\text{GS}}\rangle$ exhibits hidden antiferromagnetic order in all the three directions, i.e., $\mathscr{O}^{(\alpha)}_{\text{string}}(\Phi_{\text{GS}}) > 0$ for $\alpha = 1, 2, 3$, where the string order parameter $\mathscr{O}^{(\alpha)}_{\text{string}}(\Phi_{\text{GS}})$ is defined in (7.2.7). This implies (at least for sufficiently large L) that

$$
\langle \Phi_{\text{GS}} | \Big(\frac{\hat{\mathscr{O}}^{(\alpha)}_{\text{string}}}{L} \Big)^2 | \Phi_{\text{GS}} \rangle \ge q_\alpha, \tag{8.1.10}
$$

with L independent constants $q_\alpha > 0$ for $\alpha = 1, 2, 3$. Note that this is precisely the type of long-range order that we studied in Part I. See, in particular, (3.4.3).

We now define, exactly as in (3.4.7), the trial state in the spirit of Horsch and von der Linden [41] as

$$
|\Gamma^{(\alpha)}\rangle = \frac{\hat{\mathscr{O}}^{(\alpha)}_{\text{string}} |\Phi_{\text{GS}}\rangle}{\big\| \hat{\mathscr{O}}^{(\alpha)}_{\text{string}} |\Phi_{\text{GS}}\rangle \big\|}, \tag{8.1.11}
$$

for $\alpha = 1, 2$, and 3. Then the following analogue of Theorem 3.1 (p. 67) was proved by Koma and Tasaki [58].

Theorem 8.2 *Assume* (8.1.10) *for the ground state of the model* (8.1.1). *Then there exists three independent excited states* $|\Psi_\nu\rangle$ *with* $\nu = 1, 2, 3$, *whose energy eigenvalues* E_ν *satisfy* $E_{\text{GS}} < E_\nu \le E_{\text{GS}} + C_\nu L^{-1}$, *where* C_ν *are constants independent of* L.

We have thus established that, in the model (8.1.1), the presence of hidden antiferromagnetic order inevitably implies the existence of three low-lying energy eigenstates, i.e., the appearance of near four-fold degenerate "ground states". The argument however is not strong enough to show that the separations of the energies are exponentially small in L.

Proof Although the core of the proof is the same as that of Theorem 3.1, we need some extra care about the symmetry, which will turn out to be essential later in Sect. 8.2.3. To make the required symmetry manifest, we here prove the theorem for a general class of models including (8.1.1).

---

[11] This is easily verified by noting that the eigenvalues of $\hat{S}^{(\alpha)}_x$ are 0, ± 1. The relation is valid only for $S = 1$.

Let $\hat{H} = \sum_{x=1}^{L} \hat{h}_x$ be a Hamiltonian on the open chain $\Lambda_L = \{1, \ldots, L\}$. The local Hamiltonian $\hat{h}_x$ acts only on sites y such that $|x - y| \le r$, and satisfies $\|\hat{h}_x\| \le h_0$, where r and h_0 are L-independent constants. We do not assume translation invariance. We make an essential assumption that $(\hat{U}_\pi^{(\alpha)})^\dagger \hat{h}_x \hat{U}_\pi^{(\alpha)} = \hat{h}_x$ for each x and $\alpha = 1, 2, 3$. In other words the local Hamiltonians are invariant under the π-rotation about each of the three axes. This is nothing but $\mathbb{Z}_2 \times \mathbb{Z}_2$ symmetry, which will be discussed in Sect. 8.2.3 and play an important role throughout Sect. 8.3. We further assume that there exists a ground state $|\Phi_{\mathrm{GS}}\rangle$ such that $\hat{U}_\pi^{(\alpha)}|\Phi_{\mathrm{GS}}\rangle = |\Phi_{\mathrm{GS}}\rangle$ for $\alpha = 1, 2, 3$.[12] Note that all the assumptions are valid for the anisotropic model (8.1.1).

Recall that a key in the Proof of Theorem 3.1 is the estimate $\|[\mathcal{O}_L, [\hat{H}, \mathcal{O}_L]]\| \le (\mathrm{const.})|\mathcal{B}_L|$, which follows in a straightforward manner from (3.4.9) and (3.4.10) by using the locality of $\hat{H}$ and $\mathcal{O}_L$. Here the string order operator (8.1.8) is not local. But since the "string part" $\exp[i\pi \sum_{y=1}^{x-1} \hat{S}_y^{(\alpha)}]$ is nothing but the π-rotation on the sublattice $\{1, 2, \ldots, x-1\}$, we find that $[\hat{S}_x^{(\alpha)} \exp[i \sum_{y=1}^{x-1} \hat{S}_y^{(\alpha)}], \hat{h}_z] = 0$ if $|x-z| > r$. Note that the use of open boundary conditions is essential here. By using this, and the similar observation for double commutators, we can easily prove the key estimate $\|[\hat{\mathcal{O}}_{\mathrm{string}}^{(\alpha)}, [\hat{H}, \hat{\mathcal{O}}_{\mathrm{string}}^{(\alpha)}]]\| \le (\mathrm{const.})L$. Then, by simply repeating the argument in the Proof of Theorem 3.1, we see that $|\Gamma^{(\alpha)}\rangle$ with $\alpha = 1, 2, 3$ are low-lying states.

The theorem is proved if we can show that $|\Gamma^{(1)}\rangle$, $|\Gamma^{(2)}\rangle$, $|\Gamma^{(3)}\rangle$, and $|\Phi_{\mathrm{GS}}\rangle$ are orthogonal with each other. To see this recall that $(\hat{U}_\pi^{(\alpha)})^\dagger \hat{S}_x^{(\beta)} \hat{U}_\pi^{(\alpha)} = -\hat{S}_x^{(\beta)}$ if $\alpha \ne \beta$ (see (2.1.16)) which implies $(\hat{U}_\pi^{(\alpha)})^\dagger \hat{\mathcal{O}}_{\mathrm{string}}^{(\beta)} \hat{U}_\pi^{(\alpha)} = -\hat{\mathcal{O}}_{\mathrm{string}}^{(\beta)}$ for $\alpha \ne \beta$. Since $\hat{U}_\pi^{(\alpha)}|\Phi_{\mathrm{GS}}\rangle = |\Phi_{\mathrm{GS}}\rangle$, we find that

$$\hat{U}_\pi^{(\alpha)}|\Gamma^{(\beta)}\rangle = \begin{cases} |\Gamma^{(\beta)}\rangle & \text{if } \alpha = \beta; \\ -|\Gamma^{(\beta)}\rangle & \text{if } \alpha \ne \beta, \end{cases} \tag{8.1.12}$$

which shows the desired orthogonality. ∎

8.2 Hidden $\mathbb{Z}_2$ x $\mathbb{Z}_2$ Symmetry Breaking

We have seen that the Haldane phase of the $S = 1$ chain, which is very likely to be a nontrivial phase distinct from the trivial large-D phase, is characterized by the two exotic properties:

(H1) The ground state exhibits no long range order, but has hidden antiferromagnetic order.

(H2) The ground state of a periodic chain (or the infinite chain) is unique, but there are four near degenerate "ground states" in an open chain (which reflect the emergence of effective $S = 1/2$ states at the edges).

[12] One can weaken the assumption to $\hat{U}_\pi^{(\alpha)}|\Phi_{\mathrm{GS}}\rangle = \pm|\Phi_{\mathrm{GS}}\rangle$.

On top of these properties there is the mystery which has been around since the original discovery of Haldane's:

(H3) A system with unbroken continuous symmetry is expected to be gapless, but some $S = 1$ antiferromagnetic chains with continuous symmetry have a unique disordered ground state with a finite gap.

In 1992, Kennedy and Tasaki argued that all the three exotic properties, which should be called the Haldane phenomena, can be naturally understood as a consequence of hidden $\mathbb{Z}_2 \times \mathbb{Z}_2$ symmetry breaking [48, 49]. This picture seems to have played some roles in clarifying the nature of the Haldane phase and in developing the modern notion of symmetry protected topological phase. We shall discuss this view point in some detail.

We mainly focus on the anisotropic model with two parameters λ and D introduced in Sect. 8.2.1. By using the nonlocal unitary transformation designed to map hidden order to manifest order (Sect. 8.2.2), we develop in Sect. 8.2.3 the picture of hidden $\mathbb{Z}_2 \times \mathbb{Z}_2$ symmetry breaking. The remark at the end of the section about the role of $\mathbb{Z}_2 \times \mathbb{Z}_2$ symmetry is crucial for further development.

8.2.1 *Phase Diagram of the λ-D Model*

To discuss the picture of hidden $\mathbb{Z}_2 \times \mathbb{Z}_2$ symmetry breaking, it is illuminating to study the anisotropic $S = 1$ chain with two parameters λ and D. Consider an open chain $\Lambda_L = \{-(L/2)+1, \ldots, (L/2)-1, L/2\}$, where we numbered the sites according to our original definition (3.1.2). We study an $S = 1$ model with uniaxial anisotropy with Hamiltonian

$$\hat{H}_{\lambda,D} = \sum_{x=-(L/2)+1}^{(L/2)-1} \{\hat{S}_x^{(1)}\hat{S}_{x+1}^{(1)} + \hat{S}_x^{(2)}\hat{S}_{x+1}^{(2)} + \lambda\, \hat{S}_x^{(3)}\hat{S}_{x+1}^{(3)}\} + D \sum_{x=-(L/2)+1}^{L/2} (\hat{S}_x^{(3)})^2, \quad (8.2.1)$$

where λ is the Ising anisotropy parameter and D is the crystal field anisotropy parameter. This is the one-dimensional version of the general model (2.5.14). It reduces to the anisotropic model (8.1.1) by setting $\lambda = 1$, and to the antiferromagnetic Heisenberg model by further setting $D = 0$. Here we focus only on the region with $\lambda > 0$.

Let $|\Phi_{\mathrm{GS}}\rangle$ be a ground state of $\hat{H}_{\lambda,D}$ for given values of λ and D. We define the Néel order parameter as[13]

$$\mathcal{O}_{\text{Néel}}^{(\alpha)}(\Phi_{\mathrm{GS}}) := \lim_{y-x\uparrow\infty} \lim_{L\uparrow\infty} (-1)^{y-x} \langle \Phi_{\mathrm{GS}} | \hat{S}_x^{(\alpha)} \hat{S}_y^{(\alpha)} | \Phi_{\mathrm{GS}} \rangle, \quad (8.2.2)$$

and the string order parameter as

[13]To be rigorous the existence of the limits in (8.2.2) and (8.2.3) are not proved in general.

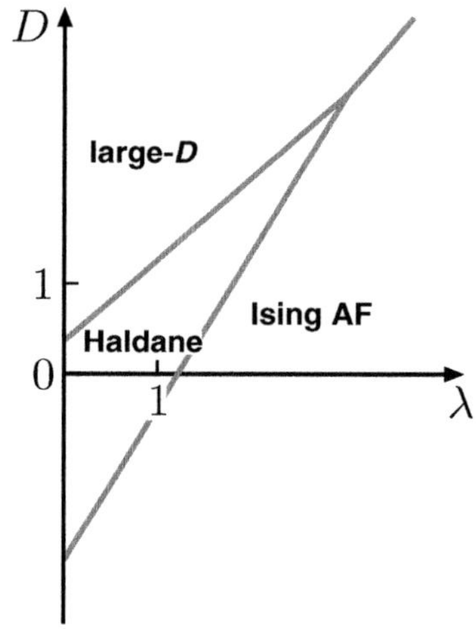

Fig. 8.6 A schematic ground state phase diagram of the anisotropic model (8.2.1). It is expected that there are three phases, the large-D phase, the Haldane phase, and the Ising AF (antiferromagnetic) phase, in the parameter region with $\lambda > 0$. As we have discussed already there is no long-range order in the large-D and the Haldane phases, but there is hidden antiferromagnetic order in the Haldane phase. There is long-range antiferromagnetic order in the 3-direction in the Ising AF phase. From the point of view of the hidden $\mathbb{Z}_2 \times \mathbb{Z}_2$ symmetry breaking, the large-D, the Ising AF, and the Haldane phases are characterized by no symmetry breaking, partial symmetry breaking, and full symmetry breaking, respectively. In the parameter region with $\lambda \leq 0$, one encounters richer phase structure with the XY and the ferromagnetic phases

$$\mathcal{O}^{(\alpha)}_{\mathrm{string}}(\Phi_{\mathrm{GS}}) := -\lim_{y-x\uparrow\infty}\lim_{L\uparrow\infty}\langle\Phi_{\mathrm{GS}}|\hat{S}^{(\alpha)}_x \exp[i\pi\sum_{u=x+1}^{y-1}\hat{S}^{(\alpha)}_u]\,\hat{S}^{(\alpha)}_y|\Phi_{\mathrm{GS}}\rangle, \tag{8.2.3}$$

which is the same as (7.2.7) except for the boundary conditions. The meanings of the two order parameters have been thoroughly discussed already. See, e.g., Sects. 3.4, 7.2.1, and 8.1.2.

Although the existence of hidden antiferromagnetic order does not imply the existence of Néel order, it is quite likely that nonzero Néel order inevitably implies nonzero hidden antiferromagnetic order. This is indeed justified by the following theorem.

Theorem 8.3 *Let* $\lambda \geq 0$ *and* $D \in \mathbb{R}$. *For* $\alpha = 1, 2, 3$, *one has*[14]

$$\left|\mathcal{O}^{(\alpha)}_{\mathrm{N\acute{e}el}}(\Phi_{\mathrm{GS}})\right| \leq \mathcal{O}^{(\alpha)}_{\mathrm{string}}(\Phi_{\mathrm{GS}}). \tag{8.2.4}$$

See [49] for the proof (which makes use of the path integral representation as in Sect. 6.3), and [47] for an extension.

[14] It is also proved in general that $\mathcal{O}^{(1)}_{\mathrm{N\acute{e}el}}(\Phi_{\mathrm{GS}}) = \mathcal{O}^{(2)}_{\mathrm{N\acute{e}el}}(\Phi_{\mathrm{GS}}) \geq 0$. See [49].

Figure 8.6 is a schematic phase diagram of the ground states of the model (8.2.1) obtained mainly numerically [8, 87, 88]. Apart from the Haldane phase and the large-D phase, which we discussed in Sect. 8.1.1, there appears the Ising AF (antiferromagnetic) phase when λ is large. The Ising AF phase is characterized by spontaneous breaking of the $\mathbb{Z}_2$ symmetry with respect to the π rotation about the 1-axis, $\hat{U}_\pi^{(1)}$. The symmetry breaking ground state exhibits long-range Néel order in the 3-direction. The phase diagram is rigorously justified (by using the cluster expansion) only for $D \gg 1$ where the model is in the large-D phase, and for $\lambda \gg 1$ where the model is in the Ising AF phase.

In terms of the two order parameters, these phases are characterized as follows.

large-D	$\mathscr{O}^{(\alpha)}_{\text{Néel}} = \mathscr{O}^{(\alpha)}_{\text{string}} = 0$ for $\alpha = 1, 2, 3$
Haldane	$\mathscr{O}^{(\alpha)}_{\text{Néel}} = 0$ for $\alpha = 1, 2, 3$, $\mathscr{O}^{(1)}_{\text{string}} = \mathscr{O}^{(2)}_{\text{string}} \neq 0$, $\mathscr{O}^{(3)}_{\text{string}} \neq 0$
Ising AF	$\mathscr{O}^{(3)}_{\text{string}} \geq \mathscr{O}^{(3)}_{\text{Néel}} > 0$, $\mathscr{O}^{(\alpha)}_{\text{Néel}} = \mathscr{O}^{(\alpha)}_{\text{string}} = 0$ for $\alpha = 1, 2$

Our goal is to understand this behavior from the viewpoint of hidden $\mathbb{Z}_2 \times \mathbb{Z}_2$ symmetry.

8.2.2 *Nonlocal Unitary Transformation*

To make notations simpler we renumber, only in this subsection, the lattice sites and let $\Lambda_L = \{1, 2 \ldots, L\}$. The central object of the present section is a non-local transformation described by the unitary operator

$$\hat{U}_{\text{KT}} = \prod_{\substack{u,v \in \Lambda_L \\ (u<v)}} \exp\bigl[i\pi \hat{S}_u^{(3)} \hat{S}_v^{(1)}\bigr]. \tag{8.2.5}$$

The unitary transformation was introduced by Kennedy and Tasaki [48, 49], and later written in the above concise (and illuminating) form by Oshikawa [78]. It is sometimes called the Kennedy–Tasaki transformation.

We first note that the operators $\exp\bigl[i\pi \hat{S}_u^{(3)} \hat{S}_v^{(1)}\bigr]$ with different u, v commute with each other. See the end of the subsection. Since we obviously have $(\exp[i\pi \hat{S}_u^{(3)} \hat{S}_v^{(1)}])^2 = \hat{1}$, we find that $(\hat{U}_{\text{KT}})^2 = \hat{1}$, and hence $\hat{U}_{\text{KT}} = \hat{U}_{\text{KT}}^\dagger$. Therefore we shall only use $\hat{U}_{\text{KT}}$ in what follows (even when $\hat{U}_{\text{KT}}^\dagger$ may be more appropriate[15]).

Let us first examine the action of $\hat{U}_{\text{KT}}$ on the standard basis state $|\Psi^{\boldsymbol{\sigma}}\rangle = \bigotimes_{x \in \Lambda_L} |\psi_x^{\sigma_x}\rangle$, where $\boldsymbol{\sigma} = (\sigma_x)_{x \in \Lambda_L}$ with $\sigma_x = 0, \pm 1$ is a spin configuration on the chain. Noting the commutativity, we can write

[15] See (A.1.17) for the transformation of operators.

$$\hat{U}_{\mathrm{KT}} = \prod_{x=2}^{L} \exp\Big[i\pi \Big(\sum_{u=1}^{x-1} \hat{S}_u^{(3)}\Big)\, \hat{S}_x^{(1)}\Big], \tag{8.2.6}$$

which implies

$$\hat{U}_{\mathrm{KT}}|\Psi^{\boldsymbol{\sigma}}\rangle = \Big\{\prod_{x=2}^{L} \exp\Big[i\pi \Big(\sum_{u=1}^{x-1} \sigma_u\Big)\, \hat{S}_x^{(1)}\Big]\Big\}|\Psi^{\boldsymbol{\sigma}}\rangle. \tag{8.2.7}$$

On the single spin Hilbert space $\mathfrak{h}_x \cong \mathbb{C}^3$, we have

$$\exp\Big[i\pi \Big(\sum_{u=1}^{x-1} \sigma_u\Big)\, \hat{S}_x^{(1)}\Big] = \begin{cases} \hat{1} & \text{if } \sum_{u=1}^{x-1} \sigma_u \text{ is even,} \\ e^{i\pi \hat{S}_x^{(1)}} = \begin{pmatrix} 0 & 0 & -1 \\ 0 & -1 & 0 \\ -1 & 0 & 0 \end{pmatrix} & \text{if } \sum_{u=1}^{x-1} \sigma_u \text{ is odd,} \end{cases} \tag{8.2.8}$$

where we used (2.1.33). Thus the spin at x is left unchanged if $\sum_{u=1}^{x-1} \sigma_u$ is even, and flipped as $|\psi_x^{\pm}\rangle \to -|\psi_x^{\mp}\rangle$, $|\psi_x^{0}\rangle \to -|\psi_x^{0}\rangle$ if $\sum_{u=1}^{x-1} \sigma_u$ is odd. This transformation is applied to all $x = 1, \ldots, L$. Therefore if we write $\hat{U}_{\mathrm{KT}}|\Psi^{\boldsymbol{\sigma}}\rangle$ as $\pm|\Psi^{\boldsymbol{\sigma}'}\rangle$ (where we are not bothered by the overall sign here), we find that the new spin configuration $\boldsymbol{\sigma}' = (\sigma'_x)_{x\in\Lambda_L}$ is given by $\sigma'_1 = \sigma_1$ and

$$\sigma'_x = (-1)^{\sum_{u=1}^{x-1} \sigma_u} \sigma_x, \tag{8.2.9}$$

for $x = 2, \ldots, L$. It is found that the map (8.2.9) converts hidden antiferromagnetic order in the original configuration $\boldsymbol{\sigma}$ into ordinary ferromagnetic order in the new configuration $\boldsymbol{\sigma}'$. For example, take the configuration $\boldsymbol{\sigma}$ we have examined in (8.1.4), which has hidden antiferromagnetic order and a local defect surrounded by the dotted oval:

$$+\ -\ +\ 0\ -\ 0\ +\ \boxed{+\ -}\ 0\ 0\ 0\ -\ 0\ +\ - \tag{8.2.10}$$

We find that the configuration $\boldsymbol{\sigma}'$ determined by the map (8.2.9) is

$$+\ +\ +\ 0\ +\ 0\ +\ -\ -\ 0\ 0\ 0\ +\ 0\ +\ + \tag{8.2.11}$$

Note that the hidden order has been converted into manifest ferromagnetic order, and the defect into a local pair of flipped spins. We thus conclude that the nonlocal unitary operator $\hat{U}_{\mathrm{KT}}$ transforms a state $|\Phi\rangle$ with hidden antiferromagnetic order into a state $\hat{U}_{\mathrm{KT}}|\Phi\rangle$ with manifest ferromagnetic order.

We next examine how $\hat{U}_{\rm KT}$ transforms operators. It can be shown (see below) that

$$\hat{U}_{\rm KT}\, \hat{S}_x^{(1)} \hat{U}_{\rm KT} = \hat{S}_x^{(1)} \exp\bigl[i\pi \sum_{v=x+1}^{L} \hat{S}_v^{(1)}\bigr], \tag{8.2.12}$$

$$\hat{U}_{\rm KT}\, \hat{S}_x^{(2)} \hat{U}_{\rm KT} = \exp\bigl[i\pi \sum_{u=1}^{x-1} \hat{S}_u^{(3)}\bigr]\, \hat{S}_x^{(2)} \exp\bigl[i\pi \sum_{v=x+1}^{L} \hat{S}_v^{(1)}\bigr], \tag{8.2.13}$$

$$\hat{U}_{\rm KT}\, \hat{S}_x^{(3)} \hat{U}_{\rm KT} = \exp\bigl[i\pi \sum_{u=1}^{x-1} \hat{S}_u^{(3)}\bigr]\, \hat{S}_x^{(3)}. \tag{8.2.14}$$

Note that the local operators $\hat{S}_x^{(\alpha)}$ are transformed into nonlocal operators with "strings".

But some local combinations of operators are transformed into a local operator. We find, for example, from (8.2.14) that

$$\begin{aligned}\hat{U}_{\rm KT}\, \hat{S}_x^{(3)} \hat{S}_{x+1}^{(3)} \hat{U}_{\rm KT} &= (\hat{U}_{\rm KT}\, \hat{S}_x^{(3)} \hat{U}_{\rm KT})(\hat{U}_{\rm KT}\, \hat{S}_{x+1}^{(3)} \hat{U}_{\rm KT})\\ &= \exp\bigl[i\pi \sum_{u=1}^{x-1} \hat{S}_u^{(3)}\bigr]\, \hat{S}_x^{(3)} \exp\bigl[i\pi \sum_{u=1}^{x} \hat{S}_u^{(3)}\bigr]\, \hat{S}_{x+1}^{(3)}\\ &= \exp\bigl[i\pi\, \hat{S}_x^{(3)}\bigr]\, \hat{S}_x^{(3)} \hat{S}_{x+1}^{(3)} = -\hat{S}_x^{(3)} \hat{S}_{x+1}^{(3)}.\end{aligned} \tag{8.2.15}$$

Note that the "strings" have been canceled. We also used the identity $\exp\bigl[i\pi\, \hat{S}_x^{(\alpha)}\bigr]\, \hat{S}_x^{(\alpha)} = -\hat{S}_x^{(\alpha)}$ for $S = 1$.[16] In this way we find that the Hamiltonian (8.2.1) is transformed into another short ranged Hamiltonian as

$$\begin{aligned}\hat{H}'_{\lambda,D} &= \hat{U}_{\rm KT}\, \hat{H}_{\lambda,D} \hat{U}_{\rm KT}\\ &= \sum_{x=1}^{L} \{-\hat{S}_x^{(1)} \hat{S}_{x+1}^{(1)} + \hat{S}_x^{(2)} \exp\bigl[i\pi(\hat{S}_x^{(3)} + \hat{S}_{x+1}^{(1)})\bigr]\, \hat{S}_{x+1}^{(2)} - \lambda\, \hat{S}_x^{(3)} \hat{S}_{x+1}^{(3)}\}\\ &\quad + D \sum_{x=1}^{L} (\hat{S}_x^{(3)})^2.\end{aligned} \tag{8.2.16}$$

It is also found from (8.2.12) and (8.2.14) that

$$\hat{U}_{\rm KT}\, \hat{S}_x^{(\alpha)} \hat{S}_y^{(\alpha)} \hat{U}_{\rm KT} = -\hat{S}_x^{(\alpha)} \exp\bigl[i\pi \sum_{u=x+1}^{y-1} \hat{S}_u^{(\alpha)}\bigr]\, \hat{S}_y^{(\alpha)}, \tag{8.2.17}$$

for any $x < y$, and $\alpha = 1$ and 3.

---

[16] See footnote 11 in p. 237.

Details about the transformation For the interested reader, we describe some details of the transformation, and derive the relations mentioned above.

We start from the commutativity of the operators $\exp[i\pi \hat{S}_u^{(3)} \hat{S}_v^{(1)}]$. It is obvious that $\left[\exp[i\pi \hat{S}_u^{(3)} \hat{S}_v^{(1)}], \exp[i\pi \hat{S}_z^{(3)} \hat{S}_w^{(1)}]\right] = 0$ when u, v, w, z are all distinct. We only need to show that $\left[\exp[i\pi \hat{S}_u^{(3)} \hat{S}_v^{(1)}], \exp[i\pi \hat{S}_v^{(3)} \hat{S}_w^{(1)}]\right] = 0$ when u, v, w are distinct. It suffices to check this for any state which is a simultaneous eigenstate of $\hat{S}_u^{(3)}$ and $\hat{S}_w^{(1)}$. Denoting the eigenvalues of $\hat{S}_u^{(3)}$ and $\hat{S}_w^{(1)}$ as $M, M' = 0, \pm 1$, the desired relation becomes $[e^{i\pi M \hat{S}_v^{(1)}}, e^{i\pi M' \hat{S}_v^{(3)}}] = 0$, which has already been shown in (2.1.25).

We next examine the transformation $\hat{U}_{\mathrm{KT}}\, \hat{S}_x^{(\alpha)} \hat{U}_{\mathrm{KT}}$, and show (8.2.12)–(8.2.14). First it is obvious that $\exp\left[i\pi \hat{S}_u^{(3)} \hat{S}_v^{(1)}\right] \hat{S}_x^{(\alpha)} \exp\left[i\pi \hat{S}_u^{(3)} \hat{S}_v^{(1)}\right] = \hat{S}_x^{(\alpha)}$ when x is distinct from u and v. Consider $\exp\left[i\pi \hat{S}_x^{(3)} \hat{S}_v^{(1)}\right] \hat{S}_x^{(\alpha)} \exp\left[i\pi \hat{S}_x^{(3)} \hat{S}_v^{(1)}\right]$ for $x \neq v$. Since this is obviously equal to $\hat{S}_x^{(\alpha)}$ when $\alpha = 3$, we assume $\alpha = 1, 2$. Working in the sector in which $\hat{S}_v^{(1)}$ has a fixed eigenvalue $M = 0, \pm 1$, and recalling (2.1.21), the operator in question becomes

$$\exp\left[i\pi M \hat{S}_x^{(3)}\right] \hat{S}_x^{(\alpha)} \exp\left[i\pi M \hat{S}_x^{(3)}\right] = \begin{cases} \hat{S}_x^{(\alpha)} & \text{if } M = 0, \\ -\hat{S}_x^{(\alpha)} & \text{if } M = \pm 1. \end{cases} \tag{8.2.18}$$

Note that the right-hand side is written in a unified manner as $\hat{S}_x^{(\alpha)} e^{i\pi M}$. We can thus express the result in an operator form as

$$\exp\left[i\pi \hat{S}_x^{(3)} \hat{S}_v^{(1)}\right] \hat{S}_x^{(\alpha)} \exp\left[i\pi \hat{S}_x^{(3)} \hat{S}_v^{(1)}\right] = \begin{cases} \hat{S}_x^{(\alpha)} \exp\left[i\pi \hat{S}_v^{(1)}\right] & \text{if } \alpha = 1, 2, \\ \hat{S}_x^{(\alpha)} & \text{if } \alpha = 3. \end{cases} \tag{8.2.19}$$

Similarly we have for $x \neq u$ that

$$\exp\left[i\pi \hat{S}_u^{(3)} \hat{S}_x^{(1)}\right] \hat{S}_x^{(\alpha)} \exp\left[i\pi \hat{S}_u^{(3)} \hat{S}_x^{(1)}\right] = \begin{cases} \hat{S}_x^{(\alpha)} & \text{if } \alpha = 1, \\ \exp\left[i\pi \hat{S}_u^{(3)}\right] \hat{S}_x^{(\alpha)} & \text{if } \alpha = 2, 3. \end{cases} \tag{8.2.20}$$

By using (8.2.19) or (8.2.20) repeatedly we get the desired (8.2.12), (8.2.13) and (8.2.14).

8.2.3 *The Picture of Hidden $\mathbb{Z}_2 \times \mathbb{Z}_2$ Symmetry Breaking*

Symmetry of the transformed Hamiltonian Let us investigate the property of the new Hamiltonian $\hat{H}'_{\lambda,D}$ obtained in (8.2.16). It is notable that $\hat{H}'_{\lambda,D}$ is still local after the nonlocal transformation $\hat{U}_{\mathrm{KT}}$. Recall that the original Hamiltonian $\hat{H}_{\lambda,D}$ is invariant under $\hat{U}_\theta^{(3)}$, the rotation about the 3-axis, by any angle θ, and also under $\hat{U}_\pi^{(1)}$ and $\hat{U}_\pi^{(2)}$, the π rotations about the 1 and the 2-axes. The Hamiltonian has continuous (as well as discrete) symmetry. But, as far as we consider only local symmetry,[17] the new Hamiltonian $\hat{H}'_{\lambda,D}$ is invariant only under $\hat{U}_\pi^{(1)}$, $\hat{U}_\pi^{(2)}$, and $\hat{U}_\pi^{(3)}$, the π rotations about the 1, 2, and 3-axes. As we have seen in Sect. 2.1, these rotations are related, e.g., by $\hat{U}_\pi^{(1)}\hat{U}_\pi^{(2)} = \hat{U}_\pi^{(3)}$, and the collection $\{\hat{1}, \hat{U}_\pi^{(1)}, \hat{U}_\pi^{(2)}, \hat{U}_\pi^{(3)}\}$ forms a representation of the group $\mathbb{Z}_2 \times \mathbb{Z}_2$. Thus the new Hamiltonian $\hat{H}'_{\lambda,D}$ only has discrete $\mathbb{Z}_2 \times \mathbb{Z}_2$ symmetry.

Let us investigate the properties of the ground states. Let $|\Phi_{\rm GS}\rangle$ be the ground state of the original Hamiltonian $\hat{H}_{\lambda,D}$ and $|\Phi'_{\rm GS}\rangle$ be the ground state of the transformed Hamiltonian $\hat{H}'_{\lambda,D} = \hat{U}_{\rm KT}\hat{H}_{\lambda,D}\hat{U}_{\rm KT}$. The two states are related by $|\Phi'_{\rm GS}\rangle = \hat{U}_{\rm KT}|\Phi_{\rm GS}\rangle$.

Hidden symmetry breaking in the Haldane phase We have seen that the Hamiltonian $\hat{H}'_{\lambda,D}$ only contains short range interactions, and has $\mathbb{Z}_2 \times \mathbb{Z}_2$ symmetry. It is also notable that interactions in the 1 and the 3 directions, i.e., $-\hat{S}_x^{(1)}\hat{S}_{x+1}^{(1)} - \lambda\,\hat{S}_x^{(3)}\hat{S}_{x+1}^{(3)}$, are the standard ferromagnetic interactions. This suggests the possibility that the corresponding ground state $|\Phi'_{\rm GS}\rangle$ exhibits long-range ferromagnetic order in the 1 and 3 directions, leading to spontaneous breakdown of the $\mathbb{Z}_2 \times \mathbb{Z}_2$ symmetry.

To make this idea precise, we define the ferromagnetic order parameter for $|\Phi'_{\rm GS}\rangle$ by[18]

$$\mathscr{O}_{\rm ferro}^{(\alpha)}(\Phi'_{\rm GS}) = \lim_{y-x\uparrow\infty}\lim_{L\uparrow\infty}\langle\Phi'_{\rm GS}|\hat{S}_x^{(\alpha)}\hat{S}_y^{(\alpha)}|\Phi'_{\rm GS}\rangle. \tag{8.2.21}$$

Our expectation is that, in a certain range of parameters λ and D, the ground state exhibits ferromagnetic long-range order in the 1 and 3 directions in the sense that

$$\mathscr{O}_{\rm ferro}^{(1)}(\Phi'_{\rm GS}) > 0 \quad\text{and}\quad \mathscr{O}_{\rm ferro}^{(3)}(\Phi'_{\rm GS}) > 0. \tag{8.2.22}$$

Note that $\mathscr{O}_{\rm ferro}^{(1)}(\Phi'_{\rm GS}) > 0$ represents breakdown of the π rotation symmetry about the 3-axis, and $\mathscr{O}_{\rm ferro}^{(3)}(\Phi'_{\rm GS}) > 0$ represents breakdown of the π rotation symmetry about the 1-axis.

By using the transformation (8.2.17), we find, for $\alpha = 1, 3$, that

$$\begin{aligned}\langle\Phi'_{\rm GS}|\hat{S}_x^{(\alpha)}\hat{S}_y^{(\alpha)}|\Phi'_{\rm GS}\rangle &= \langle\Phi_{\rm GS}|\hat{U}_{\rm KT}\,\hat{S}_x^{(\alpha)}\hat{S}_y^{(\alpha)}\hat{U}_{\rm KT}|\Phi_{\rm GS}\rangle\\ &= -\langle\Phi_{\rm GS}|\hat{S}_x^{(\alpha)}\exp\Big[i\pi\sum_{u=x+1}^{y-1}\hat{S}_u^{(\alpha)}\Big]\,\hat{S}_y^{(\alpha)}|\Phi_{\rm GS}\rangle,\end{aligned} \tag{8.2.23}$$

where the right-hand side is precisely the string correlation function (7.2.6). We thus arrive at an interesting identity

$$\mathscr{O}_{\rm ferro}^{(\alpha)}(\Phi'_{\rm GS}) = \mathscr{O}_{\rm string}^{(\alpha)}(\Phi_{\rm GS}), \tag{8.2.24}$$

[17] By local symmetry we mean invariance under a unitary operator that can be written as a product of identical unitary operators acting on single spin Hilbert spaces.

[18] In the present subsection, we renumber lattice sites again so that $\Lambda_L = \{-(L/2)+1, \ldots, (L/2)-1, L/2\}$ as in Sect. 8.2.1.

where the string order parameter $\mathscr{O}^{(\alpha)}_{\text{string}}(\Phi_{\text{GS}})$ is defined in (8.2.3) (or in (7.2.7)). Therefore the assumption (8.2.22) about the full breakdown of the $\mathbb{Z}_2 \times \mathbb{Z}_2$ symmetry in $|\Phi'_{\text{GS}}\rangle$ implies $\mathscr{O}^{(1)}_{\text{string}}(\Phi_{\text{GS}}) > 0$ and $\mathscr{O}^{(3)}_{\text{string}}(\Phi_{\text{GS}}) > 0$ for the ground state $|\Phi_{\text{GS}}\rangle$ of the original Hamiltonian $\hat{H}_{\lambda,D}$. Since we know $\mathscr{O}^{(1)}_{\text{string}}(\Phi_{\text{GS}}) = \mathscr{O}^{(2)}_{\text{string}}(\Phi_{\text{GS}})$ by symmetry, we find that $|\Phi_{\text{GS}}\rangle$ exhibits hidden antiferromagnetic order in all the directions, $\alpha = 1, 2$, and 3. This explains the first exotic property (H1) mentioned in the beginning of Sect. 8.2.

In fact the assumption of full $\mathbb{Z}_2 \times \mathbb{Z}_2$ symmetry breaking in $|\Phi'_{\text{GS}}\rangle$ also explains the properties (H2) and (H3).

Suppose that the $\mathbb{Z}_2 \times \mathbb{Z}_2$ symmetry is fully broken in the infinite volume ground states of the transformed model with Hamiltonian $\hat{H}'_{\lambda,D}$. There appear four symmetry breaking ground states since the group $\mathbb{Z}_2 \times \mathbb{Z}_2$ consists of four elements.[19] Then, in the corresponding finite chain, there should appear nearly degenerate four "ground states" (in the sense of Part I). Since the Hamiltonians $\hat{H}'_{\lambda,D} = \hat{U}_{\text{KT}} \hat{H}_{\lambda,D} \hat{U}_{\text{KT}}$ and $\hat{H}_{\lambda,D}$ have exactly the same energy eigenvalues with exactly the same degeneracies for any finite L, we conclude that $\hat{H}_{\lambda,D}$ also has four nearly degenerate "ground states".[20] The property (H2) has been explained.

To explain the property (H3), i.e., the emergence of the Haldane gap, we recall that spontaneous breakdown of discrete symmetry in ground states generally leads to a nonvanishing energy gap above the ground state energy. This is because, in such a situation, elementary excitations are basically "kinks" connecting different ground states. See the example of the quantum Ising model in Sect. 3.3. We see that, by the nonlocal transformation, the mystery (H3) turns into a situation commonly observed when discrete symmetry is spontaneously broken.

To sum, the Haldane phenomena, namely, the exotic properties (H1), (H2), and (H3) of the Haldane phase naturally follow from the assumption that the ground states of $\hat{H}'_{\lambda,D}$ fully break the $\mathbb{Z}_2 \times \mathbb{Z}_2$ symmetry. Of course this symmetry breaking is never manifest in the original model with the Hamiltonian $\hat{H}_{\lambda,D}$. We call it hidden symmetry breaking because it is revealed only by applying the nonlocal unitary transformation $\hat{U}_{\text{KT}}$.

The phase diagram The whole phase diagram of the model (8.2.1), summarized in Fig. 8.6, can also be interpreted in terms of the $\mathbb{Z}_2 \times \mathbb{Z}_2$ symmetry. When D is very large, where the configuration $\cdots 00000000 \cdots$ basically characterizes the ground state of (8.2.16), it is obvious that the ground state $|\Phi'_{\text{GS}}\rangle$ does not break any symmetry. Thus the large-D phase should be characterized by the absence of symmetry breaking.

When λ is large, where the ferromagnetic interaction in the 3-direction dominates the Hamiltonian (8.2.16), it is expected that the ground state $|\Phi'_{\text{GS}}\rangle$ exhibits ferromagnetic order only in the 3-direction, thus breaking only the $\mathbb{Z}_2$ symmetry

---

[19] The four ground states correspond to the four choices of the signs of the magnetizations $\omega'(\hat{S}^{(1)}_x) = \pm m^{(1)}$ and $\omega'(\hat{S}^{(3)}_x) = \pm m^{(3)}$, where $\omega'(\cdot)$ denotes the expectation in the ground states.

[20] Note that the exact correspondence is only valid for a finite L. In the infinite volume limit, the original model with $\hat{H}_{\lambda,D}$ has a unique ground state rather than four.

described by $\hat{U}_\pi^{(1)}$, the π rotation about the 1-axis. We thus have $\mathscr{O}_{\text{ferro}}^{(1)}(\Phi'_{\text{GS}}) = 0$ and $\mathscr{O}_{\text{ferro}}^{(3)}(\Phi'_{\text{GS}}) > 0$ for the transformed model, which is equivalent to $\mathscr{O}_{\text{string}}^{(1)}(\Phi_{\text{GS}}) = \mathscr{O}_{\text{string}}^{(2)}(\Phi_{\text{GS}}) = 0$ and $\mathscr{O}_{\text{string}}^{(3)}(\Phi_{\text{GS}}) > 0$ for the ground state $|\Phi_{\text{GS}}\rangle$ of the original Hamiltonian $\hat{H}_{\lambda,D}$. Since genuine hidden antiferromagnetic order, i.e., hidden antiferromagnetic order without any manifest long-range order, is expected to be nearly isotropic (see Sect. 8.1.2), this anisotropic behavior suggests that $|\Phi_{\text{GS}}\rangle$ possesses Néel order in the 3-direction, i.e., $\mathscr{O}_{\text{Néel}}^{(1)}(\Phi_{\text{GS}}) = \mathscr{O}_{\text{Néel}}^{(2)}(\Phi_{\text{GS}}) = 0$ and $\mathscr{O}_{\text{Néel}}^{(3)}(\Phi_{\text{GS}}) > 0$. The Ising AF phase therefore can be characterized by the partial breakdown of the $\mathbb{Z}_2 \times \mathbb{Z}_2$ symmetry.

In conclusion the three phases of the model (8.2.1) are precisely characterized by breakdown of the $\mathbb{Z}_2 \times \mathbb{Z}_2$ symmetry as follows.

	$\mathbb{Z}_2$ for $\hat{U}_\pi^{(1)}$	$\mathbb{Z}_2$ for $\hat{U}_\pi^{(3)}$
large-D	unbroken	unbroken
Ising AF	broken	unbroken
Haldane	broken	broken

Variational calculation The above picture based on the $\mathbb{Z}_2 \times \mathbb{Z}_2$ symmetry is supported by the following simple but illuminating variational calculation [48, 49]. Take $a, b, c \in \mathbb{C}$ such that $|a|^2 + |b|^2 + |c|^2 = 1$, and consider a simple tensor product state

$$|\Theta'_{a,b,c}\rangle = \bigotimes_{x \in \Lambda_L} \{a|\psi_x^0\rangle + b|\psi_x^+\rangle + c|\psi_x^-\rangle\}. \tag{8.2.25}$$

We regard $|\Theta'_{a,b,c}\rangle$ as a candidate for the ground state of the transformed Hamiltonian $\hat{H}'_{\lambda,D}$, and perform a naive variational calculation.

A straightforward calculation shows that[21]

$$\begin{aligned} E(a, b, c) &:= \frac{1}{L}\langle \Theta'_{a,b,c}|\hat{H}'_{\lambda,D}|\Theta'_{a,b,c}\rangle \\ &= -2\Re[a^2(b^2 + c^2)^*] - 2|a|^2(|b|^2 + |c|^2) \\ &\quad - \lambda(|b|^2 - |c|^2)^2 + D(|b|^2 + |c|^2). \end{aligned} \tag{8.2.26}$$

For given values of λ and D we shall minimize $E(a, b, c)$ to find the corresponding $|\Theta'_{a,b,c}\rangle$ which best approximates the ground state of $\hat{H}'_{\lambda,D}$. Then the state $|\Theta_{a,b,c}\rangle = \hat{U}_{\text{KT}}|\Theta'_{a,b,c}\rangle$ should approximate the ground state of the original Hamiltonian $\hat{H}_{\lambda,D}$.

We find that the number of solutions[22] (a, b, c) at which the variational energy $E(a, b, c)$ takes its minimum can be one, two, or four, depending on the values of λ and D. This of course means that the number of the variational ground states of $\hat{H}'_{\lambda,D}$

[21] The expression (8.2.26), which is simplified by using the condition $|a|^2 + |b|^2 + |c|^2 = 1$, is not suitable for numerical calculation. See the original work [49] for the full expression, which was used to produce Fig. 8.7.

[22] We of course exclude equivalent solutions which differ only by a phase factor.

can be one, two, or four. Figure 8.7 shows typical behavior of $E(a, b, c)$ in each of the three cases, and Fig. 8.8 is the corresponding phase diagram.

In the parameter region with $D > 4$ and $D > \lambda > 0$, we find that the minimum of $E(a, b, c)$ is attained at a single point with $a = 1$, $b = c = 0$. The resulting variational ground state is simply $|\Theta'_{1,0,0}\rangle = \bigotimes_{x\in\Lambda_L} |\psi^0_x\rangle$. It is easy to see, from (8.2.9), that the nonlocal unitary transformation $\hat{U}_{\mathrm{KT}}$ leaves this state unchanged, and we also have $|\Theta_{1,0,0}\rangle = \hat{U}_{\mathrm{KT}}|\Theta'_{1,0,0}\rangle = \bigotimes_{x\in\Lambda_L} |\psi^0_x\rangle$. We have thus obtained a somewhat crude approximation of the ground state in the large-D phase.

In the region with $2\lambda - D > 4$, $\lambda > D$, and $\lambda > 0$, the minimum of $E(a, b, c)$ is attained at two points $(0, 1, 0)$ and $(0, 0, 1)$. This means that there are two distinct variational ground states $|\Theta'_{0,1,0}\rangle = \bigotimes_{x\in\Lambda_L} |\psi^+_x\rangle$ and $|\Theta'_{0,0,1}\rangle = \bigotimes_{x\in\Lambda_L} |\psi^-_x\rangle$. We find that the corresponding approximate ground states of the original model are

$$|\Theta_{0,1,0}\rangle = \hat{U}_{\mathrm{KT}}|\Theta'_{0,1,0}\rangle = (-1)^{L/2}\bigotimes_{j=1}^{L/2}\bigl\{|\psi^+_{-(L/2)+2j-1}\rangle \otimes |\psi^-_{-(L/2)+2j}\rangle\bigr\}, \tag{8.2.27}$$

$$|\Theta_{0,0,1}\rangle = \hat{U}_{\mathrm{KT}}|\Theta'_{0,0,1}\rangle = (-1)^{L/2}\bigotimes_{j=1}^{L/2}\bigl\{|\psi^-_{-(L/2)+2j-1}\rangle \otimes |\psi^+_{-(L/2)+2j}\rangle\bigr\}. \tag{8.2.28}$$

These are nothing but the Néel state (2.5.2) which have the maximum Néel order in the 3-direction. We have recovered the Ising AF phase, where the $\mathbb{Z}_2 \times \mathbb{Z}_2$ symmetry is only partly broken.

Finally in the parameter region with $D < 4$, $2\lambda - D < 4$, and $\lambda > 0$, we find that the minimum of $E(a, b, c)$ is attained at four points $(\bar{a}, \pm\bar{b}, 0)$ and $(\bar{a}, 0, \pm\bar{b})$, where $\bar{a} = \sqrt{(4 + D - 2\lambda)/(8 - 2\lambda)}$ and $\bar{b} = \sqrt{(4 - D)/(8 - 2\lambda)}$. It is clear that the $\mathbb{Z}_2 \times \mathbb{Z}_2$ symmetry is fully broken, leading to the four degenerate variational ground states. The resulting variational ground states $|\Theta_{a,b,c}\rangle = \hat{U}_{\mathrm{KT}}|\Theta'_{a,b,c}\rangle$ of the original model, which exhibit hidden antiferromagnetic order, are no longer simple product states since the nonlocal unitary transformation generates entanglement between different sites. Interestingly the approximate ground states $|\Theta_{a,b,c}\rangle$ for $\lambda = 1$ and $D = 0$, which corresponds to the antiferromagnetic Heisenberg chain, are found to be exactly the VBS state [48, 49]. We thus see that the variational calculation becomes exact for the AKLT model.[23] As is clear from the Fig. 8.8 the present calculation does not recover the ground state phase diagram in Fig. 8.6 quantitatively. But the coincidence is quite satisfactory, given the simplicity of the approximation.

The true meaning of the $\mathbb{Z}_2 \times \mathbb{Z}_2$ symmetry We have followed Kennedy and Tasaki [48, 49], who emphasized the importance of the discrete $\mathbb{Z}_2 \times \mathbb{Z}_2$ symmetry possessed by the Hamiltonian $\hat{H}'_{\lambda,D}$ obtained by the nonlocal unitary transformation. Some time later, Pollmann, Turner, Berg, and Oshikawa [82] pointed out an important role played by the same $\mathbb{Z}_2 \times \mathbb{Z}_2$ symmetry of the original Hamiltonian $\hat{H}_{\lambda,D}$. To

[23] It is likely that the present calculation is equivalent to a variational calculation within a certain class of matrix product states. But the breakdown of the $\mathbb{Z}_2 \times \mathbb{Z}_2$ symmetry, which is apparent in the present formulation, may not be seen clearly in matrix product state calculations.

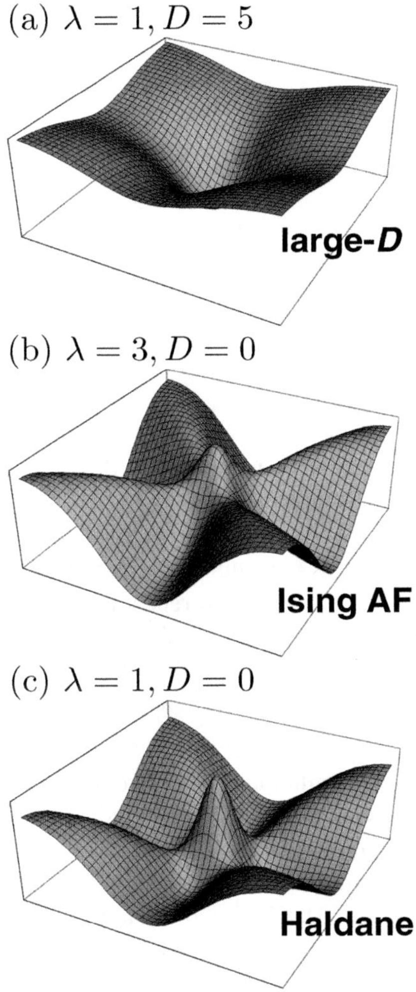

Fig. 8.7 Typical behavior of the variational energy $E(a, b, c)$ for the simple tensor product state $|\Theta'_{a,b,c}\rangle$. $E(a, b, c)$ is plotted as a function of $b/a, c/a \in [-3, 3]$ assuming $a, b, c \in \mathbb{R}$. Note the symmetry $E(a, b, c) = E(a, c, b)$. **a** When $\lambda = 1$ and $D = 5$, the energy attains the minimum at a single point $(1, 0, 0)$, which corresponds to the origin of the graph. This represents the large-D phase, where the $\mathbb{Z}_2 \times \mathbb{Z}_2$ symmetry is not broken. **b** When $\lambda = 3$ and $D = 0$, the energy attains the minimum at two points $(0, 1, 0)$ and $(0, 0, 1)$, which correspond to $b/a = \pm\infty$ and $c/a = \pm\infty$, respectively. This represents the Ising AF phase, where the $\mathbb{Z}_2 \times \mathbb{Z}_2$ symmetry is partially broken. **c** When $\lambda = 1$ and $D = 0$, which correspond to the antiferromagnetic Heisenberg chain, the energy attains the minimum at four points $(1/\sqrt{3}, \pm\sqrt{2/3}, 0)$ and $(1/\sqrt{3}, 0, \pm\sqrt{2/3})$. This represents the Haldane phase, where the $\mathbb{Z}_2 \times \mathbb{Z}_2$ symmetry is fully broken. Moreover the variational states $\hat{U}_{\mathrm{KT}}|\Theta'_{a,b,c}\rangle$ in this case are nothing but the VBS states (© Hal Tasaki 2020. All Rights Reserved)

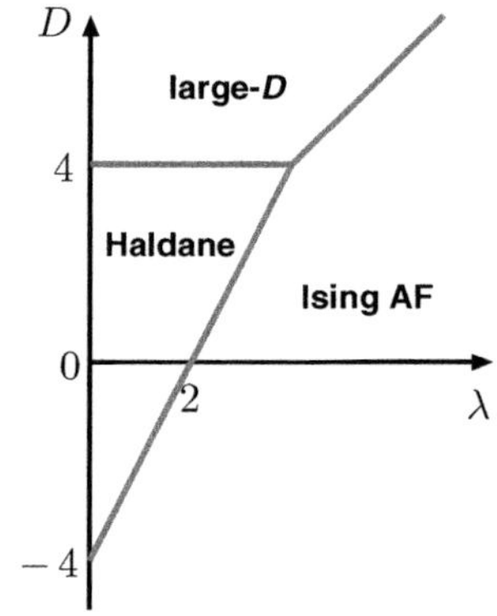

Fig. 8.8 The ground state phase diagram of the Hamiltonian (8.2.1) obtained by the variational calculation. The feature of the actual phase diagram in Fig. 8.6 is recovered qualitatively (© Hal Tasaki 2020. All Rights Reserved)

understand this point was indeed essential for the development of the notion of the symmetry protected topological phase, which we shall describe in Sect. 8.3.

The first simple but crucial observation is that the π rotation operators $\hat{U}_\pi^{(1)}$, $\hat{U}_\pi^{(2)}$ and $\hat{U}_\pi^{(3)}$, which describe the $\mathbb{Z}_2\times\mathbb{Z}_2$ symmetry, all commute with the nonlocal unitary transformation $\hat{U}_{\mathrm{KT}}$. This is easily shown by repeating the proof of the commutativity of the operators $\exp[i\pi\hat{S}_u^{(3)}\hat{S}_v^{(1)}]$ presented at the end of Sect. 8.2.2.

We then find that the invariance $(\hat{U}_\pi^{(\alpha)})^\dagger\hat{H}'_{\lambda,D}\hat{U}_\pi^{(\alpha)} = \hat{H}'_{\lambda,D}$ is equivalent to the invariance $(\hat{U}_\pi^{(\alpha)})^\dagger\hat{H}_{\lambda,D}\hat{U}_\pi^{(\alpha)} = \hat{H}_{\lambda,D}$. This means that the $\mathbb{Z}_2\times\mathbb{Z}_2$ symmetry of the transformed Hamiltonian $\hat{H}'_{\lambda,D}$ is not new symmetry acquired by $\hat{H}'_{\lambda,D}$, but symmetry inherited from the original Hamiltonian $\hat{H}_{\lambda,D}$. In other words, among various types of symmetry of $\hat{H}_{\lambda,D}$, only the $\mathbb{Z}_2\times\mathbb{Z}_2$ symmetry "passes through" the nonlocal unitary transformation and is present in $\hat{H}'_{\lambda,D}$.

But this is not the only importance of the $\mathbb{Z}_2\times\mathbb{Z}_2$ symmetry. Recall the transformation rules (8.2.12)–(8.2.14), which show that the local operator $\hat{S}_x^{(\alpha)}$ is mapped to a nonlocal operator by $\hat{U}_{\mathrm{KT}}$. Consider a local product of spin operators where the numbers of the 1, 2, and 3 components are n_1, n_2, and n_3, respectively. For example, $\hat{S}_x^{(1)}\hat{S}_{x+1}^{(2)}(\hat{S}_{x+2}^{(3)})^2$ has $n_1 = n_2 = 1$ and $n_3 = 2$. If we transform this operator using $\hat{U}_{\mathrm{KT}}$, "strings" of operators are generated according to (8.2.12)–(8.2.14). It is easy to check that the string cancels, as in (8.2.15), if and only if both $n_1 + n_2$ and $n_2 + n_3$ are even. But this condition is precisely the necessary and sufficient condition for the operator to be $\mathbb{Z}_2\times\mathbb{Z}_2$ invariant. To see this one only needs to see that $\hat{U}_\pi^{(1)}$ maps $(\hat{S}_x^{(1)}, \hat{S}_x^{(2)}, \hat{S}_x^{(3)})$ into $(\hat{S}_x^{(1)}, -\hat{S}_x^{(2)}, -\hat{S}_x^{(3)})$, and $\hat{U}_\pi^{(3)}$ maps $(\hat{S}_x^{(1)}, \hat{S}_x^{(2)}, \hat{S}_x^{(3)})$ into $(-\hat{S}_x^{(1)}, -\hat{S}_x^{(2)}, \hat{S}_x^{(3)})$. The product $\hat{S}_x^{(1)}\hat{S}_{x+1}^{(2)}\hat{S}_{x+2}^{(3)}$ with $n_1 = n_2 = n_3 = 1$ is an example of a $\mathbb{Z}_2\times\mathbb{Z}_2$ invariant operator.

Let us summarize this essential observation of Pollmann, Turner, Berg, and Oshikawa [82].

Proposition 8.4 *Let $\hat{H}$ be a Hamiltonian of an $S = 1$ open chain with short range interactions. A necessary and sufficient condition for the transformed Hamiltonian*

$\hat{H}' = \hat{U}_{\mathrm{KT}} \hat{H} \hat{U}_{\mathrm{KT}}$ *to have only short range interactions is that* $\hat{H}$ *is* $\mathbb{Z}_2 \times \mathbb{Z}_2$ *invariant, i.e.,* $(\hat{U}_\pi^{(\alpha)})^\dagger \hat{H} \hat{U}_\pi^{(\alpha)} = \hat{H}$ *for* $\alpha = 1, 2, 3$. *(In this case* $\hat{H}'$ *is also* $\mathbb{Z}_2 \times \mathbb{Z}_2$ *invariant.)*

We thus conclude that the picture of hidden $\mathbb{Z}_2 \times \mathbb{Z}_2$ symmetry breaking is effective if and only if the original Hamiltonian has $\mathbb{Z}_2 \times \mathbb{Z}_2$ symmetry.

8.3 Symmetry Protected Topological Phase

It is now understood that the Haldane phase in quantum spin chains with odd S should be understood as a symmetry protected topological (SPT) phase. Here we shall carefully discuss this new notion and related mathematical ideas for the characterization and classification of SPT phases.

We start by discussing in Sect. 8.3.1 some issues or puzzles related to the picture of the hidden $\mathbb{Z}_2 \times \mathbb{Z}_2$ symmetry breaking. In Sect. 8.3.2, we introduce the idea of symmetry protected topological phase, and see that it solves all the puzzles. We then discuss the basic idea behind the "topological" indices that characterize SPT phases in Sect. 8.3.3. We shall see that the essence of nontrivial SPT phases is in entanglement properties of the ground states. In Sect. 8.3.4, we go deeper into the theory of SPT phases, and discuss their characterization in terms of projective representations of symmetry groups. The techniques developed in Sect. 8.3.4 is used to derive an interesting no-go theorem in Sect. 8.3.5. Finally, in Sect. 8.3.6, we sketch the essence of fully rigorous index theorems for SPT phases, which essentially complete the general theory of the Haldane phase.

8.3.1 More on the Picture of the Hidden $\mathbb{Z}_2 \times \mathbb{Z}_2$ Symmetry Breaking

The success of the picture of the hidden $\mathbb{Z}_2 \times \mathbb{Z}_2$ symmetry breaking for the $S = 1$ chain may suggest that the full $\mathbb{Z}_2 \times \mathbb{Z}_2$ symmetry breaking is a universal character of integer spin antiferromagnetic chains exhibiting the Haldane phenomena. However the situation is much more complicated and fascinating, as we shall describe below.

The VBS states with higher integer S Let us first examine the properties of the VBS state with general integer S. We shall see that some properties do not fit well with the picture of hidden $\mathbb{Z}_2 \times \mathbb{Z}_2$ symmetry breaking.

Consider, for example, the $S = 2$ VBS state (see Sect. 7.3.1) on an open chain. Recall that we represent each spin with $S = 2$ by symmetrizing four spins with $S = 1/2$. Then on the site at one edge of the chain there remain two free spins:

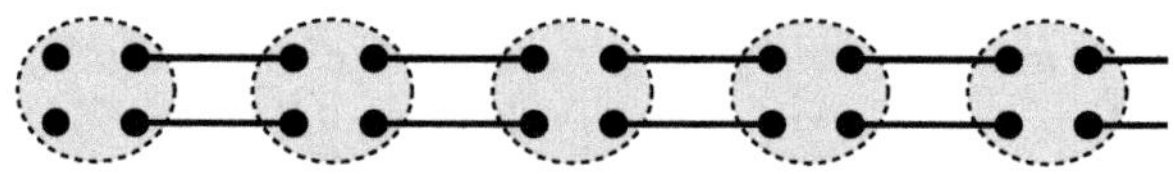

(8.3.1)

Since spins surrounded by the dotted circle must be symmetrized, these two spins form an effective single $S = 1$ spin:

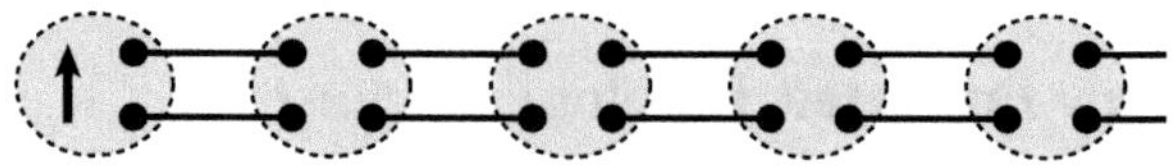

(8.3.2)

Noting that a spin with $S = 1$ leads to three-fold degeneracy, we find that the corresponding $S = 2$ AKLT model on an open chain has nine-fold degenerate ground states.

Similarly we see that the AKLT model with general integer S on an open chain has effective $S/2$ spins at each boundary, which lead to $(S + 1)^2$ fold degeneracy in the ground states. These ground states are of course the generalized VBS states. This observation does not fit well with the picture of the $\mathbb{Z}_2 \times \mathbb{Z}_2$ symmetry breaking, which only predicts four-fold degeneracy. In particular, for an even S, the degeneracy $(S + 1)^2$ of the AKLT model is not an integer multiple of four.

The problems of hidden antiferromagnetic order and related hidden $\mathbb{Z}_2 \times \mathbb{Z}_2$ symmetry breaking for the VBS state with a general integer spin S were systematically studied by Oshikawa [78]. By properly extending the nonlocal unitary transformation $\hat{U}_{\mathrm{KT}}$ to a general integer S,[24] it was observed that the transformed Hamiltonian again has $\mathbb{Z}_2 \times \mathbb{Z}_2$ symmetry. No higher symmetry, such as $\mathbb{Z}_{S+1} \times \mathbb{Z}_{S+1}$ symmetry, was found. It again holds that hidden antiferromagnetic order, which can be measured by the string order parameter, in the original VBS states corresponds to full $\mathbb{Z}_2 \times \mathbb{Z}_2$ symmetry breaking in the transformed model.

But, rather surprisingly, the hidden antiferromagnetic order was found to depend on S as

$$\mathcal{O}^{(\alpha)}_{\mathrm{string}}(\Phi^S_{\mathrm{VBS}}) \begin{cases} > 0 & \text{if } S = 1, 3, 5, \ldots, \\ = 0 & \text{if } S = 2, 4, 6, \ldots. \end{cases} \tag{8.3.3}$$

For the hidden $\mathbb{Z}_2 \times \mathbb{Z}_2$ symmetry, this implies that the symmetry is fully broken in the VBS with odd S, but it is not broken at all for the VBS state with even S. Recall that the whole story in the part II of the book started from Haldane's discovery that there is a qualitative difference in low energy properties between antiferromagnetic Heisenberg chains with an integer S and a half-odd-integer S. Remarkably Oshikawa's observation suggests that there is a further qualitative difference between chains with an odd S and even S.

As we shall see below in Sect. 8.3.2, the above observations indeed capture a universal property of spin S antiferromagnetic chains.

[24] The definition (8.2.5), which was indeed given by Oshikawa [78], readily extends to general S, while the original definition in [48, 49] was only for $S = 1$.

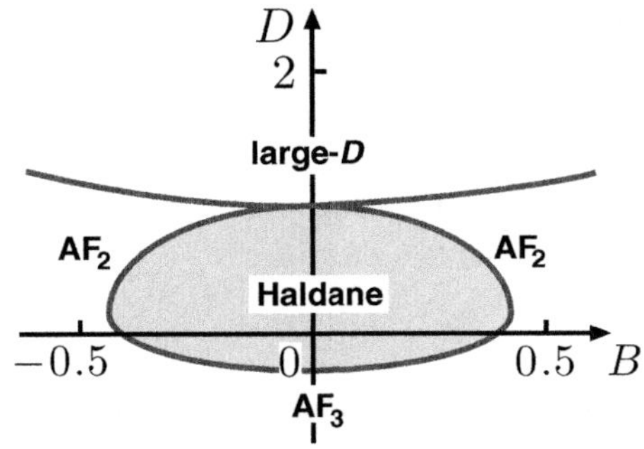

Fig. 8.9 A schematic ground state phase diagram of the Hamiltonian (8.3.4) obtained by Gu and Wen [31]. The picture of hidden antiferromagnetic order does not work for the present model, but there is a distinct phase including the Heisenberg point. Although this phase cannot be characterized neither by hidden antiferromagnetic order nor the emergence of edge states, we should identify it as the Haldane phase. Surrounding the Haldane phase, there are large-D phase and two antiferromagnetic phases, denoted as AF_2 and AF_3, where the spins point in the 2 or the 3 directions, respectively (© Hal Tasaki 2020. All Rights Reserved)

Spin chain without $\mathbb{Z}_2 \times \mathbb{Z}_2$ symmetry Gu and Wen [31] studied an $S = 1$ spin chain with Hamiltonian

$$\hat{H}_{D,B} = \sum_{x=1}^{L} \{\hat{\boldsymbol{S}}_x \cdot \hat{\boldsymbol{S}}_{x+1} + D(\hat{S}_x^{(3)})^2 + B\,\hat{S}_x^{(1)}\}, \tag{8.3.4}$$

where (as before) D is the crystal field anisotropy, and B is the uniform magnetic field in the 1-direction. The model reduces to the anisotropic model (8.1.1) by setting $B = 0$. As we have seen in Sects. 8.1.1 and 8.2.1, the present model is in the trivial large-D phase for $B = 0$ and $D > D_c \simeq 1$, and is in the nontrivial Haldane phase for $B = 0$ and $D_c > D > D'_c$. The two phases can be sharply distinguished by the fact that the hidden $\mathbb{Z}_2 \times \mathbb{Z}_2$ symmetry is not broken in the former while it is fully broken in the latter. We have also seen that the $\mathbb{Z}_2 \times \mathbb{Z}_2$ symmetry in the original Hamiltonian $\hat{H}_D$ is essential for this picture to be useful.

Now note that the term $B\,\hat{S}_x^{(1)}$ in (8.3.4) breaks the invariance under the π rotation about the 3-axis. The model only has $\mathbb{Z}_2$ symmetry with respect to the π rotation about the 1-axis. This fact, or a direct application of the transformation rule (8.2.12), implies that the Hamiltonian $\hat{U}_{KT}\hat{H}_{D,B}\hat{U}_{KT}$ is no longer short ranged for $B \neq 0$. We conclude that the picture of the hidden $\mathbb{Z}_2 \times \mathbb{Z}_2$ symmetry cannot be applied to this model with $B \neq 0$. Likewise the picture of hidden antiferromagnetic order, which is basically equivalent to the $\mathbb{Z}_2 \times \mathbb{Z}_2$ symmetry picture, does not work.

Nevertheless the ground state phase diagram obtained by the tensor-entanglement-filtering renormalization method [31] shows that the model possesses a distinct phase with a unique disordered ground state accompanied by a gap. The phase includes the Heisenberg point, and is surrounded by the disordered large-D phase and phases with long-range antiferromagnetic order. See Fig. 8.9. As we have discussed above this phase cannot be characterized by the hidden $\mathbb{Z}_2 \times \mathbb{Z}_2$ symmetry breaking. It is also known that edge spin degrees of freedom generally do not appear in this phase. It seems clear, however, that the phase should be identified as the Haldane phase.

This important example demonstrates that the Haldane phase cannot be fully characterized by hidden antiferromagnetic order, or, equivalently, by full breakdown of the hidden $\mathbb{Z}_2 \times \mathbb{Z}_2$ symmetry.[25]

8.3.2 Symmetry Protected Topological (SPT) Phase

It has become clear, especially from the works by Gu and Wen [31], and by Pollmann, Turner, Berg, and Oshikawa [81, 82], that the Haldane phase should be regarded as a nontrivial symmetry protected topological (SPT) phase, which is most clearly characterized by the presence of "entanglement protected by symmetry". In this section we introduce the notion of SPT phases, and discuss basic pictures about the characterization of the Haldane phase.

Classification of unique gapped ground states Let $\hat{H}_0$ and $\hat{H}_1$ be short ranged Hamiltonians of a quantum spin chain with spin S. We assume that both $\hat{H}_0$ and $\hat{H}_1$ have a unique ground state with a nonvanishing energy gap. We say that the two Hamiltonians $\hat{H}_0$ and $\hat{H}_1$, or the corresponding two ground states, are continuously connected if there exists a class of short ranged Hamiltonians $\hat{H}_s$ with $0 \le s \le 1$ such that $\hat{H}_s$ depends continuously on s, coincides with $\hat{H}_0$ and $\hat{H}_1$ when $s = 0$ and 1, respectively, and has a unique ground state (which is also required to depend continuously on s) with a nonvanishing energy gap for any s. In short $\hat{H}_0$ and $\hat{H}_1$ are continuously connected if there is a path of models between them that does not pass through any singular points. Since we are interested in classifying bulk properties of the ground states, the above process should be done in a suitably defined system on the infinite chain. See [4, 6] for a finer scheme of classification which focuses on chains with open boundaries.

It is now strongly believed that, if one does not impose any symmetry on Hamiltonians, all short ranged one-dimensional Hamiltonians (with a unique gapped ground state) are continuously connected with each other [15, 17, 86]. This claim has been rigorously established within matrix product states by Ogata in [72–74]. As a corollary of Theorem 1.14 in [74], it is known rigorously, for example, that the $S = 1$ VBS state (7.1.12) exhibiting the Haldane phenomena can be continuously connected to the trivial ground state of the decoupled Hamiltonian (8.1.2).[26] A path of models which connects the VBS state to a product state (which is different from the ground state of (8.1.2)) was explicitly constructed in [4]. But all these may not be too sur-

[25] In this model the Haldane phase is protected by bond-centered inversion symmetry. See Sect. 8.3.2.

[26] Such a path of models is not constructed explicitly, but can be constructed in principle by following the scheme developed in [72–74].

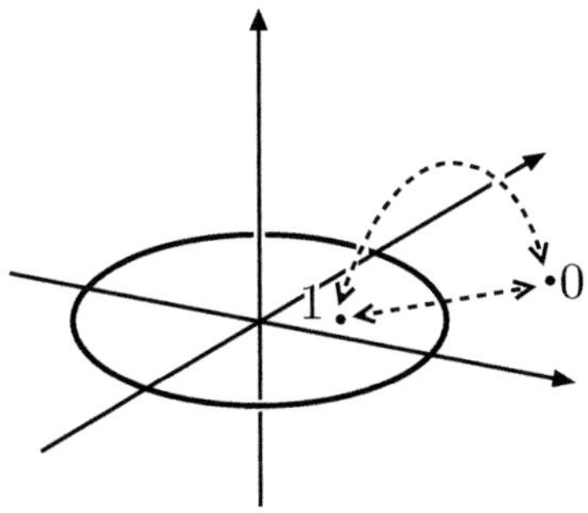

Fig. 8.10 Schematic image of a symmetry protected topological phase. The two models, which are represented by the dots 0 and 1, are continuously connected in the large parameter space. But in the restricted parameter space with certain fixed symmetry, they are separated by a definite phase boundary. It is important to note that the two phases cannot be distinguished by ordinary order parameters because we assume that there is no symmetry breaking

prising if we recall that even the solid phase in ordinary matter need to be "protected" by translation symmetry, as we remarked at the end of Sect. 3.3.[27]

The situation can be essentially different if symmetry is imposed on Hamiltonians. Suppose that the Hamiltonians $\hat{H}_0$ and $\hat{H}_1$ as above have certain common symmetry. If we require $\hat{H}_s$ with any s to also have the same symmetry, then it may happen that $\hat{H}_0$ and $\hat{H}_1$ can never be connected continuously for any choice of $\hat{H}_s$. If that is the case we regard that the two Hamiltonians $\hat{H}_0$ and $\hat{H}_1$, or the corresponding ground states, belong to different phases (protected by the imposed symmetry). In this manner, short ranged Hamiltonians with a unique gapped ground state that have certain fixed symmetry are classified into several phases. We recall that these phases cannot be distinguished by ordinary order parameters since we are assuming that there is no symmetry breaking. See Fig. 8.10.

It is clear that, no matter what symmetry is fixed, there is always one phase to which trivial tensor product states belong. This phase should be called the trivial phase. If and when a phase other than the trivial phase exists, it is called a symmetry protected topological (SPT) phase (or, more precisely, a nontrivial SPT phase).[28] Here the term "topological" indicates that the phase is not characterized by any symmetry breaking. A very important class of SPT phases are that of topological insulators [44, 101, 102].

Haldane phase for $S = 1$ as an SPT phase In 2009, after more than a quarter of a century from Haldane's discovery, Gu and Wen finally pointed out that the Haldane phase should be understood as a nontrivial symmetry protected topological phase [31]. Soon after that, Pollmann, Turner, Berg, and Oshikawa identified a complete

---

[27] See however the discussion about topological order in Sect. 8.4. The situation can be essentially different in quantum systems in two or higher dimensions.

[28] In some literature a ground state in a nontrivial SPT phase is said to possess symmetry protected topological order (SPTO).

set of symmetries required to protect the Haldane phase [81, 82]. They concluded that the Haldane phase in antiferromagnetic $S = 1$ quantum spin chains is a symmetry protected topological phase that is protected by one of the following three types of symmetry:

(S1) $\mathbb{Z}_2 \times \mathbb{Z}_2$ symmetry
(S2) time-reversal symmetry
(S3) bond-centered inversion symmetry

No other symmetry, including the translation invariance, is required.[29]

We stress that the assertion that the Haldane phase is an SPT phase is not only an interesting observation, but is a statement with a nontrivial predictive power. To see this, take the $S = 1$ chain, and let $\hat{H}_0$ be a Hamiltonian in the trivial phase, e.g, the decoupled Hamiltonian (8.1.2), and $\hat{H}_1$ be a Hamiltonian in the Haldane phase, e.g., the AKLT Hamiltonian (7.1.1). It then follows from the above assertion that, if $\hat{H}_s$ for any $0 \le s \le 1$ has one of the symmetries (S1), (S2), or (S3), then the one-parameter family of models defined by $\hat{H}_s$ must go through a phase transition at some s between 0 and 1.

We now follow Pollmann, Turner, Berg, and Oshikawa [81, 82], and discuss the mechanism by which each symmetry protects the Haldane phase. We shall present more sophisticated (and even rigorous) discussions based on "topological" $\mathbb{Z}_2$ indices in Sects. 8.3.3, 8.3.4, and 8.3.6.

(S1) $\mathbb{Z}_2 \times \mathbb{Z}_2$ *symmetry:* We require that the Hamiltonian $\hat{H}_s$ with any s satisfies $(\hat{U}_\pi^{(\alpha)})^\dagger \hat{H}_s \hat{U}_\pi^{(\alpha)} = \hat{H}_s$ for $\alpha = 1$ and 3. In fact this case has already been fully discussed in Sect. 8.2. As is summarized in Proposition 8.4 (p. 250) by Pollmann, Turner, Berg, and Oshikawa, a short ranged Hamiltonian with $\mathbb{Z}_2 \times \mathbb{Z}_2$ symmetry is transformed into another $\mathbb{Z}_2 \times \mathbb{Z}_2$ invariant short ranged Hamiltonian by the non-local transformation (8.2.5). We can then repeat the discussion in Sect. 8.2.3 for this new Hamiltonian. We simply analyze the standard $\mathbb{Z}_2 \times \mathbb{Z}_2$ symmetry breaking, which can be detected by the ferromagnetic order parameters, i.e., the magnetizations in the 1 and 3 directions. The phase with full $\mathbb{Z}_2 \times \mathbb{Z}_2$ symmetry breaking is identified with the Haldane phase in the original Hamiltonian.

(S2) *Time-reversal symmetry:* We now require that $\hat{H}_s$ satisfies $\hat{\Theta}^{-1} \hat{H}_s \hat{\Theta} = \hat{H}_s$, where the time-reversal map $\hat{\Theta}$ is defined in (2.3.16). Although we are interested in bulk properties, we shall consider an open chain and focus on one of the boundaries.[30] As we discussed in Sect. 8.1.3, an effective $S = 1/2$ degree of freedom is expected to emerge at each boundary if the model is in the Haldane phase. The two effective spins at the two edges are in general weakly coupled, but should become almost free as the chain becomes large, leading to a two-fold degeneracy at each boundary. Note also that there is no such edge degeneracy in a trivial ground state as in the large-D phase.

---

[29] As was discussed in footnote 51 in p. 216, the one-dimensional Briegel–Raussendorf state (cluster state) (7.3.16) is also in a nontrivial SPT phase protected by $\mathbb{Z}_2 \times \mathbb{Z}_2$ symmetry. See, e.g., [91].

[30] We shall see in Sect. 8.3.3 that the existence of an edges spin with half-odd-integral S implies nontrivial entanglement property in the infinite chain (without boundaries).

If we change the Hamiltonian slightly, the two-fold degeneracy is lifted in general. We argue however that, as long as the Hamiltonian preserves time-reversal symmetry and the edge spin remains, the degeneracy cannot be lifted. To see this we assume that the edge $S = 1/2$ spin is described by an effective edge Hamiltonian which inherits the time-reversal invariance of the whole Hamiltonian. But, for a single $S = 1/2$ spin, the only time-reversally invariant Hamiltonian is a constant. We see that the degeneracy cannot be lifted.[31] The only way to lift the degeneracy is to destabilize the edge spin, but this should involve a qualitative change in the bulk properties of the state. We conclude that the two-fold degeneracy (at each boundary), which characterizes the Haldane phase, is protected by time-reversal symmetry.

(S3) *Bond-centered inversion symmetry:* Consider a periodic chain $\{1, 2, \ldots, L\}$ with odd L. The bond-centered inversion is the map $x \to L - x$ on the lattice sites $x = 1, \ldots, L$, which induces the map $\boldsymbol{\sigma} = (\sigma_1, \sigma_2, \ldots, \sigma_L) \to \boldsymbol{\sigma}_{\text{inv}} = (\sigma_L, \sigma_{L-1}, \ldots, \sigma_1)$ on spin configurations. We then define the unitary operator $\hat{U}_{\text{inv}}$ by

$$\hat{U}_{\text{inv}}|\Psi^{\boldsymbol{\sigma}}\rangle = |\Psi^{\boldsymbol{\sigma}_{\text{inv}}}\rangle, \tag{8.3.5}$$

for any $\boldsymbol{\sigma}$, and require that $\hat{H}_s$ satisfies $(\hat{U}_{\text{inv}})^\dagger \hat{H}_s \hat{U}_{\text{inv}} = \hat{H}_s$ for any s. Let $|\Phi_{\text{GS}}\rangle$ be the ground state of $\hat{H}_s$ for some fixed s. If the ground state is unique, we must have $\hat{U}_{\text{inv}}|\Phi_{\text{GS}}\rangle = \sigma_{\text{inv}}|\Phi_{\text{GS}}\rangle$. By noting that $(\hat{U}_{\text{inv}})^2 = \hat{1}$, we find that the phase factor (or the parity) σ_{inv} is either 1 or -1.

Consider the VBS state $|\Phi_{\text{VBS}}\rangle$ defined by (7.1.11) and (7.1.12) on the periodic chain with L sites. Since the inversion flips the sign of each valence-bond, one readily finds that $\hat{U}_{\text{inv}}|\Phi_{\text{VBS}}\rangle = (-1)^L|\Phi_{\text{VBS}}\rangle$. See (S.63) in p. 505 for an example with $L = 3$. On the other hand, the state $|\Phi_0\rangle = \bigotimes_{x=1}^{L}|\psi_x^0\rangle$ with all 0's, which the ground state of the trivia Hamiltonian (8.1.2) in the large-D phase, clearly satisfies $\hat{U}_{\text{inv}}|\Phi_0\rangle = |\Phi_0\rangle$ for any L. Thus, as long as we work in the periodic chain with odd L, the VBS state has parity $\sigma_{\text{inv}} = -1$ and the large-D ground state has parity $\sigma_{\text{inv}} = 1$. This suggests that there is a gapless transition point in between them.

The above elementary discussion is a prototype of the characterization of SPT phases in terms of "topological" indices, and the parity σ_{inv} may be regarded as the simplest $\mathbb{Z}_2$ index. We note that the above argument works only for finite periodic chains with odd sites, while the fully rigorous theorem discussed in Sect. 8.3.6 works for infinite systems.

When an $S = 1$ antiferromagnetic spin chain is protected by symmetry (S1), the picture of hidden $\mathbb{Z}_2 \times \mathbb{Z}_2$ symmetry developed in Sect. 8.2.3 guarantees that the Haldane phase can be precisely characterized by the presence of hidden antiferromagnetic order and four-fold near degeneracy originating from edge spins.

When a model is protected by (S2), but not (S1), the hidden $\mathbb{Z}_2 \times \mathbb{Z}_2$ symmetry is not present. This means that the Haldane phase cannot be characterized by hidden antiferromagnetic order. But, as is clear from the above discussion, the effective $S = 1/2$ degrees of freedom at the boundaries are still present, protected by time-

[31] This is the simplest (and trivial) case of Kramers degeneracy. See Sect. 2.3.

reversal symmetry. Thus the emergence of edge spins can be a criterion for the Haldane phase.

Most interestingly, if a model only has symmetry (S3), neither hidden antiferromagnetic order nor edge spins can characterize the Haldane phase. Nevertheless the Haldane phase is there as a distinct nontrivial SPT phase. This is precisely the case in the model (8.3.4) of Gu and Wen. See Fig. 8.9. Even in this case the Haldane phase is distinguished by a "topological" $\mathbb{Z}_2$ index that characterizes the entanglement properties of the ground state, as we shall see in Sect. 8.3.6.

The reader might wonder whether there is a single Haldane phase protected by the three types of symmetry (S1), (S2), and (S3). It is known that the situation is more complicated. In spin chains with both $\mathbb{Z}_2 \times \mathbb{Z}_2$ symmetry and time-reversal symmetry, for example, it is known that the nontrivial phase protected by $\mathbb{Z}_2 \times \mathbb{Z}_2$ symmetry is further divided into four distinct phases [61]. See Problem 8.3.4.d in p. 274.

Antiferromagnetic spin chains with higher S Let us move on to discuss antiferromagnetic chains with general integer spin S. In Sect. 8.3.1 we have seen some results that suggest a qualitative difference between odd S and even S. It is indeed believed that, when S is an odd integer, a spin S quantum antiferromagnetic chain exhibiting the Haldane gap is in a nontrivial SPT phase protected by one of (S1), (S2), or (S3) above [81, 82]. When S is an even integer, on the other hand, it is believed that the model exhibiting the Haldane gap belongs to the trivial phase. It is continuously connected to trivial models, no matter what symmetry we impose.

Let us investigate the role of each symmetry.

(S1) $\mathbb{Z}_2 \times \mathbb{Z}_2$ *symmetry:* This is again easy. We first note that Proposition 8.4 (p. 250) about the role of the $\mathbb{Z}_2 \times \mathbb{Z}_2$ symmetry is extended (as it is) to general integral S [82]. A short ranged Hamiltonian of a spin chain with integer S which has $\mathbb{Z}_2 \times \mathbb{Z}_2$ symmetry (generated by $\hat{U}_\pi^{(1)}$ and $\hat{U}_\pi^{(3)}$) is mapped by the nonlocal unitary transformation (8.2.5) (extended by Oshikawa [78]) into another short ranged Hamiltonian with $\mathbb{Z}_2 \times \mathbb{Z}_2$ symmetry. Hidden antiferromagnetic order in the original model is mapped to ordinary ferromagnetic order in the new model. Thus, exactly as in the case with $S = 1$, hidden antiferromagnetic order (in the original model) can be a sign of a stable phase, i.e., a nontrivial SPT phase. But, as we have seen in (8.3.3), the spin S VBS state has hidden antiferromagnetic order if S is odd, but does not if S is even. This immediately implies that the spin S VBS state with odd S is in a nontrivial SPT phase protected by $\mathbb{Z}_2 \times \mathbb{Z}_2$ symmetry, while that with even S is not.

(S2) *Time-reversal symmetry:* Let us start from the spin S VBS state, and investigate the stability against a perturbation which preserves time-reversal symmetry. We again focus on one of the boundaries of a long open chain. As we saw in (8.3.1) and (8.3.2), the spin S VBS state on an open chain has effective spin $S/2$ degree of freedom at each boundary. When we perturb the Hamiltonian, it is expected that the behavior at each edge is described by an effective Hamiltonian for a single $S/2$ spin, provided that the edge degrees of freedom is stable and the chain is long enough. It is expected that this effective Hamiltonian inherits time-reversal invariance of the whole Hamiltonian.

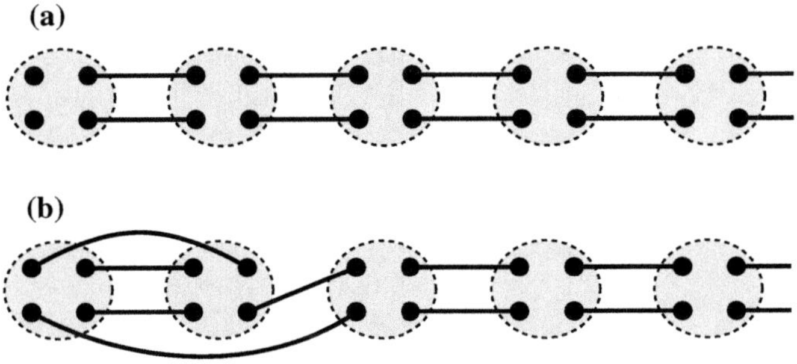

Fig. 8.11 **a** The left end of the $S = 2$ VBS state on an open chain. There are two free $S = 1/2$'s at the end which form an effective $S = 1$ edge spin. **b** By modifying the above configuration near the left end, one obtains a state consisting only of singlet pairs. This suggests that it is possible to "gap out" the edge spin degrees of freedom by a perturbation near the boundary to get a unique ground state with a gap

We then note that, when S is odd, any time-reversally symmetric Hamiltonian for spin $S/2$ must have at least doubly degenerate ground states because of Kramers degeneracy. See Sect. 2.3. The only way to lift this degeneracy is to destabilize the edge spins, but this should involve a drastic change in the bulk of the chain. We thus conclude that, when S is odd, an $S = 1/2$ edge spin degree of freedom (that corresponds to the double degeneracy) remains stable, protected by time-reversal symmetry.[32]

When S is even, on the other hand, there is no general reason that the edge spin should remain stable. In fact inspection shows that one may "gap out" the edge spin without changing the bulk property of the model. See Fig. 8.11.

(S3) *Bond-centered inversion symmetry:* The argument for $S = 1$ can be extended to general S in a straightforward manner. For the spin S VBS state $|\Phi^S_{\mathrm{VBS}}\rangle$ on the periodic chain with L sites, one readily finds that $\hat{U}_{\mathrm{inv}}|\Phi^S_{\mathrm{VBS}}\rangle = (-1)^{LS}|\Phi^S_{\mathrm{VBS}}\rangle$. When both S and L are odd, we see that $|\Phi^S_{\mathrm{VBS}}\rangle$ has odd parity, indicating that it cannot be continuously connected to the trivial state, which has even parity. When S is even, $|\Phi^S_{\mathrm{VBS}}\rangle$ always has even parity, suggesting that there is nothing protecting the VBS state.

In order to unambiguously conclude that a spin chain with even S exhibiting the Haldane gap belongs to the trivial phase, one still needs to show that its ground state can be continuously connected to trivial states. In fact a one parameter family of matrix product states for the $S = 2$ chain with all the symmetry (S1), (S2), and (S3) which interpolates between the VBS state and the trivial product state $|\Phi_0\rangle = \bigotimes_{x=1}^{L}|\psi_x^0\rangle$ without passing through any gapless states was explicitly constructed in [82]. More strikingly, it was observed numerically by Tonegawa, Okamoto, Nakano, Sakai, Nomura, and Kaburagi that the $S = 2$ antiferromagnetic Heisenberg chain

[32]By using the same argument based on the property stated in Problem 2.2.a (p. 23), we can show the stability of edge spins for models with odd S and $\mathbb{Z}_2 \times \mathbb{Z}_2$ symmetry.

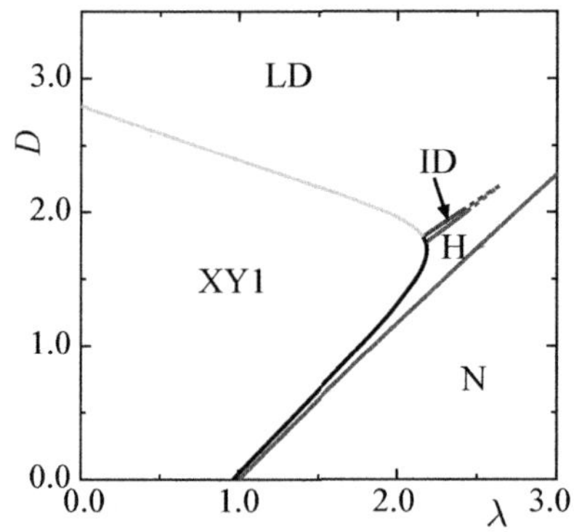

Fig. 8.12 The ground state phase diagram of the $S = 2$ chain with the anisotropic Hamiltonian (8.2.1) with two parameters λ and D. The Hamiltonian has all the symmetry (S1), (S2), and (S3). There are the large-D phase (LD), the XY phase (XY1), the Néel phase (N), the Haldane phase (H), and the intermediate D phase (ID). Remarkably the Haldane phase is continuously connected to the trivial large-D phase. It is also remarkable that although the Haldane phase is very narrow for small D, the Heisenberg point with $\lambda = 1$ and $D = 0$ is still in the Haldane phase, precisely as Haldane concluded in his original works! This indicates that the SU(2) invariance plays an essential role in stabilizing the Haldane phase. But it seems that we still do not fully understand underlying mechanisms. Finally the intermediate-D phase (ID) is a nontrivial SPT phase, whose existence was predicted by Oshikawa back in 1992 [78]. Reproduced with permission from [96]. (c) (2011) The Physical Society of Japan

(which exhibits the Haldane gap) is continuously connected to trivial models within a physically realistic parameter space with all the symmetry (S1), (S2), and (S3) [96]. Figure 8.12 shows the ground state phase diagram of the $S = 2$ chain with the anisotropic Hamiltonian (8.2.1) (with periodic boundary conditions). The Haldane "phase", indicated by "H", is clearly connected to the large-D "phase", indicated by "LD", exactly as the liquid "phase" is continuously connected to the gas "phase" in Fig. 3.5.

8.3.3 Entanglement and "Topological" Indices for SPT Phases

Different SPT phases cannot be distinguished by a local order parameter since there is no symmetry breaking by definition. Although the den Nijs–Rommelse string order parameter can characterize the Haldane phase, it is useful only for models with $\mathbb{Z}_2 \times \mathbb{Z}_2$ symmetry. The most general and clear classification of SPT phases is given by "topological" indices that characterize the nature of entanglement in the ground state. Here we use the word "topological" to indicate that these indices are invariant when the unique gapped ground state is continuously modified.

In the present section we follow Pollmann, Turner, Berg, and Oshikawa [81, 82], and discuss the basic idea of the "topological" indices in a heuristic manner.

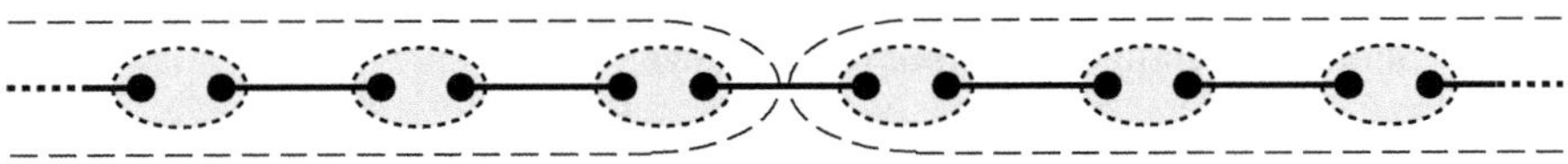

Fig. 8.13 The $S = 1$ VBS state on the infinite chain. The left and the right half-infinite chains are entangled via the spin-singlet connecting the two regions ()

In Sect. 8.3.4, we see how the idea can be made into precise definition for matrix product states. In Sect. 8.3.6, we discuss Ogata's index theorems, where the indices are rigorously defined for a general unique gapped ground state in the infinite chain.

There are some other approaches to characterize nontrivial SPT phases. The earlier work based on the Berry phase for quantum spin chains [36, 37, 39, 40] may be closely related to the approach with the "topological" indices. See also [32, 80] for a characterization of the Haldane phase in terms of general nonlocal order parameters, and [89] for a discussion based on topological field theories. The formulae for "topological" indices in these works suggest that the indices for SPT and the indices for free fermion systems such as the Chern number for quantum Hall effect (see the remark in p. 228) may be discussed on equal footing. See, e.g., (2.38) or (2.39) of [89].

Entanglement and $\mathbb{Z}_2$ indices Let us heuristically discuss the basic idea behind the "topological" indices of Pollmann, Turner, Berg, and Oshikawa's. We formally consider a spin system and its ground state on the infinite chain $\mathbb{Z}$. We imagine that the whole chain is divided into the left half-infinite chain $\{\ldots, -2, -1\}$ and the right half-infinite chain $\{0, 1, 2, \ldots\}$. The "topological" indices are designed to characterize the way the left half and the right half are entangled. This is done by examining the response of the entanglement structure of the ground state to symmetry transformations.

As an illustrative example, we take (without being mathematically careful) the $S = 1$ VBS state on the infinite chain. It is clear that the entanglement between the left half and the right half is mediated by the spin-singlet that connects the two regions. See Fig. 8.13. The VBS state is thus represented in the form of a singlet pair as

$$|\Phi_{\text{VBS}}\rangle = \frac{1}{\sqrt{2}}\bigl\{|\Phi_\uparrow\rangle_{\text{L}} \otimes |\Psi_\downarrow\rangle_{\text{R}} - |\Phi_\downarrow\rangle_{\text{L}} \otimes |\Psi_\uparrow\rangle_{\text{R}}\bigr\}, \tag{8.3.6}$$

where $|\Phi_\sigma\rangle_{\text{L}}$ and $|\Psi_\sigma\rangle_{\text{R}}$ are (formal expressions of) the states on the left half-infinite chain and the right half-infinite chain, respectively, with the edge spin $\sigma =\uparrow, \downarrow$. We should regard $|\Phi_\sigma\rangle_{\text{L}}$ and $|\Psi_\sigma\rangle_{\text{R}}$ as effective states (in a certain sense), since they are not the standard ket states. Here it is essential that $|\Phi_\sigma\rangle_{\text{L}}$ and $|\Psi_\sigma\rangle_{\text{R}}$ can be regarded as states with $S = 1/2$. These $S = 1/2$ states are reminiscent of the edge states at open ends although we are here dealing with a single infinite chain.

Let us generalize the above observation to a general unique gapped ground state of a system on the infinite chain. We write the ground state formally as $|\Phi_{\text{GS}}\rangle$. To examine the entanglement in the ground state, we write it (again formally) in the form of Schmidt decomposition (see (A.1.21)) as

$$|\Phi_{\mathrm{GS}}\rangle = \sum_j \sqrt{p_j}\, |\Phi_j\rangle_{\mathrm{L}} \otimes |\Psi_j\rangle_{\mathrm{R}}, \tag{8.3.7}$$

with $p_j > 0$ and $\sum_j p_j = 1$, where $\{|\Phi_j\rangle_{\mathrm{L}}\}_{j=1,2,\ldots}$ and $\{|\Psi_j\rangle_{\mathrm{R}}\}_{j=1,2,\ldots}$ are certain orthonormal sets of effective states in the left half and the right half. The sum over j is in general an infinite sum, but we expect that only a finite number of j's are dominant since a unique gapped ground state is known to have short-range entanglement [34]. Then the reduced density matrix for the right half-infinite chain is given by

$$\hat{\rho}_{\mathrm{R}} := \sum_j {}_{\mathrm{L}}\langle\Phi_j|\Big(|\Phi_{\mathrm{GS}}\rangle\langle\Phi_{\mathrm{GS}}|\Big)|\Phi_j\rangle_{\mathrm{L}} = \sum_j p_j |\Psi_j\rangle_{\mathrm{R}}\, {}_{\mathrm{R}}\langle\Psi_j|. \tag{8.3.8}$$

The corresponding entanglement entropy of the ground state $|\Phi_{\mathrm{GS}}\rangle$ is defined by $S_{\mathrm{LR}} := -\sum_j p_j \log p_j$. See Appendix A.1.

We first focus on models with time-reversal symmetry. The uniqueness of the ground state implies that $|\Phi_{\mathrm{GS}}\rangle$ is time-reversal invariant, and hence so is the reduced density matrix $\hat{\rho}_{\mathrm{R}}$. We thus have

$$\sum_j p_j |\Psi_j\rangle_{\mathrm{R}}\, {}_{\mathrm{R}}\langle\Psi_j| = \sum_j p_j |\Psi'_j\rangle_{\mathrm{R}}\, {}_{\mathrm{R}}\langle\Psi'_j|, \tag{8.3.9}$$

where $|\Psi'_j\rangle_{\mathrm{R}} = \hat{\Theta}|\Psi_j\rangle_{\mathrm{R}}$ is the time-reversal of $|\Psi_j\rangle_{\mathrm{R}}$. Let p be one of $p_1, p_2, \ldots$, and let $J(p) := \{j \mid p_j = p\}$ be the corresponding set of indices. Then the invariance (8.3.9) implies

$$\sum_{j\in J(p)} |\Psi_j\rangle_{\mathrm{R}}\, {}_{\mathrm{R}}\langle\Psi_j| = \sum_{j\in J(p)} |\Psi'_j\rangle_{\mathrm{R}}\, {}_{\mathrm{R}}\langle\Psi'_j|, \tag{8.3.10}$$

for each p.

We now assume that, as in the case of the VBS state, all the states $|\Phi_j\rangle_{\mathrm{L}}$ and $|\Psi_j\rangle_{\mathrm{R}}$ can be effectively regarded as states with half-odd-integer spins.[33] Then from (2.3.29), which is the essence of Kramers degeneracy, we see that $|\Psi_j\rangle_{\mathrm{R}}$ and $|\Psi'_j\rangle_{\mathrm{R}}$ are orthogonal. We then find from (8.3.10) that $|J(p)|$ must be even and hence $|J(p)| \geq 2$ for all p. This implies that the entanglement entropy $S_{\mathrm{LR}} = -\sum_j p_j \log p_j$ can never be less than $\log 2$. We have thus obtained a lower bound for the entanglement entropy from the requirement of symmetry. For the VBS state (8.3.6), we have $p_1 = p_2 = 1/2$, and hence $S_{\mathrm{LR}} = \log 2$, which saturates the lower bound. The presence of such "entanglement imposed by symmetry" is a definite sign that the state is in a nontrivial SPT phase. Note that there is no such mandatory entanglement if $|\Psi_j\rangle_{\mathrm{R}}$ are states with integer spins. Of course there is nonzero entanglement in

[33] This is what one expects when the $S = 1$ VBS state is slightly perturbed (within the Haldane phase). The $S = 1/2$ degree of freedom may couple to excitations with integer spins to form states with half-odd-integer spins.

general, but one may modify the state continuously near the origin of the chain (without breaking the symmetry) to get a state without entanglement.

Therefore the desired index should distinguish between the cases where the states $|\Psi_j\rangle_R$ have half-odd-integer spins and integer spins. The distinction is precisely given by the property of $\hat{\Theta}^2$ noted just below (2.3.17). We can define the index $\sigma_{tr} = \pm 1$ as a constant such that $\hat{\Theta}^2|\Psi_j\rangle_R = \sigma_{tr}|\Psi_j\rangle_R$ for all j.[34] A model with $\sigma_{tr} = -1$, which corresponds to half-odd-integer (effective) spins on the half-infinite chains, is in a nontrivial SPT phase, namely, the Haldane phase, while a model with $\sigma_{tr} = 1$, which corresponds to integer (effective) spins, is in the trivial phase.

Models with $\mathbb{Z}_2 \times \mathbb{Z}_2$ (or dihedral) symmetry can be treated in essentially the same manner. See Problem 8.3.3.a below. Again the index $\sigma_D = \pm 1$ should distinguish between the case where effective states on the half-infinite chains have half-odd-integer spins and the case where they have integer spins. This is attained if it holds that $\hat{U}_\pi^{(1)}\hat{U}_\pi^{(3)}|\Psi_j\rangle_R = \sigma_D\,\hat{U}_\pi^{(3)}\hat{U}_\pi^{(1)}|\Psi_j\rangle_R$ for all j. The model with $\sigma_D = -1$, which corresponds to half-odd-integer (effective) spins, is in the Haldane phase.

The treatment of models with bond-centered inversion symmetry is different, but is not difficult (at least formally). Assume that the model is invariant under the inversion map $x \to 1 - x$. We formally define the inversion transformation $\hat{U}_{inv}$ as in (8.3.5) from the map $x \to 1 - x$. Then the index $\sigma_{inv} = \pm 1$ is simply the parity defined by $\hat{U}_{inv}|\Phi_{GS}\rangle = \sigma_{inv}|\Phi_{GS}\rangle$, where we understand that the ground state admits the Schmidt decomposition (8.3.7) in terms of effective states. It is clear that a state which has $\sigma_{inv} = -1$, with the VBS state (8.3.6) being an example, must have nonzero entanglement with entanglement entropy not less than log 2.

Problem 8.3.3.a Work out the details of the case with $\mathbb{Z}_2 \times \mathbb{Z}_2$ symmetry. [solution→p.511]

Toward precise definitions of the indices The above discussion is based on the vaguely defined notion of effective states that live on the half-infinite chains, and cannot be accepted as a proper theoretical argument. We need to find ways to define and analyze the transformations of effective states.

We should stress that it is insufficient to consider ground states (of a system with integer S) on the long open chain $\{1 - \frac{L}{2}, 2 - \frac{L}{2}, \dots, \frac{L}{2}\}$ and the corresponding Schmidt decomposition (8.3.7). Although everything is well-defined in this case, we see that all the states $|\Phi_j\rangle_L$ and $|\Psi_j\rangle_R$ have integer spins by definition, and the indices σ_{tr} and σ_D always take the trivial value 1.[35] The effective states considered above must characterize how the infinite ground state (when decomposed into left and right) behaves near the origin of the chain. We shall see in Sect. 8.3.4 that such a characterization can be realized in the framework of matrix product states. In Sect. 8.3.6, we further see that such a characterization is possible with a complete generality and rigor by using sophisticated notions from the operator algebraic approach to quantum spin chains.

---

[34] Of course the same index is obtained by $\hat{\Theta}^2|\Phi_j\rangle_L = \sigma_{tr}|\Phi_j\rangle_L$.

[35] The parity σ_{inv} may be 1 or -1, but it is not easy to relate the value of the parity to the behavior of the infinite system.

8.3.4 "Topological" Indices for Matrix Product States

Pollmann, Turner, Berg, and Oshikawa [81, 82] found that, if one works within the matrix product formalism, one can directly study the transformation of the (effective) states in the half-infinite chains discussed heuristically in Sect. 8.3.3. In this section, we shall see how the "topological" indices are defined by using the powerful uniqueness theorem, Theorem 7.6 in p. 203, for injective matrix product states, and also work out some examples for $S = 1$ systems. The technique developed here is used in Sect. 8.3.5 to prove a Lieb–Schultz–Mattis type no-go theorem.

The essence of the index theory presented here is the characterization of SPT phases in terms of projective representations of the relevant symmetry on the matrices defining a matrix product state. There is an earlier work by Perez-Garcia, Wolf, Sanz, Verstraete, and Cirac [79], where essential techniques used in [81, 82] and similar ideas to make use of projective representations are discussed. The viewpoint based on projective representations, which is by now standard, further leads to a classification of possible symmetry protected topological phases, with the aid of group cohomology [15, 17, 29, 86].

Although the above mentioned works [79, 81, 82] seem to be the first physics literature to focus on projective representations (on the matrices for a matrix product state) of a symmetry group, we note that, back in 2001, Matsui developed a mathematical theory for quantum spin chains based on projective representations of group symmetry, not only for matrix product states but for more general pure states which satisfy the split property [63] (see also the discussion after Theorem 8.6 in p. 276).[36]

Unitary transformation of a matrix product state Consider a quantum spin system with spin S on the chain $\{1, \dots, L\}$, and take a translation invariant matrix product state

$$|\Phi\rangle = \sum_{\sigma} \mathrm{Tr}[\mathsf{A}^{\sigma_1} \mathsf{A}^{\sigma_2} \cdots \mathsf{A}^{\sigma_L}]\,|\Psi^{\sigma}\rangle. \tag{8.3.11}$$

This is nothing but the general MPS (7.2.36), with some $D \times D$ matrices A^{σ} where $\sigma = -S, \dots, S$. We make an essential assumption that the matrix product state is injective, i.e., the collection of matrices $(\mathsf{A}^{\sigma})_{\sigma=-S,\dots,S}$ satisfies (7.2.41) and the properties (i), (ii), (iii) stated in Theorem 7.5 in p. 202. Physically speaking, this means that the state $|\Phi\rangle$ has short range correlation, and is not a superposition of macroscopically different states.

Take (and fix) an arbitrary unitary operators $\hat{u}$ on the single spin Hilbert space $\mathfrak{h}_0 \cong \mathbb{C}^{2S+1}$, and let $\hat{u}_x$ be its identical copy acting on the Hilbert space $\mathfrak{h}_x$ for the spin at site $x = 1, \dots, L$. We consider the product $\hat{U} = \bigotimes_{x=1}^{L} \hat{u}_x$. We shall examine how the unitary transformation by $\hat{U}$ is incorporated into the matrix product representation. From the definitions, we find

[36] We also learned that the emergence of projective representations in matrix product states was realized by Fannes, Nachtergaele, and Werner in 1990s, although the work was not published. (B. Nachtergaele, private communication. See also the remark at the end of Sect. 1 of [63].)

$$
\begin{aligned}
\hat{U}|\Phi\rangle &= \sum_{\boldsymbol{\sigma}} \mathrm{Tr}[\mathsf{A}^{\sigma_1}\mathsf{A}^{\sigma_2}\cdots\mathsf{A}^{\sigma_L}]\,\hat{U}|\Psi^{\boldsymbol{\sigma}}\rangle \\
&= \sum_{\boldsymbol{\sigma},\boldsymbol{\sigma}'} \mathrm{Tr}[\mathsf{A}^{\sigma_1}\mathsf{A}^{\sigma_2}\cdots\mathsf{A}^{\sigma_L}]\,|\Psi^{\boldsymbol{\sigma}'}\rangle\Big(\prod_{x=1}^{L}\langle\psi^{\sigma'_x}|\hat{u}|\psi^{\sigma_x}\rangle\Big) \\
&= \sum_{\boldsymbol{\sigma}} \mathrm{Tr}[\tilde{\mathsf{A}}^{\sigma_1}\tilde{\mathsf{A}}^{\sigma_2}\cdots\tilde{\mathsf{A}}^{\sigma_L}]\,|\Psi^{\boldsymbol{\sigma}}\rangle,
\end{aligned}
\tag{8.3.12}
$$

where we exchanged the symbols $\boldsymbol{\sigma}$ and $\boldsymbol{\sigma}'$ in the final line, and set

$$
\tilde{\mathsf{A}}^{\sigma} = \sum_{\sigma'=-S}^{S} \langle\psi^{\sigma}|\hat{u}|\psi^{\sigma'}\rangle\mathsf{A}^{\sigma'}. \tag{8.3.13}
$$

It is easily verified that

$$
\sum_{\sigma=-S}^{S} \tilde{\mathsf{A}}^{\sigma}(\tilde{\mathsf{A}}^{\sigma})^{\dagger} = \sum_{\sigma=-S}^{S} \mathsf{A}^{\sigma}(\mathsf{A}^{\sigma})^{\dagger}, \tag{8.3.14}
$$

and that the transfer matrix defined from $\tilde{\mathsf{A}}^{\sigma}$ as in (7.2.42) is identical to that defined from A^{σ}. This implies an important fact that the collection of matrices $(\tilde{\mathsf{A}}^{\sigma})_{\sigma=-S,\ldots,S}$ is also injective.

Let us demand that $|\Phi\rangle$ is invariant under the unitary transformation $\hat{U}$, i.e.,

$$
\hat{U}|\Phi\rangle = e^{i\eta_L}|\Phi\rangle, \tag{8.3.15}
$$

for any L, with some phase factor $\eta_L \in \mathbb{R}$. Obviously the invariance is guaranteed if the new matrices in (8.3.13) are related to the original matrices by

$$
\tilde{\mathsf{A}}^{\sigma} = e^{i\eta_L/L}\,\mathsf{U}^{\dagger}\mathsf{A}^{\sigma}\mathsf{U}, \tag{8.3.16}
$$

for all σ, with a $D \times D$ unitary matrix U independent of σ and L. See the following examples, where we indeed find such a unitary matrix by inspection. What is essential here is that the converse is also true. By Theorem 7.6 (p. 203) due to Fannes, Nachtergaele, and Werner [28], we find that the invariance (8.3.15) implies the existence and the uniqueness (up to a phase) of the unitary matrix U satisfying (8.3.16).[37] (See also [79] for a closely related discussion.) This also implies, as anticipated, that the phase factor η_L can be chosen as $\eta_L = L\,\zeta$ where ζ is independent of L. By combining (8.3.16) with $\eta_L/L = \zeta$ and (8.3.13), we find the relation

$$
\sum_{\sigma'=-S}^{S} \langle\psi^{\sigma}|\hat{u}|\psi^{\sigma'}\rangle\mathsf{A}^{\sigma'} = e^{i\zeta}\,\mathsf{U}^{\dagger}\mathsf{A}^{\sigma}\mathsf{U}, \tag{8.3.17}
$$

[37]It suffices to set $\mathsf{B}^{\sigma} = e^{-i\eta_L/L}\tilde{\mathsf{A}}^{\sigma}$, and apply the theorem.

which can be solved for A^σ on the left-hand side to give

$$\mathsf{A}^\sigma = e^{i\zeta} \sum_{\sigma'=-S}^{S} \langle\psi^\sigma|\hat{u}^\dagger|\psi^{\sigma'}\rangle \mathsf{U}^\dagger \mathsf{A}^{\sigma'} \mathsf{U}. \tag{8.3.18}$$

To sum we have seen that the assumption (8.3.15) about the invariance of $|\Phi\rangle$ implies the existence of ζ and U, and the nontrivial constraint (8.3.18) on the matrices A^σ.

Before proceeding, let us argue that the above unitary matrix U precisely describes the transformation of the (effective) states that live on the half-infinite chains discussed in Sect. 8.3.3. This is most easily seen diagrammatically (without being bothered by cumbersome matrix product notations). Consider a translation invariant MPS with matrices $(\mathsf{A}^\sigma)_{\sigma=-S,\dots,S}$ on the infinite chain,[38] and represent it as

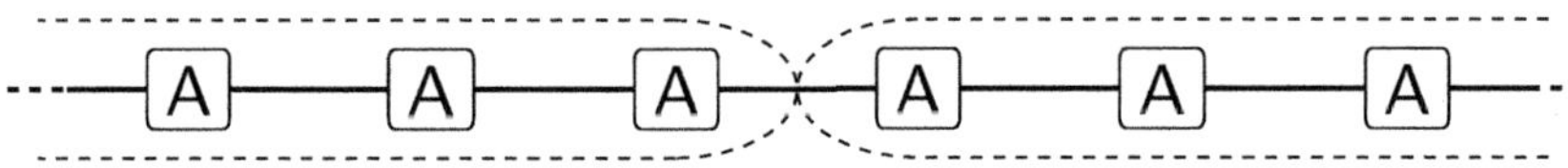

We are again interested in the way the left half and the right half are entangled. Let us apply the unitary transformation $\hat{u}_x$ on every lattice site. As in (8.3.12), we get an MPS with new matrices $\tilde{\mathsf{A}}^\sigma$, which we represent as

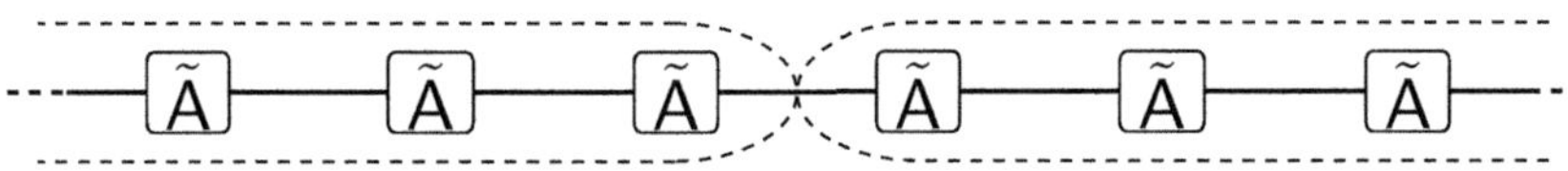

Recalling that $\tilde{\mathsf{A}}^\sigma = (\text{const.})\mathsf{U}^\dagger\mathsf{A}^\sigma\mathsf{U}$ as in (8.3.16), we can write the same state as

where we used $\mathsf{U}\mathsf{U}^\dagger = \mathsf{I}$ except at the dividing point. The diagram clearly demonstrates that the state on the right is transformed by $\mathsf{U}^\dagger$ and the state on the left is transformed by U.

Symmetry transformation of an MPS ground state Let us move onto the problem of classifying gapped unique ground states. Suppose that the model has symmetry described by a group G, which can be a finite group such as $\mathbb{Z}_2$ or $\mathbb{Z}_2 \times \mathbb{Z}_2$, or an infinite group such as U(1) or SU(2). We assume that a unitary representation $\hat{u}(\cdot)$ of G on the single-spin Hilbert space $\mathfrak{h}_0$ is given, i.e., for each $g \in G$ there is a $(2S+1) \times (2S+1)$ matrix $\hat{u}(g)$ such that $\hat{u}(e) = \hat{1}$ and $\hat{u}(g)\hat{u}(h) = \hat{u}(gh)$ for any $g, h \in G$. See Appendix A.5. For each $g \in G$, let $\hat{U}(g) = \bigotimes_{x=1}^{L} \hat{u}_x(g)$, where

[38] Although we do not go into details such an MPS on the infinite chain can be defined rigorously. In fact MPS was originally defined in [26, 27] for infinite systems.

$\hat{u}_x(g)$ is the copy of $\hat{u}(g)$ on $\mathfrak{h}_x$. Of course $\hat{U}(g)$ with $g \in G$ form a representation of G. We further assume that the model has translation invariance.[39]

We assume that the model has a unique gapped ground state $|\Phi\rangle$ that is written as an injective matrix product state (8.3.11). The uniqueness implies that the state $|\Phi\rangle$ is invariant under the unitary transformation $\hat{U}(g)$ for any $g \in G$, i.e.,

$$\hat{U}(g)|\Phi\rangle = e^{i\eta_L(g)}|\Phi\rangle \tag{8.3.19}$$

with some constants $\eta_L(g)$. This is of course exactly the same as (8.3.15). Then, by setting $\hat{u} = \hat{u}(g)$ in (8.3.18), we have

$$\mathsf{A}^\sigma = e^{i\zeta(g)} \sum_{\sigma'=-S}^{S} \langle\psi^\sigma|\hat{u}^\dagger(g)|\psi^{\sigma'}\rangle \mathsf{U}^\dagger(g)\mathsf{A}^{\sigma'}\mathsf{U}(g). \tag{8.3.20}$$

for each $g \in G$, where the constant $\zeta(g) \in \mathbb{R}$ and the $D \times D$ matrix $\mathsf{U}(g)$ depend only on g. (As usual we write $\mathsf{U}^\dagger(g)$ to mean $\{\mathsf{U}(g)\}^\dagger$.) We shall choose $\zeta(e) = 0$ and $\mathsf{U}(e) = \mathsf{I}$, which is always possible.

We now substitute (8.3.20) into the same relation with g replaced by $h \in G$ to find

$$\begin{aligned}\mathsf{A}^\sigma &= e^{i\{\zeta(g)+\zeta(h)\}} \sum_{\sigma',\sigma''} \langle\psi^\sigma|\hat{u}^\dagger(h)|\psi^{\sigma''}\rangle \mathsf{U}^\dagger(h) \langle\psi^{\sigma''}|\hat{u}^\dagger(g)|\psi^{\sigma'}\rangle \mathsf{U}^\dagger(g)\mathsf{A}^{\sigma'}\mathsf{U}(g)\mathsf{U}(h) \\ &= e^{i\{\zeta(g)+\zeta(h)\}} \sum_{\sigma'} \langle\psi^\sigma|\hat{u}^\dagger(gh))|\psi^{\sigma'}\rangle \{\mathsf{U}(g)\mathsf{U}(h)\}^\dagger \mathsf{A}^{\sigma'}\mathsf{U}(g)\mathsf{U}(h).\end{aligned} \tag{8.3.21}$$

On the other hand the relation (8.3.20) with g replaced by gh reads

$$\mathsf{A}^\sigma = e^{i\zeta(gh)} \sum_{\sigma'} \langle\psi^\sigma|\hat{u}^\dagger(gh))|\psi^{\sigma'}\rangle \mathsf{U}^\dagger(gh)\mathsf{A}^{\sigma'}\mathsf{U}(gh). \tag{8.3.22}$$

Noting that the requirement (8.3.19) (with $\eta_L(g) = L\zeta(g)$) and the property $\hat{U}(g)\hat{U}(h) = \hat{U}(gh)$ readily imply $\zeta(g) + \zeta(h) = \zeta(gh)$, we see from (8.3.21) and (8.3.22) that

$$\{\mathsf{U}(g)\mathsf{U}(h)\}^\dagger \mathsf{A}^{\sigma'}\mathsf{U}(g)\mathsf{U}(h) = \mathsf{U}^\dagger(gh)\mathsf{A}^{\sigma'}\mathsf{U}(gh), \tag{8.3.23}$$

for any $g, h \in G$ and any σ. Writing $\mathsf{W} = \mathsf{U}(g)\mathsf{U}(h)\mathsf{U}^\dagger(gh)$, this implies $[\mathsf{W}, \mathsf{A}^\sigma] = 0$ for any σ. By using the relation repeatedly we find $[\mathsf{W}, \mathsf{A}^{\sigma_1}\mathsf{A}^{\sigma_2}\ldots\mathsf{A}^{\sigma_{\ell_0}}] = 0$ for any $\sigma_1, \ldots, \sigma_{\ell_0}$. Then the property (i) in Theorem 7.5 (p. 202), which is valid because of the assumed injectivity, implies that W commutes with any $D \times D$ matrix, and hence $\mathsf{W} = \mu\mathsf{I}$ for some constant $\mu \in \mathbb{C}$. Since the unitarity of W implies $|\mu| = 1$, we can write $\mu = e^{i\phi(g,h)}$. We have thus arrived at an important relation

---

[39]Translation invariant is not necessary to protect the Haldane phase, but this assumption is necessary for the present argument based on the uniqueness theorem (Theorem 7.6 in p. 203) for injective MPS. In Ogata's index theorems discussed in Sect. 8.3.6 translation invariance is not required.

$$\mathsf{U}(g)\mathsf{U}(h) = e^{i\phi(g,h)}\mathsf{U}(gh), \tag{8.3.24}$$

with a phase factor $\phi(g, h) \in \mathbb{R}$.

If $\phi(g, h)$ is always zero, then the relation (8.3.24) says that the matrices $\mathsf{U}(g)$ with $g \in G$ form a representation of G. In general, however, the Eq. (8.3.24) allows a nonvanishing extra phase $\phi(g, h)$. In such a case, $\mathsf{U}(g)$ with $g \in G$ are said to form a projective representation of the group G. Recall that we already discussed the notion of projective representation in Sect. 2.1.

Classification of projective representations and SPT phases We have seen that one can associate a projective representation of G with any injective matrix product state $|\Phi\rangle$ with symmetry descried by G. In order to connect this observation to the problem of SPT phases, let us discuss the notion of projective representation in more detail.

We first note that some projective representations are related with each other in a trivial manner. Let $\mathsf{U}'(g) = e^{i\psi(g)}\mathsf{U}(g)$ for all $g \subset G$, where $\psi(g) \in \mathbb{R}$ is an arbitrary function on G. Then one can replace $\mathsf{U}(g)$ with $\mathsf{U}'(g)$ without changing the right-hand side of (8.3.20). The phase $\phi'(g, h)$ associated with $\mathsf{U}'(g)$ as in (8.3.24) is given by

$$\phi'(g, h) = \phi(g, h) + \psi(gh) - \psi(g) - \psi(h) \ (\mathrm{mod}\ 2\pi) \quad \text{for any } g, h \in G. \tag{8.3.25}$$

This means that two projective representations whose phase functions $\phi(\cdot, \cdot)$ and $\phi'(\cdot, \cdot)$ are related by (8.3.25) are equivalent. In particular if the phase function $\phi(\cdot, \cdot)$ of a projective representation is written as $\phi(g, h) = \psi(gh) - \psi(g) - \psi(h) \ (\mathrm{mod}\ 2\pi)$ for any $g, h \in G$ with some $\psi(\cdot)$, then the projective representation is equivalent to a genuine representation with zero phase factor. In such a case we say that the projective representation is trivial.

Recall the important fact that the unitary matrix U which satisfies (8.3.16) is determined uniquely only up to a phase factor. We therefore find that the projective representation associated with a given injective matrix product state $|\Phi\rangle$ is unique up to the equivalence discussed above. This uniqueness is essential for the characterization and the classification of symmetry protected topological phases.

The question then is whether there are nontrivial projective representations for a given group G. This is, of course, a very well understood problem in mathematics. Observe that the associativity $\{\mathsf{U}(g_1)\mathsf{U}(g_2)\}\mathsf{U}(g_3) = \mathsf{U}(g_1)\{\mathsf{U}(g_2)\mathsf{U}(g_3)\}$, which must hold for any $g_1, g_2, g_3 \in G$, along with (8.3.24), imposes a nontrivial constraint on the phase function, namely,

$$\phi(g_1, g_2) + \phi(g_1 g_2, g_3) = \phi(g_2, g_3) + \phi(g_1, g_2 g_3) \ (\mathrm{mod}\ 2\pi). \tag{8.3.26}$$

The desired classification of projective representations of G boils down to finding phase functions $\phi(\cdot, \cdot)$ satisfying (8.3.26), noting that two phase functions related by (8.3.25) should be regarded as equivalent. Mathematically such a classification is precisely given by an object called the second group cohomology[40] $\mathrm{H}^2(G, \mathrm{U}(1))$.

For the group $\mathbb{Z}_2 \times \mathbb{Z}_2$, which describes the symmetry (S1) that protects the Haldane phase, it is known that $\mathrm{H}^2(\mathbb{Z}_2 \times \mathbb{Z}_2, \mathrm{U}(1)) = \mathbb{Z}_2$, where $\mathbb{Z}_2$ is a group with two elements (see Appendix A.5). Thus there are exactly two types of projective representation of the group $\mathbb{Z}_2 \times \mathbb{Z}_2$, one trivial and one nontrivial. See Problem 8.3.4.a below for an elementary proof. One may define the "topological" index $\sigma_{\mathrm{D}} = \pm 1$ by assigning $\sigma_{\mathrm{D}} = 1$ to a trivial projective representation and $\sigma_{\mathrm{D}} = -1$ to a nontrivial projective representation. (We have indeed seen back in Sect. 2.1 that a single $S = 1/2$ spin gives a nontrivial projective representation (2.1.31) of $\mathbb{Z}_2 \times \mathbb{Z}_2$.)

We stress that inequivalent projective representations are truly distinct, and can never be connected continuously. This fact has a deep implication on the classification problem of unique gapped ground states. Suppose that there is a $\mathbb{Z}_2 \times \mathbb{Z}_2$ invariant ground state (which can be written as an injective matrix product state) that is accompanied by a nontrivial projective representation of $\mathbb{Z}_2 \times \mathbb{Z}_2$. When we modify the state continuously, keeping the injectivity and $\mathbb{Z}_2 \times \mathbb{Z}_2$ invariance, the accompanying projective representation remains to be nontrivial. This is because the projective representation is unique, and a discontinuous jump is necessary to change the equivalence class of projective representations. As long as $\mathbb{Z}_2 \times \mathbb{Z}_2$ symmetry is preserved, the only way to change the class of projective representations is to break the injectivity, and that corresponds to going through a gapless phase boundary point.

It is then expected that, in $\mathbb{Z}_2 \times \mathbb{Z}_2$ symmetric spin chains with odd integer S, the Haldane phase is characterized by a nontrivial projective representation of $\mathbb{Z}_2 \times \mathbb{Z}_2$ with $\sigma_{\mathrm{D}} = -1$, while the trivial phase is characterized by a trivial projective representation with $\sigma_{\mathrm{D}} = 1$. We shall confirm this expectation below by examining the VBS state and the large-D state for $S = 1$. The same should be true for general odd integer S. We note that the overall story about the projective representations is the same for chains with even integer S. In these models it happens that the VBS state is accompanied by a trivial projective representation of $\mathbb{Z}_2 \times \mathbb{Z}_2$, and hence belongs to the trivial phase.

This picture can be readily generalized to spin chains protected by a general symmetry group G. Trivial ground states are accompanied by a trivial projective representation, and nontrivial SPT phases correspond to nontrivial projective representations. Each inequivalent nontrivial projective representation should correspond to a distinct nontrivial SPT phase.

For example, if the symmetry group G is the cyclic group $\mathbb{Z}_n$ or $U(1)$, it is known that $\mathrm{H}^2(G, U(1))$ contains only one element. Thus the spin chains protected by such local symmetry can have only a single trivial phase.[41] It is also useful to note that $\mathrm{H}^2(\mathrm{SO}(3), \mathrm{U}(1)) = \mathbb{Z}_2$. The rotational symmetry, which is strictly larger than

[40] The reader with a rough idea about cohomology may recognize that $\psi(gh) - \psi(g) - \psi(h)$ in (8.3.25) is a coboundary, and the condition (8.3.26), which can be organized as $\phi(g_2, g_3) - \phi(g_1 g_2, g_3) + \phi(g_1, g_2 g_3) - \phi(g_1, g_2) = 0 \pmod{2\pi}$, is a cocycle condition. See, e.g., Appendices A, B of [70] for more about group cohomology.

[41] The reader might wonder why time-reversal symmetry or inversion symmetry, which are $\mathbb{Z}_2$, can protect the Haldane phase. These types of symmetry does not fall into the present scheme, since

$\mathbb{Z}_2 \times \mathbb{Z}_2$ symmetry, does not lead to richer phase structure. There can only be two phases, the Haldane phase and the trivial phase. In [70], an SU(3) extension of the AKLT model, which has $\mathbb{Z}_3 \times \mathbb{Z}_3$ symmetry, is studied. Spin chains protected by $\mathbb{Z}_3 \times \mathbb{Z}_3$ symmetry have richer phases since $\mathrm{H}^2(\mathbb{Z}_3 \times \mathbb{Z}_3, \mathrm{U}(1)) = \mathbb{Z}_3$.

Problem 8.3.4.a Suppose that four matrices $\mathsf{\rho}(e)$, $\mathsf{\rho}(a)$, $\mathsf{\rho}(b)$, and $\mathsf{\rho}(c)$ form a projective representation of the group $\mathbb{Z}_2 \times \mathbb{Z}_2 = \{e, a, b, c\}$, i.e., $\mathsf{\rho}(e) = \mathsf{I}$ and $\mathsf{\rho}(g)\mathsf{\rho}(h) = e^{i\phi(g,h)}\mathsf{\rho}(gh)$ for all $g, h \in \mathbb{Z}_2 \times \mathbb{Z}_2$ with some phase function $\phi(g, h) \in \mathbb{R}$. By using the multiplication rule (2.1.27) of $\mathbb{Z}_2 \times \mathbb{Z}_2$ and the cocycle condition (8.3.26), prove that the projective representation is either trivial or equivalent to that realized by the identity $\hat{1}$ and the π-rotation operators $\hat{u}_1$, $\hat{u}_2$, and $\hat{u}_3$ for a half-odd-integer spin as in (2.1.31). (Hint: It is convenient to introduce an equivalent projective representation $\tilde{\mathsf{\rho}}(\cdot)$ whose phase function satisfies $\tilde{\phi}(g, g) = 0$ for all $g \in \mathbb{Z}_2 \times \mathbb{Z}_2$.) [solution→p.511]

Symmetry transformations for $S = 1$ chain Let us study examples of $\mathbb{Z}_2 \times \mathbb{Z}_2$ symmetric states of the $S = 1$ chain. We denote by $\hat{S}^{(\alpha)}$ with $\alpha = 1, 2, 3$ the spin operators (3×3 matrices) for a single spin with $S = 1$. See (2.1.9). For some simple transformations, we shall explicitly write down the relation (8.3.13) which determines the transformed matrices.

First consider the unitary operator $\exp[-i\theta \hat{S}^{(3)}]$, the rotation by θ about the 3-axis. Its matrix form is readily obtained from (2.1.9) as $\exp[-i\theta \hat{S}^{(3)}] = \begin{pmatrix} e^{-i\theta} & 0 & 0 \\ 0 & 1 & 0 \\ 0 & 0 & e^{i\theta} \end{pmatrix}$.

From (8.3.13), we find that the matrices $\tilde{\mathsf{A}}^\sigma$ representing the transformed state $\hat{U}^{(3)}_\theta|\Phi\rangle$ are given by

$$\tilde{\mathsf{A}}^+ = e^{-i\theta}\mathsf{A}^+, \quad \tilde{\mathsf{A}}^0 = \mathsf{A}^0, \quad \tilde{\mathsf{A}}^- = e^{i\theta}\mathsf{A}^-. \tag{8.3.27}$$

We next concentrate on the π rotations, and use the simplified notation $\hat{u}_\alpha = \exp[-i\pi \hat{S}^{(\alpha)}]$ for $\alpha = 1, 2, 3$. Recall that we used the same notation in Sect. 2.1. Since $S = 1$ is an integer, we have $(\hat{u}_\alpha)^2 = \hat{1}$ and $[\hat{u}_\alpha, \hat{u}_\beta] = 0$. See (2.1.23) and (2.1.25). We also know that these operators satisfy the relations (2.1.29), and hence the operators $\hat{1}$, $\hat{u}_1$, $\hat{u}_2$, and $\hat{u}_3$ form a genuine representation (rather than a projective representation) of the group $\mathbb{Z}_2 \times \mathbb{Z}_2$.

Since we have already examined $\hat{u}_3$ (which is obtained by setting $\theta = \pi$ in the above), let us take a look at $\hat{u}_1$, whose matrix form is $\hat{u}_1 = \begin{pmatrix} 0 & 0 & -1 \\ 0 & -1 & 0 \\ -1 & 0 & 0 \end{pmatrix}$, as we have seen in (2.1.33). Again from (8.3.13), we see that the transformed matrices are

$$\tilde{\mathsf{A}}^+ = -\mathsf{A}^-, \quad \tilde{\mathsf{A}}^0 = -\mathsf{A}^0, \quad \tilde{\mathsf{A}}^- = -\mathsf{A}^+. \tag{8.3.28}$$

---

the former uses antiunitary operator in the representation, and the latter is not a local symmetry. We shall treat the case with time-reversal symmetry separately below.

Problem 8.3.4.b Find the relation corresponding to (8.3.27) or (8.3.28) for $\hat{u}_2$. [solution→p.512]

For a given injective matrix product state $|\Phi\rangle$, there exist matrices U_1, U_2, and U_3 that satisfy the basic relation (8.3.18) with $\hat{u}_1$, $\hat{u}_2$, and $\hat{u}_3$, respectively. Then the four matrices I, U_1, U_2, and U_3 form a projective representation of $\mathbb{Z}_2 \times \mathbb{Z}_2$.

Recall that projective representations of $\mathbb{Z}_2 \times \mathbb{Z}_2$ are classified into exactly two types, trivial ones and nontrivial ones (see Problem 8.3.4.a). We can then define the desired "topological" index $\sigma_{\mathrm{D}} = \pm 1$ by the relation

$$\mathsf{U}_\alpha \mathsf{U}_\beta = \sigma_{\mathrm{D}} \mathsf{U}_\beta \mathsf{U}_\alpha, \tag{8.3.29}$$

for any $\alpha, \beta = 1, 2, 3$ with $\alpha \neq \beta$. To see this one only needs to check that σ_{D} in (8.3.29) is invariant under the phase change $\mathsf{U}_\alpha \to e^{i\psi_\alpha}\mathsf{U}_\alpha$, and to recall that one has $\sigma_{\mathrm{D}} = -1$ in the nontrivial projective representation (2.1.31) realized by a single spin with $S = 1/2$.

$\mathbb{Z}_2 \times \mathbb{Z}_2$ symmetry: VBS state Let us take the $S = 1$ VBS state $|\Phi_{\mathrm{VBS}}\rangle$, which should belong to the nontrivial SPT phase. Recall that the VBS state is written in the matrix product form (8.3.11) with 2×2 matrices

$$\mathsf{A}^+ = \begin{pmatrix} 0 & 0 \\ -\frac{1}{\sqrt{2}} & 0 \end{pmatrix}, \quad \mathsf{A}^0 = \begin{pmatrix} \frac{1}{2} & 0 \\ 0 & -\frac{1}{2} \end{pmatrix}, \quad \mathsf{A}^- = \begin{pmatrix} 0 & \frac{1}{\sqrt{2}} \\ 0 & 0 \end{pmatrix}, \tag{8.3.30}$$

as we have seen in (7.2.12). We know that this matrix product state is injective. See footnote 31 in p. 203. Since the VBS state has SU(2) symmetry, it is of course invariant under any transformation generated by $\hat{u}_1$, $\hat{u}_2$, or $\hat{u}_3$.

As a warm up, let us examine $\exp[-i\theta\hat{S}^{(3)}]$, the θ rotation about the 3-axis. One can easily check that the matrices (8.3.27) can be written in the desired form (8.3.16) with the phase factor $e^{i\eta_L/L} = 1$ and the matrix $\mathsf{U}_3(\theta) = \begin{pmatrix} e^{-i\theta/2} & 0 \\ 0 & e^{i\theta/2} \end{pmatrix}$. It should be noted that this $\mathsf{U}_3(\theta)$ is nothing but the rotation operator $\exp[-i\theta\hat{S}^{(3)}]$, but for a spin with $S = 1/2$. By setting $\theta = \pi$, we get $\mathsf{U}_3 = \begin{pmatrix} -i & 0 \\ 0 & i \end{pmatrix}$.

We next examine $\hat{u}_1$, the π rotation about the 1-axix. One can again easily check that (8.3.28) is written in the form (8.3.16) with $e^{i\eta_L/L} = 1$ and $\mathsf{U}_1 = \begin{pmatrix} 0 & -i \\ -i & 0 \end{pmatrix}$. This matrix is again equal to $\exp[-i\pi\hat{S}^{(1)}]$ for an $S = 1/2$ spin. See (2.1.32).

One can similarly find the remaining matrix U_2 corresponding to $\hat{u}_2$. See Problem 8.3.4.c below. Then I, U_1, U_2, and U_3 form a projective representation of $\mathbb{Z}_2 \times \mathbb{Z}_2$. We readily find that this is a nontrivial projective representation by noting (from a direct computation) that $\mathsf{U}_1\mathsf{U}_3 = -\mathsf{U}_3\mathsf{U}_1$, which means $\sigma_{\mathrm{D}} = -1$. We conclude that the $S = 1$ VBS state is accompanied by a nontrivial projective representation of the group $\mathbb{Z}_2 \times \mathbb{Z}_2$, as we have expected.

Needless to say the anticommutativity $\mathsf{U}_1\mathsf{U}_3 = -\mathsf{U}_3\mathsf{U}_1$ is nothing but the relation (2.1.25) for a single $S = 1/2$. The above observation is consistent with the picture that (effective) states on the half-infinite chains have $S = 1/2$. See Sect. 8.3.3, in particular, Fig. 8.13.

Problem 8.3.4.c Find the matrix U_2 which corresponds to $\hat{u}_2$. [solution→p.512]

$\mathbb{Z}_2 \times \mathbb{Z}_2$ symmetry: MPS for the large-D phase It is useful to see an example where we encounter a trivial projective representation of $\mathbb{Z}_2 \times \mathbb{Z}_2$. The simplest (and indeed important) example is the trivial product state $\bigotimes_{x=1}^{L} |\psi_x^0\rangle$, which is the ground state of the trivial anisotropic Hamiltonian (8.1.2). The state is obviously written in an MPS form with 1×1 matrices $\mathsf{A}^{\pm} = (0)$ and $\mathsf{A}^0 = (1)$. We then see that (8.3.27), (8.3.28), and (S.88) are written in the form (8.3.16) with $\mathsf{U}_1 = \mathsf{U}_2 = \mathsf{U}_3 = \mathsf{I}$ (and suitable η_L), and hence the index is $\sigma_{\mathrm{D}} = 1$.

Let us take a look at a slightly less trivial state. Again take the $S = 1$ spin chain and consider the matrix product state (8.3.11) defined by matrices

$$\mathsf{A}^+ = \begin{pmatrix} 0 & 1 & 0 \\ 0 & 0 & a \\ 0 & 0 & 0 \end{pmatrix}, \quad \mathsf{A}^0 = \begin{pmatrix} \sqrt{2}\,a & 0 & 0 \\ 0 & 1 & 0 \\ 0 & 0 & \sqrt{2}\,a \end{pmatrix}, \quad \mathsf{A}^- = \begin{pmatrix} 0 & 0 & 0 \\ a & 0 & 0 \\ 0 & 1 & 0 \end{pmatrix}, \tag{8.3.31}$$

where the parameters $a \in \mathbb{R}$ can be anything, but we assume $0 < a \ll 1$. The set of matrices is designed to be injective.[42] The resulting matrix product state mimics the ground state of the anisotropic Hamiltonian (8.1.1) with $D \gg 1$. The reader is invited to examine the definition and confirm that the state describes a gas of pairs of $+$ and $-$, as we have seen in Sect. 8.1.2.[43] See, in particular, Fig. 8.3a.

Consider again the rotation $\exp[-i\theta\hat{S}^{(3)}]$, under which the state $|\Phi\rangle$ is invariant. The matrices resulting from the transformed state $\hat{U}_\theta^{(3)}|\Phi\rangle$ are given by (8.3.27). In this case these matrices are written in the form (8.3.16) with $e^{i\eta_L/L} = 1$ and $\mathsf{U}_3(\theta) = \begin{pmatrix} e^{i\theta} & 0 & 0 \\ 0 & 1 & 0 \\ 0 & 0 & e^{-i\theta} \end{pmatrix}$. By setting $\theta = \pi$, we have $\mathsf{U}_3 = \begin{pmatrix} -1 & 0 & 0 \\ 0 & 1 & 0 \\ 0 & 0 & -1 \end{pmatrix}$. Similarly we see that (8.3.28) is written in the form (8.3.16) with $e^{i\eta_L/L} = -1$ and $\mathsf{U}_1 = \begin{pmatrix} 0 & 0 & -1 \\ 0 & -1 & 0 \\ -1 & 0 & 0 \end{pmatrix}$. We also find by examining (S.88) that $\mathsf{U}_2 = \begin{pmatrix} 0 & 0 & 1 \\ 0 & -1 & 0 \\ 1 & 0 & 0 \end{pmatrix}$ with the phase factor $e^{i\eta_L/L} = -1$. Note that these U_1, U_2, and U_3 are the same as the matrices (2.1.33) for the π-rotations of a single $S = 1$. Obviously we have a genuine

[42]This can be checked by verifying the condition (iii) of Theorem 7.5 (p. 202) with the aid of a computer.

[43]Suppose one starts from the vector $(0, 1, 0)^{\mathrm{t}}$. Multiplication by an arbitrary number of A^0 leaves the vector as it is, representing a sequence of 0's. If we multiply this by A^+, which corresponds to insertion of a single $+$, the vector becomes $(1, 0, 0)^{\mathrm{t}}$. Multiplying n times by A^0, which means a sequence of 0's, the vector becomes $((\sqrt{2}\,a)^n, 0, 0)^{\mathrm{t}}$. Now by multiplying this by A^-, which is insertion of a single $-$, the vector returns to $(0, 2^{n/2}a^{n+1}, 0)^{\mathrm{t}}$. One can again multiply A^0 many times without changing the vector. Note that the weight of the configuration, $2^{n/2}a^{n+1}$, decreases rapidly as n increases, indicating that $+$ and $-$ form a pair and stay close to each other.

representation (or a trivial projective representation) of $\mathbb{Z}_2 \times \mathbb{Z}_2$, which means that the state is characterized by the index $\sigma_{\mathrm{D}} = 1$.

Time-reversal symmetry The Haldane phase may be protected either by $\mathbb{Z}_2 \times \mathbb{Z}_2$, time-reversal, or bond-centered inversion symmetry. The treatment of the latter two is slightly different from that of local unitary symmetry discussed above. We here discuss the case of time-reversal symmetry in detail, and leave the case of inversion symmetry to the original work of Pollmann, Turner, Berg, and Oshikawa [81].

For simplicity we only consider an injective matrix product state $|\Phi\rangle$ with $S = 1$, keeping in mind the VBS state with matrices (8.3.30) and the large-D state with (8.3.31). Let $\hat{\Theta}$ be the time-reversal map defined in (2.3.16). From the matrix product form (8.3.11) for $|\Phi\rangle$, we find that its time-reversal is written as

$$\begin{aligned}\hat{\Theta}|\Phi\rangle &= \sum_{\sigma} \hat{\Theta}\mathrm{Tr}[\mathsf{A}^{\sigma_1}\mathsf{A}^{\sigma_2} \cdots \mathsf{A}^{\sigma_L}]\,|\Psi^{\sigma}\rangle \\ &= \sum_{\sigma}\Big\{\prod_{x\in\Lambda}(-1)^{1+\sigma_x}\Big\}\Big(\mathrm{Tr}[\mathsf{A}^{-\sigma_1}\mathsf{A}^{-\sigma_2} \cdots \mathsf{A}^{-\sigma_L}]\Big)^*\,|\Psi^{\sigma}\rangle \\ &= \sum_{\sigma}\mathrm{Tr}[\tilde{\mathsf{A}}^{\sigma_1}\tilde{\mathsf{A}}^{\sigma_2} \cdots \tilde{\mathsf{A}}^{\sigma_L}]\,|\Psi^{\sigma}\rangle,\end{aligned} \tag{8.3.32}$$

with

$$\tilde{\mathsf{A}}^+ = (\mathsf{A}^-)^*, \quad \tilde{\mathsf{A}}^0 = -(\mathsf{A}^0)^*, \quad \tilde{\mathsf{A}}^- = (\mathsf{A}^+)^*, \tag{8.3.33}$$

where A^* denotes the matrix obtained by taking the complex conjugate of all the entries in A. It is not the Hermitian conjugate. One can again easily verify that $(\tilde{\mathsf{A}}^+, \tilde{\mathsf{A}}^0, \tilde{\mathsf{A}}^-)$ is injective when $(\mathsf{A}^+, \mathsf{A}^0, \mathsf{A}^-)$ is injective.

Let us now assume that $|\Phi\rangle$ has time-reversal symmetry, i.e., $\hat{\Theta}|\Phi\rangle = \pm|\Phi\rangle$ for any L, where the sign may depend on L. As before we find that a necessary and sufficient condition for the invariance is that the matrices $\tilde{\mathsf{A}}^{\sigma}$ defined in (8.3.33) is written as

$$\tilde{\mathsf{A}}^{\sigma} = e^{i\zeta}\,\mathsf{U}^{\dagger}\mathsf{A}^{\sigma}\mathsf{U}, \tag{8.3.34}$$

for any $\sigma = 0, \pm 1$, with a constant ζ and a unitary matrix U. An essential observation by Pollmann, Turner, Berg, and Oshikawa [81] is that the above unitary matrix satisfies $\mathsf{U}\mathsf{U}^* = \sigma_{\mathrm{tr}}\mathsf{I}$, where $\sigma_{\mathrm{tr}} = \pm 1$ is the "topological" index.

To see this we combine (8.3.33) and (8.3.34) to find

$$(-1)^{1+\sigma}(\mathsf{A}^{-\sigma})^* = e^{i\zeta}\mathsf{U}^{\dagger}\mathsf{A}^{\sigma}\mathsf{U}, \tag{8.3.35}$$

for any $\sigma = 0, \pm 1$. By taking the complex conjugate and letting $\sigma \to -\sigma$, we get

$$\mathsf{A}^{\sigma} = (-1)^{1+\sigma}e^{-i\zeta}(\mathsf{U}^{\dagger})^*(\mathsf{A}^{-\sigma})^*\mathsf{U}^*. \tag{8.3.36}$$

As in Sect. 2.3, we denote by K the complex conjugation map (which is not a matrix), i.e., $\mathsf{K}(v_1, \ldots, v_D)^{\mathrm{t}} = (v_1^*, \ldots, v_D^*)^{\mathrm{t}}$. By introducing the antiunitary operators

$\mathsf{V} = \mathsf{UK}$ and $\mathsf{V}^\dagger = \mathsf{KU}^\dagger$ (which satisfy $\mathsf{VV}^\dagger = \mathsf{V}^\dagger\mathsf{V} = \mathsf{I}$), (8.3.36) is written as[44]

$$\mathsf{A}^\sigma = (-1)^{1+\sigma} e^{-i\zeta}\, \mathsf{V}^\dagger \mathsf{A}^{-\sigma}\mathsf{V}. \tag{8.3.37}$$

See Appendix A.4.3 for the treatment of antiunitary operators. By substituting this relation (with σ replaced by $-\sigma$) to itself, we find

$$\begin{aligned}\mathsf{A}^\sigma &= (-1)^{1+\sigma} e^{-i\zeta}\, \mathsf{V}^\dagger\{(-1)^{1-\sigma} e^{-i\zeta}\, \mathsf{V}^\dagger \mathsf{A}^\sigma \mathsf{V}\}\,\mathsf{V}\\ &= (\mathsf{V}^\dagger)^2 \mathsf{A}^\sigma\, \mathsf{V}^2,\end{aligned} \tag{8.3.38}$$

where $\mathsf{V}^2 = \mathsf{UU}^*$ is a unitary matrix. Note that (8.3.38) can be written as $[\mathsf{V}^2, \mathsf{A}^\sigma] = 0$ for any σ. As before (see the discussion below (8.3.23)) this implies that $\mathsf{V}^2 = \mu\mathsf{I}$ for some constant $\mu \in \mathbb{C}$. To see the allowed values for μ, we note that $\mathsf{V}^2 = \mathsf{UU}^* = \mu\mathsf{I}$ is rewritten as $\mathsf{U} = \mu\mathsf{U}^{\mathrm{t}}$. By using the final relation twice, we find $\mu^2 = 1$, which means $\mu = \pm 1$.[45] We identify μ with the index σ_{tr}. Note that the characterization $\mathsf{V}^2 = \sigma_{\mathrm{tr}}\mathsf{I}$ of the index is consistent with the heuristic discussion in Sect. 8.3.3.

We now examine the two examples. For the VBS state with matrices (8.3.30), one readily finds that (8.3.34) with $\zeta = 0$ and $\mathsf{U} = \begin{pmatrix} 0 & 1 \\ -1 & 0 \end{pmatrix}$ reproduces (8.3.33). Since $\mathsf{U}^2 = -\mathsf{I}$, we see that the index takes the nontrivial value $\sigma_{\mathrm{tr}} = -1$. For the large-$D$ state with (8.3.31), we see that $\zeta = \pi$ and $\mathsf{U} = \begin{pmatrix} 0 & 0 & 1 \\ 0 & -1 & 0 \\ 1 & 0 & 0 \end{pmatrix}$ do the same. Since $\mathsf{U}^2 = \mathsf{I}$, we find $\sigma_{\mathrm{tr}} = 1$.

When an injective matrix product state has time-reversal symmetry, the "topological" index σ_{tr} is restricted to either 1 or -1, and is not allowed to vary continuously. We thus conclude that there must be a phase transition, i.e., a jump in the index, between the VBS state and the large-D state. In this case too the appearance of nontrivial index in the VBS state is a direct manifestation that the effective states on the half-infinite chains have $S = 1/2$. See Sect. 8.3.3, in particular, Fig. 8.13.

Problem 8.3.4.d Define the twisted-VBS state $|\Phi_{\text{t-VBS}}\rangle$ as a matrix product state with

$$\mathsf{A}^{\pm}_{\text{t-VBS}} = \mathsf{A}^{\pm}_{\text{VBS}}, \quad \mathsf{A}^{0}_{\text{t-VBS}} = i\, \mathsf{A}^{0}_{\text{VBS}}, \tag{8.3.39}$$

where the matrices $\mathsf{A}^\sigma_{\text{VBS}}$ are those in (8.3.30). Show that the twisted-VBS state is in the nontrivial SPT phase with respect to $\mathbb{Z}_2 \times \mathbb{Z}_2$ symmetry, but is in the trivial phase with respect to time-reversal symmetry. This means that in the class of spin chains with both $\mathbb{Z}_2 \times \mathbb{Z}_2$ symmetry and time-reversal symmetry, the VBS state and the twisted VBS state belong to distinct nontrivial phases. These are indeed two of the four nontrivial phases located by Liu, Liu, and Wen [61]. The above consideration

---

[44] This is verified by checking that $(\mathsf{U}^\dagger)^*(\mathsf{A}^{-\sigma})^*\mathsf{U}^*\boldsymbol{v} = \mathsf{KU}^\dagger\mathsf{A}^{-\sigma}\mathsf{UK}\boldsymbol{v} = \mathsf{V}^\dagger\mathsf{A}^{-\sigma}\mathsf{V}\boldsymbol{v}$ for any vector $\boldsymbol{v}$.

[45] One can derive the same result by only using antiunitary operators. Note that $\mathsf{V}^2 = \mu\mathsf{I}$ implies $\mathsf{V} = \mu\mathsf{V}^\dagger$ and hence $\mathsf{V}^\dagger = \mu\mathsf{V}$. The two relations imply $\mathsf{V} = \mu^2\mathsf{V}$, which shows $\mu^2 = 1$. We have noted that $(\alpha\mathsf{X})\dagger = \mathsf{X}^\dagger\alpha^* = \alpha\mathsf{X}^\dagger$ for any antilinear operator X and $\alpha \in \mathbb{C}$. See Appendix A.4.3.

shows that, as long as time-reversal invariance is kept (and we work within matrix product states), the VBS and the twisted VBS states cannot be continuously connected with each other. Show, on the other hand, that the two states can be continuously connected with each other when only $\mathbb{Z}_2 \times \mathbb{Z}_2$ symmetry is kept. [solution→p.512]

SPT phases in two or higher dimensions Similar characterization and classification of symmetry protected topological phases in two or higher dimensions is still a widely open problem, especially from the point of view of mathematical physics. So far the most complete picture is the one given by Chen, Gu, Liu, and Wen [18, 19], who argued that SPT phases in d-dimensional quantum spin systems are classified in terms of the group cohomology $\mathrm{H}^{d+1}(G, \mathrm{U}(1))$ of the symmetry group G.[46] See also [24, 66, 68, 83, 98, 104]. We note however that the classification is less complete compared with the corresponding result in one dimension by Pollmann, Turner, Berg, and Oshikawa. This is because there is no complete counterpart in higher dimensions of the advanced theory of matrix product states (including the approximation theorem and the uniqueness theorem).[47] (But see [68] for some results in this direction.)

In the above classification, the VBS states in two or higher dimensions discussed in Sect. 7.3.2 are not regarded as states in proper SPT phases. They are regarded to be in a "weak" SPT phases which need to be protected not only by the internal symmetry but also by translation invariance. See Sect. 10.4.1 of [105]. The same is true for the Briegel–Raussendorf state (or the cluster state) (7.3.29) in two or higher-dimensions, as was shown in [66]. See Sect. 7.3.3 for the Briegel–Raussendorf state.

The first example of a model in a genuine two-dimensional SPT phase is the CZX model constructed by Chen, Liu, and Wen [16, 105]. A simpler (and an elegant) example was then constructed by Levin and Gu [59]. Miller and Miyake [66] and Yoshida [104] showed that the generalized Briegel–Raussendorf states (or the hyper graph states) on two-dimensional lattices (see Sect. 7.3.3) provide examples of states in genuine two-dimensional SPT phases. (See also Sect. II.F of [19] where a closely related class of states is introduced earlier.) The simplest example is the generalized Briegel–Raussendorf state (or the hyper graph state) on the triangular lattice that we treated in Problem 7.3.3.c (p. 220). It is known that this state is related to the Levin-Gu state by a simple transformation. See Sect. 4.3.1 of [98].

See, e.g., [20, 43, 68] for other related approaches to SPT phases in two or higher dimensions. For an interesting recent attempt at characterizing a (weak) two-dimensional SPT phases from the viewpoint of computational power, see [21, 84].

---

[46]The theory is related to the earlier work on topological gauge theories by Dijkgraaf and Witten [23].

[47]Even on heuristic level, it is believed that the classification of SPT phases in terms of the group cohomology is incomplete, and one needs to invoke the notion of cobordism [45].

8.3.5 *Lieb–Schultz–Mattis Type Theorem Without Continuous Symmetry*

In this short section, we use the technique developed in the previous section to prove an interesting theorem, i.e., Theorem 8.7 below, which shows that the projective representation of on-site symmetry must be trivial in order for an injective matrix product state to exist. The topic is not directly related to SPT phases.

Illuminating Corollary The most important consequence of Theorem 8.7 is the following no-go theorem.

Corollary 8.5 *In a quantum spin chain with half-odd-integer S, there can be no translation invariant injective matrix product states which are $\mathbb{Z}_2 \times \mathbb{Z}_2$ invariant or time-reversally invariant.*

This statement was first presented by Chen, Gu, and Wen [17] as a part of their general classification.[48] Watanabe, Po, Vishwanath, and Zaletel [97] later showed a derivation for the case with time-reversal symmetry.[49]

As we shall see below, the Proof of Corollary 8.5 is based on the observation that the existence of an injective matrix product state is inconsistent with nontriviality of the projective representation of the on-site symmetry. As far as we know such an argument was first discovered back in 2001 by Matsui [63], who treated not only matrix product states but more general pure states with split property of systems with continuous symmetry. Recall that the original Lieb–Schultz–Mattis theorem (see Sect. 6.2) is proved by explicitly constructing low lying variational states orthogonal to the ground state. It is interesting that the two essentially different arguments lead to similar no-go theorems.

No-go theorem for a general unique gapped ground state Since a matrix product state is injective when it has exponentially decaying correlations and is not a superposition of macroscopically distinct states, the corollary suggests that the following Lieb–Schultz–Mattis-type theorem is valid. Unlike the original Lieb–Schultz–Mattis theorem (see Sect. 6.2), which makes an essential use of the U(1) symmetry, the following statement assumes only discrete symmetry.

Theorem 8.6 *Consider a quantum spin chain with half-odd-integer spin and short-ranged translation invariant Hamiltonian. We assume that the model has either $\mathbb{Z}_2 \times \mathbb{Z}_2$ symmetry or time-reversal symmetry. Then it can never be the case that the corresponding ground state is unique and accompanied by a nonzero gap.*

The "proof" of the theorem, assuming Corollary 8.5, goes as follows. Suppose that the model has a unique gapped ground state. A unique ground state must have the

---

[48]The case with $\mathbb{Z}_2 \times \mathbb{Z}_2$ symmetry follows from the general statement in Sect. V.B.4, and the case with time-reversal symmetry is treated in Sect. V.C.

[49]Not all of the results in [97] are mathematically rigorous. But we expect that the argument in Supporting Information of [97] (that includes [S1–9]) can be made rigorous by using standard techniques in matrix product states, provided that one only considers quantum spin chains.

same symmetry as the Hamiltonian, and hence should be translationally invariant and also $\mathbb{Z}_2 \times \mathbb{Z}_2$ or time-reversally invariant. Since it is known that matrix product states can approximate any translationally invariant states with short range correlation, the ground state should be approximated by a suitable matrix product state with the same symmetry. But that contradicts Corollary 8.5.

Although the above argument is plausible, there are still some essential mathematical gaps in it, which are not easy to fill. In short one must have a very good control on the approximation by (injective) matrix product states in order to justify the claim, but the currently available results [1, 27] are not enough for the purpose.

But recently Ogata and Tasaki [77] provided a fully rigorous Proof of Theorem 8.6, which does not make use of matrix product approximations. The proof instead makes use of a deep analogy between the Cuntz algebra associated with a pure state with the split property and the matrix product formulation, and is based on closely related results by Matsui [62, 63]. See also the end of Sect. 7.2.2.

Setting and the main theorem We now describe and prove the main theorem of the section.

Let G be a finite group that describes the symmetry of the model. We assume that there is a representation $s : G \to \{1, -1\}$, i.e., a function $s(g) = \pm 1$ such that $s(g)s(h) = s(gh)$ for any $g, h \in G$. (Then one inevitably has $s(e) = 1$.) We write

$$C_g[x] = \begin{cases} x & \text{if } s(g) = 1, \\ x^* & \text{if } s(g) = -1, \end{cases} \tag{8.3.40}$$

where x may be a complex number or a matrix. (As before we denote by A^* the matrix obtained by taking the complex conjugate of all the entries of a matrix A.) For each $g \in G$, take a unitary operator $\hat{u}(g)$ on the single-spin Hilbert space $\mathfrak{h}_0 \cong \mathbb{C}^{2S+1}$. We set $\hat{u}(e) = \hat{1}$. We then define

$$\hat{v}(g) = \begin{cases} \hat{u}(g) & \text{if } s(g) = 1, \\ \hat{u}(g)\hat{K} & \text{if } s(g) = -1, \end{cases} \tag{8.3.41}$$

where $\hat{K}$ is the complex conjugation map introduced in Sect. 2.3. See also Appendix A.4.3. Note that $\hat{v}(g)$ is antiunitary if $s(g) = -1$. We assume that $\hat{v}(\cdot)$ forms a projective representation of G, i.e.,

$$\hat{v}(g)\hat{v}(h) = e^{i\tilde{\phi}(g,h)}\hat{v}(gh), \tag{8.3.42}$$

for any $g, h \in G$ with some $\tilde{\phi}(g, h) \in \mathbb{R}$. Note that, unlike in Sect. 8.3.4, where we considered projective representations acting on matrices for MPS, we are here considering the projective representation on the single-spin Hilbert space. The phase factor $\tilde{\phi}(g, h)$ must satisfy the cocycle condition as in (8.3.26) (with a suitable modification from the antiunitarity), but we do not use it here.

Suppose that, as a special case of (8.3.41), we have a genuine representation $\hat{v}_0(\cdot)$ of G, i.e., $\hat{v}_0(e) = \hat{1}$ and $\hat{v}_0(g)\hat{v}_0(h) = \hat{v}_0(gh)$ for any $g, h \in G$. As before we say that a projective representation $\hat{v}(\cdot)$ is trivial if there is a function $\tilde{\psi}(g) \in \mathbb{R}$ such that $\hat{v}(g) = e^{i\tilde{\psi}(g)}\hat{v}_0(g)$ for any $g \in G$. Suppose that $\hat{v}(\cdot)$ is trivial. Noting that $\hat{v}(g)\hat{v}(h) = e^{i\{\tilde{\psi}(g)+s(g)\tilde{\psi}(h)\}}\hat{v}_0(g)\hat{v}_0(h)$, we find from (8.3.42) that

$$\tilde{\phi}(g, h) = \tilde{\psi}(g) + s(g)\tilde{\psi}(h) - \tilde{\psi}(gh) \ (\mathrm{mod}\ 2\pi). \tag{8.3.43}$$

Conversely, we readily find that any projective representation whose phase function $\tilde{\phi}(\cdot, \cdot)$ satisfies (8.3.43) is trivial.

We now consider a spin system with spin S on the chain $\{1, \ldots, L\}$, and define for each $g \in G$

$$\hat{V}(g) = \begin{cases} \bigotimes_{x=1}^{L} \hat{u}_x(g) & \text{if } s(g) = 1, \\ \left(\bigotimes_{x=1}^{L} \hat{u}_x(g)\right)\hat{K} & \text{if } s(g) = -1, \end{cases} \tag{8.3.44}$$

where $\hat{u}_x(g)$ is the copy of $\hat{u}(g)$ on $\mathfrak{h}_x$. Of course $\hat{V}(\cdot)$ gives a projective representation of G on the Hilbert space of the spin chain.

We are now ready to state the main theorem of the present section. As we mentioned already, very close statements are found in [17, 97]. The theorem was stated in the following concise form and proved by Yuji Tachikawa (private communication).[50]

Theorem 8.7 *Let* $|\Phi\rangle$ *be an injective matrix product state* (8.3.11). *If it holds with some* $\eta_L(g) \in \mathbb{R}$ *that*

$$\hat{V}(g)|\Phi\rangle = e^{i\eta_L(g)}|\Phi\rangle, \tag{8.3.45}$$

for any $g \in G$ *and any* L, *then the projective representation* $\hat{v}(\cdot)$ *is trivial.*

Let us see how Corollary 8.5 is obtained from the theorem. To treat the case with $\mathbb{Z}_2 \times \mathbb{Z}_2$ symmetry, we set $G = \mathbb{Z}_2 \times \mathbb{Z}_2$, and take the standard unitary projective representation given by $\hat{1}$, $\hat{u}_1$, $\hat{u}_2$, and $\hat{u}_3$ as in (2.1.29). As we have seen in (2.1.31) this projective representation is inevitably nontrivial if S is a half-odd-integer. This means that there can be no $\mathbb{Z}_2 \times \mathbb{Z}_2$ invariant injective MPS. To treat time reversal symmetry, we set $G = \mathbb{Z}_2 = \{e, a\}$ and take the projective representation $\hat{v}(e) = \hat{1}$ and $\hat{v}(a) = \hat{\Theta} = \hat{u}_2\hat{K}$. Note that we have set $s(e) = 1$ and $s(a) = -1$. As we have seen just below (2.3.9), it holds that $(\hat{v}(a))^2 = \hat{\Theta}^2 = -\hat{1}$ if S is a half-odd-integer. This again shows that the projective representation is nontrivial.

Proof of Theorem 8.7 Exactly as in (8.3.12) or (8.3.32), we write

$$\hat{V}(g)|\Phi\rangle = \sum_{\sigma} \mathrm{Tr}[\tilde{\mathsf{A}}_g^{\sigma_1}\tilde{\mathsf{A}}_g^{\sigma_2} \cdots \tilde{\mathsf{A}}_g^{\sigma_L}]\,|\Psi^{\sigma}\rangle, \tag{8.3.46}$$

---

[50] There is a generalization of Theorem 8.6 which is parallel to Theorem 8.7 (Yuji Tachikawa, private communication). The generalized theorem can be proved by combining the idea in the following proof and the technique in [77].

with

$$\tilde{\mathsf{A}}_g^\sigma = \sum_{\sigma'=-S}^{S} \langle\psi^\sigma|\hat{u}(g)|\psi^{\sigma'}\rangle\, C_g[\mathsf{A}^{\sigma'}]. \tag{8.3.47}$$

One can show for each $g \in G$ that the collection of matrices $(\tilde{\mathsf{A}}_g^\sigma)_{\sigma=-S,\dots,S}$ is injective. As in (8.3.16), the assumed invariance (8.3.45) and the uniqueness theorem (Theorem 7.6 in p. 203) imply that, for each $g \in G$, there are a constant $\zeta(g) \in \mathbb{R}$ and a unitary matrix $\mathsf{U}(g)$ such that

$$\tilde{\mathsf{A}}_g^\sigma = e^{i\zeta(g)}\, \mathsf{U}^\dagger(g)\mathsf{A}^\sigma\mathsf{U}(g), \tag{8.3.48}$$

for all $\sigma = -S, \dots, S$. With (8.3.48), this implies

$$\mathsf{A}^\sigma = C_g\Big[e^{i\zeta(g)} \sum_{\sigma'=-S}^{S} \langle\psi^\sigma|\hat{u}^\dagger(g)|\psi^{\sigma'}\rangle\mathsf{U}^\dagger(g)\mathsf{A}^{\sigma'}\mathsf{U}(g)\Big]. \tag{8.3.49}$$

For g with $s(g) = -1$, let us set $\hat{v}^\dagger(g) = \hat{K}\hat{u}^\dagger(g)$, $\mathsf{V}(g) = \mathsf{U}(g)\mathsf{K}$, and $\mathsf{V}^\dagger(g) = \mathsf{K}\mathsf{U}^\dagger(g)$. (See Appendix A.4.3 for the convention of conjugate of antilinear operators.) For g with $s(g) = 1$, where we have $\hat{v}^\dagger(g) = \hat{u}^\dagger(g)$ by definition, we set $\mathsf{V}(g) = \mathsf{U}(g)$ (and of course $\mathsf{V}^\dagger(g) = \mathsf{U}^\dagger(g)$). Then (8.3.49) is written as

$$\mathsf{A}^\sigma = e^{is(g)\zeta(g)} \sum_{\sigma'=-S}^{S} \langle\psi^\sigma|\hat{v}^\dagger(g)|\psi^{\sigma'}\rangle\mathsf{V}^\dagger(g)\mathsf{A}^{\sigma'}\mathsf{V}(g), \tag{8.3.50}$$

which is the basic relation for the proof.[51]

We now substitute (8.3.50) into the same relation with g replaced by h to get

$$\begin{aligned}
\mathsf{A}^\sigma &= e^{is(h)\zeta(h)} \sum_{\sigma'=-S}^{S} \langle\psi^\sigma|\hat{v}^\dagger(h)|\psi^{\sigma'}\rangle\mathsf{V}^\dagger(h) \\
&\qquad \Big\{e^{is(g)\zeta(g)} \sum_{\sigma''=-S}^{S} \langle\psi^{\sigma'}|\hat{v}^\dagger(g)|\psi^{\sigma''}\rangle\mathsf{V}^\dagger(g)\mathsf{A}^{\sigma''}\mathsf{V}(g)\Big\}\mathsf{V}(h) \\
&= e^{i\{s(h)\zeta(h)+s(h)s(g)\zeta(g)\}} \sum_{\sigma',\sigma''=-S}^{S} \langle\psi^\sigma|\hat{v}^\dagger(h)|\psi^{\sigma'}\rangle\, C_h\big[\langle\psi^{\sigma'}|\hat{v}^\dagger(g)|\psi^{\sigma''}\rangle\big] \\
&\qquad \mathsf{V}^\dagger(h)\mathsf{V}^\dagger(g)\mathsf{A}^{\sigma''}\mathsf{V}(g)\mathsf{V}(h) \\
&= e^{i\{s(h)\zeta(h)+s(h)s(g)\zeta(g)\}} \sum_{\sigma'=-S}^{S} \langle\psi^\sigma|\hat{v}^\dagger(h)\hat{v}^\dagger(g)|\psi^{\sigma'}\rangle\, \mathsf{V}^\dagger(h)\mathsf{V}^\dagger(g)\mathsf{A}^{\sigma'}\mathsf{V}(g)\mathsf{V}(h)
\end{aligned}$$

[51] Here $\hat{K}$ contained in $\hat{v}^\dagger(g)$ acts only on the single-spin Hilbert space. In other words $\langle\psi^\sigma|\hat{v}^\dagger(g)|\psi^{\sigma'}\rangle$ is a mere complex number.

$$= e^{i\{s(h)\zeta(h)+s(h)s(g)\zeta(g)-s(gh)\tilde{\phi}(g,h)\}} \sum_{\sigma'=-S}^{S} \langle\psi^{\sigma}|\hat{v}^{\dagger}(gh)|\psi^{\sigma'}\rangle \, \mathsf{V}^{\dagger}(h)\mathsf{V}^{\dagger}(g)\mathsf{A}^{\sigma'}\mathsf{V}(g)\mathsf{V}(h). \tag{8.3.51}$$

To get the final expression, we noted that (8.3.42) implies

$$\hat{v}^{\dagger}(h)\hat{v}^{\dagger}(g) = \hat{v}^{\dagger}(gh)\, e^{-i\tilde{\phi}(g,h)} = e^{-i\, s(g,h)\tilde{\phi}(g,h)}\, \hat{v}^{\dagger}(gh). \tag{8.3.52}$$

The same relation (8.3.50) with g replaced by gh reads

$$\mathsf{A}^{\sigma} = e^{is(gh)\zeta(gh)} \sum_{\sigma'=-S}^{S} \langle\psi^{\sigma}|\hat{v}^{\dagger}(gh)|\psi^{\sigma'}\rangle \mathsf{V}^{\dagger}(gh)\mathsf{A}^{\sigma'}\mathsf{V}(gh). \tag{8.3.53}$$

Comparing (8.3.51) and (8.3.53), we see, for each pair $g, h \in G$, that $\mathsf{V}^{\dagger}(h)\mathsf{V}^{\dagger}(g)\mathsf{A}^{\sigma}\mathsf{V}(g)\mathsf{V}(h) = c\,\mathsf{V}^{\dagger}(gh)\mathsf{A}^{\sigma}\mathsf{V}(gh)$ with a constant c such that $|c| = 1$ for all σ. But the injectivity implies that $c = 1$.[52] Then comparing the phase factors of (8.3.51) and (8.3.53) we see that

$$s(h)\zeta(h) + s(h)s(g)\zeta(g) - s(gh)\tilde{\phi}(g,h) = s(gh)\zeta(gh) \pmod{2\pi}, \tag{8.3.54}$$

which reduces to the condition (8.3.43) with $\tilde{\psi}(g) = \zeta(g)$. We thus see that the projective representation $\hat{v}(\cdot)$ is trivial.■

8.3.6 *Rigorous Index Theorems for SPT Phases*

All the discussions on symmetry protected topological phases so far have been either not rigorous or restricted to matrix product states (including the VBS state). Very recently (that was indeed during the preparation of the present book) we have seen essential progress of fully rigorous theories which directly deals with SPT phases in quantum spin chains [75, 76, 94]. In particular Ogata's sophisticated index theorems based on operator algebraic formalism provide complete characterization of SPT phases in quantum spin chains [75, 76].[53]

Here we start by discussing a prototypical problem of a topological phase transition in quantum spin chains in order to illustrate Tasaki's (elementary) rigorous

---

[52] We have $\mathsf{W}^{\dagger}\mathsf{A}^{\sigma}\mathsf{W} = c\mathsf{A}^{\sigma}$ for all σ where $\mathsf{W} = \mathsf{V}(g)\mathsf{V}(h)\mathsf{V}^{\dagger}(gh)$ is a unitary matrix. We then find $\mathsf{W}^{\dagger}\mathsf{A}^{\sigma_1}\cdots\mathsf{A}^{\sigma_{\ell_0}}\mathsf{W} = c^{\ell_0}\mathsf{A}^{\sigma_1}\cdots\mathsf{A}^{\sigma_{\ell_0}}$ for all $\sigma_1, \ldots, \sigma_{\ell_0}$. Then the property (i) in Theorem 7.5 (p. 202) implies $\mathsf{W}^{\dagger}\mathsf{M}\mathsf{W} = c^{\ell_0}\mathsf{M}$ for any $D \times D$ matrix M. This is possible only when $\mathsf{W} = t\,\mathsf{I}$ with $|t| = 1$. Then $\mathsf{W}^{\dagger}\mathsf{A}^{\sigma}\mathsf{W} = c\mathsf{A}^{\sigma}$ implies $c = 1$.

[53] There have been very few examples of well-defined indices for infinite volume states of fully interacting quantum many-body systems. Another example is the excess spin operator defined in [5]. But it has not yet been used to analyze phase transitions.

result and the limitation of previous discussions. We then attempt at reviewing the essence of Ogata's index theorems without assuming the knowledge of the operator algebraic approach to quantum spin systems.

Prototypical Problem To examine our understanding, let us focus on a prototypical problem of a topological phase transition between a nontrivial SPT phase and a trivial one. Let $s \in [0, 1]$ be a parameter, and consider an $S = 1$ chain with the Hamiltonian

$$\hat{H}_s = s \sum_{x=1}^{L} \{\hat{\boldsymbol{S}}_x \cdot \hat{\boldsymbol{S}}_{x+1} + \frac{1}{3}(\hat{\boldsymbol{S}}_x \cdot \hat{\boldsymbol{S}}_{x+1})^2\} + (1-s) \sum_{x=1}^{L} (\hat{S}_x^{(3)})^2, \tag{8.3.55}$$

which linearly interpolates the trivial Hamiltonian (8.1.2) with the ground state $|\Phi_0\rangle = \bigotimes_{x=1}^{L} |\psi_x^0\rangle$ for $s = 0$, and the AKLT Hamiltonian (7.1.1) with the VBS ground state $|\Phi_{\rm VBS}\rangle$ of (7.1.12) for $s = 1$. Clearly the model is in the trivial large-D phase when $s = 0$ and in the Haldane phase when $s = 1$. Also note that the Hamiltonian (8.3.55) with any s has any of the three types of symmetry (S1), (S2), and (S3) necessary to protect the Haldane phase (see p. 256). Then, according to what we know about symmetry protected topological phases, there should be a topological phase transition at an intermediate value of s. More precisely we expect that there is a single critical value s at which the Hamiltonian (8.3.55) becomes gapless as in Fig. 8.1 (p. 227). In fact we now have the following rigorous result.

Theorem 8.8 *The model* (8.3.55) *in its infinite volume limit undergoes a phase transition in the sense that there exists an intermediate value of s at which the model is either gapless, has more than one ground states, or exhibits discontinuity in the ground state expectation value of a local operator.*

The theorem was proved by Tasaki in 2018 as a corollary of his index theorem [94]. See below. Ogata's general index theorems [75, 76] lead to a similar (slightly weaker) conclusion for the model (8.3.55).

In view of sophisticated theories of SPT phases that we have reviewed, it might be rather surprising that the existence of a phase transition had remained unproven until 2018. But the difficulty of the proof comes precisely from the topological nature of the phase transition, which inhibits one from using standard methods in statistical mechanics. Let us briefly see why the arguments we have seen in the present chapter fail to provide a rigorous proof of a phase transition.

The simplest argument which suggests a phase transition in the model (8.3.55) is the change in parity. For the model on a periodic chain with an odd number of sites, we have seen that the trivial state $|\Phi_0\rangle$ and the VBS state $|\Phi_{\rm VBS}\rangle$ have even and odd parities, respectively, with respect to the inversion centered at a single bond. See p. 257. This indeed proves the existence of a level crossing at intermediate s for these models. But, of course, the existence of a level crossing does not necessarily imply a phase transition in the infinite volume limit. The situation is even trickier since there seems to be no such level crossings in a chain with an even number of sites.

We have seen in Sect. 8.3.4 that there are indices defined by Pollmann, Turner, Berg and Oshikawa that characterize projective representations of the symmetry. This of course provides a sophisticated support for the existence of a phase transition. There are three indices (corresponding to the three types of symmetry) which take 1 and -1 in the ground states $|\Phi_0\rangle$ and $|\Phi_{\text{VBS}}\rangle$, respectively. Then there should be a phase transition associated with a jump in an index. A major drawback of this approach is that these indices are well defined only for injective matrix product states. The condition of injectivity sets a strong limitation on the possible form of matrix product states.[54] We also note that, although any unique gapped ground state can be well approximated by matrix product states [1], the approximation is not precise enough to derive a definite conclusion about phase transitions in full ground states.

Probably the most promising argument was the picture of hidden $\mathbb{Z}_2 \times \mathbb{Z}_2$ symmetry breaking discussed in Sect. 8.2. After the application of the nonlocal unitary transformation (8.2.5) to the model (8.3.55), we find that the number of infinite volume ground states for $s = 0$ and 1 become one and four, respectively. Then, almost by definition, the infinite volume limit of the transformed model must exhibit a phase transition. This rigorous result strongly suggests that the original model too undergoes a phase transition. But we still do not have any proof in this direction. This is because the original model always has a unique ground state, and we do not know anything about the nature of the phase transition in the transformed model (except for the change in the number of ground states).

Tasaki's Proof of Theorem 8.8 is based on an elementary index theorem. By making use of an earlier observation by Nakamura and Todo [71], Tasaki defined a $\mathbb{Z}_2$ index[55] for a unique gapped ground state in spin chains with certain symmetry. By proving the invariance of the index under continuous modification of the model, the existence of a phase transition between the trivial phase and the Haldane phase was proved. The types of symmetry assumed in the theory are precisely (S1), (S2), and (S3) (see p. 256) identified by Pollmann, Turner, Berg, and Oshikawa [81, 82], but with an additional U(1) symmetry. The need of extra U(1) symmetry shows the limitation of this elementary approach. Ogata's index theorems, on the other hand, do not require any symmetry other than (S1), (S2), or (S3).

Basic setup for Ogata's index theorems Let us now turn to a brief review of Ogata's general index theorems [75, 76].[56] In the index theory of Pollmann, Turner, Berg, and Oshikawa [81, 82], which we fully discussed in Sect. 8.3.4, it was essential that a projective representation of $\mathbb{Z}_2 \times \mathbb{Z}_2$ or time-reversal symmetry emerges when one examines the transformation of a matrix product state. It was observed that one gets a nontrivial projective representation in the VBS state, and a trivial one in the trivial states. As we have noted in Sect. 8.3.3, the appearance of a nontrivial projective

[54] Note that one can never represent $|\Phi_0\rangle$ and $|\Phi_{\text{VBS}}\rangle$ as injective matrix product states with the same inner dimension D.

[55] The index is essentially the sign of the expectation value of the Lieb–Schultz–Mattis U(1) twist operator (6.2.2).

[56] See [69] for an extension.

representation is related to the fact that effective states on the half-infinite chains have half-odd-integral spins.

It is natural that one tries to develop a similar theory for spin chains with integer S whose ground states are not necessarily matrix product states. As long as one works within the ordinary language of spin systems, however, there seems to be no chance of encountering a nontrivial projective representation. This is because all that one has are quantum spins with integer S, which always lead to a genuine representation (or, equivalently, a trivial projective representation) of $\mathbb{Z}_2 \times \mathbb{Z}_2$ or time-reversal symmetry, as we saw back in Sect. 2.1. Ogata found a way to overcome this difficulty by making use of notions from the operator algebraic formulation of quantum spin systems. Unfortunately to go through Ogata's theory in detail is much beyond the limitation of the present book (and the ability of the present author). We shall here try to illustrate the main results without going into technical details.

We start by describing the standard operator algebraic formulation of quantum spin systems on the infinite chain. Note that we have discussed the same formulation briefly in Sect. 4.3.1. The reader is recommended to take a look Appendix A.7, which is a brief review of the formulation, when necessary. For simplicity we shall here restrict ourselves to $S = 1$ quantum spin systems on the infinite chain $\mathbb{Z}$.

For $x \in \mathbb{Z}$ and $\alpha = 1, 2, 3$, we denote by $\hat{S}_x^{(\alpha)}$ the spin operator. We denote by $\mathfrak{A}_{\mathrm{loc}}$ the set of all local operators on the chain, i.e., polynomials of the spin operators $\hat{S}_x^{(\alpha)}$. The C*-algebra $\mathfrak{A} := \overline{\mathfrak{A}_{\mathrm{loc}}}$ of the spin chain is the completion of $\mathfrak{A}_{\mathrm{loc}}$ with respect to the operator norm. Similarly we denote by $\mathfrak{A}_{\mathrm{L}}$ and $\mathfrak{A}_{\mathrm{R}}$, respectively, the C*-algebras constructed from operators on the left half-infinite chain $\{\ldots, -2, -1, 0\}$ and the right half-infinite chain $\{1, 2, 3 \ldots\}$.

Fix finite constants $R > 0$ and $h_0 > 0$. For each $x \in \mathbb{Z}$ let $\hat{h}_x \in \mathfrak{A}_{\mathrm{loc}}$ be the local Hamiltonian that depends only on $\hat{S}_y^{(\alpha)}$ such that $|x - y| \le R$ and satisfies $\|\hat{h}_x\| \le h_0$. We do not require translation invariance. We write the total Hamiltonian formally as $\hat{H} = \sum_{x \in \mathbb{Z}} \hat{h}_x$ (although the sum is not well-defined).

A state ρ of the spin chain on $\mathbb{Z}$ is a linear function from $\mathfrak{A}$ to $\mathbb{C}$ which satisfies $\rho(\hat{1}) = 1$ and $\rho(\hat{A}^\dagger \hat{A}) \ge 0$ for any $\hat{A} \in \mathfrak{A}$. Physically $\rho(\hat{A})$ is the expectation value of the operator $\hat{A}$.

We assume that the model with the Hamiltonian $\hat{H}$ has a unique ground state $\omega(\cdot)$ with a nonzero energy gap. See Definitions A.25 and A.27 for the precise definitions. In many cases the ground state $\omega(\cdot)$ is constructed by the limit

$$\omega(\hat{A}) = \lim_{L \uparrow \infty} \langle \Phi_{\mathrm{GS},L} | \hat{A} | \Phi_{\mathrm{GS},L} \rangle, \tag{8.3.56}$$

for any $A \in \mathfrak{A}_{\mathrm{loc}}$, where $|\Phi_{\mathrm{GS},L}\rangle$ is the ground state (for certain boundary conditions) on the finite chain $\Lambda_L = \{-\frac{L}{2} + 1, -\frac{L}{2} + 2, \ldots, \frac{L}{2} - 1, \frac{L}{2}\}$. See the discussion after (4.3.7) for the existence of the limit.

Given a C^*-algebra and a state on it, one can go through a standard procedure known as the Gelfand–Naimark–Segal (GNS) construction to define a representation of the C^*-algebra. See Appendix A.7. Applying the construction to $\mathfrak{A}$ and the ground state $\omega(\cdot)$, one gets a separable Hilbert space $\mathscr{H}$, its element Ω, and a representation

π (i.e., a linear map from $\mathfrak{A}$ to $B(\mathcal{H})$, the set of all bounded operators on $\mathcal{H}$) such that

$$\omega(\hat{A}) = \langle \Omega, \pi(\hat{A})\Omega \rangle \tag{8.3.57}$$

for any $\hat{A} \in \mathfrak{A}$. Physically speaking the Hilbert space $\mathcal{H}$ consists of the given ground state and all states obtained by locally perturbing it.

By using Hastings' result on the area law [34], Matsui [64] proved that the unique gapped ground state $\omega(\cdot)$ has a property called the split property. Then, from the fact that $\omega(\cdot)$ is a pure state with the split property, it follows that the above Hilbert space and the representation are decomposed as $\mathcal{H} = \tilde{\mathcal{H}}_{\mathrm{L}} \otimes \tilde{\mathcal{H}}_{\mathrm{R}}$ and $\pi = \tilde{\pi}_{\mathrm{L}} \otimes \tilde{\pi}_{\mathrm{R}}$, where $\tilde{\pi}_{\mathrm{L}} : \mathfrak{A}_{\mathrm{L}} \to B(\tilde{\mathcal{H}}_{\mathrm{L}})$ and $\tilde{\pi}_{\mathrm{R}} : \mathfrak{A}_{\mathrm{R}} \to B(\tilde{\mathcal{H}}_{\mathrm{R}})$ are representations of the C* algebras $\mathfrak{A}_{\mathrm{L}}$ and $\mathfrak{A}_{\mathrm{R}}$, respectively. See, e.g., [76]. This seemingly mathematically abstract construction is essential for the characterization of the Haldane phase. In this setup Ogata considered $\mathbb{Z}_2$ indices for models with either the $\mathbb{Z}_2 \times \mathbb{Z}_2$, the time-reversal, or the bond-centered inversion symmetry, as we shall see below.[57]

Let us make some more comments on the Hilbert spaces $\tilde{\mathcal{H}}_{\mathrm{L}}$ and $\tilde{\mathcal{H}}_{\mathrm{R}}$. (One may skip this comment and jump to the discussion of the indices.) Starting from the C*-algebra $\mathfrak{A}_{\mathrm{R}}$, and the state $\omega|_{\mathfrak{A}_{\mathrm{R}}}$ (which is the restriction of the ground state ω onto $\mathfrak{A}_{\mathrm{R}}$), one can go through the GNS construction to define the Hilbert space $\mathcal{H}_{\mathrm{R}}$ and the representation π_{R} of $\mathfrak{A}_{\mathrm{R}}$. It is indeed essential that the Hilbert spaces $\mathcal{H}_{\mathrm{R}}$ and $\tilde{\mathcal{H}}_{\mathrm{R}}$ are in general different; the space $\mathcal{H}_{\mathrm{R}}$ can be "bigger" than $\tilde{\mathcal{H}}_{\mathrm{R}}$ in some sense.[58] To relate the two spaces, we first note that $\pi_{\mathrm{R}}(\mathfrak{A}_{\mathrm{R}})$, which is contained in $B(\mathcal{H}_{\mathrm{R}})$, is a representation of $\mathfrak{A}_{\mathrm{R}}$ and is itself a C^*-algebra. There is a well-know procedure to enlarge a given representation of C^*-algebra (by taking bicommutant) to make it into another mathematically natural object known as von Neumann algebra. In our case, $\pi_{\mathrm{R}}(\mathfrak{A}_{\mathrm{R}})''$ is the von Neumann algebra thus obtained. We shall not try to explain what bicommutant $''$ is, but note that $\pi_{\mathrm{R}}(\mathfrak{A}_{\mathrm{R}})''$ is the closure of $\pi_{\mathrm{R}}(\mathfrak{A}_{\mathrm{R}})$ with respect to the weak topology, and hence $\pi_{\mathrm{R}}(\mathfrak{A}_{\mathrm{R}}) \subset \pi_{\mathrm{R}}(\mathfrak{A}_{\mathrm{R}})'' \subset B(\mathcal{H}_{\mathrm{R}})$.

When the state $\omega(\cdot)$ is pure and has the split property (which is the case here) the von Neumann algebra $\pi_{\mathrm{R}}(\mathfrak{A}_{\mathrm{R}})''$ becomes a type-I factor, which is the most well-behaved type among von Neumann algebras. It is known that a type-I factor is isomorphic to the set of all bounded operators on a certain Hilbert space. This new Hilbert space can be identified with $\tilde{\mathcal{H}}_{\mathrm{R}}$ that we saw above. Thus our von Neumann algebra $\pi_{\mathrm{R}}(\mathfrak{A}_{\mathrm{R}})''$ is identified with $B(\tilde{\mathcal{H}}_{\mathrm{R}})$. The two representations of $\mathfrak{A}_{\mathrm{R}}$ are related by $\tilde{\pi}_{\mathrm{R}} = \iota_{\mathrm{R}} \circ \pi_{\mathrm{R}}$ where $\iota_{\mathrm{R}} : \pi_{\mathrm{R}}(\mathfrak{A}_{\mathrm{R}})'' \to B(\tilde{\mathcal{H}}_{\mathrm{R}})$ is the isomorphism. Of course exactly the same is true for the Hilbert spaces $\tilde{\mathcal{H}}_{\mathrm{L}}$ and $\mathcal{H}_{\mathrm{L}}$.

$\mathbb{Z}_2 \times \mathbb{Z}_2$ **symmetry** Let us first discuss Ogata's index theorem for models with $\mathbb{Z}_2 \times \mathbb{Z}_2$ symmetry.[59] We define a linear $*$-automorphism $\mathcal{R}_\alpha$ (see the beginning of

[57] In fact Ogata's theory covers much more general symmetry transformations.

[58] When $\mathcal{H}_{\mathrm{R}}$ and $\tilde{\mathcal{H}}_{\mathrm{R}}$ are different, it follows that there must be nonzero entanglement between the left and the right half of the chain. This is an indication that the ground state is in the nontrivial symmetry protected topological phase.

[59] The original paper [75] mainly deals with the case of time-reversal symmetry, but the proof is essentially the same for $\mathbb{Z}_2 \times \mathbb{Z}_2$ (and other on-site) symmetry.

Appendix A.6 for the definition of linear $*$-automorphisms), with $\alpha = 1, 2, 3$ by

$$\mathscr{R}_\alpha(\hat{S}_x^{(\beta)}) = \begin{cases} \hat{S}_x^{(\beta)} & \text{if } \beta = \alpha, \\ -\hat{S}_x^{(\beta)} & \text{if } \beta \neq \alpha, \end{cases} \tag{8.3.58}$$

for any $x \in \mathbb{Z}$ and $\beta = 1, 2, 3$. The automorphism $\mathscr{R}_\alpha$ represents the π rotation about the α-axis. We assume that the Hamiltonian is $\mathbb{Z}_2 \times \mathbb{Z}_2$ invariant in the sense that $\mathscr{R}_\alpha(\hat{h}_x) = \hat{h}_x$ for any $\alpha = 1, 2, 3$ and any $x \in \mathbb{Z}$. We (of course) assume that the ground state, $\omega(\cdot)$, of $\hat{H}$ is unique and accompanied by a nonzero gap. The uniqueness implies the $\mathbb{Z}_2 \times \mathbb{Z}_2$ invariance, i.e.,

$$\omega(\mathscr{R}_\alpha(\hat{A})) = \omega(\hat{A}), \tag{8.3.59}$$

for any $\hat{A} \in \mathfrak{A}$ and $\alpha = 1, 2, 3$.

It then follows that there is a projective representation of the group $\mathbb{Z}_2 \times \mathbb{Z}_2$ on the space $\tilde{\mathscr{H}}_\mathrm{R}$ (or $\tilde{\mathscr{H}}_\mathrm{L}$) associated with the automorphism $\mathscr{R}_\alpha$. More precisely it can be shown that there exist (essentially unique) unitary operators $\tilde{U}_1$, $\tilde{U}_2$, and $\tilde{U}_3$ on $\tilde{\mathscr{H}}_\mathrm{R}$ such that

$$\tilde{\pi}_\mathrm{R}(\mathscr{R}_\alpha(\hat{A})) = \tilde{U}_\alpha\, \tilde{\pi}_\mathrm{R}(\hat{A})\, \tilde{U}_\alpha^\dagger, \tag{8.3.60}$$

for any $\hat{A} \in \mathfrak{A}_\mathrm{R}$ and $\alpha = 1, 2, 3$. Note that (8.3.60) is essentially the Wigner's theorem, Theorem A.21 in p. 483. We see that $\hat{1}$, $\tilde{U}_1$, $\tilde{U}_2$, and $\tilde{U}_3$ form a projective representation of the $\mathbb{Z}_2 \times \mathbb{Z}_2$ group. It then follows that the unitary operators satisfy

$$\tilde{U}_\alpha \tilde{U}_\beta = \sigma_\mathrm{D}\, \tilde{U}_\beta \tilde{U}_\alpha, \tag{8.3.61}$$

for any $\alpha \neq \beta$ with a fixed constant $\sigma_\mathrm{D} = \pm 1$. The case with $\sigma_\mathrm{D} = 1$ corresponds to a trivial projective representation and that with $\sigma_\mathrm{D} = -1$ to a nontrivial projective representation. Ogata further shows that the ground state with $\sigma_\mathrm{D} = -1$ must have entanglement entropy (between the left and the right half chains) not less than $\log 2$, exactly as Pollmann, Turner, Berg, and Oshikawa [81, 82] concluded for MPS. See also Sect. 8.3.3.

The factor σ_D is the index of Ogata's. It is remarkable that the index can be defined only by considering the action of $\mathbb{Z}_2 \times \mathbb{Z}_2$ transformation on the Hilbert space $\tilde{\mathscr{H}}_\mathrm{R}$. The construction is natural and straightforward, provided that one understands essential physics and knows proper mathematics.

It is then shown that the index σ_D is equal to 1 and -1 in a trivial product ground state and in the VBS state, respectively. One might say, rather intuitively, that the Hilbert space $\tilde{\mathscr{H}}_\mathrm{R}$ contains the information about the emergent $S = 1/2$ degrees of freedom that appears at the edge of the half-infinite chain when the ground state is in the Haldane phase. See Sect. 8.3.3, in particular, Fig. 8.13.

By using the techniques of quasi-adiabatic continuation developed by Hastings [35] and Bachmann, Michalakis, Nachtergaele, and Sims [6], it is further proved that the index σ_D is invariant when the unique gapped ground state is deformed in a certain differentiable manner, keeping $\mathbb{Z}_2 \times \mathbb{Z}_2$ symmetry. This implies that one must go through a certain singularity when smoothly connecting (via models with $\mathbb{Z}_2 \times \mathbb{Z}_2$ symmetry) two unique gapped ground states that have different σ_{D}. This leads to a slightly weaker version of Theorem 8.8 in p. 281.

Time-reversal symmetry The case with time-reversal symmetry is treated in an analogous manner. We define an antilinear $*$-automorphism Ξ for time-reversal by

$$\Xi(\hat{S}_x^{(\alpha)}) = -\hat{S}_x^{(\alpha)}, \tag{8.3.62}$$

for any $x \in \mathbb{Z}$ and $\alpha = 1, 2, 3$. See Sect. 2.3, especially (2.3.23) and footnote 24 in p. 29, and Appendix A.6.2. We assume that the Hamiltonian $\hat{H}$ is time-reversally invariant, i.e., $\Xi(\hat{h}_x) = \hat{h}_x$ for any $x \in \mathbb{Z}$, and has a unique gapped ground state $\omega(\cdot)$. The ground state is then time-reversally invariant, i.e., $\omega(\Xi(\hat{A})) = \omega(\hat{A}^\dagger)$ for any $\hat{A} \in \mathfrak{A}$.

It can be then proved that there exists (essentially unique) antiunitary operator $\tilde{V}$ (representing time-reversal) on $\tilde{\mathscr{H}}_{\mathrm{R}}$ such that

$$\tilde{\pi}_{\mathrm{R}}(\Xi(\hat{A})) = \tilde{V}\, \tilde{\pi}_{\mathrm{R}}(\hat{A})\, \tilde{V}^\dagger, \tag{8.3.63}$$

for any $\hat{A} \in \mathfrak{A}_{\mathrm{R}}$. It is also proved that $\tilde{V}^2 = \sigma_{\mathrm{tr}} \hat{1}$ with $\sigma_{\mathrm{tr}} = \pm 1$. The "topological" index σ_{tr} takes the nontrivial value -1 in the VBS state and the trivial value 1 in a product state. The results about the entanglement entropy and the invariance of the index are essentially the same as above.

Inversion symmetry The case with inversion symmetry, which is treated separately in Ogata's second paper on the subject [76], requires a different treatment since one must deal with a global symmetry. We define a linear $*$-automorphism Γ for bond-centered inversion by

$$\Gamma(\hat{S}_x^{(\alpha)}) = \hat{S}_{1-x}^{(\alpha)}, \tag{8.3.64}$$

for any $x \in \mathbb{Z}$ and $\alpha = 1, 2, 3$. We denote by $\tilde{\Gamma}$ a natural map from $\mathfrak{A}_{\mathrm{L}}$ to $\mathfrak{A}_{\mathrm{R}}$ induced by Γ. We assume that the Hamiltonian $\hat{H}$ is invariant[60] under Γ, and has a unique gapped ground state $\omega(\cdot)$. The ground state is inversion invariant, i.e., $\omega(\Gamma(\hat{A})) = \omega(\hat{A})$ for any $\hat{A} \in \mathfrak{A}$.

Because of the inversion symmetry, we see that the two Hilbert spaces $\tilde{\mathscr{H}}_{\mathrm{L}}$ and $\tilde{\mathscr{H}}_{\mathrm{R}}$ are identical, and so are the two representations $\tilde{\pi}_{\mathrm{L}}$ and $\tilde{\pi}_{\mathrm{R}}$. By making use of these identifications we can rewrite the previous GNS representation $(\mathscr{H}, \pi, \Omega)$ to that on the Hilbert space $\tilde{\mathscr{H}}_{\mathrm{R}} \otimes \tilde{\mathscr{H}}_{\mathrm{R}}$ with an element $\tilde{\Omega}$. The representation is given by $\tilde{\pi} = (\tilde{\pi}_{\mathrm{R}} \circ \tilde{\Gamma}) \otimes \tilde{\pi}_{\mathrm{R}}$.

Then it can be shown that there is a unique unitary operator $\tilde{U}$ (representing the inversion) on the Hilbert space $\tilde{\mathscr{H}}_{\mathrm{R}} \otimes \tilde{\mathscr{H}}_{\mathrm{R}}$ which satisfies

$$\tilde{U}\tilde{\pi}(\hat{A})\tilde{\Omega} = \tilde{\pi}(\Gamma(\hat{A}))\tilde{\Omega} \tag{8.3.65}$$

[60]It is convenient to write the Hamiltonian as $\hat{H} = \sum_x \hat{h}_{x,x+1}$. Then the invariance means $\Gamma(\hat{h}_{x,x+1}) = \hat{h}_{-x,1-x}$ for any x.

for any $\hat{A} \in \mathfrak{A}$. Note, in particular, that $\tilde{U}\tilde{\Omega} = \tilde{\Omega}$. It is further proved that there is a unique index $\sigma_{\mathrm{inv}} = \pm 1$ such that

$$\tilde{U}(\eta \otimes \xi) = \sigma_{\mathrm{inv}}(\xi \otimes \eta) \tag{8.3.66}$$

for any $\eta, \xi \in \tilde{\mathscr{H}}_{\mathrm{R}}$. The index takes the trivial value $\sigma_{\mathrm{inv}} = 1$ for a product ground state. It is also proved that one has $\sigma_{\mathrm{inv}} = -1$ for the VBS state, but, rather interestingly, the proof is far from being easy. There are basically the same results as above about the entanglement entropy and the invariance of the index.

Needless to say the index σ_{inv} corresponds to the parity that we discussed back in Sect. 8.3.3. It should be noted however that it is impossible to define the parity of a ground state on the infinite chain. Instead of looking at the parity of the ground state directly, Ogata here examines how local excitations above the ground state respond to the inversion transformation.[61]

Discussion We have seen that there are well-defined "topological" indices that characterize a general unique gapped ground state of a spin system on the infinite chain. The existence and the stability of the indices require only one of the three classes of symmetry (S1), (S2), or (S3) identified by Pollmann, Turner, Berg, and Oshikawa (see p. 256). No other symmetry, including the translation symmetry, is required. After about 35 years from Haldane's seminal discovery, the picture that the Haldane phase for odd integer S is in a nontrivial SPT phase has been finally justified with a complete mathematical rigor.[62] Since Ogata's index theorem in [75] extends to any on-site symmetry, it is fair to say that the abstract theory of symmetry protected topological phases for quantum spin chains has also been completed.[63] For the present author, it was an unexpected surprise to witness such rapid and essential progress (during the final stage of the writing!) on one of the main topics of the book.

It is true that Ogata's theories are based on too heavy a mathematical machinery. For most physicists (including the present author) it is impossible to fully understand the theories and apply or extend them to other situations. Then a natural question is whether it is possible to develop elementary rigorous theories for SPT phases that attain the same generality as Ogata's theories. As we have already emphasized in Sect. 8.3.3, however, such a theory should properly deal with "effective states" that live on the half-infinite chains. For the moment it seems that Ogata's approach, which makes full use of the properties of pure split states, is the only mathematically natural way to proceed.[64] We finally stress that to develop concrete mathematically rigorous theories for various "topological" phases in higher dimensions is still a widely open and promising field of research.

---

[61]The invariance $\tilde{U}\tilde{\Omega} = \tilde{\Omega}$ roughly means that here one takes the convention that the ground state always has an even "parity".

[62]The invariance theorems may be improved, but this is mostly a technical problem.

[63]But recall that there is still no proof that the $S = 1$ Heisenberg antiferromagnetic chain is in the Haldane phase.

[64]The excess spin operator defined in [5] is more elementary, but we still do not know how to construct a general and powerful theory out of it.

8.4 Topological Order in Kitaev's Toric Code Model

Before ending Part II of the book, we give a brief account on the exciting topic of topological order. In particular we give an introductory review of Kitaev's toric code model, a prototypical model exhibiting topological order in its ground states.

In Sect. 3.3, we studied the phase transition in the quantum Ising model, and found that exact $\mathbb{Z}_2$ symmetry is necessary for the existence of the ordered phase. See, in particular, Fig. 3.4. We further argued in the same section that even the very familiar solid phase should be protected by translation symmetry (which is inherent in nature). See Fig. 3.5. When we encountered the notion of symmetry protected topological (SPT) phase in Sect. 8.3.2, we again saw that the presence of suitable symmetry is necessary to have well defined "topological" phases. One might be then tempted to conclude that certain symmetry must exists for the notion of a phase to be meaningful. This conclusion is probably valid in a world governed solely by classical physics, but not in our world, which is based on quantum physics.

In a certain quantum system at zero temperature (i.e., in its ground states), there may exist a distinct nontrivial phase which need not be protected by any symmetry. In such a phase, the ground states do not break any symmetry and are accompanied by a nonzero energy gap. Although there is no apparent order, the phase is characterized by robust "topological order". When one smoothly modifies (short ranged) Hamiltonian in an arbitrary manner, without respecting any symmetry, the phase with topological order can never be smoothly connected (without passing through a gapless model) to a trivial ground state, such as that of a non-interacting spin system. We can say that a phase with topological order, unlike symmetry protected topological phases, is intrinsically stable.

The notion of topological order was gradually developed from late 1980s through the study of novel quantum phenomena, and made explicit by Wen [99]. See also [100–102, 105] and many references therein. The fractional quantum Hall effect is an example of phenomena where topological order plays fundamental roles.

It is strongly believed that topological order can take place only in two or higher dimensions. In one dimension any unique gapped ground state is believe to be smoothly connected to a trivial ground state [15, 17] as we have already mentioned in the beginning of Sect. 8.3.2. See also [93] for a theorem related to the absence of topological order in one dimension.

To give a precise and general characterization of topological order is not easy. It is a common understanding that a key to characterize topological phases is the nature of entanglement; ground states in symmetry protected topological phase have only short range entanglement, while topological order is maintained by long range entanglement [100–102, 105]. Here, rather than trying to discuss general aspects of topological order, we shall present an introductory review of a prototypical model exhibiting topological order, i.e., Kitaev's toric code model. It is a very simple mathematical model,[65] whose ground states have zero correlation length. Nevertheless the

[65]Kitaev also found a more realistic spin model on the hexagonal lattice, called the Kitaev model, which reduces to the toric code model in the anisotropic limit [53]. Materials which realize the Kitaev model have also been proposed. See [7, 42, 56] and references therein.

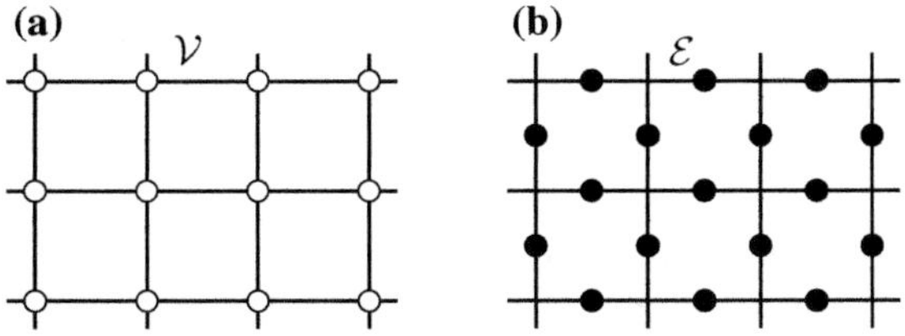

Fig. 8.14 **a** The set of usual sites of the square lattice is denoted as $\mathscr{V}$. **b** The set of sites taken at the center of each bond is denoted as $\mathscr{E}$. Spins live on sites in $\mathscr{E}$ in the toric code model

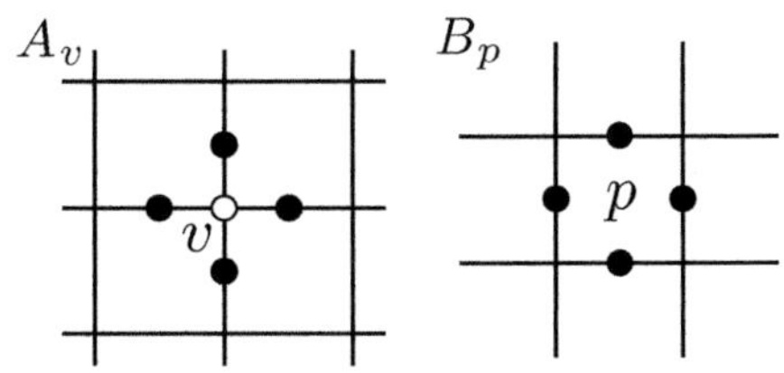

Fig. 8.15 The sets A_v and B_p. Black circles denote the elements of each set

model beautifully illustrates the essence of topological order. The present problem is not directly related to the Haldane phenomena, but we believe it is worth discussing in detail. See [60] for a generalization of the toric code model known as the string-net model.

Kitaev's toric code model and its ground state The toric code model was introduced and solved by Kitaev [51, 52] in connection with quantum computation. We will discuss the relevance of the model to quantum information near the end of this section. See [50, 105] for aspects of the toric code model which we do not cover, and [55] for background in quantum computation.

We start by describing the lattice structure used only for this model. Let $\mathscr{V}$ denote the set of sites of the $L \times L$ square lattice with periodic boundary conditions. We add a new site at the center of each bond, and denote by $\mathscr{E}$ the set of all such sites. See Fig. 8.14. We finally denote by $\mathscr{P}$ the set of all plaquettes (unit squares) of the square lattice.

For each $v \in \mathscr{V}$, we let $A_v \subset \mathscr{E}$ be the set of four sites in $\mathscr{E}$ on the bonds including v. Likewise, for each $p \in \mathscr{P}$, we let $B_p \subset \mathscr{E}$ be the set of four sites in $\mathscr{E}$ which are on the plaquette p. See Fig. 8.15.

We consider an $S = 1/2$ quantum spin system where each spin lives on a site in the set $\mathscr{E}$.[66] The Pauli matrices (see (2.1.8)) of the spin at site $x \in \mathscr{E}$ are denoted as $\hat{\sigma}_x^{(\alpha)} = 2\hat{S}_x^{(\alpha)}$ with $\alpha = 1, 2, 3$. The Hamiltonian of the toric code model is

$$\hat{H}_{\text{tc}} := -\sum_{v\in\mathscr{V}} \prod_{x\in A_v} \hat{\sigma}_x^{(3)} - \sum_{p\in\mathscr{P}} \prod_{x\in B_p} \hat{\sigma}_x^{(1)}. \tag{8.4.1}$$

[66] Equivalently one can think that spins live on bonds of the square lattice as in the lattice gauge theory.

We first note that different terms in the Hamiltonian commute with each other. That $[\prod_{x\in A_v}\hat{\sigma}_x^{(3)},\prod_{x\in A_{v'}}\hat{\sigma}_x^{(3)}]=0$ and $[\prod_{x\in B_p}\hat{\sigma}_x^{(1)},\prod_{x\in B_{p'}}\hat{\sigma}_x^{(1)}]=0$ is obvious. To see that $[\prod_{x\in A_v}\hat{\sigma}_x^{(3)},\prod_{x\in B_p}\hat{\sigma}_x^{(1)}]=0$ for any v and p, note that $A_v\cap B_p$ may contain no sites or two sites. Since $\hat{\sigma}_x^{(1)}\hat{\sigma}_x^{(3)}=-\hat{\sigma}_x^{(3)}\hat{\sigma}_x^{(1)}$, we have the desired commutativity.

Thus each term in the Hamiltonian (8.4.1) can be diagonalized simultaneously. Since both $\prod_{x\in A_v}\hat{\sigma}_x^{(3)}$ and $\prod_{x\in B_p}\hat{\sigma}_x^{(1)}$ clearly have eigenvalues ±1, we find that the energy eigenvalue can only take integral values. This implies that ground states of (8.4.1) are inevitably accompanied by an energy gap (which is at least 2). We also find that if there is a state $|\Phi_{\rm GS}\rangle$ such that

$$\Big(\prod_{x\in A_v}\hat{\sigma}_x^{(3)}\Big)|\Phi_{\rm GS}\rangle=|\Phi_{\rm GS}\rangle, \tag{8.4.2}$$

for any $v\in\mathscr{V}$, and

$$\Big(\prod_{x\in B_p}\hat{\sigma}_x^{(1)}\Big)|\Phi_{\rm GS}\rangle=|\Phi_{\rm GS}\rangle, \tag{8.4.3}$$

for any $p\in\mathscr{P}$, then $|\Phi_{\rm GS}\rangle$ is an exact ground state of (8.4.1). We shall see that such states do exist, and precisely characterize them.

For notational convenience, we write (only in this section) the basis states $|\psi^{\uparrow}\rangle$ and $|\psi^{\downarrow}\rangle$ of a single $S=1/2$ spin as $|\psi^{+1}\rangle$ and $|\psi^{-1}\rangle$, respectively, and represent spin configurations on $\mathscr{E}$ by $\boldsymbol{\tau}=(\tau_x)_{x\in\mathscr{E}}$, where $\tau_x=\pm1$. The corresponding basis state for the whole system is written as $|\Psi^{\boldsymbol{\tau}}\rangle:=\bigotimes_{x\in\mathscr{E}}|\psi_x^{\tau_x}\rangle$.

The first condition (8.4.2) is easy to treat. Since $(\prod_{x\in A_v}\hat{\sigma}_x^{(3)})|\Psi^{\boldsymbol{\tau}}\rangle=(\prod_{x\in A_v}\tau_x)|\Psi^{\boldsymbol{\tau}}\rangle$, we see that only those configurations $\boldsymbol{\tau}$ with $\prod_{x\in A_v}\tau_x=1$ for any $v\in\mathscr{V}$ can contribute to ground states. We denote by $\mathscr{T}_0$ the set of all such configurations.

Let us examine the second condition (8.4.3). Recalling that $\hat{\sigma}^{(1)}|\psi^{\tau}\rangle=|\psi^{-\tau}\rangle$ for $\tau=\pm1$, we have

$$\Big(\prod_{x\in B_p}\hat{\sigma}_x^{(1)}\Big)\bigotimes_{x\in B_p}|\psi_x^{\tau_x}\rangle=\bigotimes_{x\in B_p}|\psi_x^{-\tau_x}\rangle, \tag{8.4.4}$$

for any spin configurations on the plaquette. This immediately implies that

$$\Big(\prod_{x\in B_p}\hat{\sigma}_x^{(1)}\Big)\Big(\bigotimes_{x\in B_p}|\psi_x^{\tau_x}\rangle+\bigotimes_{x\in B_p}|\psi_x^{-\tau_x}\rangle\Big)=\Big(\bigotimes_{x\in B_p}|\psi_x^{\tau_x}\rangle+\bigotimes_{x\in B_p}|\psi_x^{-\tau_x}\rangle\Big), \tag{8.4.5}$$

or, for the whole configuration that

$$\Big(\prod_{x\in B_p}\hat{\sigma}_x^{(1)}\Big)\Big(|\Psi^{\boldsymbol{\tau}}\rangle+|\Psi^{F_p[\boldsymbol{\tau}]}\rangle\Big)=\Big(|\Psi^{\boldsymbol{\tau}}\rangle+|\Psi^{F_p[\boldsymbol{\tau}]}\rangle\Big). \tag{8.4.6}$$

Here $F_p[\boldsymbol{\tau}]$ is the spin configuration obtained from $\boldsymbol{\tau}$ by flipping the spins on the plaquette p. More precisely, if we write $\boldsymbol{\tau}'=F_p[\boldsymbol{\tau}]$, then

$$\tau'_x = \begin{cases} -\tau_x & \text{if } x \in B_p, \\ \tau_x & \text{otherwise.} \end{cases} \tag{8.4.7}$$

It is crucial to note that the configuration space $\mathcal{T}_0$ is invariant under the spin flip operation $F_p[\cdot]$, i.e., $F_p[\boldsymbol{\tau}] \in \mathcal{T}_0$ if and only if $\boldsymbol{\tau} \in \mathcal{T}_0$. We express the invariance as $F_p[\mathcal{T}_0] = \mathcal{T}_0$. The invariance implies that the conditions (8.4.2) and (8.4.3) can be consistent. In particular it implies that the superposition

$$|\Phi_{\text{GS}}\rangle = \frac{1}{\sqrt{|\mathcal{T}_0|}} \sum_{\boldsymbol{\tau} \in \mathcal{T}_0} |\Psi^{\boldsymbol{\tau}}\rangle \tag{8.4.8}$$

is an exact ground state of the Hamiltonian (8.4.1). This may be intuitively clear, but let us give a proof. The condition (8.4.2) is satisfied since only configurations from $\mathcal{T}_0$ are used. To check the condition (8.4.3), we note that (8.4.8) is rewritten by using the invariance $\sum_{\boldsymbol{\tau} \in \mathcal{T}_0} |\Psi^{\boldsymbol{\tau}}\rangle = \sum_{\boldsymbol{\tau} \in \mathcal{T}_0} |\Psi^{F_p[\boldsymbol{\tau}]}\rangle$ as

$$|\Phi_{\text{GS}}\rangle = \frac{1}{2\sqrt{|\mathcal{T}_0|}} \sum_{\boldsymbol{\tau} \in \mathcal{T}_0} \Big(|\Psi^{\boldsymbol{\tau}}\rangle + |\Psi^{F_p[\boldsymbol{\tau}]}\rangle \Big), \tag{8.4.9}$$

for any $p \in \mathcal{P}$. From (8.4.6), we see that the desired condition (8.4.3) is satisfied.

Relation with the Briegel–Raussendorf state Before proceeding to a more exciting part, we discuss an interesting relation between the Briegel–Raussendorf state (or the cluster state) studied in Sect. 7.3.3 and the ground state (8.4.8) of the toric code model. Consider an $S = 1/2$ quantum spin system on the larger lattice $\mathcal{E} \cup \mathcal{V}$, which is known as the decorated square lattice or the Lieb lattice (see Fig. 10.2). The basis states of the spin at $v \in \mathcal{V}$ are denoted as $|\tilde{\psi}_v^{\pm 1}\rangle$, a spin configuration on $\mathcal{V}$ as $\tilde{\boldsymbol{\tau}} = (\tilde{\tau}_v)_{v \in \mathcal{V}}$ with $\tilde{\tau}_v = \pm 1$, and the corresponding states on the whole $\mathcal{V}$ as $|\tilde{\Psi}^{\tilde{\boldsymbol{\tau}}}\rangle = \bigotimes_{v \in \mathcal{V}} |\tilde{\psi}_v^{\tilde{\tau}_v}\rangle$. We consider the state (7.3.33) on $\mathcal{E} \cup \mathcal{V}$, where $\mathcal{C}$ is taken as the set of neighboring pairs of sites from $\mathcal{E}$ and $\mathcal{V}$. We then get the Briegel–Raussendorf state which is written explicitly as

$$|\Xi_{\mathcal{C}}\rangle = \sum_{\boldsymbol{\tau}, \tilde{\boldsymbol{\tau}}} \Big\{ \prod_{v \in \mathcal{V}} \prod_{x \in A_v} s(\tau_x, \tilde{\tau}_v) \Big\} |\Psi^{\boldsymbol{\tau}}\rangle \otimes |\tilde{\Psi}^{\tilde{\boldsymbol{\tau}}}\rangle, \tag{8.4.10}$$

where $\boldsymbol{\tau}$ and $\tilde{\boldsymbol{\tau}}$ are summed over all spin configurations. The phase factor is given by $s(1, 1) = s(1, -1) = s(-1, 1) = 1$ and $s(-1, -1) = -1$ as in (7.3.15).

We also consider a trivial product state on $\mathcal{V}$ defined by

$$|\tilde{\Phi}_{\rightarrow}\rangle = \bigotimes_{v \in \mathcal{V}} \big(|\tilde{\psi}_v^{+1}\rangle + |\tilde{\psi}_v^{-1}\rangle \big) = \sum_{\tilde{\boldsymbol{\tau}}'} |\tilde{\Psi}^{\tilde{\boldsymbol{\tau}}'}\rangle. \tag{8.4.11}$$

Let us take the Briegel–Raussendorf state (8.4.10), and restrict part of it (on $\mathscr{V}$) to the product state (8.4.11).[67] We then get a state on $\mathscr{E}$, which is

$$\begin{aligned}\langle\tilde{\Phi}_{\rightarrow}|\Xi_{\mathscr{C}}\rangle &= \sum_{\tilde{\tau}',\tau,\tilde{\tau}}\Big\{\prod_{v\in\mathscr{V}}\prod_{x\in A_v} s(\tau_x,\tilde{\tau}_v)\Big\}|\Psi^{\tau}\rangle\,\langle\tilde{\Psi}^{\tilde{\tau}'}|\tilde{\Psi}^{\tilde{\tau}}\rangle \\ &= \sum_{\tau,\tilde{\tau}}\Big\{\prod_{v\in\mathscr{V}}\prod_{x\in A_v} s(\tau_x,\tilde{\tau}_v)\Big\}|\Psi^{\tau}\rangle \\ &= \sum_{\tau}\Big\{\prod_{v\in\mathscr{V}}\sum_{\tilde{\tau}_v=\pm 1}\prod_{x\in A_v} s(\tau_x,\tilde{\tau}_v)\Big\}|\Psi^{\tau}\rangle \\ &= \sum_{\tau}\Big\{\prod_{v\in\mathscr{V}}\Big[1+\prod_{x\in A_v} s(\tau_x,-1)\Big]\Big\}|\Psi^{\tau}\rangle,\end{aligned}\tag{8.4.12}$$

where we noted that $s(\tau,1)=1$. Also noting that $s(\tau,-1)=\tau$, we find

$$1+\prod_{x\in A_v} s(\tau_x,-1) = 1+\prod_{x\in A_v}\tau_x = \begin{cases}2 & \text{if } \prod_{x\in A_v}\tau_x = 1,\\ 0 & \text{if } \prod_{x\in A_v}\tau_x = -1.\end{cases}\tag{8.4.13}$$

We thus see that the factor $\{\prod_{v\in\mathscr{V}}[1+\prod_{x\in A_v} s(\tau_x,-1)]\}$ in the final expression of (8.4.12) is equal to $2^{|\mathscr{V}|}$ if $\boldsymbol{\tau}\in\mathscr{T}_0$ and is zero otherwise. This means, remarkably, that the state (8.4.12) is nothing but a constant multiple of the ground state (8.4.8). It is interesting that the Briegel–Raussendorf state, which is in a (weak) symmetry protected topological phase and do not possess topological order, is "upgraded" into the nontrivial ground state of the toric code model by the above simple restriction.

Topological degeneracy of ground states The exact ground state (8.4.8) is not the unique ground state of the toric code Hamiltonian (8.4.1). Interestingly ground states of the model exhibit nontrivial degeneracy which reflects the global topology of the lattice.

Recall that the spin-flip invariance of the configuration space $\mathscr{T}_0$ played and essential role in the proof that (8.4.8) is a ground state. If there exists a subset $\mathscr{T}'\subset\mathscr{T}_0$ which satisfies $F_p[\mathscr{T}']=\mathscr{T}'$ for any $p\in\mathscr{P}$, then the state $|\mathscr{T}'|^{-1/2}\sum_{\tau\in\mathscr{T}'}|\Psi^{\tau}\rangle$ is also a ground state. In fact we shall see below that $\mathscr{T}_0$ is decomposed into a disjoint union

$$\mathscr{T}_0=\mathscr{T}_{0,0}\cup\mathscr{T}_{1,0}\cup\mathscr{T}_{0,1}\cup\mathscr{T}_{1,1},\tag{8.4.14}$$

with each $\mathscr{T}_{\nu,\nu'}$ satisfying $F_p[\mathscr{T}_{\nu,\nu'}]=\mathscr{T}_{\nu,\nu'}$ for any $p\in\mathscr{P}$. Thus the toric code model (8.4.1) with periodic boundary conditions have four ground states

$$|\Phi_{\rm GS}^{\nu,\nu'}\rangle=\frac{1}{\sqrt{|\mathscr{T}_{\nu,\nu'}|}}\sum_{\tau\in\mathscr{T}_{\nu,\nu'}}|\Psi^{\tau}\rangle,\tag{8.4.15}$$

[67] More precisely we project the state (8.4.10) onto the subspace which consists of states $|\Phi\rangle\otimes|\tilde{\Phi}_{\rightarrow}\rangle$ with arbitrary $|\Phi\rangle$.

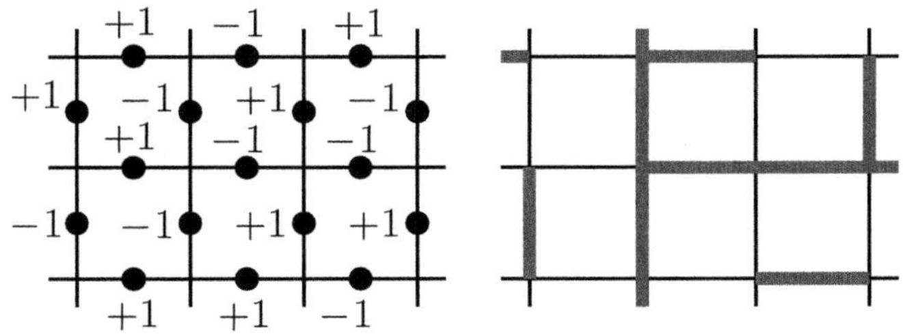

Fig. 8.16 A spin configuration and the corresponding graph. Note that this configuration does not belong to $\mathscr{T}_0$ since it contains configurations not listed in (8.4.16) (© Hal Tasaki 2020. All Rights Reserved)

with $\nu, \nu' = 0, 1$.

To see how the degeneracy emerges it is useful to introduce a graphic notation. For a given spin configuration $\boldsymbol{\tau} = (\tau_x)_{x\in\mathscr{E}}$, we draw a bond with spin $\tau_x = -1$ with a thick line, and leave a bond with $\tau_x = +1$ as it is. In this manner we get a graph of thick lines on the square lattice. Clearly there is a one-to-one correspondence between a configuration and a graph. See Fig. 8.16.

The necessary condition (8.4.2), which is equivalent to $\prod_{x\in A_v} \tau_x = 1$, requires that an even number of x in A_v to have $\tau_x = -1$. In the language of graphs this means that there must be an even number of thick lines attached to the site v. The condition of course must be satisfied for all sites in $\mathscr{V}$. To be explicit allowed patterns around a single site are the following:

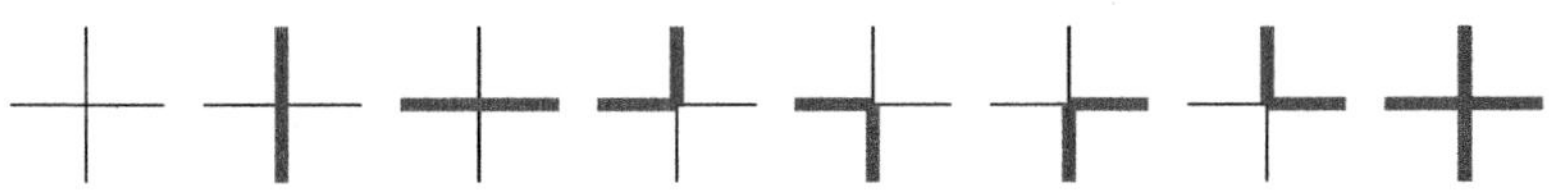

(8.4.16)

By combining these patterns, we get a graph in which thick lines form closed loops. The set of configurations $\mathscr{T}_0$ precisely corresponds to the set of all graphs consisting of closed loops.

We move onto the second necessary condition (8.4.3), which leads to the requirement of invariance $F_p[\mathscr{T}'] = \mathscr{T}'$ for a set of configurations $\mathscr{T}'$. To understand this condition graphically we note that the action of $F_p[\cdot]$ on a configuration corresponds to switching thick line and thin line on each bond on the plaquette p. The possible changes in a graph on a plaquette are the following:

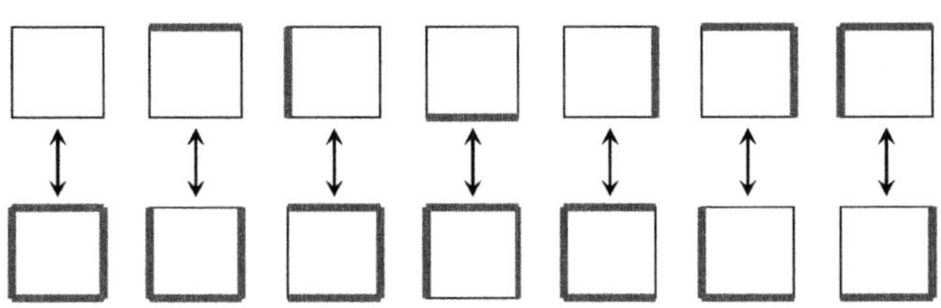

(8.4.17)

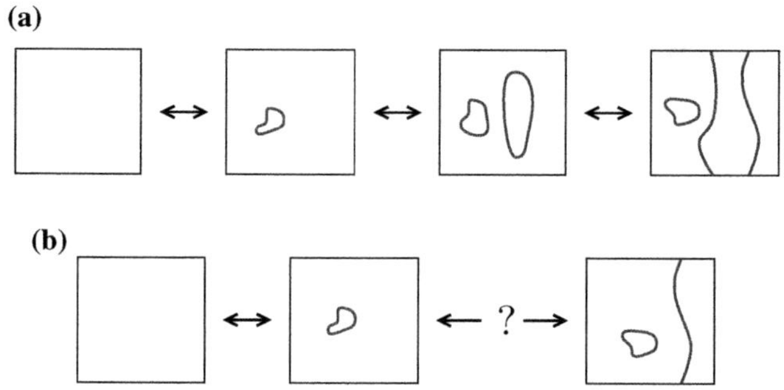

Fig. 8.17 **a** Graphs can be modified almost freely by repeatedly using the rules (8.4.17). **b** But it is impossible to start from the empty graph and reach the graph on the right, which contains a loop wrapping around the lattice once

Thus the problem of finding an invariant $\mathscr{T}'$ reduces to finding a set of graphs (of closed loops) which is invariant under the modifications (8.4.17).

By repeatedly using the rules (8.4.17), one can create a loop, modify the shape of a loop, split a loop into two, etc. Certainly a variety of graphs can be generated, starting, e.g., from the empty configuration (which corresponds to the spin configuration with $\tau_x = +1$ for all $x \in \mathscr{E}$). See Fig. 8.17a. But there are graphs that can never be generated in this manner. For example a graph that contains a single loop wrapping around the lattice once in the vertical direction is never obtained. See Fig. 8.17b.

To see this more systematically draw a horizontal line which passes through the center of bonds as Fig. 8.18, and consider, in a given graph, the number of thick lines which cross the line. When a graph is modified according to (8.4.17), the number may change, but only by 2. This means that the parity of the number of crossing is conserved under the local modifications (8.4.17).[68] Clearly the same is true for the number of crossing through a vertical line. In this way, graphs of closed loops are classified into one of the four categories labeled by the indices (ν, ν'), where $\nu = 0, 1$ is the parity of the number of crossing through the horizontal line, and $\nu' = 0, 1$ is that for the vertical line. Since graphs and configurations are in one-to-one correspondence, we have shown that the configuration space $\mathscr{T}_0$ is decomposed as (8.4.14), and hence ground states are four-fold degenerate. See Problem 8.4.b below for more rigor.

Note that the use of periodic boundary conditions is essential for the emergence of the four-fold degeneracy. If we use open boundary conditions (with a Hamiltonian properly modified at boundaries) we simply get a unique ground state. The degeneracy of the ground states reflect the global topology of the lattice.

It is interesting that the four ground states $|\Phi_{\mathrm{GS}}^{\nu,\nu'}\rangle$ with $\nu, \nu' = 0, 1$ differ only in a global manner, and hence cannot be distinguished by the expectation value of any

[68]Note also that, for a given graph, the parity is invariant on the location of the horizontal line. (In fact the line can be replaced by a curve which wraps around the lattice once in the horizontal direction.)

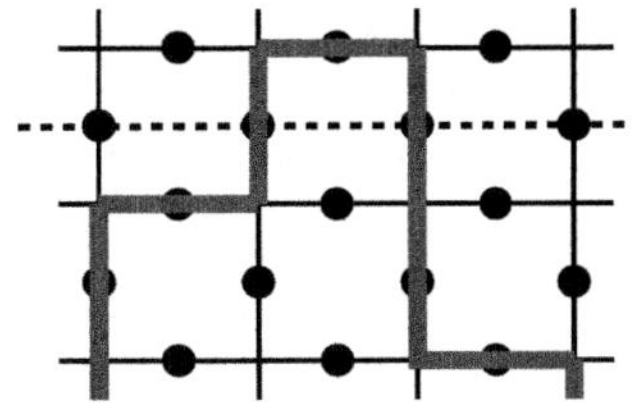

Fig. 8.18 The parity of the number of thick lines that cross the dotted line is invariant under the local modifications (8.4.17) (© Hal Tasaki 2020. All Rights Reserved)

local operator. This in particular means that the ground state of the toric code model on the infinite lattice (in the sense discussed in Sect. 4.3 or defined in Definition A.25 in p. 488) is unique. See [14] and references therein for further discussion of the toric code and related models in infinite volume.

If we denote by $\hat{Z}_{\mathrm{hor}}$ the product of $\hat{\sigma}_x^{(3)}$ for all x along the horizontal line we considered above (see Fig. 8.17b), one finds that $\hat{Z}_{\mathrm{hor}}|\Phi_{\mathrm{GS}}^{\nu,\nu'}\rangle = (-1)^{\nu}|\Phi_{\mathrm{GS}}^{\nu,\nu'}\rangle$. Similarly for the operator $\hat{Z}_{\mathrm{ver}}$ defined in the same manner for the vertical line, one has $\hat{Z}_{\mathrm{ver}}|\Phi_{\mathrm{GS}}^{\nu,\nu'}\rangle = (-1)^{\nu'}|\Phi_{\mathrm{GS}}^{\nu,\nu'}\rangle$. Thus the four ground states are clearly distinguished by these nonlocal operators.

Problem 8.4.a It is convenient to define the operators

$$\hat{A}_v := \prod_{x\in A_v} \hat{\sigma}_x^{(3)}, \quad \hat{B}_p := \prod_{x\in B_p} \hat{\sigma}_x^{(1)}. \tag{8.4.18}$$

Let $|\Phi_0\rangle = \{\prod_{p\in\mathscr{P}}(\hat{1}+\hat{B}_p)\}|\Phi_\uparrow\rangle$, where $|\Phi_\uparrow\rangle = \bigotimes_{x\in\mathscr{E}} |\psi_x^{+1}\rangle$. Show that $|\Phi_0\rangle$ is a ground state of the toric code model (8.4.1). How is this ground state related to the ground states we have defined? [solution→p.513]

Problem 8.4.b To show that there are only four ground states, one still needs to prove that all graphs with common parity indices (ν, ν') are connected by local modifications (8.4.17). Prove this fact (if the reader is rigorously minded enough to think it needs to be proved). [solution→p.513]

Stability of the topological order We now move onto the problem of stability, which touches the essence of topological order. Let us start from an important and nontrivial theorem proved by Bravyi, Hastings, and Michalakis [11]. See also [9, 65]. Consider the perturbed toric code Hamiltonian

$$\hat{H}_\varepsilon = \hat{H}_{\mathrm{tc}} + \varepsilon \sum_{x\in\mathscr{E}} \hat{V}_x, \tag{8.4.19}$$

where $\hat{V}_x$ is a function of $\hat{\sigma}_y^{(\alpha)}$ with y such that $|y-x| \le r$, where the range r is a fixed constant. We require that $\|\hat{V}_x\| \le 1$ for any x. We do not assume translation invariance. Then the following is proved in [9, 11, 65].

Theorem 8.9 *There exists a constant $\varepsilon_0 > 0$, and for any $|\varepsilon| \le \varepsilon_0$ and any L, the Hamiltonian* (8.4.19) *has four near degenerate "ground states" whose energies are separated from other energy eigenvalues by a nonzero constant independent of L.*

The theorem states that, not only the gap, but the "topological" four-fold degeneracy of the ground states remains stable when the Hamiltonian is perturbed in an arbitrary manner. It is worth comparing the situation with that in the Haldane phase. In the $S = 1$ antiferromagnetic Heisenberg chain or in the $S = 1$ AKLT model, we also encountered a nontrivial (near) four-fold degeneracy in the ground states. See Sects. 7.2.3 and 8.1.3. The degeneracy was also "topological" in the sense that it was present in open chains but not in periodic chains. The degeneracy, however, is not robust against general perturbations. It is immediately lifted when a perturbation that breaks time-reversal and $\mathbb{Z}_2 \times \mathbb{Z}_2$ symmetry, such as the uniform magnetic field, is added. The degeneracy has to be protected by either time-reversal symmetry or $\mathbb{Z}_2 \times \mathbb{Z}_2$ symmetry, as we have discussed in Sect. 8.3.2. On the contrary the degeneracy in the toric code model need not be protected by any symmetry. It is instead protected by the global topology of the model.

The robustness of the degeneracy immediately implies the existence of a stable quantum phase which we mentioned in the beginning of the section. Suppose that we start from the toric code Hamiltonian (8.4.1) and smoothly modify it into a trivial Hamiltonian, e.g., $\hat{H}_{\text{trivial}} = -\sum_{x\in\mathscr{E}} \hat{\sigma}_x^{(1)}$. We know rigorously that the four-fold near degeneracy remains when the perturbation is small enough, while models close to the trivial one have a unique ground state with a gap. This implies that the dimension of the degeneracy must change from four to one (or, more generally, four to something else) when the perturbation is strong enough, and hence there must be a phase transition.

Let us briefly (and heuristically) discuss the mechanism behind the stability. For the existence of the topological order it is essential that relevant states can be described in terms of configurations of closed loops. The order would be destroyed if loops were cut and made into open strings. To see how this is prevented, suppose that a perturbation (such as $\hat{V}_x = \hat{\sigma}_x^{(1)}$) acts onto a configuration in $\mathscr{T}_0$, and cuts a loop. But such a configuration cannot last forever because it costs extra energy coming from the first term in the Hamiltonian (8.4.1). Thus the configuration eventually returns to one in $\mathscr{T}_0$ after some transitions. This causes a local recombination of loops, but does not change the global topology of the loop configurations. In other words in order to change the parity index (ν, ν') it is necessary to make a global modification of configuration, which requires infinite order of perturbation.

The robust degeneracy is indeed essential for Kitaev's original motivation in quantum computation. When building a quantum computer, one needs to prepare a system of qubits which can stably store superpositions of distinct states.[69] By using the four ground states $|\Phi_{0,0}\rangle$, $|\Phi_{1,0}\rangle$, $|\Phi_{0,1}\rangle$, and $|\Phi_{1,1}\rangle$ of the toric code model, one can represent two bits of classical information, and also their superpositions. The stability

[69] A qubit is mathematically equivalent to a spin $S = 1/2$. The basis states $|\psi^{\uparrow}\rangle$ and $|\psi^{\downarrow}\rangle$ of a qubit are written as $|0\rangle$ and $|1\rangle$, respectively, to stress that a qubit is a quantum analogue of a bit (in information theory) which can be 0 or 1.

of the degeneracy suggests that the qubits constructed in this manner are tolerant to perturbation from external systems. The name toric code stands for a quantum memory that makes use of the geometry of a torus. It is an example of similar quantum memories generally called surface codes.[70]

Unfortunately the foregoing discussion about the stability applies only to ground states. It is known that the topological order in the toric code model is unstable to thermal fluctuation. This is most naively seen from the behavior of the correlation function of the loop operator $\hat{Z}_{\rm hor}$. Let $\langle\cdots\rangle_{\beta,L}$ denote the expectation in the equilibrium state at inverse temperature β of the toric code model defined on the $L \times L$ square lattice. It is obvious from symmetry that $\langle\hat{Z}_{\rm hor}\rangle_{\beta,L} = 0$. This equality corresponds to (3.2.10) for the classical Ising model, and carries no information about possible order in the thermal equilibrium state. If the topological order remains at nonzero temperatures, it is likely that one sees some indication in the behavior of corresponding correlation functions. See (3.2.5) or (3.2.6) for the Ising model. Let $\hat{Z}'_{\rm hor}$ be the loop operator defined with respect to the horizontal line parallel to and separated by distance ℓ from the original horizontal line used to define $\hat{Z}_{\rm hor}$. Then it is not hard to show that

$$\langle\hat{Z}_{\rm hor}\hat{Z}'_{\rm hor}\rangle_{\beta,L} \simeq (\tanh\beta)^{\ell L}, \tag{8.4.20}$$

for any β. See Problem 8.4.c for the derivation and the precise expression. When we fix L and ℓ, and let $\beta \uparrow \infty$, the correlation converges to 1, recovering the perfect topological order in the ground states. When we fix β, on the other hand, the correlation decays exponentially in ℓL for any $\beta < \infty$. This suggests that the topological order is absent for any nonzero temperatures. See [12] for detailed discussion. The situation is more delicate in three dimensions [13]. From the view point of quantum information, the instability of topological order against thermal fluctuation corresponds to non-feasibility of stable quantum memories at nonzero temperatures. See [10, 103] and references therein.[71]

Problem 8.4.c Derive (8.4.20). Although this is not difficult, one needs some experience in statistical mechanics. See the footnote if the reader needs a hint.[72] [solution→p.513]

---

[70] At the time of writing we have seen a realization of a quantum computer of moderate size, with 53 qubits [2]. So far these qubits have been used as "noisy qubits" to perform not completely reliable quantum computation. They may be also used, probably more importantly, as "physical qubits" (which correspond to spins in the toric code model) to realize a small number of stable "logical qubits" (which correspond to the degenerate ground states). The goal of the latter approach is to build fault-tolerant quantum computers. See footnote 71 below. (I am sure that this footnote will quickly become obsolete.)

[71] It is possible (at least theoretically) to perform active error corrections to build qubits (based on the toric code (or a related) model) that are fault-tolerant, i.e., work reliably under thermal fluctuation and other disturbance (or imperfection). See, e.g., [30].

[72] If $\hat{A}$ is an operator such that $(\hat{A})^2 = \hat{1}$, then it holds that $e^{\beta\hat{A}} = \cosh\beta\,(\hat{1} + \hat{A}\tanh\beta)$. Try using this relation for $\hat{A}_v$ and $\hat{B}_p$ defined in (8.4.18).

References

1. I. Arad, Z. Landau, U. Vazirani, T. Vidick, Rigorous RG algorithms and area laws for low energy eigenstates in 1D. Commun. Math. Phys. **356**, 65–105 (2017), arXiv:1602.08828
2. F. Arute, K. Arya, R. Babbush, et al., Quantum supremacy using a programmable superconducting processor. Nature **574**, 505–120 (2019), https://www.nature.com/articles/s41586-019-1666-5
3. J.E. Avron, R. Seiler, B. Simon, Homotopy and quantization in condensed matter physics. Phys. Rev. Lett. **51**, 51–53 (1983)
4. S. Bachmann, B. Nachtergaele, Product vacua with boundary states and the classification of gapped phases. Commun. Math. Phys. **329**, 509–544 (2014), arXiv:1212.3718
5. S. Bachmann, B. Nachtergaele, On gapped phases with a continuous symmetry and boundary operators. J. Stat. Phys. **154**, 91–112 (2014), arXiv:1307.0716
6. S. Bachmann, S. Michalakis, B. Nachtergaele, R. Sims, Automorphic equivalence within gapped phases of quantum lattice systems. Commun. Math. Phys. **309**, 835–871 (2012), arXiv:1102.0842
7. A. Banerjee, J. Yan, J. Knolle, C.A. Bridges, M.B. Stone, M.D. Lumsden, D.G. Mandrus, D.A. Tennant, R. Moessner, S.E. Nagler, Neutron tomography of magnetic Majorana fermions in a proximate quantum spin liquid. Science **356**, 1055–1059 (2017), arXiv:1609.00103
8. R. Botet, R. Jullien, M. Kolb, Finite-size-scaling study of the spin-1 Heisenberg-Ising chain with uniaxial anisotropy. Phys. Rev. B **28**, 3914 (1983)
9. S. Bravyi, M.B. Hastings, A short proof of stability of topological order under local perturbations. Commun. Math. Phys. **307**, 609–627 (2011), arXiv:1001.4363
10. S. Bravyi, B. Terhal, A no-go theorem for a two-dimensional self-correcting quantum memory based on stabilizer codes. New J. Phys. **11**, 043029 (2009). https://doi.org/10.1088/1367-2630/11/4/043029
11. S. Bravyi, M.B. Hastings, S. Michalakis, Topological quantum order: stability under local perturbations. J. Math. Phys. **51**, 093512 (2010), arXiv:1001.0344
12. C. Castelnovo, C. Chamon, Entanglement and topological entropy of the toric code at finite temperature. Phys. Rev. B **76**, 184442 (2007), arXiv:0704.3616
13. C. Castelnovo, C. Chamon, Topological order in a three-dimensional toric code at finite temperature. Phys. Rev. B **78**, 155120 (2008), arXiv:0804.3591
14. M. Cha, P. Naaijkens, B. Nachtergaele, The complete set of infinite volume ground states for Kitaev's abelian quantum double models. Commun. Math. Phys. **357**, 125–157 (2018), arXiv:1608.04449
15. X. Chen, Z.-C. Gu, X.-G. Wen, Complete classification of 1D gapped quantum phases in interacting spin systems. Phys. Rev. B **84**, 235128 (2011), arXiv:1103.3323
16. X. Chen, Z.-X. Liu, X.-G. Wen, Two-dimensional symmetry-protected topological orders and their protected gapless edge excitations. Phys. Rev. B **84**, 235141 (2011), arXiv:1106.4752
17. X. Chen, Z.-C. Gu, X.-G. Wen, Classification of gapped symmetric phases in one-dimensional spin systems. Phys. Rev. B **83**, 035107 (2011), arXiv:1008.3745
18. X. Chen, Z.-C. Gu, Z.-X. Liu, X.-G. Wen, Symmetry-protected topological orders in interacting bosonic systems. Science **338**, 1604–1606 (2012), arXiv:1301.0861
19. X. Chen, Z.-C. Gu, Z.-X. Liu, X.-G. Wen, Symmetry protected topological orders and the group cohomology of their symmetry group. Phys. Rev. B **87**, 155114 (2013), arXiv:1106.4772
20. X. Chen, Y.-M. Lu, A. Vishwanath, Symmetry-protected topological phases from decorated domain walls. Nat. Commun. **5**, 3507 (2014), https://www.nature.com/articles/ncomms4507
21. A.K. Daniel, R.N. Alexander, A. Miyake, Computational universality of symmetry-protected topologically ordered cluster phases on 2D archimedean lattices, preprint (2019), arXiv:1907.13279
22. M. den Nijs, K. Rommelse, Preroughening transitions in crystal surfaces and valence-bond phases in quantum spin chains. Phys. Rev. B **40**, 4709 (1989)

23. R. Dijkgraaf, E. Witten, Topological gauge theories and group cohomology. Commun. Math. Phys. **129**, 393–429 (1990), https://projecteuclid.org/euclid.cmp/1104180750
24. D.V. Else, C. Nayak, Classifying symmetry-protected topological phases through the anomalous action of the symmetry on the edge. Phys. Rev. B **90**, 235137 (2014), arXiv:1409.5436
25. M. Endres, M. Cheneau, T. Fukuhara, C. Weitenberg, P. Schauß, C. Gross, L. Mazza, M.C. Bañuls, L. Pollet, I. Bloch, S. Kuhr, Observation of correlated particle-hole pairs and string order in low-dimensional Mott insulators. Science **334**, 200 (2011), arXiv:1108.3317
26. M. Fannes, B. Nachtergaele, R.F. Werner, Valence bond states on quantum spin chains as ground states with spectral gap. J. Phys. A Math. Gen. **24**, L185–L190 (1991)
27. M. Fannes, B. Nachtergaele, R.F. Werner, Finitely correlated states on quantum spin chains. Commun. Math. Phys. **144**, 443–490 (1992), https://projecteuclid.org/euclid.cmp/1104249404
28. M. Fannes, B. Nachtergaele, R.F. Werner, Finitely correlated pure states. J. Funct. Anal. **120**, 511–534 (1994), ftp://ace2.ma.utexas.edu/pub/mp_arc/converted/92/92-132.ps.gz
29. L. Fidkowski, A. Kitaev, Topological phases of fermions in one dimension. Phys. Rev. B **83**, 075103 (2011), arXiv:1008.4138
30. K. Fujii, *Quantum Computation with Topological Codes: from Qubit to Topological Fault-Tolerance*. SpringerBriefs in Mathematical Physics, vol. 8 (Springer, Berlin, 2015), arXiv:1504.01444
31. Z.-C. Gu, X.-G. Wen, Tensor-entanglement-filtering renormalization approach and symmetry-protected topological order. Phys. Rev. B **80**, 155131 (2009), arXiv:0903.1069
32. J. Haegeman, D. Perez-Garcia, I. Cirac, N. Schuch, An order parameter for symmetry-protected phases in one dimension. Phys. Rev. Lett. **109**, 050402 (2012), arXiv:1201.4174
33. M. Hagiwara, K. Katsumata, I. Affleck, B.I. Halperin, J.P. Renard, Observation of $S = 1/2$ degrees of freedom in an $S = 1$ linear-chain Heisenberg antiferromagnet. Phys. Rev. Lett. **65**, 3181 (1990)
34. M. Hastings, An area law for one-dimensional quantum systems. J. Stat. Mech. P08024 (2007), arXiv:0705.2024
35. M.B. Hastings, Quasi-adiabatic continuation for disordered systems: applications to correlations, Lieb-Schultz-Mattis, and Hall conductance, preprint (2010), arXiv:1001.5280
36. Y. Hatsugai, Quantized Berry phases as a local order parameter of a quantum liquid. J. Phys. Soc. Jpn. **75**, 123601 (2006), arXiv:cond-mat/0603230
37. Y. Hatsugai, Quantized Berry phases for a local characterization of spin liquids in frustrated spin systems. J. Phys.: Condens. Matter **19**, 145209 (2007), arXiv:cond-mat/0607024
38. Y. Hatsugai, M. Kohmoto, Numerical study of the hidden antiferromagnetic order in the Haldane phase. Phys. Rev. B **44**, 11789–11794 (1991)
39. T. Hirano, H. Katsura, Y. Hatsugai, Topological classification of gapped spin chains: quantized Berry phase as a local order parameter. Phys. Rev. B **77**, 094431 (2008), arXiv:0710.4198
40. T. Hirano, H. Katsura, Y. Hatsugai, Degeneracy and consistency condition for Berry phases: gap closing under a local gauge twist. Phys. Rev. B **78**, 054431 (2008), arXiv:0803.3185
41. P. Horsch, W. von der Linden, Spin-correlations and low lying excited states of the spin-1/2 Heisenberg antiferromagnet on a square lattice. Z. Phys. B **72**, 181–193 (1988)
42. G. Jackeli, G. Khaliullin, Mott insulators in the strong spin-orbit coupling limit: from Heisenberg to a quantum compass and kitaev models. Phys. Rev. Lett. **102**, 017205 (2009), arXiv:0809.4658
43. S. Jiang, Y. Ran, Anyon condensation and a generic tensor-network construction for symmetry protected topological phases. Phys. Rev. B **95**, 125107 (2017), arXiv:1611.07652
44. C.L. Kane, E.J. Mele, Quantum spin Hall effect in graphene. Phys. Rev. Lett. **95**, 226801 (2005), arXiv:cond-mat/0411737
45. A. Kapustin, Bosonic topological insulators and paramagnets: a view from cobordisms, preprint (2014), arXiv:1404.6659
46. T. Kennedy, Exact diagonalization of open spin 1 chains. J. Phys.: Condens. Matter **2**, 5737–5745 (1990)

47. T. Kennedy, Non-positive matrix elements for Hamiltonians of spin-1 chains. J. Phys.: Condens. Matter **6**, 8015–8022 (1994)
48. T. Kennedy, H. Tasaki, Hidden $\mathbb{Z}_2 \times \mathbb{Z}_2$ symmetry breaking in Haldane-gap antiferromagnets. Phys. Rev. B **45**, 304–307 (1992)
49. T. Kennedy, H. Tasaki, Hidden symmetry breaking and the Haldane phase in $S = 1$ quantum spin chains. Commun. Math. Phys. **147**, 431–484 (1992), https://projecteuclid.org/euclid.cmp/1104250747
50. A. Kitaev, C. Laumann, Topological phases and quantum computation. Lectures given by Kitaev at the 2008 Les Houches Summer School "Exact methods in low-dimensional physics and quantum computing", arXiv:0904.2771
51. A.Yu. Kitaev, Quantum computations: algorithms and error correction. Russ. Math. Surv. **52**, 1191–1249 (1997). https://doi.org/10.1070/RM1997v052n06ABEH002155
52. A.Yu. Kitaev, Fault-tolerant quantum computation by anyons. Ann. Phys. **303**, 2–30 (2003), arXiv:quant-ph/9707021
53. A.Yu. Kitaev, Anyons in an exactly solved model and beyond. Ann. Phys. **321**, 2–111 (2006), arXiv:cond-mat/0506438
54. A.Yu. Kitaev, Periodic table for topological insulators and superconductors. AIP Conf. Proc. **1134**, 22 (2009), arXiv:0901.2686
55. A.Yu. Kitaev, A.H. Shen, M.N. Vyalyi, *Classical and Quantum Computation*. Graduate Studies in Mathematics, vol. 47 (American Mathematical Society, Providence, 2002)
56. K. Kitagawa, T. Takayama, Y. Matsumoto, A. Kato, R. Takano, Y. Kishimoto, S. Bette, R. Dinnebier, G. Jackeli, H. Takagi, A spin-orbital-entangled quantum liquid on a honeycomb lattice. Nature **554**, 341–345 (2018)
57. M. Kohmoto, Topological invariant and the quantization of the Hall conductance. Ann. Phys. **160**, 343–354 (1985)
58. T. Koma, H. Tasaki, Symmetry breaking and finite-size effects in quantum many-body systems. J. Stat. Phys. **76**, 745–803 (1994) (The version in the arXiv is more complete than the published version.), arXiv:cond-mat/9708132
59. M. Levin, Z.-C. Gu, Braiding statistics approach to symmetry-protected topological phases. Phys. Rev. B **86**, 115109 (2012), arXiv:1202.3120
60. M. Levin, X.-G. Wen, String-net condensation: a physical mechanism for topological phases. Phys. Rev. B **71**, 045110 (2005), arXiv:cond-mat/0404617
61. Z.-X. Liu, M. Liu, X.-G. Wen, Gapped quantum phases for the $S = 1$ spin chain with D_{2h} symmetry. Phys. Rev. B **84**, 075135 (2011), arXiv:1101.5680
62. T. Matsui, A characterization of finitely correlated pure states. Infin. Dimens. Anal. Quantum Probab. **1**, 647–661 (1998)
63. T. Matsui, The split property and the symmetry breaking of the quantum spin chain. Commun. Math. Phys. **218**, 393–416 (2001)
64. T. Matsui, Boundedness of entanglement entropy and split property of quantum spin chains. Rev. Math. Phys. 1350017 (2013), arXiv:1109.5778
65. S. Michalakis, J.P. Zwolak, Stability of frustration-free Hamiltonians. Commun. Math. Phys. **322**, 277–302 (2013), arXiv:1109.1588
66. J. Miller, A. Miyake, Hierarchy of universal entanglement in 2D measurement-based quantum computation. npj Quantum Inf. **2**, 16036 (2016), https://www.nature.com/articles/npjqi201636
67. S. Miyashita, S. Yamamoto, Effects of edges in $S = 1$ Heisenberg antiferromagnetic chains. Phys. Rev. B **48**, 913–919 (1993)
68. A. Molnar, Y. Ge, N. Schuch, J.I. Cirac, A generalization of the injectivity condition for projected entangled pair states, preprint (2017), arXiv:1706.07329
69. A. Moon, Automorphic equivalence preserves the split property, preprint (2019), arXiv:1903.00944
70. T. Morimoto, H. Ueda, T. Momoi, A. Furusaki, $\mathbb{Z}_3$ symmetry-protected topological phases in the SU(3) AKLT model. Phys. Rev. B **90**, 235111 (2014), arXiv:1409.1659

71. M. Nakamura, S. Todo, Order parameter to characterize valence-bond-solid states in quantum spin chains. Phys. Rev. Lett. **89**, 077204 (2002), arXiv:cond-mat/0112377
72. Y. Ogata, A class of asymmetric gapped Hamiltonians on quantum spin chains and its characterization I. Commun. Math. Phys. **348**, 847–895 (2016), arXiv:1510.07753
73. Y. Ogata, A class of asymmetric gapped Hamiltonians on quantum spin chains and its characterization II. Commun. Math. Phys. **348**, 897–957 (2016), arXiv:1510.07751
74. Y. Ogata, A class of asymmetric gapped Hamiltonians on quantum spin chains and its characterization III. Commun. Math. Phys. **352**, 1205–1263 (2017), arXiv:1606.05508
75. Y. Ogata, A $\mathbb{Z}_2$-index of symmetry protected topological phases with time reversal symmetry for quantum spin chains, preprint (2018), arXiv:1810.01045
76. Y. Ogata, A $\mathbb{Z}_2$-index of symmetry protected topological phases with reflection symmetry for quantum spin chains, preprint (2019), arXiv:1904.01669.pdf
77. Y. Ogata, H. Tasaki, Lieb-Schultz-Mattis type theorems for quantum spin chains without continuous symmetry. Commun. Math. Phys. **372**, 951–962 (2019), arXiv:1808.08740
78. M. Oshikawa, Hidden $\mathbb{Z}_2 \times \mathbb{Z}_2$ symmetry in quantum spin chains with arbitrary integer spin. J. Phys. Condens. Matter **4**, 7469 (1992)
79. D. Perez-Garcia, M.M. Wolf, M. Sanz, F. Verstraete, J.I. Cirac, String order and symmetries in quantum spin lattices. Phys. Rev. Lett. **100**, 167202 (2008), arXiv:0802.0447
80. F. Pollmann, A.M. Turner, Detection of symmetry-protected topological phases in one dimension. Phys. Rev. B **86**, 125441 (2012), arXiv:1204.0704
81. F. Pollmann, A.M. Turner, E. Berg, M. Oshikawa, Entanglement spectrum of a topological phase in one dimension. Phys. Rev. B **81**, 064439 (2010), arXiv:0910.1811
82. F. Pollmann, A.M. Turner, E. Berg, M. Oshikawa, Symmetry protection of topological phases in one-dimensional quantum spin systems. Phys. Rev. B **85**, 075125 (2012), arXiv:0909.4059
83. H. Poulsen-Nautrup, T.-C. Wei, Symmetry-protected topologically ordered states for universal quantum computation. Phys. Rev. A **92**, 052309 (2015), arXiv:1509.02947
84. R. Raussendorf, C. Okay, D.-S. Wang, D.T. Stephen, H. Poulsen-Nautrup, A computationally universal phase of quantum matter. Phys. Rev. Lett. **122**, 090501 (2019), arXiv:1803.00095
85. A.P. Schnyder, A. Ryu, A. Furusaki, A.W.W. Ludwig, Classification of topological insulators and superconductors in three spatial dimensions. Phys. Rev. B **78**, 195125 (2008), arXiv:0803.2786
86. N. Schuch, D. Perez-Garcia, I. Cirac, Classifying quantum phases using matrix product states and PEPS. Phys. Rev. B **84**, 165139 (2011), arXiv:1010.3732
87. H.J. Schulz, Phase diagrams and correlation exponents for quantum spin chains of arbitrary spin quantum number. Phys. Rev. B **34**, 6372 (1986)
88. H.J. Schulz, T. Ziman, Finite-length calculation of η and phase diagrams of quantum spin chains. Phys. Rev. B **33**, 6545 (1986)
89. K. Shiozaki, S. Ryu, Matrix product states and equivariant topological field theories for bosonic symmetry-protected topological phases in (1+1) dimensions. J. High Energy Phys. **2017**, 100 (2017). https://doi.org/10.1007/JHEP04(2017)100
90. B. Simon, Holonomy, the quantum adiabatic theorem, and Berry's phase. Phys. Rev. Lett. **51**, 2167–2170 (1983)
91. W. Son, L. Amico, V. Vedral, Topological order in 1D cluster state protected by symmetry. Quantum Inf. Process. **11**, 1961–1968 (2012), arXiv:1111.7173
92. H. Tasaki, Quantum liquid in antiferromagnetic chains: a stochastic geometric approach to the Haldane gap. Phys. Rev. Lett. **66**, 798–801 (1991). Errata: Phys. Rev. Lett. **75**, 354 (1995)
93. H. Tasaki, Lieb-Schultz-Mattis theorem with a local twist for general one-dimensional quantum systems. J. Stat. Phys. **170**, 653–671 (2018), arXiv:1708.05186
94. H. Tasaki, Topological phase transition and $\mathbb{Z}_2$ index for $S = 1$ quantum spin chains. Phys. Rev. Lett. **121**, 140604 (2018), arXiv:1804.04337
95. D.J. Thouless, M. Kohmoto, M.P. Nightingale, M. den Nijs, Quantized Hall conductance in a two-dimensional periodic potential. Phys. Rev. Lett. **49**, 405–408 (1982). https://doi.org/10.1103/PhysRevLett.49.405

96. T. Tonegawa, K. Okamoto, H. Nakano, T. Sakai, K. Nomura, M. Kaburagi, Haldane, large-D, and intermediate-D states in an $S = 2$ quantum spin chain with on-site and XXZ anisotropies. J. Phys. Soc. Jpn. **80**, 043001 (2011), arXiv:1011.6568
97. H. Watanabe, H.C. Po, A. Vishwanath, M.P. Zaletel, Filling constraints for spin-orbit coupled insulators in symmorphic and nonsymmorphic crystals. Proc. Natl. Acad. Sci. U.S.A. **112**, 14551–14556 (2015), http://www.pnas.org/content/112/47/14551
98. T.-C. Wei, Quantum spin models for measurement-based quantum computation. Adv. Phys.: X **1**, 1461026 (2018). https://doi.org/10.1080/23746149.2018.1461026
99. X.-G. Wen, Topological orders in rigid states. Int. J. Mod. Phys. B **04**, 239 (1990), http://dao.mit.edu/~wen/pub/topo.pdf
100. X.-G. Wen, *Quantum Field Theory of Many-Body Systems: from the Origin of Sound to an Origin of Light and Electrons* (Oxford University Press, Oxford, 2007)
101. X.-G. Wen, Colloquium: zoo of quantum-topological phases of matter. Rev. Mod. Phys. **89**, 41004 (2017), arXiv:1610.03911
102. X.-G. Wen, Choreographed entanglement dances: topological states of quantum matter. Science **363**, 834 (2019), http://science.sciencemag.org/content/363/6429/eaal3099
103. B. Yoshida, Feasibility of self-correcting quantum memory and thermal stability of topological order. Ann. Phys. **326**, 2566–2633 (2011), arXiv:1103.1885
104. B. Yoshida, Topological phases with generalized global symmetries. Phys. Rev. B **93**, 155131 (2016), arXiv:1508.03468
105. B. Zeng, X. Chen, D.-L. Zhou, X.-G. Wen, *Quantum Information Meets Quantum Matter: from Quantum Entanglement to Topological Phase in Many-Body Systems*. Quantum Science and Technology (Springer, Berlin, 2019), arXiv:1508.02595

Part III
Hubbard Model and the Origin of Magnetism

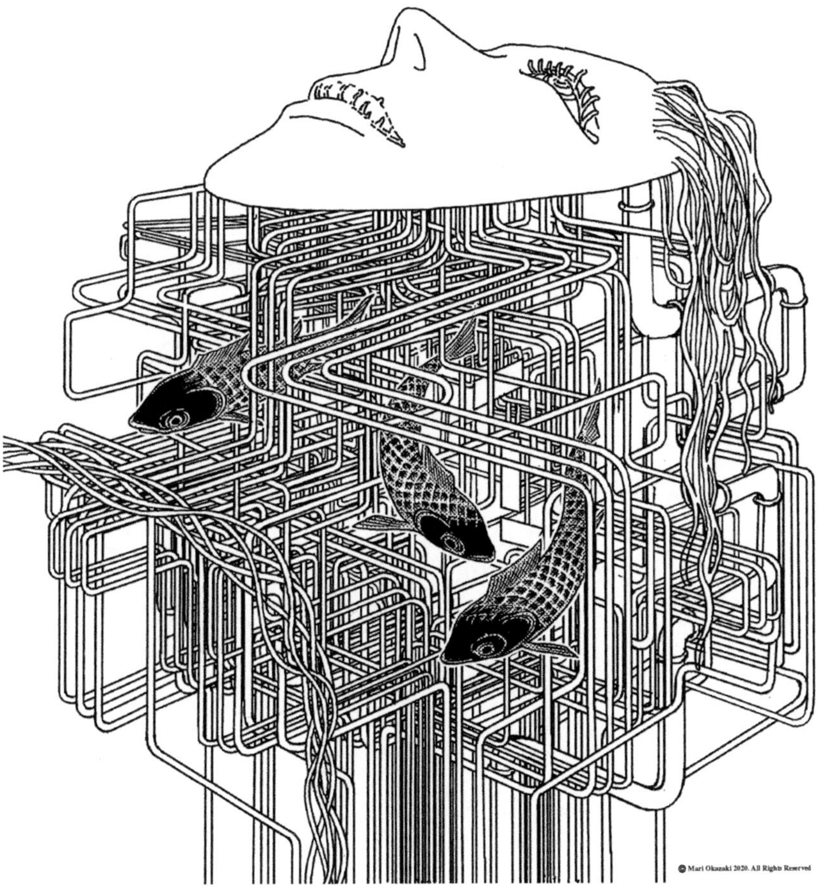

In the final part of the present book, we study the Hubbard model, a standard idealized model for strongly interacting itinerant electrons in a solid. The model allows us to focus on intricate "competition" between the wave-like nature and the particle-like nature of electrons, which leads to various nontrivial and unexpected phenomena.

We start from an introduction to the model and discuss its basic properties carefully. We then focus on the important class of models at half-filling, and discuss fundamental theorems by Lieb. The theorems shed light on the origin of antiferromagnetism, ferrimagnetism, and superconductivity. We then discuss a variety of rigorous results on the generation of ferromagnetic order. To understand the origin of ferromagnetism was historically the main motivation for the introduction of the Hubbard model.

Chapter 9
Introduction to the Hubbard Model

Here we introduce the Hubbard model, without assuming advanced knowledge on the theoretical treatment of quantum many-particles systems. In Sect. 9.2, we start from the wave function description of multiple electrons, and introduce the Fock space representation (also known as the "second quantization" formalism), which is standard in modern condensed matter physics. We then give a precise definition of the Hubbard model, and discuss its basic features in Sect. 9.3.

9.1 What is the Hubbard Model?

In the preceding chapters, we have mostly studied quantum spin systems, where each electron (which carries a spin $S = 1/2$) is assumed to be strictly localized at an atomic site, and spin degrees of freedom of nearby electrons interact directly with each other. In Part III of the book, we shall take one step further down to the microscopic level, and study the origin of the interaction between localized spins. The basic idea is that the interaction between spins does not come directly from fundamental laws of physics,[1] but is generated effectively from interplay between quantum mechanical motion of multiple electrons and the strong Coulomb interaction between them.

This means that we need to treat a theoretical model for interacting electrons in a solid. The many-body Schrödinger equation of electrons in a potential generated by a crystal is an obvious candidate, but such a model is formidably difficult to study (even numerically). The Hubbard model, a minimum model which takes into account quantum mechanical motion of electrons and repulsive interaction between them, may be the right problem for us. Although being too simple as a realistic model in condensed matter physics, the Hubbard model is no doubt one of the most important models in modern theoretical physics. In spite of its simple definition, the model is believed to exhibit various interesting phenomena including metal-insulator transition, antiferromagnetism, ferrimagnetism, ferromagnetism, Tomonaga–Luttinger

[1] The magnetic dipole interaction is too weak to explain observed magnetism.

H. Tasaki, *Physics and Mathematics of Quantum Many-Body Systems*,
Graduate Texts in Physics, https://doi.org/10.1007/978-3-030-41265-4_9

liquid, and superconductivity. Theoretical studies over the years have also revealed that to understand various properties of the Hubbard model is generally extremely difficult. We believe that in course of getting deeper understanding of nontrivial properties of the Hubbard model, we will learn many new techniques, concepts, and ways of thinking, both physical and mathematical.

The importance of the Hubbard model may be best understood from the point of view of universality that we have discussed in Sect. 1.1. The model may not be realistic, but we nevertheless believe that nontrivial phenomena and mechanisms found in the idealized model can also be found in other systems in the same "universality class". The universality class is expected to be large and rich enough so that it contains various realistic strongly interacting electron systems with complicated details which are ignored in the idealized model. We also note that experimental techniques to construct ultracold fermion systems which "simulate" the Hubbard models are rapidly developing. See [2, 13, 14].

In the following chapters, we discuss selected mathematically rigorous results about the ground states of various versions of the Hubbard model, placing main emphasis on the origin of magnetic ordering. See [10, 18] for reviews of rigorous results which cover topics not treated here, and [19] for a more detailed review focused on ferromagnetism in the Hubbard model. We do not cover the rich field, originated by Lieb and Wu [11], of the exact solutions of the one-dimensional Hubbard model. We recommend, e.g., [3, 12]. The reader interested in the connection of the Hubbard model to condensed matter physics is suggested to study Chap. 4 of [4]. See also [6] for a very readable introduction to the Hubbard model.

Hubbard model and the origin of ferromagnetism It may be interesting and enlightening to briefly discuss the history of the Hubbard model in connection to the research on the origin of ferromagnetism.

The origin of strong ferromagnetic order observed in some materials has been a mystery in physical science for a long time. Since quantum mechanical many-electron systems without interactions universally exhibit paramagnetism, the origin of ferromagnetism should be sought in electron-electron interactions. In most solids, however, the dominant part of the interaction between electrons is the Coulomb interaction, which is perfectly spin-independent. We are thus faced with a very interesting and fundamental problem to determine whether spin-independent interactions in an itinerant electron system can be the origin of ferromagnetic ordering. This problem is important not only because ferromagnetism is a very common (and useful) phenomenon, but also because it focuses on a fundamental role played by nonlinear interactions in many-body quantum mechanical systems.

It was Heisenberg [7] who first realized that ferromagnetism is an intrinsically quantum mechanical phenomenon. In Heisenberg's approach to ferromagnetism, one starts from the picture that each electron (relevant to magnetism) is almost localized at an atomic orbit. By treating the effect of the Coulomb interaction and overlap between nearby atomic orbits in a perturbative manner as in the Heitler–London theory, Heisenberg concluded that there appears an "exchange interaction" between nearby electronic spins. The exchange interaction determines magnetic properties of the system.

In a different approach to the problem of ferromagnetism, which was originated by Bloch [1], one starts from the quantum mechanical free electron gas, in which electrons are in plane-wave like states. One then treats the effect of the Coulomb interaction perturbatively, and tries to find instabilities against certain magnetic ordering. When combined with the Hartree–Fock approximation (or a mean-field theory), this approach leads to the picture that there is an instability toward ferromagnetism when the density of states at the fermi energy and the Coulomb interaction are sufficiently large.

In spite of a considerable number of attempts to improve these ideas, neither of these two approaches has yet produced a truly convincing explanation about the origin of ferromagnetism.

A modern version of the problem of the origin of ferromagnetism was formulated by Kanamori [9], Gutzwiller [5], and Hubbard [8] in the 1960s,[2] who studied the simple tight-binding model of electrons with on-site Coulomb interaction, now called the Hubbard model. When there is no electron-electron interaction, the model exhibits paramagnetism as an inevitable consequence of the Pauli exclusion principle. Kanamori, Gutzwiller, and Hubbard asked, among other things, whether the paramagnetism found for a non-interacting system can be converted into ferromagnetism when there is sufficiently large Coulomb interaction. This is a concrete formulation of the fundamental problem that we discussed above.

It is worth noting that the on-site Coulomb interaction itself is completely independent of electronic spins, and it does not favor any magnetic ordering. One does not find any term in the Hubbard Hamiltonian which explicitly favors ferromagnetism (or any other ordering). See (9.3.17), (9.3.28), and (9.3.34). Our theoretical goal is to show that magnetic ordering arises as a consequence of a subtle interplay between the kinetic motion of electrons and the short-ranged Coulomb interaction.

The problem of ferromagnetism in the Hubbard model was studied intensively mainly by various heuristic methods. But by now there have been some rigorous results by Nagaoka, Lieb, Mileke, and Tasaki, which convincingly show that certain versions of the model can describe stable ferromagnetism. We shall discuss these results in Sect. 10.2 and in Chap. 11.

9.2 Tight-Binding Description of Electrons in a Solid

Before introducing the Hubbard model, we discuss in detail the tight-binding description of electrons in a solid. In this section we focus on the Hilbert space and basic operators, leaving the Hamiltonian to the next Sect. 9.3.

We introduce the conventional wave function formalism in Sect. 9.2.1, and the Fock space representation (or the "second quantization" formalism), which we use throughout Part III of the book, in Sect. 9.2.3. To bridge these sections, we discuss

[2] A similar formulation was given earlier, for example, in [17].

the creation and annihilation operators within the framework of wave functions in Sect. 9.2.2.

9.2.1 Wave Functions for Electrons

Single electron Let the lattice Λ be a finite set of sites $x, y, \ldots$, as in Sect. 2.2. In the tight-binding description of electrons in a solid, which is a kind of low-energy effective theory, each site in Λ represents an atom in the solid. One then declares that electrons can live only on lattice sites, and can hop from one site to another. Thus a quantum mechanical state of an electron is fully specified by the amplitude $\varphi(x, \sigma) \in \mathbb{C}$ for each site $x \in \Lambda$ and spin $\sigma = \uparrow, \downarrow$.

The idea behind this description is that the actual state of the electron is described by the wave function

$$\phi_\sigma(\boldsymbol{r}) = \sum_{x\in\Lambda} \varphi(x, \sigma)\, w_x(\boldsymbol{r}), \tag{9.2.1}$$

where $\boldsymbol{r} \in \mathbb{R}^3$, and $w_x(\boldsymbol{r})$ denotes the (fixed) wave function of one relevant electronic orbit[3] around the atom x. Of course, an atom has more than one orbit, but we here assume that electrons in other orbits do not play significant roles in determining low-energy properties of the system, and may be "forgotten" in approximate treatments. See Fig. 9.1.

Let us write $\tilde{\Lambda} := \Lambda \times \{\uparrow, \downarrow\}$, and denote its elements as $u, v, w, \ldots \in \tilde{\Lambda}$. Since $u = (x, \sigma)$, the above amplitude, or the wave function, is written as $\varphi(u)$. Let us collectively denote the wave function as $\boldsymbol{\varphi} = (\varphi(u))_{u\in\tilde{\Lambda}}$. We here avoid using the bra-ket notation, which is reserved for the Fock space representation introduced in Sect. 9.2.3 The single-electron Hilbert space $\tilde{\mathfrak{h}}$ is the set of all such $\boldsymbol{\varphi}$. For any $\boldsymbol{\varphi} \in \tilde{\mathfrak{h}}$ and $\boldsymbol{\psi} = (\psi(u))_{u\in\tilde{\Lambda}} \in \tilde{\mathfrak{h}}$, their inner product is

$$\langle \boldsymbol{\varphi}, \boldsymbol{\psi} \rangle := \sum_{u\in\tilde{\Lambda}} \{\varphi(u)\}^* \,\psi(u). \tag{9.2.2}$$

Clearly $\tilde{\mathfrak{h}} \cong \mathbb{C}^{|\tilde{\Lambda}|}$, or, if the reader is familiar with the Hilbert space theory, $\tilde{\mathfrak{h}} \cong \ell^2(\tilde{\Lambda}; \mathbb{C})$. We reserve the symbol $\mathfrak{h}$ for the Hilbert space of a particle on Λ without spin, which consists of wave functions $\boldsymbol{\varphi} = (\varphi(x))_{x\in\Lambda}$.

Multiple electrons The Hilbert space of N distinguishable particles is given by the tensor product $\underbrace{\tilde{\mathfrak{h}} \otimes \cdots \otimes \tilde{\mathfrak{h}}}_{N}$, whose element is an N particle wave function $\boldsymbol{\Phi} = (\Phi(u_1, \ldots, u_N))_{u_1,\ldots,u_N\in\tilde{\Lambda}}$ with arbitrary $\Phi(u_1, \ldots, u_N) \in \mathbb{C}$. Here we are labeling the particles as $1, 2, \ldots, N$, and denoting by u_j the coordinate of the j-th particle. The inner product of $\boldsymbol{\Phi}$ and $\boldsymbol{\Psi} = (\Psi(u_1, \ldots, u_N))_{u_1,\ldots,u_N\in\tilde{\Lambda}}$ is

[3] We also assume that the orbit is nondegenerate. This idealization may not be realistic in general.

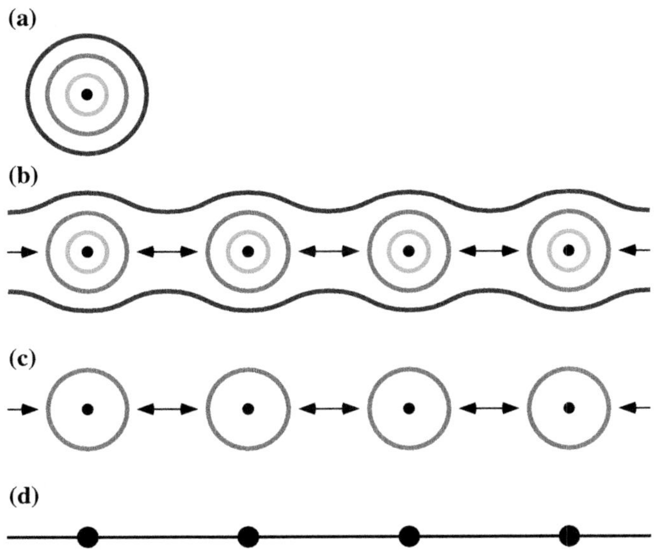

Fig. 9.1 Highly schematic figures which explain the philosophy of the tight-binding description. **a** A single atom which has multiple electrons in different orbits. **b** When atoms come together to form a solid, electrons in the black orbits become itinerant, while those in the light gray orbits are still localized at the original atomic sites. Electrons in the gray orbits are mostly localized around the atomic sites, but tunnel to nearby gray orbits with non-negligible probabilities. **c** We only consider the electrons in the gray orbits, which are expected to play essential roles in determining various aspects of low-energy physics of the system. **d** If the gray orbit is non-degenerate, we get a lattice model in which electrons live on lattice sites and hop from one site to another. In the special case where each lattice site is occupied by a single electron, the situation reduces to that of quantum spin systems. See Sect. 4.3.2 of [4] for a more careful discussion

$$\langle \boldsymbol{\Phi}, \boldsymbol{\Psi} \rangle := \sum_{u_1,\ldots,u_N \in \tilde{\Lambda}} \{\Phi(u_1, \ldots, u_N)\}^* \, \Psi(u_1, \ldots, u_N). \tag{9.2.3}$$

But this description is not suited for electrons, which are indistinguishable with each other. Since electrons are fermions, any wave function must be antisymmetric with respect to the change of the labels of electrons as,[4] e.g., $\Phi(u_1, u_2) = -\Phi(u_2, u_1)$, or, in general,

$$\Phi(u_1, \ldots, u_N) = (-1)^p \Phi(u_{p(1)}, \ldots, u_{p(N)}). \tag{9.2.4}$$

[4]The common expression "the wave function must be antisymmetric under particle exchange" can be confusing since it might suggest that particles are literally exchanged through a physical process. One should realize that we are here dealing with the description of an instantaneous state of the system; the change of the labels is not a physical process.

Here p is an arbitrary permutation of $\{1, \ldots, N\}$, and $(-1)^p = \pm 1$ denotes its parity. Thus the Hilbert space for (the wave functions of) N electrons on the lattice Λ is the antisymmetric tensor product

$$\mathcal{H}_N^{\mathrm{wf}} := \mathsf{P}_{\mathrm{as}} \underbrace{\tilde{\mathfrak{h}} \otimes \cdots \otimes \tilde{\mathfrak{h}}}_{N}, \tag{9.2.5}$$

where P_{as} is the projection onto totally antisymmetric wave functions, i.e.,

$$(\mathsf{P}_{\mathrm{as}}\boldsymbol{\Phi})(u_1, \ldots, u_N) = \frac{1}{N!} \sum_p (-1)^p \Phi(u_{p(1)}, \ldots, u_{p(N)}), \tag{9.2.6}$$

with the sum running over all the $N!$ permutations.[5] The inner product of $\boldsymbol{\Phi}, \boldsymbol{\Psi} \in \mathcal{H}_N^{\mathrm{wf}}$ is still given by (9.2.3). It will turn out that the electron number N takes an integral value between 0 and $2|\Lambda|$ since there are no totally antisymmetric wave functions with $N > 2|\Lambda|$. The space $\mathcal{H}_1^{\mathrm{wf}}$ is of course $\tilde{\mathfrak{h}}$ itself, and the space $\mathcal{H}_0^{\mathrm{wf}}$ should be identified with $\mathbb{C}$.

9.2.2 *Creation and Annihilation Operators in the Wave Function Formalism*

In order to give a connection between the wave function formalism in Sect. 9.2.1 and the Fock space representation in Sect. 9.2.3, we here construct and study creation and annihilation operators for fermions within the language of wave functions.[6] The quick reader can skip this subsection and jump to Sect. 9.2.3. We do not make use of the material here in the rest of the book.

Creation operator For any $\boldsymbol{\psi} \in \tilde{\mathfrak{h}}$ and $N = 1, 2, \ldots$, we wish to define the creation operator $\mathsf{C}^\dagger(\boldsymbol{\psi}) : \mathcal{H}_{N-1}^{\mathrm{wf}} \to \mathcal{H}_N^{\mathrm{wf}}$, which "adds" the state $\boldsymbol{\psi}$ to an arbitrary $N-1$ particle state $\boldsymbol{\Phi}$ to generate a new N particle state $\mathsf{C}^\dagger(\boldsymbol{\psi})\boldsymbol{\Phi}$.

When $N = 2$ a natural definition of such an operator is

$$(\mathsf{C}^\dagger(\boldsymbol{\psi})\boldsymbol{\varphi})(u_1, u_2) = \frac{1}{\sqrt{2}}\big\{\psi(u_1)\varphi(u_2) - \psi(u_2)\varphi(u_1)\big\}. \tag{9.2.7}$$

Note that $\psi(u_1)\varphi(u_2) - \psi(u_2)\varphi(u_1)$ is the only antisymmetric two particle wave function that one can generate from $\psi(u)$ and $\varphi(u)$. The norm of the new state is readily found to be

---

[5] We denote in general by $(\boldsymbol{\Phi})(u_1, \ldots, u_N)$ the component $\Phi(u_1, \ldots, u_N)$ of a wave function $\boldsymbol{\Phi}$.

[6] There is a corresponding pedagogical account by the present author on the Fock space representation in ordinary quantum mechanics of fermions or bosons [20].

$$\|\mathsf{C}^\dagger(\boldsymbol{\psi})\boldsymbol{\varphi}\|^2 = \frac{1}{2}\sum_{u_1,u_2\in\tilde{\Lambda}}\{\psi(u_1)\varphi(u_2)-\psi(u_2)\varphi(u_1)\}^*\{\psi(u_1)\varphi(u_2)-\psi(u_2)\varphi(u_1)\}$$
$$= \|\boldsymbol{\psi}\|^2\,\|\boldsymbol{\varphi}\|^2 - |\langle\boldsymbol{\psi},\boldsymbol{\varphi}\rangle|^2. \tag{9.2.8}$$

Suppose that $\|\boldsymbol{\psi}\| = \|\boldsymbol{\varphi}\| = 1$. Then the new state $\mathsf{C}^\dagger(\boldsymbol{\psi})\boldsymbol{\varphi}$ is normalized only when $\langle\boldsymbol{\psi},\boldsymbol{\varphi}\rangle = 0$. We even have $\mathsf{C}^\dagger(\boldsymbol{\psi})\boldsymbol{\varphi} = 0$ when $\boldsymbol{\psi} = e^{i\theta}\boldsymbol{\varphi}$. But this is natural since "adding" the same state is prohibited by the antisymmetry, a mathematical expression of the Pauli exclusion principle.

Let us consider the most straightforward extension of (9.2.7) to general N. For an arbitrary $\boldsymbol{\Phi} \in \mathscr{H}_{N-1}^{\mathrm{wf}}$, we define

$$(\mathsf{C}^\dagger(\boldsymbol{\psi})\boldsymbol{\Phi})(u_1,\ldots,u_N) := \frac{1}{\sqrt{N}}\sum_{j=1}^{N}(-1)^{j+1}\,\psi(u_j)\,\Phi(u_1,\ldots,\breve{u}_j,\ldots,u_N), \tag{9.2.9}$$

for any $u_1,\ldots,u_N \in \tilde{\Lambda}$. Here and in what follows we denote by $u_1,\ldots,\breve{u}_j,\ldots,u_N$ the sequence without u_j, i.e., $u_1,\ldots,u_{j-1},u_{j+1},\ldots,u_N$. Thus (9.2.9) with $N=3$ reads

$$(\mathsf{C}^\dagger(\boldsymbol{\psi})\boldsymbol{\Phi})(u_1,u_2,u_3)$$
$$= \frac{1}{\sqrt{3}}\big\{\psi(u_1)\,\Phi(u_2,u_3) - \psi(u_2)\,\Phi(u_1,u_3) + \psi(u_3)\,\Phi(u_1,u_2)\big\}. \tag{9.2.10}$$

Note that the right-hand side of (9.2.9) or (9.2.10) is indeed antisymmetric in $u_1,\ldots,u_N$. For $N=1$ and $\boldsymbol{\Phi} = 1 \in \mathscr{H}_0^{\mathrm{wf}}$, the definition (9.2.9) gives

$$(\mathsf{C}^\dagger(\boldsymbol{\psi})1)(u) = \psi(u). \tag{9.2.11}$$

It is also obvious that $\mathsf{C}^\dagger(\boldsymbol{\psi})$ is linear in $\boldsymbol{\psi}$, i.e.,

$$\mathsf{C}^\dagger(\alpha\boldsymbol{\psi} + \beta\boldsymbol{\xi}) = \alpha\,\mathsf{C}^\dagger(\boldsymbol{\psi}) + \beta\,\mathsf{C}^\dagger(\boldsymbol{\xi}), \tag{9.2.12}$$

for any $\boldsymbol{\psi}, \boldsymbol{\xi} \in \tilde{\mathfrak{h}}$ and $\alpha, \beta \in \mathbb{C}$.

Annihilation operator We define the annihilation operator $\mathsf{C}(\boldsymbol{\psi}) : \mathscr{H}_N^{\mathrm{wf}} \to \mathscr{H}_{N-1}^{\mathrm{wf}}$ for any $N = 1, 2, \ldots$ by $\mathsf{C}(\boldsymbol{\psi}) := \{\mathsf{C}^\dagger(\boldsymbol{\psi})\}^\dagger$. It should be the operator that "removes" the state $\boldsymbol{\psi}$ from an arbitrary N particle state $\boldsymbol{\Phi}$ to generate a new $N-1$ particle state $\mathsf{C}(\boldsymbol{\psi})\boldsymbol{\Phi}$. Let us determine the action of $\mathsf{C}(\boldsymbol{\psi})$ from the identity

$$\langle\mathsf{C}^\dagger(\boldsymbol{\psi})\boldsymbol{\Xi}, \boldsymbol{\Phi}\rangle = \langle\boldsymbol{\Xi}, \mathsf{C}(\boldsymbol{\psi})\boldsymbol{\Phi}\rangle, \tag{9.2.13}$$

for arbitrary $\boldsymbol{\Xi} \in \mathscr{H}_{N-1}^{\mathrm{wf}}$ and $\boldsymbol{\Phi} \in \mathscr{H}_N^{\mathrm{wf}}$ with $N = 1, 2, \ldots$. From (9.2.9), we find

$$\langle \mathsf{C}^\dagger(\boldsymbol{\psi})\Xi, \boldsymbol{\Phi}\rangle$$
$$= \sum_{u_1,\ldots,u_N} \frac{1}{\sqrt{N}} \sum_{j=1}^{N} (-1)^{j+1} \big\{\psi(u_j)\, \Xi(u_1,\ldots,\breve{u}_j,\ldots,u_N)\big\}^* \, \Phi(u_1,\ldots,u_N)$$

Let us fix j, and write the sequence $(u_1,\ldots,\breve{u}_j,\ldots,u_N)$ as $(w_1,\ldots,w_{N-1})$ (or, in other words, set $w_i = u_i$ for $i < j$ and $w_{i-1} = u_i$ for $i > j$). Then we have $\Phi(u_1,\ldots,u_N) = \Phi(w_1,\ldots,w_{j-1},u_j,w_j,\ldots,w_{N-1}) = (-1)^{j+1}\Phi(u_j,w_1,\ldots,w_{N-1})$, where we used the antisymmetry (9.2.4) to get the final identity. By rewriting u_j as v, we see that the summands with different j produce the same expression, and hence

$$= \sum_{w_1,\ldots,w_{N-1}} \{\Xi(w_1,\ldots,w_{N-1})\}^* \sqrt{N} \sum_{v} \{\psi(v)\}^* \, \Phi(v,w_1,\ldots,w_{N-1}). \quad (9.2.14)$$

By comparing this expression with the right-hand side of (9.2.13), we find for any $\boldsymbol{\Phi} \in \mathscr{H}_N^{\mathrm{wf}}$ with $N = 1, 2, \ldots$ that

$$(\mathsf{C}(\boldsymbol{\psi})\boldsymbol{\Phi})(u_1,\ldots,u_{N-1}) = \sqrt{N} \sum_{v\in\tilde{\Lambda}} \{\psi(v)\}^* \, \Phi(v,u_1,\ldots,u_{N-1}), \quad (9.2.15)$$

which is the desired expression of the annihilation operator. The right-hand side of (9.2.15) is obviously antisymmetric in $u_1,\ldots,u_{N-1}$. We also define

$$\mathsf{C}(\boldsymbol{\psi})\alpha = 0, \quad (9.2.16)$$

for any $\alpha \in \mathscr{H}_0^{\mathrm{wf}} \cong \mathbb{C}$ since $\mathsf{C}^\dagger(\boldsymbol{\psi})\boldsymbol{\Phi} \notin \mathscr{H}_0^{\mathrm{wf}}$ for all possible $\boldsymbol{\Phi}$.

Anticommutation relations We shall derive the anticommutation relations (9.2.19), (9.2.20), and (9.2.26), which provide the essential characterization of the creation and annihilation operators. Let $\boldsymbol{\varphi}, \boldsymbol{\psi} \in \tilde{\mathfrak{h}}$. For any $\Xi \in \mathscr{H}_N^{\mathrm{wf}}$ with $N = 2, 3, \ldots$, we find by using (9.2.15) twice that

$$\begin{aligned}(\mathsf{C}(\boldsymbol{\varphi})\mathsf{C}(\boldsymbol{\psi})\Xi)(u_1,\ldots,u_{N-2}) &= \sqrt{N-1} \sum_{v\in\tilde{\Lambda}} \{\varphi(v)\}^* (\mathsf{C}(\boldsymbol{\psi})\Xi)(v,u_1,\ldots,u_{N-2}) \\ &= \sqrt{N(N-1)} \sum_{v,w\in\tilde{\Lambda}} \{\varphi(v)\,\psi(w)\}^* \, \Xi(w,v,u_1,\ldots,u_{N-2}),\end{aligned} \quad (9.2.17)$$

and

$$(\mathsf{C}(\boldsymbol{\psi})\mathsf{C}(\boldsymbol{\varphi})\Xi)(u_1,\ldots,u_{N-2}) = \sqrt{N(N-1)} \sum_{v,w\in\tilde{\Lambda}} \{\psi(w)\,\varphi(v)\}^* \, \Xi(v,w,u_1,\ldots,u_{N-2}). \quad (9.2.18)$$

Noting that $\Xi(v, w, u_1, \ldots, u_{N-2}) = -\Xi(w, v, u_1, \ldots, u_{N-2})$, we find $\mathsf{C}(\boldsymbol{\varphi})\,\mathsf{C}(\boldsymbol{\psi})\Xi = -\mathsf{C}(\boldsymbol{\psi})\mathsf{C}(\boldsymbol{\varphi})\Xi$. Since Ξ is arbitrary this leads to the anticommutation relation for the annihilation operators

$$\{\mathsf{C}(\boldsymbol{\varphi}), \mathsf{C}(\boldsymbol{\psi})\} = 0, \tag{9.2.19}$$

for any $\boldsymbol{\varphi}, \boldsymbol{\psi} \in \tilde{\mathfrak{h}}$, where the anticommutator is defined as $\{\hat{A}, \hat{B}\} := \hat{A}\hat{B} + \hat{B}\hat{A}$ for any operators $\hat{A}$ and $\hat{B}$. By taking the adjoint, we also find for the creation operators that

$$\{\mathsf{C}^\dagger(\boldsymbol{\varphi}), \mathsf{C}^\dagger(\boldsymbol{\psi})\} = 0. \tag{9.2.20}$$

To evaluate the anticommutator $\{\mathsf{C}(\boldsymbol{\varphi}), \mathsf{C}^\dagger(\boldsymbol{\psi})\}$ is interesting but a little complicated. We encourage the reader to write down the case with $N = 2$ explicitly. Take an arbitrary $\Xi \in \mathscr{H}_N^{\mathrm{wf}}$ with $N = 1, 2, \ldots$. First it is easy to see from (9.2.9) and (9.2.15) that

$$\begin{aligned}
&(\mathsf{C}^\dagger(\boldsymbol{\psi})\mathsf{C}(\boldsymbol{\varphi})\Xi)(u_1, \ldots, u_N) \\
&\quad = \frac{1}{\sqrt{N}} \sum_{j=1}^{N} (-1)^{j+1} \psi(u_j) (\mathsf{C}(\boldsymbol{\varphi})\Xi)(u_1, \ldots, \breve{u}_j, \ldots, u_N) \\
&\quad = \sum_{j=1}^{N} (-1)^{j+1} \psi(u_j) \sum_{v} \{\varphi(v)\}^* \, \Xi(v, u_1, \ldots, \breve{u}_j, \ldots, u_N).
\end{aligned} \tag{9.2.21}$$

Next we simply use (9.2.15) to observe that

$$(\mathsf{C}(\boldsymbol{\varphi})\mathsf{C}^\dagger(\boldsymbol{\psi})\Xi)(u_1, \ldots, u_N) = \sqrt{N+1} \sum_{v} \{\varphi(v)\}^* (\mathsf{C}^\dagger(\boldsymbol{\psi})\Xi)(v, u_1, \ldots, u_N). \tag{9.2.22}$$

Then by using (9.2.9) we see that

$$\begin{aligned}
&(\mathsf{C}^\dagger(\boldsymbol{\psi})\Xi)(v, u_1, \ldots, u_N) \\
&\quad = \frac{1}{\sqrt{N+1}} \Bigl\{ \psi(v)\, \Xi(u_1, \ldots, u_N) + \sum_{j=1}^{N} (-1)^j \psi(u_j)\, \Xi(v, u_1, \ldots, \breve{u}_j, \ldots, u_N) \Bigr\}.
\end{aligned} \tag{9.2.23}$$

Note that j in this expression corresponds to $j - 1$ in (9.2.9). Substituting this back to (9.2.22), we get

$$(\mathsf{C}(\boldsymbol{\varphi})\mathsf{C}^\dagger(\boldsymbol{\psi})\Xi)(u_1,\ldots,u_N)$$
$$=\sum_v \{\varphi(v)\}^* \Big\{\psi(v)\,\Xi(u_1,\ldots,u_N)+\sum_{j=1}^{N}(-1)^j\psi(u_j)\,\Xi(v,u_1,\ldots,\breve{u}_j,\ldots,u_N)\Big\}. \tag{9.2.24}$$

By summing (9.2.21) and (9.2.24), we finally obtain

$$(\mathsf{C}(\boldsymbol{\varphi})\mathsf{C}^\dagger(\boldsymbol{\psi})\Xi+\mathsf{C}^\dagger(\boldsymbol{\psi})\mathsf{C}(\boldsymbol{\varphi})\Xi)(u_1,\ldots,u_N)=\langle\boldsymbol{\varphi},\boldsymbol{\psi}\rangle\,\Xi(u_1,\ldots,u_N), \tag{9.2.25}$$

for any $\boldsymbol{\Xi}$. This leads to the anticommutation relation

$$\{\mathsf{C}(\boldsymbol{\varphi}),\mathsf{C}^\dagger(\boldsymbol{\psi})\}=\langle\boldsymbol{\varphi},\boldsymbol{\psi}\rangle, \tag{9.2.26}$$

for any $\boldsymbol{\varphi},\boldsymbol{\psi}\in\tilde{\mathfrak{h}}$.

9.2.3 *The Fock Space Representation*

We now introduce the Fock space representation for tight-binding electron systems, which we shall use throughout Part III of the present book. The Fock space representation is also known as the "second quantization" formalism, but one should note that it is strictly equivalent to the conventional wave function formalism (for a many-particle system) discussed in Sects. 9.2.1 and 9.2.2. The unfortunate and misleading name "second quantization" simply reflects the (constructive) confusion in the early history.

As we have already explained, one can skip Sect. 9.2.2 and start reading this subsection after Sect. 9.2.1. Everything is explained in a logically consistent manner. The relation to the formalism in Sect. 9.2.2 is discussed at the end of the subsection.

Basic fermion operators With each lattice site $x\in\Lambda$ and spin index $\sigma=\uparrow,\downarrow$, we associate a fermion operator $\hat{c}_{x,\sigma}$. One can freely take the conjugate,[7] products, and linear combinations (with complex coefficients) of these operators (and the identity operator) to get new operators. We require that these operators satisfy the anticommutation relations

$$\{\hat{c}^\dagger_{x,\sigma},\hat{c}^\dagger_{y,\tau}\}=\big\{\hat{c}_{x,\sigma},\hat{c}_{y,\tau}\big\}=0, \tag{9.2.27}$$

and

$$\{\hat{c}_{x,\sigma},\hat{c}^\dagger_{y,\tau}\}=\delta_{x,y}\delta_{\sigma,\tau}, \tag{9.2.28}$$

for any $x,y\in\Lambda$ and $\sigma,\tau=\uparrow,\downarrow$, where the anticommutator is defined as $\{\hat{A},\hat{B}\}=\hat{A}\hat{B}+\hat{B}\hat{A}$. Note that (9.2.27) implies $(\hat{c}_{x,\sigma})^2=(\hat{c}^\dagger_{x,\sigma})^2=0$.

[7]We write $\hat{c}^\dagger_{x,\sigma}$ instead of $(\hat{c}_{x,\sigma})^\dagger$.

Physically speaking, $\hat{c}_{x,\sigma}$ and $\hat{c}^{\dagger}_{x,\sigma}$ are the annihilation and creation operators, respectively, of an electron at site x with spin σ. The corresponding number operator is defined as

$$\hat{n}_{x,\sigma} = \hat{c}^{\dagger}_{x,\sigma}\hat{c}_{x,\sigma}. \tag{9.2.29}$$

From (9.2.27) and (9.2.28), we find that

$$(\hat{n}_{x,\sigma})^2 = \hat{c}^{\dagger}_{x,\sigma}\hat{c}_{x,\sigma}\hat{c}^{\dagger}_{x,\sigma}\hat{c}_{x,\sigma} = \hat{c}^{\dagger}_{x,\sigma}(1 - \hat{c}^{\dagger}_{x,\sigma}\hat{c}_{x,\sigma})\hat{c}_{x,\sigma} = \hat{c}^{\dagger}_{x,\sigma}\hat{c}_{x,\sigma} = \hat{n}_{x,\sigma}. \tag{9.2.30}$$

Thus we obtain $\hat{n}_{x,\sigma}(1 - \hat{n}_{x,\sigma}) = 0$, which implies that $\hat{n}_{x,\sigma}$ can only have eigenvalues 0 and 1. This is another mathematical realization of the Pauli exclusion principle. It is also easy to see that

$$[\hat{n}_{x,\sigma}, \hat{c}^{\dagger}_{y,\tau}] = \delta_{x,y}\delta_{\sigma,\tau}\,\hat{c}^{\dagger}_{x,\sigma}, \quad [\hat{n}_{x,\sigma}, \hat{c}_{y,\tau}] = -\delta_{x,y}\delta_{\sigma,\tau}\,\hat{c}_{x,\sigma}, \tag{9.2.31}$$

and

$$[\hat{n}_{x,\sigma}, \hat{n}_{y,\tau}] = 0, \tag{9.2.32}$$

for any $x, y \in \Lambda$ and $\sigma, \tau = \uparrow, \downarrow$. We further define the number operator for site x as $\hat{n}_x = \hat{n}_{x,\uparrow} + \hat{n}_{x,\downarrow}$, and the total number operator as

$$\hat{N} := \sum_{x\in\Lambda} \hat{n}_x = \sum_{\substack{x\in\Lambda\\ \sigma=\uparrow,\downarrow}} \hat{n}_{x,\sigma}. \tag{9.2.33}$$

Hilbert space We can now construct the Hilbert space for many-electron problems. We start from a single normalized state $|\Phi_{\mathrm{vac}}\rangle$, which represents the (fictitious) state with no electrons in the lattice. This property is mathematically represented as

$$\hat{c}_{x,\sigma}|\Phi_{\mathrm{vac}}\rangle = 0 \quad \text{for any } x \in \Lambda \text{ and } \sigma = \uparrow, \downarrow. \tag{9.2.34}$$

We fix a positive integer N such that $1 \le N \le 2|\Lambda|$, which is the total number of electrons in the system. It is useful to express the electron number in terms of the filling factor $\nu = N/(2|\Lambda|)$, which takes a value in the range $0 < \nu \le 1$. We usually fix ν and make N and $|\Lambda|$ large. Take N arbitrary sites $x_1, x_2, \ldots, x_N \in \Lambda$ (with possible overlaps) and spin indices $\sigma_1, \sigma_2, \ldots, \sigma_N = \uparrow, \downarrow$, and define the state

$$|\Psi_{(x_1,\sigma_1),\ldots,(x_N,\sigma_N)}\rangle = \hat{c}^{\dagger}_{x_1,\sigma_1}\cdots\hat{c}^{\dagger}_{x_N,\sigma_N}|\Phi_{\mathrm{vac}}\rangle. \tag{9.2.35}$$

We interpret (9.2.35) as the state in which there is an electron at site x_i with spin σ_i for $i = 1, \ldots, N$. We allow all possible states of the form (9.2.35), but make proper identification according to the anticommutation relations (9.2.27). In particular, the state (9.2.35) is vanishing whenever $(x_i, \sigma_i) = (x_j, \sigma_j)$ for some $i \neq j$. This is noth-

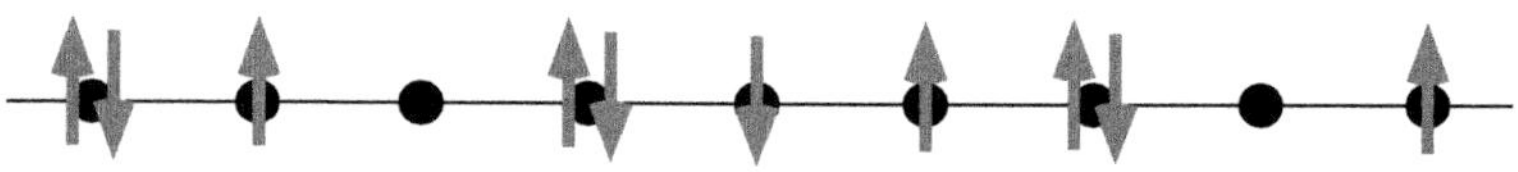

Fig. 9.2 An allowed configuration in a tight-binding model with a single orbit per site. Each site in the lattice can be either empty, singly occupied by an electron with up or down spin, or doubly occupied by electrons with opposite spins. If we consider the Hubbard type interaction, we get an extra energy $U_x \geq 0$ whenever two electrons occupy a single site x. See (9.3.28). In the simpler interaction (9.3.29) with uniform repulsive energy U, the interaction energy for the above configuration is $3U$

ing but the Pauli exclusion principle.[8] In other words, each site in the lattice can be either empty, singly occupied by an electron with up or down spin, or doubly occupied by electrons with opposite spins. See Fig. 9.2.

The Hilbert space $\mathscr{H}_N$ for the given electron number N consists of all possible linear combinations of the basis states (9.2.35). The inner product of any two states is determined by the basic relation $\langle\Phi_{\mathrm{vac}}|\Phi_{\mathrm{vac}}\rangle = 1$, the anticommutation relations (9.2.27), (9.2.28), and linearity.

Suppose that $|\Psi_{(x_1,\sigma_1),\ldots,(x_N,\sigma_N)}\rangle$ of (9.2.35) is nonvanishing. Then repeatedly using the commutation relation (9.2.31), and noting that $\hat{n}_{x,\sigma}|\Phi_{\mathrm{vac}}\rangle = 0$, one finds

$$\hat{n}_{x,\sigma}|\Psi_{(x_1,\sigma_1),\ldots,(x_N,\sigma_N)}\rangle = \begin{cases} |\Psi_{(x_1,\sigma_1),\ldots,(x_N,\sigma_N)}\rangle & \text{if } (x_i,\sigma_i) = (x,\sigma) \text{for some } i; \\ 0 & \text{otherwise,} \end{cases} \tag{9.2.36}$$

where we used the general relation (A.2.2) for commutators. This shows that, as expected, $\hat{n}_{x,\sigma}$ counts the number of electrons at x with spin σ. From this we find that $\hat{N}\,|\Psi_{(x_1,\sigma_1),\ldots,(x_N,\sigma_N)}\rangle = N\,|\Psi_{(x_1,\sigma_1),\ldots,(x_N,\sigma_N)}\rangle$.

The Hilbert space that contains all possible electron numbers

$$\mathscr{F} := \mathscr{H}_0 \oplus \mathscr{H}_1 \oplus \cdots \oplus \mathscr{H}_{2|\Lambda|}, \tag{9.2.37}$$

where we set $\mathscr{H}_0 \cong \mathbb{C}$, is know as the Fock space. A general element of $\mathscr{F}$ is $\sum_{N=0}^{2|\Lambda|}|\Phi^{(N)}\rangle$ with arbitrary $|\Phi^{(N)}\rangle \in \mathscr{H}_N$. Note that the operators $\hat{c}_{x,\sigma}$, $\hat{c}^\dagger_{x,\sigma}$, and $\hat{n}_{x,\sigma}$ can be regarded as operators that act on the Fock space $\mathscr{F}$. This is the reason that we call the present formalism the Fock space representation. In the present chapter, however, we exclusively work within the Hilbert space $\mathscr{H}_N$ with a fixed electron number, and do not make use of the Fock space.[9] This is sufficient since the number of electrons is strictly conserved.

Spin operators Since electrons carry spin $S = 1/2$, we can associate the spin operator $\hat{\mathbf{S}}_x = (\hat{S}^{(1)}_x, \hat{S}^{(2)}_x, \hat{S}^{(3)}_x)$ with each site $x \in \Lambda$. To define the spin operators, write the components of spin matrices (2.1.7) for $S = 1/2$ as $S^{(\alpha)}_{\sigma,\tau}$, where

[8]If we set $N > 2|\Lambda|$, then the state (9.2.35) always vanishes.

[9]Recall that in Sect. 5.3 we used a subspace (5.3.1) the Fock space for bosons to discuss symmetry breaking associated with Bose–Einstein condensation.

$\alpha = 1, 2, 3$ and $\sigma, \tau = \uparrow, \downarrow$. To be explicit, nonvanishing components are given by $S^{(1)}_{\uparrow,\downarrow} = S^{(1)}_{\downarrow,\uparrow} = 1/2$, $S^{(2)}_{\uparrow,\downarrow} = -S^{(2)}_{\downarrow,\uparrow} = -i/2$, and $S^{(3)}_{\uparrow,\uparrow} = -S^{(3)}_{\downarrow,\downarrow} = 1/2$. We then define

$$\hat{S}^{(\alpha)}_x := \sum_{\sigma,\tau=\uparrow,\downarrow} \hat{c}^\dagger_{x,\sigma}\, S^{(\alpha)}_{\sigma,\tau}\, \hat{c}_{x,\tau}, \tag{9.2.38}$$

for each $x \in \Lambda$. It is evident from (9.2.34) that $\hat{S}^{(\alpha)}_x |\Phi_{\rm vac}\rangle = 0$.

Let us show that these operators are indeed spin operators in the sense that they satisfy the commutation relations (2.1.1) for quantum mechanical angular momenta. Let $(\alpha, \beta, \gamma) = (1, 2, 3), (2, 3, 1)$, or $(3, 1, 2)$ so that we expect $[\hat{S}^{(\alpha)}_x, \hat{S}^{(\beta)}_x] = i\, \hat{S}^{(\gamma)}_x$. Noting that $S^{(\alpha)}_{\sigma,\tau}$ are mere complex coefficients, we find from (9.2.38) that

$$[\hat{S}^{(\alpha)}_x, \hat{S}^{(\beta)}_x] = \sum_{\sigma,\tau,\mu,\nu=\uparrow,\downarrow} S^{(\alpha)}_{\sigma,\tau}\, S^{(\beta)}_{\mu,\nu}\, [\hat{c}^\dagger_{x,\sigma}\hat{c}_{x,\tau}, \hat{c}^\dagger_{x,\mu}\hat{c}_{x,\nu}]. \tag{9.2.39}$$

By an explicit calculation using (9.2.27) and (9.2.28), one finds

$$[\hat{c}^\dagger_{x,\sigma}\hat{c}_{x,\tau}, \hat{c}^\dagger_{x,\mu}\hat{c}_{x,\nu}] = \delta_{\tau,\mu}\, \hat{c}^\dagger_{x,\sigma}\hat{c}_{x,\nu} - \delta_{\sigma,\nu}\hat{c}^\dagger_{x,\mu}\hat{c}_{x,\tau}. \tag{9.2.40}$$

See the solution to the Problem 9.2.3.b below. We therefore get

$$\begin{aligned}
[\hat{S}^{(\alpha)}_x, \hat{S}^{(\beta)}_x] &= \sum_{\sigma,\tau,\nu=\uparrow,\downarrow} S^{(\alpha)}_{\sigma,\tau}\, S^{(\beta)}_{\tau,\nu}\, \hat{c}^\dagger_{x,\sigma}\hat{c}_{x,\nu} - \sum_{\sigma,\tau,\mu=\uparrow,\downarrow} S^{(\alpha)}_{\sigma,\tau}\, S^{(\beta)}_{\mu,\sigma}\, \hat{c}^\dagger_{x,\mu}\hat{c}_{x,\tau} \\
&= \sum_{\sigma,\tau,\mu=\uparrow,\downarrow} (S^{(\alpha)}_{\sigma,\tau}\, S^{(\beta)}_{\tau,\mu} - S^{(\beta)}_{\sigma,\tau}\, S^{(\alpha)}_{\tau,\mu})\hat{c}^\dagger_{x,\sigma}\hat{c}_{x,\mu} \\
&= \sum_{\sigma,\mu=\uparrow,\downarrow} i\, S^{(\gamma)}_{\sigma,\mu}\, \hat{c}^\dagger_{x,\sigma}\hat{c}_{x,\mu} = i\, \hat{S}^{(\gamma)}_x,
\end{aligned} \tag{9.2.41}$$

which is the desired commutation relation.

Since it is obvious that

$$[\hat{S}^{(\alpha)}_x, \hat{S}^{(\beta)}_y] = 0, \tag{9.2.42}$$

for any α and β when $x \neq y$, we see that the total spin operator

$$\hat{\boldsymbol{S}}_{\rm tot} := \sum_{x\in\Lambda} \hat{\boldsymbol{S}}_x \tag{9.2.43}$$

also satisfies the commutation relation (2.1.1), and hence is a quantum mechanical angular momentum. This fact will be of essential importance when we discuss magnetism in the Hubbard model. As usual, we denote the eigenvalue of $(\hat{\boldsymbol{S}}_{\rm tot})^2$ as $S_{\rm tot}(S_{\rm tot}+1)$, where $S_{\rm tot} = 0, 1, \ldots, N/2$ if N is even, and $S_{\rm tot} = 1/2, 3/2, \ldots, N/2$ if N is odd.

By using the explicit form of spin matrices, one finds from (9.2.38) that

$$\hat{S}^{+}_{x}=\hat{c}^{\dagger}_{x,\uparrow}\hat{c}_{x,\downarrow},\quad \hat{S}^{-}_{x}=\hat{c}^{\dagger}_{x,\downarrow}\hat{c}_{x,\uparrow},\quad \hat{S}^{(3)}_{x}=\frac{\hat{n}_{x,\uparrow}-\hat{n}_{x,\downarrow}}{2},\tag{9.2.44}$$

where $\hat{S}^{\pm}_{x}=\hat{S}^{(1)}_{x}\pm i\hat{S}^{(2)}_{x}$. See Problem 9.2.3.a below. If we (tentatively) define $|\psi^{\sigma}_{x}\rangle=\hat{c}^{\dagger}_{x,\sigma}|\Phi_{\rm vac}\rangle$ for $\sigma=\uparrow,\downarrow$, one finds from the anticommutation relations (9.2.27) and (9.2.28) that the operators (9.2.44) precisely recover the expected rules (2.1.4) and (2.1.5) for a quantum spin with $S=1/2$. For the total spin operators $\hat{S}^{(\alpha)}_{\rm tot}:=\sum_{x\in\Lambda}\hat{S}^{(\alpha)}_{x}$ and $\hat{S}^{\pm}_{\rm tot}:=\hat{S}^{(1)}_{\rm tot}\pm i\hat{S}^{(2)}_{\rm tot}$, we of course have

$$\hat{S}^{+}_{\rm tot}=\sum_{x\in\Lambda}\hat{c}^{\dagger}_{x,\uparrow}\hat{c}_{x,\downarrow},\quad \hat{S}^{-}_{\rm tot}=\sum_{x\in\Lambda}\hat{c}^{\dagger}_{x,\downarrow}\hat{c}_{x,\uparrow},\quad \hat{S}^{(3)}_{\rm tot}=\frac{1}{2}\sum_{x\in\Lambda}(\hat{n}_{x,\uparrow}-\hat{n}_{x,\downarrow}).\tag{9.2.45}$$

Problem 9.2.3.a Write down $\hat{S}^{(1)}_{x}$, $\hat{S}^{(2)}_{x}$, and $\hat{S}^{(3)}_{x}$ explicitly by using the definition (9.2.38), and confirm the expressions (9.2.44). Express $(\hat{\boldsymbol{S}}_{x})^2=(\hat{S}^{(1)}_{x})^2+(\hat{S}^{(2)}_{x})^2+(\hat{S}^{(3)}_{x})^2$ in terms of the number operator $\hat{n}_x$, and discuss the implication of the expression. [solution→p.514]

General fermion operators Let $\boldsymbol{\varphi}=(\varphi(x,\sigma))_{x\in\Lambda,\sigma=\uparrow,\downarrow}\in\tilde{\mathfrak{h}}$ be an arbitrary state (wave function) of a singe electron. (See Sect. 9.2.1.) We define the corresponding creation and annihilation operators by

$$\hat{C}^{\dagger}(\boldsymbol{\varphi})=\sum_{\substack{x\in\Lambda\\ \sigma=\uparrow,\downarrow}}\varphi(x,\sigma)\,\hat{c}^{\dagger}_{x,\sigma},\quad \hat{C}(\boldsymbol{\varphi})=\sum_{\substack{x\in\Lambda\\ \sigma=\uparrow,\downarrow}}\{\varphi(x,\sigma)\}^{*}\,\hat{c}_{x,\sigma}.\tag{9.2.46}$$

Clearly $\hat{C}^{\dagger}(\boldsymbol{\varphi})$ is linear in $\boldsymbol{\varphi}$, i.e., $\hat{C}^{\dagger}(\alpha\boldsymbol{\varphi}+\beta\boldsymbol{\psi})=\alpha\,\hat{C}^{\dagger}(\boldsymbol{\varphi})+\beta\,\hat{C}^{\dagger}(\boldsymbol{\psi})$ for any $\boldsymbol{\varphi},\boldsymbol{\psi}\in\tilde{\mathfrak{h}}$ and $\alpha,\beta\in\mathbb{C}$. From the basic anticommutation relations (9.2.27) and (9.2.28), we see that these operators also satisfy the anticommutation relations

$$\{\hat{C}(\boldsymbol{\varphi}),\hat{C}(\boldsymbol{\psi})\}=\{\hat{C}^{\dagger}(\boldsymbol{\varphi}),\hat{C}^{\dagger}(\boldsymbol{\psi})\}=0,\tag{9.2.47}$$

$$\{\hat{C}(\boldsymbol{\varphi}),\hat{C}^{\dagger}(\boldsymbol{\psi})\}=\langle\boldsymbol{\varphi},\boldsymbol{\psi}\rangle,\tag{9.2.48}$$

for any $\boldsymbol{\varphi},\boldsymbol{\psi}\in\tilde{\mathfrak{h}}$. Again (9.2.47) implies $(\hat{C}(\boldsymbol{\varphi}))^2=(\hat{C}^{\dagger}(\boldsymbol{\varphi}))^2=0$.

Let A be an operator which acts on the single-electron Hilbert space $\tilde{\mathfrak{h}}$. It can be regarded as a $2|\Lambda|\times 2|\Lambda|$ matrix $\mathsf{A}=(a_{(x,\sigma),(y,\tau)})_{x,y\in\Lambda,\sigma,\tau=\uparrow,\downarrow}$. We define the corresponding bilinear form of fermion operators by

$$\hat{B}(\mathsf{A})=\sum_{\substack{x,y\in\Lambda\\ \sigma,\tau=\uparrow,\downarrow}}\hat{c}^{\dagger}_{x,\sigma}\,a_{(x,\sigma),(y,\tau)}\,\hat{c}_{y,\tau},\tag{9.2.49}$$

which is sometimes called the "second quantization" of the operator A. By using the anticommutation relations (9.2.27) and (9.2.28), one easily verifies the commutation relation

$$[\hat{B}(\mathsf{A}), \hat{C}^\dagger(\boldsymbol{\varphi})] = \hat{C}^\dagger(\mathsf{A}\boldsymbol{\varphi}), \tag{9.2.50}$$

which sheds light on the relation between the single-particle quantum mechanics and the Fock space representation. This relation will be useful later. A similar commutation relation for two bilinear forms

$$[\hat{B}(\mathsf{A}), \hat{B}(\mathsf{B})] = \hat{B}([\mathsf{A}, \mathsf{B}]) \tag{9.2.51}$$

is also easy to prove.

Problem 9.2.3.b Verify the relations (9.2.50) and (9.2.51). [solution→p.514]

Slater determinant state Let $\boldsymbol{\varphi}^{(1)}, \ldots, \boldsymbol{\varphi}^{(N)} \in \tilde{\mathfrak{h}}$ be arbitrary single-electron states (wave functions), and consider the N electron state of the form

$$|\Phi\rangle = \hat{C}^\dagger(\boldsymbol{\varphi}^{(1)}) \cdots \hat{C}^\dagger(\boldsymbol{\varphi}^{(N)})|\Phi_{\text{vac}}\rangle. \tag{9.2.52}$$

This state is known as the Slater determinant state,[10] and plays an important role in the theory of many fermions. Roughly speaking the Slater determinant state (9.2.52) describes the state in which N electrons occupy each of the single-electron states $\boldsymbol{\varphi}^{(1)}, \ldots, \boldsymbol{\varphi}^{(N)}$.

For arbitrary $\boldsymbol{\psi}^{(1)}, \ldots, \boldsymbol{\psi}^{(N)} \in \tilde{\mathfrak{h}}$, define another Slater determinant state $|\Psi\rangle = \hat{C}^\dagger(\boldsymbol{\psi}^{(1)}) \cdots \hat{C}^\dagger(\boldsymbol{\psi}^{(N)})|\Phi_{\text{vac}}\rangle$. We have the following useful lemma.

Lemma 9.1 *The inner product of the two Slater determinant states $|\Phi\rangle$ and $|\Psi\rangle$ is given by*

$$\begin{aligned}\langle\Phi|\Psi\rangle &= \begin{vmatrix} \langle\boldsymbol{\varphi}^{(1)}, \boldsymbol{\psi}^{(1)}\rangle & \langle\boldsymbol{\varphi}^{(1)}, \boldsymbol{\psi}^{(2)}\rangle & \cdots & \langle\boldsymbol{\varphi}^{(1)}, \boldsymbol{\psi}^{(N)}\rangle \\ \langle\boldsymbol{\varphi}^{(2)}, \boldsymbol{\psi}^{(1)}\rangle & \langle\boldsymbol{\varphi}^{(2)}, \boldsymbol{\psi}^{(2)}\rangle & \cdots & \langle\boldsymbol{\varphi}^{(2)}, \boldsymbol{\psi}^{(N)}\rangle \\ \vdots & \vdots & \ddots & \vdots \\ \langle\boldsymbol{\varphi}^{(N)}, \boldsymbol{\psi}^{(1)}\rangle & \langle\boldsymbol{\varphi}^{(N)}, \boldsymbol{\psi}^{(2)}\rangle & \cdots & \langle\boldsymbol{\varphi}^{(N)}, \boldsymbol{\psi}^{(N)}\rangle \end{vmatrix} \\ &= \sum_p (-1)^p \prod_{j=1}^N \langle\boldsymbol{\varphi}^{(j)}, \boldsymbol{\psi}^{(p(j))}\rangle, \end{aligned} \tag{9.2.53}$$

where p is summed over all permutations of $\{1, \ldots, N\}$. Here $|\mathsf{A}|$ denotes the determinant of an $N \times N$ matrix A.

Proof The relation is shown by noting that

$$\langle\Phi|\Psi\rangle = \langle\Phi_{\text{vac}}|\hat{C}(\boldsymbol{\varphi}^{(N)}) \cdots \hat{C}(\boldsymbol{\varphi}^{(1)})\hat{C}^\dagger(\boldsymbol{\psi}^{(1)}) \cdots \hat{C}^\dagger(\boldsymbol{\psi}^{(N)})|\Phi_{\text{vac}}\rangle, \tag{9.2.54}$$

and repeatedly using (9.2.48) and $\hat{C}(\boldsymbol{\varphi})|\Phi_{\text{vac}}\rangle = 0$. First observe that

[10]The wave function corresponding to this state is $(N!)^{-1/2}\sum_p(-1)^p\varphi^{(1)}(u_{p(1)})\cdots\varphi^{(N)}(u_{p(N)})$, where p is summed over all permutations. The sum has a precise form of a determinant.

$$\hat{C}(\boldsymbol{\varphi}^{(1)})\hat{C}^\dagger(\boldsymbol{\psi}^{(1)})\cdots\hat{C}^\dagger(\boldsymbol{\psi}^{(N)})|\Phi_{\text{vac}}\rangle$$
$$=\sum_{j=1}^{N}(-1)^{j-1}\langle\boldsymbol{\varphi}^{(1)},\boldsymbol{\psi}^{(j)}\rangle\,\hat{C}^\dagger(\boldsymbol{\psi}^{(1)})\cdots\hat{C}^\dagger(\boldsymbol{\psi}^{(j-1)})\hat{C}^\dagger(\boldsymbol{\psi}^{(j+1)})\cdots\hat{C}^\dagger(\boldsymbol{\psi}^{(N)})|\Phi_{\text{vac}}\rangle. \tag{9.2.55}$$

where $\hat{C}^\dagger(\boldsymbol{\psi}_j)$ is missing in the summand. This procedure can be repeated until all $\hat{C}^\dagger$ and $\hat{C}$ are gone. We easily find for general N that

$$\hat{C}(\boldsymbol{\varphi}^{(N)})\cdots\hat{C}(\boldsymbol{\varphi}^{(1)})\hat{C}^\dagger(\boldsymbol{\psi}^{(1)})\cdots\hat{C}^\dagger(\boldsymbol{\psi}^{(N)})|\Phi_{\text{vac}}\rangle=\sum_{p}\eta(p)\prod_{j=1}^{N}\langle\boldsymbol{\varphi}^{(j)},\boldsymbol{\psi}^{(p(j))}\rangle|\Phi_{\text{vac}}\rangle, \tag{9.2.56}$$

and hence

$$\langle\Phi|\Psi\rangle=\sum_{p}\eta(p)\prod_{j=1}^{N}\langle\boldsymbol{\varphi}^{(j)},\boldsymbol{\psi}^{(p(j))}\rangle, \tag{9.2.57}$$

where p is summed over all permutations, and $\eta(p)=\pm1$ is a certain sign factor that depends only on p. We need to show that $\eta(p)=(-1)^p$. This can be done by examining the sign appearing in relations like (9.2.55), but there is an easier trick. Fix a permutation p_0, and note that the expression (9.2.54) and the anticommutation relation (9.2.47) imply

$$\langle\Phi|\Psi\rangle=(-1)^{p_0}\,\langle\Phi_{\text{vac}}|\hat{C}(\boldsymbol{\varphi}^{(N)})\cdots\hat{C}(\boldsymbol{\varphi}^{(1)})\hat{C}^\dagger(\boldsymbol{\psi}^{(p_0(1))})\cdots\hat{C}^\dagger(\boldsymbol{\psi}^{(p_0(N))})|\Phi_{\text{vac}}\rangle. \tag{9.2.58}$$

Exactly as before the right-hand side can be rewritten as in (9.2.57), but we know for sure that the result contains the term $(-1)^{p_0}\prod_{j=1}^{N}\langle\boldsymbol{\varphi}^{(j)},\boldsymbol{\psi}^{(p_0(j))}\rangle$, which has the desired sign factor. Since p_0 is arbitrary we have shown that $\eta(p)=(-1)^p$. ∎

We now state the following lemma, which provides an essential characterization of Slater determinant states.

Lemma 9.2 *The Slater determinant state (9.2.52) is nonvanishing if and only if the N states $\boldsymbol{\varphi}^{(1)},\ldots,\boldsymbol{\varphi}^{(N)}$ are linearly independent.*

Proof We note that (9.2.53) implies $\langle\Phi|\Phi\rangle=|\mathsf{G}|$, where G is the Gram matrix associated with $\boldsymbol{\varphi}^{(1)},\ldots,\boldsymbol{\varphi}^{(N)}$, i.e., an $N\times N$ matrix defined by $(\mathsf{G})_{j,k}=\langle\boldsymbol{\varphi}^{(j)},\boldsymbol{\varphi}^{(k)}\rangle$. It is a well known theorem in linear algebra that determinant $|\mathsf{G}|$ is nonzero if and only if $\boldsymbol{\varphi}^{(1)},\ldots,\boldsymbol{\varphi}^{(N)}$ are linearly independent. ∎

The following lemma is also interesting. It says that the Slater determinant state (9.2.52) is completely determined by the subspace spanned by the states $\boldsymbol{\varphi}^{(1)},\ldots,\boldsymbol{\varphi}^{(N)}$.

Lemma 9.3 *Suppose that two sets $\{\boldsymbol{\varphi}^{(1)},\ldots,\boldsymbol{\varphi}^{(N)}\}$ and $\{\boldsymbol{\psi}^{(1)},\ldots,\boldsymbol{\psi}^{(N)}\}$ of states in $\mathfrak{h}$ span the same N-dimensional subspace of $\mathfrak{h}$. Then there is a nonvanishing constant $\alpha\in\mathbb{C}$, and we have*

$$\hat{C}^\dagger(\boldsymbol{\varphi}^{(1)})\cdots\hat{C}^\dagger(\boldsymbol{\varphi}^{(N)})|\Phi_{\text{vac}}\rangle = \alpha\,\hat{C}^\dagger(\boldsymbol{\psi}^{(1)})\cdots\hat{C}^\dagger(\boldsymbol{\psi}^{(N)})|\Phi_{\text{vac}}\rangle. \tag{9.2.59}$$

Proof From the assumption we have $\boldsymbol{\varphi}^{(j)} = \sum_{k=1}^{N} \beta_{j,k}\,\boldsymbol{\psi}^{(k)}$ with some $\beta_{j,k} \in \mathbb{C}$ for any $j = 1, \ldots, N$. Thus we get

$$\hat{C}^\dagger(\boldsymbol{\varphi}^{(1)})\cdots\hat{C}^\dagger(\boldsymbol{\varphi}^{(N)}) = \sum_{k_1,\ldots,k_N=1}^{N} \beta_{1,k_1}\cdots\beta_{N,k_N}\,\hat{C}^\dagger(\boldsymbol{\psi}^{(k_1)})\cdots\hat{C}^\dagger(\boldsymbol{\psi}^{(k_N)}). \tag{9.2.60}$$

Since $\{\hat{C}^\dagger(\boldsymbol{\psi}^{(k)})\}^2 = 0$, the product $\hat{C}^\dagger(\boldsymbol{\psi}^{(k_1)})\cdots\hat{C}^\dagger(\boldsymbol{\psi}^{(k_N)})$ can be either vanishing or equal to $\pm\hat{C}^\dagger(\boldsymbol{\psi}^{(1)})\cdots\hat{C}^\dagger(\boldsymbol{\psi}^{(N)})$, we thus get

$$\hat{C}^\dagger(\boldsymbol{\varphi}^{(1)})\cdots\hat{C}^\dagger(\boldsymbol{\varphi}^{(N)}) = \alpha\,\hat{C}^\dagger(\boldsymbol{\psi}^{(1)})\cdots\hat{C}^\dagger(\boldsymbol{\psi}^{(N)}). \tag{9.2.61}$$

Note that α cannot be zero since the left-hand side is nonvanishing because of the previous Lemma 9.2. ∎

Spin dependent operators It is also convenient to define fermion operators which explicitly depend on the spin index $\sigma =\uparrow,\downarrow$. Let us denote by $\mathfrak{h}$ the Hilbert space of a single spinless particle on Λ. More precisely $\mathfrak{h} \cong \mathbb{C}^{|\Lambda|}$ consists of wave functions of the form $\boldsymbol{\varphi} = (\varphi(x))_{x\in\Lambda}$. For any $\boldsymbol{\varphi} \in \mathfrak{h}$ and $\sigma =\uparrow,\downarrow$, we define, as in (9.2.46), the corresponding creation and annihilation operators by

$$\hat{C}^\dagger_\sigma(\boldsymbol{\varphi}) = \sum_{x\in\Lambda}\varphi(x)\,\hat{c}^\dagger_{x,\sigma}, \quad \hat{C}_\sigma(\boldsymbol{\varphi}) = \sum_{x\in\Lambda}\{\varphi(x)\}^*\,\hat{c}_{x,\sigma}. \tag{9.2.62}$$

The anticommutation relations, which immediately follow from (9.2.27) and (9.2.28), are

$$\{\hat{C}_\sigma(\boldsymbol{\varphi}), \hat{C}_\tau(\boldsymbol{\psi})\} = \{\hat{C}^\dagger{}_\sigma(\boldsymbol{\varphi}), \hat{C}^\dagger{}_\tau(\boldsymbol{\psi})\} = 0, \tag{9.2.63}$$

$$\{\hat{C}_\sigma(\boldsymbol{\varphi}), \hat{C}^\dagger{}_\tau(\boldsymbol{\psi})\} = \langle\boldsymbol{\varphi}, \boldsymbol{\psi}\rangle\,\delta_{\sigma,\tau}, \tag{9.2.64}$$

for any $\boldsymbol{\varphi}, \boldsymbol{\psi} \in \mathfrak{h}$ and $\sigma, \tau =\uparrow,\downarrow$.

The following lemma is elementary but will be used frequently.

Lemma 9.4 (General basis of $\mathscr{H}_N$) *Let $\boldsymbol{\varphi}^{(1)}, \ldots, \boldsymbol{\varphi}^{(|\Lambda|)}$ be $|\Lambda|$ arbitrary states in $\mathfrak{h}$ which are linearly independent (but not necessarily orthogonal with each other). Then the states*

$$|\Gamma_{S_\uparrow,S_\downarrow}\rangle = \left(\prod_{j\in S_\uparrow}\hat{C}^\dagger_\uparrow(\boldsymbol{\varphi}^{(j)})\right)\left(\prod_{j\in S_\downarrow}\hat{C}^\dagger_\downarrow(\boldsymbol{\varphi}^{(j)})\right)|\Phi_{\text{vac}}\rangle \tag{9.2.65}$$

with arbitrary subsets $S_\uparrow$ and $S_\downarrow$ of $\{1, 2, \ldots, |\Lambda|\}$ such that $|S_\uparrow| + |S_\downarrow| = N$ span the Hilbert space $\mathscr{H}_N$.

Proof Since the $|\Lambda|$ vectors are linearly independent, we have $\delta_{x,y} = \sum_{j=1}^{|\Lambda|} a_{x,j} \varphi_y^{(j)}$ with some regular matrix $(a_{x,j})_{x\in\Lambda, j=1,\dots,|\Lambda|}$. Then we find from the definition (9.2.62) that

$$\hat{c}_{x,\sigma}^{\dagger} = \sum_{j=1}^{|\Lambda|} a_{x,j}\, \hat{C}_{\sigma}^{\dagger}(\boldsymbol{\varphi}^{(j)}), \tag{9.2.66}$$

which means that the basis state (9.2.35) can be written as a linear combination of the states (9.2.65). ∎

From the expression (9.2.45) of the total spin operators, the basic anticommutation relations (9.2.27), (9.2.28), and the commutation relations (9.2.31), we find

$$\begin{aligned} &[\hat{S}_{\rm tot}^{+}, \hat{c}_{x,\uparrow}^{\dagger}] = 0, && [\hat{S}_{\rm tot}^{-}, \hat{c}_{x,\uparrow}^{\dagger}] = \hat{c}_{x,\downarrow}^{\dagger}, && [\hat{S}_{\rm tot}^{(3)}, \hat{c}_{x,\uparrow}^{\dagger}] = \frac{1}{2}\hat{c}_{x,\uparrow}^{\dagger}, \\ &[\hat{S}_{\rm tot}^{+}, \hat{c}_{x,\downarrow}^{\dagger}] = \hat{c}_{x,\uparrow}^{\dagger}, && [\hat{S}_{\rm tot}^{-}, \hat{c}_{x,\downarrow}^{\dagger}] = 0, && [\hat{S}_{\rm tot}^{(3)}, \hat{c}_{x,\downarrow}^{\dagger}] = -\frac{1}{2}\hat{c}_{x,\downarrow}^{\dagger}. \end{aligned} \tag{9.2.67}$$

This automatically implies for any $\boldsymbol{\varphi} \in \mathfrak{h}$ that

$$\begin{aligned} &\big[\hat{S}_{\rm tot}^{+}, \hat{C}^{\dagger}{}_{\uparrow}(\boldsymbol{\varphi})\big] = 0, && \big[\hat{S}_{\rm tot}^{-}, \hat{C}^{\dagger}{}_{\uparrow}(\boldsymbol{\varphi})\big] = \hat{C}^{\dagger}{}_{\downarrow}(\boldsymbol{\varphi}), && \big[\hat{S}_{\rm tot}^{(3)}, \hat{C}^{\dagger}{}_{\uparrow}(\boldsymbol{\varphi})\big] = \frac{1}{2}\hat{C}^{\dagger}{}_{\uparrow}(\boldsymbol{\varphi}), \\ &\big[\hat{S}_{\rm tot}^{+}, \hat{C}^{\dagger}{}_{\downarrow}(\boldsymbol{\varphi})\big] = \hat{C}^{\dagger}{}_{\uparrow}(\boldsymbol{\varphi}), && \big[\hat{S}_{\rm tot}^{-}, \hat{C}^{\dagger}{}_{\downarrow}(\boldsymbol{\varphi})\big] = 0, && \big[\hat{S}_{\rm tot}^{(3)}, \hat{C}^{\dagger}{}_{\downarrow}(\boldsymbol{\varphi})\big] = -\frac{1}{2}\hat{C}^{\dagger}{}_{\downarrow}(\boldsymbol{\varphi}). \end{aligned} \tag{9.2.68}$$

One then finds

$$\begin{aligned} \big[\hat{S}_{\rm tot}^{+}, \hat{C}^{\dagger}{}_{\uparrow}(\boldsymbol{\varphi})\,\hat{C}^{\dagger}{}_{\downarrow}(\boldsymbol{\varphi})\big] &= \big[\hat{S}_{\rm tot}^{+}, \hat{C}^{\dagger}{}_{\uparrow}(\boldsymbol{\varphi})\big]\,\hat{C}^{\dagger}{}_{\downarrow}(\boldsymbol{\varphi}) + \hat{C}^{\dagger}{}_{\uparrow}(\boldsymbol{\varphi})\,\big[\hat{S}_{\rm tot}^{+}, \hat{C}^{\dagger}{}_{\downarrow}(\boldsymbol{\varphi})\big] \\ &= \{\hat{C}^{\dagger}{}_{\uparrow}(\boldsymbol{\varphi})\}^2 = 0, \end{aligned} \tag{9.2.69}$$

and similarly that $\big[\hat{S}_{\rm tot}^{-}, \hat{C}^{\dagger}{}_{\uparrow}(\boldsymbol{\varphi})\,\hat{C}^{\dagger}{}_{\downarrow}(\boldsymbol{\varphi})\big] = 0$. It is also easily checked that $\big[\hat{S}_{\rm tot}^{(3)}, \hat{C}^{\dagger}{}_{\uparrow}(\boldsymbol{\varphi})\,\hat{C}^{\dagger}{}_{\downarrow}(\boldsymbol{\varphi})\big] = 0$. We have thus confirmed that, for any $\boldsymbol{\varphi} \in \mathfrak{h}$,

$$\big[\hat{S}_{\rm tot}^{(\alpha)}, \hat{C}^{\dagger}{}_{\uparrow}(\boldsymbol{\varphi})\,\hat{C}^{\dagger}{}_{\downarrow}(\boldsymbol{\varphi})\big] = 0, \tag{9.2.70}$$

for $\alpha = 1, 2, 3$, i.e., the operator $\hat{C}^{\dagger}{}_{\uparrow}(\boldsymbol{\varphi})\,\hat{C}^{\dagger}{}_{\downarrow}(\boldsymbol{\varphi})$ is SU(2) invariant. This is a mathematical expression of the well known fact that, when a single state $\boldsymbol{\varphi} \in \mathfrak{h}$ is occupied by two electrons with opposite spins, the total spin state must be a spin-singlet.[11] (See Appendix A.3.3.)

The commutation relations (9.2.70) immediately imply

[11] There is a shorter (but abstract) proof of the SU(2) invariance based on Lemma 9.3. The two-electron state created by $\hat{C}^{\dagger}{}_{\uparrow}(\boldsymbol{\varphi})\,\hat{C}^{\dagger}{}_{\downarrow}(\boldsymbol{\varphi})$ corresponds to the two-dimensional subspace of $\tilde{\mathfrak{h}}$ spanned by $\varphi(x)\,\delta_{\sigma,\uparrow}$ and $\varphi(x)\,\delta_{\sigma,\downarrow}$, which is $\{\boldsymbol{\varphi}\} \times \mathbb{C}^2$. Since the subspace is invariant under the action of SU(2) (i.e., the spin rotation acting on $\mathbb{C}^2$), the state is also invariant.

$$\big[(\hat{S}_{\mathrm{tot}})^2, \hat{C}^\dagger{}_\uparrow(\varphi)\, \hat{C}^\dagger{}_\downarrow(\varphi)\big] = 0, \tag{9.2.71}$$

where $(\hat{S}_{\mathrm{tot}})^2 = (\hat{S}^{(1)}_{\mathrm{tot}})^2 + (\hat{S}^{(2)}_{\mathrm{tot}})^2 + (\hat{S}^{(3)}_{\mathrm{tot}})^2$. Since $(\hat{S}_{\mathrm{tot}})^2|\Phi_{\mathrm{vac}}\rangle = 0$, we readily see

$$(\hat{S}_{\mathrm{tot}})^2\, \hat{C}^\dagger{}_\uparrow(\varphi)\, \hat{C}^\dagger{}_\downarrow(\varphi)|\Phi_{\mathrm{vac}}\rangle = 0, \tag{9.2.72}$$

i.e., the state $\hat{C}^\dagger{}_\uparrow(\varphi)\, \hat{C}^\dagger{}_\downarrow(\varphi)|\Phi_{\mathrm{vac}}\rangle$ is a spin-singlet, as we have discussed above.

For an operator (i.e., a $|\Lambda| \times |\Lambda|$ matrix) A which acts on $\mathfrak{h}$, we define, for $\sigma = \uparrow, \downarrow$,

$$\hat{B}_\sigma(\mathsf{A}) = \sum_{x,y\in\Lambda} \hat{c}^\dagger_{x,\sigma}\, (\mathsf{A})_{x,y}\, \hat{c}_{y,\sigma}, \tag{9.2.73}$$

which is nothing but the second quantization (9.2.49) restricted to a single value of σ. Then we can show, as in (9.2.50),

$$[\hat{B}_\sigma(\mathsf{A}), \hat{C}^\dagger{}_\tau(\varphi)] = \delta_{\sigma,\tau}\, \hat{C}^\dagger_\sigma(\mathsf{A}\varphi), \tag{9.2.74}$$

for any $\varphi \in \mathfrak{h}$ and $\sigma, \tau = \uparrow, \downarrow$.

Problem 9.2.3.c Consider a two-electron state

$$|\xi\rangle = \sum_{x,y\in\Lambda} \xi(x,y)\, \hat{c}^\dagger_{x,\uparrow}\, \hat{c}^\dagger_{y,\downarrow}|\Phi_{\mathrm{vac}}\rangle, \tag{9.2.75}$$

where $\xi(x, y)$ is a complex valued function on $\Lambda \times \Lambda$. Compute $(\hat{S}_{\mathrm{tot}})^2|\xi\rangle$, assuming that $\xi(\cdot,\cdot)$ is symmetric, i.e., $\xi(x, y) = \xi(y, x)$, or antisymmetric, i.e., $\xi(x, y) = -\xi(y, x)$. [solution→p.515]

The solution of the above problem might appear to be inconsistent with the fact that, in a system of two spins with $S = 1/2$, the states $|\Phi_{1,0}\rangle = (|\uparrow\rangle_1|\downarrow\rangle_2 + |\downarrow\rangle_1|\uparrow\rangle_2)/\sqrt{2}$ and $|\Phi_{0,0}\rangle = (|\uparrow\rangle_1|\downarrow\rangle_2 - |\downarrow\rangle_1|\uparrow\rangle_2)/\sqrt{2}$ have $S_{\mathrm{tot}} = 1$ and $S_{\mathrm{tot}} = 0$, respectively. See (A.3.22) and (A.3.23). To see that these results are consistent, note that the standard representation in spin systems is recovered if fermion operators are ordered according to the sites. Therefore one of the spin triplet is written as

$$|\Phi_{1,0}\rangle = \frac{1}{\sqrt{2}}(\hat{c}^\dagger_{1,\uparrow}\hat{c}^\dagger_{2,\downarrow} + \hat{c}^\dagger_{1,\downarrow}\hat{c}^\dagger_{2,\uparrow})|\Phi_{\mathrm{vac}}\rangle. \tag{9.2.76}$$

If we order the fermion operators according the spin index, this becomes $|\Phi_{1,0}\rangle = (\hat{c}^\dagger_{1,\uparrow}\hat{c}^\dagger_{2,\downarrow} - \hat{c}^\dagger_{2,\uparrow}\hat{c}^\dagger_{1,\downarrow})|\Phi_{\mathrm{vac}}\rangle/\sqrt{2}$, which is written in the form (9.2.75) with antisymmetric $\xi(x, y)$. Similarly the spin singlet is written as

$$|\Phi_{0,0}\rangle = \frac{1}{\sqrt{2}}(\hat{c}^\dagger_{1,\uparrow}\hat{c}^\dagger_{2,\downarrow} - \hat{c}^\dagger_{1,\downarrow}\hat{c}^\dagger_{2,\uparrow})|\Phi_{\mathrm{vac}}\rangle = \frac{1}{\sqrt{2}}(\hat{c}^\dagger_{1,\uparrow}\hat{c}^\dagger_{2,\downarrow} + \hat{c}^\dagger_{2,\uparrow}\hat{c}^\dagger_{1,\downarrow})|\Phi_{\mathrm{vac}}\rangle. \tag{9.2.77}$$

Equivalence to the wave function formalism Since we have completely described the Fock space representation (or the "second quantization" formalism), let us see that it is equivalent to the wave function formalism discussed in Sects. 9.2.1 and 9.2.2. This is indeed almost trivial. The reader who skipped Sect. 9.2.2 should skip this part as well.

The first essential observation is that the creation and annihilation operators $\mathsf{C}^\dagger(\boldsymbol{\varphi})$ and $\mathsf{C}(\boldsymbol{\varphi})$ constructed in the wave function formalism and the same operators $\hat{C}^\dagger(\boldsymbol{\varphi})$ and $\hat{C}(\boldsymbol{\varphi})$ introduced in the Fock space representation should be identified. Indeed both the operators have the same linear (or antil-inear) dependence on the single-electron state $\boldsymbol{\varphi} \in \tilde{\mathfrak{h}}$, and satisfy exactly the same anticommutation relations, namely, (9.2.19), (9.2.20), and (9.2.26) for $\mathsf{C}^\dagger(\boldsymbol{\varphi})$, $\mathsf{C}(\boldsymbol{\varphi})$, and (9.2.47) and (9.2.48) for $\hat{C}^\dagger(\boldsymbol{\varphi})$, $\hat{C}(\boldsymbol{\varphi})$.

The second essential observation is that the vacuum state $|\Phi_{\rm vac}\rangle \in \mathscr{H}_0$ in the Fock space representation can be identified with $1 \in \mathscr{H}_0^{\rm wf}$ in the wave function formalism. Since the requirement (9.2.34) for the vacuum is written as $\hat{C}(\boldsymbol{\varphi})|\Phi_{\rm vac}\rangle = 0$ for any $\boldsymbol{\varphi} \in \tilde{\mathfrak{h}}$, it precisely corresponds to the relation (9.2.16). Then by applying creation operators and taking linear combinations, we can extend this identification to $\mathscr{H}_N^{\rm wf}$ and $\mathscr{H}_N$ with any $N \le 2|\Lambda|$.

9.3 Definition of the Hubbard Model

We shall now define the Hubbard model. We start from the single-electron problem in Sect. 9.3.1, and treat many-electron problem in Sect. 9.3.2. We also summarize some useful transformations for the Hubbard Hamiltonian in Sect. 9.3.3.

9.3.1 Single-Electron Schrödinger Equation

Let us return to the tight-binding description of a single electron that we introduced in Sect. 9.2.1, and discuss the corresponding Schrödinger equation.

The most general Schrödinger equation for an electron on the lattice Λ is

$$\sum_{\substack{y\in\Lambda\\ \tau=\uparrow,\downarrow}} \tilde{t}_{(x,\sigma),(y,\tau)}\,\varphi(y,\tau) = \varepsilon\,\varphi(x,\sigma) \quad \text{for any } x \in \Lambda \text{ and } \sigma = \uparrow,\downarrow, \tag{9.3.1}$$

where the generalized hopping amplitude $(\tilde{t}_{(x,\sigma),(y,\tau)})_{x,y\in\Lambda,\sigma,\tau=\uparrow,\downarrow}$ satisfies $\tilde{t}_{(x,\sigma),(y,\tau)} = (\tilde{t}_{(y,\tau),(x,\sigma)})^* \in \mathbb{C}$. The generalized hopping amplitude should (in principle) be determined from the electronic structure of the material, but is sometimes simply given by hand. Here ε and $\varphi(x,\sigma)$ are the energy eigenvalue and the energy eigenstate (wave function) to be determined. The generalized hopping amplitude describes several different physical objects. The diagonal term $\tilde{t}_{(x,\sigma),(x,\sigma)}$ is the (possibly σ dependent) on-site potential. The term $\tilde{t}_{(x,\sigma),(y,\sigma)}$ with $x \neq y$ represents (again, possibly σ dependent) hopping of electron from y to x, while $\tilde{t}_{(x,\sigma),(y,\tau)}$ with $\sigma \neq \tau$ represents a spin-flip (possibly accompanied by hopping).

In the present book, we restrict ourselves to a smaller class of models where $\tilde{t}_{(x,\sigma),(y,\tau)}$ is given by

$$\tilde{t}_{(x,\sigma),(y,\tau)} = \begin{cases} t_{x,y} & \text{if } \sigma = \tau, \\ 0 & \text{if } \sigma \neq \tau, \end{cases} \tag{9.3.2}$$

with the hopping amplitude $(t_{x,y})_{x,y\in\Lambda}$ satisfying $t_{x,y} = (t_{y,x})^* \in \mathbb{C}$. This means that we are excluding spin-flip processes, and spin-dependence in on-site potential or hopping. Physically speaking, we are assuming that the spin-orbit coupling is negligible.

With (9.3.2), the Schrödinger equation (9.3.1) decouples into two identical equations for $\sigma =\uparrow$ and $\downarrow$. It suffices to consider the Schrödinger equation for a spinless particle[12]

$$\sum_{y\in\Lambda} t_{x,y}\,\varphi(y) = \varepsilon\,\varphi(x) \quad \text{for any } x \in \Lambda. \tag{9.3.3}$$

If we define a $|\Lambda| \times |\Lambda|$ hopping matrix T by $(\mathsf{T})_{x,y} = t_{x,y}$, the Schrödinger equation (9.3.3) is compactly written as

$$\mathsf{T}\boldsymbol{\varphi} = \varepsilon\,\boldsymbol{\varphi}, \tag{9.3.4}$$

where $\boldsymbol{\varphi} = (\varphi(x))_{x\in\Lambda} \in \mathfrak{h}$ can be regarded as a $|\Lambda|$ dimensional (column) vector.

Since T is Hermitian, the eigenvalue equation (9.3.4) has $|\Lambda|$ real eigenvalues $\varepsilon_1, \ldots, \varepsilon_{|\Lambda|}$. The corresponding eigenstate $\boldsymbol{\psi}^{(j)} = (\psi_x^{(j)})_{x\in\Lambda}$ satisfies

$$\mathsf{T}\boldsymbol{\psi}^{(j)} = \varepsilon_j\boldsymbol{\psi}^{(j)}, \tag{9.3.5}$$

for each $j = 1, \ldots, |\Lambda|$. We can assume (if necessary) that the energy eigenstates form an orthonormal basis, i.e., $\langle\boldsymbol{\psi}^{(j)}, \boldsymbol{\psi}^{(k)}\rangle = \delta_{j,k}$ for any $j, k = 1, \ldots, |\Lambda|$.

Example Let us take a look at the simplest example. Consider the d-dimensional hypercubic lattice Λ_L with L^d sites and the corresponding set of bonds $\mathscr{B}_L$ defined in (3.1.2) and (3.1.3), respectively. We set

$$t_{x,y} = \begin{cases} -t & \text{if } \{x, y\} \in \mathscr{B}_L, \\ 0 & \text{otherwise}, \end{cases} \tag{9.3.6}$$

where $t \in \mathbb{R}$ is the hopping amplitude. Then the corresponding Schrödinger equation (9.3.3) becomes

$$-t\sum_{y\in\mathscr{N}(x)} \varphi(y) = \varepsilon\,\varphi(x) \quad \text{for any} x \in \Lambda_L, \tag{9.3.7}$$

[12]It is also common to put a minus sign in front of $t_{x,y}$.

where $\mathscr{N}(x) = \{y \mid \{x, y\} \in \mathscr{B}_L\}$ is the set of sites neighboring to x.

It is easily found that the energy eigenstates and eigenvalues are parameterized by the wave number vector $\boldsymbol{k} = (k_1, \ldots, k_d) \in \mathscr{K}_L$, where the set $\mathscr{K}_L$ is defined in (4.1.17). For each $\boldsymbol{k} \in \mathscr{K}_L$, one has $\mathsf{T}\boldsymbol{\psi}^{(\boldsymbol{k})} = \varepsilon(\boldsymbol{k})\,\boldsymbol{\psi}^{(\boldsymbol{k})}$ with

$$\psi_x^{(\boldsymbol{k})} = L^{-d/2} \exp\Bigl[i \sum_{j=1}^{d} k_j x_j\Bigr], \tag{9.3.8}$$

where we wrote $x = (x_1, \ldots, x_d)$, and

$$\varepsilon(\boldsymbol{k}) = -2t \sum_{j=1}^{d} \cos k_j. \tag{9.3.9}$$

In this simplest example, the energy eigenstates (9.3.8) are of the form of the "plane-wave", and energy eigenvalues (9.3.9) obey the simple cosine dispersion relation.

Band structure in a translation invariant system The energy spectrum in the above model can be regarded as the simplest example of an energy band in a single-electron problem with translation invariance. We shall here discuss the general theory of the band structure in a translation invariant model.

Let $\mathscr{U}$ be a finite set that we call the unit cell. We consider a decorated hypercubic lattice $\Lambda = \Lambda_L \times \mathscr{U}$, where Λ_L is again the $L \times \cdots \times L$ hypercubic lattice in (3.1.2). We shall fix the unit cell $\mathscr{U}$ and let L become large. A site $x \in \Lambda$ is thus written as $x = (r, u)$ with $r \in \Lambda_L$ and $u \in \mathscr{U}$. For a site $x = (r, u)$ we define its translation by $s \in \mathbb{Z}^d$ as $x + s = (r + s, u)$. Here, and in what follows, we use periodic boundary conditions for the hypercubic lattice Λ_L.

Suppose that we have a hopping matrix $\mathsf{T} = (t_{x,y})_{x,y\in\Lambda}$ on Λ which is translation invariant, i.e., $t_{x+s,y+s} = t_{x,y}$ for any $x, y \in \Lambda$ and $s \in \mathbb{Z}^d$. Because of the translation invariance, one can assume that the energy eigenstate satisfying the Schrödinger equation (9.3.3) is written as[13]

$$\varphi(r, u) = e^{i\boldsymbol{k}\cdot r}\xi_u, \tag{9.3.10}$$

where $\boldsymbol{k} = (k_1, \ldots, k_d)$ is an element of $\mathscr{K}_L$ defined in (4.1.17). We here wrote $r = (r_1, \ldots, r_d)$ and $\boldsymbol{k} \cdot r = \sum_{j=1}^{d} k_j r_j$.

Substituting (9.3.10) into (9.3.3), we get

$$\sum_{s\in\Lambda_L,\, v\in\mathscr{U}} t_{(r,u),(s,v)}\, e^{i\boldsymbol{k}\cdot s}\xi_v = \varepsilon\, e^{i\boldsymbol{k}\cdot r}\xi_u, \tag{9.3.11}$$

for any $(r, u) \in \Lambda$. By using the translation invariance of $t_{x,y}$ this is rewritten as

[13]To see this it suffices to note that the translation operator and the hopping matrix T commute with each other, and hence are simultaneously diagonalizable.

$$\sum_{v \in \mathscr{U}} S^{(\boldsymbol{k})}_{u,v}\, \xi_v = \varepsilon\, \xi_u, \tag{9.3.12}$$

for each $\boldsymbol{k} \in \mathscr{K}_L$, where the effective hopping is defined by

$$S^{(\boldsymbol{k})}_{u,v} = \sum_{s \in \Lambda_L} t_{(o,u),(s,v)}\, e^{i\boldsymbol{k}\cdot s}, \tag{9.3.13}$$

with $o = (0, \ldots, 0) \in \Lambda_L$. We note that the above observation corresponds to the Bloch theorem for the Schrödinger equation in a periodic potential.

Since the effective hopping matrix $\mathsf{S}^{(\boldsymbol{k})} = (S^{(\boldsymbol{k})}_{u,v})_{u,v\in\mathscr{U}}$ (with a fixed $\boldsymbol{k} \in \mathscr{K}_L$) is Hermitian, it has $|\mathscr{U}|$ (possibly degenerate) eigenvalues, which we denote as $\varepsilon_\mu(\boldsymbol{k})$ with $\mu = 1, \ldots, |\mathscr{U}|$. We can assume that each $\varepsilon_\mu(\boldsymbol{k})$ depends continuously on $\boldsymbol{k} \in [-\pi, \pi]^d$. This means that the whole spectrum (i.e., the set of eigenvalues) of the hopping matrix T is written as $\cup_{\mu=1}^{|\mathscr{U}|} \mathrm{Spec}_\mu$ with $\mathrm{Spec}_\mu = \{\varepsilon_\mu(\boldsymbol{k}) \,|\, \boldsymbol{k} \in \mathscr{K}_L\}$, where Spec_μ is known as the μ-th energy band. The eigenvalue $\varepsilon_\mu(\boldsymbol{k})$ as a function of $\boldsymbol{k}$ is called the dispersion relation of the μ-th band.

Simple model with two bands We shall examine a simple one-dimensional model with a nontrivial band structure. See Problem 11.3.1.a (p. 399) for a more complicated example with flat bands.

Let L be an even integer, and take the one-dimensional lattice $\Lambda = \{1, \frac{3}{2}, 2, \ldots, L, L + \frac{1}{2}, \}$ with periodic boundary conditions (where we identify $L + 1$ with 1), and set

$$t_{x,y} = \begin{cases} -t & \text{if } |x - y| = 1/2, \\ V & \text{if } x = y \text{ is an integer}, \\ 0 & \text{otherwise}, \end{cases} \tag{9.3.14}$$

with $t, V \in \mathbb{R}$. We have on-site potential V on every other sites.

Here we can set $\Lambda_L = \{1, 2, \ldots, L\}$, $\mathscr{U} = \{0, \frac{1}{2}\}$ and express any $x \in \Lambda$ as $x = r + u$ with $r \in \Lambda_L$ and $u \in \mathscr{U}$. For $k \in \mathscr{K}_L = \{(2\pi/L)n \,|\, n = -(L/2) + 1, \ldots, L/2\}$, the effective hopping matrix (9.3.13) is given by

$$\mathsf{S}^{(k)} = \begin{pmatrix} V & -t(1 + e^{-ik}) \\ -t(1 + e^{ik}) & 0 \end{pmatrix}. \tag{9.3.15}$$

By solving the eigenvalue equation, one finds that the present model has two bands with dispersion relations

$$\varepsilon_\pm(k) = \frac{V \pm \sqrt{V^2 + \{4t\cos(k/2)\}^2}}{2}. \tag{9.3.16}$$

9.3.2 Hamiltonian of the Hubbard Model

We are now ready to write down the Hamiltonian of the Hubbard model. Let us start from the many-electron model without interaction, and then add the interaction Hamiltonian.

Non-interacting system We consider a model of many electrons on the lattice Λ which undergoes quantum mechanical motion described by the hopping amplitude $\mathsf{T} = (t_{x,y})_{x,y\in\Lambda}$. We still neglect interactions between electrons. Then the Hamiltonian is given by

$$\hat{H}_{\text{hop}} := \sum_{\sigma=\uparrow,\downarrow} \hat{B}_\sigma(\mathsf{T}) = \sum_{\substack{x,y\in\Lambda \\ \sigma=\uparrow,\downarrow}} t_{x,y}\,\hat{c}^\dagger_{x,\sigma}\hat{c}_{y,\sigma}, \tag{9.3.17}$$

where $\hat{B}_\sigma(\cdot)$ is defined in (9.2.73).

To see that (9.3.17) describes the desired hopping, take a single-electron state $\hat{C}^\dagger_\sigma(\boldsymbol{\varphi})|\Phi_{\text{vac}}\rangle$ with $\boldsymbol{\varphi} \in \mathfrak{h}$, and consider the Schrödinger equation

$$\hat{H}_{\text{hop}}\hat{C}^\dagger_\sigma(\boldsymbol{\varphi})|\Phi_{\text{vac}}\rangle = \varepsilon\,\hat{C}^\dagger_\sigma(\boldsymbol{\varphi})|\Phi_{\text{vac}}\rangle. \tag{9.3.18}$$

Noting that $\hat{B}_\sigma(\mathsf{T})|\Phi_{\text{vac}}\rangle = 0$, the left-hand side of (9.3.18) is evaluated as

$$\begin{aligned}\hat{H}_{\text{hop}}\hat{C}^\dagger_\sigma(\boldsymbol{\varphi})|\Phi_{\text{vac}}\rangle &= \sum_{\tau=\uparrow,\downarrow} \hat{B}_\tau(\mathsf{T})\,\hat{C}^\dagger_\sigma(\boldsymbol{\varphi})|\Phi_{\text{vac}}\rangle \\ &= [\hat{B}_\sigma(\mathsf{T}), \hat{C}^\dagger_\sigma(\boldsymbol{\varphi})]\,|\Phi_{\text{vac}}\rangle = \hat{C}^\dagger_\sigma(\mathsf{T}\boldsymbol{\varphi})|\Phi_{\text{vac}}\rangle,\end{aligned} \tag{9.3.19}$$

where we used the commutation relation (9.2.74). Since this should be equal to the right-hand side of (9.3.18), we get $\mathsf{T}\boldsymbol{\varphi} = \varepsilon\,\boldsymbol{\varphi}$, which is nothing but the single-electron Schrödinger equation (9.3.4).

It also follows from the commutation relations (9.2.31) that $[\hat{H}_{\text{hop}}, \hat{N}] = 0$. This is obvious since hopping never changes the electron number.

Assuming that we have obtained the energy eigenstates and eigenvalues of the single-electron Schrödinger equation (9.3.3) or (9.3.4), it is easy to diagonalize the many-body Hamiltonian (9.3.17). For $j = 1, 2, \ldots, |\Lambda|$, let $\boldsymbol{\psi}^{(j)}$ be the eigenstate satisfying (9.3.5) with the energy eigenvalue ε_j. Let us assume (for convenience) that $\{\boldsymbol{\psi}^{(j)}\}_{j=1,\ldots,|\Lambda|}$ forms an orthonormal basis. Define the new fermion operator by

$$\hat{a}^\dagger_{j,\sigma} = \hat{C}^\dagger_\sigma(\boldsymbol{\psi}^{(j)}), \tag{9.3.20}$$

for $j = 1, 2, \ldots, |\Lambda|$ and $\sigma = \uparrow, \downarrow$. From the commutation relation (9.2.74), we see

$$[\hat{H}_{\text{hop}}, \hat{a}^\dagger_{j,\sigma}] = [\hat{B}_\sigma(\mathsf{T}), \hat{C}^\dagger_\sigma(\boldsymbol{\psi}^{(j)})] = \hat{C}^\dagger_\sigma[\mathsf{T}\boldsymbol{\psi}^{(j)}] = \varepsilon_j\,\hat{a}^\dagger_{j,\sigma}. \tag{9.3.21}$$

Let $S_\uparrow$ and $S_\downarrow$ be arbitrary subsets of $\{1, 2, \ldots, |\Lambda|\}$ such that $|S_\uparrow| + |S_\downarrow| = N$, and define

$$|\Psi_{S_\uparrow,S_\downarrow}\rangle = \Bigl(\prod_{j\in S_\uparrow} \hat{a}^\dagger_{j,\uparrow}\Bigr)\Bigl(\prod_{j\in S_\downarrow} \hat{a}^\dagger_{j,\downarrow}\Bigr)|\Phi_{\rm vac}\rangle. \tag{9.3.22}$$

Lemma 9.4 shows that these states form a basis of $\mathscr{H}_N$. By repeatedly using the commutation relation (9.3.21), and noting that $\hat{H}_{\rm hop}|\Phi_{\rm vac}\rangle = 0$, one finds

$$\hat{H}_{\rm hop}|\Psi_{S_\uparrow,S_\downarrow}\rangle = \Bigl(\sum_{j\in S_\uparrow} \varepsilon_j + \sum_{j\in S_\downarrow} \varepsilon_j\Bigr)|\Psi_{S_\uparrow,S_\downarrow}\rangle, \tag{9.3.23}$$

where we used the relation (A.2.2) for commutators. We have thus diagonalized the hopping Hamiltonian $\hat{H}_{\rm hop}$. See also Problem 9.3.2.a below.

By choosing subsets $S_\uparrow$, $S_\downarrow$ which minimize the energy eigenvalue $\sum_{j\in S_\uparrow} \varepsilon_j + \sum_{j\in S_\downarrow} \varepsilon_j$, we get ground state(s) of the present non-interacting model. In particular, if the single-electron energy eigenvalues are nondegenerate and the electron number N is even, the ground state of $\hat{H}_{\rm hop}$ is unique and written as

$$|\Phi_{\rm GS}\rangle = \Bigl(\prod_{j=1}^{N/2} \hat{a}^\dagger_{j,\uparrow}\hat{a}^\dagger_{j,\downarrow}\Bigr)|\Phi_{\rm vac}\rangle. \tag{9.3.24}$$

We here ordered the energy eigenvalues as $\varepsilon_j < \varepsilon_{j+1}$ for $j = 1, \ldots, |\Lambda| - 1$. The ground state (9.3.24) is nothing but the state obtained by "filling up" the low energy levels with both up and down spin electrons, as one learns in elementary quantum mechanics (Fig. 9.3). Recall that, as we have seen in (9.2.71), the operator $\hat{a}^\dagger_{j,\uparrow}\hat{a}^\dagger_{j,\downarrow}$ creates a spin-singlet of two electrons occupying the same state $\psi^{(j)}$. In particular (9.2.71) and $(\hat{\boldsymbol{S}}_{\rm tot})^2|\Phi_{\rm vac}\rangle = 0$ immediately implies $(\hat{\boldsymbol{S}}_{\rm tot})^2|\Phi_{\rm GS}\rangle = 0$, and hence $S_{\rm tot} = 0$, for the ground state (9.3.24). We see that the ground state has no spin at all; it exhibits paramagnetism.

In the two examples in Sect. 9.3.1, we have seen that the single-electron wave function of the energy eigenstate behaves as "waves" as in (9.3.8) or (9.3.10). This is indeed a common feature if the hopping amplitude has translation invariance. The fact that the Hamiltonian $\hat{H}_{\rm hop}$ is diagonalized in the basis (9.3.22) suggests that the electrons generally behave as "waves" in non-interacting many-electron systems.

Problem 9.3.2.a Show that the operators $\hat{a}_{j,\sigma}$ satisfy anticommutation relations

$$\{\hat{a}_{j,\sigma}, \hat{a}_{k,\tau}\} = \{\hat{a}^\dagger_{j,\sigma}, \hat{a}^\dagger_{k,\tau}\} = 0, \tag{9.3.25}$$

$$\{\hat{a}_{j,\sigma}, \hat{a}^\dagger_{k,\tau}\} = \delta_{j,k}\,\delta_{\sigma,\tau}, \tag{9.3.26}$$

for any $j, k = 1, \ldots, |\Lambda|$ and $\sigma, \tau = \uparrow, \downarrow$. Note that these relations have exactly the same forms as the basic anticommutation relations (9.2.27) and (9.2.28). Then show that the hopping Hamiltonian (9.3.17) is written in a diagonalized form

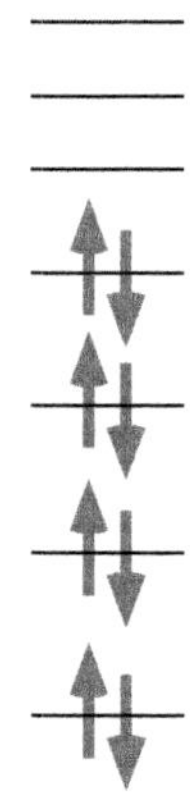

Fig. 9.3 Schematic picture of the ground state of a non-interacting many-electron system. The lowest $N/2$ single-electron energy levels are "filled" by both up spin and down spin electrons. The state naturally exhibits paramagnetism known as Pauli paramagnetism (© Hal Tasaki 2020. All Rights Reserved)

$$\hat{H}_{\text{hop}} = \sum_{\sigma=\uparrow,\downarrow} \sum_{j=1}^{|\Lambda|} \varepsilon_j \, \hat{\tilde{n}}_{j,\sigma}, \tag{9.3.27}$$

where $\hat{\tilde{n}}_{j,\sigma} = \hat{a}^{\dagger}_{j,\sigma}\hat{a}_{j,\sigma}$ is the number operator for the single-electron state $\boldsymbol{\psi}^{(j)}$. (Hint: The completeness condition $\sum_{j=1}^{|\Lambda|} \psi^{(j)}_z (\psi^{(j)}_y)^* = \delta_{z,y}$ is useful.) Show (9.3.23) by using the expression (9.3.27). [solution→p.515]

Interaction Hamiltonian It is believed that interactions between electrons in a solid mainly come from the electrostatic Coulomb force. Although the Coulomb force in vacuum is long-ranged, we consider an extremely short-range interaction which acts only when two electrons occupy the same site (i.e., the same atomic orbit). One crude justification of such a short-range (Coulomb) interaction comes from the observation that the Coulomb force should be most dominant when two electrons approach within the minimum possible distance. In a slightly more sophisticated justification, one argues that the long range Coulomb force is screened (in a certain self-consistent manner) by other electrons including those in different orbital states which we had decided to forget. Our point of view, based on the philosophy of universality, is that models with artificial short-range interactions are worth studying because they are among the minimum models which can be studied to elicit universal properties of strongly interacting electron systems.

The Hamiltonian for such short-range Coulomb repulsion is given by

$$\hat{H}_{\text{int}} := \sum_{x\in\Lambda} U_x \, \hat{n}_{x,\uparrow}\hat{n}_{x,\downarrow}, \tag{9.3.28}$$

where $U_x \geq 0$ is the energy from the repulsive interaction at site x. From the (trivial) commutation relation (9.2.32), we see that $[\hat{H}_{\text{int}}, \hat{N}] = 0$, which is obvious. It is also common to study the following simplified interaction Hamiltonian where the repulsive energy is set $U_x = U \geq 0$ for all x:

$$\hat{H}_{\text{int}} = U \sum_{x\in\Lambda} \hat{n}_{x,\uparrow}\hat{n}_{x,\downarrow} \tag{9.3.29}$$

See Fig. 9.2.

Note that the expression (9.3.28) is already diagonalized, and it is straightforward to write down eigenstates of $\hat{H}_{\text{int}}$. Let $\Gamma_\uparrow$ and $\Gamma_\downarrow$ be arbitrary subsets of the lattice Λ such that $|\Gamma_\uparrow| + |\Gamma_\downarrow| = N$, and define

$$|\Xi_{\Gamma_\uparrow,\Gamma_\downarrow}\rangle = \Bigl(\prod_{x\in\Gamma_\uparrow} \hat{c}^\dagger_{x,\uparrow}\Bigr)\Bigl(\prod_{x\in\Gamma_\downarrow} \hat{c}^\dagger_{x,\downarrow}\Bigr)|\Phi_{\text{vac}}\rangle. \tag{9.3.30}$$

Clearly these states form a basis of $\mathscr{H}_N$. By recalling (A.2.2), and repeatedly using the commutation relation $[\hat{n}_{x,\sigma}, \hat{c}^\dagger_{y,\tau}] = \delta_{x,y}\delta_{\sigma,\tau}\,\hat{c}^\dagger_{y,\tau}$, we find for $\sigma, \tau = \uparrow, \downarrow$ that

$$\Bigl[\hat{n}_{x,\sigma}, \Bigl(\prod_{y\in\Gamma_\tau} \hat{c}^\dagger_{y,\tau}\Bigr)\Bigr] = \begin{cases} \Bigl(\prod_{y\in\Gamma_\tau} \hat{c}^\dagger_{y,\tau}\Bigr) & \text{if } \sigma = \tau \text{ and } x \in \Gamma_\tau, \\ 0 & \text{otherwise.} \end{cases} \tag{9.3.31}$$

We thus have

$$\Bigl[\hat{n}_{x,\uparrow}\hat{n}_{x,\downarrow}, \Bigl(\prod_{y\in\Gamma_\uparrow} \hat{c}^\dagger_{y,\uparrow}\Bigr)\Bigl(\prod_{y\in\Gamma_\downarrow} \hat{c}^\dagger_{y,\downarrow}\Bigr)\Bigr] = \begin{cases} \Bigl(\prod_{y\in\Gamma_\uparrow} \hat{c}^\dagger_{y,\uparrow}\Bigr)\Bigl(\prod_{y\in\Gamma_\downarrow} \hat{c}^\dagger_{y,\downarrow}\Bigr) & \text{if } x \in \Gamma_\uparrow \cap \Gamma_\downarrow, \\ \Bigl(\prod_{y\in\Gamma_\uparrow} \hat{c}^\dagger_{y,\uparrow}\Bigr)\Bigl(\prod_{y\in\Gamma_\downarrow} \hat{c}^\dagger_{y,\downarrow}\Bigr)\hat{n}_{x,\uparrow} & \text{if } x \in \Gamma_\downarrow \backslash \Gamma_\uparrow, \\ \Bigl(\prod_{y\in\Gamma_\uparrow} \hat{c}^\dagger_{y,\uparrow}\Bigr)\Bigl(\prod_{y\in\Gamma_\downarrow} \hat{c}^\dagger_{y,\downarrow}\Bigr)\hat{n}_{x,\downarrow} & \text{if } x \in \Gamma_\uparrow \backslash \Gamma_\downarrow, \\ 0 & \text{otherwise.} \end{cases} \tag{9.3.32}$$

With this commutation relation and $\hat{n}_{x,\sigma}|\Phi_{\text{vac}}\rangle = 0$, we readily find

$$\hat{H}_{\text{int}}|\Xi_{\Gamma_\uparrow,\Gamma_\downarrow}\rangle = \Bigl(\sum_{x\in\Gamma_\uparrow\cap\Gamma_\downarrow} U_x\Bigr)\,|\Xi_{\Gamma_\uparrow,\Gamma_\downarrow}\rangle, \tag{9.3.33}$$

which indeed is an obvious result, since the interaction (9.3.28) essentially picks up doubly occupied sites.

Then the ground state of $\hat{H}_{\text{int}}$ for a given electron number N can be constructed by choosing subsets $\Gamma_\uparrow$, $\Gamma_\downarrow$ that minimize the energy eigenvalue $\sum_{x\in\Gamma_\uparrow\cap\Gamma_\downarrow} U_x$. When $N \le |\Lambda|$, one can always choose $\Gamma_\uparrow$ and $\Gamma_\downarrow$ such that $\Gamma_\uparrow \cap \Gamma_\downarrow = \emptyset$. In this case, the ground state energy is 0, and the ground states are highly degenerate.

We have thus observed the interaction Hamiltonian $\hat{H}_{\text{int}}$ is most naturally treated if we regard electrons as (classical) "particles" which live on lattice sites.

The Hubbard model The Hubbard model is a tight-binding electron model in which electrons hop around the lattice and interact with each other through short-range repulsive interactions. The full Hamiltonian of the Hubbard model is simply

$$\hat{H} = \hat{H}_{\mathrm{hop}} + \hat{H}_{\mathrm{int}}, \tag{9.3.34}$$

where, of course, $\hat{H}_{\mathrm{hop}}$ is the hopping Hamiltonian (9.3.17) and $\hat{H}_{\mathrm{int}}$ is the interaction Hamiltonian (9.3.28) or (9.3.29). Apparently we have $[\hat{H}, \hat{N}] = [\hat{H}_{\mathrm{hop}}, \hat{N}] + [\hat{H}_{\mathrm{int}}, \hat{N}] = 0$. We can thus look for eigenstates of $\hat{H}$ in $\mathscr{H}_N$ with a fixed electron number N.

We have already seen that each of $\hat{H}_{\mathrm{hop}}$ or $\hat{H}_{\mathrm{int}}$ can be easily diagonalized. We have also observed, however, that electrons behave as "waves" in $\hat{H}_{\mathrm{hop}}$, while they behave as "particles" in $\hat{H}_{\mathrm{int}}$. How do they behave in a system whose Hamiltonian is the sum of these totally different Hamiltonians? This is indeed a fascinating problem which is deeply related to the wave-particle dualism in quantum physics. We might say that many of the important models in quantum many-body problems, including the φ^4 quantum field theory, the Kondo problem, and, of course, the Hubbard model are minimum models which take into account both the wave-like nature and the particle-like nature (through point-like nonlinear interactions) of matter.

From a technical point of view, the wave-particle dualism implies that the Hamiltonians $\hat{H}_{\mathrm{int}}$ and $\hat{H}_{\mathrm{hop}}$ do not commute with each other. Although each Hamiltonian is diagonalizable, it is highly nontrivial (or impossible) to find the properties of their sum. Of course, mathematical difficulty does not automatically guarantee that the model is worth studying. A truly exciting characteristic of the Hubbard model is that, though the Hamiltonian $\hat{H}_{\mathrm{hop}}$ or $\hat{H}_{\mathrm{int}}$ does not favor any nontrivial order, their sum $\hat{H} = \hat{H}_{\mathrm{hop}} + \hat{H}_{\mathrm{int}}$ is believed to generate various types of nontrivial order including antiferromagnetism, ferromagnetism, and superconductivity.

When we sum up the two innocent Hamiltonians $\hat{H}_{\mathrm{hop}}$ and $\hat{H}_{\mathrm{int}}$, competition between their wave-like and particle-like characters takes place, and one gets various interesting "physics". In the subsequent chapters we shall (partially) confirm this fascinating and challenging scenario from the point of view of mathematical physics.

9.3.3 *Transformations and Symmetry of the Hubbard Hamiltonian*

Let us discuss some transformations of the Hubbard Hamiltonian, which will be useful in latter chapters. The quick reader can only check the statement about the SU(2) invariance, proceed to the next chapter, and come back to study the remaining material when necessary.

SU(2) invariance Since we have defined the total spin operator $\hat{\mathbf{S}}_{\mathrm{tot}}$ in (9.2.43), we can define the operators for uniform rotation of the spin degrees of freedom by

$\hat{U}_\theta^{(\alpha)} = e^{-i\theta\hat{S}_{\rm tot}^{(\alpha)}}$, which is exactly the same as (2.2.11) for quantum spin systems. We then claim that the Hubbard Hamiltonian (9.3.34) is SU(2) invariant, i.e., invariant under any rotation as in (2.4.3). Let us show this important property.

An operator $\hat{A}$ is SU(2) invariant when $[\hat{A}, \hat{S}_{\rm tot}^{(\alpha)}] = 0$ for $\alpha = 1, 2, 3$. This is of course equivalent to

$$[\hat{A}, \hat{S}_{\rm tot}^{\pm}] = 0 \quad \text{and} \quad [\hat{A}, \hat{S}_{\rm tot}^{(3)}] = 0. \tag{9.3.35}$$

We shall demonstrate (9.3.35) for operators that constitute the Hubbard Hamiltonian. By properly modifying the commutation relations (9.2.67), we get

$$[\hat{c}_{x,\uparrow}^\dagger, \hat{S}_{\rm tot}^+] = 0, \quad [\hat{c}_{x,\downarrow}^\dagger, \hat{S}_{\rm tot}^+] = -\hat{c}_{x,\uparrow}^\dagger, \quad [\hat{c}_{x,\uparrow}, \hat{S}_{\rm tot}^+] = \hat{c}_{x,\downarrow}, \quad [\hat{c}_{x,\downarrow}, \hat{S}_{\rm tot}^+] = 0. \tag{9.3.36}$$

We then find $[\hat{c}_{x,\uparrow}^\dagger\hat{c}_{y,\uparrow}, \hat{S}_{\rm tot}^+] = \hat{c}_{x,\uparrow}^\dagger\hat{c}_{y,\downarrow}$ and $[\hat{c}_{x,\downarrow}^\dagger\hat{c}_{y,\downarrow}, \hat{S}_{\rm tot}^+] = -\hat{c}_{x,\uparrow}^\dagger\hat{c}_{y,\downarrow}$, which implies $[\sum_{\sigma=\uparrow,\downarrow}\hat{c}_{x,\sigma}^\dagger\hat{c}_{y,\sigma}, \hat{S}_{\rm tot}^+] = 0$. Similarly we can show that $[\sum_{\sigma=\uparrow,\downarrow}\hat{c}_{x,\sigma}^\dagger\hat{c}_{y,\sigma}, \hat{S}_{\rm tot}^-] = 0$. On the other hand it easily follows from the commutation relations (9.2.31) that $[\hat{c}_{x,\sigma}^\dagger\hat{c}_{y,\sigma}, \hat{S}_{\rm tot}^{(3)}] = 0$ for any x, y and σ. We have thus found that the operator $\sum_{\sigma=\uparrow,\downarrow}\hat{c}_{x,\sigma}^\dagger\hat{c}_{y,\sigma}$ with any x, y is SU(2) invariant. This in particular means that $\hat{H}_{\rm hop}$ is SU(2) invariant, and also $\hat{n}_x = \sum_{\sigma=\uparrow,\downarrow}\hat{c}_{x,\sigma}^\dagger\hat{c}_{x,\sigma}$ is SU(2) invariant for each x. The latter implies the SU(2) invariance of $\hat{H}_{\rm int}$ because $\hat{n}_{x,\uparrow}\hat{n}_{x,\downarrow} = \{(\hat{n}_x)^2 - \hat{n}_x\}/2$.

Gauge transformation Take an arbitrary phase factor $\theta_x \in \mathbb{R}$ for each $x \in \Lambda$, and write $\boldsymbol{\theta} = (\theta_x)_{x\in\Lambda}$. We define the unitary operator $\hat{U}_{\boldsymbol{\theta}}^{\rm g} : \mathcal{H}_N \to \mathcal{H}_N$ for the gauge transformation as

$$\hat{U}_{\boldsymbol{\theta}}^{\rm g} := \prod_{x\in\Lambda} e^{-i\theta_x\hat{n}_x}. \tag{9.3.37}$$

Note that the unitary operator only changes the quantum mechanical U(1) phase.

By examining the action on the basis state (9.2.35), one easily verifies that[14]

$$(\hat{U}_{\boldsymbol{\theta}}^{\rm g})^\dagger\,\hat{c}_{x,\sigma}^\dagger\,\hat{U}_{\boldsymbol{\theta}}^{\rm g} = e^{i\theta_x}\,\hat{c}_{x,\sigma}^\dagger, \quad (\hat{U}_{\boldsymbol{\theta}}^{\rm g})^\dagger\,\hat{c}_{x,\sigma}\,\hat{U}_{\boldsymbol{\theta}}^{\rm g} = e^{-i\theta_x}\,\hat{c}_{x,\sigma}. \tag{9.3.38}$$

We also see directly from the definition (9.3.37) that $(\hat{U}_{\boldsymbol{\theta}}^{\rm g})^\dagger\,\hat{n}_{x,\sigma}\,\hat{U}_{\boldsymbol{\theta}}^{\rm g} = \hat{n}_{x,\sigma}$. This means $(\hat{U}_{\boldsymbol{\theta}}^{\rm g})^\dagger\,\hat{H}_{\rm int}\,\hat{U}_{\boldsymbol{\theta}}^{\rm g} = \hat{H}_{\rm int}$, i.e., the interaction Hamiltonian (9.3.28) is invariant under any gauge transformation.[15] But the hopping Hamiltonian (9.3.17) is transformed as

---

[14] It is common not to define the unitary operator explicitly and only declare the transformation of operators: $\hat{c}_{x,\sigma}^\dagger \to e^{i\theta_x}\,\hat{c}_{x,\sigma}^\dagger$ and $\hat{c}_{x,\sigma} \to e^{-i\theta_x}\,\hat{c}_{x,\sigma}$. Although the existence of the unitary operator is obivous in the present case, it is known in general that a transformation of operators satisfying certain conditions guarantees (through Wigner's theorem) that there is a corresponding unitary operator. See Appendix A.6.

[15] Note that we are here using the first convention of transformations of operators as in (A.1.16). When transforming Hamiltonians, however, it is often more convenient to use the second convention (A.1.17).

$$(\hat{U}^{\mathrm{g}}_{\boldsymbol{\theta}})^\dagger\, \hat{H}_{\mathrm{hop}}\, \hat{U}^{\mathrm{g}}_{\boldsymbol{\theta}} = \sum_{\substack{x,y\in\Lambda\\ \sigma=\uparrow,\downarrow}} e^{i(\theta_x-\theta_y)}\, t_{x,y}\, \hat{c}^\dagger_{x,\sigma}\hat{c}_{y,\sigma}. \tag{9.3.39}$$

This means that phase of the hopping amplitude $t_{x,y}$ can be modified as $t_{x,y} \to e^{i(\theta_x-\theta_y)}\, t_{x,y}$.[16]

To see a useful application, suppose that the hopping amplitude is bipartite, i.e., one can decompose the lattice as $\Lambda = A \cup B$ with $A \cap B = \emptyset$ so that $t_{x,y} \neq 0$ only when $x \in A$, $y \in B$ or $x \in B$, $y \in A$. Then one can realize the sign change $t_{x,y} \to -t_{x,y}$ by the gauge transformation with $\boldsymbol{\theta}$ such that $\theta_x = 0$ for $x \in A$ and $\theta_x = \pi$ for $x \in B$.

Particle-hole transformation For each $x \in \Lambda$ and $\sigma = \uparrow, \downarrow$, let $\hat{U}^{\mathrm{ph}}_{x,\sigma} := \hat{c}_{x,\sigma} - \hat{c}^\dagger_{x,\sigma}$. Note that this is a unitary operator since $(\hat{U}^{\mathrm{ph}}_{x,\sigma})^\dagger\, \hat{U}^{\mathrm{ph}}_{x,\sigma} = \hat{c}^\dagger_{x,\sigma}\hat{c}_{x,\sigma} + \hat{c}_{x,\sigma}\hat{c}^\dagger_{x,\sigma} = 1$. It is also easy to check from the anticommutation relations (9.2.27), (9.2.28) that

$$(\hat{U}^{\mathrm{ph}}_{x,\sigma})^\dagger\, \hat{c}_{x,\sigma}\, \hat{U}^{\mathrm{ph}}_{x,\sigma} = -\hat{c}^\dagger_{x,\sigma}, \quad (\hat{U}^{\mathrm{ph}}_{x,\sigma})^\dagger\, \hat{c}^\dagger_{x,\sigma}\, \hat{U}^{\mathrm{ph}}_{x,\sigma} = -\hat{c}_{x,\sigma}. \tag{9.3.40}$$

We then define the unitary operator $\hat{U}^{\mathrm{ph}} : \mathscr{H}_N \to \mathscr{H}_{2|\Lambda|-N}$ for particle-hole transformation by

$$\hat{U}^{\mathrm{ph}} := \prod_{x\in\Lambda} (\hat{U}^{\mathrm{ph}}_{x,\uparrow}\, \hat{U}^{\mathrm{ph}}_{x,\downarrow}). \tag{9.3.41}$$

Note that this definition is independent of the ordering of the product since $\hat{U}^{\mathrm{ph}}_{x,\uparrow}\, \hat{U}^{\mathrm{ph}}_{x,\downarrow}$ with different x commute. Noting, e.g., that $(\hat{U}^{\mathrm{ph}}_{x,\downarrow})^\dagger (\hat{U}^{\mathrm{ph}}_{x,\uparrow})^\dagger\, \hat{c}_{x,\uparrow} \hat{U}^{\mathrm{ph}}_{x,\uparrow}\, \hat{U}^{\mathrm{ph}}_{x,\downarrow} = -(\hat{U}^{\mathrm{ph}}_{x,\downarrow})^\dagger\, \hat{c}^\dagger_{x,\uparrow}\, \hat{U}^{\mathrm{ph}}_{x,\downarrow} = \hat{c}^\dagger_{x,\uparrow} (\hat{U}^{\mathrm{ph}}_{x,\downarrow})^\dagger\, \hat{U}^{\mathrm{ph}}_{x,\downarrow} = \hat{c}^\dagger_{x,\uparrow}$, one finds

$$(\hat{U}^{\mathrm{ph}})^\dagger\, \hat{c}_{x,\sigma}\, \hat{U}^{\mathrm{ph}} = \hat{c}^\dagger_{x,\sigma}, \quad (\hat{U}^{\mathrm{ph}})^\dagger\, \hat{c}^\dagger_{x,\sigma}\, \hat{U}^{\mathrm{ph}} = \hat{c}_{x,\sigma}, \tag{9.3.42}$$

for any $x \in \Lambda$ and $\sigma = \uparrow, \downarrow$. The transformation switches the roles of the creation and annihilation operators, thus replacing a particle with a "hole".[17]

Since

$$(\hat{U}^{\mathrm{ph}})^\dagger\, \hat{c}^\dagger_{x,\sigma}\hat{c}_{y,\sigma}\, \hat{U}^{\mathrm{ph}} = \hat{c}_{x,\sigma}\hat{c}^\dagger_{y,\sigma} = -\hat{c}^\dagger_{y,\sigma}\hat{c}_{x,\sigma}, \tag{9.3.43}$$

for any $x \neq y$, and

$$(\hat{U}^{\mathrm{ph}})^\dagger\, \hat{n}_{x,\sigma}\, \hat{U}^{\mathrm{ph}} = \hat{c}_{x,\sigma}\hat{c}^\dagger_{x,\sigma} = 1 - \hat{c}^\dagger_{x,\sigma}\hat{c}_{x,\sigma} = 1 - \hat{n}_{x,\sigma}, \tag{9.3.44}$$

we see that the hopping Hamiltonian (9.3.17) transforms as

---

[16] Note that for any distinct sites $x_1, \ldots, x_n$, the product $t_{x_1,x_2}\, t_{x_2,x_3} \cdots t_{x_{n-1},x_n}\, t_{x_n,x_1}$ does not change under the gauge transformation.

[17] It is common only to declare the transformation $\hat{c}^\dagger_{x,\sigma} \to \hat{c}_{x,\sigma}$, $\hat{c}_{x,\sigma} \to \hat{c}^\dagger_{x,\sigma}$. See footnote 14.

$$(\hat{U}^{\text{ph}})^\dagger\, \hat{H}_{\text{hop}}\, \hat{U}^{\text{ph}} = -\sum_{\substack{x,y\in\Lambda\\ \sigma=\uparrow,\downarrow}} (t_{x,y})^*\, \hat{c}^\dagger_{x,\sigma}\hat{c}_{y,\sigma} + 2\sum_{x,\in\Lambda} t_{x,x}, \tag{9.3.45}$$

where we noted that $t_{y,x} = (t_{x,y})^*$. Thus, apart from the additional constant (which only shifts the energy), the transformation $t_{x,y} \to -(t_{x,y})^*$ is realized.

From (9.3.44), we see that the interaction Hamiltonian (9.3.28) transforms as

$$(\hat{U}^{\text{ph}})^\dagger\, \hat{H}_{\text{int}}\, \hat{U}^{\text{ph}} = \sum_x U_x(1-\hat{n}_{x,\uparrow})(1-\hat{n}_{x,\downarrow}) = \hat{H}_{\text{int}} - \sum_x U_x\hat{n}_x + \sum_x U_x. \tag{9.3.46}$$

For the simpler interaction Hamiltonian (9.3.29) with $U_x = U$, the right-hand side becomes $\hat{H}_{\text{int}} - U\hat{N} + |\Lambda|\,U$, which means that the interaction is essentially invariant under the particle-hole transformation. Sometimes one uses the symmetric version of the interaction Hamiltonian

$$\hat{H}'_{\text{int}} := \sum_x U_x\Big(\hat{n}_{x,\uparrow} - \frac{1}{2}\Big)\Big(\hat{n}_{x,\downarrow} - \frac{1}{2}\Big), \tag{9.3.47}$$

which is strictly invariant under the hole-particle transformation.

Problem 9.3.3.a Consider the combination of the particle-hole transformation and the gauge transformation, $\hat{U}^{\text{ph},\theta}_{x,\sigma} = e^{-i\theta\hat{n}_{x,\sigma}}(\hat{c}_{x,\sigma} - \hat{c}^\dagger_{x,\sigma})$ for $\theta \in \mathbb{R}$. What are $(\hat{U}^{\text{ph},\theta}_{x,\sigma})^\dagger\, \hat{c}_{x,\sigma}\, \hat{U}^{\text{ph},\theta}_{x,\sigma}$ and $(\hat{U}^{\text{ph},\theta}_{x,\sigma})^\dagger\, \hat{c}^\dagger_{x,\sigma}\, \hat{U}^{\text{ph},\theta}_{x,\sigma}$? [solution→p.516]

Shiba transformation Let us finally discuss a useful transformation, often called the Shiba transformation,[18] which is a combination of the particle-hole transformation and a gauge transformation, both applied only to up-spin electrons. (Compare the above Problem 9.3.3.a and the definition (9.3.48).) For a certain class of models, the transformation switches the sign of the interaction energy U_x, thus converting the (standard) repulsive Hubbard model into an attractive Hubbard model, and vice versa.

Let us assume that the hopping amplitude $(t_{x,y})_{x,y\in\Lambda}$ is bipartite, i.e., one can decompose the lattice as $\Lambda = A \cup B$ with $A \cap B = \emptyset$ so that $t_{x,y} \neq 0$ only when $x \in A$, $y \in B$ or $x \in B$, $y \in A$. This in particular means that the on-site potential $t_{x,x}$ is vanishing. We also make a crucial assumption that $t_{x,y} \in \mathbb{R}$ for any x and y.

For each $x \in \Lambda$, let

$$\hat{U}^{\text{S}}_x := \begin{cases} \hat{c}_{x,\uparrow} - \hat{c}^\dagger_{x,\uparrow} & \text{if } x \in A;\\ e^{-i\pi\hat{n}_{x,\uparrow}}(\hat{c}_{x,\uparrow} - \hat{c}^\dagger_{x,\uparrow}) = \hat{c}_{x,\uparrow} + \hat{c}^\dagger_{x,\uparrow} & \text{if } x \in B, \end{cases} \tag{9.3.48}$$

which is again unitary. We easily see

[18]The transformation appears in Shiba's paper [15]. According to [3], it was first used by Lieb and Wu in [11].

$$(\hat{U}_x^{\mathrm{S}})^\dagger\,\hat{c}_{x,\uparrow}\,\hat{U}_x^{\mathrm{S}}=-(-1)^x\hat{c}_{x,\uparrow}^\dagger,\quad(\hat{U}_x^{\mathrm{S}})^\dagger\,\hat{c}_{x,\uparrow}^\dagger\,\hat{U}_x^{\mathrm{S}}=-(-1)^x\hat{c}_{x,\uparrow},\tag{9.3.49}$$

where $(-1)^x=1$ if $x\in A$ and $(-1)^x=-1$ if $x\in B$.

The Shiba transformation is defined by the unitary operator

$$\hat{U}^{\mathrm{S}}:=\prod_{x\in\Lambda}\hat{U}_x^{\mathrm{S}}.\tag{9.3.50}$$

Here we have fixed an arbitrary ordering of sites in Λ. From (9.3.49), one finds

$$\begin{aligned}(\hat{U}^{\mathrm{S}})^\dagger\,\hat{c}_{x,\uparrow}^\dagger\,\hat{U}^{\mathrm{S}}&=(-1)^{|\Lambda|}(-1)^x\,\hat{c}_{x,\uparrow},\quad(\hat{U}^{\mathrm{S}})^\dagger\,\hat{c}_{x,\uparrow}\,\hat{U}^{\mathrm{S}}=(-1)^{|\Lambda|}(-1)^x\,\hat{c}_{x,\uparrow}^\dagger,\\(\hat{U}^{\mathrm{S}})^\dagger\,\hat{c}_{x,\downarrow}^\dagger\,\hat{U}^{\mathrm{S}}&=(-1)^{|\Lambda|}\hat{c}_{x,\downarrow}^\dagger,\quad(\hat{U}^{\mathrm{S}})^\dagger\,\hat{c}_{x,\downarrow}\,\hat{U}^{\mathrm{S}}=(-1)^{|\Lambda|}\hat{c}_{x,\downarrow},\end{aligned}\tag{9.3.51}$$

Then the change of the signs of hopping as in (9.3.43) is precisely canceled by the change from the gauge transformation, and one finds, rather remarkably, that

$$(\hat{U}^{\mathrm{S}})^\dagger\,\hat{H}_{\mathrm{hop}}\,\hat{U}^{\mathrm{S}}=\hat{H}_{\mathrm{hop}},\tag{9.3.52}$$

where we recalled that $t_{x,y}$ are real, and hence $t_{x,y}=t_{y,x}$. The interaction Hamiltonian, on the other hand, is affected only by the partial particle-hole transformation, and changes as

$$(\hat{U}^{\mathrm{S}})^\dagger\,\hat{H}_{\mathrm{int}}\,\hat{U}^{\mathrm{S}}=\sum_{x\in\Lambda}U_x(1-\hat{n}_{x,\uparrow})\hat{n}_{x,\downarrow}=-\hat{H}_{\mathrm{int}}+\sum_{x\in\Lambda}U_x\hat{n}_{x,\downarrow}.\tag{9.3.53}$$

The sign of the interaction has changed, and there also appeared (unwanted) on-site potential term only for down-spin electrons. When $U_x=U$ as in (9.3.29), this extra term becomes $U\sum_{x\in\Lambda}\hat{n}_{x,\downarrow}$. This is basically innocent since $\sum_{x\in\Lambda}\hat{n}_{x,\downarrow}$ is a conserved quantity.

Another useful trick is to use symmetric interaction Hamiltonian $\hat{H}'_{\mathrm{int}}$ of (9.3.47). In this case the partial particle-hole transformation brings $\hat{n}_{x,\uparrow}-(1/2)$ into $-\{\hat{n}_{x,\uparrow}-(1/2)\}$, and the interaction changes the sign. The whole Hubbard Hamiltonian transforms as

$$(\hat{U}^{\mathrm{S}})^\dagger\,(\hat{H}_{\mathrm{hop}}+\hat{H}'_{\mathrm{int}})\,\hat{U}^{\mathrm{S}}=\hat{H}_{\mathrm{hop}}-\hat{H}'_{\mathrm{int}},\tag{9.3.54}$$

and hence the model with repulsive interaction $U_x\ge0$ is mapped to the model with attractive interaction $-U_x\le0$. One must note however that the Shiba transformation changes the number of up electrons $N_\uparrow$ into $|\Lambda|-N_\uparrow$. Only when $N_\uparrow$ happens to be $|\Lambda|/2$, which is the case in half-filled models discussed in Chap. 10, the transformation preserves the Hilbert space.

Problem 9.3.3.b For $x\in\Lambda$ and $\alpha=1,2,3$, define the η-paring operators (or the pseudo spin operators) by $\hat{\eta}_x^{(\alpha)}:=(\hat{U}^{\mathrm{S}})^\dagger\,\hat{S}_x^{(\alpha)}\,\hat{U}^{\mathrm{S}}$. They obviously satisfy the commutation relations $[\hat{\eta}_x^{(\alpha)},\hat{\eta}_x^{(\beta)}]=i\sum_{\gamma=1,2,3}\varepsilon_{\alpha\beta\gamma}\,\hat{\eta}_x^{(\gamma)}$, as in (2.1.1). Exactly as for spin operators, we define $\hat{\eta}_x^{\pm}:=\hat{\eta}_x^{(1)}\pm i\hat{\eta}_x^{(2)}$, $\hat{\eta}_{\mathrm{tot}}^{(\alpha)}:=\sum_{x\in\Lambda}\hat{\eta}_x^{(\alpha)}$, and $\hat{\eta}_{\mathrm{tot}}^{\pm}:=\sum_{x\in\Lambda}\hat{\eta}_x^{\pm}$.

Express $\hat{\eta}_x^{\pm}$ and $\hat{\eta}_x^{(3)}$ in terms of the fermion operators and the number operators. Show in general that $[\hat{\eta}_{\rm tot}^{(\alpha)}, \hat{S}_{\rm tot}^{(\beta)}] = 0$ for any α and β. Let us write the Hubbard Hamiltonian which appears in (9.3.54) as $\hat{H}' = \hat{H}_{\rm hop} + \hat{H}'_{\rm int}$. Show that $[\hat{H}', \hat{S}_{\rm tot}^{(\alpha)}] = 0$ and $[\hat{H}', \hat{\eta}_{\rm tot}^{(\alpha)}] = 0$ for $\alpha = 1, 2, 3$, which implies that the Hamiltonian $\hat{H}'$ has six independent conserved quantities.[19] [solution→p.516]

9.4 Bosonic Hubbard Model

Here we discuss the formulation of the bosonic version of the Hubbard model, which we studied in Chap. 5 in connection to the phenomenon of Bose–Einstein condensation. We will be brief, mainly stressing the difference between the original fermionic version and the bosonic version.

Wave function The bosonic Hubbard model is mainly regarded as a model of cold atoms trapped in an optical lattice. Bosonic atoms may have an integer (hyperfine) spin $S = 0, 1, 2, \ldots$. Thus the coordinate for a single particle is $u = (x, \sigma) \in \tilde{\Lambda} = \Lambda \times \{-S, \ldots, S\}$. The most standard model studied in Chap. 5 corresponds to $S = 0$, where $\tilde{\Lambda} = \Lambda$. The single particle Hilbert space $\mathfrak{h}$ consists of wave functions of the form $\boldsymbol{\varphi} = (\varphi(u))_{u\in\tilde{\Lambda}}$, exactly as in the case of electrons.

An N-particle wave function is again $\boldsymbol{\Phi} = (\Phi(u_1, \ldots, u_N))_{u_1,\ldots,u_N\in\tilde{\Lambda}}$, but should satisfy

$$\Phi(u_1, \ldots, u_N) = \Phi(u_{p(1)}, \ldots, u_{p(N)}), \tag{9.4.1}$$

for any permutation p. The wave function is symmetric with respect to the change of the particle labels. Thus the Hilbert space of the wave functions for N particles is

$$\mathscr{H}_N^{\rm wf} := \mathsf{P}_{\rm sym}\, \underbrace{\mathfrak{h} \otimes \cdots \otimes \mathfrak{h}}_{N}, \tag{9.4.2}$$

where the projection onto the symmetric part is defined by $(\mathsf{P}_{\rm sym}\boldsymbol{\Phi})(u_1, \ldots, u_N) = (N!)^{-1} \sum_p \Phi(u_{p(1)}, \ldots, u_{p(N)})$. In models of bosons, the total particle number N can be any nonnegative integer.

Creation and annihilation operators The construction of the creation and annihilation operators within the wave function formalism is essentially the same as that in Sect. 9.2.2.[20] Basically all the formulae and derivations are valid as they are if one replaces the sign factor $(-1)^{j+1}$ by 1.

Thus the definition (9.2.9) of the creation operator becomes

[19]The Hamiltonian has $SO(4) \cong SU(2) \times SU(2)/\mathbb{Z}_2$ symmetry, as was first noted in [21]. For details see, e.g., Sect. 2.2.5 of [3] or Sect. 1 of [16].

[20]See footnote 6 in p. 310.

$$(\mathsf{A}^\dagger(\boldsymbol{\psi})\boldsymbol{\Phi})(u_1,\ldots,u_N) := \frac{1}{\sqrt{N}}\sum_{j=1}^{N}\psi(u_j)\,\Phi(u_1,\ldots,\breve{u}_j,\ldots,u_N), \tag{9.4.3}$$

and consequently the action of the annihilation operator is

$$(\mathsf{A}(\boldsymbol{\psi})\boldsymbol{\Phi})(u_1,\ldots,u_{N-1}) = \sqrt{N}\sum_{v\in\tilde{\Lambda}}\{\psi(v)\}^*\,\Phi(v,u_1,\ldots,u_{N-1}), \tag{9.4.4}$$

which is the same as (9.2.15). Repeating the proofs of the anticommutation relations (9.2.19), (9.2.20), and (9.2.26), one instead gets commutation relations

$$[\mathsf{A}(\boldsymbol{\varphi}),\mathsf{A}(\boldsymbol{\psi})]=0,\quad [\mathsf{A}^\dagger(\boldsymbol{\varphi}),\mathsf{A}^\dagger(\boldsymbol{\psi})]=0,\quad [\mathsf{A}(\boldsymbol{\varphi}),\mathsf{A}^\dagger(\boldsymbol{\psi})]=\langle\boldsymbol{\varphi},\boldsymbol{\psi}\rangle. \tag{9.4.5}$$

Fock space representation The Fock space representation for bosons is again basically the same as that for fermions developed in Sect. 9.2.3. For each $(x,\sigma)\in\tilde{\Lambda}$, we define the creation operator $\hat{a}^\dagger_{x,\sigma}$ and the annihilation operator $\hat{a}_{x,\sigma}$ which satisfy the commutation relations

$$[\hat{a}^\dagger_{x,\sigma},\hat{a}^\dagger_{y,\tau}]=[\hat{a}_{x,\sigma},\hat{a}_{y,\tau}]=0,\quad [\hat{a}_{x,\sigma},\hat{a}^\dagger_{y,\tau}]=\delta_{x,y}\delta_{\sigma,\tau}. \tag{9.4.6}$$

We have already seen the spinless version of the relations in (5.1.1). Now the number operator $\hat{n}_{x,\sigma}=\hat{a}^\dagger_{x,\sigma}\hat{a}_{x,\sigma}$ takes any nonnegative integer as its eigenvalue. The operators

$$\hat{A}^\dagger(\boldsymbol{\varphi}) := \sum_{u\in\tilde{\Lambda}}\varphi(u)\,\hat{a}^\dagger_u,\quad \hat{A}(\boldsymbol{\varphi}) := \sum_{u\in\tilde{\Lambda}}\{\varphi(u)\}^*\,\hat{a}_u, \tag{9.4.7}$$

for any $\boldsymbol{\varphi}\in\tilde{\mathfrak{h}}$, recover the commutation relations (9.4.5), and can be identified with the operators $\mathsf{A}^\dagger(\boldsymbol{\varphi})$ and $\mathsf{A}(\boldsymbol{\varphi})$ constructed within the wave function formalism.

The Hilbert space for N particles is spanned by states of the form

$$\hat{a}^\dagger_{x_1,\sigma_1}\cdots\hat{a}^\dagger_{x_N,\sigma_N}|\Phi_{\mathrm{vac}}\rangle, \tag{9.4.8}$$

where $|\Phi_{\mathrm{vac}}\rangle$ again satisfies $\hat{a}_{x,\sigma}|\Phi_{\mathrm{vac}}\rangle=0$ for any $(x,\sigma)\in\tilde{\Lambda}$. Now there can be any overlaps within the sequence $(x_1,\sigma_1),\ldots,(x_N,\sigma_N)$, and the state (9.4.8) is always nonvanishing.

The rest of the construction is essentially parallel to the case of electrons. One should check that the hopping part of the Hamiltonian (5.1.3) is nothing but the "second quantization" of the standard Schrödinger equation on Λ_L with the hopping (9.3.6).

References

1. F. Bloch, Bemerkung zur Elektronentheorie des Ferromagnetismus und der elektrischen Leitfähigkeit. Z. Phys. **57**, 545 (1929)
2. C.S. Chiu, G. Ji, A. Mazurenko, D. Greif, M. Greiner, Quantum state engineering of a Hubbard system with ultracold fermions. Phys. Rev. Lett. **120**, 243201 (2018), arXiv:1712.07114
3. F.H.L. Essler, H. Frahm, F. Göhmann, A. Klümper, V.E. Korepin, *The One-Dimensional Hubbard Model* (Cambridge University Press, Cambridge, 2010)
4. P. Fazekas, *Lecture Notes on Electron Correlation and Magnetism*. Series in Modern Condensed Matter Physics, vol. 5 (World Scientific, Singapore, 1999)
5. M.C. Gutzwiller, Effect of correlation on the ferromagnetism of transition metals. Phys. Rev. Lett. **10**, 159 (1963)
6. B. Hayes, Hip-hop physics. Am. Sci. **97**, 438 (2009), https://www.americanscientist.org/article/hip-hop-physics#
7. W.J. Heisenberg, Zur theorie des ferromagnetismus. Z. Phys. **49**, 619–636 (1928)
8. J. Hubbard, Electron correlations in narrow energy bands. Proc. Roy. Soc. (London) **A276**, 238 (1963)
9. J. Kanamori, Electron correlation and ferromagnetism of transition metals. Prog. Theor. Phys. **30**, 275 (1963), https://academic.oup.com/ptp/article/30/3/275/1865799
10. E.H. Lieb, The Hubbard model: some rigorous results and open problems, in *Proceedings of the XIth International Congress of Mathematical Physics*, Paris, 1994, D. Iagolnitzer ed., pp. 392–412. International Press (1995), arXiv:cond-mat/9311033
11. E.H. Lieb, F.Y. Wu, Absence of Mott transition in an exact solution of the short-range, one-band model in one dimension. Phys. Rev. Lett. **20**, 1445 (1968); Erratum: Phys. Rev. Lett. **21**, 192 (1968)
12. E.H. Lieb, F.Y. Wu, *The one-dimensional Hubbard model: a reminiscence*. Phys. A **321**, 1–27 (2003), arXiv:cond-mat/0207529
13. A. Mazurenko, C.S. Chiu, G. Ji, M.F. Parsons, M. Kanász-Nagy, R. Schmidt, F. Grusdt, E. Demler, D. Greif, M. Greiner, A cold-atom Fermi-Hubbard antiferromagnet. Nature **545**, 462–466 (2017), arXiv:1612.08436
14. M.F. Parsons, A. Mazurenko, C.S. Chiu, G. Ji, D. Greif, M. Greiner, Site-resolved measurement of the spin-correlation function in the Fermi-Hubbard model. Science **353**, 1253–1256 (2016), http://science.sciencemag.org/content/353/6305/1253
15. H. Shiba, Thermodynamic properties of the one-dimensional half-filled-band Hubbard model. II: application of the grand canonical method. Prog. Theor. Phys. **48**, 2171–2186 (1972), https://academic.oup.com/ptp/article/48/6/2171/1857344
16. M. Shiroishi, H. Ujino, M. Wadati, SO(4) symmetry of the transfer matrix for the one-dimensional hubbard model. J. Phys. A: Math. Gen. **31**, 2341–2358 (1998), arXiv:cond-mat/9708064
17. J.C. Slater, H. Statz, G.F. Koster, A two-electron example of ferromagnetism. Phys. Rev. **91**, 1323 (1953)
18. H. Tasaki, The Hubbard model: an introduction and selected rigorous results. J. Phys. Condens. Matter **10**, 4353 (1998), arXiv:cond-mat/9512169
19. H. Tasaki, From Nagaoka's ferromagnetism to flat-band ferromagnetism and beyond – An introduction to ferromagnetism in the Hubbard model. Prog. Theor. Phys. **99**, 489–548 (1998), arXiv:cond-mat/9712219
20. H. Tasaki, Introduction to the "second quantization" formalism for non-relativistic quantum mechanics: a possible substitution for Sections 6.7 and 6.8 of Feynman's "Statistical Mechanics", unpublished note (2018), arxiv:1812.10732
21. C.N. Yang, S.C. Zhang, SO_4 symmetry in a hubbard model. Mod. Phys. Lett. **B4**, 759–776 (1990). Reprinted in M. Rasetti (ed.) "The Hubbard Model: Recent Results" (World Scientific, 1991) and A. Montorsi (ed.) "The Hubbard Model: A reprint volume" (World Scientific, 1992), https://www.worldscientific.com/doi/abs/10.1142/S021797929100050X

Chapter 10
Half-Filled Models: Lieb's Theorems and the Origin of Antiferromagnetism and Ferrimagnetism

A lattice electron system in which the electron number N is identical to the number of sites $|\Lambda|$ is said to be half-filled because the maximum possible value of N is $2|\Lambda|$. Half-filled models represent physically natural situations since a system becomes half-filled if each atom contributes one electron to the system of electrons in consideration. In the present chapter we discuss the theorem of fundamental significance due to Lieb, which sheds light on the origin of antiferromagnetism and ferrimagnetism in interacting itinerant electron systems at half-filling. We start the chapter by a short section (Sect. 10.1) with a formal perturbative argument about the relation between the half-filled Hubbard model and the antiferromagnetic Heisenberg model. Then in Sect. 10.2, we present a thorough discussion of Lieb's theorems. We shall state main claims, discuss important applications, and give complete and hopefully readable proofs.

10.1 Formal Perturbation Theory for Strong Interaction

In this preparatory section we shall see, from a non-rigorous perturbative argument, that low energy properties of the Hubbard model at half-filling resemble those of the antiferromagnetic Heisenberg model.

Similar discussion already appears in the old work of Anderson [1]. See Sect. 5.1 of [6], Appendix 2.A of [4], and Sect. 3.2 of [2] for modern treatments. For rigorous perturbation theory in the similar spirit, see, e.g., [3] and references therein.

Toy model with two electrons on two sites As a warm up, let us study possibly the simplest nontrivial Hubbard model, namely, the model with two electrons on the lattice $\Lambda = \{1, 2\}$ with two sites. The Hamiltonian (9.3.34) reads

$$\hat{H} = t \sum_{\sigma=\uparrow,\downarrow} (\hat{c}^\dagger_{1,\sigma}\hat{c}_{2,\sigma} + \hat{c}^\dagger_{2,\sigma}\hat{c}_{1,\sigma}) + U\,(\hat{n}_{1,\uparrow}\hat{n}_{1,\downarrow} + \hat{n}_{2,\uparrow}\hat{n}_{2,\downarrow}), \qquad (10.1.1)$$

H. Tasaki, *Physics and Mathematics of Quantum Many-Body Systems*, Graduate Texts in Physics, https://doi.org/10.1007/978-3-030-41265-4_10

where $t \in \mathbb{R}$. We shall examine the eigenstates, in particular the ground state, of (10.1.1).

First suppose that both the electrons have up spin. Since two up-spin electrons cannot occupy a single site (i.e., $(\hat{c}^\dagger_{x,\uparrow})^2 = 0$), the only possible state is $|\Phi_{1,1}\rangle = \hat{c}^\dagger_{1,\uparrow}\hat{c}^\dagger_{2,\uparrow}|\Phi_{\rm vac}\rangle$, which clearly satisfies $\hat{H}|\Phi_{1,1}\rangle = 0$. We also see that $\hat{H}|\Phi_{1,-1}\rangle = 0$, where $|\Phi_{1,-1}\rangle = \hat{c}^\dagger_{1,\downarrow}\hat{c}^\dagger_{2,\downarrow}|\Phi_{\rm vac}\rangle$. Note that both the states have[1] $S_{\rm tot} = 1$, and hence are two elements of a spin triplet. The remaining one of the triplet should be a superposition of $\hat{c}^\dagger_{1,\uparrow}\hat{c}^\dagger_{2,\downarrow}|\Phi_{\rm vac}\rangle$ and $\hat{c}^\dagger_{1,\downarrow}\hat{c}^\dagger_{2,\uparrow}|\Phi_{\rm vac}\rangle$. By using the commutation relation

$$[\hat{c}^\dagger_{x,\sigma}\hat{c}_{y,\sigma}, \hat{c}^\dagger_{z,\tau}] = \delta_{y,z}\delta_{\sigma,\tau}\,\hat{c}^\dagger_{x,\sigma}, \tag{10.1.2}$$

one easily finds that $|\Phi_{1,0}\rangle = \frac{1}{\sqrt{2}}(\hat{c}^\dagger_{1,\uparrow}\hat{c}^\dagger_{2,\downarrow} + \hat{c}^\dagger_{1,\downarrow}\hat{c}^\dagger_{2,\uparrow})|\Phi_{\rm vac}\rangle$ satisfies $\hat{H}|\Phi_{1,0}\rangle = 0$. As we have seen in (9.2.76), the state has total spin $S_{\rm tot} = 1$. We have thus found a triplet of energy eigenstates which have zero energy.

Let us examine energy eigenstates with $S_{\rm tot} = 0$. Since these states are orthogonal to $|\Phi_{1,0}\rangle$, we can write them as

$$|\Phi\rangle = \bigl\{\alpha(\hat{c}^\dagger_{1,\uparrow}\hat{c}^\dagger_{2,\downarrow} - \hat{c}^\dagger_{1,\downarrow}\hat{c}^\dagger_{2,\uparrow}) + \beta\,\hat{c}^\dagger_{1,\uparrow}\hat{c}^\dagger_{1,\downarrow} + \gamma\,\hat{c}^\dagger_{2,\uparrow}\hat{c}^\dagger_{2,\downarrow}\bigr\}|\Phi_{\rm vac}\rangle, \tag{10.1.3}$$

where $\alpha, \beta, \gamma \in \mathbb{C}$ are coefficients to be determined from the Schrödinger equation. Again by using (10.1.2), we find

$$\begin{aligned}\hat{H}|\Phi\rangle = \bigl\{2t\alpha(\hat{c}^\dagger_{1,\uparrow}\hat{c}^\dagger_{1,\downarrow} + \hat{c}^\dagger_{2,\uparrow}\hat{c}^\dagger_{2,\downarrow}) + t(\beta+\gamma)(\hat{c}^\dagger_{1,\uparrow}\hat{c}^\dagger_{2,\downarrow} - \hat{c}^\dagger_{1,\downarrow}\hat{c}^\dagger_{2,\uparrow})\\ + U\beta\,\hat{c}^\dagger_{1,\uparrow}\hat{c}^\dagger_{1,\downarrow} + U\gamma\,\hat{c}^\dagger_{2,\uparrow}\hat{c}^\dagger_{2,\downarrow}\bigr\}|\Phi_{\rm vac}\rangle,\end{aligned} \tag{10.1.4}$$

which should be equal to $E|\Phi\rangle$. One then easily finds that there are three energy eigenstates. Especially the one with negative energy eigenvalue $E_{\rm GS} = \{U - \sqrt{U^2 + 16t^2}\}/2$ gives the ground state, which is characterized by $\beta = \gamma$ and $\alpha = 2t\beta/E_{\rm GS}$. We note that $E_{\rm GS} \simeq -4t^2/U$ if $|t| \ll U$. We have thus found that the ground state is unique and has total spin $S_{\rm tot} = 0$. This suggests that antiferromagnetic interaction between the two spins are generated.

Problem 10.1.a Find the remaining two energy eigenvalues. [solution→p.516].

Perturbative analysis Let us now turn to the general Hubbard model on lattice Λ with Hamiltonian (9.3.34), which is the sum of the general hopping Hamiltonian (9.3.17) with hopping amplitude $t_{x,y} = (t_{y,x})^* \in \mathbb{C}$ and the general interaction Hamiltonian (9.3.28) with on-site interaction $U_x > 0$. We assume that the system is at half-filling, i.e., the electron number is $N = |\Lambda|$. We focus on the situation where the interaction energy is much larger than the hopping matrix elements. We shall be heuristic, and treat the interaction Hamiltonian $\hat{H}_{\rm int}$ as the unperturbed Hamiltonian, and the hopping Hamiltonian $\hat{H}_{\rm hop}$ as a perturbation.

---

[1]Recall that the eigenvalue of $(\hat{\boldsymbol{S}}_{\rm tot})^2$ is denoted as $S_{\rm tot}(S_{\rm tot}+1)$.

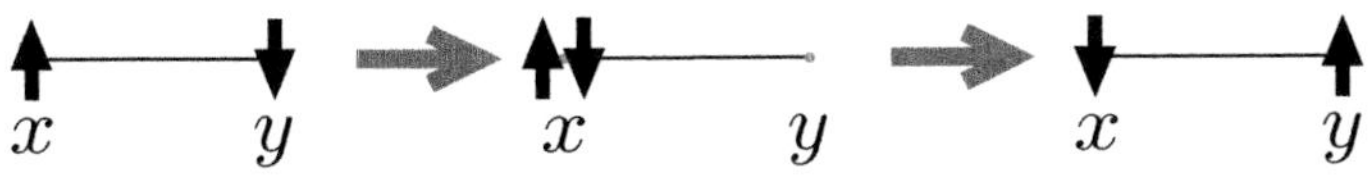

Fig. 10.1 When electrons hop twice, spins on sites x and y may be exchanged. This process generates the super-exchange interaction (also known as the kinetic exchange interaction), which is the origin of antiferromagnetism in the half-filled Hubbard model. ()

We first examine the unperturbed Hamiltonian $\hat{H}_{\mathrm{int}}$, assuming that $U_x > 0$ for all $x \in \Lambda$. The problem is indeed solved in Sect. 9.3.2, and we already know the eigenstates (9.3.30) and the eigenvalues (9.3.33). In particular we have seen that a ground state with zero energy is realized whenever the two subsets $\Gamma_\uparrow, \Gamma_\downarrow \subset \Lambda$ are taken to satisfy $\Gamma_\uparrow \cap \Gamma_\downarrow = \emptyset$ so that there are no doubly occupied sites. Now in the half-filled model, where we have $|\Gamma_\uparrow| + |\Gamma_\downarrow| = |\Lambda|$, the condition $\Gamma_\uparrow \cap \Gamma_\downarrow = \emptyset$ automatically implies that $\Gamma_\uparrow \cup \Gamma_\downarrow = \Lambda$. Since this means that each site is occupied either by an up-spin or down-spin electron, any ground state $|\Xi_{\Gamma_\uparrow,\Gamma_\downarrow}\rangle$ is rewritten (apart from a factor ± 1) in the form

$$|\Psi^{\boldsymbol{\sigma}}\rangle = \Bigl(\prod_{x \in \Lambda} c^\dagger_{x,\sigma_x}\Bigr)|\Phi_{\mathrm{vac}}\rangle, \tag{10.1.5}$$

where we ordered the sites in Λ in an arbitrary (but fixed) manner, and assume that the product is taken according to the ordering. Here $\boldsymbol{\sigma} = (\sigma_x)_{x\in\Lambda}$ with $\sigma_x = \uparrow, \downarrow$ is a spin configuration. Note that the state precisely corresponds to the basic basis state (2.2.1) for quantum spin systems. Since any spin configuration $\boldsymbol{\sigma}$ is allowed, we find that the ground states (10.1.5) are $2^{|\Lambda|}$ fold degenerate.

Let us examine the effect of the perturbation $\hat{H}_{\mathrm{hop}}$. For simplicity, we assume that $t_{x,x} = 0$ for any $x \in \Lambda$.[2] By operating $\hat{H}_{\mathrm{hop}}$ once onto $|\Psi^{\boldsymbol{\sigma}}\rangle$, an electron hops, and we inevitably get a state with one vacant site and one doubly occupied site. The resulting state is not a ground state of $\hat{H}_{\mathrm{int}}$. We thus find that the lowest order contribution from this perturbation theory comes from the second order.

Figure 10.1 shows a process that is taken into account in the second order perturbation theory. The electron at site y hops to site x with the transition amplitude $t_{x,y}$, resulting in a new state with extra energy U_x. Then one of the two electrons at site x hops back to site y, and the system returns to one of the ground states. In this process, spins at the sites x and y may be exchanged as Fig. 10.1 shows. Note that when the sites x and y have the same spin to begin with, the hopping between these sites is prohibited by the Pauli principle. This means that the above second order perturbation process lowers the energy of states in which the spins at sites x and y are pointing in the opposite direction (or, more precisely, the states in which the total spin is zero). We thus get an effective antiferromagnetic interaction between

[2]It is easy to remove this assumption since the diagonal part $\sum_{x,\sigma} t_{x,x}\,\hat{c}^\dagger_{x,\sigma}\hat{c}_{x,\sigma} = \sum_x t_{x,x}(\hat{n}_{x,\uparrow} + \hat{n}_{x,\downarrow})$ only shifts the energy of the unperturbed ground states by a constant $\sum_x t_{x,x}$, and slightly modifies the second order perturbation.

electronic spins, which is known as the super-exchange interaction or the kinetic exchange interaction.

Let us treat this perturbation theory more systematically. We denote by $\mathscr{H}_0$ the space spanned by $|\Psi^{\boldsymbol{\sigma}}\rangle$ with all possible spin configuration $\boldsymbol{\sigma}$, and by $\hat{P}_0$ the orthogonal projection onto $\mathscr{H}_0$. We still assume $t_{x,x} = 0$ and $U_x > 0$ for all $x \in \Lambda$. The fact that the first order perturbation has no contribution can be expressed as $\hat{P}_0 \, \hat{H}_{\text{hop}} \, \hat{P}_0 = 0$.

As we shall see at the end of the subsection, the effect of the second order perturbation is described by the effective spin Hamiltonian

$$\hat{H}_{\text{spin}} = -\hat{P}_0 \, \hat{H}_{\text{hop}} (\hat{H}_{\text{int}})^{-1} \hat{H}_{\text{hop}} \, \hat{P}_0, \tag{10.1.6}$$

which acts on the space $\mathscr{H}_0$. See (10.1.20). Although the interaction Hamiltonian $\hat{H}_{\text{int}}$ is not invertible, we are regarding it as an operator acting on the space orthogonal to $\mathscr{H}_0$. Then all the eigenvalues of $\hat{H}_{\text{int}}$ are positive (see (9.3.33)) and the inverse $(\hat{H}_{\text{int}})^{-1}$ is well defined.

Let us examine the precise form of the effective Hamiltonian (10.1.6). For any $|\Phi\rangle \in \mathscr{H}_0$, we have

$$(\hat{H}_{\text{int}})^{-1} \hat{H}_{\text{hop}} |\Phi\rangle = (\hat{H}_{\text{int}})^{-1} \sum_{\substack{x,y \in \Lambda \\ \sigma = \uparrow,\downarrow}} t_{x,y} \, \hat{c}^\dagger_{x,\sigma} \hat{c}_{y,\sigma} |\Phi\rangle = \sum_{\substack{x,y \in \Lambda \\ \sigma = \uparrow,\downarrow}} \frac{t_{x,y}}{U_x} \, \hat{c}^\dagger_{x,\sigma} \hat{c}_{y,\sigma} |\Phi\rangle, \tag{10.1.7}$$

because the site x is doubly occupied in the state $\hat{c}^\dagger_{x,\sigma} \hat{c}_{y,\sigma} |\Phi\rangle$ (if the state is nonzero). We thus find

$$\hat{P}_0 \, \hat{H}_{\text{hop}} (\hat{H}_{\text{int}})^{-1} \hat{H}_{\text{hop}} \, \hat{P}_0 = \sum_{\substack{x,y \in \Lambda \\ \sigma,\tau = \uparrow,\downarrow}} \frac{t_{y,x} \, t_{x,y}}{U_x} \, \hat{c}^\dagger_{y,\tau} \hat{c}_{x,\tau} \hat{c}^\dagger_{x,\sigma} \hat{c}_{y,\sigma} \, \hat{P}_0, \tag{10.1.8}$$

where we noted that one electron must hop from x to y to resolve the double occupancy so that the excited state returns to the subspace $\mathscr{H}_0$. For any $x \neq y$, we find from the anticommutation relations (9.2.27), (9.2.28) that

$$\begin{aligned}\sum_{\sigma,\tau = \uparrow,\downarrow} \hat{c}^\dagger_{y,\tau} \hat{c}_{x,\tau} \hat{c}^\dagger_{x,\sigma} \hat{c}_{y,\sigma} &= \sum_{\sigma = \uparrow,\downarrow} \hat{c}^\dagger_{y,\sigma} \hat{c}_{y,\sigma} - \sum_{\sigma,\tau = \uparrow,\downarrow} \hat{c}^\dagger_{y,\tau} \hat{c}_{y,\sigma} \hat{c}^\dagger_{x,\sigma} \hat{c}_{x,\tau} \\ &= \hat{n}_y - (\hat{S}^+_y \hat{S}^-_x + \hat{S}^-_y \hat{S}^+_x) - \sum_{\sigma = \uparrow,\downarrow} \hat{n}_{x,\sigma} \hat{n}_{y,\sigma},\end{aligned} \tag{10.1.9}$$

where we evaluated the terms with $\sigma \neq \tau$ and $\sigma = \tau$ separately, and used (9.2.44) and (9.2.29). One can also verify that $\sum_{\sigma=\uparrow,\downarrow} \hat{n}_{x,\sigma} \hat{n}_{y,\sigma} = 2(\hat{S}^{(3)}_x \hat{S}^{(3)}_y + \frac{1}{4} \hat{n}_x \hat{n}_y)$. By also recalling that $\hat{S}^+_y \hat{S}^-_x + \hat{S}^-_y \hat{S}^+_x = 2(\hat{S}^{(1)}_x \hat{S}^{(1)}_y + \hat{S}^{(2)}_x \hat{S}^{(2)}_y)$, and noting that $\hat{n}_x \hat{P}_0 = \hat{P}_0$, we finally see that

$$\hat{H}_{\text{spin}} = \sum_{x,y\in\Lambda} \frac{|t_{x,y}|^2}{U_x} 2\Big(\hat{\boldsymbol{S}}_x \cdot \hat{\boldsymbol{S}}_y - \frac{1}{4}\Big)\hat{P}_0. \tag{10.1.10}$$

By defining the set of bonds by $\mathscr{B} = \{\{x, y\} \,|\, t_{x,y} \neq 0\}$ (where we identify $\{x, y\}$ and $\{y, x\}$) and the exchange interaction by

$$J_{x,y} = 2\Big(\frac{|t_{x,y}|^2}{U_x} + \frac{|t_{x,y}|^2}{U_y}\Big), \tag{10.1.11}$$

we can write (10.1.10) as

$$\hat{H}_{\text{spin}} = \sum_{\{x,y\}\in\mathscr{B}} J_{x,y}\Big(\hat{\boldsymbol{S}}_x \cdot \hat{\boldsymbol{S}}_y - \frac{1}{4}\Big)\hat{P}_0, \tag{10.1.12}$$

which is nothing but the Hamiltonian (2.5.1) (with nonuniform exchange interactions) of the $S = 1/2$ antiferromagnetic Heisenberg model. This observation suggests that the low-energy behavior of the half-filled Hubbard model is well described by the antiferromagnetic Heisenberg model when U_x are much larger than $t_{x,y}$.

When applied to the two site model studied in the beginning of the section, the effective Hamiltonian (10.1.12) yields the energy eigenvalues 0 and $-4t^2/U$ since $\hat{\boldsymbol{S}}_x \cdot \hat{\boldsymbol{S}}_y$ has eigenvalues $1/4$ and $-3/4$ (see (2.2.19)). This is consistent with the exact calculation.

Perturbation theory We shall treat the second order perturbation theory for degenerate ground states in a general setting. We derive the expression (10.1.20) of the effective Hamiltonian, and also state a lemma which will be used later.

Let $\mathscr{H}$ be a finite dimensional Hilbert space. We consider a Hamiltonian $\hat{H}_0$ with D_0-fold degenerate ground states. We assume that the ground state energy is zero, and any other energy eigenvalue is not less than $E_{\text{gap}} > 0$. We denote by $\mathscr{H}_0$ the space spanned by all the ground states of $\hat{H}_0$, and by $\mathscr{H}_\perp$ its orthogonal complement. We let $\hat{P}_0$ be the orthogonal projection onto $\mathscr{H}_0$.

We consider the perturbed Hamiltonian

$$\hat{H}(\lambda) = \hat{H}_0 + \lambda\hat{V}, \tag{10.1.13}$$

where $\lambda \in \mathbb{R}$ is a small parameter, and $\hat{V}$ is an arbitrary perturbation Hamiltonian such that $\hat{V}|\Phi\rangle \in \mathscr{H}_\perp$ for any $|\Phi\rangle \in \mathscr{H}_0$. This property implies that there is no first-order correction to the ground states, and one should look at the second order perturbation.

Let us write the exact Schrödinger equation for $\hat{H}(\lambda)$ as

$$(\hat{H}_0 + \lambda\hat{V})(|\Phi\rangle + |\Gamma\rangle) = E\,(|\Phi\rangle + |\Gamma\rangle), \tag{10.1.14}$$

where $|\Phi\rangle \in \mathscr{H}_0$ and $|\Gamma\rangle \in \mathscr{H}_\perp$. Operating $\hat{P}_0$ to (10.1.14) and recalling that $\hat{H}_0|\Phi\rangle = 0$ and $\hat{H}_0|\Gamma\rangle \in \mathscr{H}_\perp$, we get

$$\lambda \hat{P}_0 \hat{V} |\Gamma\rangle = E\,|\Phi\rangle. \tag{10.1.15}$$

Operating $(\hat{1} - \hat{P}_0)$ to (10.1.14), we find

$$\lambda \hat{V}|\Phi\rangle + \hat{H}_0|\Gamma\rangle + \lambda(\hat{1} - \hat{P}_0)\hat{V}|\Gamma\rangle = E\,|\Gamma\rangle, \tag{10.1.16}$$

which implies

$$\lambda \hat{V}|\Phi\rangle = \bigl\{-\hat{H}_0 + E + \lambda(\hat{P}_0 - \hat{1})\hat{V}\bigr\}|\Gamma\rangle \simeq -\hat{H}_0|\Gamma\rangle. \tag{10.1.17}$$

Here we assumed that E and $\lambda\hat{V}$ are much smaller than the energy gap E_{gap}. We then find

$$|\Gamma\rangle \simeq -\lambda \hat{H}_0^{-1}\hat{V}|\Phi\rangle, \tag{10.1.18}$$

where we noted that $\hat{H}_0^{-1}$ is well-defined in the space $\mathscr{H}_\perp$. Substituting (10.1.18) into (10.1.15), we find that the Schrödinger equation (10.1.14) is approximately rewritten as

$$-\lambda^2 \hat{P}_0\,\hat{V}\,\hat{H}_0^{-1}\hat{V}|\Phi\rangle \simeq E|\Phi\rangle. \tag{10.1.19}$$

This means that, within the second order perturbation theory, low-energy eigenstates of $\hat{H}(\lambda)$ are determined by the effective Hamiltonian

$$\hat{H}_{\mathrm{eff}} = -\hat{P}_0\,\hat{V}\,\hat{H}_0^{-1}\hat{V}\,\hat{P}_0, \tag{10.1.20}$$

which is an operator on the Hilbert space $\mathscr{H}_0$. The effective spin Hamiltonian in (10.1.6) is give by $\hat{H}_{\mathrm{spin}} = \lambda^2\hat{H}_{\mathrm{eff}}$.

Although the above discussion is not meant to be rigorous, one can derive some (weak) rigorous statements from the above argument. The following lemma, for example, will be useful when we prove Lieb's theorem for the repulsive Hubbard model in Sect. 10.2.2.

Lemma 10.1 *Suppose that the effective Hamiltonian (10.1.20) is well-defined, and its ground state $|\Phi_{\text{eff-GS}}\rangle$ is unique. Then the ground state of $\hat{H}(\lambda)$ is unique for sufficiently small λ, and converges to $|\Phi_{\text{eff-GS}}\rangle$ as $\lambda \to 0$.*

It may be useful to check that the ground state of the two-site Hubbard model studied in the beginning of the present section indeed converges to that of the two-site antiferromagnetic Heisenberg model.

Proof By continuity there are exactly D_0 independent eigenstates of $\hat{H}(\lambda)$ whose eigenvalues converge to zero as $\lambda \to 0$. We can also assume that each of these D_0 eigenstates depends continuously on λ. Let us denote these eigenstates as $|\Xi_j(\lambda)\rangle$ with $j = 1, \ldots, D_0$. We thus have $\hat{H}(\lambda)|\Xi_j(\lambda)\rangle = E_j(\lambda)|\Xi_j(\lambda)\rangle$ with

$\lim_{\lambda\to 0} E_j(\lambda) = 0$. Note that we must have $|\Xi_j(0)\rangle \in \mathscr{H}_0$ for all j. Setting $|\Phi\rangle = \hat{P}_0|\Xi_j(\lambda)\rangle$ and $|\Gamma\rangle = (\hat{1} - \hat{P}_0)|\Xi_j(\lambda)\rangle$, we find from (10.1.15) and (10.1.17) that

$$\hat{P}_0\,\hat{V}\frac{1}{-\hat{H}_0 + E_j(\lambda) + \lambda(\hat{P}_0 - \hat{1})\hat{V}}\hat{V}\,\hat{P}_0|\Xi_j(\lambda)\rangle = \frac{E_j(\lambda)}{\lambda^2}\hat{P}_0|\Xi_j(\lambda)\rangle. \quad (10.1.21)$$

We here noted that $-\hat{H}_0 + E_j(\lambda) + \lambda(\hat{P}_0 - \hat{1})\hat{V}$ is invertible in $\mathscr{H}_\perp$ when λ is sufficiently small. The left-hand side of (10.1.21) clearly converges to $\hat{H}_{\rm eff}|\Xi_j(0)\rangle$ as $\lambda \to 0$. Since the right-hand side should also have a well-defined $\lambda \to 0$ limit, we get

$$\hat{H}_{\rm eff}|\Xi_j(0)\rangle = \varepsilon_j|\Xi_j(0)\rangle, \quad (10.1.22)$$

with $\varepsilon_j = \lim_{\lambda\to 0} E_j(\lambda)/\lambda^2$. We have found that $|\Xi_j(0)\rangle$ is an eigenstate of $\hat{H}_{\rm eff}$. Because the states $|\Xi_j(0)\rangle$ with $j = 1, \ldots, D_0$ are linearly independent, one of them must coincide with $|\Phi_{\text{eff-GS}}\rangle$. By relabeling j if necessary, we therefore have $|\Xi_1(0)\rangle = (\text{const.})|\Phi_{\text{eff-GS}}\rangle$. Thus ε_1 is the ground state energy of $\hat{H}_{\rm eff}$. Recalling that $\varepsilon_j > \varepsilon_1$ for any $j = 2, \ldots, D_0$ by assumption, and $\varepsilon_j = \lim_{\lambda\to 0} E_j(\lambda)/\lambda^2$, we find that $|\Xi_1(\lambda)\rangle$ is the unique ground state of $\hat{H}(\lambda)$ for sufficiently small λ. ∎

10.2 Lieb's Theorems

In 1989, Lieb proved two theorems for the Hubbard model, which are of fundamental importance in the theory of quantum many-body systems [10]. One of the theorems provides, among other things, partial support to the conjecture about the similarity of the half-filled Hubbard model and the antiferromagnetic Heisenberg model that we discussed in the previous section. The theorems also lead to rigorous examples of itinerant electron systems exhibiting ferrimagnetism or superconductivity.

In the present section, we shall give a thorough discussion of Lieb's theorems and their applications. We start from the basic theorem for the attractive Hubbard model (Sect. 10.2.1) and then proceed to a more useful theorem for the repulsive Hubbard model (Sect. 10.2.2). We then discuss in Sect. 10.2.3 two important applications of the theorems, namely to ferrimagnetism and to superconductivity. In Sect. 10.2.4, we present detailed proofs of the main theorems presented in Sect. 10.2.1. It is likely that the proof of the Lieb theorem is vaguely regarded as something very difficult to comprehend. By going through our exposition, which is intended to be as elementary and careful as possible, the reader will find that the proof makes use only of elementary linear algebra (nothing more than the diagonalization of Hermitian matrices). We encourage the reader to go through the proof to appreciate the beautiful and highly unexpected argument, known as the spin-space reflection positivity, devised by Lieb to prove the theorem.[3] In Sect. 10.2.5, we briefly discuss some extensions of Lieb's theorems.

10.2.1 Lieb's Theorem for the Attractive Hubbard Model

In this subsection, we focus on a very different model, namely, the attractive Hubbard model with an even number of electrons, and describe basic theorems. We shall give complete and detailed proofs of the theorems in Sect. 10.2.4. In the next subsection we shall see that these theorems can be turned into important theorems for the standard repulsive Hubbard model at half-filling.

The model covered here is surprisingly general. The hopping Hamiltonian is

$$\hat{H}_{\text{hop}} = \sum_{\substack{x,y\in\Lambda \\ \sigma=\uparrow,\downarrow}} t_{x,y}\,\hat{c}^\dagger_{x,\sigma}\hat{c}_{y,\sigma}, \tag{10.2.1}$$

which is exactly the same as (9.3.17). Here the hopping amplitude is quite arbitrary except for the two requirements that $t_{x,y} = t_{y,x} \in \mathbb{R}$ and that the whole lattice Λ is connected through nonvanishing $t_{x,y}$.[4] Note, in particular, that arbitrary on-site potential $t_{x,x} \in \mathbb{R}$ is allowed. We consider the attractive on-site interaction

$$\hat{H}_{\text{att-int}} = -\sum_{x\in\Lambda} U_x\,\hat{n}_{x,\uparrow}\hat{n}_{x,\downarrow}, \tag{10.2.2}$$

with arbitrary $U_x > 0$.

The following theorem of Lieb's is the basis of the whole theories in the present chapter.

Theorem 10.2 (Lieb's theorem) *Consider the attractive Hubbard model with the Hamiltonian* $\hat{H} = \hat{H}_{\text{hop}} + \hat{H}_{\text{att-int}}$, *and let the electron number* N *be an arbitrary even integer such that* $0 < N \le 2|\Lambda|$. *Then the ground state is unique and has* $S_{\text{tot}} = 0$.

That the ground state has $S_{\text{tot}} = 0$ may not be surprising since the attractive interaction (10.2.2) favors doubly occupied sites. In particular, the ground states of the model in the limit $U_x \uparrow \infty$ for all $x \in \Lambda$ are

$$|\Theta_A\rangle = \Bigl(\prod_{x\in A} \hat{c}^\dagger_{x,\uparrow}\hat{c}^\dagger_{x,\downarrow}\Bigr)|\Phi_{\text{vac}}\rangle, \tag{10.2.3}$$

---

[3]The Proof of Theorem 4.1 (p. 75), which establishes the existence of long-range order in the antiferromagnetic Heisenberg model, is also based on the reflection positivity method. But the proof of Lieb's theorem is much simpler.

[4]More precisely this means that, for any $x, y \in \Lambda$ such that $x \ne y$, there exists a finite sequence $z_1, z_2, \ldots, z_n \in \Lambda$ with the properties $z_1 = x$, $z_n = y$, and $t_{z_j,z_{j+1}} \ne 0$ for $j = 1, \ldots, n-1$.

where A is an arbitrary subset of Λ with $|A| = N/2$. As we have seen in (9.2.71), $\hat{c}^\dagger_{x,\uparrow}\hat{c}^\dagger_{x,\downarrow}$ creates a spin-singlet pair, which has $S_{\text{tot}} = 0$, on site x. Note however that the ground states in this case is highly degenerate (unless $N = 2|\Lambda|$) since $|\Theta_A\rangle$ with any A is a ground state. The uniqueness guaranteed by Lieb's theorem is a consequence of finiteness of U_x, and is a result of subtle interplay between the interaction $\hat{H}_{\text{att-int}}$ and the hopping $\hat{H}_{\text{hop}}$.[5]

In the opposite limit with $U_x = 0$ for all $x \in \Lambda$, the statement of the theorem seems to follow from the elementary observation that the state (9.3.24), which apparently has $S_{\text{tot}} = 0$, is an exact ground state. But we should note that the ground state may not be unique when some of the eigenvalues of the hopping matrix $\mathsf{T} = (t_{x,y})_{x,y\in\Lambda}$ are degenerate. It is in fact possible that some simple hopping matrix T has a highly degenerate eigenvalue. See Sects. 10.2.3 and 11.3. In such a case, the uniqueness stated in the proof is a result of nontrivial effect caused by the interaction.

The method developed by Lieb to prove Theorem 10.2 can be extended to study certain correlation functions. The following theorem is due to Tian [23]. See also [20, 25].

Theorem 10.3 *Let $|\Phi_{\text{GS}}\rangle$ be the unique ground state of $\hat{H} = \hat{H}_{\text{hop}} + \hat{H}_{\text{att-int}}$ with even N such that $0 < N \le 2|\Lambda|$. For any $x, y \in \Lambda$, we have*

$$\langle\Phi_{\text{GS}}|\hat{c}^\dagger_{x,\uparrow}\hat{c}^\dagger_{x,\downarrow}\hat{c}_{y,\downarrow}\hat{c}_{y,\uparrow}|\Phi_{\text{GS}}\rangle > 0. \tag{10.2.4}$$

Note that the correlation function $\langle\Phi_{\text{GS}}|\hat{c}^\dagger_{x,\uparrow}\hat{c}^\dagger_{x,\downarrow}\hat{c}_{y,\downarrow}\hat{c}_{y,\uparrow}|\Phi_{\text{GS}}\rangle$ is a measure of off-diagonal long-range order (ODLRO). Exactly as Bose–Einstein condensation is characterized by the behavior of the correlation function $\langle\Phi_{\text{GS}}|\hat{a}^\dagger_x\hat{a}_y|\Phi_{\text{GS}}\rangle$ as in (5.2.1), the standard superconductivity can be detected by this correlation function. We note however that Theorem 10.3 alone has no direct implication on the presence or absence of ODLRO. In Theorem 10.8 (p. 359), we shall see that the existence of ODLRO can be proved in some cases.

10.2.2 Lieb's Theorem for the Repulsive Hubbard Model

Lieb's theorem is most naturally stated for the attractive Hubbard model, but it can be converted into a theorem for the standard repulsive Hubbard model by using the Shiba transformation discussed in Sect. 9.3.3. Although this conversion is possible only for a special class of models, the resulting theorem turns out to be of profound importance in the study of strongly interacting electron systems.

Here we shall first state the theorems, and then describe how they can be obtained from the theorems for the attractive Hubbard model.

[5] It may happen however that the model exhibits phase separation and does not have a unique ground state in the thermodynamic limit.

Main statements We consider the Hubbard model on a finite lattice Λ. We assume that the hopping amplitude $(t_{x,y})_{x,y\in\Lambda}$ is bipartite, i.e., the lattice is decomposed into two sublattices as $\Lambda = A \cup B$ with $A \cap B = \emptyset$, and $t_{x,y}$ is nonvanishing only when $x \in A, y \in B$ or $x \in B, y \in A$. This means that electrons can hop only between different sublattices. We also assume that $t_{x,y} = t_{y,x} \in \mathbb{R}$ as before, and that the whole lattice Λ is connected through nonvanishing $t_{x,y}$. Then the hopping Hamiltonian $\hat{H}_{\text{hop}}$ is defined by (10.2.1) or (9.3.17).

We can treat two types of interactions. One is the standard interaction (9.3.29) with uniform repulsive energy $U > 0$:

$$\hat{H}_{\text{int}} = U \sum_{x\in\Lambda} \hat{n}_{x,\uparrow}\hat{n}_{x,\downarrow} \tag{10.2.5}$$

The other is the symmetric form (9.3.47) with local repulsive energy $U_x > 0$ for any $x \in \Lambda$:

$$\hat{H}'_{\text{int}} := \sum_{x\in\Lambda} U_x \Big(\hat{n}_{x,\uparrow} - \frac{1}{2}\Big)\Big(\hat{n}_{x,\downarrow} - \frac{1}{2}\Big) \tag{10.2.6}$$

We can now state Lieb's theorem for the repulsive Hubbard model at half-filling [10].

Theorem 10.4 (Lieb's theorem) *Consider the Hubbard model with $\hat{H} = \hat{H}_{\text{hop}} + \hat{H}_{\text{int}}$ or $\hat{H}' = \hat{H}_{\text{hop}} + \hat{H}'_{\text{int}}$, and let the electron number be $N = |\Lambda|$. Then any ground state has total spin $S_{\text{tot}} = \big||A| - |B|\big|/2$. The ground states are exactly $2S_{\text{tot}} + 1 = \big||A| - |B|\big| + 1$ fold degenerate.*

Note that the $2S_{\text{tot}} + 1$ fold degeneracy is the trivial (and mandatory) degeneracy that comes from the SU(2) invariance of the model. See Theorem A.16 in p. 473. The theorem therefore states that the ground states are unique apart from the unavoidable spin degeneracy.

The total spin S_{tot} of the ground state determined in the theorem is exactly the same as that of the ground state(s) of the $S = 1/2$ antiferromagnetic Heisenberg model on the same lattice. See the Lieb–Mattis theorem (Theorem 2.3 in p. 42). We should stress however that the straightforward Perron–Frobenius argument used in the proof of the Lieb–Mattis theorem does not apply to the Hubbard model (except in one dimension). This is not only a technical difficulty, but is a consequence of the essential fact that quantum mechanical processes allowed in the Hubbard model are in general much richer and more complex than those in the Heisenberg model. The proof of Lieb's theorem indeed makes use of a kind of Perron–Frobenius argument, but should be combined with a very clever mathematical idea called the spin-space reflection positivity. See Sect. 10.2.4.

For the standard uniform interaction (10.2.5), Lieb's theorem is valid for any value of Coulomb repulsion U, only provided that it is positive. It is quite likely that low-energy properties of the Hubbard model are drastically different in the weak coupling region with small U and in the strong coupling region with large U. It is remarkable that a single proof of Lieb's covers the whole range of $U > 0$ and clarifies the basic properties of the ground states.

Needless to say, however, the knowledge about the total spin of the ground states does not completely determine the nature of the ground states. When two sublattices have the same number of sites, for example, one finds from the theorem that the finite volume ground state is unique and has $S_{\mathrm{tot}} = 0$. We have already seen in the previous parts of the present book that ground states with $S_{\mathrm{tot}} = 0$ can exhibit drastically different properties.

Corresponding to Theorem 10.3, we have the following result, which was first noted by Shen, Qiu, and Tian [20]. See also [18, 25].

Theorem 10.5 *Under the same condition as Theorem 10.4, we have for any ground state* $|\Phi_{\mathrm{GS}}\rangle$ *and for* $x, y \in \Lambda$ *that*

$$\langle\Phi_{\mathrm{GS}}|(\hat{S}_x^{(1)}\hat{S}_y^{(1)} + \hat{S}_x^{(2)}\hat{S}_y^{(2)})|\Phi_{\mathrm{GS}}\rangle \begin{cases} > 0 & \text{when } x, y \in A \text{ or } x, y \in B, \\ < 0 & \text{when } x \in A, y \in B \text{ or } x \in B, y \in A. \end{cases} \tag{10.2.7}$$

In the special case with $|A| = |B|$, we know that the ground state $|\Phi_{\mathrm{GS}}\rangle$ is unique and is SU(2) invariant. Then (10.2.7) implies

$$\langle\Phi_{\mathrm{GS}}|\hat{\mathbf{S}}_x \cdot \hat{\mathbf{S}}_y|\Phi_{\mathrm{GS}}\rangle \begin{cases} > 0 & \text{when } x, y \in A \text{ or } x, y \in B, \\ < 0 & \text{when } x \in A, y \in B \text{ or } x \in B, y \in A. \end{cases} \tag{10.2.8}$$

The correlation inequalities (10.2.7), (10.2.8) also suggest the similarity with the antiferromagnetic Heisenberg model. In fact (10.2.8) is exactly the same as the inequality (2.5.7), and (10.2.7) is the same as more general inequality (4.1.15).

As is stated in Theorem 4.1 in p. 75, the $S = 1/2$ antiferromagnetic Heisenberg model on the cubic lattice, for example, exhibits antiferromagnetic long-range order in the ground state. It is likely that the same is true for the half-filled Hubbard model on the same lattice with sufficiently large U. But, for the moment, there are no methods or ideas which are useful in proving this conjecture. To extend the powerful method of the (spatial) reflection positivity used to prove Theorem 4.1 seems hopeless.

Proof Let us prove Theorems 10.4 and 10.5, assuming Theorems 10.2 and 10.3, respectively, for the attractive models. Although Theorem 10.4 is essentially a corollary of Theorem 10.2, there are some nontrivial steps in the proof.

Proof of Theorem 10.4, Assuming Theorem 10.2 We first note that the model with the uniform interaction (10.2.5) can be regarded as a special case of the model with the symmetric interaction (10.2.6). To see this we set $U_x = U$ for all $x \in \Lambda$ in (10.2.6) to find that

$$\hat{H}'_{\mathrm{int}} = U\sum_{x\in\Lambda}\hat{n}_{x,\uparrow}\hat{n}_{x,\downarrow} - \frac{U}{2}\sum_{x\in\Lambda}(\hat{n}_{x,\uparrow} + \hat{n}_{x,\downarrow}) + \frac{U}{4}|\Lambda| = \hat{H}_{\mathrm{int}} - \frac{U}{2}N + \frac{U}{4}|\Lambda|, \tag{10.2.9}$$

where the right-hand is essentially the uniform interaction (10.2.5). We can thus concentrate on the symmetric interaction (10.2.6). Our Hubbard Hamiltonian with repulsive interaction is $\hat{H}^{\mathrm{rep}} = \hat{H}_{\mathrm{hop}} + \hat{H}'_{\mathrm{int}}$ with arbitrary $U_x > 0$.

We assume for the moment that $N = |\Lambda|$ is even. Let $\mathscr{H}_{N/2,N/2}$ be the subspace of $\mathscr{H}_N$ consisting of states with exactly $N/2$ up-spin electrons and $N/2$ down-spin electrons. In other words, $\mathscr{H}_{N/2,N/2}$ consists of all possible linear combinations of the basis states (9.2.35) which satisfy $\sum_{j=1}^{N} \sigma_j = 0$. Since the Hamiltonian $\hat{H}^{\mathrm{rep}} = \hat{H}_{\mathrm{hop}} + \hat{H}'_{\mathrm{int}}$ is SU(2) invariant (see Sect. 9.3.3), Theorem A.17 (p. 473) guarantees that one finds all the energy eigenvalues in the subspace with $S^{(3)}_{\mathrm{tot}} = 0$, which is nothing but $\mathscr{H}_{N/2,N/2}$. We shall thus work within $\mathscr{H}_{N/2,N/2}$.[6]

We now use the Shiba transformation $\hat{U}^{\mathrm{S}}$ discussed at the end of Sect. 9.3.3 to convert the repulsive Hubbard model into the attractive Hubbard model. Recall that the Shiba transformation changes the electron numbers as $N_\uparrow \to |\Lambda| - N_\uparrow$ and $N_\downarrow \to N_\downarrow$. Thus the transformation leaves the subspace $\mathscr{H}_{N/2,N/2}$ invariant. As we have seen in (9.3.54), the Shiba transformation maps the Hamiltonian $\hat{H}^{\mathrm{rep}} = \hat{H}_{\mathrm{hop}} + \hat{H}'_{\mathrm{int}}$ into[7]

$$\hat{H}^{\mathrm{att}} := \hat{U}^{\mathrm{S}} \hat{H}^{\mathrm{rep}} (\hat{U}^{\mathrm{S}})^\dagger = \hat{H}_{\mathrm{hop}} - \hat{H}'_{\mathrm{int}}. \tag{10.2.10}$$

Note that

$$-\hat{H}'_{\mathrm{int}} = \hat{H}_{\mathrm{att\text{-}int}} + \sum_{x\in\Lambda} \frac{U_x}{2}(\hat{n}_{x,\uparrow} + \hat{n}_{x,\downarrow}) + \frac{1}{4}\sum_{x\in\Lambda} U_x, \tag{10.2.11}$$

where the attractive interaction $\hat{H}_{\mathrm{att\text{-}int}}$ is defined in (10.2.2). Therefore, by including the on-site potential $(U_x/2)(\hat{n}_{x,\uparrow} + \hat{n}_{x,\downarrow})$ into the hopping Hamiltonian (and omitting the additive constant), we see that $\hat{H}^{\mathrm{att}} = \hat{H}_{\mathrm{hop}} - \hat{H}'_{\mathrm{int}}$ has a precise form to which Theorem 10.2 applies. We find that the ground state of $\hat{H}^{\mathrm{att}}$ (in the whole Hilbert space $\mathscr{H}_N$) is unique and has $S_{\mathrm{tot}} = 0$, and hence is in $\mathscr{H}_{N/2,N/2}$. We find in particular that the ground state of $\hat{H}^{\mathrm{att}}$ in $\mathscr{H}_{N/2,N/2}$ is unique. We thus conclude that the ground state of the repulsive Hubbard Hamiltonian $\hat{H}^{\mathrm{rep}}$ within $\mathscr{H}_{N/2,N/2}$ is unique. Let us call it $|\Phi^{(0)}_{\mathrm{GS}}\rangle$.

Let us see how the preceding argument should be modified when $N = |\Lambda|$ is odd. Theorem A.17 (p. 473) now allows us to work within $\mathscr{H}_{(N+1)/2,(N-1)/2}$, the subspace with $(N+1)/2$ up-spin electrons and $(N-1)/2$ down-spin electrons. The Shiba transformation maps this subspace into $\mathscr{H}_{(N-1)/2,(N-1)/2}$. Note that Theorem 10.2 again implies that the ground state of $\hat{H}^{\mathrm{att}}$ with $N-1$ electrons is unique and is in $\mathscr{H}_{(N-1)/2,(N-1)/2}$. We thus see that the ground state of $\hat{H}^{\mathrm{rep}}$ in $\mathscr{H}_{(N+1)/2,(N-1)/2}$ is unique. We call it also $|\Phi^{(0)}_{\mathrm{GS}}\rangle$.

It remains to determine the total spin of $|\Phi^{(0)}_{\mathrm{GS}}\rangle$. This can be done in several manners, but we here follow the argument in the original proof of Lieb's [10]. We treat the cases with even and odd N at the same time.

---

[6]Recall that we used the same strategy in the proof of the Marshall–Lieb–Mattis theorem (Theorem 2.2 in p. 39). See p. 40.

[7]Note that we are here using the second convention of transformations of operators as in (A.1.17).

The strategy is as follows. Think about varying the parameters of the model continuously without violating the conditions of the theorem. Then the ground state (within $\mathscr{H}_{N/2,N/2}$ or $\mathscr{H}_{(N+1)/2,(N-1)/2}$) remains unique, and changes continuously. It is crucial to note that the total spin S_{tot} of the ground state does not change in this process.[8] We then make use of the perturbative analysis in Sect. 10.1 (in a rigorous manner) to connect the ground state of the Hubbard model to that of the antiferromagnetic Heisenberg. The foregoing argument shows that S_{tot} of the ground state of the Hubbard model is the same as S_{tot} of the ground state of the antiferromagnetic Heisenberg model.[9] But the latter is already known from the Lieb–Mattis theorem (Theorem 2.3 in p. 42).

To be precise, we make the following continuous changes of the parameters. For each $x \in \Lambda$, we change U_x to a common constant $U > 0$. For each pair $x, y \in \Lambda$ such that $x \neq y$, we change $t_{x,y}$ to a constant $\lambda > 0$ if $t_{x,y}$ is positive, and to $-\lambda$ if $t_{x,y}$ is negative. We do not modify $t_{x,y}$ which is zero to begin with. In this manner we get a Hubbard model with uniform repulsive interaction U and hopping amplitudes $\pm\lambda$. Let us imagine that $|\lambda| \ll U$, and apply the perturbation theory in Sect. 10.1, treating the hopping Hamiltonian as a perturbation. Then, exactly as in (10.1.12), we see that the effective Hamiltonian defined in (10.1.6) is given by

$$\hat{H}_{\text{eff}} = \frac{4\lambda^2}{U} \sum_{\{x,y\}\in\mathscr{B}} \left(\hat{\boldsymbol{S}}_x \cdot \hat{\boldsymbol{S}}_y - \frac{1}{4}\right), \tag{10.2.12}$$

which is nothing but the Hamiltonian of the $S = 1/2$ antiferromagnetic Heisenberg model on Λ. The set of bonds is given by $\mathscr{B} = \{\{x, y\} \mid t_{x,y} \neq 0\}$. The assumptions about the connectivity and bipartiteness of the hopping amplitude imply that the lattice $(\Lambda, \mathscr{B})$ is connected and bipartite. Then we can apply the Lieb–Mattis theorem (Theorem 2.3 in p. 42) to conclude that the ground state of (10.2.12) (in the sector with $S_{\text{tot}}^{(3)} = 0$ or $S_{\text{tot}}^{(3)} = 1/2$) is unique and has $S_{\text{tot}} = \big||A| - |B|\big|/2$. We now recall that Lemma 10.1 shows that, when we let $\lambda \to 0$, the ground state (within $\mathscr{H}_{N/2,N/2}$ or $\mathscr{H}_{(N+1)/2,(N-1)/2}$) of the Hubbard model converges to that of the Heisenberg model (10.2.12). From the invariance of S_{tot} (which we have discussed above), we conclude that the ground state $|\Phi_{\text{GS}}^{(0)}\rangle$ of the Hubbard model also has $S_{\text{tot}} = \big||A| - |B|\big|/2$.

The degeneracy of the ground states in the whole Hilbert space $\mathscr{H}_N$ readily follows from Theorem A.16 (p. 473). ∎

Proof of Theorem 10.5, Assuming Theorem 10.3 This is easy. Take a ground state $|\Phi_{\text{GS}}^{\text{rep}}\rangle$ of the repulsive Hamiltonian $\hat{H}^{\text{rep}}$ in $\mathscr{H}_{N_1,N_2}$, the subspace with N_1 up-spin electrons and N_2 down-spin electrons, where $N_1 + N_2 = |\Lambda|$. The Shiba transformation maps this subspace into $\mathscr{H}_{|\Lambda|-N_1,N_2} = \mathscr{H}_{N_2,N_2}$. Obviously $|\Phi_{\text{GS}}^{\text{att}}\rangle = \hat{U}^{\text{S}}|\Phi_{\text{GS}}^{\text{rep}}\rangle$

---

[8] Recall that the ground state is a simultaneous eigenstate of the Hamiltonian and $(\hat{\boldsymbol{S}}_{\text{tot}})^2$. By continuity the eigenvalue of the latter, which is $S_{\text{tot}}(S_{\text{tot}} + 1)$, cannot vary.

[9] It is essential that the spin operators $\hat{S}_x^{(\alpha)}$ defined for the Hubbard model, when restricted onto the subspace without any doubly occupied sites, exactly coincide with that of the quantum spin system.

is the ground state of the attractive Hamiltonian $\hat{H}^{\text{att}} = \hat{U}^{\text{S}} \hat{H}^{\text{rep}} (\hat{U}^{\text{S}})^{\dagger}$ in the subspace $\mathscr{H}_{N_2,N_2}$. Note also that the ground state correlation functions are transformed as $\langle \Phi_{\text{GS}}^{\text{rep}} | \hat{A} | \Phi_{\text{GS}}^{\text{rep}} \rangle = \langle \Phi_{\text{GS}}^{\text{att}} | \hat{U}^{\text{S}} \hat{A} (\hat{U}^{\text{S}})^{\dagger} | \Phi_{\text{GS}}^{\text{att}} \rangle$. If we abbreviate the relation $\hat{U}^{\text{S}} \hat{A} (\hat{U}^{\text{S}})^{\dagger} = \hat{A}'$ as $\hat{A} \to \hat{A}'$, we find that the Shiba transformation (9.3.51) maps the spin operators as

$$\hat{S}_x^+ = \hat{c}_{x,\uparrow}^{\dagger} \hat{c}_{x,\downarrow} \to (-1)^x \hat{c}_{x,\uparrow} \hat{c}_{x,\downarrow}, \quad \hat{S}_x^- = \hat{c}_{x,\downarrow}^{\dagger} \hat{c}_{x,\uparrow} \to (-1)^x \hat{c}_{x,\downarrow}^{\dagger} \hat{c}_{x,\uparrow}^{\dagger}, \tag{10.2.13}$$

where $(-1)^x = 1$ if $x \in A$ and $(-1)^x = -1$ if $x \in B$. This implies that

$$(-1)^x (-1)^y \langle \Phi_{\text{GS}}^{\text{rep}} | (\hat{S}_x^+ \hat{S}_y^-) | \Phi_{\text{GS}}^{\text{rep}} \rangle = \langle \Phi_{\text{GS}}^{\text{att}} | \hat{c}_{y,\uparrow}^{\dagger} \hat{c}_{y,\downarrow}^{\dagger} \hat{c}_{x,\downarrow} \hat{c}_{x,\uparrow} | \Phi_{\text{GS}}^{\text{att}} \rangle. \tag{10.2.14}$$

Then the desired inequalities (10.2.7) for $|\Phi_{\text{GS}}^{\text{rep}}\rangle$ follow from (10.2.4) for $|\Phi_{\text{GS}}^{\text{att}}\rangle$. ∎

10.2.3 Lieb's Ferrimagnetism

When applied to models defined on asymmetric bipartite lattices, Lieb's theorems yield nontrivial and beautiful results of considerable physical significance. For the repulsive Hubbard models at half-filling, we find from Theorem 10.4 that there is a large class of models which exhibit ferromagnetism, or, more precisely, ferrimagnetism [10]. This is among the most influential achievements in modern mathematical physics, and is also directly related to the original motivation for the study of the Hubbard model, namely, to understand the origin of ferromagnetism. See Sect. 9.1. For the attractive Hubbard model, one can use Theorem 10.2 to show that a class of models exhibits superconductivity, or, more precisely, off-diagonal long-range order that characterizes condensation of electron pairs [19]. Although these models with attractive interaction may not be realistic as they are, it is remarkable that there are many simple models which can be proved to exhibit superconductivity.

As in the previous section we consider a lattice Λ which is decomposed into two sublattices as $\Lambda = A \cup B$. We are interested in the situation where the numbers of sites in the two sublattices differ macroscopically. We assume, without loss of generality, that $|A| > |B|$. There are many examples in which we have

$$|A| - |B| = a\,|\Lambda|, \tag{10.2.15}$$

with a constant a such that $0 < a < 1$.

A typical example is the decorated square lattice in Fig. 10.2, where the sublattice A consists of sites indicated by white dots, and the sublattice B consists of sites indicated by black dots. When the sublattice B is identical to the $L \times L$ square lattice, we have $|A| = 2L^2$, $|B| = L^2$, and $|\Lambda| = |A| + |B| = 3L^2$. We thus have (10.2.15) with $a = 1/3$. The lattice was called the CuO lattice in the 1980s (when the present author entered the field), but recently it is mostly called the Lieb lattice. This is

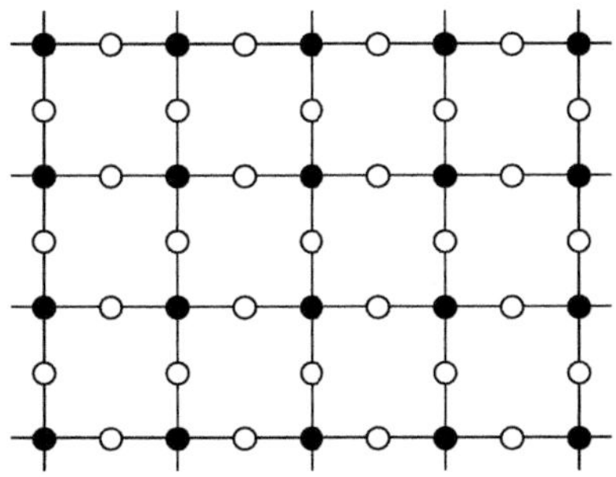

Fig. 10.2 An example of a bipartite lattice in which the numbers of sites in the two sublattices are macroscopically different. Lieb's theorems imply that the repulsive Hubbard model at half-filling on this lattice exhibits ferrimagnetism, and the attractive Hubbard model with filling factor ν in the range $1/3 < \nu < 2/3$ exhibits superconductivity. The lattice used to be called the CuO lattice, but now is mostly known as the Lieb lattice.

because the lattice was studied as a typical example exhibiting Lieb's ferrimagnetism that we shall discuss now.[10]

Ferrimagnetism Consider the repulsive Hubbard model at half-filling defined on a bipartite lattice, and assume that the conditions for Theorem 10.4 are satisfied. Then the theorem shows that ground states of the model have $S_{\text{tot}} = (|A| - |B|)/2$, and are $|A| - |B| + 1$ fold degenerate. When (10.2.15) holds, we have $S_{\text{tot}} = (a/2)|\Lambda|$, which means that the ground states have bulk magnetization. In other words the system behaves as a macroscopic magnet in its ground states. As we have discussed in Sect. 9.1, to understand the origin of ferromagnetic order observed in nature was a major motivation to study the Hubbard model. Lieb's theorem provided historically the first natural and rigorous example of ferromagnetism (in a broad sense) in itinerant electron systems.[11]

The origin of the bulk magnetic moment is intuitively understandable in the models with large interaction U. For large U, the perturbative analysis in Sect. 10.1 suggests that the ground state of the Hubbard model is similar to that of the antiferromagnetic Heisenberg model. We then naturally expect that the Hubbard model exhibits ferrimagnetism, i.e., antiferromagnetism where the magnetic moments on the two sublattices do not balance. We have already discussed such a phenomenon for the Heisenberg model at the end of Sect. 4.1.

In fact Theorem 4.4 (p. 78) about the existence of ferrimagnetic long-range order in the Heisenberg model is readily extended to the Hubbard model. In order to characterize ferrimagnetic long-range order, we use the SU(2) invariant operator

[10]See, e.g., [22] for interesting attempts to realize the Lieb lattice by using cold atoms in an optical trap.

[11]Ferromagnetism in the Hubbard model due to Nagaoka, which takes place in a singular situation, was already known. See Sect. 11.2.

$$(\hat{\boldsymbol{O}}_L)^2 := \sum_{x,y\in\Lambda} (-1)^x(-1)^y\, \hat{\boldsymbol{S}}_x \cdot \hat{\boldsymbol{S}}_y, \tag{10.2.16}$$

which is the same as (4.1.12). Then by using Lieb's theorem (Theorem 10.4) instead of the Lieb–Mattis theorem (Theorem 2.3 in p. 42), and the inequality (10.2.7) instead of (4.1.15), we can prove the following theorem of Shen, Qiu, and Tian's [20] exactly in the same manner as Theorem 4.4.

Theorem 10.6 *Under the same condition as Theorem 10.4 (p. 350), we have*

$$\langle \Phi_{\rm GS}|(\hat{\boldsymbol{O}}_L)^2|\Phi_{\rm GS}\rangle \geq \left(\frac{|A|-|B|}{2}\right)^2, \tag{10.2.17}$$

for any ground state $|\Phi_{\rm GS}\rangle$. *The left-hand side is independent of the choice of the ground state.*

Flat band in the single-electron problem Consider the same repulsive Hubbard model that satisfies the conditions for Theorem 10.4, now in the opposite limit where the interactions U_x are small or vanishing. In Sect. 9.3.2 (see, in particular, (9.3.24) and Fig. 9.3), we observed that noninteracting systems universally exhibit paramagnetism. This observation might appear to contradict Lieb's ferrimagnetism, but one should recall that the conclusion in Sect. 9.3.2 was based on the assumption that the single-electron energy eigenvalues are non-degenerate.

As the following simple proposition shows, there is an inevitable degeneracy in the single-electron energy spectrum in the corresponding model on an asymmetric bipartite lattice.

Proposition 10.7 *Suppose that the hopping amplitude* $\mathsf{T} = (t_{x,y})_{x,y\in\Lambda}$ *is bipartite (see the beginning of Sect.* 10.2.2 *for the definition). We assume* $|A| \geq |B|$. *We denote by* ε_j *with* $j = 1, \ldots, |\Lambda|$ *the single-electron energy eigenvalues determined by* (9.3.5). *We assume that they are ordered so that* $\varepsilon_j \leq \varepsilon_{j+1}$. *Then the energy eigenvalues have the symmetry* $\varepsilon_{|\Lambda|-j} = -\varepsilon_j$ *for any* $j = 1, \ldots, |\Lambda|$. *Moreover there are at least* $|A| - |B|$ *energy eigenvalues which are zero. This means that* $\varepsilon_j = 0$ *for all* j *such that* $j \in I_0 := \{|B|+1, |B|+2, \ldots, |A|\}$.

Proof Take any energy eigenstate $\boldsymbol{\varphi} = (\varphi(x))_{x\in\Lambda} \in \mathfrak{h} \cong \mathbb{C}^{|\Lambda|}$ with eigenvalue ε. We thus have $\sum_{y\in\Lambda} t_{x,y}\varphi(y) = \varepsilon\varphi(x)$. Define $\tilde{\boldsymbol{\varphi}} = (\tilde{\varphi}(x))_{x\in\Lambda}$ by $\tilde{\varphi}(x) = \varphi(x)$ if $x \in A$ and $\tilde{\varphi}(x) = -\varphi(x)$ if $x \in B$. Then it clear holds that $\sum_{y\in\Lambda} t_{x,y}\tilde{\varphi}(y) = -\varepsilon\tilde{\varphi}(x)$ for any $x \in \Lambda$. This proves the symmetry $\varepsilon_{|\Lambda|-j} = -\varepsilon_j$.

We now assume $|A| > |B|$. Let $\mathfrak{h}_A$ (resp. $\mathfrak{h}_B$) be the $|A|$ dimensional (resp. $|B|$ dimensional) subspace of the single-electron Hilbert space $\mathfrak{h}$ that consists of states $\boldsymbol{\varphi} = (\varphi(x))_{x\in\Lambda}$ such that $\varphi(x) = 0$ unless $x \in A$ (resp. $x \in B$). Bipartiteness implies that T maps $\mathfrak{h}_A$ to $\mathfrak{h}_B$. Since the dimension of $\mathfrak{h}_A$ is larger than that of $\mathfrak{h}_B$, there must be a $|A| - |B|$ dimensional subspace of $\mathfrak{h}_A$ that is mapped to 0. We thus see that there are $|A| - |B|$ independent zero-energy eigenstates which have nonvanishing components only on the A sublattice. ■

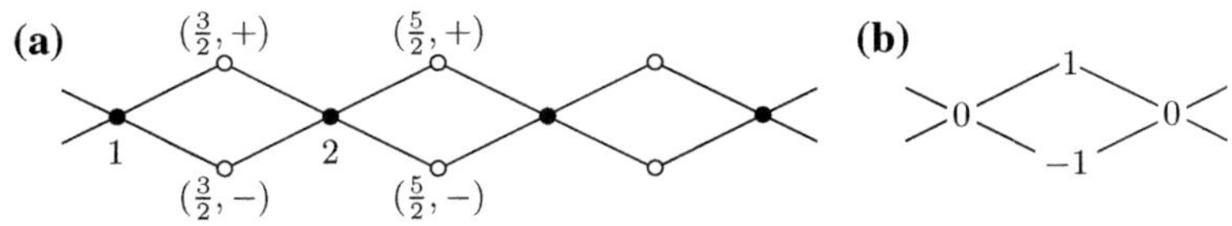

Fig. 10.3 **a** The diamond chain is a simple example which satisfies (10.2.15) with $a = 1/3$. **b** The state localized at the two sites with half-odd-integer x is an exact zero-energy eigenstate of the single-electron Schrödinger equation (10.2.18). All such states span the flat band. (© Hal Tasaki 2020. All Rights Reserved)

When the model has translation invariance, and hence the notion of energy bands is meaningful (see Sect. 9.3.1), the highly degenerate zero-energy eigenstates are said to form a flat band (or a dispersion-less band).[12]

To see a simple example, consider the diamond chain in Fig. 10.3, which is constructed as follows. Let $x = \frac{1}{2}, 1, \frac{3}{2}, 2, \ldots, L - \frac{1}{2}, L$. We assume that each integer x represents a single site in the chain. For each half-odd-integer x, we assume that there are two sites $(x, +)$ and $(x, -)$. In this way we get a chain with $3L$ sites, which clearly has a bipartite structure. We then assume that there is a constant hopping amplitude $t \in \mathbb{R}$ between the neighboring sites as in Fig. 10.3a. The corresponding single-electron Schrödinger equation (9.3.4) reads

$$t \sum_{\nu=\pm} \bigl\{\varphi((x - \tfrac{1}{2}, \nu)) + \varphi((x + \tfrac{1}{2}, \nu))\bigr\} = \varepsilon\, \varphi(x) \quad \text{for } x = 1, 2, \ldots, L,$$
$$t\bigl\{\varphi(x - \tfrac{1}{2}) + \varphi(x + \tfrac{1}{2})\bigr\} = \varepsilon\, \varphi((x, \nu)) \quad \text{for } x = \tfrac{1}{2}, \tfrac{3}{2}, \ldots, L - \tfrac{1}{2} \text{ and } \nu = \pm, \tag{10.2.18}$$

where we use periodic boundary conditions. One readily finds that there are $2L$ energy eigenstates $\boldsymbol{\psi}^{(k)}$ given by

$$\begin{cases} \psi^{(k)}(x) = \sqrt{2}\, e^{ikx} & \text{for } x = 1, 2, \ldots, L, \\ \psi^{(k)}((x, \nu)) = e^{ikx} & \text{for } x = \frac{1}{2}, \frac{3}{2}, \ldots, L - \frac{1}{2} \text{ and } \nu = \pm, \end{cases} \tag{10.2.19}$$

where $k = (2\pi/L)n$ with $n = 1, \ldots, 2L$, and the corresponding energy eigenvalues are $\varepsilon(k) = 2\sqrt{2}\, t \cos(k/2)$. In addition there are exactly L energy eigenstates with zero energy, which are localized at the two sites with half-odd-integer x.[13] The state has only nonvanishing components $+1$ on $(x, +)$ and -1 on $(x, -)$. See Fig. 10.3b. These L states span the flat band.

We now go back to the general situation described in Proposition 10.7, and examine the ground states of the system without any interactions. It is crucial to understand the peculiar structure of the single-electron energy eigenvalues. According to

---

[12] That some tight-binding electron models have flat bands had been known much before. See, e.g., [28].

[13] In many cases one finds eigenstates within a flat band which are strictly localized on a finite number of sites, as was noted, e.g., in [8, 21]. See also Sect. 11.3.1.

Proposition 10.7, there are $|B|$ energy eigenstates with non-positive energy eigenvalues, which are in general expected to form normal dispersive bands. Then there are $|A| - |B|$ eigenstates with zero energy, and another set of $|B|$ eigenstates with nonnegative energy eigenvalues. Note that $|B| + (|A| - |B|) + |B|$ is indeed equal to the total number of energy eigenstates, i.e., $|A| + |B| = |\Lambda|$. Now, in the ground states of non-interacting system, the $|B|$ states with non-positive energy should be filled with both up-spin and down-spin electrons. Since this requires $2|B|$ electrons, there remain $N - 2|B| = |A| - |B|$ electrons. (Recall that we are at the half-filling, i.e., $N = |\Lambda| = |A| + |B|$.) This number is exactly the same as the number of zero-energy states. This means that the flat band at the center of the single-electron spectrum is exactly half-filled. It seems to be a general rule that a half-filled flat band (or a half-filled nearly flat band) has strong tendency to generate ferromagnetic order. See Sects. 11.3 and 11.4, in particular the discussion at the end of Sect. 11.3.1.

As in (9.3.20), we denote by $\hat{a}^\dagger_{j,\sigma}$ the creation operator of an electron with spin σ in the single-electron energy eigenstate which corresponds to ε_j. As in Proposition 10.7, we denote by $I_0 = \{|B| + 1, \ldots, |A|\}$ the set of indices j which correspond to the zero-energy eigenstates. For any two subsets $S_\uparrow$ and $S_\downarrow$ of I_0 such that $|S_\uparrow| + |S_\downarrow| = |A| - |B|$, define

$$|\Theta_{S_\uparrow,S_\downarrow}\rangle = \Bigl(\prod_{j\in S_\uparrow} \hat{a}^\dagger_{j,\uparrow}\Bigr)\Bigl(\prod_{j\in S_\downarrow} \hat{a}^\dagger_{j,\downarrow}\Bigr)\Bigl(\prod_{j=1}^{|B|} \hat{a}^\dagger_{j,\uparrow}\hat{a}^\dagger_{j,\downarrow}\Bigr)|\Phi_{\mathrm{vac}}\rangle, \qquad (10.2.20)$$

which is an N electron state. Since $\varepsilon_j = 0$ for any $j \in I_0$, we see that $|\Theta_{S_\uparrow,S_\downarrow}\rangle$ for any choice of $S_\uparrow$ and $S_\downarrow$ are ground states of the non-interacting Hamiltonian $\hat{H}_{\mathrm{hop}}$. This should be compared with the unique ground state (9.3.24) for the model with nondegenerate ε_j. We have thus seen that the ground states for $U = 0$ (or $U_x = 0$ for all x) are highly degenerate. In particular the total spin can take any values between 0 and $(|A| - |B|)/2$. According to Lieb's theorem (Theorem 10.4) any small amount of repulsive interaction lifts the degeneracy, and only the states with maximum total spin remain as the ground states. This is the reason that Lieb's ferrimagnetism takes place for any nonzero values of U_x.

Problem 10.2.3.a Consider the Lieb lattice in Fig. 10.2 with periodic boundary conditions. We assume that the B sublattice is identical to the $L \times L$ square lattice with even L. Assume that there is a constant hopping amplitude $t \in \mathbb{R}$ between each pair of neighboring sites, i.e., sites connected by a bond in Fig. 10.2. What are zero-energy eigenstates of the corresponding single-electron Schrödinger equation (9.3.4)? (I guarantee that the problem is not as trivial as the reader might first think. I myself have enjoyed solving it.) [solution→p.517]

Superconductivity In 1993, Shen and Qiu [19] found that Lieb's theorem for the attractive Hubbard model, Theorem 10.2 in p. 348, may be combined with the clever observation by Yang and Zhang [29] to show the existence of off-diagonal long-range order relevant to superconductivity in a class of attractive Hubbard models on asymmetric bipartite lattices. We believe that this is an important result, at least from a theoretical point of view, since there are not so many models in which superconductivity can be rigorously established. It is also important that their result is valid in a finite range of filling factor including the half-filling.

Let us describe and prove the result of Shen and Qiu. We consider a lattice Λ which is decomposed into two sublattices as $\Lambda = A \cup B$, where $|A| \geq |B|$. We assume that the hopping amplitude $(t_{x,y})_{x,y\in\Lambda}$ is bipartite, i.e., $t_{x,y}$ is nonvanishing only when $x \in A$, $y \in B$ or $x \in B$, $y \in A$. We further assume that $t_{x,y} = t_{y,x} \in \mathbb{R}$, and that the whole lattice Λ is connected through nonvanishing $t_{x,y}$. We then consider the standard hopping Hamiltonian (10.2.1) or (9.3.17). For the interaction, we use the following symmetric form

$$\hat{H}'_{\text{att-int}} := -\sum_x U_x\Big(\hat{n}_{x,\uparrow} - \frac{1}{2}\Big)\Big(\hat{n}_{x,\downarrow} - \frac{1}{2}\Big), \tag{10.2.21}$$

where we assume $U_x > 0$ for all $x \in \Lambda$. Then the following was proved in [19].

Theorem 10.8 (Lieb–Shen–Qiu superconductivity) *Consider the attractive Hubbard model defined as above, and let electron number N be even and satisfy $2|B| \leq N \leq 2|A|$. Then the unique ground state $|\Phi_{\text{GS}}\rangle$ of $\hat{H}_{\text{hop}} + \hat{H}'_{\text{att-int}}$ satisfies*

$$\langle\Phi_{\text{GS}}|\Big(\sum_{x\in\Lambda}\hat{c}_{x,\downarrow}\hat{c}_{x,\uparrow}\Big)^{\dagger}\Big(\sum_{x\in\Lambda}\hat{c}_{x,\downarrow}\hat{c}_{x,\uparrow}\Big)|\Phi_{\text{GS}}\rangle \geq \Big(|A| - \frac{N}{2}\Big)\Big(\frac{N}{2} - |B|\Big). \tag{10.2.22}$$

Recall that the uniqueness of the ground state is a direct consequence of the original theorem of Lieb's (Theorem 10.2 in p. 348).

Suppose that (10.2.15) is satisfied with $0 < a < 1$. Denoting the filling factor as $\nu = N/(2|\Lambda|)$, the inequality (10.2.22) is written as

$$\frac{1}{|\Lambda|^2}\langle\Phi_{\text{GS}}|\Big(\sum_{x\in\Lambda}\hat{c}_{x,\downarrow}\hat{c}_{x,\uparrow}\Big)^{\dagger}\Big(\sum_{x\in\Lambda}\hat{c}_{x,\downarrow}\hat{c}_{x,\uparrow}\Big)|\Phi_{\text{GS}}\rangle \geq \Big(\frac{1+a}{2} - \nu\Big)\Big(\nu - \frac{1-a}{2}\Big), \tag{10.2.23}$$

which shows that the left-hand side is strictly positive for any $|\Lambda|$ provided that the filling factor is in the range $(1-a)/2 < \nu < (1+a)/2$. This proves that the ground states have nonzero off-diagonal long-range order that indicates condensation of fermion pairs. We stress that this is a standard criterion for characterizing superconductivity (without letting electrical current go through the system).

In a weakly interacting system, the above range of filling factor can be naturally understood in terms of the structure of the single-electron spectrum characterized by Proposition 10.7. When the $|B|$ energy eigenstates with non-positive energy are fully filled by up-spin and down-spin electrons, the filling factor is $\nu = 2|B|/(2|\Lambda|) = (1-a)/2$. When both the non-positive energy eigenstates and the $|A| - |B|$ zero-energy eigenstates (i.e., the flat band) are completely filled, then the filling factor is $\nu = \{2|B| + 2(|A| - |B|)\}/(2|\Lambda|) = (1+a)/2$. Thus the condition $(1-a)/2 < \nu < (1+a)/2$ means that the flat band is partially filled.

In this case too the ground states of the non-interacting system are given by (10.2.20) with $S_\uparrow$ and $S_\downarrow$ being any subsets of I_0 such that $|S_\uparrow| + |S_\downarrow| = N - 2|B|$. Again the ground states are highly degenerate. It is understood that any amount of attractive interaction lifts the degeneracy and we end up with a unique ground state that has $S_{\rm tot}=0$, and, more over, exhibits superconducting off-diagonal long-range order. We should stress however that the Lieb-Shen-Qui superconductivity continues to strongly interacting regime, where the above picture based on single-electron energy eigenstates becomes meaningless.

Proof of Theorem 10.8 From Theorem 10.2 we know that the unique ground state $|\Phi_{\rm GS}\rangle$ is in $\mathscr{H}_{N/2,N/2}$, i.e., the space with $N/2$ up-spin electrons and $N/2$ down-spin electrons. By the Shiba transformation (9.3.51), the Hamiltonian $\hat{H}_{\rm hop}+\hat{H}'_{\rm att\text{-}int}$ is mapped to that of the repulsive Hubbard model $\hat{H}_{\rm hop}+\hat{H}'_{\rm int}$, where the symmetric repulsive interaction is defined in (10.2.6) or (9.3.47). Recall, however, that the Shiba transformation changes the number of up-spin and down-spin electrons from $N/2$ and $N/2$, respectively, to $|\Lambda|-(N/2)$ and $N/2$, respectively. Thus the mapped state $|\Phi'_{\rm GS}\rangle$ is the unique ground state of the repulsive Hamiltonian $\hat{H}_{\rm hop}+\hat{H}'_{\rm int}$ in $\mathscr{H}_{|\Lambda|-(N/2),N/2}$, i.e., the space of states with $|\Lambda|-(N/2)$ up-spin electrons and $N/2$ down-spin electrons. Note that the total number of electrons is $|\Lambda|$. We also see that $|\Phi'_{\rm GS}\rangle$ has $S^{(3)}_{\rm tot}=(|\Lambda|-N)/2$. As long as $|S^{(3)}_{\rm tot}|\le(|A|-|B|)/2$, which is equivalent to $2|B|\le N\le 2|A|$, the same argument as in the Proof of Theorem 10.4 (p. 350), i.e., the mapping to the corresponding antiferromagnetic Heisenberg model, shows that $|\Phi'_{\rm GS}\rangle$ has total spin $S_{\rm tot}=(|A|-|B|)/2$.

We now recall that by the Shiba transformation the operators relevant to the ODLRO is transformed as

$$\begin{aligned}\hat{c}^\dagger_{x,\uparrow}\hat{c}^\dagger_{x,\downarrow}\hat{c}_{y,\downarrow}\hat{c}_{y,\uparrow} &\to (-1)^x(-1)^y\hat{S}^-_x\hat{S}^+_y\\ &=(-1)^x(-1)^y(\hat{S}^{(1)}_x\hat{S}^{(1)}_y+\hat{S}^{(2)}_x\hat{S}^{(2)}_y)-\delta_{x,y}\,\hat{S}^{(3)}_x.\end{aligned}\tag{10.2.24}$$

Then the inequality (10.2.4) implies that

$$\begin{aligned}0&\le\langle\Phi_{\rm GS}|\hat{c}^\dagger_{x,\uparrow}\hat{c}^\dagger_{x,\downarrow}\hat{c}_{y,\downarrow}\hat{c}_{y,\uparrow}|\Phi_{\rm GS}\rangle\\ &=(-1)^x(-1)^y\langle\Phi'_{\rm GS}|(\hat{S}^{(1)}_x\hat{S}^{(1)}_y+\hat{S}^{(2)}_x\hat{S}^{(2)}_y)|\Phi'_{\rm GS}\rangle,\end{aligned}\tag{10.2.25}$$

whenever $x\ne y$. We then observe that

$$\begin{aligned}&\langle\Phi_{\rm GS}|\Bigl(\sum_{x\in\Lambda}\hat{c}_{x,\downarrow}\hat{c}_{x,\uparrow}\Bigr)^\dagger\Bigl(\sum_{x\in\Lambda}\hat{c}_{x,\downarrow}\hat{c}_{x,\uparrow}\Bigr)|\Phi_{\rm GS}\rangle\\ &\quad=\langle\Phi'_{\rm GS}|\sum_{\alpha=1,2}\Bigl(\sum_{x\in\Lambda}(-1)^x\hat{S}^{(\alpha)}_x\Bigr)^2|\Phi'_{\rm GS}\rangle-\langle\Phi'_{\rm GS}|\Bigl(\sum_{x\in\Lambda}\hat{S}^{(3)}_x\Bigr)|\Phi'_{\rm GS}\rangle\\ &\quad\ge\langle\Phi'_{\rm GS}|\sum_{\alpha=1,2}\Bigl(\sum_{x\in\Lambda}\hat{S}^{(\alpha)}_x\Bigr)^2|\Phi'_{\rm GS}\rangle-\langle\Phi'_{\rm GS}|\Bigl(\sum_{x\in\Lambda}\hat{S}^{(3)}_x\Bigr)|\Phi'_{\rm GS}\rangle\\ &\quad=\langle\Phi'_{\rm GS}|\{(\hat{S}^{(1)}_{\rm tot})^2+(\hat{S}^{(2)}_{\rm tot})^2\}|\Phi'_{\rm GS}\rangle-\langle\Phi'_{\rm GS}|\hat{S}^{(3)}_{\rm tot}|\Phi'_{\rm GS}\rangle\\ &\quad=S_{\rm tot}(S_{\rm tot}+1)-(S^{(3)}_{\rm tot})^2+S^{(3)}_{\rm tot}\\ &\quad=\Bigl(|A|-\frac{N}{2}\Bigr)\Bigl(\frac{N}{2}+1-|B|\Bigr)\ge\Bigl(|A|-\frac{N}{2}\Bigr)\Bigl(\frac{N}{2}-|B|\Bigr),\end{aligned}\tag{10.2.26}$$

where we used (10.2.25) to get the lower bound in the third line. In the fifth line we wrote $S_{\rm tot} = (|A| - |B|)/2$ and $S^{(3)}_{\rm tot} = (|\Lambda| - N)/2 = (|A| + |B| - N)/2$. ∎

10.2.4 Proofs of Theorems 10.2 and 10.3

Let us now prove Lieb's theorem for the attractive Hubbard model, Theorem 10.2 in p. 348, which is the core of the whole theories discussed in the present chapter. As we have already declared, we here try to be as elementary and clear as possible. Many technical details which are compressed in the original proof and other expositions are carefully explained by only using elementary mathematics.[14] We here basically follow the original proof of Lieb's in [10], but also make use of some ideas from [16, 18, 25]. We also prove Theorem 10.3 (p. 349), which is easy after seeing the proof of the main theorem.

The essential ingredient of the proof is the method of spin-space reflection positivity, which should be compared with the spatial reflection positivity method employed in Sect. 4.1.2 to prove the existence of long-range order in the ground state of the anti-ferromagnetic Heisenberg model. The two arguments of course share many aspects in common, but the present spin-space version is much simpler mainly because of the better symmetry of the model.

Recall that the electron number N is an arbitrary even integer such that $0 < N \le 2|\Lambda|$. As before we shall work in the subspace $\mathscr{H}_{N/2,N/2}$, which consists of all states with $N/2$ up-spin electrons and $N/2$ down-spin electrons. This is justified by the SU(2) invariance of the Hamiltonian and Theorem A.17 (p. 473), which guarantee that one can find all the energy eigenvalues (of the system with N electrons) in $\mathscr{H}_{N/2,N/2}$.

Matrix representation of states We start by introducing a description of states which is essential for the method of spin space reflection positivity. Let $\mathscr{S} = \{A \mid A \subset \Lambda, |A| = N/2\}$ be the set of all subsets of Λ containing $N/2$ sites. As in (9.3.30), we define the basis state of $\mathscr{H}_{N/2,N/2}$ by

$$|\Xi_{A,B}\rangle := \Bigl(\prod_{x\in A} \hat{c}^\dagger_{x,\uparrow}\Bigr)\Bigl(\prod_{x\in B} \hat{c}^\dagger_{x,\downarrow}\Bigr)|\Phi_{\rm vac}\rangle, \tag{10.2.27}$$

for any $A, B \in \mathscr{S}$. Here, and in what follows, we fix an ordering of sites in Λ, and assume that products always respect the ordering.

For any $A, B \in \mathscr{S}$, let

$$T_{A,B} := \langle\Phi_{\rm vac}|\Bigl(\prod_{x\in A} \hat{c}^\dagger_{x,\sigma}\Bigr)^\dagger \hat{H}_{\rm hop}\Bigl(\prod_{x\in B} \hat{c}^\dagger_{x,\sigma}\Bigr)|\Phi_{\rm vac}\rangle. \tag{10.2.28}$$

[14] It might be interesting to compare the present proof with that of the Perron–Frobenius theorem in Appendix A.4.1. See also an extension of the Perron–Frobenius theorem in [5].

We see that $T_{A,B}$ is real, independent of $\sigma =\uparrow, \downarrow$, and symmetric, i.e., $T_{A,B} = T_{B,A}$. Note that (10.2.27) and (10.2.28) imply

$$\langle \Xi_{A',B'} | \hat{H}_{\rm hop} | \Xi_{A,B} \rangle = T_{A',A}\, \delta_{B',B} + \delta_{A',A}\, T_{B',B}. \tag{10.2.29}$$

We then see that

$$\begin{aligned} \hat{H}_{\rm hop} | \Xi_{A,B} \rangle &= \sum_{A',B' \in \mathcal{S}} | \Xi_{A',B'} \rangle \langle \Xi_{A',B'} | \hat{H}_{\rm hop} | \Xi_{A,B} \rangle \\ &= \sum_{A' \in \mathcal{S}} T_{A',A} | \Xi_{A',B} \rangle + \sum_{B' \in \mathcal{S}} T_{B,B'} | \Xi_{A,B'} \rangle, \end{aligned} \tag{10.2.30}$$

where we used the symmetry $T_{B',B} = T_{B,B'}$ for later convenience. As for the interaction Hamiltonian (10.2.2), we immediately see that

$$\hat{H}_{\text{att-int}} | \Xi_{A,B} \rangle = -\sum_{x \in \Lambda} U_x\, I_A^{(x)} I_B^{(x)} | \Xi_{A,B} \rangle, \tag{10.2.31}$$

where we set

$$I_A^{(x)} = \begin{cases} 1 & x \in A, \\ 0 & x \notin A, \end{cases} \tag{10.2.32}$$

for any $x \in \Lambda$ and $A \in \mathcal{S}$.

Let $\mathsf{W} = (W_{A,B})_{A,B \in \mathcal{S}}$ (with $W_{A,B} \in \mathbb{C}$) be a $|\mathcal{S}| \times |\mathcal{S}|$ matrix whose entries are indexed by elements of $\mathcal{S}$. For any such matrix W, we define

$$|\Gamma(\mathsf{W})\rangle = \sum_{A,B \in \mathcal{S}} W_{A,B} | \Xi_{A,B} \rangle. \tag{10.2.33}$$

It is clear that any state in $\mathcal{H}_{N/2,N/2}$ can be written in this form. Note also that, for any matrices W and W',

$$\langle \Gamma(\mathsf{W}) | \Gamma(\mathsf{W}') \rangle = \sum_{A,B \in \mathcal{S}} (W_{A,B})^*\, W'_{A,B} = \mathrm{Tr}[\mathsf{W}^\dagger\, \mathsf{W}']. \tag{10.2.34}$$

We remark that the representation (10.2.33) should be compared with the matrix representation of the ground state (4.1.70), which played an essential role in the spatial reflection positivity method.

Suppose that $|\Gamma(\mathsf{W})\rangle$ happens to be an eigenstate of $\hat{H} = \hat{H}_{\rm hop} + \hat{H}_{\text{att-int}}$, i.e., it satisfies the Schrödinger equation $\hat{H} | \Gamma(\mathsf{W}) \rangle = E | \Gamma(\mathsf{W}) \rangle$. By using (10.2.30) and (10.2.31), the left-hand side is written as

$$\hat{H}|\Gamma(\mathsf{W})\rangle = \sum_{A,B,A'\in\mathcal{S}} T_{A',A} W_{A,B}|\Xi_{A',B}\rangle + \sum_{A,B,B'\in\mathcal{S}} W_{A,B} T_{B,B'}|\Xi_{A,B'}\rangle$$
$$- \sum_{A,B\in\mathcal{S}} \sum_{x\in\Lambda} U_x\, I_A^{(x)} W_{A,B}\, I_B^{(x)} |\Xi_{A,B}\rangle. \tag{10.2.35}$$

Since this should be equal to $E \sum_{A,B\in\mathcal{S}} W_{A,B}|\Xi_{A,B}\rangle$, we find

$$\sum_{A'\in\mathcal{S}} T_{A,A'} W_{A',B} + \sum_{B'\in\mathcal{S}} W_{A,B'} T_{B',B} - \sum_{x\in\Lambda} U_x\, I_A^{(x)} W_{A,B}\, I_B^{(x)} = E\, W_{A,B}, \tag{10.2.36}$$

for any $A, B \in \mathcal{S}$. By introducing matrices T and $\mathsf{I}^{(x)}$ by $(\mathsf{T})_{A,B} = T_{A,B}$ and $(\mathsf{I}^{(x)})_{A,B} = \delta_{A,B}\, I_A^{(x)}$, the Schrödinger equation (10.2.36) is written in a remarkably compact manner as

$$\mathsf{TW} + \mathsf{WT} - \sum_{x\in\Lambda} U_x\, \mathsf{I}^{(x)}\mathsf{W}\mathsf{I}^{(x)} = E\,\mathsf{W}. \tag{10.2.37}$$

Since T and $\mathsf{I}^{(x)}$ are Hermitian, we find

$$\mathsf{TW}^\dagger + \mathsf{W}^\dagger\mathsf{T} - \sum_{x\in\Lambda} U_x\, \mathsf{I}^{(x)}\mathsf{W}^\dagger\mathsf{I}^{(x)} = E\,\mathsf{W}^\dagger, \tag{10.2.38}$$

and hence $|\Gamma(\mathsf{W}^\dagger)\rangle$ is also an energy eigenstate with energy E. This represents the symmetry between up-spin electrons and down-spin electrons. Let us define Hermitian matrices $\mathsf{W}_+ = \mathsf{W} + \mathsf{W}^\dagger$ and $\mathsf{W}_- = (\mathsf{W} - \mathsf{W}^\dagger)i$. By linearity we see that $|\Gamma(\mathsf{W}_\pm)\rangle$, if nonvanishing, are also energy eigenstates with energy E. Note that at least one of W_+ and W_- is nonzero. We thus conclude that it is possible to list all (independent) energy eigenstates in the form $|\Gamma(\mathsf{W})\rangle$ with Hermitian W.

Basic Lemma and the Proof of Theorem 10.2 We are now ready to state the following basic lemma. The lemma is proved in the rest of the present subsection through several steps.

Lemma 10.9 *Suppose that a ground state (in the subspace $\mathcal{H}_{N/2,N/2}$) of $\hat{H} = \hat{H}_{\text{hop}} + \hat{H}_{\text{att-int}}$ is written as $|\Phi_{\text{GS}}\rangle = |\Gamma(\mathsf{W}_{\text{GS}})\rangle$ with a Hermitian matrix W_{GS}. Then W_{GS} must be either positive definite or negative definite.*[15]

Let us prove Lieb' theorem, Theorem 10.2 in p. 348, by assuming the above lemma. We first prove the uniqueness of the ground state within $\mathcal{H}_{N/2,N/2}$. Suppose

[15] A Hermitian matrix M is said to be positive definite if all the eigenvalues are positive, or, equivalently, $\boldsymbol{v}^\dagger\mathsf{M}\boldsymbol{v} > 0$ for any nonzero vector $\boldsymbol{v}$. A Hermitian matrix M is said to be negative definite if $-\mathsf{M}$ is positive definite. A Hermitian matrix M is said to be positive semidefinite (or nonnegative) if all the eigenvalues are nonnegative, or, equivalently, $\boldsymbol{v}^\dagger\mathsf{M}\boldsymbol{v} \geq 0$ for any vector $\boldsymbol{v}$.

We use the following standard notation here and in what follows: $\boldsymbol{v} = (v_A)_{A\in\mathcal{S}}$ denotes a column vector, and $\boldsymbol{v}^\dagger$ a row vector. The product $\boldsymbol{u}^\dagger\boldsymbol{v} = \sum_{A\in\mathcal{S}} (u_A)^* v_A$ is a scalar, and $\boldsymbol{v}\boldsymbol{u}^\dagger$ is a matrix such that $(\boldsymbol{v}\boldsymbol{u}^\dagger)_{A,B} = v_A(u_B)^*$. Recall that $\mathrm{Tr}[\boldsymbol{v}\boldsymbol{u}^\dagger] = \boldsymbol{u}^\dagger\boldsymbol{v}$.

that the ground states within $\mathscr{H}_{N/2,N/2}$ are degenerate. Then one can take mutually orthogonal ground states $|\Phi_{\mathrm{GS}}\rangle = |\Gamma(\mathsf{W}_{\mathrm{GS}})\rangle$ and $|\Phi'_{\mathrm{GS}}\rangle = |\Gamma(\mathsf{W}'_{\mathrm{GS}})\rangle$ where W_{GS} and $\mathsf{W}'_{\mathrm{GS}}$ are positive definite. But the formula (10.2.34) implies $\langle\Phi'_{\mathrm{GS}}|\Phi_{\mathrm{GS}}\rangle = \mathrm{Tr}[\mathsf{W}'_{\mathrm{GS}}\mathsf{W}_{\mathrm{GS}}] > 0$, which is a contradiction.[16]

We next prove that the unique ground state has $S_{\mathrm{tot}} = 0$. We first recall that the uniqueness implies that the ground state is an eigenstate of $(\hat{\boldsymbol{S}}_{\mathrm{tot}})^2$. Then note that we have $(\mathsf{W}_{\mathrm{GS}})_{A,A} \neq 0$ for any $A \in \mathscr{S}$ since W_{GS} is either positive or negative definite. This means that the ground state $|\Gamma(\mathsf{W}_{\mathrm{GS}})\rangle$ contains the component $|\Xi_{A,A}\rangle$, which has $S_{\mathrm{tot}} = 0$.

Since the unique ground state $|\Phi_{\mathrm{GS}}\rangle$ in $\mathscr{H}_{N/2,N/2}$ (which is indeed the subspace with $S^{(3)}_{\mathrm{tot}} = 0$) has $S_{\mathrm{tot}} = 0$, we find from Theorem A.16 (p. 473) that $|\Phi_{\mathrm{GS}}\rangle$ is indeed the unique ground state in the whole Hilbert space.

Spin-space reflection positivity An essential key of the Proof of Lemma 10.9 is the bound (10.2.43), which is a major consequence of the spin-space reflection positivity. For an arbitrary Hermitian W such that $\mathrm{Tr}[\mathsf{W}^2] = 1$ (and hence $\langle\Gamma(\mathsf{W})|\Gamma(\mathsf{W})\rangle = 1$), we write the corresponding energy expectation value as $\mathscr{E}(\mathsf{W}) = \langle\Gamma(\mathsf{W})|\hat{H}|\Gamma(\mathsf{W})\rangle$. By recalling (10.2.35), we find

$$\begin{aligned}\mathscr{E}(\mathsf{W}) &= \sum_{A,B,A'\in\mathscr{S}} (W_{A',B})^* T_{A',A} W_{A,B} + \sum_{A,B,B'\in\mathscr{S}} (W_{A,B'})^* W_{A,B} T_{B,B'} \\ &\quad - \sum_{x\in\Lambda} U_x \sum_{A,B\in\mathscr{S}} (W_{A,B})^* I_A^{(x)} W_{A,B}\, I_B^{(x)} \\ &= 2\,\mathrm{Tr}[\mathsf{W}^2\mathsf{T}] - \sum_{x\in\Lambda} U_x\ \mathrm{Tr}[\mathsf{W}\mathsf{I}^{(x)}\mathsf{W}\mathsf{I}^{(x)}],\end{aligned} \tag{10.2.39}$$

where we noted that $\mathsf{W} = \mathsf{W}^\dagger$. We write W in its spectral decomposition form[17]

$$\mathsf{W} = \sum_{j=1}^{|\mathscr{S}|} \lambda_j\, \boldsymbol{v}_j \boldsymbol{v}_j^\dagger, \tag{10.2.40}$$

where $\boldsymbol{v}_j$ is the eigenvector of W with eigenvalue $\lambda_j \in \mathbb{R}$. We have chosen the eigenvectors so that $\{\boldsymbol{v}_j\}_{j=1,\ldots,|\mathscr{S}|}$ forms an orthonormal basis. Substituting this into (10.2.39), we get

---

[16]Here we used the following easy lemma. **Lemma** Let M and M' be $D \times D$ matrices. If M is positive definite and M' is positive semidefinite but nonvanishing, then $\mathrm{Tr}[\mathsf{M}\mathsf{M}'] > 0$. ***Proof*** Take the orthonormal basis $\{\boldsymbol{u}_j\}_{j=1,\ldots,D}$ which consists of the eigenvectors of M', i.e., $\mathsf{M}'\boldsymbol{u}_j = m_j\boldsymbol{u}_j$ with $m_j \geq 0$ for all j and $m_j > 0$ for some j. Then $\mathrm{Tr}[\mathsf{M}\mathsf{M}'] = \sum_{j=1}^{D} m_j(\boldsymbol{u}_j^\dagger \mathsf{M}\boldsymbol{u}_j) > 0$.

[17]See footnote 15 for the notation.

$$
\begin{aligned}
\mathscr{E}(\mathsf{W}) =& 2\sum_{j=1}^{|\mathscr{S}|}(\lambda_j)^2 \operatorname{Tr}[\boldsymbol{v}_j\boldsymbol{v}_j^\dagger \mathsf{T}] - \sum_{x\in\Lambda} U_x \sum_{j,k=1}^{|\mathscr{S}|} \lambda_j\lambda_k \operatorname{Tr}[\boldsymbol{v}_j\boldsymbol{v}_j^\dagger \mathsf{I}^{(x)}\boldsymbol{v}_k\boldsymbol{v}_k^\dagger \mathsf{I}^{(x)}] \\
=& 2\sum_{j=1}^{|\mathscr{S}|}(\lambda_j)^2 (\boldsymbol{v}_j^\dagger \mathsf{T}\boldsymbol{v}_j) - \sum_{x\in\Lambda} U_x \sum_{j,k=1}^{|\mathscr{S}|} \lambda_j\lambda_k |\boldsymbol{v}_j^\dagger \mathsf{I}^{(x)}\boldsymbol{v}_k|^2.
\end{aligned}
\tag{10.2.41}
$$

With the same λ_j and $\boldsymbol{v}_j$ as in (10.2.40), we define

$$
|\mathsf{W}| := \sum_{j=1}^{|\mathscr{S}|} |\lambda_j|\, \boldsymbol{v}_j\boldsymbol{v}_j^\dagger, \tag{10.2.42}
$$

which is a positive semidefinite matrix. Note also that $\operatorname{Tr}[\mathsf{W}^2] = \sum_{j=1}^{|\mathscr{S}|}(\lambda_j)^2 = \operatorname{Tr}[|\mathsf{W}|^2]$. We thus see $\operatorname{Tr}[|\mathsf{W}|^2] = 1$, and hence $|\Gamma(|\mathsf{W}|)\rangle$ is normalized. Noting that $U_x > 0$ and $\lambda_j \in \mathbb{R}$, we see that the right-hand side of (10.2.41) never increases when we replace λ_j by $|\lambda_j|$. This is the spin-space reflection positivity argument. We therefore get the essential inequality:

$$
\mathscr{E}(\mathsf{W}) \geq \mathscr{E}(|\mathsf{W}|) \tag{10.2.43}
$$

Note that the sign of U_x is essential for the bound (10.2.43). In other words, only the attractive Hubbard model has (spin-space) reflection positivity.

We now suppose that a ground state $|\Phi_{\mathrm{GS}}\rangle$ is written as $|\Gamma(\mathsf{W}_{\mathrm{GS}})\rangle$ with Hermitian W_{GS}. Then since (10.2.43) implies $E_{\mathrm{GS}} = \mathscr{E}(\mathsf{W}_{\mathrm{GS}}) \geq \mathscr{E}(|\mathsf{W}_{\mathrm{GS}}|)$, we find that $|\Gamma(|\mathsf{W}_{\mathrm{GS}}|)\rangle$ is also a ground state. Note that this conclusion is obtained without using the fact that Λ is connected via nonvanishing $t_{x,y}$.

Implication of connectivity To complete the Proof of Lemma 10.9, we state the following lemma.

Lemma 10.10 *Let* R *be an* $|\mathscr{S}| \times |\mathscr{S}|$ *positive semidefinite matrix which satisfies*

$$
\mathsf{T}\mathsf{R} + \mathsf{R}\mathsf{T} - \sum_{x\in\Lambda} U_x\, \mathsf{I}^{(x)}\mathsf{R}\mathsf{I}^{(x)} = E\mathsf{R}, \tag{10.2.44}
$$

for some E. *Then* R *must be either positive definite or zero.*

Note that (10.2.44) is nothing but the Schrödinger equation (10.2.37). We shall prove the lemma below.

By using Lemma 10.10, we shall now prove Lemma 10.9, namely, that any Hermitian matrix that represents a ground state must be positive or negative definite. Suppose that a ground state $|\Phi_{\mathrm{GS}}\rangle$ is written as $|\Gamma(\mathsf{W}_{\mathrm{GS}})\rangle$ with Hermitian W_{GS}. We have just seen from the spin-space reflection positivity argument that $|\mathsf{W}_{\mathrm{GS}}|$ also represents a ground state. Let us set $\mathsf{R} = |\mathsf{W}_{\mathrm{GS}}|$. Since R represents a ground state it satisfies the Schrödinger equation (10.2.37) or (10.2.44) with $E = E_{\mathrm{GS}}$. We thus conclude from the lemma that $\mathsf{R} = |\mathsf{W}_{\mathrm{GS}}|$ is positive definite because it is is clearly nonzero.

Let us next set

$$\mathsf{R} = |\mathsf{W}_{\mathrm{GS}}| - \mathsf{W}_{\mathrm{GS}} = \sum_{j=1}^{|\mathscr{S}|} (|\lambda_j| - \lambda_j)\, \boldsymbol{v}_j \boldsymbol{v}_j^{\dagger}, \tag{10.2.45}$$

which is positive semidefinite by definition. Since W_{GS} satisfies the Schrödinger equation (10.2.37) with $E = E_{\mathrm{GS}}$, this R also satisfies (10.2.44). We conclude that R is either zero or positive definite. When $\mathsf{R} = 0$, we immediately see from (10.2.45) that W_{GS} equals $|\mathsf{W}_{\mathrm{GS}}|$ and hence is positive definite. When R is positive definite, (10.2.45) implies $\lambda_j < 0$ for all j and hence W_{GS} is negative definite (and we have $\mathsf{W}_{\mathrm{GS}} = -|\mathsf{W}_{\mathrm{GS}}|$). Lemma 10.9 has been proved.

Proof of Lemma 10.10 Let the kernel of R be $\ker \mathsf{R} := \{\boldsymbol{v} \mid \mathsf{R}\boldsymbol{v} = \boldsymbol{0}\}$. Take an arbitrary $\boldsymbol{v} \in \ker \mathsf{R}$, and "sandwich" (10.2.44) by $\boldsymbol{v}^{\dagger}$ and $\boldsymbol{v}$. Since $\mathsf{R}\boldsymbol{v} = \boldsymbol{0}$ and $\boldsymbol{v}^{\dagger}\mathsf{R} = \boldsymbol{0}$, we find

$$\sum_{x \in \Lambda} U_x\, \boldsymbol{v}^{\dagger} \mathsf{I}^{(x)} \mathsf{R} \mathsf{I}^{(x)} \boldsymbol{v} = 0. \tag{10.2.46}$$

Noting that $U_x > 0$ and R is positive semidefinite, we find $\boldsymbol{v}^{\dagger}\mathsf{I}^{(x)}\mathsf{R}\mathsf{I}^{(x)}\boldsymbol{v} = 0$ and hence[18] $\mathsf{R}\mathsf{I}^{(x)}\boldsymbol{v} = 0$ for each $x \in \Lambda$. We have thus found that $\boldsymbol{v} \in \ker \mathsf{R}$ implies $\mathsf{I}^{(x)}\boldsymbol{v} \in \ker \mathsf{R}$ for any x. For any $A \in \mathscr{S}$, let $\mathsf{I}^{(A)} = \prod_{x \in A} \mathsf{I}^{(x)}$, which is the projection matrix onto the unit vector $\boldsymbol{\delta}^{(A)}$ defined by $(\boldsymbol{\delta}^{(A)})_B = \delta_{A,B}$. From the above observation, we see that

$$\boldsymbol{v} \in \ker \mathsf{R} \implies \mathsf{I}^{(A)}\boldsymbol{v} \in \ker \mathsf{R}, \tag{10.2.47}$$

for any $A \in \mathscr{S}$. Again take an arbitrary $\boldsymbol{v} \in \ker \mathsf{R}$, and let the whole (10.2.44) act on $\boldsymbol{v}$. Since we now know $\mathsf{R}\mathsf{I}^{(x)}\boldsymbol{v} = \boldsymbol{0}$, we get $\mathsf{R}\mathsf{T}\boldsymbol{v} = \boldsymbol{0}$. We thus see

$$\boldsymbol{v} \in \ker \mathsf{R} \implies \mathsf{T}\boldsymbol{v} \in \ker \mathsf{R}. \tag{10.2.48}$$

Noting that $\mathsf{T}\boldsymbol{\delta}^{(A)} = \sum_{B \in \mathscr{S}} T_{B,A}\, \boldsymbol{\delta}^{(B)}$ for any $A \in \mathscr{S}$, we see that $\mathsf{I}^{(B)}\mathsf{T}\boldsymbol{\delta}^{(A)} \neq \boldsymbol{0}$ if $T_{B,A} \neq 0$. Therefore, if $\boldsymbol{\delta}^{(A)} \in \ker \mathsf{R}$ and $T_{B,A} \neq 0$, then one sees from (10.2.47) and (10.2.48) that $\mathsf{I}^{(B)}\mathsf{T}\boldsymbol{\delta}^{(A)} \in \ker \mathsf{R}$, and hence $\boldsymbol{\delta}^{(B)} \in \ker \mathsf{R}$. We have thus found an important relation

$$\boldsymbol{\delta}^{(A)} \in \ker \mathsf{R} \implies \boldsymbol{\delta}^{(B)} \in \ker \mathsf{R} \text{ for any } B \in \mathscr{S} \text{ such that } T_{B,A} \neq 0. \tag{10.2.49}$$

Now, assume that R is not positive definite, and hence there exists at least one nonzero vector $\boldsymbol{v}_0$ in $\ker \mathsf{R}$. The vector $\boldsymbol{v}_0$ has at least one nonzero component. Let's say that the A_0 component of $\boldsymbol{v}_0$ is nonvanishing, which is equivalent to $\mathsf{I}^{(A_0)}\boldsymbol{v}_0 \neq \boldsymbol{0}$.

[18] Since $\boldsymbol{v}^{\dagger}\mathsf{I}^{(x)}\mathsf{R}\mathsf{I}^{(x)}\boldsymbol{v} = |\sqrt{\mathsf{R}}\,\mathsf{I}^{(x)}\boldsymbol{v}|^2$, we find $\sqrt{\mathsf{R}}\,\mathsf{I}^{(x)}\boldsymbol{v} = \boldsymbol{0}$.

Since (10.2.47) implies $\mathsf{I}^{(A_0)}\boldsymbol{v}_0 \in \ker \mathsf{R}$, we see that $\boldsymbol{\delta}^{(A_0)} \in \ker \mathsf{R}$. Let $A_1, \ldots, A_n \in \mathscr{S}$ be such that $T_{A_i,A_0} \neq 0$. Then we find from (10.2.49) that $\boldsymbol{\delta}^{(A_i)} \in \ker \mathsf{R}$ for $i = 1, \ldots, n$. This procedure can be repeated to conclude $\boldsymbol{\delta}^{(A)} \in \ker \mathsf{R}$ for any $A \in \mathscr{S}$. Here we noted that the whole configuration space $\mathscr{S}$ is connected by nonzero $T_{B,A}$ because the whole lattice Λ is connected by nonzero $t_{x,y}$.[19] We thus find that $\ker \mathsf{R}$ is the whole space, and R is zero. ∎

Proof of Theorem 10.3 We finish the present section by proving Theorem 10.3 (p. 349) about the correlation function. For any $x, y \in \Lambda$ such that $x \neq y$ and any $A, B \in \mathscr{S}$, let

$$S^{x,y}_{A,B} = \langle \Phi_{\mathrm{vac}}| \Big(\prod_{z\in A} \hat{c}^\dagger_{z,\sigma}\Big)^\dagger \hat{c}^\dagger_{x,\sigma}\hat{c}_{y,\sigma} \Big(\prod_{z\in B} \hat{c}^\dagger_{z,\sigma}\Big)|\Phi_{\mathrm{vac}}\rangle, \tag{10.2.50}$$

which is real and independent of $\sigma =\uparrow, \downarrow$. Unlike $T_{A,B}$ defined in (10.2.28), $S^{x,y}_{A,B}$ is not symmetric in A and B. Noting that $\langle \Xi_{A',B'}|\hat{c}^\dagger_{x,\uparrow}\hat{c}_{y,\uparrow}\hat{c}^\dagger_{x,\downarrow}\hat{c}_{y,\downarrow}|\Xi_{A,B}\rangle = S^{x,y}_{A',A}\,S^{x,y}_{B',B}$, we find for any Hermitian W that

$$\begin{aligned}
\langle \Gamma(\mathsf{W})|\hat{c}^\dagger_{x,\uparrow}\hat{c}^\dagger_{x,\downarrow}\hat{c}_{y,\downarrow}\hat{c}_{y,\uparrow}|\Gamma(\mathsf{W})\rangle &= \langle \Gamma(\mathsf{W})|\hat{c}^\dagger_{x,\uparrow}\hat{c}_{y,\uparrow}\hat{c}^\dagger_{x,\downarrow}\hat{c}_{y,\downarrow}|\Gamma(\mathsf{W})\rangle \\
&= \sum_{A,B,A',B'\in\mathscr{S}} (W_{A',B'})^*\, S^{x,y}_{A',A}\, S^{x,y}_{B',B}\, W_{A,B} \\
&= \mathrm{Tr}[\mathsf{W}\mathsf{S}^{x,y}\,\mathsf{W}(\mathsf{S}^{x,y})^\dagger],
\end{aligned} \tag{10.2.51}$$

where we defined $(\mathsf{S}^{x,y})_{A,B} = S^{x,y}_{A,B}$. We now take the unique ground state $|\Phi_{\mathrm{GS}}\rangle$, and write it as $|\Phi_{\mathrm{GS}}\rangle = |\Gamma(\mathsf{W}_{\mathrm{GS}})\rangle$ with positive definite W_{GS}. Then we have $\langle \Phi_{\mathrm{GS}}|\hat{c}^\dagger_{x,\uparrow}\hat{c}^\dagger_{x,\downarrow}\hat{c}_{y,\downarrow}\hat{c}_{y,\uparrow}|\Phi_{\mathrm{GS}}\rangle = \mathrm{Tr}[\mathsf{W}_{\mathrm{GS}}\,\mathsf{S}^{x,y}\,\mathsf{W}_{\mathrm{GS}}(\mathsf{S}^{x,y})^\dagger] > 0$ from the lemma in footnote 16 (p. 364) because $\mathsf{S}^{x,y}\,\mathsf{W}_{\mathrm{GS}}(\mathsf{S}^{x,y})^\dagger$ is positive semidefinite and nonvanishing.

10.2.5 Extensions and Other Rigorous Results

There have been various extensions of Lieb's theorems to other situations and models. Let us briefly discuss an extension to finite temperatures, and very briefly mention other results.

A theorem which can be regarded as a finite temperature version of Lieb's theorem was proved by Kubo and Kishi [9]. It deals with the charge susceptibility and the on-site pairing (superconducting) susceptibility for the equilibrium state of the repulsive Hubbard model at half-filling. We define the thermodynamic function J

---

[19] To be rigorous we need to show that the connectivity of the lattice Λ implies the connectivity of the configuration space $\mathscr{S}$. This was indeed done in the more complicated case of spin systems in p. 41, and that proof applies to the present case. (Consider the case with $S = 1/2$, and identify the set of sites with $\sigma = 1/2$ with $A \subset \Lambda$.) For a direct proof for this simpler situation, see, e.g., (the first half of) the Proof of Lemma in [7].

corresponding to the grand canonical ensemble by

$$J(\beta, \mu, (\gamma_x)_{x\in\Lambda}, (\eta_x)_{x\in\Lambda})$$
$$= -\frac{1}{\beta}\log \mathrm{Tr}\exp\Bigl[-\beta\Bigl(\hat{H} - \mu\hat{N} - \sum_{\substack{x\in\Lambda \\ \sigma=\uparrow,\downarrow}} \gamma_x n_{x,\sigma} - \sum_{x\in\Lambda} \eta_x(\hat{c}^\dagger_{x,\uparrow}\hat{c}^\dagger_{x,\downarrow} + \hat{c}_{x,\downarrow}\hat{c}_{x,\uparrow})\Bigr)\Bigr], \tag{10.2.52}$$

where β and μ are the inverse temperature and the chemical potential, respectively, and the trace is taken over the Hilbert spaces with all possible electron numbers. We added to the Hamiltonian two terms with fictitious external fields $(\gamma_x)_{x\in\Lambda}$ and $(\eta_x)_{x\in\Lambda}$ to test for the possible charge ordering and superconducting ordering, respectively.

We define the charge susceptibility χ^{c} and the on-site pairing susceptibility χ^{p} by

$$\chi^{\mathrm{c}}_{\mathbf{q}}(\beta, \mu) = -\frac{\partial}{\partial\tilde{\gamma}_{\mathbf{q}}}\frac{\partial}{\partial\tilde{\gamma}_{-\mathbf{q}}} J(\beta, \mu, (\gamma_x), (\eta_x))\Bigr|_{(\gamma_x)=(\eta_x)=0} \geq 0, \tag{10.2.53}$$

and

$$\chi^{\mathrm{p}}_{\mathbf{q}}(\beta, \mu) = -\frac{\partial}{\partial\tilde{\eta}_{\mathbf{q}}}\frac{\partial}{\partial\tilde{\eta}_{-\mathbf{q}}} J(\beta, \mu, (\gamma_x), (\eta_x))\Bigr|_{(\gamma_x)=(\eta_x)=0} \geq 0. \tag{10.2.54}$$

The Fourier transformation of the external fields are

$$\tilde{\gamma}_{\mathbf{q}} = |\Lambda|^{-1/2}\sum_{x\in\Lambda} \gamma_x\, e^{i\mathbf{q}\cdot x}, \quad \tilde{\eta}_{\mathbf{q}} = |\Lambda|^{-1/2}\sum_{x\in\Lambda} \eta_x\, e^{i\mathbf{q}\cdot x}, \tag{10.2.55}$$

where $\mathbf{q}$ is a wave number vector corresponding to the lattice Λ (which we assume to have a periodic structure).

Then the following theorem was proved by Kubo and Kishi [9].

Theorem 10.11 *Consider the Hubbard model with uniform repulsive interaction* (10.2.5) *with* $U > 0$. *We assume that the conditions for Theorem 10.4 (p. 350) (except for that of the electron number) are satisfied. Then for any* $\beta > 0$ *and for any wave number vector* $\mathbf{q}$, *we have*

$$\chi^{\mathrm{c}}_{\mathbf{q}}(\beta, U/2) \leq \frac{1}{U}, \quad \text{and} \quad \chi^{\mathrm{p}}_{\mathbf{q}}(\beta, U/2) \leq \frac{2}{U}. \tag{10.2.56}$$

Note that the choice $\mu = U/2$ corresponds to the half-filling. The theorem states that the charge and the on-site paring susceptibilities for any wave vector $\mathbf{q}$ are finite in the half-filled model at finite temperatures. This means that the model does not exhibit any CDW ordering or superconducting ordering.

Lieb's theorems and the corresponding finite-temperature theorem were extended to various models. For extensions to the periodic Anderson model, see, e.g., [17, 24,

27], for extensions to the Kondo lattice model, see, e.g., [16, 17], and for extensions to systems with electrons coupled to phonons, see [7]. Review articles on these topics such as [18, 25, 26] may be useful. See also [15] for a unified extension of Lieb's theorem to the Hubbard model coupled to phonons or photons.

There are some interesting and nontrivial rigorous results (which are mainly due to Lieb and his collaborators) on related models of itinerant electron at half-filling. See [12] for the uniform density theorem, [11, 14] for the flux phase problem, and [13] for the stability of the Peierls instability.

References

1. P.W. Anderson, New approach to the theory of superexchange interactions. Phys. Rev. **115**, 2–13 (1959)
2. A. Auerbach, *Interacting Electrons and Quantum Magnetism* (Springer, New York, 1994)
3. N. Datta, R. Fernández, and J. Fröhlich, Effective hamiltonians and phase diagrams for tight-binding models. J. Stat. Phys. **96**, 545–611 (1999), arxiv:math-ph/9809007
4. F.H.L. Essler, H. Frahm, F. Göhmann, A. Klümper, V.E. Korepin, *The One-Dimensional Hubbard Model* (Cambridge University Press, Cambridge, 2010)
5. W.G. Faris, Invariant cones and uniqueness of the ground state for fermion systems. J. Math. Phys. **13**, 1285 (1972)
6. P. Fazekas, *Lecture Notes on Electron Correlation and Magnetism*. Series in Modern Condensed Matter Physics, vol. 5 (World Scientific, Singapore, 1999)
7. J.K. Freericks, E.H. Lieb, Ground state of a general electron-phonon Hamiltonian is a spin singlet. Phys. Rev. B **51**, 2812–2821 (1995), arxiv:cond-mat/9407026
8. M. Kohmoto, B. Sutherland, Electronic states on a penrose lattice. Phys. Rev. Lett. **56**, 2740 (1986)
9. K. Kubo, K. Kishi, Rigorous bounds on the susceptibilities of the Hubbard model. Phys. Rev. B **41**, 4866 (1990)
10. E.H. Lieb, Two theorems on the Hubbard model. Phys. Rev. Lett. **62**, 1201–1204 (1989)
11. E.H. Lieb, The flux phase of the half-filled band. Phys. Rev. Lett. **73**, 2158 (1994)
12. E.H. Lieb, M. Loss, R.J. McCann, Uniform density theorem for the Hubbard model. J. Math. Phys. **34**, 891 (1993)
13. E.H. Lieb, B, Nachtergaele, The stability of the Peierls instability for ring-shaped molecules. Phys. Rev. B **51**, 4777–4791(1995), arxiv:cond-mat/9410100v1
14. N. Macris, B. Nachtergalel, On the flux phase conjecture at half-filling: an improved proof. J. Stat. Phys. **85** 745–761 (1996), arxiv:cond-mat/9604043
15. T. Miyao, Stability of ferromagnetism in many-electron systems. J. Stat. Phys. **176**, 1211–1271 (2019), arxiv:1712.05529
16. S.-Q. Shen, Total spin and antiferromagnetic correlation in the Kondo model. Phys. Rev. B **53**, 14252 (1996)
17. S.-Q. Shen, Applications of reflection positivity in strongly correlated electron systems. Phys. Rev. B **54**, 4397 (1996)
18. S.-Q. Shen, Strongly correlated electron systems: spin-reflection positivity and some rigorous results. Int. J. Mod. Phys. B **12**, 709–779 (1998)
19. S.-Q. Shen, Z.-M. Qiu, Exact demonstration of off-diagonal long-range order in the ground state of a Hubbard model. Phys. Rev. Lett. **71**, 4238 (1993)
20. S.-Q. Shen, Z.-M. Qiu, G.-S. Tian, Ferrimagnetic long-range order of the Hubbard model. Phys. Rev. Lett. **72**, 1280 (1994)
21. B. Sutherland, Localization of electronic wave functions due to topology. Phys. Rev. B **34**, 5208–5211 (1986)

22. S. Taie, H. Ozawa, T. Ichinose, T. Nishio, A. Nakajima, Y. Takahashi, Coherent driving and freezing of bosonic matter wave in an optical Lieb lattice. Sci. Adv. **1**, e150085 (2015), http://advances.sciencemag.org/content/1/10/e1500854
23. G.-S. Tian, Rigorous theorems on off-diagonal long-range order in the negative-U Hubbard model. Phys. Rev. B **45**, 3145–3148 (1992)
24. G.-S. Tian, Antiferromagnetic correlation in the half-filled strongly correlated electron models at nonzero temperature: a rigorous result. Phys. Rev. B **63**, 224413 (2001)
25. G.-S. Tian, Lieb's spin-reflection-positivity method and its applications to strongly correlated electron systems. J. Stat. Phys. **116**, 629–680 (2004)
26. H. Tsunetsugu, M. Sigrist, K. Ueda, The ground-state phase diagram of the one-dimensional Kondo lattice model. Rev. Mod. Phys. **69**, 809–863 (1997)
27. K. Ueda, H. Tsunetsugu, M. Sigrist, Singlet ground state of the periodic Anderson model at half filling: a rigorous result. Phys. Rev. Lett. **68**, 1030 (1992)
28. D. Weaire, M.F. Thorpe, Electronic properties of an amorphous solid. i: a simple tight-binding theory. Phys. Rev. B **4**, 2508–2520 (1971)
29. C.N. Yang, S.C. Zhang, SO_4 symmetry in a Hubbard model. Mod. Phys. Lett. **B4**, 759–776 (1990). Reprinted in M. Rasetti (ed.) *The Hubbard Model: Recent Results* (World Scientific, Singapore, 1991), and A. Montorsi (ed.) *The Hubbard Model: A Reprint Volume* (World Scientific, Singapore, 1992). https://www.worldscientific.com/doi/abs/10.1142/S021797929100050X

Chapter 11
The Origin of Ferromagnetism

In the present chapter we focus on ferromagnetism (or, more precisely, saturated ferromagnetism) in the ground states of the Hubbard model. Recalling that both the non-interacting models and the non-hopping models exhibit paramagnetism, we see that ferromagnetism can be generated only through nontrivial "competition" between wave-like nature and particle-like nature of electrons. After discussing basic properties of saturated ferromagnetism in the Hubbard model in Sect. 11.1, we present some rigorous results which establish that the ground states of certain versions of the Hubbard models exhibit ferromagnetism. They include Nagaoka's ferromagnetism for systems with infinitely large U (Sect. 11.2), flat-band ferromagnetism by Mielke and by Tasaki (Sect. 11.3), and ferromagnetism in nearly-flat-band Hubbard model due to Tasaki (Sect. 11.4). The final result is of particular importance since it deals with a situation where ferromagnetism is intrinsically a non-perturbative phenomenon, and above mentioned competition plays an essential role. We end the chapter by briefly discussing the fascinating but extremely difficult problem of metallic ferromagnetism in Sect. 11.5.

11.1 Basic Properties of Ferromagnetism

We shall now focus on the emergence of ferromagnetism in several versions of the Hubbard model. As we discussed in Sect. 9.1, to study the origin of ferromagnetism was the main motivation for the introduction of the Hubbard model. Let us remark at this point that ferromagnetism, despite being quite remarkable and interesting phenomenon, is indeed not ubiquitous. We are very much familiar with ferromagnetism in iron, but most metals are paramagnets. The examples of Hubbard models exhibiting ferromagnetism that we treat in Sects. 11.3 and 11.4 describe insulators, but in reality ferromagnetic insulators do exist but are very rare. We might say that ferromagnetism is a phenomenon that is robust enough to be observed in nature, but is found only in relatively small ranges in the "parameter space".

In this preparatory section, we discuss some basic properties of saturated ferromagnetism, to which we simply refer as ferromagnetism throughout the chapter. In Sect. 11.1.1, we give a precise definition of ferromagnetism, and state a basic propo-

H. Tasaki, *Physics and Mathematics of Quantum Many-Body Systems*,
Graduate Texts in Physics, https://doi.org/10.1007/978-3-030-41265-4_11

sition which characterizes the nature of ferromagnetic ground states. In Sect. 11.1.2, we present two elementary theorems which show that ferromagnetism is impossible in certain situations. These theorems should make clear that ferromagnetism is indeed a non-perturbative phenomenon in general. In Sect. 11.1.3, we introduce and discuss a simple toy model, which illustrates some essential properties of ferromagnetism in the Hubbard model. Finally in Sect. 11.1.4, we briefly discuss the Stoner criterion, which suggests that ferromagnetism should be sought in models with large (single-electron) density of states and/or large Coulomb interaction.

Let us also make a brief comment on the problem of ferromagnetism at nonzero temperatures. It has been shown rigorously that any Hubbard model with short range hopping in one or two dimensions does not exhibit any magnetic ordering at nonzero temperatures [12, 21]. The mechanism behind this statement is essentially the same as that for quantum spin systems, which we have discussed thoroughly in Sect. 4.4.3. In fact the Proofs of Theorems 4.24 and 4.25 (p. 124) can be applied to the Hubbard model (or a more general lattice electron model) almost as they are.[1] Therefore we must treat models in (at least) three dimensions in order to have ferromagnetism stable at nonzero temperatures. We expect that ferromagnetism in three dimensional versions of some of the models treated in this chapter survives at finite temperatures, but there are no rigorous results in this direction. As we have noted in Sect. 4.4.4, the existence of ferromagnetic order in the quantum ferromagnetic Heisenberg model at low enough temperatures is not yet proved [8]. The corresponding problem in the Hubbard model must be much harder.

11.1.1 Definition and Basic Theorem

We consider the standard Hubbard Hamiltonian

$$\hat{H} = \hat{H}_{\text{hop}} + \hat{H}_{\text{int}}, \tag{11.1.1}$$

with

$$\hat{H}_{\text{hop}} = \sum_{\substack{x,y\in\Lambda \\ \sigma=\uparrow,\downarrow}} t_{x,y}\, \hat{c}^\dagger_{x,\sigma}\hat{c}_{y,\sigma}, \quad \hat{H}_{\text{int}} = U\sum_{x\in\Lambda} \hat{n}_{x,\uparrow}\hat{n}_{x,\downarrow}, \tag{11.1.2}$$

which are (9.3.17) and (9.3.29), respectively. We can treat the interaction Hamiltonian (9.3.28) with site dependent interaction U_x in most cases, but shall stick to the above simpler form.

In the present chapter we study the strongest form of ferromagnetism, namely saturated ferromagnetism, which is defined as follows. We write $S_{\max} := N/2$.

[1]One uses the same complex rotation (4.4.26), regarding the spin operator as written in terms of the fermion operators.

Definition 11.1 (*Saturated ferromagnetism*) Consider the Hubbard model with Hamiltonian (11.1.1) with N electrons. We say that the model exhibits ferromagnetism (or, more precisely, saturated ferromagnetism) if any ground state $|\Phi_{\mathrm{GS}}\rangle$ has the maximum possible total spin $S_{\mathrm{tot}} = S_{\max}$, i.e., $(\hat{\mathbf{S}}_{\mathrm{tot}})^2|\Phi_{\mathrm{GS}}\rangle = S_{\max}(S_{\max}+1)|\Phi_{\mathrm{GS}}\rangle$.

We should note that saturated ferromagnetism is indeed a very special form of ferromagnetism. In general we expect partial ferromagnetism where S_{tot} of the ground states is macroscopically large but the ratio $S_{\mathrm{tot}}/S_{\max}$ is strictly less than unity.[2] In the present book, we however concentrate only on saturated ferromagnetism, and refer to it simply as ferromagnetism. This is mainly because the study of partial ferromagnetism is so difficult that there are almost no rigorous results to be discussed. Theoretical study of (saturated) ferromagnetism in the Hubbard model is relatively easier because of the property stated in the following Proposition 11.2.

As in Sect. 9.3.2, we denote by $\hat{a}^\dagger_{j,\sigma}$ the creation operator of the single-electron energy eigenstate $\boldsymbol{\psi}^{(j)}$ with energy eigenvalue ε_j, where $j = 1, 2, \ldots, |\Lambda|$. It satisfies the commutation relation $[\hat{H}_{\mathrm{hop}}, \hat{a}^\dagger_{j,\sigma}] = \varepsilon_j \hat{a}^\dagger_{j,\sigma}$, which is (9.3.21). We have seen that the eigenstates of $\hat{H}_{\mathrm{hop}}$ are explicitly written as (9.3.22). Let us define

$$E_{\mathrm{ferro}} := \min_{\substack{\Gamma\subset\{1,\ldots,|\Lambda|\} \\ (|\Gamma|=N)}} \sum_{j\in\Gamma} \varepsilon_j, \tag{11.1.3}$$

which is the ground state energy of the system of N spin-less fermions with non-interacting Hamiltonian $\hat{H}_{\mathrm{hop}}$.

Proposition 11.2 *Suppose that a Hubbard model exhibits (saturated) ferromagnetism as in Definition 11.1. Then for any* $\Gamma \subset \{1, \ldots, |\Lambda|\}$ *such that* $|\Gamma| = N$ *and* $\sum_{j\in\Gamma} \varepsilon_j = E_{\mathrm{ferro}}$, *and for any* $M = -S_{\max}, S_{\max}+1, \ldots, S_{\max}$, *the state*

$$|\Phi^{\Gamma,M}_{\mathrm{GS}}\rangle := (\hat{S}^-_{\mathrm{tot}})^{S_{\max}-M}\Bigl(\prod_{j\in\Gamma} \hat{a}^\dagger_{j,\uparrow}\Bigr)|\Phi_{\mathrm{vac}}\rangle, \tag{11.1.4}$$

is a ground state of the Hubbard model. Moreover any ground state is a linear combination of these ground states.

This proposition shows that the ground states of a Hubbard model simplifies considerably when the model exhibits saturated ferromagnetism. To see the essence of the proof, consider a subspace of states that consists only of up-spin electrons. Since the interaction Hamiltonian $\hat{H}_{\mathrm{int}}$ always gives zero when acting on such states, the Hubbard model, when restricted on this subspace, reduces to a non-interacting electron model. The energy eigenstates then take the simple Slater determinant form (9.3.22) with $S_\downarrow = \emptyset$. In particular, the lowest energy state in the subspace is given by (11.1.4) with $M = S_{\max}$.

---

[2] Lieb's example in Sect. 10.2.3 satisfies this criterion, but it should better be interpreted as ferrimagnetism rather than partial ferromagnetism.

This drastic simplification is a result of the particular form of the Hubbard interaction, where only electrons on the same site interact. We should note however that the appearance of ferromagnetism is a highly nontrivial phenomenon, which must be caused by the interaction effect.

Proof of Proposition 11.2 The proposition relies essentially on the SU(2) invariance of the Hubbard model. As in Appendix A.3, we denote by $\mathscr{H}_{J,M}$ the space of states with $(\hat{\boldsymbol{S}}_{\text{tot}})^2|\Phi\rangle = J(J+1)|\Phi\rangle$ and $\hat{S}^{(3)}_{\text{tot}}|\Phi\rangle = M|\Phi\rangle$. (Here we set $\hat{J} = \hat{\boldsymbol{S}}_{\text{tot}}$.)

We first note that there is a ground state in[3] $\mathscr{H}_{S_{\max},S_{\max}}$, i.e., the space of states consisting of N up-spin electrons. Then we see from the above discussion that $|\Phi^{\Gamma,S_{\max}}_{\text{GS}}\rangle$, which is the lowest energy state within $\mathscr{H}_{S_{\max},S_{\max}}$, must be a ground state of the Hubbard model. Note that Lemma A.14 (p. 471) implies that $|\Phi^{\Gamma,M}_{\text{GS}}\rangle$ is nonzero for $M = -S_{\max}, S_{\max}+1, \ldots, S_{\max}$. Since $[\hat{S}^{-}_{\text{tot}}, \hat{H}] = 0$, we see that $|\Phi^{\Gamma,M}_{\text{GS}}\rangle$ is a ground state for any M.

To see that any ground state is a linear combination of the states (11.1.4), take an arbitrary ground state $|\Phi_{\text{GS}}\rangle$, and decompose it as $|\Phi_{\text{GS}}\rangle = \sum_{M=-N/2}^{N/2} |\Phi^{M}_{\text{GS}}\rangle$, where $|\Phi^{M}_{\text{GS}}\rangle \in \mathscr{H}_{S_{\max},M}$. Note that each $|\Phi^{M}_{\text{GS}}\rangle$, if nonzero, is a ground state. For each nonzero $|\Phi^{M}_{\text{GS}}\rangle$, let $|\Phi^{M,\uparrow}_{\text{GS}}\rangle = (\hat{S}^{+}_{\text{tot}})^{S_{\max}-M}|\Phi^{M}_{\text{GS}}\rangle$, which is nonzero by Lemma A.14. Since $|\Phi^{M,\uparrow}_{\text{GS}}\rangle$ is a ground state and belongs to $\mathscr{H}_{S_{\max},S_{\max}}$, it is written as a linear combination of the states (11.1.4) with $M = S_{\max}$. Since $|\Phi^{M}_{\text{GS}}\rangle = (\text{const.})(\hat{S}^{-}_{\text{tot}})^{S_{\max}-M}|\Phi^{M,\uparrow}_{\text{GS}}\rangle$, we see that $|\Phi^{M}_{\text{GS}}\rangle$ is also a linear combination of states (11.1.4). ∎

Suppose that the single-electron energy eigenstates (which are ordered so that $\varepsilon_j \le \varepsilon_{j+1}$) satisfy $\varepsilon_N < \varepsilon_{N+1}$. Then the choice of Γ in the ground state (11.1.4) is unique, and we must set $\Gamma = \{1, 2, \ldots, N\}$. In such a case the ground states are exactly $2S_{\max} + 1 = N + 1$ fold degenerate.

11.1.2 *Instability of Ferromagnetism*

To see that ferromagnetism is indded a delicate phenomenon, we discuss some results which show that the Hubbard model with certain conditions does not exhibit ferromagnetism.

We have seen in Sect. 9.3.2 that the non-interacting model, whose Hamiltonian is $\hat{H}_{\text{hop}}$, in general exhibits Pauli paramagnetism as in (9.3.24). See, in particular, Fig. 9.3. In the same section, we have also seen that the non-hopping model, whose Hamiltonian is $\hat{H}_{\text{int}}$, exhibits a simple paramagnetism. Neither $\hat{H}_{\text{hop}}$ nor $\hat{H}_{\text{int}}$ favors ferromagnetism.

We next consider the situation in which the Coulomb interaction U is nonzero but small. One can easily prove the following theorem which states that there cannot be (saturated) ferromagnetism when U is small, unless the single-electron energy

---

[3]***Proof*** Since $\hat{H}$ and $\hat{S}^{(3)}_{\text{tot}}$ are simultaneously diagonalizable, there is a ground state in $\mathscr{H}_{S_{\max},M_0}$ for some M_0. Then Theorem A.16 (p. 473) implies that there is a ground state in $\mathscr{H}_{S_{\max},M}$ for each M.

eigenvalues ε_j are highly degenerate. We again denote by $\hat{a}^\dagger_{j,\sigma}$ the creation operator of the single-electron energy eigenstate with eigenvalue ε_j, where $j = 1, 2, \ldots, |\Lambda|$. We also assume that the energy eigenvalues are ordered as $\varepsilon_j \le \varepsilon_{j+1}$.

Theorem 11.3 (Impossibility of ferromagnetism for small U) *Suppose* $0 \le U < \varepsilon_N - \varepsilon_1$. *Then the ground state of the Hubbard model does not have* $S_{\text{tot}} = S_{\max}$, *i.e., the model does not exhibit (saturated) ferromagnetism.*

Note that $\varepsilon_N - \varepsilon_1$, which is known as the fermi energy, is usually independent of the system size when the filling factor $N/(2|\Lambda|)$ is fixed.

Proof One of the lowest energy states with $S_{\text{tot}} = S_{\max}$ is given by

$$|\Phi^\uparrow_{\text{GS}}\rangle = \Bigl(\prod_{j=1}^{N} \hat{a}^\dagger_{j,\uparrow}\Bigr)|\Phi_{\text{vac}}\rangle, \tag{11.1.5}$$

whose energy is $E_{\text{ferro}} = \sum_{j=1}^{N} \varepsilon_j$. Consider a normalized trial state

$$|\Psi\rangle = \hat{a}^\dagger_{1,\downarrow}\Bigl(\prod_{j=1}^{N-1} \hat{a}^\dagger_{j,\uparrow}\Bigr)|\Phi_{\text{vac}}\rangle, \tag{11.1.6}$$

which is obtained from (11.1.5) by removing the up-spin electron which has the highest energy and then adding a down-spin electron which has the lowest energy. Noting the $SU(2)$ invariance of $\hat{a}^\dagger_{1,\downarrow}\hat{a}^\dagger_{1,\uparrow}$ as shown in (9.2.70) and (9.2.71), one finds that $|\Psi\rangle$ has $S_{\text{tot}} = S_{\max} - 1$. We want to evaluate the energy expectation value $\langle\Psi|\hat{H}|\Psi\rangle$. For the kinetic energy, we readily see that $\langle\Psi|\hat{H}_{\text{hop}}|\Psi\rangle = \varepsilon_1 + \sum_{j=1}^{N-1} \varepsilon_j$. For $\hat{H}_{\text{int}}$, we note that the trivial inequality $\hat{n}_{x,\uparrow} \le 1$ implies $\hat{H}_{\text{int}} \le U \sum_{x\in\Lambda} \hat{n}_{x,\downarrow}$ to get $\langle\Psi|\hat{H}_{\text{int}}|\Psi\rangle \le \langle\Psi|U\sum_{x\in\Lambda} \hat{n}_{x,\downarrow}|\Psi\rangle = U$. We therefore have

$$\langle\Psi|\hat{H}|\Psi\rangle - E_{\text{ferro}} \le \varepsilon_1 - \varepsilon_N + U < 0. \tag{11.1.7}$$

From the variational principle, we see that E_{ferro} is not the ground state energy of $\hat{H}$. ∎

Although the above theorem ensures that the ground state cannot be ferromagnetic, it does not provide any information about the nature of the true ground state of the model. To study the latter explicitly is in general a very difficult problem.

Let us also discuss the situation in which the interaction may be large but the density of electrons is very low. It is expected that the chance of electrons to collide with each other in this case becomes very small. It is likely that the model is close to an ideal gas, and there is no ferromagnetism.

This naive guess is justified for "healthy" models in dimensions three (or higher). The dimensionality of the lattice is taken into account by assuming that there are positive constants[4] c, n_0, ρ_0, and d, and the single electron energy eigenvalues satisfy

$$\varepsilon_n - \varepsilon_1 \geq c \left(\frac{n - n_0}{|\Lambda|} \right)^{2/d}, \tag{11.1.8}$$

for any n such that $n \geq n_0$ and $n/|\Lambda| \leq \rho_0$. Note that the right-hand side represents the n dependence of energy levels in the standard quantum mechanical system of a single particle in d-dimensions. Then we have the following theorem due to Pieri, Daul, Baeriswyl, Dzierzawa, and Fazekas [46].

Theorem 11.4 (Impossibility of ferromagnetism at low densities) *Take a $\hat{H}_{\text{hop}}$ which has translation invariance and satisfies (11.1.8) with positive c, n_0, ρ_0, and $d > 2$. Then there exists a constant $\rho_1 > 0$, and the corresponding Hubbard model does not exhibit (saturated) ferromagnetism for any $U \geq 0$ if $N/|\Lambda| \leq \rho_1$ holds.*

Outline of Proof The naive trial state (11.1.6) does not work for large U. We follow [52], and consider the Roth state [49]

$$|\tilde{\Psi}\rangle = \hat{P}_0 |\Psi\rangle, \tag{11.1.9}$$

where $|\Psi\rangle$ is defined in (11.1.6), and

$$\hat{P}_0 = \prod_{x \in \Lambda} (1 - \hat{n}_{x,\uparrow} \hat{n}_{x,\downarrow}) \tag{11.1.10}$$

is the orthogonal projection (called the Gutzwiller projection) onto the space with no doubly occupied sites. Because of the projection, the state (11.1.9) minimizes the Coulomb interaction as $\hat{H}_{\text{int}} |\Psi\rangle = 0$. Thus we only need to evaluate the expectation value of $\hat{H}_{\text{hop}}$. After a slightly complicated calculation whose details can be found in Appendix F of [73], we find

$$\frac{\langle \widetilde{\Psi} | H | \widetilde{\Psi} \rangle}{\langle \widetilde{\Psi} | \widetilde{\Psi} \rangle} - E_{\text{ferro}} \leq \varepsilon_1 - \varepsilon_N + c' \rho, \tag{11.1.11}$$

where $\rho = N/|\Lambda|$ is the electron density and $c' > 0$ is a constant. From the assumption (11.1.8), we find that the right-hand side becomes strictly negative for sufficiently small ρ provided that $d > 2$. ∎

That we have a restriction on dimensionality in Theorem 11.4 is not merely technical. In a one-dimensional system, moving electrons must eventually collide with each other for an obvious geometric reason. Thus a one-dimensional model cannot be regarded as close to an ideal gas, no matter how low the electron density is. We do not know whether the inapplicability of the theorem to two-dimensional systems is physically meaningful or not.

---

[4]The constants n_0 and d represent the degeneracy of the single-electron ground states and the dimension of the system, respectively.

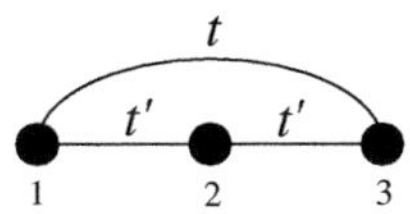

Fig. 11.1 The lattice and the hopping amplitude of the toy model. By considering the system with two electrons on this lattice, we can observe some very important aspects of ferromagnetism in the Hubbard model ()

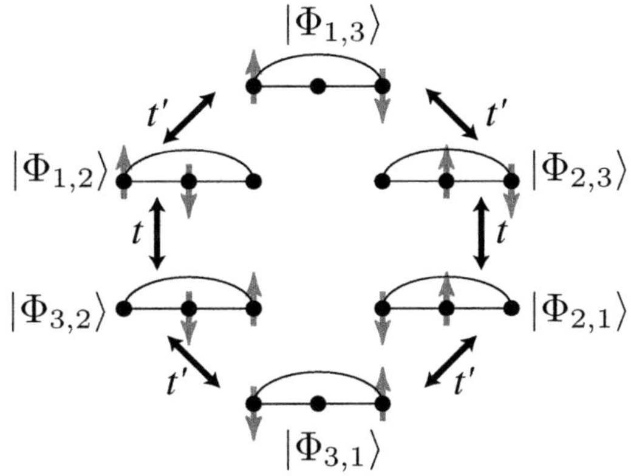

Fig. 11.2 Allowed states and transition amplitude in the toy model with $U = \infty$, which leads to the ring exchange process. The total spin of the ground states can be easily read off from this diagram ()

11.1.3 Toy Model with Two Electrons

As a starting point of our study of ferromagnetism, we consider a very simple toy model with two electrons. Interestingly, some essential features of ferromagnetism found in many-electron systems (that we will discuss in the following sections) are already present in this model.

The smallest possible Hubbard model which is away from half-filling is that with two electrons on a lattice with three sites. Consider the lattice $\Lambda = \{1, 2, 3\}$, and put one electron with $\sigma = \uparrow$ and one with $\sigma = \downarrow$. The hopping matrix is defined by $t_{1,2} = t_{2,3} = t'$, and $t_{1,3} = t$. Note that there are two kinds of hoppings t and t'. Since the sign of t' can be changed by the gauge transformation $\hat{c}_{2,\sigma} \to -\hat{c}_{2,\sigma}$ (see Sect. 9.3.3), we shall fix $t' > 0$. Figure 11.1 shows the lattice and the hopping amplitude. For simplicity, we assume there is only one kind of interaction, and set $U_1 = U_2 = U_3 = U \geq 0$. We have $S_{\max} = 1$ because $N = 2$. Therefore we can say that there appears ferromagnetism if the ground state has $S_{\text{tot}} = 1$, i.e., if it is a spin-triplet.

Let us take the limit $U \uparrow \infty$, in which the effect of the interaction becomes most drastic, and consider only those states with finite energies. This is equivalent to considering only states in which two electrons never occupy a same site.[5] There are six basis states which satisfy the constraint, and they can be written as $|\Phi_{x,y}\rangle = \hat{c}^\dagger_{x,\uparrow}\hat{c}^\dagger_{y,\downarrow}|\Phi_{\rm vac}\rangle$ where $x, y = 1, 2, 3$, and $x \neq y$. Transition amplitudes between these states are shown in Fig. 11.2. We find that the problem is equivalent to that of a quantum mechanical particle hopping around on a ring consisting of six sites. The basic structure of the ground state can be determined by the standard Perron–Frobenius sign convention.[6] By also taking into account the symmetry between the basis states, the ground state for $t < 0$ is written as

$$|\Phi_{\rm GS}^{(t<0)}\rangle = |\Phi_{1,2}\rangle + |\Phi_{3,2}\rangle - \alpha(t,t')|\Phi_{3,1}\rangle + |\Phi_{2,1}\rangle + |\Phi_{2,3}\rangle - \alpha(t,t')|\Phi_{1,3}\rangle, \tag{11.1.12}$$

and that for $t > 0$ as

$$|\Phi_{\rm GS}^{(t>0)}\rangle = |\Phi_{1,2}\rangle - |\Phi_{3,2}\rangle + \beta(t,t')|\Phi_{3,1}\rangle - |\Phi_{2,1}\rangle + |\Phi_{2,3}\rangle - \beta(t,t')|\Phi_{1,3}\rangle, \tag{11.1.13}$$

where $\alpha(t,t')$ and $\beta(t,t')$ are positive functions of t and t'.

To find the total spin of these states, it suffices to concentrate on two lattice sites, say sites 1 and 2, and note that $|\Phi_{\rm GS}^{(t<0)}\rangle = |\Phi_{1,2}\rangle + |\Phi_{2,1}\rangle + \cdots$, and $|\Phi_{\rm GS}^{(t>0)}\rangle = |\Phi_{1,2}\rangle - |\Phi_{2,1}\rangle + \cdots$. It immediately follows that $|\Phi_{\rm GS}^{(t<0)}\rangle$ has $S_{\rm tot} = 0$, and $|\Phi_{\rm GS}^{(t>0)}\rangle$ has $S_{\rm tot} = 1$. See (9.2.76) and (9.2.77). We see that ferromagnetic coupling is generated when t is positive.

It will be useful to see the mechanism that generates the ferromagnetism in this simple model. Note that the states $|\Phi_{1,2}\rangle$ and $|\Phi_{2,1}\rangle$ can be found in the upper left and and the lower right, respectively, in the diagram of Fig. 11.2. By starting from $|\Phi_{1,2}\rangle$ and following the possible transitions, one reaches the state $|\Phi_{2,1}\rangle$. In other words, electrons hop around in the lattice, and the spins on sites 1 and 2 are "exchanged." When $t > 0$, the quantum mechanical amplitude associated with the exchange process generates the superposition of the two states which precisely yields ferromagnetism. The process leading to ferromagnetic coupling may be called the ring exchange.

Let us briefly look at the cases with finite U. Let $E_{\min}(S_{\rm tot})$ be the lowest energy among the states which have total spin $S_{\rm tot}$. In Fig. 11.3, we plotted $E_{\min}(0)$ and $E_{\min}(1)$ for the toy model with $t = t'/2 > 0$ as functions of $U \geq 0$. As is suggested by the result in the $U \uparrow \infty$ limit, we have ferromagnetism in the sense that $E_{\min}(0) > E_{\min}(1)$ when U is sufficiently large. A level crossing takes place at finite U, and the system is no longer ferromagnetic for small U. Even in the simplest toy model, ferromagnetism is a nonperturbative phenomenon which takes place only when U is sufficiently large.

The only exception is the case with $t = t' > 0$. See Fig. 11.4. For this parameter value, the ground states are degenerate and have both $S_{\rm tot} = 0$ and 1 when $U = 0$. Ferromagnetic state is the only ground state for any $U > 0$.

---

[5]See the Proof of Theorem 11.5 in Sect. 11.2 for the treatment of the $U \uparrow \infty$ limit.

[6]If the transition amplitude between two states is negative (*resp.*, positive), one superposes the two states with the same (*resp.*, opposite) signs.

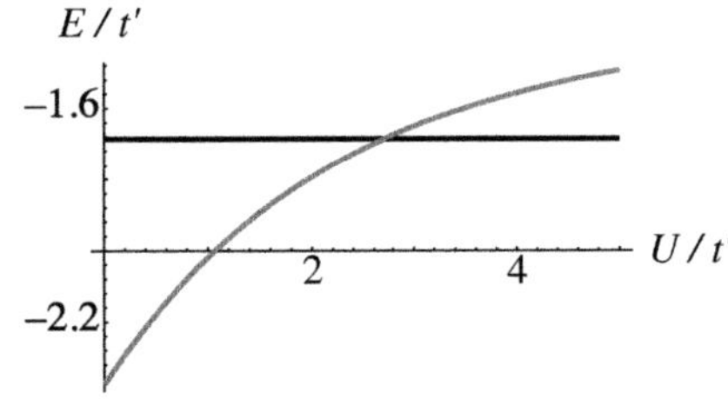

Fig. 11.3 The U dependence of $E_{\mathrm{min}}(0)$ (gray curve) and $E_{\mathrm{min}}(1)$ (black line) in the toy model with $t = t'/2 > 0$. We have ferromagnetism in the sense that $E_{\mathrm{min}}(0) > E_{\mathrm{min}}(1)$ when U is sufficiently large. We find that ferromagnetism is a nonperturbative phenomenon (© Hal Tasaki 2020. All Rights Reserved)

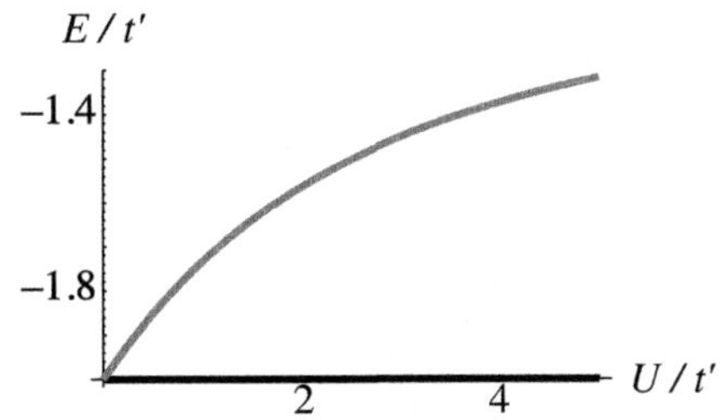

Fig. 11.4 The U dependence of $E_{\mathrm{min}}(0)$ (gray curve) and $E_{\mathrm{min}}(1)$ (black line) in the toy model with $t = t' > 0$. Only for this special parameter, we have ferromagnetism $E_{\mathrm{min}}(0) > E_{\mathrm{min}}(1)$ for any value of $U > 0$. One can regard this case as the simplest example of the flat-band ferromagnetism that we will discuss in Sect. 11.3 (© Hal Tasaki 2020. All Rights Reserved)

Figures 11.3 and 11.4 show that the energy $E_{\mathrm{min}}(1)$ of ferromagnetic states is independent of U. This is because ferromagnetic states do not feel the on-site Coulomb interaction, as we have seen in the Proof of Proposition 11.2.

11.1.4 Stoner Criterion

Before closing the present preparatory section, we briefly discuss a simple condition, known as the Stoner criterion, for ferromagnetism in interacting electron systems. It is based on a very crude mean-field type approximation, and represents a condition for instability of Pauli paramagnetism against ferromagnetic ordering. Although the Stoner criterion should not be regarded as a reliable condition, it gives us a rough idea about the roles played by the interaction U and the density of states D_{F} at the Fermi energy.

Consider the Hubbard model in the space with $N_\uparrow$ up-spin electrons and $N_\downarrow$ down-spin electrons, where $N_\uparrow + N_\downarrow = N$ is fixed. We shall make a bold approximation to the interaction Hamiltonian, and replace $\hat{n}_{x,\uparrow}$ and $\hat{n}_{x,\downarrow}$ by their averages as

$$\hat{H}_{\text{int}} = U \sum_{x \in \Lambda} \hat{n}_{x,\uparrow} \hat{n}_{x,\downarrow} \;\rightarrow\; U \sum_{x \in \Lambda} \langle \hat{n}_{x,\uparrow} \rangle \langle \hat{n}_{x,\downarrow} \rangle = U\,|\Lambda| \frac{N_\uparrow}{|\Lambda|} \frac{N_\downarrow}{|\Lambda|} = U \frac{N_\uparrow N_\downarrow}{|\Lambda|}. \tag{11.1.14}$$

We again denote by $\hat{a}^\dagger_{j,\sigma}$ the creation operator of the the single-electron energy eigenstate with ε_j, where $j = 1, 2, \ldots, |\Lambda|$ and $\varepsilon_j \le \varepsilon_{j+1}$. Then the state with the lowest kinetic energy with given $N_\uparrow$ and $N_\downarrow$ is

$$|\Phi^{N_\uparrow, N_\downarrow}_{MF}\rangle = \Bigl(\prod_{j=1}^{N_\uparrow} \hat{a}^\dagger_{j,\uparrow}\Bigr)\Bigl(\prod_{j=1}^{N_\downarrow} \hat{a}^\dagger_{j,\downarrow}\Bigr)|\Phi_{\text{vac}}\rangle. \tag{11.1.15}$$

The total energy of this state is thus given by

$$E_{\text{MF}}(N_\uparrow, N_\downarrow) = \sum_{j=1}^{N_\uparrow} \varepsilon_j + \sum_{j=1}^{N_\downarrow} \varepsilon_j + U \frac{N_\uparrow N_\downarrow}{|\Lambda|}. \tag{11.1.16}$$

Let us compare the energies of the paramagnetic state $|\Phi^{N/2,N/2}_{MF}\rangle$ and the state $|\Phi^{(N/2)+n,(N/2)-n}_{MF}\rangle$ obtained by flipping n spins from down to up. Note that the latter state has $S_{\text{tot}} = n$. From (11.1.16), we find that the energy difference is

$$\begin{aligned} \Delta E &= E_{\text{MF}}\Bigl(\frac{N}{2} + n, \frac{N}{2} - n\Bigr) - E_{\text{MF}}\Bigl(\frac{N}{2}, \frac{N}{2}\Bigr) \\ &= \sum_{\ell=1}^{n} (\varepsilon_{(N/2)+\ell} - \varepsilon_{(N/2)-n+\ell}) - U \frac{n^2}{|\Lambda|} \\ &\simeq n^2 \Delta\varepsilon - U \frac{n^2}{|\Lambda|}. \end{aligned} \tag{11.1.17}$$

We here approximated $\varepsilon_{(N/2)+\ell} - \varepsilon_{(N/2)-n+\ell} \simeq n\,\Delta\varepsilon$, where $\Delta\varepsilon$ is the average spacing of the single-electron energy eigenvalues near the Fermi energy $\varepsilon_{N/2}$. In a standard system, the energy level spacing behaves as $\Delta\varepsilon \simeq (|\Lambda|\, D_{\text{F}})^{-1}$ with a size independent quantity D_{F} known as the (single-electron) density of states (per volume) at the Fermi energy. We thus find

$$\Delta E \simeq \frac{n^2}{|\Lambda|\, D_{\text{F}}} (1 - D_{\text{F}}\, U). \tag{11.1.18}$$

This estimate implies that the paramagnetic state with $n = 0$ is unstable towards increasing n if

$$D_{\text{F}}\, U \gtrsim 1, \tag{11.1.19}$$

which is the Stoner criterion.

As is clear from the derivation, which is not justified in any reasonable limits, the criterion is far from being reliable. There are many examples in which $D_{\text{F}}\, U$ is extremely large but ferromagnetism is not observed. Nevertheless the Stoner criterion provides us with a rough guiding principle that ferromagnetism should be sought in models with large D_F and/or large U.

11.2 Nagaoka's Ferromagnetism

In the present section we discuss Nagaoka's ferromagnetism, which historically was the first rigorous example of ferromagnetism in the Hubbard model. In short Nagaoka's theorem [42] states that some Hubbard models exhibit ferromagnetism when the Coulomb repulsion U is infinitely large, and the number of electrons is one less than the number of the lattice sites (i.e., $N = |\Lambda| - 1$). Given the general fact that a half-filled system with $N = |\Lambda|$, where Lieb's theorem apply, never shows ferromagnetism, this is a rather striking result. It demonstrates that strongly interacting electron systems can exhibit very rich and sometimes surprising properties. Note also that the necessity to take $U \uparrow \infty$ is roughly compatible with the Stoner criterion (11.1.19).

When $U = \infty$ and $N = |\Lambda| - 1$, states with finite energies have no doubly occupied sites, and there is exactly one empty site, which we call the "hole." The basic mechanism of Nagaoka's ferromagnetism is that the hole hops around the lattice and generates a suitable linear combination of the basis states, in such a way that the resulting state exhibits ferromagnetism. It is remarkable that the motion of a single hole can turn (expected) antiferromagnetism at exact half-filling into ferromagnetism.[7] See Fig. 11.5. Thouless also discussed a similar mechanism in a different context [75]. Note that the ferromagnetism observed in the toy model studied in Sect. 11.1.3 can be regarded as the simplest version of Nagaoka's ferromagnetism. See, in particular, Fig. 11.2.

Here we give a complete proof of Nagaoka's theorem in its most generalized form due to Tasaki [66]. We believe that the reader will find the general proof [66, 73], which is a refinement of that by Nagaoka [42], straightforward and simple. In Sect. 11.2.1, we state and prove a weak version of Nagaoka's theorem. Then we shall see in Sect. 11.2.2 that this theorem can be easily strengthened to give the full theorem of Nagaoka's if we properly formulate the notion of connectivity of configurations.

Although Nagaoka's ferromagnetism is of considerable interest from theoretical and mathematical points of view, it is now understood that the ferromagnetic state obtained through this mechanism is rather pathological and has little to do with realistic ferromagnetism. There have been many studies which indicate that Nagaoka's ferromagnetism is destroyed when one relaxes the extreme conditions necessary for

[7] In Japanese institutions it often happens that the presence of a single foreign participant in a seminar room makes everybody shift from Japanese to English. I used to mention this phenomenon in the introduction to talks about Nagaoka's ferromagnetism. I remember one seminar at RIMS in Kyoto, where I had a perfect situation for this joke; the topic was about my refinement of Nagaoka's theorem, there was exactly one foreign participant at the seminar, and among many Japanese participants was Yosuke Nagaoka!

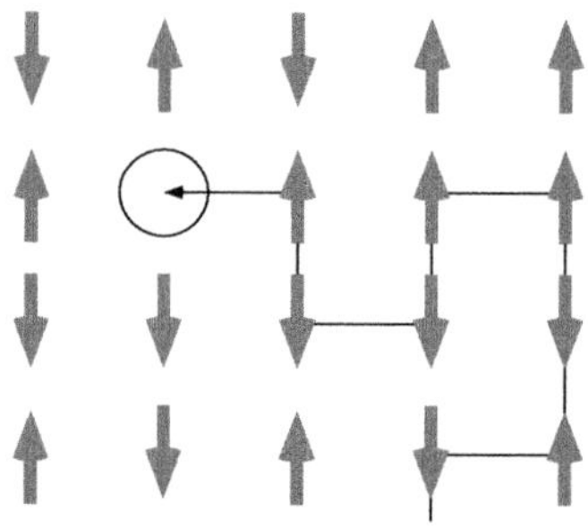

Fig. 11.5 Schematic picture of the origin of Nagaoka's ferromagnetism. When the hole hops around the lattice, the spin configuration changes. For a model with $t_{x,y} \geq 0$, the hole motion produces a precise linear combination of various spin configurations, which leads to a ferromagnetic state (© Hal Tasaki 2020. All Rights Reserved)

the theorem to be valid. We shall briefly summarize some of such results about instability of Nagaoka's ferromagnetism in Sect. 11.2.3.

11.2.1 Weak Version of Nagaoka's Theorem

We consider the standard Hubbard Hamiltonian $\hat{H} = \hat{H}_{\text{hop}} + \hat{H}_{\text{int}}$ with

$$\hat{H}_{\text{hop}} = \sum_{\substack{x,y\in\Lambda \\ \sigma=\uparrow,\downarrow}} t_{x,y}\, \hat{c}^{\dagger}_{x,\sigma}\hat{c}_{y,\sigma}, \quad \hat{H}_{\text{int}} = U\sum_{x\in\Lambda} \hat{n}_{x,\uparrow}\hat{n}_{x,\downarrow}. \tag{11.2.1}$$

Then we have the following.

Theorem 11.5 (Weak version of Nagaoka's theorem) *Assume that the hopping amplitude satisfies $t_{x,y} = t_{y,x} \geq 0$ for any $x, y \in \Lambda$, and consider the Hubbard model with $N = |\Lambda| - 1$ in the limit $U \uparrow \infty$. Then among the ground states there are at least $(2S_{\text{tot}} + 1)$ states with total spin $S_{\text{tot}} = S_{\text{max}} (= N/2)$.*

Note that the theorem does not establish the existence of ferromagnetism since it does not state that the ferromagnetic states are the only ground states. In fact the condition of the theorem is valid even in the trivial non-hopping model with $t_{x,y} = 0$ for all $x, y \in \Lambda$, which exhibits paramagnetism as we have seen in Sect. 9.3.2.

Proof of Theorem 11.5 We shall prove the theorem, following [66, 73]. Let $\mathcal{H}_N^{\text{hc}}$ be the subspace (of the whole Hilbert space $\mathcal{H}_N$ with N electrons) which consists of states without any doubly occupied sites.[8] Here the script "hc" stands for "hardcore".

[8]To be precise, $\mathcal{H}_N^{\text{hc}}$ is spanned by the basis states (9.2.35) with the constraint that $x_j \neq x_k$ when $j \neq k$.

Note that any $|\Phi\rangle \in \mathscr{H}_N^{\rm hc}$ satisfies $\hat{H}_{\rm int}|\Phi\rangle = 0$. We denote by $\hat{P}_{\rm hc}$ the orthogonal projection onto $\mathscr{H}_N^{\rm hc}$.

We are interested in the ground state of the Hubbard Hamiltonian $\hat{H} = \hat{H}_{\rm hop} + \hat{H}_{\rm int}$ in the limit $U \uparrow \infty$. The model in the limit is described by the effective Hamiltonian

$$\hat{H}_{\rm eff} := \hat{P}_{\rm hc}\,\hat{H}\,\hat{P}_{\rm hc}, \tag{11.2.2}$$

which is an operator on the hardcore subspace $\mathscr{H}_N^{\rm hc}$. This fact should be obvious, but is proved in Theorem A.12 (p. 470). In particular the desired ground states of $\hat{H}$ in the $U \uparrow \infty$ limit are ground states of $\hat{H}_{\rm eff}$.

We follow Nagaoka [42], and prepare a basis for the Hilbert space $\mathscr{H}_N^{\rm hc}$, which is most convenient for the present proof. Our basis states are specified by the position $x \in \Lambda$ of the hole and the spin configuration $\boldsymbol{\sigma} = (\sigma_y)_{y\in\Lambda\backslash\{x\}} \in \mathscr{S}_{\Lambda\backslash\{x\}}$ in $\Lambda\backslash\{x\}$, which is the rest of the lattice. (For any subset $\Lambda' \subset \Lambda$, we denote by $\mathscr{S}_{\Lambda'}$ the set of all spin configurations $(\sigma_x)_{x\in\Lambda'}$ with $\sigma_x = \uparrow, \downarrow$.) We define

$$|\Phi_{x,\boldsymbol{\sigma}}\rangle := \hat{c}_{x,\uparrow}\left(\prod_{y\in\Lambda}\hat{c}^\dagger_{y,\sigma'_y}\right)|\Phi_{\rm vac}\rangle = \hat{c}_{x,\downarrow}\left(\prod_{y\in\Lambda}\hat{c}^\dagger_{y,\sigma''_y}\right)|\Phi_{\rm vac}\rangle, \tag{11.2.3}$$

where the product is taken over all the sites in Λ with an arbitrary but fixed order. We gave two equivalent expressions for $|\Phi_{x,\boldsymbol{\sigma}}\rangle$. The spin configurations $(\sigma'_y)_{y\in\Lambda}$ and $(\sigma''_y)_{y\in\Lambda}$ are almost the same as $\boldsymbol{\sigma} = (\sigma_y)_{y\in\Lambda\backslash\{x\}}$, and defined by $\sigma'_y = \sigma''_y = \sigma_y$ for all $y \in \Lambda\backslash\{x\}$. As for the missing site x, we set $\sigma'_x = \uparrow$ and $\sigma''_x = \downarrow$. Thus, in the definition (11.2.3), we are simply supplying an electron at the hole site x, and then annihilating the electron by $\hat{c}_{x,\uparrow}$ or $\hat{c}_{x,\downarrow}$. This may appear as redundant, but, in this way, we get precise fermion signs appropriate for our purpose.

We wish to examine the action of $\hat{H}_{\rm eff} = \hat{P}_{\rm hc}\,\hat{H}_{\rm hop}\hat{P}_{\rm hc}$ on the states (11.2.3). In order to get a nonvanishing contribution, an electron must hop into the hole site x from one of the sites connected to x via nonzero $t_{x,z}$. We therefore examine the action of $\sum_{\sigma=\uparrow,\downarrow}\hat{c}^\dagger_{x,\sigma}\hat{c}_{z,\sigma} = -\sum_{\sigma=\uparrow,\downarrow}\hat{c}_{z,\sigma}\hat{c}^\dagger_{x,\sigma}$ on $|\Phi_{x,\boldsymbol{\sigma}}\rangle$. Using the two equivalent expressions (11.2.3) for $|\Phi_{x,\boldsymbol{\sigma}}\rangle$, we get

$$\begin{aligned}\sum_{\sigma=\uparrow,\downarrow}\hat{c}^\dagger_{x,\sigma}c_{z,\sigma}|\Phi_{x,\boldsymbol{\sigma}}\rangle &= -\hat{c}_{z,\uparrow}\hat{n}_{x,\uparrow}\Bigl(\prod_{y\in\Lambda}\hat{c}^\dagger_{y,\sigma'_y}\Bigr)|\Phi_{\rm vac}\rangle - \hat{c}_{z,\downarrow}\hat{n}_{x,\downarrow}\Bigl(\prod_{y\in\Lambda}\hat{c}^\dagger_{y,\sigma''_y}\Bigr)|\Phi_{\rm vac}\rangle\\ &= -\hat{c}_{z,\uparrow}\Bigl(\prod_{y\in\Lambda}\hat{c}^\dagger_{y,\sigma'_y}\Bigr)|\Phi_{\rm vac}\rangle - \hat{c}_{z,\downarrow}\Bigl(\prod_{y\in\Lambda}\hat{c}^\dagger_{y,\sigma''_y}\Bigr)|\Phi_{\rm vac}\rangle\\ &= -|\Phi_{z,\boldsymbol{\sigma}_{z\to x}}\rangle,\end{aligned} \tag{11.2.4}$$

where $\boldsymbol{\sigma}_{z\to x} \in \mathscr{S}_{\Lambda\backslash z}$ is the new spin configuration on $\Lambda\backslash z$ obtained from $\boldsymbol{\sigma}$ by moving σ_z to the site x. Note that, in the second and the third expressions in (11.2.4), only one of the two terms survive depending on the value of σ_z. Therefore the matrix elements of the effective Hamiltonian $\hat{H}_{\rm eff}$ are given by

$$\langle \Phi_{y,\tau}|\hat{H}_{\text{eff}}|\Phi_{x,\sigma}\rangle = \begin{cases} -t_{x,y} & \text{if } \tau = \sigma_{y\to x} \\ 0 & \text{otherwise.} \end{cases} \tag{11.2.5}$$

Let $|\Phi_{\text{GS}}\rangle$ be an arbitrary normalized ground state of $\hat{H}_{\text{eff}}$, and expand it using the above basis as

$$|\Phi_{\text{GS}}\rangle = \sum_{x\in\Lambda} \sum_{\sigma\in\mathcal{S}_{\Lambda\backslash\{x\}}} \varphi(x,\sigma)|\Phi_{x,\sigma}\rangle. \tag{11.2.6}$$

Since the matrix elements (11.2.5) of $\hat{H}_{\text{eff}}$ are real, we can assume that the coefficients $\varphi(x,\sigma)$ are real.[9] For each $x \in \Lambda$ we define

$$\xi_x = \Bigl(\sum_{\sigma\in\mathcal{S}_{\Lambda\backslash\{x\}}} \{\varphi(x,\sigma)\}^2\Bigr)^{1/2}. \tag{11.2.7}$$

Let the corresponding normalized ferromagnetic state be

$$|\Phi_{\uparrow}\rangle := \sum_{x\in\Lambda} \xi_x |\Phi_{x,(\uparrow)}\rangle, \tag{11.2.8}$$

where $(\uparrow)$ denotes the spin configuration with all spins up.

By using (11.2.6) and (11.2.5), we find

$$\begin{aligned}
\langle \Phi_{\text{GS}}|\hat{H}_{\text{eff}}|\Phi_{\text{GS}}\rangle &= \sum_{x,y\in\Lambda} \sum_{\substack{\sigma\in\mathcal{S}_{\Lambda\backslash\{x\}} \\ \tau\in\mathcal{S}_{\Lambda\backslash\{y\}}}} \varphi(y,\tau)\varphi(x,\sigma)\langle \Phi_{y,\tau}|\hat{H}_{\text{eff}}|\Phi_{x,\sigma}\rangle \\
&= -\sum_{x,y\in\Lambda} t_{x,y} \sum_{\sigma\in\mathcal{S}_{\Lambda\backslash\{x\}}} \varphi(y,\sigma_{y\to x})\varphi(x,\sigma) \\
&\geq -\sum_{x,y\in\Lambda} t_{x,y} \Bigl(\sum_{\sigma\in\mathcal{S}_{\Lambda\backslash\{x\}}} \{\varphi(y,\sigma_{y\to x})\}^2\Bigr)^{1/2} \Bigl(\sum_{\sigma\in\mathcal{S}_{\Lambda\backslash\{x\}}} \{\varphi(x,\sigma)\}^2\Bigr)^{1/2} \\
&= -\sum_{x,y\in\Lambda} t_{x,y}\, \xi_y \xi_x = \langle \Phi_{\uparrow}|\hat{H}_{\text{eff}}|\Phi_{\uparrow}\rangle,
\end{aligned} \tag{11.2.9}$$

where we used the Schwarz inequality (with the assumption $t_{x,y} \geq 0$) to get the third line. This bound shows that $|\Phi_{\uparrow}\rangle$ is also a ground state. ∎

---

[9] If this is not the case, we redefine $ia\{\varphi(x,\sigma) - \varphi(x,\sigma)^*\}$ as $\varphi(x,\sigma)$, where $a \in \mathbb{R}$ is a normalization factor. The corresponding $|\Phi_{\text{GS}}\rangle$ is also a ground state.

11.2.2 Nagaoka's Theorem and the Connectivity Condition

We now state a stronger theorem, namely, Nagaoka's theorem in its most general form [42, 66, 73]. Under an additional condition (which are verified in many situations), we prove that the model exhibits ferromagnetism in the sense of Definition 11.1, i.e., ferromagnetic ground states are the only possible ground states. See [20] for an extension of Nagaoka's theorem to the SU(n) version of the Hubbard model, and [41] for a unified extension to the Hubbard model coupled to phonons or photons.

The key is the Perron–Frobenius theorem, Theorem A.18 in p. 475. To apply the theorem to the present problem, we identify the matrix M with the matrix representation (11.2.5) of the effective Hamiltonian in the basis states (11.2.3) with a fixed $S_{\mathrm{tot}}^{(3)} = \sum_{y\in\Lambda\backslash\{x\}} \sigma_y$. Because of (11.2.5) and the assumption $t_{x,y} \geq 0$, the condition (i) (for Theorem A.18) on the non-positivity of off-diagonal matrix elements is satisfied. The condition (ii) on the connectivity, on the other hand, is not always valid. This motivates us to introduce the following connectivity condition.

Definition 11.6 (*Connectivity condition*) A Hubbard model with $U = \infty$ and $N = |\Lambda| - 1$ is said to satisfy the connectivity condition if all the basis states $|\Phi_{x,\sigma}\rangle$ with common $S_{\mathrm{tot}}^{(3)} = \sum_{y\in\Lambda\backslash\{x\}} \sigma_y$ are connected (in the sense of the Perron–Frobenius theorem) with each other through nonvanishing matrix elements of $\hat{H}_{\mathrm{eff}}$.

Note that the connectivity condition is a condition for hopping amplitude $(t_{x,y})_{x,y\in\Lambda}$ (or the graph defined by the hopping amplitude). As we shall see below, the connectivity condition is known to be satisfied in the Hubbard model with nearest neighbor hopping on essentially any standard lattices in two or higher dimensions.

Consider a model which satisfies the connectivity condition. Then we can readily apply the Perron–Frobenius theorem to see that the ground state in each subspace with a fixed $S_{\mathrm{tot}}^{(3)}$ is unique. Then Theorem 11.5 implies that this ground state must be ferromagnetic. We have thus proved the following.

Theorem 11.7 (Nagaoka's theorem) *Consider an arbitrary Hubbard model with* $t_{x,y} \geq 0$ *for any* $x, y \in \Lambda$, $N = |\Lambda| - 1$, *and* $U = \infty$, *and further assume that the model satisfies the connectivity condition. Then the ground states have total spin* $S_{\mathrm{tot}} = S_{\max} (= N/2)$, *and are non-degenerate apart from the trivial* $(2S_{\max} + 1)$*-fold degeneracy.*

In contrast to the weak theorem (Theorem 11.5), the above theorem states that all the ground states are ferromagnetic. This improvement is essential if one is interested in the emergence of ferromagnetism.

Since the theorem asserts the non-degeneracy of the ferromagnetic ground states for $U = \infty$, the continuity of energy eigenvalues in U implies that the statement of the theorem is valid also for sufficiently large but finite U. However, we have no meaningful estimates on how large U should be. When U is finite it is expected that antiferromagnetic interaction is generated by the mechanism (which does not involve the hole) discussed in Sect. 10.1. Then it is very likely that U must grow indefinitely as $|\Lambda|$ is increased in order to maintain ferromagnetism.

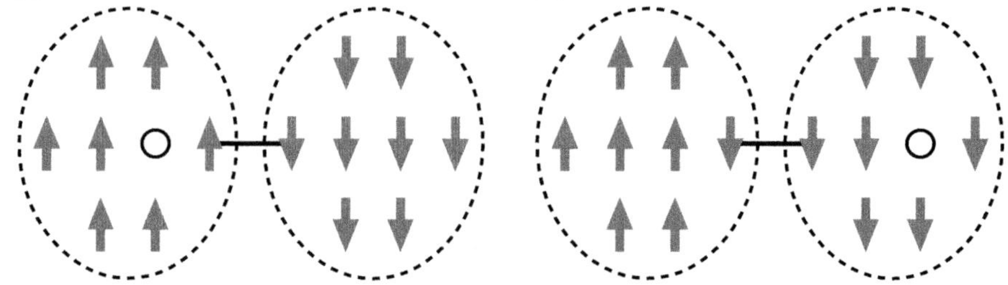

Fig. 11.6 A lattice which is connected but not biconnected. The lattice can be decomposed into two parts (within which one may have any connections) connected by a single bond. Suppose that one starts from a configuration (**a**), where the left part contains only up-spin electrons and a hole, while the right part contains only down-spin electrons. When the hole moves to the right part, one gets a configuration (**b**). If the hole go backs to the left part, one again gets a configuration as in **a**. It is impossible to carry an up-spin electron to the right part. The connectivity condition is violated (© Hal Tasaki 2020. All Rights Reserved)

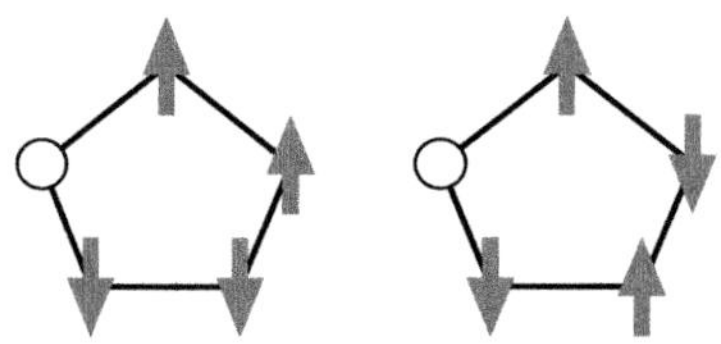

Fig. 11.7 The connectivity condition does not hold on a pentagon. It is interesting to compare the situation with those in a triangle (Fig. 11.8) and a square (Fig. 11.9) (© Hal Tasaki 2020. All Rights Reserved)

Connectivity condition In order to make Theorem 11.7 meaningful, we still need to verify the connectivity condition for some concrete models. Let $\mathscr{B}$ be the set of bonds $\{x, y\} = \{y, x\}$ such that $x \neq y$ and $t_{x,y} \neq 0$. We shall examine the validity of the connectivity condition for the lattice (or the graph) $(\Lambda, \mathscr{B})$.

Let us first discuss a necessary and sufficient condition for the connectivity condition [4]. Obviously the connectivity condition does not always hold. If $(\Lambda, \mathscr{B})$ is a one-dimensional lattice with open boundary conditions, for example, it is easy to see that there is no way of modifying spin configurations by the motion of a hole. Generalizing this observation, one sees that the connectivity condition does not hold if the lattice $(\Lambda, \mathscr{B})$ is not biconnected.[10] See Fig. 11.6. Another counter example is $(\Lambda, \mathscr{B})$ which forms a hexagon. It is obvious that one can never connect the two configurations depicted in Fig. 11.7 by moving the hole.

[10]A lattice (or a graph) is biconnected (or non-separable) if and only if one cannot make it disconnected by removing a single site.

In fact it is known that these two examples essentially represent all cases where the connectivity condition fails. By making use of a graph theoretic result by Wilson [81] on the "15 puzzle" problem,[11] Bobrow, Stubis, and Li recently derived the following necessary and sufficient condition for the connectivity [4].

Theorem 11.8 (Necessary and sufficient condition for the connectivity) *A lattice* $(\Lambda, \mathcal{B})$ *satisfies the connectivity condition if and only if it is biconnected and it is not a simple loop (i.e., periodic chain) with more than four sites.*

We thus see that the connectivity condition is satisfied in essentially any standard lattice in two or higher dimensions. We leave the Proof of Theorem 11.8 to the original papers [4, 81].

Let us next state a simple sufficient condition for the connectivity, which can be proved easily.[12] The sufficient condition is implicit in Nagaoka's original work [42].

We introduce some terminology to state the condition. By a loop of length m, we mean an ordered set $(x_1, \ldots, x_m)$ of sites in Λ such that $\{x_i, x_{i+1}\} \in \mathcal{B}$ for all $i = 1, \ldots, m-1$, and $\{x_m, x_1\} \in \mathcal{B}$. We say that a pair $\{x, y\}$ of lattice sites (which may not be in $\mathcal{B}$) is an exchange bond if (E1) x and y belong to a common loop of length three or four, and (E2) the whole lattice remains connected (through bonds in $\mathcal{B}$) when both the sites x and y are removed.

Lemma 11.9 (A sufficient condition for the connectivity) *If the whole lattice is connected by exchange bonds, then the model satisfies the connectivity condition.*

Note that all bonds in $\mathcal{B}$ are also exchange bonds in triangular, square, simple cubic, fcc, or bcc lattices. These lattices trivially satisfy the above sufficient condition. The hexagonal lattice does not satisfy the sufficient condition, but it does satisfy the connectivity condition because of Theorem 11.8.

Proof of Lemma 11.9 Suppose that we are given an arbitrary configuration of $N = |\Lambda| - 1$ electrons on Λ. Our goal is to show that we can get an arbitrary configuration with the same $S^{(3)}_{\mathrm{tot}}$ by moving the single hole along bonds in $\mathcal{B}$.

Let $\{x, y\}$ be an exchange bond. We show below that we can exchange the spins at sites x and y without changing the configuration outside $\{x, y\}$. Since the whole lattice is connected via exchange bonds, this means we can generate any permutation of spin configurations by successive exchanges on the exchange bonds.[13] This proves the connectivity condition.

We now prove the desired property of exchange bonds. Let $\{x, y\}$ be an exchange bond, and assume that x and y are occupied by electrons with opposite spins. We first bring the hole (by successive hops outside $\{x, y\}$) to a site other than x or y on the loop (of length three or four) that contains both x and y. Next we let the hole move

---

[11]Those who read Japanese might enjoy a short article by Nagaoka entitled "The 15 Puzzle" [43].

[12]The sufficient condition is, in a sense, physical since it only makes use of basic exchange processes caused by local motions of the hole. But, after all, we should note that Nagaoka's ferromagnetism itself is not quite physical.

[13]For a proof of this property, see "Proof of the property (iii)" in p. 41.

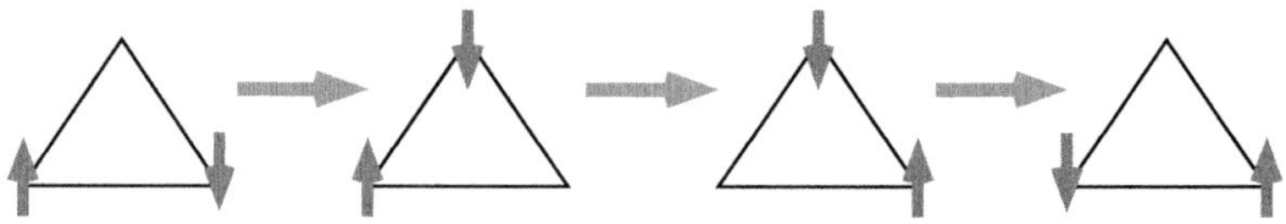

Fig. 11.8 The two spins are exchanged when the hole hops around the loop once. This is the same as Fig. 11.2 for the toy model discussed in Sect. 11.1.3 (© Hal Tasaki 2020. All Rights Reserved)

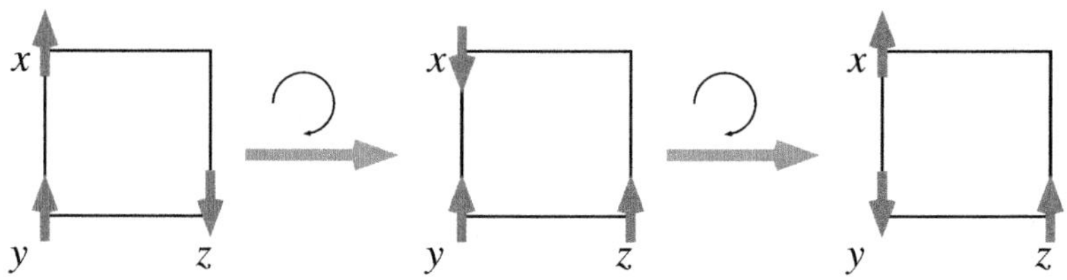

Fig. 11.9 The spins at x and z are exchanged when the hole hops around the loop once. The spins at y and z are exchanged when the hole hops around the loop twice (or once in the opposite orientation) (© Hal Tasaki 2020. All Rights Reserved)

along the loop until the spins at x and y are exchanged. In a loop of length three, this is realized after the hole goes around the loop once, as in Fig. 11.8. In a loop of length four, we have to move the hole along the loop once or twice, depending on the spin configuration, as in Fig. 11.9. Note that the exchange on a length four loop is possible because electronic spins take (only) two values.[14] Finally, we bring the hole back to the original location by following the same path as before backwards. We then recover exactly the same spin configuration, except on sites x and y. ∎

11.2.3 Instability of Nagaoka's Ferromagnetism

There is no doubt that Nagaoka's ferromagnetism is nontrivial and interesting. We have seen that completely spin-independent on-site Coulomb interaction and quantum mechanics for many fermions together can generate strong order in electron spins. It is also clear, on the other hand, that the mechanism that generates ferromagnetism is rather singular; the only motion allowed in the whole system is that of the single hole, and this motion generates the ferromagnetic order in the whole system. One expects that the resulting ferromagnetic state is physically pathological.

The pathology is clearly seen in the properties of low energy excitations. Kusakabe and Aoki studied spin-wave excitations above the ground state and found unusual behaviors [25]. Most notably it was shown that the width of the energy band (of the spin-wave excitation) shrinks to zero as the system size is increased. This result also

[14]This point is relevant to the extension of Nagaoka's ferromagnetism to the SU(n) Hubbard model [4, 20].

suggests that the Nagaoka's ferromagnetism does not emerge in a model with large but finite U which is independent of the system size. See also [3].

The pathology may be removed if the Nagaoka's ferromagnetism can be extended to models which have multiple holes with nonzero density. If each hole contributes in aligning spins in a finite region then it might be possible that healthy ferromagnetism in the whole system is generated. Unfortunately, the Perron–Frobenius argument which works for the one-hole case fails even for models with two holes. In fact it has been shown, e.g., in [7, 56, 76] that ferromagnetism does not emerge when there are two holes.

There is a considerable number of rigorous works (including that in Nagaoka's original work [42]) which establish that saturated ferromagnetism does not take place when one relaxes the strict conditions necessary for Nagaoka's theorem. See, for example, [14, 15, 52, 54]. Most of these works are essentially based on variational arguments where one constructs sophisticated variational states which have lower energies than the ferromagnetic state. See also [27, 47, 48] for results which indicate the instability of Nagaoka's ferromagnetism. We also recommend Sect. 8.4 of [10], which contains extensive discussion about the instability of Nagaoka's ferromagnetism. As far as we know, there are no rigorous results about the stability of Nagaoka's ferromagnetism in the Hubbard model.

11.3 Flat-Band Ferromagnetism

In the present section, we discuss versions of the Hubbard model which exhibit a type of ferromagnetism known as flat-band ferromagnetism. Flat-band ferromagnetism takes place in particularly designed systems in which the single-electron ground states have macroscopic degeneracy. Recall that Nagaoka's ferromagnetism satisfies the Stoner criterion $D_{\mathrm{F}}\, U \gtrsim 1$ because of the large Coulomb interaction U. In flat-band ferromagnetism, the criterion is satisfied for the opposite reason that the density of states D_{F} is infinitely large. Unlike Nagaoka's ferromagnetism, which is likely to be unstable against perturbation, flat-band ferromagnetism is believed to be stable against various perturbation. There are indeed some rigorous results about stability, as we shall discuss in Sect. 11.4.

The first class of examples of flat-band ferromagnetism was discovered in 1991 by Mielke [34], whose elegant construction makes use of graph theoretic notions. Later, Tasaki [67], who had been independently working on a similar problem, found a different class of models which exhibit flat-band ferromagnetism.[15]

We start by discussing Tasaki's model in detail in Sect. 11.3.1 since the theory is more elementary. We emphasize that the proof of the emergence of ferromagnetism in this model clearly illustrates the mechanism by which ferromagnetic exchange interaction is generated from the Coulomb repulsion and quantum mechanics for fermions. Then we describe in Sect. 11.3.2 the class of flat-band models discovered

[15] It was very early days of arXiv, and neither Mielke nor myself were posting papers. I learned about Mielke's work from my colleague some time after it was published in the journal. I was at that time working on a draft of my paper [67].

by Mielke, whose construction is based on the notion of line graphs. After discussing a general method for constructing models with flat-bands in Sect. 11.3.3, we finally discuss the general theory of flat-band ferromagnetism due to Mielke in Sect. 11.3.4. We note that the topics treated in Sects. 11.3.3 and 11.3.4 are somewhat advanced. The reader may skip these subsections and proceed directly to Sect. 11.4.

After the works of Mielke and Tasaki, various versions of the Hubbard model that exhibit flat-band ferromagnetism have been found.[16] See, e.g., [6, 13, 29, 31, 32, 51, 58, 60, 77]. We should always keep in mind that the Hubbard model is a highly simplified theoretical model. To find the implications of results for the Hubbard model in realistic many-electron systems is an extremely difficult but a challenging problem. See, e.g., [1, 11, 17, 26, 28, 44, 55] for some interesting proposals about experimental realization of flat-band ferromagnetism.

11.3.1 Tasaki's Flat-Band Ferromagnetism

We shall discuss Tasaki's flat-band ferromagnetism in detail. We define the simplest version of the model, give a complete proof, and discuss the band structure and the mechanism of the ferromagnetism. We also describe a general procedure, called the cell construction, to define a class of models that exhibit flat-band ferromagnetism.

The model on the decorated hypercubic lattice and the main theorem Let $\mathscr{E}$ be the set of sites in the d-dimensional $L \times L$ hypercubic lattice with unit lattice spacing and periodic boundary conditions. It is exactly the same as Λ_L defined in (3.1.2). We take a new site in the middle of each bond (i.e., a pair of sites separated by distance 1) of the hypercubic lattice, and denote by $\mathscr{I}$ the collection of all such sites. We study the Hubbard model on the decorated hypercubic lattice $\Lambda = \mathscr{E} \cup \mathscr{I}$. Sites in $\mathscr{E}$ are called external sites, and in $\mathscr{I}$ are called internal sites, for a reason to become clear when we discuss the general construction at the end of the section. One may regard the lattice structure as crudely mimicking that of a metallic oxide, where sites in $\mathscr{E}$ and $\mathscr{I}$ are identified with metallic atoms and oxygen atoms, respectively. The two-dimensional version of Λ is the same as the Lieb lattice. See Fig. 10.2 in p. 355 and Fig. 11.10.

Let us define two types of localized states in the single-electron Hilbert space $\mathfrak{h} \cong \mathbb{C}^{|\Lambda|}$. Let $\nu > 0$ be a fixed parameter. For each $p \in \mathscr{E}$, we define $\boldsymbol{\alpha}_p = (\alpha_p(x))_{x\in\Lambda} \in \mathfrak{h}$ by

$$\alpha_p(x) = \begin{cases} 1 & \text{if } x = p, \\ -\nu & \text{if } |x-p| = 1/2, \\ 0 & \text{otherwise,} \end{cases} \tag{11.3.1}$$

[16] We believe it fair to say that rigorous results preceded numerical works in the study of ferromagnetism in the Hubbard model. There also appeared numerical works in various versions of the Hubbard model, but we shall not try to list them.

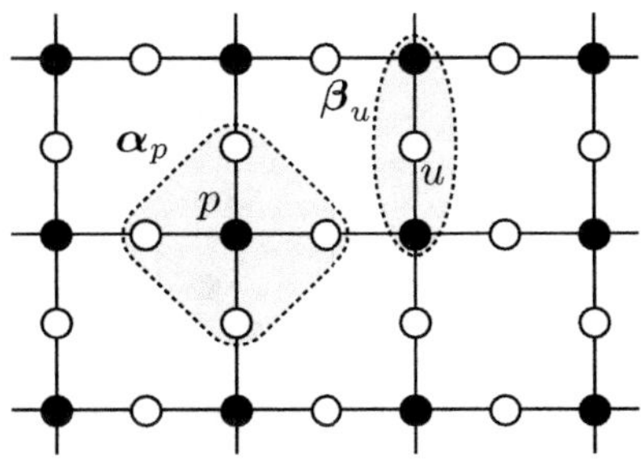

Fig. 11.10 The decorated hypercubic lattice with $d = 2$, which is the same as the Lieb lattice (Fig. 10.2). Black dots represent sites in $\mathscr{E}$ (which may be identified with metallic atoms) and white dots represent sites in $\mathscr{I}$ (which may be identified with oxygen atoms). Single-electron states $\boldsymbol{\alpha}_p$ and $\boldsymbol{\beta}_u$ are localized around sites $p \in \mathscr{E}$ and $u \in \mathscr{I}$, respectively (© Hal Tasaki 2020. All Rights Reserved)

and, for each $u \in \mathscr{I}$, we define $\boldsymbol{\beta}_u = (\beta_u(x))_{x \in \Lambda} \in \mathfrak{h}$ by

$$\beta_u(x) = \begin{cases} 1 & \text{if } x = u, \\ \nu & \text{if } |x - u| = 1/2, \\ 0 & \text{otherwise.} \end{cases} \tag{11.3.2}$$

See Fig. 11.10. Although ν is completely arbitrary here, it is useful to imagine that $0 < \nu \ll 1$. Then $\boldsymbol{\alpha}_p$ represents a state in which the electron is mostly localized at a metallic site p with small components on the neighboring sites, and likewise $\boldsymbol{\beta}_u$ represents a state nearly localized at an oxygen site u. We will see that the ferromagnetic ground states consist only of electrons in the $\boldsymbol{\alpha}$ states as in (11.3.9).

Note that $\langle \boldsymbol{\alpha}_p, \boldsymbol{\beta}_u \rangle = 0$ for any $p \in \mathscr{E}$ and $u \in \mathscr{I}$. The following lemma, which should be expected, is essential.

Lemma 11.10 $\{\boldsymbol{\alpha}_p\}_{p \in \mathscr{E}} \cup \{\boldsymbol{\beta}_u\}_{u \in \mathscr{I}}$ *is a basis of* $\mathfrak{h}$.

Proof $\boldsymbol{\alpha}_p$ with $p \in \mathscr{E}$ are linearly independent since only $\boldsymbol{\alpha}_p$ has nonzero component on site p. Likewise $\boldsymbol{\beta}_u$ with $u \in \mathscr{I}$ are linearly independent. Noting that $\{\boldsymbol{\alpha}_p\}_{p \in \mathscr{E}}$ and $\{\boldsymbol{\beta}_u\}_{u \in \mathscr{I}}$ span mutually orthogonal spaces with dimensions $|\mathscr{E}|$ and $|\mathscr{I}|$, respectively, and that the dimension of $\mathfrak{h}$ is $|\Lambda| = |\mathscr{E}| + |\mathscr{I}|$, we see that $\{\boldsymbol{\alpha}_p\}_{p \in \mathscr{E}} \cup \{\boldsymbol{\beta}_u\}_{u \in \mathscr{I}}$ forms a basis of $\mathfrak{h}$. ∎

Let us define fermion operators corresponding to $\boldsymbol{\alpha}_p$ and $\boldsymbol{\beta}_u$ by

$$\hat{a}_{p,\sigma} := \hat{C}_\sigma(\boldsymbol{\alpha}_p) = \hat{c}_{p,\sigma} - \nu \sum_{\substack{u \in \mathscr{I} \\ (|u-p|=1/2)}} \hat{c}_{u,\sigma}, \tag{11.3.3}$$

for $p \in \mathscr{E}$, and

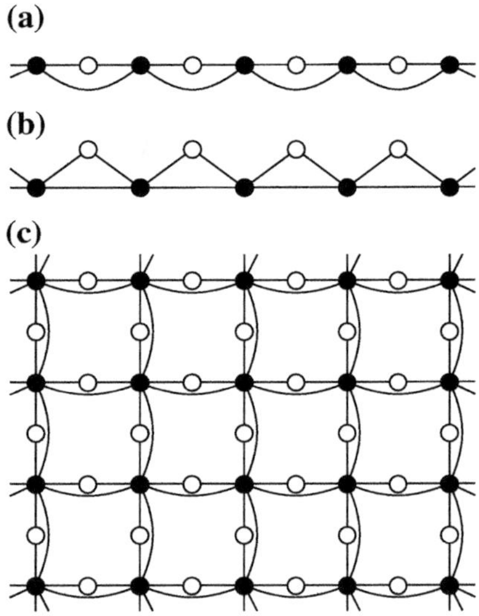

Fig. 11.11 Hopping in the Tasaki model in **a** $d = 1$ and **c** $d = 2$. Note that the lattice for $d = 1$ can be modified (without changing the connection) into the Delta chain in **b**. In addition to the hopping amplitude indicated by the lines, we have on-site potential which is fine tuned. See (11.3.22). Note that on each bond we have a structure with three sites that resembles the toy model for ferromagnetism described in Fig. 11.1 (p. 377) ()

$$\hat{b}_{u,\sigma} := \hat{C}_\sigma(\boldsymbol{\beta}_u) = \hat{c}_{u,\sigma} + \nu \sum_{\substack{p\in\mathscr{E} \\ (|p-u|=1/2)}} \hat{c}_{p,\sigma}, \tag{11.3.4}$$

for $u \in \mathscr{I}$. We then consider the Hubbard model with the Hamiltonian $\hat{H} = \hat{H}_{\text{hop}} + \hat{H}_{\text{int}}$, where the hopping Hamiltonian is defined as

$$\hat{H}_{\text{hop}} = t \sum_{\substack{u\in\mathscr{I} \\ \sigma=\uparrow,\downarrow}} \hat{b}^\dagger_{u,\sigma}\hat{b}_{u,\sigma}, \tag{11.3.5}$$

with $t > 0$, and the interaction Hamiltonian as

$$\hat{H}_{\text{int}} = U \sum_{x\in\Lambda} \hat{n}_{x,\uparrow}\hat{n}_{x,\downarrow}. \tag{11.3.6}$$

Note that our hopping Hamiltonian has a particular form while the interaction Hamiltonian is standard. Recalling the definition (11.3.4) of $\hat{b}_{u,\sigma}$, one sees that $\hat{H}_{\text{hop}}$ contains hopping between a pair of sites in $\mathscr{E}$ separated by distance 1 as well as a pair of sites in $\mathscr{E}$ and $\mathscr{I}$ separated by distance 1/2 as in Fig. 11.11. See also (11.3.22) below for an explicit expression of the hopping amplitude.

The following theorem in [40, 67] establishes the emergence of ferromagnetism in the present model.

Theorem 11.11 (Tasaki's flat-band ferromagnetism) *Consider the above Hubbard model with electron number $N = |\mathscr{E}| = L^d$. Then for any $\nu > 0$, $t > 0$, and $U > 0$,* *the ground states have $S_{\text{tot}} = S_{\max} = N/2$, and are unique apart from the trivial $2S_{\max} + 1 = N + 1$ fold degeneracy.*

Before proving the theorem, let us construct ground states of the model. Note that $\langle \boldsymbol{\alpha}_p, \boldsymbol{\beta}_u \rangle = 0$ and (9.2.64) imply

$$\{\hat{b}_{u,\sigma}, \hat{a}^\dagger_{p,\tau}\} = 0, \tag{11.3.7}$$

for any $u \in \mathscr{I}$, $p \in \mathscr{E}$, and $\sigma, \tau = \uparrow, \downarrow$. This in particular means $[\hat{b}^\dagger_{u,\sigma}\hat{b}_{u,\sigma}, \hat{a}^\dagger_{p,\tau}] = 0$, and hence

$$\hat{H}_{\text{hop}} |\Phi_{\boldsymbol{\alpha}\text{ all}\uparrow}\rangle = 0, \tag{11.3.8}$$

where

$$|\Phi_{\boldsymbol{\alpha}\text{ all}\uparrow}\rangle = \Bigl(\prod_{p\in\mathscr{E}} \hat{a}^\dagger_{p,\uparrow}\Bigr)|\Phi_{\text{vac}}\rangle \tag{11.3.9}$$

is a ferromagnetic state where all the $\boldsymbol{\alpha}$ states are filled with up-spin electrons. It is also obvious that $\hat{H}_{\text{int}}|\Phi_{\boldsymbol{\alpha}\text{ all}\uparrow}\rangle = 0$, and we find that $\hat{H}|\Phi_{\boldsymbol{\alpha}\text{ all}\uparrow}\rangle = 0$. Noting that $\hat{H}_{\text{hop}} \ge 0$, $\hat{H}_{\text{int}} \ge 0$, and hence $\hat{H} \ge 0$ (see Appendix A.2.3 for inequalities for self-adjoint operators), we see that the ground state energy of $\hat{H}$ is zero, and the ferromagnetic state $|\Phi_{\boldsymbol{\alpha}\text{ all}\uparrow}\rangle$ is a ground state. This is indeed a straightforward consequence of the (artificial) definition (11.3.5). The main point of Theorem 11.11 is that $|\Phi_{\boldsymbol{\alpha}\text{ all}\uparrow}\rangle$ is essentially the unique ground state. Since only trivial $2S_{\text{tot}} + 1$ fold degeneracy is allowed by the theorem, we see that the ground states are written as

$$|\Phi^{(M)}_{\text{GS}}\rangle = (\hat{S}^-_{\text{tot}})^{(N/2)-M}|\Phi_{\boldsymbol{\alpha}\text{ all}\uparrow}\rangle \tag{11.3.10}$$

with $M = -S_{\max}, -S_{\max}+1, \ldots, S_{\max}$.

Proof of Theorem 11.11 We shall prove that any ground state must be a linear combination of the ground states (11.3.10). Here each step in the proof highlights essential physical mechanism for ferromagnetism. We shall make clear the physical meaning of each argument.

Frustration free properties: Let $|\Phi_{\text{GS}}\rangle$ be an arbitrary ground state of $\hat{H}$ with $N = |\mathscr{E}|$ electrons. Since the ground state energy is zero, we have $\hat{H}|\Phi_{\text{GS}}\rangle = 0$, and hence $\hat{H}_{\text{hop}}|\Phi_{\text{GS}}\rangle = 0$ and $\hat{H}_{\text{int}}|\Phi_{\text{GS}}\rangle = 0$. Recalling the definitions (11.3.5) and (11.3.6), we find from Lemma A.10 (p. 469) about frustration-free Hamiltonians that $\hat{b}^\dagger_{u,\sigma}\hat{b}_{u,\sigma}|\Phi_{\text{GS}}\rangle = 0$ for each u, σ, and $\hat{n}_{x,\uparrow}\hat{n}_{x,\downarrow}|\Phi_{\text{GS}}\rangle = 0$ for each x. We further see from Lemma A.11 (p. 469) about positive semidefinite operators that

$$\hat{b}_{u,\sigma}|\Phi_{\text{GS}}\rangle = 0 \text{ for any } u \in \mathscr{I} \text{ and } \sigma = \uparrow, \downarrow, \tag{11.3.11}$$

and

$$\hat{c}_{x,\downarrow}\hat{c}_{x,\uparrow}|\Phi_{\text{GS}}\rangle = 0 \text{ for any } x \in \Lambda, \tag{11.3.12}$$

where we noted that $\hat{n}_{x,\uparrow}\hat{n}_{x,\downarrow} = (\hat{c}_{x,\downarrow}\hat{c}_{x,\uparrow})^\dagger\, \hat{c}_{x,\downarrow}\hat{c}_{x,\uparrow}$. The conditions (11.3.11) and (11.3.12), which we shall call the zero-energy conditions, will turn out to be quite useful.

Restriction to the lowest band: From Lemma 9.4 (p. 321) about general basis of the Hubbard model and Lemma 11.10 above, we see that our ground state is (indeed, any state is) expanded as

$$|\Phi_{\mathrm{GS}}\rangle = \sum_{\substack{A_\uparrow, A_\downarrow \subset \mathcal{E} \\ B_\uparrow, B_\downarrow \subset \mathcal{I} \\ (|A_\uparrow|+|A_\downarrow|+|B_\uparrow|+|B_\downarrow|=N)}} f(A_\uparrow, A_\downarrow, B_\uparrow, B_\downarrow)\Big(\prod_{p\in A_\uparrow} \hat{a}^\dagger_{p,\uparrow}\Big)\Big(\prod_{p\in A_\downarrow} \hat{a}^\dagger_{p,\downarrow}\Big)\Big(\prod_{u\in B_\uparrow} \hat{b}^\dagger_{u,\uparrow}\Big)\Big(\prod_{u\in B_\downarrow} \hat{b}^\dagger_{u,\downarrow}\Big)|\Phi_{\mathrm{vac}}\rangle, \tag{11.3.13}$$

where $f(A_\uparrow, A_\downarrow, B_\uparrow, B_\downarrow)$ is a certain coefficient. It should be clear that the zero-energy condition (11.3.11) shows that $|\Phi_{\mathrm{GS}}\rangle$ does not contain any $\hat{b}^\dagger_{u,\sigma}$, i.e., one has $f(A_\uparrow, A_\downarrow, B_\uparrow, B_\downarrow) = 0$ unless $B_\uparrow = B_\downarrow = \emptyset$.[17] Thus the ground state is written as

$$|\Phi_{\mathrm{GS}}\rangle = \sum_{\substack{A_\uparrow, A_\downarrow \subset \mathcal{E} \\ (|A_\uparrow|+|A_\downarrow|=N)}} g(A_\uparrow, A_\downarrow)\Big(\prod_{p\in A_\uparrow} \hat{a}^\dagger_{p,\uparrow}\Big)\Big(\prod_{p\in A_\downarrow} \hat{a}^\dagger_{p,\downarrow}\Big)|\Phi_{\mathrm{vac}}\rangle, \tag{11.3.14}$$

where $g(A_\uparrow, A_\downarrow) = f(A_\uparrow, A_\downarrow, \emptyset, \emptyset)$.

As we shall see in the next part, the states created by the $\hat{a}^\dagger$ operators belong to the lowest flat-band of the single-electron spectrum. That we can discuss the low energy properties of the model within the space spanned by the lowest band is of essential importance for the rest of the proof (and in the physical picture of nearly-flat-band ferromagnetism).

Repulsion in the state space: We next make use of the zero-energy condition (11.3.12), which represents the effect of local Coulomb repulsion. Note that $[\hat{c}_{p,\downarrow}\hat{c}_{p,\uparrow}, \hat{a}^\dagger_{q,\sigma}] = 0$ for any $p, q \in \mathcal{E}$ such that $p \neq q$, and that

$$\hat{c}_{p,\downarrow}\hat{c}_{p,\uparrow}\hat{a}^\dagger_{p,\uparrow}\hat{a}^\dagger_{p,\downarrow}|\Phi_{\mathrm{vac}}\rangle = |\Phi_{\mathrm{vac}}\rangle, \tag{11.3.15}$$

for any $p \in \mathcal{E}$. These relations imply that the condition (11.3.12) with $x = p \in \mathcal{E}$ cannot be valid if the ground state (11.3.14) contains $\hat{a}^\dagger_{p,\uparrow}\hat{a}^\dagger_{p,\downarrow}$. This means that one has $g(A_\uparrow, A_\downarrow) = 0$ whenever $A_\uparrow \cap A_\downarrow \neq \emptyset$. We have thus found, not too surprisingly, that repulsion in the real space leads to repulsion in the state space (described by $\hat{a}^\dagger$ operators). Of course this is not the only effect of the repulsive interaction, as we shall see below.

---

[17] This is in fact not entirely obvious because $\boldsymbol{\beta}_u$ with $u \in \mathcal{I}$ are not mutually orthogonal. Let us give a careful proof. Define the Gramm matrix G by $(\mathsf{G})_{u,v} = \langle \boldsymbol{\beta}_u, \boldsymbol{\beta}_v\rangle = \{\hat{b}_{u,\sigma}, \hat{b}^\dagger_{v,\sigma}\}$ for $u, v \in \mathcal{I}$. The linear independence of $\{\boldsymbol{\beta}_u\}_{u\in\mathcal{I}}$ implies that G is invertible. We define the dual operators by $\hat{b}'_{u,\sigma} = \sum_{v\in\mathcal{I}} (\mathsf{G}^{-1})_{u,v}\hat{b}_{v,\sigma}$, which clearly satisfy $\{\hat{b}'_{u,\sigma}, \hat{b}^\dagger_{v,\tau}\} = \delta_{u,v}\delta_{\sigma,\tau}$. We also have $\{\hat{b}'_{u,\sigma}, \hat{a}^\dagger_{p,\tau}\} = 0$ for any $u \in \mathcal{I}$ and $p \in \mathcal{E}$. Since (11.3.11) implies $\hat{b}'_{u,\sigma}|\Phi_{\mathrm{GS}}\rangle = 0$ for any $u \in \mathcal{I}$ and $\sigma =\uparrow, \downarrow$, we get the desired property.

Since $|A_\uparrow| + |A_\downarrow| = N = |\mathscr{E}|$, we find that $A_\uparrow \cap A_\downarrow = \emptyset$ implies $A_\uparrow \cup A_\downarrow = \mathscr{E}$. We therefore see that the ground state (11.3.14) is further rewritten as

$$|\Phi_{\text{GS}}\rangle = \sum_{\boldsymbol{\sigma}} C(\boldsymbol{\sigma})\Big(\prod_{p\in\mathscr{E}} \hat{a}^\dagger_{p,\sigma_p}\Big)|\Phi_{\text{vac}}\rangle, \tag{11.3.16}$$

where $\boldsymbol{\sigma} = (\sigma_p)_{p\in\mathscr{E}}$ (with $\sigma_p =\uparrow, \downarrow$) is summed over all possible spin configurations on $\mathscr{E}$, and $C(\boldsymbol{\sigma})$ is a coefficient determined from $g(A_\uparrow, A_\downarrow)$. We have fixed an arbitrary ordering of the elements of $\mathscr{E}$, and assume that any product respects the ordering. The expression (11.3.16) may be called the spin system representation.

Ferromagnetic exchange interaction: Let us examine the implication of the zero-energy condition (11.3.12) when x is in $\mathscr{I}$. Take $p, q \in \mathscr{E}$ separated by distance 1, and let $u \in \mathscr{I}$ be the unique site in between them. Then we see from the definition (11.3.3) that

$$\hat{c}_{u,\downarrow}\hat{c}_{u,\uparrow}\hat{a}^\dagger_{p,\sigma}\hat{a}^\dagger_{q,\tau}|\Phi_{\text{vac}}\rangle = \begin{cases} |\Phi_{\text{vac}}\rangle & \text{if } (\sigma,\tau) = (\uparrow,\downarrow), \\ -|\Phi_{\text{vac}}\rangle & \text{if } (\sigma,\tau) = (\downarrow,\uparrow), \\ 0 & \text{if } (\sigma,\tau) = (\uparrow,\uparrow), (\downarrow,\downarrow). \end{cases} \tag{11.3.17}$$

Note that the sign changes when the spins are exchanged because of the fermionic anticommuation relation.

For the same p, q, let us rewrite the spin system representation (11.3.16) as

$$|\Phi_{\text{GS}}\rangle = \sum_{\boldsymbol{\sigma}} \operatorname{sgn}(p,q)\, C(\boldsymbol{\sigma})\, \hat{a}^\dagger_{p,\sigma_p}\hat{a}^\dagger_{q,\sigma_q}\Big(\prod_{p'\in\mathscr{E}\backslash\{p,q\}} \hat{a}^\dagger_{p',\sigma_{p'}}\Big)|\Phi_{\text{vac}}\rangle, \tag{11.3.18}$$

where $\operatorname{sgn}(p,q) = \pm 1$ is a fermionic sign factor which depends only on p and q. Noting that $\hat{c}_{u,\downarrow}\hat{c}_{u,\uparrow}$ commutes with $(\prod_{p'\in\mathscr{E}\backslash\{p,q\}} \hat{a}^\dagger_{p',\sigma_{p'}})$, we see from (11.3.17) that

$$\begin{aligned} &\hat{c}_{u,\downarrow}\hat{c}_{u,\uparrow}|\Phi_{\text{GS}}\rangle \\ &\quad= \sum_{\boldsymbol{\sigma}'} \operatorname{sgn}(p,q)\,\{C\big((\uparrow,\downarrow,\boldsymbol{\sigma}')\big) - C\big((\downarrow,\uparrow,\boldsymbol{\sigma}')\big)\}\Big(\prod_{p'\in\mathscr{E}\backslash\{p,q\}} \hat{a}^\dagger_{p',\sigma_{p'}}\Big)|\Phi_{\text{vac}}\rangle. \end{aligned} \tag{11.3.19}$$

We here wrote $\boldsymbol{\sigma}$ as $(\sigma_p, \sigma_q, \boldsymbol{\sigma}')$, where $\boldsymbol{\sigma}'$ is the spin configuration on $\mathscr{E}\backslash\{p,q\}$. Since the states in the sum are linearly independent, we see that the zero-energy condition $\hat{c}_{u,\downarrow}\hat{c}_{u,\uparrow}|\Phi_{\text{GS}}\rangle = 0$ implies $C\big((\uparrow,\downarrow,\boldsymbol{\sigma}')\big) = C\big((\downarrow,\uparrow,\boldsymbol{\sigma}')\big)$ for any $\boldsymbol{\sigma}'$. In other words, we have found

$$C(\boldsymbol{\sigma}) = C(\boldsymbol{\sigma}_{p\leftrightarrow q}), \tag{11.3.20}$$

for any spin configuration $\boldsymbol{\sigma}$. Here $\boldsymbol{\sigma}_{p\leftrightarrow q}$ is the new spin configuration obtained from $\boldsymbol{\sigma}$ by swapping σ_p and σ_q. This is the exchange interaction that generates the ferromagnetism in the present model. Note that it is crucial here that the two $\boldsymbol{\alpha}$ states on the neighboring $\mathscr{E}$ sites share a common site in between them. See Fig. 11.12. We

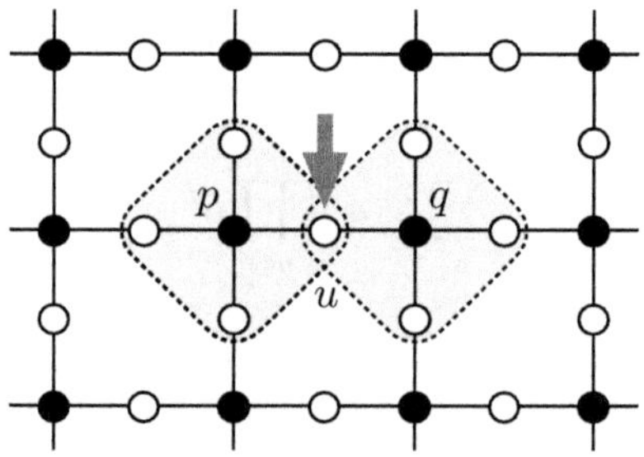

Fig. 11.12 A schematic picture of the mechanism which generates the exchange interaction (11.3.20). The two states $\boldsymbol{\alpha}_p$ and $\boldsymbol{\alpha}_q$ on the adjacent sites $p, q \in \mathscr{E}$ shares a common site $u \in \mathscr{I}$. When we take into account the effect of on-site repulsion at site u, we find that the spins localized around p and q, respectively, interact ferromagnetically. This is essentially the same as the direct exchange mechanism discussed by Heisenberg [16]

also note that this mechanism is essentially the same as Heisenberg's direct exchange interaction, which was originally discussed within a perturbative theory [16]. See also Sect. 11.4.1, in particular Fig. 11.19.

By using (11.3.20) repeatedly we see that $C(\boldsymbol{\sigma}) = C(\boldsymbol{\sigma}')$ whenever $\sum_{p\in\mathscr{E}} \sigma_p = \sum_{p\in\mathscr{E}} \sigma'_p$. This is because all the spin configurations with common $S^{(3)}_{\mathrm{tot}}$ are connected through swapping of spins on neighboring sites.[18] We thus find that the ground state within the space of fixed $S^{(3)}_{\mathrm{tot}}$ is unique. Since we already know that there is a ferromagnetic ground state (11.3.10) in this space, we see that the unique ground state has $S_{\mathrm{tot}} = S_{\mathrm{max}}$.[19]

We have thus proved that the ground state (11.3.16) is linear combination of the states (11.3.10).

Hopping amplitude, the band structure, and the role of the flat band We have represented the hopping Hamiltonian compactly as in (11.3.5) by using the $\hat{b}$ operators. By using the definition (11.3.4) of $\hat{b}_{u,\sigma}$, one can write (11.3.5) in the standard form

$$\hat{H}_{\mathrm{hop}} = \sum_{\substack{x,y\in\Lambda \\ \sigma=\uparrow,\downarrow}} t_{x,y}\, \hat{c}^\dagger_{x,\sigma} \hat{c}_{y,\sigma}, \tag{11.3.21}$$

with hopping amplitude given by

[18] This should be obvious, but see "Proof of the property (iii)" in p. 41 for a rigorous proof.

[19] By recalling (2.4.11), one can directly show that the superposition of basis states (with a common $S^{(3)}_{\mathrm{tot}}$) with equal weights leads to a ferromagnetic state.

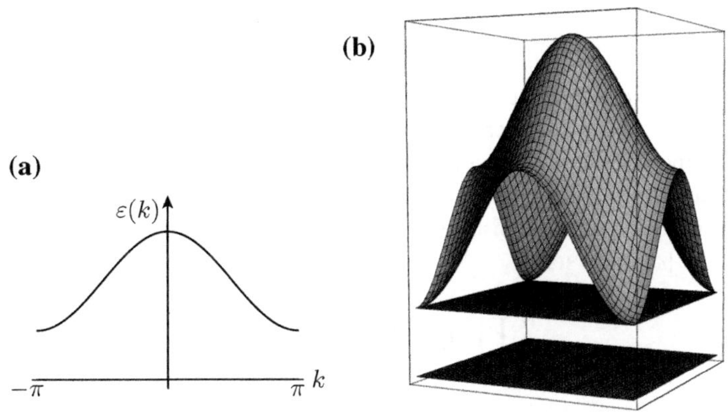

Fig. 11.13 The dispersion relations (11.3.23) of the Tasaki model on the decorated d-dimensional hypercubic lattice. **a** There are two bands in the one-dimensional model. A completely flat lower band is separated by a nonzero gap from the upper band with a cosine dispersion relation. **b** There are two flat-bands and one band with a cosine dispersion in the two-dimensional model. The lowest band is again separated by a nonzero gap from the rest of the spectrum. (The horizontal axes represent $k_1, k_2 \in [-\pi, \pi]$ and the vertical axis represents $\varepsilon(k_1, k_2)$.) (© Hal Tasaki 2020. All Rights Reserved)

$$t_{x,y} = \begin{cases} \nu t & \text{if } |x-y| = 1/2, \\ \nu^2 t & \text{if } x, y \in \mathscr{E} \text{ and } |x-y| = 1, \\ t & \text{if } x = y \in \mathscr{I}, \\ 2d\nu^2 t & \text{if } x = y \in \mathscr{E} \\ 0 & \text{otherwise.} \end{cases} \tag{11.3.22}$$

See Fig. 11.11.

One can then solve the corresponding single-electron Schrödinger equation (9.3.3) to find the band structure. It is found that the model has $d+1$ bands with dispersion relations

$$\varepsilon_\mu(k) = \begin{cases} 0 & \mu = 1, \\ t & \mu = 2, \ldots, d, \\ t + 2\nu^2 t \sum_{j=1}^d (1 + \cos k_j) & \mu = d+1, \end{cases} \tag{11.3.23}$$

where $k \in \mathscr{K}_L$ is a wave vector defined in (4.1.17). See Fig. 11.13. Rather pathologically, d bands are flat (or dispersion-less) and only the highest band has a cosine dispersion. See below for the derivation.

Clearly the lowest flat-band is most important for the present study of ferromagnetism. There are $|\mathscr{K}_L| = L^d$ states with zero energy, and are separated from other single-electron energy eigenstates by the energy gap $t > 0$. In fact the presence of the degenerate zero-energy states is obvious because the definition (11.3.5) and the anticommutation relation (11.3.7) readily imply

$$\hat{H}_{\text{hop}}\,\hat{a}^\dagger_{p,\sigma}|\Phi_{\text{vac}}\rangle = 0, \tag{11.3.24}$$

for any $p \in \mathscr{E}$. Note that the number of the zero-energy states $\hat{a}^\dagger_{p,\sigma}|\Phi_{\text{vac}}\rangle$, which is $|\mathscr{E}|$, is the same as the degeneracy of the flat-band.

The presence of the macroscopic degeneracy in the single-electron ground states leads to a pathological degeneracy in the many-electron problem without interaction. To see this, note that, in the non-interacting model with Hamiltonian $\hat{H}_{\text{hop}}$ and electron number $N = |\mathscr{E}|$, any state

$$\Bigl(\prod_{p\in A_\uparrow}\hat{a}^\dagger_{p,\uparrow}\Bigr)\Bigl(\prod_{p\in A_\downarrow}\hat{a}^\dagger_{p,\downarrow}\Bigr)|\Phi_{\text{vac}}\rangle, \tag{11.3.25}$$

which appeared in (11.3.14), with arbitrary subsets $A_\uparrow, A_\downarrow \subset \mathscr{E}$ with $|A_\uparrow| + |A_\downarrow| = N$ is an exact ground state with energy zero. The ground states are highly degenerate, and, in particular, can take any total spin S_{tot} ranging from 0 to $S_{\max} = N/2$. This is a kind of paramagnetism. Then the role of U in the interacting model is to lift the degeneracy and to select those states with maximum S_{tot} as the only ground states. This is why Theorem 11.11 is valid for any positive value of U (rather than for U larger than a certain critical value). In this sense one can say that, in the flat-band ferromagnetism, there is no true competition (see Sect. 9.1) between the hopping Hamiltonian and the interaction Hamiltonian. We shall come back to this issue in Sect. 11.4.

Recall also that we have set the electron number N identical to the number of sites in $\mathscr{E}$. This choice enabled us to write down the spin system representation (11.3.16), in which each site in $\mathscr{E}$ (i.e., metallic atom) carries a spin. In the language of single-electron band structure, the choice $N = |\mathscr{E}|$ corresponds to the half-filling of the lowest flat-band. This feature is common for other examples of flat-band ferromagnetism, including that by Mielke. Very interestingly we observed in Sect. 10.2.3 that Lieb's ferrimagnetism also takes place when the flat-band (in the middle of the spectrum) is half-filled.

We finally comment on the derivation of the dispersion relations (11.3.23) for $\mu \geq 2$, i.e., the higher bands. The derivation is easy for $d = 1$. It is convenient to set $\mathscr{E} = \{1, 2, \ldots, L\}$ and $\mathscr{I} = \{\frac{1}{2}, \frac{3}{2}, \ldots, L - \frac{1}{2}\}$. Note that the definition (11.3.4) implies the anticommutation relation

$$\{\hat{b}_{v,\tau}, \hat{b}^\dagger_{u,\sigma}\} = \begin{cases} (1 + 2\nu^2)t\,\delta_{\sigma,\tau} & \text{if } v = u, \\ \nu^2 t\,\delta_{\sigma,\tau} & \text{if } |v - u| = 1, \\ 0 & \text{otherwise.} \end{cases} \tag{11.3.26}$$

We then find for any $\varphi_u \in \mathbb{C}$ (with $u \in \mathscr{I}$) that

$$\hat{H}_{\text{hop}} \sum_{u\in\mathscr{I}} \varphi_u\,\hat{b}^\dagger_{u,\sigma}|\Phi_{\text{vac}}\rangle = t \sum_{u\in\mathscr{I}} \varphi_u\bigl\{(1 + 2\nu^2)\hat{b}^\dagger_{u,\sigma} + \nu^2(\hat{b}^\dagger_{u+1,\sigma} + \hat{b}^\dagger_{u-1,\sigma})\bigr\}|\Phi_{\text{vac}}\rangle. \tag{11.3.27}$$

This means that the Schrödinger equation $\hat{H}_{\text{hop}} \sum_{u \in \mathscr{I}} \varphi_u \, \hat{b}^\dagger_{u,\sigma} |\Phi_{\text{vac}}\rangle =$ $\varepsilon \sum_{u \in \mathscr{I}} \varphi_u \, \hat{b}^\dagger_{u,\sigma} |\Phi_{\text{vac}}\rangle$ reduces to

$$\nu^2 t(\varphi_{u+1} + \varphi_{u-1}) + (1 + 2\nu^2)t \, \varphi_u = \varepsilon \varphi_u, \tag{11.3.28}$$

which is essentially the same as the simplest tight-binding Schrödinger equation (9.3.7). One readily finds that the energy eigenvalues are given by $\varepsilon(k) = t + 2\nu^2 t(1 + \cos k)$, which is the desired dispersion relation for $\mu = 2$ in (11.3.23). The case with $d \geq 2$ may also be treated using the $\hat{b}^\dagger$ operators, but a straightforward calculation also works as we see in the following problem.

Problem 11.3.1.a Derive the dispersion relations (11.3.23) for general d by computing the effective hopping matrix given by (9.3.13) explicitly and then obtaining its eigenvalues. [solution→p.517]

The origin of the ferromagnetism Let us briefly discuss how the origin of the ferromagnetism in the present model can be understood intuitively. We shall see that at least three different physical pictures are possible.

First, as we have already discussed in the proof, our proof suggests that the "first order" effect of the Coulomb interaction at internal sites is relevant for ferromagnetic order. One may say that the spins of electrons nearly localized at sites in $\mathscr{E}$ (i.e., metallic sites) align with each other because of the direct exchange interaction. In this sense the model may be regarded as realizing Heisenberg's original picture of ferromagnetism. See also Sect. 11.4.1, in particular Fig. 11.19.

Secondly, one may view the present model as an assembly of the toy model with three sites discussed Sect. 11.1.3. One can then imagine that the ring exchange process in Fig. 11.1 is taking place in each part of the system and maintaining global ferromagnetism.

The third interpretation, which is most heuristic, is based on the band picture. The model has a flat-band at the bottom of the single-electron spectrum, which is exactly half-filled because we set $N = |\mathscr{E}|$. As we have discussed above, the model with $U = 0$ has highly degenerate ground states with all possible S_{tot}. The Stoner criterion (see Sect. 11.1.4) then suggests that the degeneracy is lifted and one gets ferromagnetism for any $U > 0$ because we have $U D_{\text{F}} = \infty$ for any nonzero U.

Cell construction Before closing this long subsection, we briefly discuss a general procedure, called the cell construction [73], to define models which have a flat-band and exhibit flat-band ferromagnetism. We note that, before Tasaki's work [67], Brandt and Giesekus [5] had proposed a class of Hubbard models which have a very similar cell structure. The Brandt-Giesekus model has interesting exact ground states with RVB (resonating valence-bond) structure, and exhibits paramagnetism. See also [68, 69].[20] In [59], Tanaka uses a similar construction to define a class of Hubbard models in which flat-band ferromagnetism is proved with a more sophisticated argument.

[20] In [68] we conjectured that a version of the Brandt-Giesekus model may exhibit superconductivity. We now believe that this is (unfortunately) not the case.

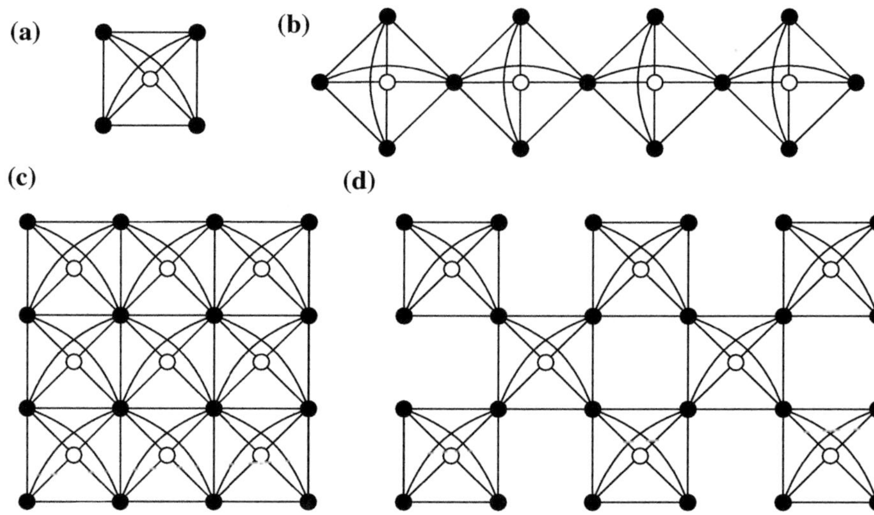

Fig. 11.14 **a** A cell with one internal site and four external sites, and **b** one dimensional and **c, d** two dimensional lattices constructed by assembling it ()

By a cell, we mean a set of site $C = \{u, p_1, \ldots, p_n\}$, where u is called the internal site, and $p_1, \ldots, p_n$ with $n \geq 2$ are called the external sites. We construct our lattice Λ by assembling a finite number of cells (which may or may not be identical), and identifying some external sites from different cells to regard them as a single site in Λ. There can be external sites which remain unidentified with other sites. We do not make such identifications for the internal sites. We assume that Λ thus constructed is connected. The lattice Λ is naturally decomposed as $\Lambda = \mathscr{E} \cup \mathscr{I}$, where $\mathscr{E}$ is the set of external sites and $\mathscr{I}$ the set of internal sites. For each $u \in \mathscr{I}$, we denote by C_u the unique cell to which u belongs.

We consider an electron system on the lattice Λ as usual, and define

$$\hat{a}_{p,\sigma} = \hat{c}_{p,\sigma} - \nu \sum_{\substack{u \in \mathscr{I} \\ (C_u \ni p)}} \hat{c}_{u,\sigma}, \tag{11.3.29}$$

for $p \in \mathscr{E}$, and

$$\hat{b}_{u,\sigma} = \hat{c}_{u,\sigma} + \nu \sum_{p \in C_u \cap \mathscr{E}} \hat{c}_{p,\sigma}, \tag{11.3.30}$$

for $u \in \mathscr{I}$, where $\nu > 0$ is a constant. Then the Hubbard Hamiltonian $\hat{H} = \hat{H}_{\text{hop}} + \hat{H}_{\text{int}}$ is defined as before by (11.3.5) and (11.3.6). Then Theorem 11.11 (p. 392) is valid as it is. The proof is not very much different, but see [73] for details.

Note that the models we have mainly studied in this subsection are constructed from cells with one internal site and two external sites. In particular the Delta chain

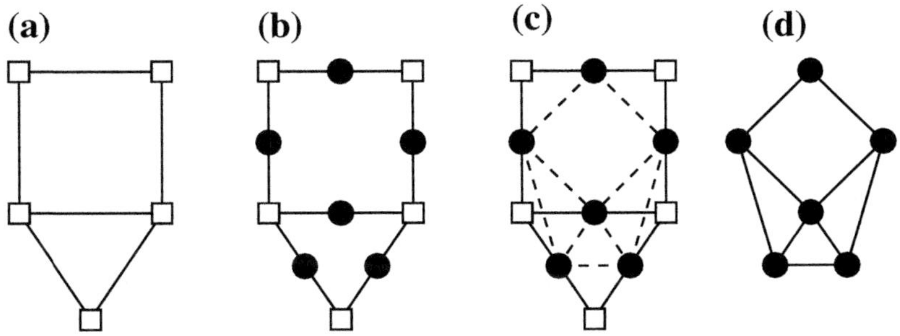

Fig. 11.15 The procedure to construct the line graph from a given connected graph. **a** The original graph $(\tilde{\Lambda}, \tilde{\mathcal{B}})$. **b** One adds a new site (drawn as a black dot) at the middle of each bond. The collection of such sites are denoted as Λ. **c** One connects two black dots by a new bond if they are directly connected to a common original site in $\tilde{\Lambda}$ (white square). **d** Leaving only the black dots and the bonds in between them, one gets the line graph $(\Lambda, \mathcal{B})$ (© Hal Tasaki 2020. All Rights Reserved)

in Fig. 11.11b obviously consists of the triangular cell 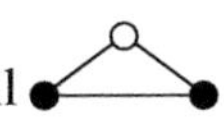. Figure 11.14 shows a cell with four external sites, and some lattices in one and two dimensions that can be constructed by assembling it.

11.3.2 Mielke's Flat-Band Ferromagnetism

We now discuss the Mielke model, which was the first example of flat-band ferromagnetism. To define the model we start by introducing the graph theoretic notion of line graph. Let $(\tilde{\Lambda}, \tilde{\mathcal{B}})$ be a general lattice (or graph), where $\tilde{\Lambda}$ denotes the set of sites (or vertices) $\alpha, \beta \ldots$, and $\tilde{\mathcal{B}}$ is the set of bonds (or edges) $\{\alpha, \beta\} = \{\beta, \alpha\}$ where $\alpha, \beta \in \tilde{\Lambda}$ and $\alpha \neq \beta$. We assume that the lattice $(\tilde{\Lambda}, \tilde{\mathcal{B}})$ is connected.[21] Starting from $(\tilde{\Lambda}, \tilde{\mathcal{B}})$, we shall construct a new lattice $(\Lambda, \mathcal{B})$, which is called the line graph of $(\tilde{\Lambda}, \tilde{\mathcal{B}})$. We take a new site at the middle of each bond $\{\alpha, \beta\} \in \tilde{\mathcal{B}}$, and let Λ be the collection of all such sites. Then we define the set of bonds $\mathcal{B}$ by declaring that $\{x, y\} \in \mathcal{B}$ if and only if $x, y \in \Lambda$, $x \neq y$, and x and y are directly connected to (at least) one common site $\alpha \in \tilde{\Lambda}$. See Fig. 11.15. As is shown in Fig. 11.16, the checkerboard lattice and the kagomé lattice[22] are the line graphs of the square lattice and the hexagonal lattice, respectively.

Let us consider a tight-biding electron system on the lattice Λ with the hopping Hamiltonian

---

[21] See footnote 28 in p. 33 for the definition of connectedness.

[22] "Kagomé" is a Japanese word that means the mesh of woven bamboo. Thus "me" is pronounced like "mesh" (without "sh").

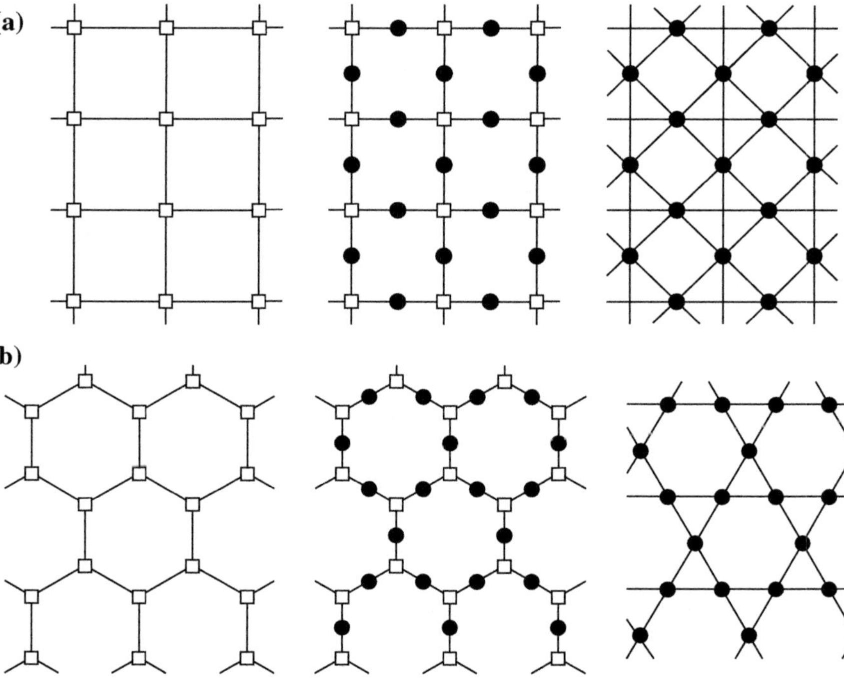

Fig. 11.16 **a** The line graph of the square lattice (left) is the checkerboard lattice (right). **b** The line graph of the hexagonal lattice (left) is the kagomé lattice (right) (© Hal Tasaki 2020. All Rights Reserved)

$$\hat{H}_{\text{hop}} = t \sum_{\substack{\{x,y\} \in \mathscr{B} \\ \sigma=\uparrow,\downarrow}} \hat{c}^\dagger_{x,\sigma} \hat{c}_{y,\sigma} + 2t \sum_{x \in \Lambda} \hat{n}_x, \tag{11.3.31}$$

with $t > 0$. The second term is added only to make the ground state energy zero. Note that there are only constant hopping between neighboring sites, and no fine tuning as in the Tasaki model (see Sect. 11.3.1) is necessary. The corresponding single-electron Schrödinger equation (9.3.3) is

$$2t\varphi(x) + t \sum_{\substack{y \in \Lambda \\ (\{x,y\} \in \mathscr{B})}} \varphi(y) = \varepsilon\, \varphi(x) \quad \text{for any } x \in \Lambda. \tag{11.3.32}$$

Let us define $D(\tilde{\Lambda}, \widetilde{\mathscr{B}}) := |\widetilde{\mathscr{B}}| - |\tilde{\Lambda}| + 1 = |\Lambda| - |\tilde{\Lambda}| + 1$ if the original lattice $(\tilde{\Lambda}, \widetilde{\mathscr{B}})$ is bipartite, and $D(\tilde{\Lambda}, \widetilde{\mathscr{B}}) := |\widetilde{\mathscr{B}}| - |\tilde{\Lambda}| = |\Lambda| - |\tilde{\Lambda}|$ if $(\tilde{\Lambda}, \widetilde{\mathscr{B}})$ is not bipartite.[23] Then the following preliminary but essential result was proved by Mielke [33].

[23] See p. 37 and Fig. 2.1 for the definition of bipartiteness.

Theorem 11.12 (Flat-band in a general line graph) *The single-electron Schrödinger equation (11.3.32) has exactly $D(\tilde{\Lambda}, \widetilde{\mathscr{B}})$ zero eigenstates with energy zero. All the other energy eigenvalues are strictly positive.*

We shall prove the theorem in Sect. 11.3.3. Thus the tight-binding electron model on a translation invariant line graph has a flat-band (or flat-bands) at the bottom of the single-electron energy spectrum. For example, the model on the d-dimensional version of the checkerboard lattice (Fig. 11.16a) with $d \geq 2$, which is the line graph of the standard d-dimensional hypercubic lattice, has d bands with the dispersion relations

$$\varepsilon_\mu(k) = \begin{cases} 0 & \mu = 1, \dots, d-1 \\ 2t \sum_{j=1}^{d}(1 + \cos k_j) & \mu = d. \end{cases} \tag{11.3.33}$$

There are $d-1$ bands with zero energy, and one band with the cosine dispersion. Note that there is no energy gap between the flat bands and the dispersive band. The gapless nature is a generic feature of the tight-binding model on the line graph of a standard bipartite lattice, as we shall show in Sect. 11.3.3. The tight-binding model on a line graph also has strictly localized single-electron energy eigenstate with zero energy. See Problem 11.3.2.b below.

We note that the dispersion relations (11.3.33) are essentially the same as those of the higher d bands of the Tasaki model show in (11.3.23). See Fig. 11.13. This is not an accident, as the following problem shows.

Problem 11.3.2.a Examine the relation between the Tasaki model in Sect. 11.3.1 and the Mielke model on the d-dimensional checkerboard lattice. Then derive the dispersion relations (11.3.33) by using the dispersion relation (11.3.23) of the Tasaki model. (Hint: Generalize the analysis that include (11.3.26), (11.3.27), and (11.3.28) for $d = 1$ to higher dimensions.) In Sect. 11.3.3, we also show that (11.3.33) is easily obtained from (9.3.9), the dispersion relation for the simplest model on the d-dimensional hypercubic lattice. [solution→p.518]

Problem 11.3.2.b Find localized zero-energy single-electron states for the checkerboard lattice and the kagomé lattice (Fig. 11.16). [solution→p.519]

We now consider the Hubbard model on Λ with the Hamiltonian $\hat{H} = \hat{H}_{\text{hop}} + \hat{H}_{\text{int}}$, where the hopping Hamiltonian is (11.3.31) and the interaction Hamiltonian is the most standard (11.3.6) or (9.3.29).

In 1991, Mielke [34, 35] proved the following general theorem, which rigorously established, for the first time, the existence of versions of the Hubbard model which exhibit saturated ferromagnetism for finite values of the interaction U. We state it without a proof.

Theorem 11.13 (Mielke's flat-band ferromagnetism) *Assume that the original lattice $(\tilde{\Lambda}, \widetilde{\mathscr{B}})$ is biconnected,*[24] *and consider the above Hubbard model with electron*

[24]A lattice (or a graph) is biconnected (or two-fold connected) if and only if one cannot make it disconnected by removing a single site.

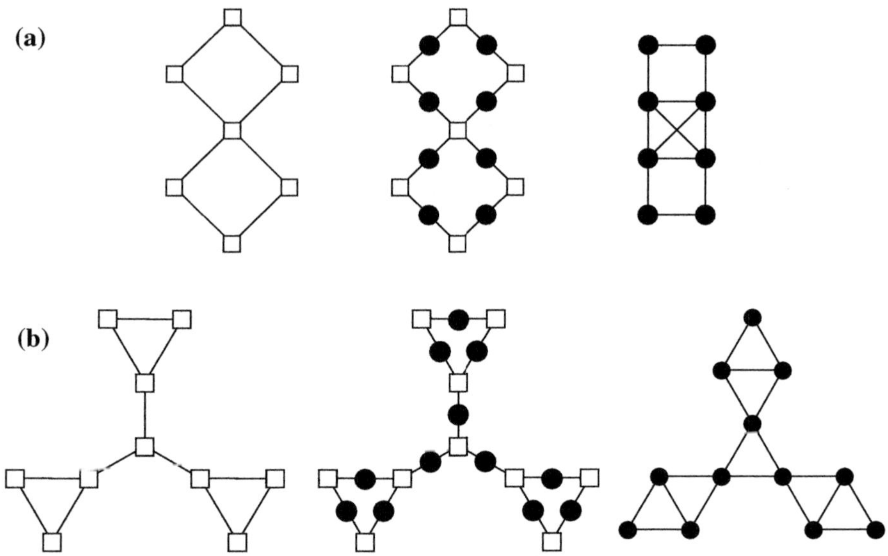

Fig. 11.17 Two lattices (whose sites are drawn as white squares) which are connected but not biconnected. Both the lattices have $D(\tilde{\Lambda}, \widetilde{\mathscr{B}}) = 2$. We consider the Hubbard model with $N = 2$ on the corresponding line graphs (whose sites are drawn as big black dots). It is found that the model on **a** does not exhibit ferromagnetism, while that on **(b)** does. See Problem 11.3.2.c. The examples are due to Akinori Tanaka (private communication)

number $N = D(\tilde{\Lambda}, \widetilde{\mathscr{B}})$. *Then, for any* $t > 0$ *and* $U > 0$, *the ground states have* $S_{\mathrm{tot}} = S_{\max} = N/2$, *and are unique apart from the trivial* $2S_{\max} + 1 = N + 1$ *fold degeneracy.*

Exactly as in the Tasaki model, ferromagnetism takes place when the lowest flat-bands (or, more precisely, the space of highly degenerate single-electron ground states) are half-filled. The condition of biconnectedness is satisfied in many lattices in two or higher dimensions including the checkerboard lattice and the kagomé lattice in Fig. 11.16. Mathematically speaking, biconnectedness of $(\tilde{\Lambda}, \widetilde{\mathscr{B}})$ is not a necessary condition for the Hubbard model on the line graph $(\Lambda, \mathscr{B})$ to exhibit ferromagnetism. See Fig. 11.17 and Problem 11.3.2.c below.

In Sect. 11.3.1, we have seen that one may use at least three different pictures to intuitively understand Tasaki's flat-band ferromagnetism. Note that the first picture based on the direct exchange interaction between almost localized electrons does not apply to the present model, where all the lattice sites are identical. The second picture based on the ring exchange process (see Fig. 11.2) applies as it is to the model on the kogomé lattice (Fig. 11.16b). To apply the picture to the model on the checkerboard lattice (Fig. 11.16a), we note that here the role of the triangle is played by the square in which all the four sites are directly connected.[25] The third picture based on the Stoner criterion needs not be modified.

Problem 11.3.2.c Prove the fact stated in the caption for Fig. 11.17. This is done by first finding two linearly independent zero energy states of the corresponding single-electron Schrödinger equation, which is easy for (a) and rather nontrivial for (b). [solution→p.519]

11.3.3 Construction of Tight-Binding Models with Flat-Bands

In this section we discuss a general method for constructing tight-binding models with flat-bands. The general method allows us to understand the origin of the flat-bands in Mielke's and Tasaki's models (and even those in Lieb's model) in a unified manner.[26] We also prove Theorem 11.12 stated above, and derive some properties about the single-electron energy eigenvalues in Mileke's model. As we noted before the present and the next subsections are somewhat advanced, and may be skipped.

In the present section, we only focus on properties of the single-electron Schrödinger equation (9.3.3) or (9.3.4) corresponding to a certain hopping matrix T. The argument in this section is more or less standard, and is essentially the same as that used by Mielke in his seminal paper [33]. We here follow a general formulation in [18].

General construction Consider two sets of sites $\tilde{\Lambda}$ and Λ, whose elements are denoted as $\alpha, \beta, \ldots \in \tilde{\Lambda}$ and $x, y, \ldots \in \Lambda$. Examples include $\tilde{\Lambda}$ and Λ that appeared in the construction of line graphs. See Figs. 11.15 and 11.16.

Consider a $|\tilde{\Lambda}| \times |\Lambda|$ matrix $\mathsf{S} = (s_{\alpha,x})_{\alpha\in\tilde{\Lambda},x\in\Lambda}$ with arbitrary entries $s_{\alpha,x} \in \mathbb{C}$. We then define hopping matrices on $\tilde{\Lambda}$ and Λ by

$$\widetilde{\mathsf{T}} := \mathsf{S}\,\mathsf{S}^\dagger \quad \text{and} \quad \mathsf{T} := \mathsf{S}^\dagger\,\mathsf{S}, \tag{11.3.34}$$

respectively. More explicitly they are written as $\widetilde{\mathsf{T}} = (\tilde{t}_{\alpha,\beta})_{\alpha,\beta\in\tilde{\Lambda}}$ and $\mathsf{T} = (t_{x,y})_{x,y\in\Lambda}$ with

$$\tilde{t}_{\alpha,\beta} = \sum_{x\in\Lambda} s_{\alpha,x}\,(s_{\beta,x})^* \quad \text{and} \quad t_{x,y} = \sum_{\alpha\in\tilde{\Lambda}} (s_{\alpha,x})^*\,s_{\alpha,y}. \tag{11.3.35}$$

Clearly both $\widetilde{\mathsf{T}}$ and T are nonnegative matrices. We further have the following crucial property.

Lemma 11.14 *Hopping matrices $\widetilde{\mathsf{T}}$ and T have exactly identical positive eigenvalues with identical multiplicities.*

---

[25] A lattice in which all the sites are connected is called a complete graph.

[26] We note that this is by no means the unique general method to construct tight-binding models with flat-bands.

Proof Assume that $\mathsf{T}\boldsymbol{\varphi} = \lambda\boldsymbol{\varphi}$ with $\lambda > 0$, where $\boldsymbol{\varphi} = (\varphi(x))_{x\in\Lambda}$ is a nonzero $|\Lambda|$-dimensional vector (or a wave function on Λ). We define $\widetilde{\boldsymbol{\varphi}} = \mathsf{S}\boldsymbol{\varphi}$, where $\widetilde{\boldsymbol{\varphi}} = (\widetilde{\varphi}(\alpha))_{\alpha\in\tilde{\Lambda}}$ is a $|\tilde{\Lambda}|$-dimensional vector (or a wave function on $\tilde{\Lambda}$). We first note that $\langle\widetilde{\boldsymbol{\varphi}}, \widetilde{\boldsymbol{\varphi}}\rangle = \langle\boldsymbol{\varphi}, \mathsf{S}^\dagger\,\mathsf{S}\boldsymbol{\varphi}\rangle = \langle\boldsymbol{\varphi}, \mathsf{T}\boldsymbol{\varphi}\rangle = \lambda\langle\boldsymbol{\varphi}, \boldsymbol{\varphi}\rangle > 0$, and hence $\widetilde{\boldsymbol{\varphi}}$ is nonzero. We then observe that $\widetilde{\mathsf{T}}\widetilde{\boldsymbol{\varphi}} = \mathsf{S}\,\mathsf{S}^\dagger\,\mathsf{S}\boldsymbol{\varphi} = \mathsf{S}\,\mathsf{T}\boldsymbol{\varphi} = \lambda\mathsf{S}\boldsymbol{\varphi} = \lambda\widetilde{\boldsymbol{\varphi}}$, and hence $\widetilde{\boldsymbol{\varphi}}$ is an eigenvector of $\widetilde{\mathsf{T}}$ with eigenvalue λ. Finally suppose that there is a nonzero vector $\boldsymbol{\varphi}'$ orthogonal to $\boldsymbol{\varphi}$ which also satisfies $\mathsf{T}\boldsymbol{\varphi}' = \lambda\boldsymbol{\varphi}'$ with the same λ, and define $\widetilde{\boldsymbol{\varphi}}' = \mathsf{S}\boldsymbol{\varphi}'$. We then see that $\widetilde{\boldsymbol{\varphi}}$ and $\widetilde{\boldsymbol{\varphi}}'$ are orthogonal because $\langle\widetilde{\boldsymbol{\varphi}}', \widetilde{\boldsymbol{\varphi}}\rangle = \langle\boldsymbol{\varphi}', \mathsf{S}^\dagger\,\mathsf{S}\boldsymbol{\varphi}\rangle = \lambda\langle\boldsymbol{\varphi}', \boldsymbol{\varphi}\rangle = 0$. By repeating the same argument switching the roles of $\tilde{\Lambda}$ and Λ, we get the desired statement. ∎

Let us assume $|\tilde{\Lambda}| < |\Lambda|$, and see how flat bands emerge. We first note that the number of independent eigenvectors of $\widetilde{\mathsf{T}}$ with positive eigenvalues cannot exceed the dimension $|\tilde{\Lambda}|$. We thus find from Lemma 11.14 that the hopping matrix T must have at least $|\Lambda| - |\tilde{\Lambda}|$ independent eigenvectors with eigenvalue zero. We conclude that the tight-binding Schrödinger equation with the hopping matrix T has flat bands provided that $|\Lambda| - |\tilde{\Lambda}|$ is proportional to the system size (and the system has translation invariance).

Relation to Lieb's flat-band models Consider the composite lattice $\Lambda^{\mathrm{tot}} = \Lambda \cup \tilde{\Lambda}$. (Examples are found in Fig. 11.15b, the figures at the middle of Figs. 11.16, and 11.18.) With the matrix S given above, we define the hopping matrix $\mathsf{T}^{\mathrm{tot}} = (t^{\mathrm{tot}}_{u,v})_{u,v\in\Lambda^{\mathrm{tot}}}$ as

$$t^{\mathrm{tot}}_{u,v} = \begin{cases} s_{u,v} & \text{if } u \in \tilde{\Lambda},\, v \in \Lambda, \\ (s_{v,u})^* & \text{if } u \in \Lambda,\, v \in \tilde{\Lambda}, \\ 0 & \text{otherwise}, \end{cases} \tag{11.3.36}$$

or, in the matrix form, as

$$\mathsf{T}^{\mathrm{tot}} = \begin{pmatrix} \mathsf{0} & \mathsf{S}^\dagger \\ \mathsf{S} & \mathsf{0} \end{pmatrix}. \tag{11.3.37}$$

The matrix $\mathsf{T}^{\mathrm{tot}}$ describes bipartite hopping amplitude on the lattice Λ^{tot}. Note that the situation is similar to that in Sect. 10.2.3, where we discussed Lieb's ferrimagnetism.

Let $\boldsymbol{\varphi} = (\varphi(x))_{x\in\Lambda}$ be such that $\mathsf{T}\boldsymbol{\varphi} = \mathbf{0}$. Then it also holds (from Lemma A.11 in p. 469) that $\mathsf{S}\boldsymbol{\varphi} = \mathbf{0}$, and hence $\mathsf{T}^{\mathrm{tot}}(\boldsymbol{\varphi} \oplus \mathbf{0}) = \mathbf{0}$. We thus see that a zero-energy eigenstate of the hopping matrix T on Λ always corresponds to a zero-energy eigenstate of $\mathsf{T}^{\mathrm{tot}}$ on Λ^{tot}. In other words the flat-band (at the bottom of the spectrum) for T correspond to a flat-band (in the middle of the spectrum) for $\mathsf{T}^{\mathrm{tot}}$. Recall that the latter played a central role in Sect. 10.2.3. See also Proposition 10.7 (p. 356).

Mielke's flat-band model We now apply the present general formalism to Mielke's model on line graphs. As in Sect. 11.3.2, let $(\tilde{\Lambda}, \tilde{\mathscr{B}})$ be an arbitrary connected lattice, and $(\Lambda, \mathscr{B})$ be the corresponding line graph. We define the matrix $\mathsf{S} = (s_{\alpha,x})_{\alpha\in\tilde{\Lambda},x\in\Lambda}$ by

$$s_{\alpha,x} = \begin{cases} \sqrt{t} & \text{if } \alpha \in \tilde{\Lambda} \text{ and } x \in \Lambda \text{ are directly connected,} \\ 0 & \text{otherwise,} \end{cases} \tag{11.3.38}$$

with $t > 0$. Then we immediately see from (11.3.35) that

$$t_{x,y} = \begin{cases} 2t & \text{if } x = y, \\ t & \text{if } \{x, y\} \in \mathscr{B}, \\ 0 & \text{otherwise,} \end{cases} \tag{11.3.39}$$

which precisely corresponds to the hopping Hamiltonian (11.3.31) of the Mielke model.

It is also useful to consider the hopping matrix $\widetilde{\mathsf{T}} = (\tilde{t}_{\alpha,\beta})_{\alpha,\beta\in\tilde{\Lambda}}$ on the original lattice $\tilde{\Lambda}$. Again from (11.3.35), we find

$$\tilde{t}_{\alpha,\beta} = \begin{cases} z(\alpha)\, t & \text{if } \alpha = \beta, \\ t & \text{if } \{\alpha, \beta\} \in \widetilde{\mathscr{B}}, \\ 0 & \text{otherwise,} \end{cases} \tag{11.3.40}$$

where $z(\alpha)$ is the number of $\beta \in \tilde{\Lambda}$ such that $\{\alpha, \beta\} \in \widetilde{\mathscr{B}}$. Let $\widetilde{\boldsymbol{\varphi}} = (\widetilde{\varphi}(\alpha))_{\alpha\in\tilde{\Lambda}}$ be a normalized eigenstate of $\widetilde{\mathsf{T}}$ with eigenvalue λ. Since $\tilde{t}_{\alpha,\beta} \in \mathbb{R}$, we can assume $\widetilde{\varphi}(\alpha) \in \mathbb{R}$. Then we see from (11.3.40) that

$$\lambda = \langle \widetilde{\boldsymbol{\varphi}}, \widetilde{\mathsf{T}}\widetilde{\boldsymbol{\varphi}} \rangle = \sum_{\alpha,\beta\in\tilde{\Lambda}} \widetilde{\varphi}(\alpha)\, \tilde{t}_{\alpha,\beta} \widetilde{\varphi}(\beta) = t \sum_{\{\alpha,\beta\}\in\widetilde{\mathscr{B}}} (\widetilde{\varphi}(\alpha) + \widetilde{\varphi}(\beta))^2. \tag{11.3.41}$$

See (2.4.14) and Problem 2.4.d (p. 35) for essentially the same relation. Let us examine whether $\widetilde{\mathsf{T}}$ has zero as an eigenvalue. By setting $\lambda = 0$ in (11.3.41), we see that $\widetilde{\varphi}(\alpha) = -\widetilde{\varphi}(\beta)$ for any $\{\alpha, \beta\} \in \widetilde{\mathscr{B}}$. Since $(\tilde{\Lambda}, \widetilde{\mathscr{B}})$ is connected, this condition is satisfied by a unique $\widetilde{\boldsymbol{\varphi}}$ only when $(\tilde{\Lambda}, \widetilde{\mathscr{B}})$ is bipartite. We thus find that $\widetilde{\mathsf{T}}$ has exactly one eigenstate with eigenvalue zero when $(\tilde{\Lambda}, \widetilde{\mathscr{B}})$ is bipartite, and does not have zero as its eigenvalue when $(\tilde{\Lambda}, \widetilde{\mathscr{B}})$ is not bipartite. Equivalently, the number of independent eigenstates with positive eigenvalues of $\widetilde{\mathsf{T}}$ is $|\tilde{\Lambda}| - 1$ when $(\tilde{\Lambda}, \widetilde{\mathscr{B}})$ is bipartite, and is $|\tilde{\Lambda}|$ when $(\tilde{\Lambda}, \widetilde{\mathscr{B}})$ is not bipartite. This observation, with Lemma 11.14, proves Theorem 11.12 about the number of zero-energy single-electron states in Mielke's flat-band model, whose hopping matrix is T.

From Lemma 11.14, we further see that nonzero eigenvalues of T are exactly the same as nonzero eigenvalues of $\widetilde{\mathsf{T}}$, which are in general much easier to study. For example let $(\tilde{\Lambda}, \widetilde{\mathscr{B}})$ be the d-dimensional hypercubic lattice with nearest neighbor bonds. Then $(\Lambda, \mathscr{B})$ is the d-dimensional checkerboard lattice in Fig. 11.16a. Then, instead of solving the Schrödinger equation on the checkerboard lattice, which is rather complicated, one only needs to solve the equation for the simple hypercubic lattice. This is easy, and is indeed already done in (9.3.9). We thus readily obtain the dispersion relation for $\varepsilon_d(k)$ in (11.3.33), by switching the sign of t and shifting the

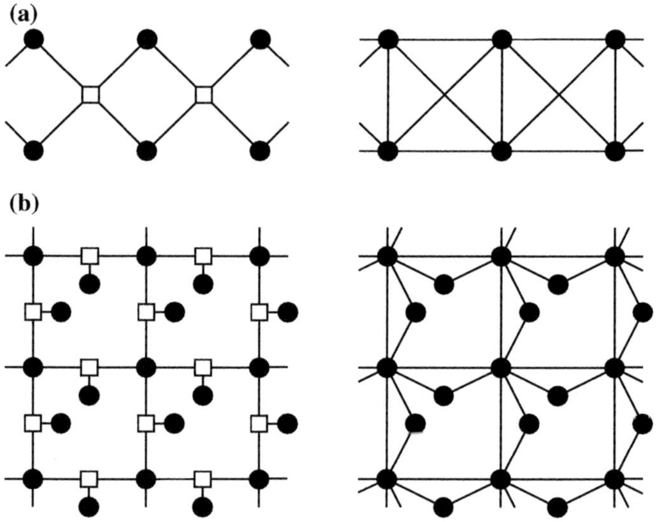

Fig. 11.18 On the left, we show the lattice $\Lambda \cup \tilde{\Lambda}$, where white squares denote sites in $\tilde{\Lambda}$ and black dots denote sites in Λ. The lines correspond to nonvanishing $s_{\alpha,x}$. On the right we show the resulting lattice $(\Lambda, \mathscr{B})$, on which we have an electron model with flat-bands (or, more precisely, highly degenerate single-electron ground states). **a** The diamond chain $\Lambda \cup \tilde{\Lambda}$ reduces to the crossed ladder. **b** Starting from a decorated square lattice with dangling bonds, one gets the decorated square lattice for the Tasaki model in Fig. 11.11c (© Hal Tasaki 2020. All Rights Reserved)

energy by $2dt$ in (9.3.9). Likewise the band structure of the model on the kagomé lattice can be determined from the band structure of the tight-binding model on the hexagonal lattice.

Let us finally comment on the gapless nature of the single-electron spectrum in the Mielke model. Assume that $(\tilde{\Lambda}, \tilde{\mathscr{B}})$ is a general connected bipartite lattice. Then one can transform the hopping matrix $\tilde{\mathsf{T}}$ as $\mathsf{U}\tilde{\mathsf{T}}\mathsf{U} = -t\boldsymbol{\Delta}$, where $\boldsymbol{\Delta}$ is the lattice Laplacian defined in (2.4.13). Here U is the unitary matrix for gauge transformation defined by $(\mathsf{U})_{\alpha,\beta} = (-1)^{\alpha}\,\delta_{\alpha,\beta}$, where the sign factor $(-1)^{\alpha} = \pm 1$ distinguishes the two sublattices (see, e.g., Sect. 9.3.3). It is known that, for most translation invariant lattices with periodic boundary conditions, the Laplacian has an almost continuous band of nonpositive eigenvalues including zero. This means that the hopping matrix T has a continuum of positive eigenvalues just above the highly degenerate eigenvalue zero. One also sees that when $(\tilde{\Lambda}, \tilde{\mathscr{B}})$ is a non-bipartite lattice like the triangular lattice, then there is a nonzero gap above the eigenvalue zero in the spectrum of the corresponding T on the line graph.

Other models The above construction certainly works for models not related to line graphs. If one starts from the lattice $\Lambda \cup \tilde{\Lambda}$ as in the left of Fig. 11.18a, then one obtains the crossed ladder as in the right of Fig. 11.18a. This corresponds to the flat-band model studied in [60], where, interestingly, the existence of saturated ferromagnetism is proved when there are one more electrons than the half-filling of the flat-band.

Consider the lattice in the left of Fig. 11.18b, which has dangling bonds. The resulting lattice $(\Lambda, \mathscr{B})$ depicted in the right of Fig. 11.18b is nothing but the decorated square lattice of the Tasaki model. See Fig. 11.11c. Furthermore by setting $s_{\alpha,x} = \sqrt{t}$ for dangling bonds $\{\alpha, x\}$, and $s_{\alpha,x} = \sqrt{t}\,\nu$ for other bonds, the hopping matrix T precisely recovers the hopping amplitude (11.3.22) of the Tasaki model. (This construction appear in [18].)

11.3.4 General Theory of Flat-Band Ferromagnetism

In the present section we discuss the general theory of flat-band ferromagnetism developed by Mielke in 1993 [36, 38]. We treat a general Hubbard model with a flat lowest band (or, more precisely, degenerate single-electron ground states), and prove Theorem 11.15, which states a compact necessary and sufficient condition for the emergence of saturated ferromagnetism when the number of electrons is equal to the degree of degeneracy. Although it is usually not necessary to resort to the general condition when working on concrete models, it is important to see that there is a clear necessary and sufficient condition. In the proof we shall see another important necessary and sufficient condition stated in Theorem 11.17, which is also due to Mielke.

Setting and main theorem Let Λ be a finite lattice. As before we denote by $\mathfrak{h} = \{\boldsymbol{\varphi} = (\varphi(x))_{x\in\Lambda} \mid \varphi(x) \in \mathbb{C}\}$ the single-electron Hilbert space for Λ. We consider a general hopping matrix $\mathsf{T} = (t_{x,y})_{x,y\in\Lambda}$ with complex hopping amplitude $t_{x,y} \in \mathbb{C}$ such that $\mathsf{T}^\dagger = \mathsf{T}$ (i.e., $(t_{x,y})^* = t_{y,x}$ for any $x, y, \in \Lambda$) and $\mathsf{T} \geq 0$. Let $\mathfrak{h}_0 := \ker \mathsf{T}$ be the subspace which consists of $\boldsymbol{\varphi} \in \mathfrak{h}$ such that $\mathsf{T}\boldsymbol{\varphi} = \mathbf{0}$. We assume that $\mathfrak{h}_0$ is not empty, and write $D_0 := \dim \mathfrak{h}_0$. We denote by P_0 the orthogonal projection matrix onto the space $\mathfrak{h}_0$, and define $\Lambda_0 := \{x \in \Lambda \mid (\mathsf{P}_0)_{x,x} \neq 0\}$.

We consider a Hubbard model on Λ with the standard Hamiltonian $\hat{H} = \hat{H}_{\mathrm{hop}} + \hat{H}_{\mathrm{int}}$, where $\hat{H}_{\mathrm{hop}}$ is the hopping Hamiltonian (9.3.17) corresponding to the hopping matrix T, and $\hat{H}_{\mathrm{int}}$ is the interaction Hamiltonian (11.3.6) or (9.3.29) with arbitrary $U > 0$. As in the Mielke model and the Tasaki model, we set the electron number as $N = D_0$, which corresponds to the exact half-filling of the flat-band (or, more precisely, the space of degenerate single-electron ground states). The following theorem of Mieleke's [36, 38] is the main result of the present section.

Theorem 11.15 *The above Hubbard model exhibits ferromagnetism, i.e., has $N+1$ fold degenerate ground states with $S_{\mathrm{tot}} = N/2$, if and only if the $|\Lambda_0| \times |\Lambda_0|$ matrix $((\mathsf{P}_0)_{x,y})_{x,y\in\Lambda_0}$ is irreducible.*

To be precise we say that the matrix $((\mathsf{P}_0)_{x,y})_{x,y\in\Lambda_0}$ is reducible if and only if Λ_0 is decomposed as $\Lambda_0 = \Lambda_1 \cup \Lambda_2$ with $\Lambda_1 \cap \Lambda_2 = \emptyset$, $\Lambda_1 \neq \emptyset$, and $\Lambda_2 \neq \emptyset$, and it holds that $(\mathsf{P}_0)_{x,y} = 0$ for any $x \in \Lambda_1$ and $y \in \Lambda_2$. We say $((\mathsf{P}_0)_{x,y})_{x,y\in\Lambda_0}$ is irreducible when it is not reducible.

The theorem provides us with a simple and elegant criterion, which is based only on the subspace $\mathfrak{h}_0$, for the emergence of flat-band ferromagnetism. However, as we have mentioned already, it rarely happens that one really needs to invoke this condition to establish the existence of ferromagnetism. An example in which the above condition is applied can be found in [19]. Needless to say the existence of a flat-band does not necessarily imply ferromagnetism. Mielke constructed flat-band models with reducible $((\mathsf{P}_0)_{x,y})_{x,y\in\Lambda_0}$, which do not exhibit ferromagnetism [39].

One can represent the matrix element $(\mathsf{P}_0)_{x,y}$ by using an off-diagonal correlation function; it is indeed the way Mielke first formulated the theorem. Let $|\Phi_{\text{ferro GS}}\rangle$ be the unique lowest-energy state within the states consisting only of N up-spin electrons. Then one has

$$(\mathsf{P}_0)_{x,y} = \langle\Phi_{\text{ferro GS}}|\hat{c}^\dagger_{y,\uparrow}\hat{c}_{x,\uparrow}|\Phi_{\text{ferro GS}}\rangle. \tag{11.3.42}$$

To see this let $\{\boldsymbol{\psi}_j\}_{j=1,\ldots,D_0}$ be an arbitrary orthonormal basis of the zero-energy space $\mathfrak{h}_0$. We then have

$$|\Phi_{\text{ferro GS}}\rangle = \Bigl(\prod_{j=1}^{D_0}\hat{C}^\dagger_\uparrow(\boldsymbol{\psi}_j)\Bigr)|\Phi_{\text{vac}}\rangle. \tag{11.3.43}$$

This, with the anticommutation relation (9.2.64), implies

$$\langle\Phi_{\text{ferro GS}}|\hat{c}^\dagger_{y,\uparrow}\hat{c}_{x,\uparrow}|\Phi_{\text{ferro GS}}\rangle = \sum_{j=1}^{D_0}\psi_j(x)\{\psi_j(y)\}^*, \tag{11.3.44}$$

where the right-hand side is nothing but $(\mathsf{P}_0)_{x,y}$.

Proof of the Main Theorem We shall give a complete and detailed Proof of Theorem 11.15.[27] In Theorem 11.17 below, we state another necessary and sufficient condition for flat-band ferromagnetism.

We start by proving the following lemma due to Mielke [36, 38], which we believe to be of essential importance in the study of flat-band systems.[28]

Lemma 11.16 *Assume that the hopping matrix* T *satisfies* $\mathsf{T}^\dagger = \mathsf{T}$ *and* $\mathsf{T} \geq 0$, *and let* $\mathfrak{h}_0 = \ker\mathsf{T}$ *and* $D_0 = \dim\mathfrak{h}_0$. *Then one can take a subset* $I \subset \Lambda$ *with* $|I| = D_0$ *and a basis* $\{\boldsymbol{\mu}_z\}_{z\in I}$ *of* $\mathfrak{h}_0$. *Here, for each* $z \in I$, *the basis state* $\boldsymbol{\mu}_z = (\mu_z(x))_{x\in\Lambda}$ *satisfies* $\mu_z(z) \neq 0$ *and* $\mu_z(z') = 0$ *for any* $z' \in I\backslash\{z\}$.

This means that, among the basis states, $\boldsymbol{\mu}_z$ is characterized as the only one with a nonzero component on the index site z. Recall that the Tasaki model (Sect. 11.3.1) is

[27] A large part of the present proof is due to Akinori Tanaka (private communication).

[28] Unfortunately the lemma, as far as we know, has not been applied much to concrete problems. It is possible that the lemma will be useful when one studies totally different aspects of flat-band systems.

designed so that the basis $\{\boldsymbol{\alpha}_p\}_{p\in\mathcal{E}}$ in (11.3.1) or (11.3.29) precisely has this property. An essential point of the above Lemma is that such a basis can always be taken in a flat-band system. We note that the basis states $\boldsymbol{\mu}_z$ may not be local in general.

Proof The present proof is a nice example which shows that elementary linear algebra sometimes enables us to prove nontrivial (and useful) results quite easily. Let A be a general square matrix with rank r. It is known that there exists an $r \times r$ submatrix of A with nonzero determinant, and that any $(r+1) \times (r+1)$ submatrix of A has determinant zero.[29] Since rank $\mathsf{T} = |\Lambda| - D_0$ by assumption, there exists a subset $\Lambda' \subset \Lambda$ with $|\Lambda'| = |\Lambda| - D_0$ such that the submatrix $(t_{x,y})_{x,y\in\Lambda'}$ has nonzero determinant. We set $I = \Lambda\backslash\Lambda'$. Take any $z \in I$. Then the $(|\Lambda'|+1) \times (|\Lambda'|+1)$ matrix $(t_{x,y})_{x,y\in\Lambda'\cup\{z\}}$ has determinant zero, and hence has zero as an eigenvalue. Let $\tilde{\boldsymbol{\mu}}_z = (\tilde{\mu}_z(x))_{x\in\Lambda'\cup\{z\}}$ be the corresponding eigenvector. We note that $\tilde{\mu}_z(z) \neq 0$ since, otherwise, $(t_{x,y})_{x,y\in\Lambda'}$ would have zero as an eigenvalue. Let us define a $|\Lambda|$-dimensional vector $\boldsymbol{\mu}_z = (\mu_z(x))_{x\in\Lambda}$ by

$$\mu_z(x) = \begin{cases} \tilde{\mu}_z(x) & \text{if } x \in \Lambda' \cup \{z\}, \\ 0 & \text{otherwise.} \end{cases} \tag{11.3.45}$$

We then have $\langle \boldsymbol{\mu}_z, \mathsf{T}\boldsymbol{\mu}_z \rangle = \sum_{x,y\in\Lambda'\cup\{z\}} (\tilde{\mu}_z(x))^* t_{x,y}\, \tilde{\mu}_z(y) = 0$. Since $\mathsf{T} \geq 0$, we find from the variational principle that $\mathsf{T}\boldsymbol{\mu}_z = \mathbf{0}$. Since $\boldsymbol{\mu}_z$ with $z \in I$ are obviously linearly independent, $\{\boldsymbol{\mu}_z\}_{z\in I}$ is a basis of $\mathfrak{h}_0$. ∎

For any $z, z' \in I$, we write $\boldsymbol{\mu}_z \sim \boldsymbol{\mu}_{z'}$ (and say that $\boldsymbol{\mu}_z$ and $\boldsymbol{\mu}_{z'}$ are directly connected) if there is $x \in \Lambda$ such that $\mu_z(x) \neq 0$ and $\mu_{z'}(x) \neq 0$. The whole basis $\{\boldsymbol{\mu}_z\}_{z\in I}$ is said to be connected if, for any $z, z' \in I$, there is a sequence $z_0, \ldots, z_n$ such that $z_0 = z$, $z_n = z'$, and $\boldsymbol{\mu}_{z_{j-1}} \sim \boldsymbol{\mu}_{z_j}$ for all $j = 1, \ldots, n$. We then have the following necessary and sufficient condition for flat-band ferromagnetism, which is also due to Mielke.

Theorem 11.17 *The above Hubbard model exhibits ferromagnetism, i.e., has $N+1$ fold degenerate ground states with $S_{\text{tot}} = N/2$, if and only if the basis $\{\boldsymbol{\mu}_z\}_{z\in I}$ is connected.*

Note that the theorem also implies that whether a basis (characterized by Lemma 11.16) is connected or not does not depend on the choice of the index set I or the basis. It is a property determined solely by the structure of the zero-energy state $\mathfrak{h}_0$.

Proof The proof is essentially the same as that of Theorem 11.11 (p. 392) for the Tasaki model. We shall not repeat the detailed argument, and shall be sketchy. For each $z \in I$, we define the fermion operator $\hat{a}^\dagger_{z,\sigma} = \hat{C}^\dagger_\sigma(\boldsymbol{\mu}_z)$.

[29]This characterization of r is known as the determinantal rank. Its equivalence to the standard characterization in terms of the dimension of ker A is a well-known theorem.

Let $|\Phi_{\rm GS}\rangle$ be an arbitrary ground state with $N = D_0$ electrons. Then, exactly as in (11.3.14), the condition $\hat{H}_{\rm hop}|\Phi_{\rm GS}\rangle = 0$ implies that the ground state is written only in terms of the $\hat{a}^\dagger$ operators as

$$|\Phi_{\rm GS}\rangle = \sum_{\substack{I_\uparrow, I_\downarrow \subset I \\ (|I_\uparrow|+|I_\downarrow|=D_0)}} g(I_\uparrow, I_\downarrow)\Bigl(\prod_{z\in I_\uparrow} \hat{a}^\dagger_{z,\uparrow}\Bigr)\Bigl(\prod_{z\in I_\downarrow} \hat{a}^\dagger_{z,\downarrow}\Bigr)|\Phi_{\rm vac}\rangle. \tag{11.3.46}$$

Next, by using the zero-energy condition $\hat{c}_{z,\downarrow}\hat{c}_{z,\uparrow}|\Phi_{\rm GS}\rangle = 0$ (which follows from $\hat{H}_{\rm int}|\Phi_{\rm GS}\rangle = 0$) for $z \in I$, we find that there can be no double occupancies of the $\hat{a}^\dagger$ states, and the ground state admits, as in (11.3.16), the spin system representation

$$|\Phi_{\rm GS}\rangle = \sum_{\boldsymbol{\sigma}} C(\boldsymbol{\sigma})\Bigl(\prod_{z\in I} \hat{a}^\dagger_{z,\sigma_z}\Bigr)|\Phi_{\rm vac}\rangle, \tag{11.3.47}$$

where $\boldsymbol{\sigma} = (\sigma_z)_{z\in I}$ with $\sigma_z = \uparrow, \downarrow$ is a spin configuration on I. Finally, take $x \in \Lambda\backslash I = \Lambda'$. We then find by inspection that

$$\hat{c}_{x,\downarrow}\hat{c}_{x,\uparrow}|\Phi_{\rm GS}\rangle = \sum_{\boldsymbol{\sigma}} \sum_{\substack{z_1,z_2\in I \\ (z_1<z_2)}} \mathrm{sgn}(z_1, z_2)\{C(\boldsymbol{\sigma}) - C(\boldsymbol{\sigma}_{z_1\leftrightarrow z_2})\}\times$$
$$\times\, \mu_{z_1}(x)\,\mu_{z_2}(x)\Bigl(\prod_{z\in I\backslash\{z_1,z_2\}} \hat{a}^\dagger_{z,\sigma_z}\Bigr)|\Phi_{\rm vac}\rangle. \tag{11.3.48}$$

See (11.3.19). Since $(\prod_{z\in I\backslash\{z_1,z_2\}} \hat{a}^\dagger_{z,\sigma_z})|\Phi_{\rm vac}\rangle$ with different (z_1, z_2) are linearly independent, we find that the zero-energy condition $\hat{c}_{x,\downarrow}\hat{c}_{x,\uparrow}|\Phi_{\rm GS}\rangle = 0$ for $x \in \Lambda\backslash I$ implies $C(\boldsymbol{\sigma}){=}C(\boldsymbol{\sigma}_{z_1\leftrightarrow z_2})$ for any $z_1, z_2{\in}I$ such that $z_1 \neq z_2$ and $\mu_{z_1}(x)\,\mu_{z_2}(x) \neq 0$. Recalling the definition that two basis states are directly connected, we finally find

$$C(\boldsymbol{\sigma}) = C(\boldsymbol{\sigma}_{z_1\leftrightarrow z_2}) \quad \text{if} \quad \boldsymbol{\mu}_{z_1} \sim \boldsymbol{\mu}_{z_2}. \tag{11.3.49}$$

We see that the spins associated with directly connected basis states are coupled ferromagnetically. Since these are the conditions which completely determine the ground state, we find that there can be only ferromagnetic ground states if the whole basis is connected, and there can be non-ferromagnetic ground states if it is not connected. ∎

As far as its mathematical structure is concerned, Theorem 11.17 may be regarded a straightforward generalization of Theorem 11.11 (p. 392). We should stress however that the Proof of Theorem 11.11 is after all an ad hoc argument devised for a very specific model, while it is guaranteed by Lemma 11.16 that Theorem 11.17 applies to any model that exhibits flat-band ferromagnetism. In this sense, we should understand that Lemma 11.16 and Theorem 11.17 together form an essential statement that characterizes flat-band ferromagnetism. Lemma 11.16 and Theorem 11.17 are useful in proving the emergence of flat-band ferromagnetism in concrete models. See, e.g.,

[19, 58, 59], where the uniqueness of ferromagnetic ground states is proved by using similar ideas.

We are now ready to prove the main result of the present section.

Proof of Theorem 11.15, given Theorem 11.17 We only need to prove the equivalence of irreducibility of $((\mathsf{P}_0)_{x,y})_{x,y\in\Lambda_0}$ and connectivity of the basis $\{\boldsymbol{\mu}_z\}_{z\in I}$. Note that the set $\Lambda_0 = \{x \in \Lambda \mid (\mathsf{P}_0)_{x,x} \neq 0\}$ is also written as[30] $\Lambda_0 = \{x \in \Lambda \mid \mu_z(x) \neq 0 \text{ for some } z \in I\}$.

We first show that non-connectivity of the basis $\{\boldsymbol{\mu}_z\}_{z\in I}$ implies reducibility of $((\mathsf{P}_0)_{x,y})_{x,y\in\Lambda_0}$. This is easy.

Let I and $\{\boldsymbol{\mu}_z\}_{z\in I}$ be arbitrary subset and basis characterized by Lemma 11.16. Assume that the index set I is decomposed as $I = I_1 \cup I_2$ with $I_1 \cap I_2 = \emptyset$, $I_1 \neq \emptyset$, and $I_2 \neq \emptyset$, in such a way that $\boldsymbol{\mu}_{z_1} \not\sim \boldsymbol{\mu}_{z_2}$ for any $z_1 \in I_1$ and $z_2 \in I_2$. For $j = 1, 2$, let $\Lambda_j = \{x \in \Lambda \mid \mu_z(x) \neq 0 \text{ for some } z \in I_j\}$. This leads to a decomposition $\Lambda_0 = \Lambda_1 \cup \Lambda_2$ with $\Lambda_1 \cap \Lambda_2 = \emptyset$. Note that both Λ_1 and Λ_2 are nonempty because $\Lambda_j \supset I_j \neq \emptyset$.

Now, from Lemma 9.3 in p. 320 about Slater determinant states, we find that the ferromagnetic ground state (11.3.43) can also be written as

$$|\Phi_{\text{ferro GS}}\rangle = c\Big(\prod_{z\in I} \hat{a}^\dagger_{z,\uparrow}\Big)|\Phi_{\text{vac}}\rangle, \tag{11.3.50}$$

where $\hat{a}^\dagger_{z,\sigma} = \hat{C}^\dagger_\sigma(\boldsymbol{\mu}_z)$ as before and $c \in \mathbb{C}$ is a normalization constant. We define $\hat{N}_1 := \sum_{x\in\Lambda_1} \hat{n}_x$. We see from the commutation relation (9.2.31) that

$$[\hat{N}_1, \hat{a}^\dagger_{z,\uparrow}] = \begin{cases} \hat{a}^\dagger_{z,\uparrow} & \text{if } z \in I_1 \\ 0 & \text{if } z \in I_2. \end{cases} \tag{11.3.51}$$

Take arbitrary $x \in \Lambda_1$ and $y \in \Lambda_2$, and let $|\Psi_x\rangle = \hat{c}_{x,\uparrow}|\Phi_{\text{ferro GS}}\rangle$ and $|\Psi_y\rangle = \hat{c}_{y,\uparrow}|\Phi_{\text{ferro GS}}\rangle$. From (11.3.51), it obviously holds that $\hat{N}_1|\Psi_x\rangle = (|I_1| - 1)|\Psi_x\rangle$ and $\hat{N}_1|\Psi_y\rangle = |I_1||\Psi_y\rangle$, which implies $(\mathsf{P}_0)_{x,y} = \langle\Phi_{\text{ferro GS}}|\hat{c}^\dagger_{y,\uparrow}\hat{c}_{x,\uparrow}|\Phi_{\text{ferro GS}}\rangle = \langle\Psi_y|\Psi_x\rangle = 0$.

We next prove that reducibility of $((\mathsf{P}_0)_{x,y})_{x,y\in\Lambda_0}$ implies non-connectivity of the basis $\{\boldsymbol{\mu}_z\}_{z\in I}$.

Assume that Λ_0 is decomposed as $\Lambda_0 = \Lambda_1 \cup \Lambda_2$ with $\Lambda_1 \cap \Lambda_2 = \emptyset$, $\Lambda_1 \neq \emptyset$, and $\Lambda_2 \neq \emptyset$, in such a way that $(\mathsf{P}_0)_{x,y} = 0$ for any $x \in \Lambda_1$ and $y \in \Lambda_2$. We then decompose the projection matrix as $\mathsf{P}_0 = \mathsf{P}_1 + \mathsf{P}_2$, where

---

[30] This may be obvious, but let us see a proof. If $\mu_z(x) = 0$ for some $x \in \Lambda$ and all $z \in I$, then it is obvious that $\psi_j(x) = 0$ for all $j = 1, \ldots, D_0$, where $\{\boldsymbol{\psi}_j\}_{j=1,\ldots,D_0}$ is an orthonormal basis of $\mathfrak{h}_0$. (See (11.3.44).) Thus $(\mathsf{P}_0)_{x,x} = \sum_{j=1}^{D_0}(\psi_j(x))^*\psi_j(x) = 0$. Next, suppose that $\mu_z(x) \neq 0$ for some $x \in \Lambda$ and some $z \in I$. Let us denote by P_z the projection matrix onto the unit vector proportional to $\boldsymbol{\mu}_z$. Then since $\mathsf{P}_0 \geq \mathsf{P}_z$, we have $(\mathsf{P}_0)_{x,x} \geq (\mathsf{P}_z)_{x,x} \neq 0$.

$$(\mathsf{P}_j)_{x,y} = \begin{cases} (\mathsf{P}_0)_{x,y} & \text{if } x, y \in \Lambda_j, \\ 0 & \text{otherwise,} \end{cases} \tag{11.3.52}$$

for $j = 1, 2$. Note that $\mathsf{P}_1\mathsf{P}_2 = \mathsf{P}_2\mathsf{P}_1 = \mathsf{0}$, and both P_1 and P_2 are projection matrices.

Let I and $\{\boldsymbol{\mu}_z\}_{z\in I}$ be arbitrary subset and basis characterized by Lemma 11.16. We then claim that $\mathsf{P}_1\boldsymbol{\mu}_z = \boldsymbol{\mu}_z$, $\mathsf{P}_2\boldsymbol{\mu}_z = \mathbf{0}$ if $z \in \Lambda_1$, and $\mathsf{P}_1\boldsymbol{\mu}_z = \mathbf{0}$, $\mathsf{P}_2\boldsymbol{\mu}_z = \boldsymbol{\mu}_z$ if $z \in \Lambda_2$. This clearly implies that $\boldsymbol{\mu}_{z_1} \not\sim \boldsymbol{\mu}_{z_2}$ for any $z_1 \in \Lambda_1 \cap I$ and $z_2 \in \Lambda_2 \cap I$, and hence the basis $\{\boldsymbol{\mu}_z\}_{z\in I}$ is not connected.

To show the claim, let $j(x) = 1$ if $x \in \Lambda_1$ and $j(x) = 2$ if $x \in \Lambda_2$. Noting that $(\mathsf{P}_1\boldsymbol{\varphi})(x) = 0$ if $x \in \Lambda_2$ and $(\mathsf{P}_2\boldsymbol{\varphi})(x) = 0$ if $x \in \Lambda_1$ for any $\boldsymbol{\varphi} \in \mathfrak{h}$, we have

$$(\mathsf{P}_{j(z)}\,\boldsymbol{\mu}_z)(x) = \begin{cases} ((\mathsf{P}_1 + \mathsf{P}_2)\boldsymbol{\mu}_z)(x) = \mu_z(x) & \text{if } x \in \Lambda_{j(z)}, \\ 0 & \text{otherwise,} \end{cases} \tag{11.3.53}$$

for any $z \in I$. Since P_0 and P_j $(j = 1, 2)$ commute, we also have $\mathsf{P}_0\mathsf{P}_{j(z)}\,\boldsymbol{\mu}_z = \mathsf{P}_{j(z)}\mathsf{P}_0\,\boldsymbol{\mu}_z = \mathsf{P}_{j(z)}\,\boldsymbol{\mu}_z$, which implies $\mathsf{P}_{j(z)}\,\boldsymbol{\mu}_z \in \mathfrak{h}_0$. We can then expand $\mathsf{P}_{j(z)}\,\boldsymbol{\mu}_z$ in terms of the basis $\{\boldsymbol{\mu}_z\}_{z\in I}$ as $\mathsf{P}_{j(z)}\,\boldsymbol{\mu}_z = \sum_{z'\in I} \gamma_{z'}\boldsymbol{\mu}_{z'}$. Note that (11.3.53) with $z \in \Lambda_{j(z)}$ implies $(\mathsf{P}_{j(z)}\,\boldsymbol{\mu}_z)(z) = \mu_z(z)$ and $(\mathsf{P}_{j(z)}\,\boldsymbol{\mu}_z)(z') = 0$ for $z' \in I\backslash\{z\}$. This uniquely determines the coefficients as $\gamma_{z'} = \delta_{z,z'}$. ∎

11.4 Ferromagnetism in Non-singular Hubbard Models

In the foregoing sections, we discussed two essentially different classes of models which exhibit ferromagnetism. In Sect. 11.2 we have seen that some Hubbard models with $U = \infty$ and exactly one hole exhibit ferromagnetism, i.e., Nagaoka's ferromagnetism. In Sect. 11.3, we have seen that special Hubbard models with a flat lowest band exhibit ferromagnetism for any $U > 0$ when the flat-band is half-filled. One sees that the Stoner criterion $D_{\mathrm{F}}\,U \gtrsim 1$ (see Sect. 11.1.4) is satisfied because $U = \infty$ in Nagaoka's ferromagnetism, and because $D_{\mathrm{F}} = \infty$ in the flat-band ferromagnetism. It is certainly desirable to have models which exhibit ferromagnetism and in which both D_{F} and U are finite. We shall see in the present section that one gets such non-singular Hubbard models by perturbing models exhibiting flat-band ferromagnetism, especially the Tasaki model (Sect. 11.3.1).

Let us recall that, in Lieb's ferrimagnetism (Sect. 10.2.3), Mielke's flat-band ferromagnetism (Sect. 11.3.2), and in Tasaki's flat-band ferromagnetism (Sect. 11.3.1), ferrimagnetism or ferromagnetism emerges for any value of $U > 0$. As we have discussed before, this is because the corresponding models with $U = 0$ have highly degenerate ground states with a wide range of values of S_{tot}, and the only role of the interaction U is to lift the degeneracy and to select those states with the largest S_{tot} as the only ground states. We have emphasized in Sect. 9.1 that a theoretical challenge in the Hubbard model is to control the competition between the hopping

Hamiltonian, which leads to the wave-like nature of electrons, and the interaction Hamiltonian, which leads to the particle-like nature of electrons. But we should say that there is no true such competition in those models with a flat-band which exhibit magnetic ordering for any $U > 0$.

In order to study more realistic and challenging situations without pathological degeneracy, and with true competition between wave-like and particle-like natures, we shall focus on models obtained by adding a translation invariant perturbation to the hopping Hamiltonian of the flat-band models. For a generic perturbation, the degeneracy in the single-electron ground states is lifted, and the lowest band becomes dispersive. The essential question is whether ferromagnetism realized in the flat-band models is stable against such perturbations. Note that Theorem 11.3 in p. 375 shows that a Hubbard model without a flat-band never exhibits ferromagnetism when the interaction $U > 0$ is too small. Thus the question is whether a perturbed model with a near-flat-band exhibits ferromagnetism when U is sufficiently large. Note that possible ferromagnetism is a purely non-perturbative phenomenon in such a situation.

It was conjectured in [40, 67] that Tasaki's flat-band ferromagnetism is stable against small perturbations to the hopping Hamiltonian. Kusakabe and Aoki [23, 24] made a systematic study on the stability of flat-band ferromagnetism. They argued that the flat-band models possess spin-wave excitations which have healthy dispersions, and this fact guarantees the robustness of the flat-band ferromagnetism. They also found numerical evidences for the stability of ferromagnetism in a model with a nearly-flat-band model. See also [45] for a further study on the stability. Later, as we shall discuss in detail below, the local stability for general perturbations and the global stability for a special class of perturbations were proved by Tasaki in [70, 71] and in [72, 74], respectively.

In the present section we focus on the stability of Tasaki's flat-band ferromagnetism against perturbations to the hopping Hamiltonian. In Sect. 11.4.1, we describe a non-rigorous perturbative argument, which should shed light on the physical mechanism of stability and instability. Then in Sect. 11.4.2 we treat models obtained by adding general translation invariant perturbations to the flat-band models, and discuss rigorous results on the local stability of ferromagnetism and on low energy excitations. In Sect. 11.4.3, we introduce a class of models with nearly-flat-band obtained by adding special perturbations to Tasaki's flat-band models, and prove that the models exhibit ferromagnetism provided that the Coulomb interaction U is sufficiently large. The results in Sects. 11.4.2 and 11.4.3 are among the most advanced rigorous results on ferromagnetism in itinerant electron systems.

11.4.1 Wannier State Perturbation Theory

We start with a preliminary heuristic analysis of a model obtained by adding a simple perturbation to the Tasaki model in Sect. 11.3.1. We shall see from a non-rigorous perturbative argument that the ferromagnetism found for the flat-band model survives when the band is nearly flat and the interaction U is large enough.

Model and the band structure As a simple illustrative example we shall consider a model obtained by shifting the on-site potential in the one-dimensional version of the Tasaki model. As in Sect. 11.3.1, we set $\mathscr{E} = \{1, 2, \ldots, L\}$, $\mathscr{I} = \{\frac{1}{2}, \frac{3}{2}, \ldots, L - \frac{1}{2}\}$, and $\Lambda = \mathscr{E} \cup \mathscr{I}$ with periodic boundary conditions. We then consider the hopping matrix $\mathsf{T} = (t_{x,y})_{x,y\in\Lambda}$ with

$$t_{x,y} = \begin{cases} \nu t & \text{if } |x-y| = 1/2, \\ \nu^2 t & \text{if } x, y \in \mathscr{E} \text{ and } |x-y| = 1, \\ (1+\gamma)\,t & \text{if } x = y \in \mathscr{I}, \\ 2\nu^2 t & \text{if } x = y \in \mathscr{E} \\ 0 & \text{otherwise,} \end{cases} \tag{11.4.1}$$

where $t > 0$ and $\nu > 0$. Here $\gamma \ge 0$[31] is a new parameter which shifts the on-site potential for sites in $\mathscr{I}$. The hopping amplitude (11.4.1) reduces to (11.3.22) when $\gamma = 0$. See also Fig. 11.11 in p. 392. In the present section we shall assume that $0 < \nu \ll 1$.

By solving the single-electron Schrödinger equation corresponding to (11.4.1) as we explained in Sect. 9.3.1, one finds that there are two bands with dispersion relations

$$\varepsilon_\pm(k) = \frac{t}{2}\Bigl\{1 + \gamma + C(k) \pm \sqrt{\{1+\gamma+C(k)\}^2 - 4\gamma\, C(k)}\Bigr\}, \tag{11.4.2}$$

where $k \in (-\pi, \pi]$, and $C(k) = 2\nu^2(1 + \cos k)$. Note that (11.4.2) recovers the dispersion relations (11.3.23) with a flat-band when $\gamma = 0$. Since we assumed $0 < \nu \ll 1$, it is useful to expand (11.4.2) in ν as

$$\varepsilon_+(k) = t\Bigl\{1 + \gamma + \frac{2\nu^2}{1+\gamma}(1+\cos k)\Bigr\} + O(\nu^4), \tag{11.4.3}$$

$$\varepsilon_-(k) = \frac{2\nu^2\gamma}{1+\gamma}\, t\,(1+\cos k) + O(\nu^4). \tag{11.4.4}$$

The lower band is nearly-flat when $\gamma > 0$ is small.

Wannier state basis It is convenient for our heuristic discussion to describe single-electron states in terms of the Wannier state basis, which is a standard tool for analyzing multi-band systems. In the present model, the Wannier state basis is an orthonormal basis $\{\boldsymbol{\omega}_y\}_{y\in\Lambda}$, where each basis state $\boldsymbol{\omega}_y = (\omega_y(x))_{x\in\Lambda}$ is exponentially localized at site y. The collection $\{\boldsymbol{\omega}_p\}_{p\in\mathscr{E}}$ spans the lower band, and $\{\boldsymbol{\omega}_u\}_{u\in\mathscr{I}}$ spans the upper band. Wannier states are usually defined through Fourier transformation,

[31] In fact we only need to assume that γ is sufficiently away from -1.

but in this case we find by inspection (see footnote 32 below) that

$$\omega_p(x) = \Bigl(1 + \frac{2\nu^2}{(1+\gamma)^2}\Bigr)^{-1/2} \times \begin{cases} 1 + O(\nu^3) & \text{if } x = p, \\ -\dfrac{\nu}{1+\gamma} + O(\nu^3) & \text{if } |x - p| = \frac{1}{2}, \\ -\dfrac{1}{2}\Bigl(\dfrac{\nu}{1+\gamma}\Bigr)^2 + O(\nu^3) & \text{if } |x - p| = 1, \\ O(\nu^3) & \text{otherwise,} \end{cases} \tag{11.4.5}$$

for any $p \in \mathcal{E}$. We see that the state $\boldsymbol{\omega}_p$ is sharply localized at site p because $0 < \nu \ll 1$. It should be noted that $\boldsymbol{\omega}_p$ is different from $\boldsymbol{\alpha}_p$ defined in (11.3.1) even when $\gamma = 0$. The basis states $\boldsymbol{\alpha}_p$ are strictly localized at three sites, but are not orthogonal with each other. The Wannier states $\boldsymbol{\omega}_p$ are only exponentially localized, but are orthogonal with each other.

That $\{\boldsymbol{\omega}_p\}_{p\in\mathcal{E}}$ spans the lower band can be seen from the relation

$$\mathsf{T}\boldsymbol{\omega}_p = \tau\,\{2\boldsymbol{\omega}_p + \boldsymbol{\omega}_{p-1} + \boldsymbol{\omega}_{p+1}\} + O(\nu^3), \tag{11.4.6}$$

for any $p \in \mathcal{E}$, where the effective hopping amplitude is

$$\tau = \frac{\nu^2\gamma}{1+\gamma}\,t + O(\nu^3). \tag{11.4.7}$$

The relations (11.4.6) and (11.4.7) can be checked by a straightforward calculation using (11.4.1) and (11.4.5).[32] Note that this hopping amplitude τ is consistent with the dispersion relation (11.4.4) of the lower band.

We define fermion operators corresponding to the Wannier states as

$$\hat{d}^\dagger_{x,\sigma} := \hat{C}^\dagger_\sigma(\boldsymbol{\omega}_x). \tag{11.4.8}$$

Since the Wannier state basis is orthonormal, we find from (9.2.64) that

$$\{\hat{d}^\dagger_{x,\sigma}, \hat{d}_{y,\tau}\} = \delta_{x,y}\delta_{\sigma,\tau}, \tag{11.4.9}$$

for any $x, y \in \Lambda$ and $\sigma, \tau = \uparrow, \downarrow$.

Problem 11.4.1.a Find, to the second order in ν, the Wannier state $\boldsymbol{\omega}_u$ for the upper band, and the corresponding relation like (11.4.6). This is indeed easy when (11.4.5) is given. [solution→p.519]

First order perturbation We consider the Hubbard model on the one-dimensional lattice $\Lambda = \{\frac{1}{2}, 1, \ldots, L - \frac{1}{2}, L\}$ with $N = L$ electrons and Hamiltonian $\hat{H} = \hat{H}_{\text{hop}} + \hat{H}_{\text{int}}$. The hopping Hamiltonian $\hat{H}_{\text{hop}}$ is defined by the standard formula (9.3.17)

---

[32] In fact we obtained (11.4.5) by demanding that $\langle\boldsymbol{\omega}_p, \boldsymbol{\omega}_{p+1}\rangle = 0$ and that $\mathsf{T}\boldsymbol{\omega}_p = \tau_0\,\boldsymbol{\omega}_p + \tau_1\{\boldsymbol{\omega}_{p-1} + \boldsymbol{\omega}_{p+1}\} + O(\nu^3)$ holds for some τ_0 and τ_1.

with the hopping amplitude (11.4.1), and $\hat{H}_{\text{int}}$ is the standard interaction Hamiltonian (11.3.6) or (9.3.29) with $U > 0$. We decompose the interaction Hamiltonian as $\hat{H}_{\text{int}} = \hat{H}_{\text{int}}^{\mathcal{E}} + \hat{H}_{\text{int}}^{\mathcal{I}}$, where

$$\hat{H}_{\text{int}}^{\mathcal{E}} = U \sum_{p\in\mathcal{E}} \hat{n}_{p,\uparrow}\hat{n}_{p,\downarrow}, \quad \hat{H}_{\text{int}}^{\mathcal{I}} = U \sum_{u\in\mathcal{I}} \hat{n}_{u,\uparrow}\hat{n}_{u,\downarrow}. \tag{11.4.10}$$

We shall now regard $\hat{H}_0 = \hat{H}_{\text{hop}} + \hat{H}_{\text{int}}^{\mathcal{E}}$ as the unperturbed Hamiltonian, and develop a non-rigorous perturbation theory. We first see from the assumption $0 < \nu \ll 1$ that the band gap between the two bands, which is roughly $(1+\gamma)\,t$, is much larger than the band width of the lower band, which is $4\nu^2 t\,\gamma/(1+\gamma)$. See (11.4.3) and (11.4.4). It might be reasonable to assume that low energy states of the N electron system are approximately described only in terms of the $\hat{d}_{p,\sigma}^{\dagger}$ operators with $p \in \mathcal{E}$. We next assume that the Coulomb interaction energy U is much larger than the band width $4\nu^2 t\,\gamma/(1+\gamma)$, so that double occupancy of $\hat{d}^{\dagger}$ states is inhibited. We then find that (near) degenerate ground states of the unperturbed Hamiltonian $\hat{H}_0$ are given by

$$|\Omega^{\boldsymbol{\sigma}}\rangle = \Bigl(\prod_{p\in\mathcal{E}} \hat{d}_{p,\sigma_p}^{\dagger}\Bigr)|\Phi_{\text{vac}}\rangle, \tag{11.4.11}$$

where $\boldsymbol{\sigma} = (\sigma_1, \ldots, \sigma_L)$ with $\sigma_p = \uparrow, \downarrow$ is a spin configuration on $\mathcal{E}$. Note that the anticommutation relations (11.4.9) imply

$$\langle\Omega^{\boldsymbol{\sigma}}|\Omega^{\boldsymbol{\sigma}'}\rangle = \delta_{\boldsymbol{\sigma},\boldsymbol{\sigma}'} \tag{11.4.12}$$

for any spin configurations $\boldsymbol{\sigma}$ and $\boldsymbol{\sigma}'$. This simple relation makes the following calculations easy. Although the foregoing discussion is very similar to the Proof of Theorem 11.11, one should note that here we have been making several uncontrolled approximations.

Since we have encountered (near) degenerate unperturbed ground states, we shall follow the standard procedure, and examine how the degeneracy is lifted by the first order perturbation theory. We thus need to evaluate matrix elements

$$\langle\Omega^{\boldsymbol{\sigma}}|\hat{H}_{\text{int}}^{\mathcal{I}}|\Omega^{\boldsymbol{\sigma}'}\rangle = U \sum_{u\in\mathcal{I}} \langle\Omega^{\boldsymbol{\sigma}}|\hat{n}_{u,\uparrow}\hat{n}_{u,\downarrow}|\Omega^{\boldsymbol{\sigma}'}\rangle, \tag{11.4.13}$$

for any spin configurations $\boldsymbol{\sigma}$ and $\boldsymbol{\sigma}'$. Since $\hat{n}_{u,\uparrow}\hat{n}_{u,\downarrow} = (\hat{c}_{u,\downarrow}\hat{c}_{u,\uparrow})^{\dagger}\,\hat{c}_{u,\downarrow}\hat{c}_{u,\uparrow}$, we see that $\langle\Omega^{\boldsymbol{\sigma}}|\hat{n}_{u,\uparrow}\hat{n}_{u,\downarrow}|\Omega^{\boldsymbol{\sigma}'}\rangle$ is the inner product of $\hat{c}_{u,\downarrow}\hat{c}_{u,\uparrow}|\Omega^{\boldsymbol{\sigma}}\rangle$ and $\hat{c}_{u,\downarrow}\hat{c}_{u,\uparrow}|\Omega^{\boldsymbol{\sigma}'}\rangle$. From (11.4.11), we have

$$\hat{c}_{u,\downarrow}\hat{c}_{u,\uparrow}|\Omega^{\sigma}\rangle \simeq \begin{cases} (\omega_{1/2})^2\Bigl(\prod_{p\in\mathcal{E}\backslash\{u-\frac{1}{2},u+\frac{1}{2}\}} \hat{d}^\dagger_{p,\sigma_p}\Bigr)|\Phi_{\text{vac}}\rangle & \text{if } (\sigma_{u-\frac{1}{2}},\sigma_{u+\frac{1}{2}})=(\uparrow,\downarrow),\\ -(\omega_{1/2})^2\Bigl(\prod_{p\in\mathcal{E}\backslash\{u-\frac{1}{2},u+\frac{1}{2}\}} \hat{d}^\dagger_{p,\sigma_p}\Bigr)|\Phi_{\text{vac}}\rangle & \text{if } (\sigma_{u-\frac{1}{2}},\sigma_{u+\frac{1}{2}})=(\downarrow,\uparrow),\\ 0 & \text{otherwise,} \end{cases} \tag{11.4.14}$$

where $\omega_{1/2} = \omega_p(p \pm \frac{1}{2}) \simeq -\nu/(1+\gamma)$. Here, and in what follows, near equality $\simeq$ indicates that we omit contributions of $O(\nu^3)$. We thus find that

$$\langle\Omega^{\sigma}|\hat{n}_{u,\uparrow}\hat{n}_{u,\downarrow}|\Omega^{\sigma'}\rangle \simeq \begin{cases} (\omega_{1/2})^4 & \text{if } \sigma_{u-\frac{1}{2}} \neq \sigma_{u+\frac{1}{2}} \text{ and } \boldsymbol{\sigma}' = \boldsymbol{\sigma},\\ -(\omega_{1/2})^4 & \text{if } \sigma_{u-\frac{1}{2}} \neq \sigma_{u+\frac{1}{2}} \text{ and } \boldsymbol{\sigma}' = \boldsymbol{\sigma}_{u-\frac{1}{2}\leftrightarrow u+\frac{1}{2}},\\ 0 & \text{otherwise.} \end{cases} \tag{11.4.15}$$

As before, $\boldsymbol{\sigma}_{u-\frac{1}{2}\leftrightarrow u+\frac{1}{2}}$ denotes the configuration obtained from $\boldsymbol{\sigma}$ by swapping $\sigma_{u-\frac{1}{2}}$ and $\sigma_{u+\frac{1}{2}}$.

We now recall that, in a system of two spins with $S = 1/2$, one has

$$\langle\Psi^{(\sigma_1,\sigma_2)}|\Bigl(\frac{1}{2} - 2\,\hat{\mathbf{S}}_1\cdot\hat{\mathbf{S}}_2\Bigr)|\Psi^{(\sigma_1',\sigma_2')}\rangle = \begin{cases} 1 & \text{if } \sigma_1 \neq \sigma_2 \text{ and } (\sigma_1',\sigma_2') = (\sigma_1,\sigma_2),\\ -1 & \text{if } \sigma_1 \neq \sigma_2 \text{ and } (\sigma_1',\sigma_2') = (\sigma_2,\sigma_1),\\ 0 & \text{otherwise,} \end{cases} \tag{11.4.16}$$

where $\sigma_1, \sigma_2 = \uparrow, \downarrow$. See (2.2.16). By comparing (11.4.15) and (11.4.16), we find that the effect of the first order perturbation is described by the effective spin Hamiltonian

$$\hat{H}^{(1)}_{\text{eff}} = J_1 \sum_{p=1}^{L} \Bigl(\frac{1}{4} - \hat{\hat{\mathbf{S}}}_p \cdot \hat{\hat{\mathbf{S}}}_{p+1}\Bigr), \tag{11.4.17}$$

where $\hat{\hat{\mathbf{S}}}_p$ is the spin operator corresponding to the Wannier state $\boldsymbol{\omega}_p$, defined by replacing $\hat{c}_{x,\sigma}$ in (9.2.38) with $\hat{d}_{p,\sigma}$. The exchange interaction constant is given by

$$J_1 = 2(\omega_{1/2})^4 U \simeq \frac{2\nu^4}{(1+\gamma)^4} U, \tag{11.4.18}$$

and is always positive. The effective Hamiltonian (11.4.17) is nothing but that of the ferromagnetic Heisenberg model. We note that this argument is essentially the same as Heisenberg's perturbative derivation of the direct exchange interaction [16]. Although the above perturbation theory is rather ad hoc, we know from rigorous bounds for the spin-wave excitation energy that (11.4.17) and (11.4.18) are quantitatively reliable, provided that $\gamma \ll 1$. See Sect. 11.4.2, in particular, (11.4.36).

"Second order" perturbation The above perturbation theory suggests that the present Hubbard model, whose effective Hamiltonian (11.4.17) is ferromagnetic, universally exhibits ferromagnetism, no matter whether the lower band is flat (i.e.,

$\gamma = 0$) or non-flat (i.e., $\gamma \neq 0$). Furthermore, by examining the argument, we see that it is only essential for the emergence of the ferromagnetic interaction that the Wannier states on neighboring $\mathscr{E}$ sites share a common $\mathscr{I}$ site in between them. Should we then conclude that one always gets ferromagnetism in the Hubbard model on the lattice with two sublattices, such as our $\Lambda = \mathscr{E} \cup \mathscr{I}$?

Of course this is not the case; we are definitely overestimating the tendency toward ferromagnetism, as Heisenberg was doing in his seminal work on the direct exchange interaction [16]. In the present model we also have to look at the "second order" effect, which may become comparable to the above first order effect. The contribution we have been overlooking comes from the effective hopping process (11.4.6) for the Wannier states of the lower band. Indeed if we concentrate only on states formed by $\hat{d}^\dagger_{p,\sigma}$ with $p \in \mathscr{E}$ (which are in the lower band), the situation is exactly the same as that in Sect. 10.1, where we examined the second order perturbation effect in the half-filled Hubbard model. We have seen that the super-exchange interaction emerges, and the low energy properties of the model are described by the antiferromagnetic Heisenberg model (10.1.12). Here we only need to replace the hopping amplitude $t_{x,y}$, which appears in the expression (10.1.11) of the exchange interaction constant, by the effective hopping amplitude τ in (11.4.7). We thus see that the effective Hamiltonian from the "second order" perturbation is[33]

$$\hat{H}^{(2)}_{\text{eff}} = J_2 \sum_{p=1}^{L} \Big(\hat{\tilde{\mathbf{S}}}_p \cdot \hat{\tilde{\mathbf{S}}}_{p+1} - \frac{1}{4} \Big), \tag{11.4.19}$$

with

$$J_2 \simeq \frac{4\tau^2}{U} \simeq \frac{4\gamma^2\nu^2}{(1+\gamma)^2} \frac{t^2}{U}. \tag{11.4.20}$$

Since $J_2 \geq 0$, we always get antiferromagnetic interaction from this process.

Stability and instability of ferromagnetism We have seen that there are two essentially different mechanisms which generate effective Heisenberg-type interaction between electronic spins in the present Hubbard model. The direct exchange interaction, which comes from the first order perturbation, leads to ferromagnetism, and the super-exchange interaction, which comes from the "second order" process, leads to antiferromagnetism. See Fig. 11.19. The overall effective spin Hamiltonian

$$\hat{H}_{\text{eff}} = (J_1 - J_2) \sum_{p=1}^{L} \Big(\frac{1}{4} - \hat{\tilde{\mathbf{S}}}_p \cdot \hat{\tilde{\mathbf{S}}}_{p+1} \Big), \tag{11.4.21}$$

where $J_1 > 0$ and $J_2 \geq 0$, is ferromagnetic if $J_1 - J_2 > 0$ and is antiferromagnetic if $J_1 - J_2 < 0$. For the flat-band model with $\gamma = 0$, we have $J_2 = 0$, and the model always exhibits ferromagnetism. This conclusion is consistent with the rigorous

[33] This is not really the second order perturbation which corresponds to the first order that we saw above. Here we are treating τ as a perturbation.

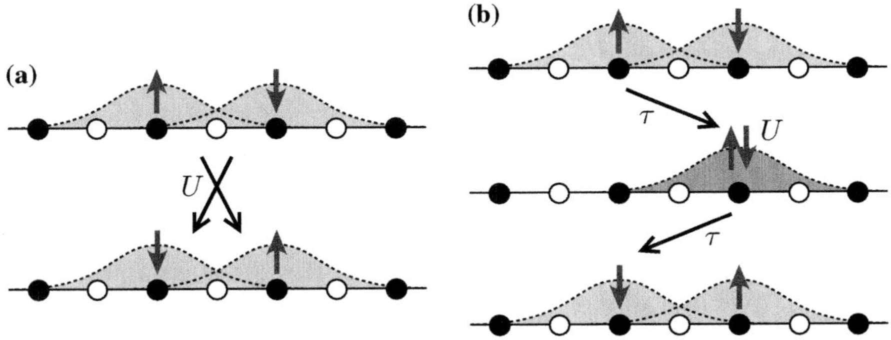

Fig. 11.19 The two exchange mechanisms in the nearly-flat-band model found from the non-rigorous Wannier state perturbation theory. **a** The direct exchange interaction due to Heisenberg leads to ferromagnetism. **b** The super-exchange interaction due to Anderson leads to antiferromagnetism. The magnetic property of the model is determined by the competition between the two effects (© Hal Tasaki 2020. All Rights Reserved)

result, namely, Theorem 11.11 in p. 392. If $\gamma > 0$, we find by comparing (11.4.18) for J_1 and (11.4.20) for J_2 that $J_1 - J_2$ is negative for sufficiently small $U > 0$. We expect that the model exhibits antiferromagnetism. As we increase U (with γ fixed) the sign of $J_1 - J_2$ changes at a critical value of U, which is estimated as

$$U_c \simeq \frac{\sqrt{2}\gamma(1+\gamma)t}{\nu}. \tag{11.4.22}$$

We expect that the model exhibits ferromagnetism when U is larger than U_c. In this sense ferromagnetism in a non-flat band model is a purely non-perturbative phenomenon. In conclusion the perturbative analysis in the present section suggests that the model exhibits ferromagnetism when the lower band is nearly flat, and the interaction is large enough.

11.4.2 Local Stability and the Spin Wave Excitation

We shall discuss the local stability of ferromagnetism in models obtained by adding an arbitrary translation invariant perturbation to the hopping Hamiltonian of the Tasaki model [70, 71]. We also discuss rigorous upper and lower bounds for the spin-wave excitation energy. These rigorous results provide strong support to the belief (or the conjecture) that the nearly-flat-band models exhibit physically realistic ferromagnetism.

Local stability theorem We consider the Hubbard model on the d-dimensional decorated hypercubic lattice $\Lambda = \mathscr{E} \cup \mathscr{I}$ as in Sect. 11.3.1. See, in particular, Figs. 11.10

and 11.11.[34] We again fix the electron number as $N = |\mathcal{E}| = L^d$. The Hamiltonian is $\hat{H} = \hat{H}_{\text{hop}} + \hat{H}_{\text{int}}$, with $\hat{H}_{\text{int}}$ being the standard interaction Hamiltonian (11.3.6). The hopping Hamiltonian in the present model is

$$\hat{H}_{\text{hop}} = t \sum_{\substack{u \in \mathcal{I} \\ \sigma=\uparrow,\downarrow}} \hat{b}^\dagger_{u,\sigma} \hat{b}_{u,\sigma} + \zeta \sum_{\substack{x,y \in \Lambda \\ \sigma=\uparrow,\downarrow}} t'_{x,y} \hat{c}^\dagger_{x,\sigma} \hat{c}_{x,\sigma}, \tag{11.4.23}$$

with $t > 0$, where $\hat{b}^\dagger$ operator is defined in (11.3.4). Here $\zeta \in \mathbb{R}$ is the perturbation parameter. Note that, when $\zeta = 0$, (11.4.23) reduces to the hopping Hamiltonian (11.3.5) for the flat-band model. The new hopping amplitude is real symmetric, i.e., $t'_{x,y} = t'_{y,x} \in \mathbb{R}$, and satisfies the translation invariance

$$t'_{x+z,y+z} = t'_{x,y}, \tag{11.4.24}$$

for any $z \in \mathbb{Z}^d$ (where we use periodic boundary conditions), and the summability

$$\sum_{y \in \Lambda} |t'_{x,y}| \le t, \quad \sum_{y \in \Lambda} |x - y|\,|t'_{x,y}| \le tR, \tag{11.4.25}$$

for any $x \in \Lambda$ with a constant $R > 0$, which may be interpreted as the range of the hopping. We thus allow any translation invariant hopping (which decays sufficiently fast) as a perturbation. The lowest band in the corresponding single-electron problem is generally no longer flat when $\zeta \neq 0$. The model with (11.4.1) studied in Sect. 11.4.1 is an example.

To state the first theorem, we define the minimum energy among the states with total spin S as

$$E_{\min}(S) := \min\bigl\{\langle\Phi|\hat{H}|\Phi\rangle \bigm| (\hat{S}_{\text{tot}})^2|\Phi\rangle = S(S+1)|\Phi\rangle,\ \|\Phi\| = 1\bigr\}, \tag{11.4.26}$$

where $S = 0, 1, \ldots, S_{\max}$. The ground state energy is written as $E_{\text{GS}} = \min_S E_{\min}(S)$. The model exhibits (saturated) ferromagnetism if and only if $E_{\min}(S_{\max}) < E_{\min}(S)$ for any S such that $S \neq S_{\max}$. It is not (yet) possible to prove that the model exhibits ferromagnetism, but the following local stability theorem was proved by Tasaki [70, 71]. The proof is based on an elementary but rigorous perturbation theory.

Theorem 11.18 (Local stability of ferromagnetic ground states) *There are constants ν_0, η_0, and ξ_0 which depend only on the dimension d and the range R. If the model parameters satisfy*

$$0 < \nu \le \nu_0, \quad |\zeta| \le \nu^3 \eta_0, \tag{11.4.27}$$

and

$$U \ge \frac{\xi_0\, t}{\nu^2} |\zeta|, \tag{11.4.28}$$

[34] In [71] general models obtained by the cell construction (see the end of Sect. 11.3.1) are treated.

then we have

$$E_{\min}(S_{\max}) < E_{\min}(S_{\max} - 1). \tag{11.4.29}$$

Note that $E_{\min}(S_{\max} - 1)$ is the minimum energy in the subspace where a single spin is flipped from the ferromagnetic state and coupled antiferromagnetically to the rest of the spins. Thus the inequality (11.4.29) shows that ferromagnetic ground states are stable against a single-spin flip. Such a local stability is a necessary but not a sufficient condition for the emergence (or the global stability) of ferromagnetism. Physically speaking, however, it is strongly believed (based on numerous examples) that the local stability generally implies the global stability in the case of ferromagnetism. See [37, 38] for related rigorous results. In this sense Theorem 11.18 provides a strong support for the emergence of saturated ferromagnetism in the perturbed model with a nearly-flat band.

The most important condition for the above local stability theorem is (11.4.28), which requires the Coulomb interaction to be sufficiently large. This is natural since we never have ferromagnetism in a generic model with $\zeta \neq 0$ if $U > 0$ is too small, as is stated in Theorem 11.3 in p. 375. We can say that the above local stability theorem establishes a truly nonperturbative result in which the "competition" between hopping and interaction is properly taken into account.

Let us stress that the problem of stability against a single-spin flip is already a highly nontrivial many-body problem. The restriction to the sector with $S_{\text{tot}} = S_{\max} - 1$ does not reduce the problem to that of a single-particle (such as a magnon) since there are plenty of spaces for the electrons to move around. Moreover there is no hope of expressing general energy eigenstates as Slater determinant states since there are both up-spin and down-spin electrons interacting via local Coulomb repulsion.

Spin-wave excitation energy By using the same rigorous perturbation theory, rigorous upper and lower bounds for the spin-wave excitation energies in the above model was obtained in [70, 71].

For $z \in \mathbb{Z}^d$, we define the translation operator $\hat{\tau}_z : \mathcal{H}_N \to \mathcal{H}_N$ by

$$\hat{\tau}_z\, \hat{c}^\dagger_{x_1,\sigma_1} \hat{c}^\dagger_{x_2,\sigma_2} \cdots \hat{c}^\dagger_{x_N,\sigma_N} |\Phi_{\text{vac}}\rangle = \hat{c}^\dagger_{x_1+z,\sigma_1} \hat{c}^\dagger_{x_2+z,\sigma_2} \cdots \hat{c}^\dagger_{x_N+z,\sigma_N} |\Phi_{\text{vac}}\rangle \tag{11.4.30}$$

for any $x_1, \ldots, x_N \in \Lambda$ and $\sigma_1, \ldots, \sigma_N = \uparrow, \downarrow$, and by assuming the linearity. We use periodic boundary conditions. The operator $\hat{\tau}_z$ is unitary, and its eigenvalues are written as $e^{-ik\cdot z}$ with $k \in \mathcal{K}_L$. Here $\mathcal{K}_L$ is the k-space defined in (4.1.17), and $k \cdot z = \sum_{\mu=1}^{d} k_\mu z_\mu$ with $k = (k_1, \ldots, k_d)$ and $z = (z_1, \ldots, z_d)$. Since both $\hat{H}$ and $\hat{S}^{(3)}_{\text{tot}}$ are translation invariant, we can look for simultaneous eigenstates of $\hat{H}$, $\hat{S}^{(3)}_{\text{tot}}$, and $\hat{\tau}_z$.[35] We then let $E_{\text{SW}}(k)$ be the smallest energy eigenvalue among the states satisfying $\hat{S}^{(3)}_{\text{tot}}|\Phi\rangle = (S_{\max} - 1)|\Phi\rangle$ and $\hat{\tau}_z|\Phi\rangle = e^{-ik\cdot z}|\Phi\rangle$ for any $z \in \mathbb{Z}^d$. This is the most general definition of the energy of the elementary spin-wave excitation with momentum k.

[35] $\hat{A}$ is translation invariant if $(\hat{\tau}_z)^{-1}\hat{A}\hat{\tau}_z = \hat{A}$, which means $[\hat{\tau}_z, \hat{A}] = 0$.

Theorem 11.19 (Bounds on the spin-wave excitation energy) *There are constants* ν_1, η_1, ξ_1, *and* ξ_2 *which depend only on the dimension d and the range R. If the model parameters satisfy*

$$0 < \nu \le \nu_1, \quad |\zeta| \le \nu^2 \eta_1, \tag{11.4.31}$$

and

$$\frac{\xi_1 t}{\nu^2}|\zeta| \le U \le \xi_2 t \nu, \tag{11.4.32}$$

then it holds that

$$F_1\, 2\nu^4 U \sum_{\mu=1}^{d}(1 - \cos k_\mu) \ge E_{\mathrm{SW}}(k) - E_{\min}(S_{\max}) \ge F_2\, 2\nu^4 U \sum_{\mu=1}^{d}(1 - \cos k_\mu), \tag{11.4.33}$$

for any $k \in \mathcal{K}_L$. *Here the constants* F_1 *and* F_2 *are written as*

$$F_1 = 1 + a_1 \nu + a_2 \frac{|\zeta|}{\nu^3} + \frac{a_3 t}{\nu^4 U}|\zeta|^2, \tag{11.4.34}$$

$$F_2 = 1 - b_1 \nu - b_2 \frac{|\zeta|}{\nu^2} - \frac{b_3 t}{\nu^4 U}|\zeta|^2, \tag{11.4.35}$$

with positive constants a_1, a_2, a_3, b_1, b_2, *and* b_3.

Therefore when ν is small, and $|\zeta|$ is much smaller (and hence the lowest band is nearly-flat), we have an almost precise estimate

$$E_{\mathrm{SW}}(k) - E_{\mathrm{GS}} \simeq 2\nu^4 U \sum_{\mu=1}^{d}(1 - \cos k_\mu), \tag{11.4.36}$$

about the spin-wave excitation energies. We recall that this dispersion relation is what one expects for the elementary magnon excitation in a ferromagnetic Heisenberg model on the hypercubic lattice $\mathcal{E}$ with the exchange interaction $J_{\mathrm{eff}} = 2\nu^4 U$. See Sect. 2.4, in particular, (2.4.17). This partially justifies the estimate (11.4.18) for the ferromagnetic coupling in the Wannier state perturbation theory as well as the conclusion by Kusakabe and Aoki that the flat-band models have normal spin-wave excitations [23, 24]. (Note that Theorem 11.19 applies also to the flat-band model.) Theorem 11.19 thus suggests that ferromagnetism realized in Tasaki's (near-)flat-band models is "healthy" from a physical point of view.

About the proofs We are not going to prove Theorems 11.18 or 11.19 in the present book. The proofs are rather involved, as can be seen from the fact that the paper [71] has 119 pages. We nevertheless stress that the essence of the proofs is quite elementary. We encourage the interested reader to take a look at (at least) Sects. 1 and 2 of [71], which can be read as a self-contained account of the rigorous perturbation method used in this work.

The basic strategy of the proofs is somewhat similar to the Wannier state perturbation theory discussed in Sect. 11.4.1. We represent the Hubbard model by using a basis consisting of exponentially localized states, and then control all the matrix elements rigorously. This is of course a formidable task in general, but is barely doable if we restrict ourselves to the subspace with $N-1$ up-spin electrons and one down-spin electrons, and make full use of the translation invariance. We need to face a tough many-body problem even within this subspace, but the problem becomes much easier because the Coulomb energy $\hat{H}_{\text{int}}$ is at most U in the subspace.

11.4.3 Ferromagnetism in Non-singular Hubbard Models

We finally discuss the result due to Tasaki [72, 74] on global stability of ferromagnetism in a Hubbard model with a nearly-flat lowest band, obtained by adding a particular perturbation to the flat-band model studied in Sect. 11.3.1. As we have already emphasized, this work established the emergence of ferromagnetism in a Hubbard model without any singularity, i.e., a model with finite density of states D_{F} and finite Coulomb interaction U. It is also important that the result is essentially non-perturbative in the sense that it can never be valid if the interaction $U > 0$ is too small.

The proof, which we shall describe in detail below, is again based on the properties of frustration-free Hamiltonians. In this case, however, the construction of frustration-free local Hamiltonians is highly nontrivial; one must take a local Hamiltonian that acts on $4d+1$ sites to make the argument work. See Fig. 11.23.

As far as we know, this strategy, which was proposed in [72], is still essentially the only method to prove the existence of ferromagnetism in a Hubbard model with a non-flat lowest band. There have been extensions of the method to the Hubbard model with a certain multi-band structure [62], on the kagomé lattice [65], on three coupled chains [77], on planar line graphs [30], and on certain three dimensional lattices [58]. See also [57] for an extension to the SU(n) Hubbard model.

The model and the main result We consider the Hubbard model on the d-dimensional decorated hypercubic lattice $\Lambda = \mathscr{E} \cup \mathscr{I}$ as in Sect. 11.3.1. See, in particular, Figs. 11.10 and 11.11. We again make use of the operators $\hat{a}_{p,\sigma}$ and $\hat{b}_{u,\sigma}$ defined in (11.3.3) and (11.3.4), respectively. Our Hamiltonian is $\hat{H} = \hat{H}_{\text{hop}} + \hat{H}_{\text{int}}$, where

$$\hat{H}_{\text{int}} = U \sum_{x \in \Lambda} \hat{n}_{x,\uparrow} \hat{n}_{x,\downarrow}, \tag{11.4.37}$$

with $U > 0$, is the standard interaction Hamiltonian (9.3.29). We consider a particular hopping Hamiltonian

$$\hat{H}_{\text{hop}} = t \sum_{\substack{u \in \mathscr{I} \\ \sigma=\uparrow,\downarrow}} \hat{b}^{\dagger}_{u,\sigma} \hat{b}_{u,\sigma} - s \sum_{\substack{p \in \Lambda \\ \sigma=\uparrow,\downarrow}} \hat{a}^{\dagger}_{p,\sigma} \hat{a}_{p,\sigma}, \tag{11.4.38}$$

with two parameters $t > 0$ and $s > 0$. Note that the model reduces to the flat-band model of Sect. 11.3.1 if we set $s = 0$. Note also that this is a special case of the hopping Hamiltonian (11.4.23) of the perturbed model with a nearly-flat band treated in Sect. 11.4.2. Again the hopping Hamiltonian (11.4.38) is written in the standard form

$$\hat{H}_{\text{hop}} = \sum_{\substack{x,y\in\Lambda \\ \sigma=\uparrow,\downarrow}} t_{x,y}\, \hat{c}^{\dagger}_{x,\sigma} \hat{c}_{y,\sigma}, \tag{11.4.39}$$

with admittedly complicated hopping amplitude given by

$$t_{x,y} = \begin{cases} \nu(t+s) & \text{if } |x-y| = 1/2, \\ \nu^2 t & \text{if } x, y \in \mathscr{E} \text{ and } |x-y| = 1, \\ -\nu^2 s & \text{if } x \neq y \in \mathscr{I} \text{ and } \exists p \in \mathscr{E} \text{ s.t. } |x-p| = |y-p| = 1/2, \\ t - 2\nu^2 s & \text{if } x = y \in \mathscr{I}, \\ 2d\nu^2 t - s & \text{if } x = y \in \mathscr{E} \\ 0 & \text{otherwise.} \end{cases} \tag{11.4.40}$$

See Fig. 11.20. By solving the corresponding single-electron Schrödinger equation (9.3.3), one finds the dispersion relations of $d+1$ bands

$$\varepsilon_\mu(k) = \begin{cases} -s - 2\nu^2 s \sum_{j=1}^{d} (1+\cos k_j) & \mu = 1, \\ t & \mu = 2, \ldots, d, \\ t + 2\nu^2 t \sum_{j=1}^{d} (1+\cos k_j) & \mu = d+1, \end{cases} \tag{11.4.41}$$

where $k \in \mathscr{K}_L$ is a wave number vector defined in (4.1.17). See Fig. 11.21. The dispersion relations should be compared with the corresponding (11.3.23) for the flat-band model. One sees that the flat lowest band has become dispersive because of the nonzero hopping parameter s. This dispersion relation is indeed easily understood. From the anticommutation relations $\{\hat{b}_{u,\sigma}, \hat{a}^{\dagger}_{p,\tau}\} = 0$, which is (11.3.7), and

$$\{\hat{a}_{p,\sigma}, \hat{a}^{\dagger}_{q,\tau}\} = \begin{cases} (1+2d\nu^2)\,\delta_{\sigma,\tau} & \text{if } p = q, \\ \nu^2\,\delta_{\sigma,\tau} & \text{if } |p-q| = 1, \\ 0 & \text{otherwise} \end{cases} \tag{11.4.42}$$

one finds

$$\hat{H}_{\text{hop}}\, \hat{a}^{\dagger}_{p,\sigma} |\Phi_{\text{vac}}\rangle = -\Big\{ s\, \hat{a}^{\dagger}_{p,\sigma} + \nu^2 s \Big(2d\, \hat{a}^{\dagger}_{p,\sigma} + \sum_{\substack{q\in\mathscr{E} \\ (|q-p|=1)}} \hat{a}^{\dagger}_{q,\sigma} \Big) \Big\} |\Phi_{\text{vac}}\rangle, \tag{11.4.43}$$

which represents nearest neighbor hopping on the d-dimensional hypercubic lattice $\mathscr{E}$. The dispersion relations for the remaining d bands are exactly the same as (11.3.23)

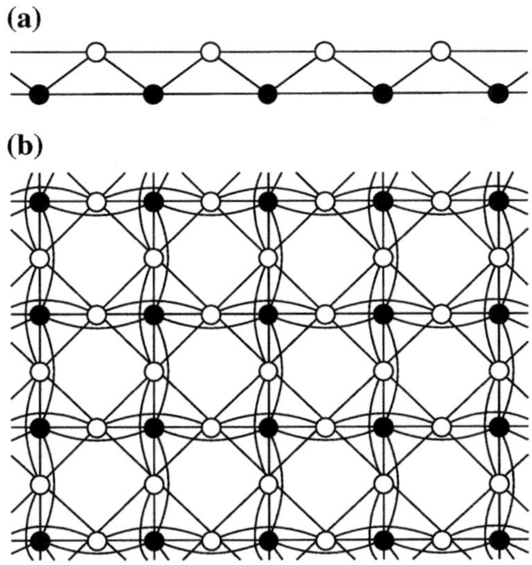

Fig. 11.20 Hopping in the non-singular model studied in Sect. 11.4.3 for **a** $d = 1$ and **b** $d = 2$. There are additional short range hopping compared with Fig. 11.11 (p. 392) for the flat-band models (© Hal Tasaki 2020. All Rights Reserved)

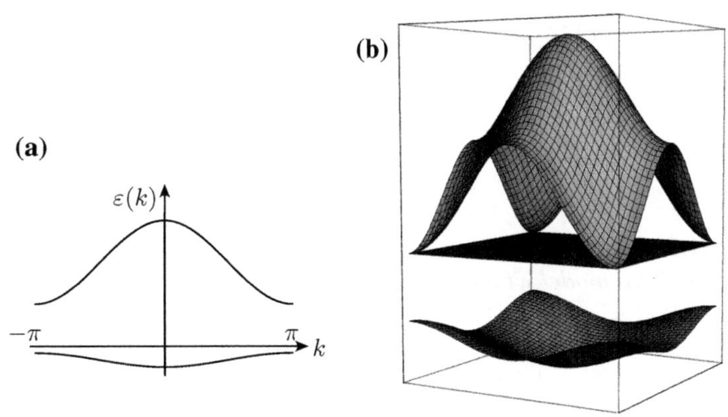

Fig. 11.21 The dispersion relations (11.4.41) of the non-singular model on the decorated d-dimensional hypercubic lattice, which should be compared with Fig. 11.13 (p. 397) for the flat-band models. **a** The two bands in the one dimensional model are now both dispersive. **b** In the two dimensional model, the lowest and the highest bands are dispersive, while the band in the middle is still flat. But the middle band has almost nothing to do with the properties of the ground states. (The horizontal axes represent $k_1, k_2 \in [-\pi, \pi]$ and the vertical axis represent $\varepsilon(k_1, k_2)$.) (© Hal Tasaki 2020. All Rights Reserved)

for the flat-band model. This is because the hopping Hamiltonian (11.4.38) does not connect $\hat{a}^\dagger$ states and $\hat{b}^\dagger$ states. (Of course the interaction Hamiltonian (11.4.37) in general mixes $\hat{a}^\dagger$ states and $\hat{b}^\dagger$ states.)

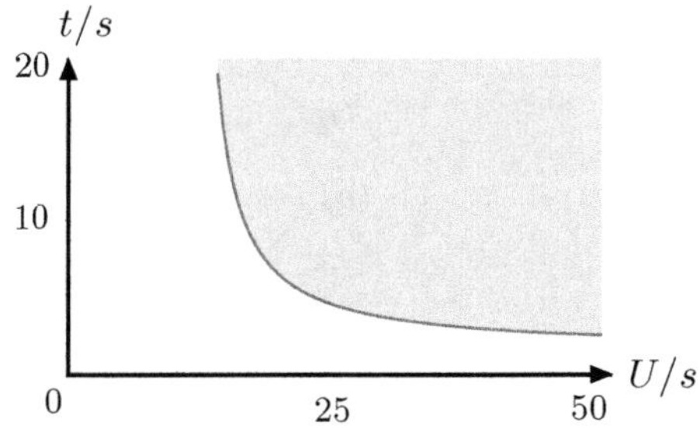

Fig. 11.22 The region in which the present Hubbard model with $d = 1$ and $\nu = 1/\sqrt{2}$ is proved to exhibit ferromagnetism. It is sufficient to chose $t/s \geq 4.5$ when $U/s = 25$, and $t/s \geq 2.6$ when $U/s = 50$ (© Hal Tasaki 2020. All Rights Reserved)

One might have noticed that both the highest and the lowest bands in (11.4.41) have perfect cosine dispersions. This is a peculiar feature of the present model, and is essentially related to the tractability of the model.[36] In a general multi-band model, one encounters dispersion relations which are not a linear functions of trigonometric functions. See, e.g., (9.3.16).

Another peculiar feature in the dispersion relations (11.4.41) is that, for $d \geq 2$, there still remain $(d-1)$ flat bands with energy t. We stress however that these flat bands have almost nothing to do with the properties of the ground states. If one does not want to have a flat-band, then it is easy to perturb the model and lift the degeneracy in such a manner that the main theorem (Theorem 11.20 below) is still valid. See Sect. 6.2 of [74].

To state the theorem, let us define, for $d \geq 2$,

$$\nu_c(d) := \sqrt{\frac{2d+1+\sqrt{4d^2+12d-7}}{4(d-1)}}. \tag{11.4.44}$$

Theorem 11.20 (Ferromagnetism in the non-singular Hubbard model) *Consider the above Hubbard model with electron number $N = |\mathcal{E}| = L^d$. Assume that $\nu > 0$ for $d = 1$, and $0 < \nu < \nu_c(d)$ for $d \geq 2$. Then for sufficiently large $t/s > 0$ and $U/s > 0$, the ground states have $S_{\text{tot}} = S_{\max} = N/2$, and are unique apart from the trivial $2S_{\max} + 1 = N + 1$ fold degeneracy.*

How large t/s and U/s should be depends only on d and ν, and not on the system size L. The theorem establishes the emergence of ferromagnetism in nearly-flat-band models, where one has $t/s \gg 1$, but it is valid even when the lower band is far from being flat. In the model with $d = 1$ and $\nu = 1/\sqrt{2}$, for example,[37] it suffices to take $t/s \geq 4.5$ when $U/s = 25$, and $t/s \geq 2.6$ when $U/s = 50$. See Fig. 11.22.

Recall that we argued in Sect. 11.4.2 that the flat-band model is stable against perturbation when $\nu \ll 1$, i.e., when $\hat{a}^\dagger$ states are almost localized at a site in $\mathcal{E}$. The

[36] We expect (or hope) that the model represents a specially tractable class of tight-binding systems, and may play the role analogous to that played by matrix product states (or the VBS state) in quantum spin chains.

[37] When $d = 1$ and $\nu = 1/\sqrt{2}$, one finds from (11.4.40) that $t_{x,x} = t - s$ for all $x \in \Lambda$. Thus the model is equivalent to that with $t_{x,x} = 0$ for all $x \in \Lambda$.

above theorem clearly covers the region where such a picture does not apply; we have $\nu_c \sim 1$ in general ($\nu_c \simeq 1.64$ for $d = 2$ and $\nu_c \simeq 1.37$ for $d = 3$), and there is no upper limit for $\nu > 0$ in $d = 1$.

Since our hopping Hamiltonian (11.4.38) is a special case of the hopping Hamiltonian (11.4.23) treated in Sect. 11.4.2, Theorem 11.19 in p. 424 on the spin-wave excitations also applies to the present model provided that $0 < \nu \ll 1, t/s \gg 1$, and U is in a suitable range. In particular, we have rigorous upper and lower bounds for the spin-wave excitation energy, which imply

$$E_{\mathrm{SW}}(k) - E_{\mathrm{GS}} \simeq 2\nu^4 U \sum_{\mu=1}^{d} (1 - \cos k_\mu), \tag{11.4.45}$$

where $E_{\mathrm{SW}}(k)$ is defined in p. 424. Thus for the range of ν, t, s, U, and d where both Theorems 11.19 and 11.20 are applicable, we have established rigorously that the ground states of the model exhibit saturated ferromagnetism, and the spin-wave excitations above the ground states have normal dispersion relations corresponding to the exchange interaction $\nu^4 U$. Starting from models of itinerant electrons, we have rigorously shown that their low energy properties perfectly coincide with those of physically natural insulating ferromagnets.

Reduction to a finite system In the rest of the present subsection, we give a complete Proof of Theorem 11.20. We start from a fundamental lemma, Lemma 11.21 below, which reduces our problem to that of a Hubbard model on a small lattice with $4d + 1$ sites. Although the basic idea of the lemma and its proof follows that in the original work [72], we here present a simplified version due to Hosho Katsura (private communication). See also [57].

The starting point is the decomposition of the Hamiltonian

$$\hat{H} = -L^d(1 + 2d\nu^2)\, s + \lambda\, \hat{H}_{\text{flat}} + \sum_{p \in \mathscr{E}} \hat{h}_p, \tag{11.4.46}$$

where

$$\hat{H}_{\text{flat}} = \sum_{\substack{u \in \mathscr{I} \\ \sigma = \uparrow, \downarrow}} \hat{b}^\dagger_{u,\sigma} \hat{b}_{u,\sigma} + \sum_{x \in \Lambda} \hat{n}_{x,\uparrow} \hat{n}_{x,\downarrow} \tag{11.4.47}$$

is nothing but the Hamiltonian of Tasaki's flat-band model studied in Sect. 11.3.1, with two parameters set as $t = U = 1$. We defined, for each $p \in \mathscr{E}$, the local Hamiltonian

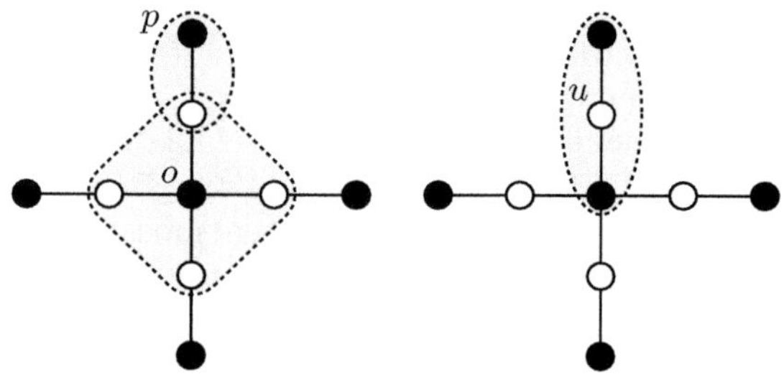

Fig. 11.23 The nine-site lattice on which the local Hamiltonian $\hat{h}_o$ for $d = 2$ acts. We have also depicted the states generated by $\hat{a}_o^\dagger$ and $\hat{a}_p^\dagger$ (left), and by $\hat{b}_u^\dagger$ (right) (© Hal Tasaki 2020. All Rights Reserved)

$$\hat{h}_p = (1 + 2d\nu^2)\, s - s \sum_{\sigma=\uparrow,\downarrow} \hat{a}_{p,\sigma}^\dagger \hat{a}_{p,\sigma} + \frac{t-\lambda}{2} \sum_{\substack{u\in\mathcal{I} \\ (|u-p|=1/2) \\ \sigma=\uparrow,\downarrow}} \hat{b}_{u,\sigma}^\dagger \hat{b}_{u,\sigma}$$
$$+ (1-\kappa)(U-\lambda)\hat{n}_{p,\uparrow}\hat{n}_{p,\downarrow} + \frac{U-\lambda}{2} \sum_{\substack{u\in\mathcal{I} \\ (|u-p|=1/2)}} \hat{n}_{u,\uparrow}\hat{n}_{u,\downarrow}$$
$$+ \frac{\kappa}{2d}(U-\lambda) \sum_{\substack{q\in\mathcal{E} \\ (|q-p|=1)}} \hat{n}_{q,\uparrow}\hat{n}_{q,\downarrow}. \tag{11.4.48}$$

We have introduced two new parameters λ and κ, which satisfy $0 < \lambda < \min\{t, U\}$ and $0 \le \kappa < 1$. Note that the local Hamiltonian $\hat{h}_p$ acts nontrivially only on the site p and its $4d$ neighbors. See Fig. 11.23. The expression (11.4.48) looks complicated, but the construction is simple; we included the $\hat{a}_{p,\sigma}^\dagger \hat{a}_{p,\sigma}$ part into $\hat{h}_p$, divided the $\hat{b}_{u,\sigma}^\dagger \hat{b}_{u,\sigma}$ into two and included in $\hat{h}_p$ with two neighboring p, and decomposed the Coulomb interaction so that there is nonzero interaction $\hat{n}_{x,\uparrow}\hat{n}_{x,\downarrow}$ on each site. Although this local lattice (on which $\hat{h}_p$ acts) may seem to be rather large, we know that this is the minim size necessary for our proof to work.

We recall that Theorem 11.11 in p. 392 for the flat-band ferromagnetism shows that the ground states of $\hat{H}_{\text{flat}}$ with $N = L^d$ electrons have maximum total spin and are exactly $N + 1$ fold degenerate. Among the ground states is the all-up state

$$|\Phi_{\boldsymbol{\alpha}\,\text{all}\,\uparrow}\rangle = \Bigl(\prod_{p\in\mathcal{E}} \hat{a}_{p,\uparrow}^\dagger\Bigr)|\Phi_{\text{vac}}\rangle, \tag{11.4.49}$$

which is (11.3.9). It is crucial to note that $\hat{h}_p|\Phi_{\boldsymbol{\alpha}\,\text{all}\,\uparrow}\rangle = 0$ for all $p \in \mathcal{E}$. To see this, one notes that the Coulomb interaction obviously vanishes in an all-up state, and also that $\hat{b}_{u,\sigma}^\dagger \hat{b}_{u,\sigma}|\Phi_{\boldsymbol{\alpha}\,\text{all}}\ \uparrow\rangle = 0$ because of the anticommutation relations (11.3.7). For the action of $\hat{a}_{p,\sigma}^\dagger \hat{a}_{p,\sigma}$, we use the anticommutation relations (11.4.42) to note $\hat{a}_{p,\uparrow}^\dagger \hat{a}_{p,\uparrow} = (1 + 2d\nu^2) - \hat{a}_{p,\uparrow}\hat{a}_{p,\uparrow}^\dagger$. We then see that $\hat{a}_{p,\uparrow}^\dagger \hat{a}_{p,\uparrow}|\Phi_{\boldsymbol{\alpha}\,\text{all}}\ \uparrow\rangle = (1 + 2d\nu^2)|\Phi_{\boldsymbol{\alpha}\,\text{all}\,\uparrow}\rangle$ since $(\hat{a}_{p,\uparrow}^\dagger)^2 = 0$. Since we clearly have $\hat{a}_{p,\downarrow}^\dagger \hat{a}_{p,\downarrow}|\Phi_{\boldsymbol{\alpha}\,\text{all}\,\uparrow}\rangle = 0$, we find $(\sum_{\sigma=\uparrow,\downarrow} \hat{a}_{p,\sigma}^\dagger \hat{a}_{p,\sigma})|\Phi_{\boldsymbol{\alpha}\,\text{all}}\ \uparrow\rangle = (1 + 2d\nu^2)|\Phi_{\boldsymbol{\alpha}\,\text{all}\,\uparrow}\rangle$.

Recall that Theorem 11.11 shows that any ground state of $\hat{H}_{\text{flat}}$ is obtained by applying an SU(2) rotation to $|\Phi_{\alpha\,\text{all}\uparrow}\rangle$. Since $\hat{h}_p$ is SU(2) invariant, this means that

$$\hat{h}_p|\Phi_{\text{GS}}^{\text{flat}}\rangle = 0, \tag{11.4.50}$$

for any ground state $|\Phi_{\text{GS}}^{\text{flat}}\rangle$ of $\hat{H}_{\text{flat}}$ with $N = L^d$, and for any $p \in \mathcal{E}$.

Lemma 11.21 *Consider the above Hubbard model with Hamiltonian $\hat{H}$ and electron number $N = |\mathcal{E}| = L^d$. If $\hat{h}_p \geq 0$ for all $p \in \mathcal{E}$, then the ground states have $S_{\text{tot}} = S_{\max} = N/2$, and are unique apart from the trivial $2S_{\max} + 1 = N + 1$ fold degeneracy.*

Proof Let $|\Phi_{\text{GS}}^{\text{flat}}\rangle$ be an arbitrary ground state with $N = |\mathcal{E}| = L^d$ of $\hat{H}_{\text{flat}}$ defined in (11.4.47). From (11.4.50) and $\hat{h}_p \geq 0$, we see that $|\Phi_{\text{GS}}^{\text{flat}}\rangle$ is a simultaneous ground state of $\hat{H}_{\text{flat}}$ and $\hat{h}_p$ for all $p \in \mathcal{E}$. This shows that the ground state energy of $\hat{H}$ is $-L^d(1 + 2d\nu^2)\,s$, and that any ground state should be a simultaneous ground state of $\hat{H}_{\text{flat}}$ and $\hat{h}_p$ for all $p \in \mathcal{E}$. (We have used Lemma A.10 in p. 469.) Then the desired uniqueness follows from Theorem 11.11 for the flat-band model. ∎

It should be noted that $\hat{h}_p$ is here regarded as an operator on the whole Hilbert space. This means that the number of electrons within the support of $\hat{h}_p$ (i.e., the $4d + 1$ sites on which $\hat{h}_p$ acts nontrivially) is not fixed. It can be anything between 0 and $2(4d + 1)$.

As is clear from the short proof, the above lemma is trivial. What is less trivial is the fact that $\hat{h}_p$ has zero eigenvalue as in (11.4.50), and what is essentially nontrivial is that there indeed exists a range of model parameters in which $\hat{h}_p$ is nonnegative.

Since $\hat{h}_p$ is after all a finite dimensional matrix, whether $\hat{h}_p \geq 0$ or not for given values of parameters may be determined numerically. In fact the region of parameters for $d = 1$ depicted in Fig. 11.22 was obtained in the original work [72] by numerically diagonalizing $\hat{h}_p$ with $\kappa = 0$.[38] Shen also performed such a computer-aided proof for higher dimensional models [53]. See also [30].

We can of course establish the nonnegativity of $\hat{h}_p$ without resorting to a computer. The following lemma will be proved below.

Lemma 11.22 *Assume that $\nu > 0$ if $d = 1$, and $0 < \nu < \nu_{\text{c}}(d)$ if $d \geq 2$, where $\nu_{\text{c}}(d)$ is defined in (11.4.44). Then for sufficiently large $t/s > 0$ and $U/s > 0$, one has $\hat{h}_p \geq 0$. (We take λ and κ to be proportional to s.)*

Proof of Theorem 11.20 *(p. 428), **given Lemmas*** 11.21 ***and*** 11.22: By simply combining the two lemmas, we get the desired main theorem. ∎

[38] One needs to set $\kappa > 0$ in Lemma 11.22. Numerical results for $d = 1$ indicate that one should set $\kappa = 0$ to get the largest range in the parameter space $(\nu, t/s, U/s)$ in which the existence of ferromagnetism is provable. (Kensuke Tamura and Hosho Katsura, private communication. See also [57].)

Why does the proof work? At this stage we wish to briefly discuss the reason why the present strategy of the proof works, especially because we ourselves still do not completely understand the reason.

The basic strategy is to first establish that ferromagnetism is realized in each small system described by the local Hamiltonian $\hat{h}_p$, and then see (by using the result for the flat-band model) that the local ferromagnetic ground states can be consistently combined to give global ferromagnetic ground states. It is rather unexpected that such a method works in an itinerant electron system with dispersive bands, since one expects that electrons behave as "waves" which are extended over the whole lattice. It is quite likely that our method relies essentially on the fact that the model behaves as insulating ferromagnet, and also on certain specific features of our model.

It seems also clear that we here have a situation similar to that in matrix product states (Sect. 7.2.2) or tensor network states (p. 213). Recall that, for a given injective matrix product state, one can always construct a parent Hamiltonian. The parent Hamiltonian is a sum of local Hamiltonians, chosen in such a way that the given MPS is a simultaneous ground state of all the local Hamiltonians. In other words, the model is frustration-free. As is noted in footnote 34 in p. 203, one has to choose the support of the local Hamiltonian large enough for the above construction to work. This is analogous to the present proof, where we needed to introduce local Hamiltonians that act on $4d + 1$ sites in order to realize a frustration-free situation. We should recall, however, that the present model is gapless and the ground states are highly degenerate because of the SU(2) symmetry. It may be interesting to understand the similarity and difference between our model and matrix product states or tensor network states, and look for the possibility of generalized theories.

Proof of Lemma 11.22 We shall prove Lemma 11.22, which is a core of our result on ferromagnetism in non-singular Hubbard models. The lemma was proved in [74] in a more general setting.[39] We here present a new transparent proof (which works only for the present model) due to Hosho Katsura and Akinori Tanaka (private communication). See also [57].

Since the system has translation invariance, it suffices to examine the local Hamiltonian $\hat{h}_o \geq 0$, where $o = (0, \dots, 0)$ is the origin of $\mathbb{Z}^d$. We here regard $\hat{h}_o$ as an operator of the electron system on a small lattice Λ_o with $4d + 1$ sites. The lattice is written as $\Lambda_o = \{o\} \cup \mathcal{E}_o \cup \mathcal{I}_o$, where $\mathcal{E}_o$ consists of $2d$ sites of the form $(0, \dots, 0, \pm 1, 0, \dots, 0)$, and $\mathcal{I}_o$ consists of $2d$ sites of the form $(0, \dots, 0, \pm 1/2, 0, \dots, 0)$. See Fig. 11.23. Then the local Hamiltonian is

[39] In [74], general models obtained by the cell construction (see the end of Sect. 11.3.1) are also treated. The Proof of Lemma 6.1 in [74] (which corresponds to our Lemma 11.23) is highly technical. We still do not know any simplifications of the proof for models other than the simplest models that we are studying here.

$$\begin{aligned}\hat{h}_o =&(1+2dv^2)\,s - s\sum_{\sigma=\uparrow,\downarrow}\hat{a}^\dagger_{o,\sigma}\hat{a}_{o,\sigma} + \frac{t-\lambda}{2}\sum_{\substack{u\in\mathscr{I}_o\\ \sigma=\uparrow,\downarrow}}\hat{b}^\dagger_{u,\sigma}\hat{b}_{u,\sigma}\\ &+(1-\kappa)(U-\lambda)\hat{n}_{o,\uparrow}\hat{n}_{o,\downarrow}+\frac{U-\lambda}{2}\sum_{u\in\mathscr{I}_o}\hat{n}_{u,\uparrow}\hat{n}_{u,\downarrow}\\ &+\frac{\kappa}{2d}(U-\lambda)\sum_{p\in\mathscr{E}_o}\hat{n}_{p,\uparrow}\hat{n}_{p,\downarrow},\end{aligned} \tag{11.4.51}$$

where we assume $0<\kappa<1$ here. The fermion operators are defined by

$$\hat{a}_{o,\sigma}=\hat{c}_{o,\sigma}-\nu\sum_{u\in\mathscr{I}_o}\hat{c}_{u,\sigma}, \tag{11.4.52}$$

and

$$\hat{b}_{u,\sigma}=\hat{c}_{u,\sigma}+\nu(\hat{c}_{o,\sigma}+\hat{c}_{2u,\sigma}), \tag{11.4.53}$$

for each $u\in\mathscr{I}_o$. Note that, by definition, $2u$ denotes a site in $\mathscr{E}_o$. These definitions are the same as previous (11.3.3) and (11.3.4). It is convenient to also define for each $p\in\mathscr{E}_o$ the operator

$$\hat{a}_{p,\sigma}:=\frac{1}{\sqrt{1+\nu^2}}(\hat{c}_{p,\sigma}-\nu\,\hat{c}_{p/2,\sigma}), \tag{11.4.54}$$

where $p/2$ is a site in $\mathscr{I}_o$. (Note that this is different from (11.3.3). We have also normalized the operator for later convenience.) See Fig. 11.23. Clearly the $4d+1$ single-electron states corresponding to the above $4d+1$ operators are linearly independent, and thus form a basis for the single-electron Hilbert space on Λ_o. We can thus represent any many-electron state on Λ_o by using $\hat{a}^\dagger$ and $\hat{b}^\dagger$ operators. We finally note that the fermion operators satisfy the anticommutation relations

$$\{\hat{a}_{o,\sigma},\hat{a}^\dagger_{o,\tau}\}=(1+2d\nu^2)\,\delta_{\sigma,\tau}\,,\quad \{\hat{a}_{o,\sigma},\hat{a}^\dagger_{p,\tau}\}=\frac{\nu^2}{\sqrt{1+\nu^2}}\,\delta_{\sigma,\tau}\,, \tag{11.4.55}$$

$$\{\hat{a}_{p,\sigma},\hat{a}^\dagger_{q,\tau}\}=\delta_{p,q}\,\delta_{\sigma,\tau}\,, \tag{11.4.56}$$

for any $p,q\in\mathscr{E}_o$, and $\sigma,\tau=\uparrow,\downarrow$, and

$$\{\hat{b}_{u,\sigma},\hat{a}^\dagger_{p,\tau}\}=0, \tag{11.4.57}$$

for any $u\in\mathscr{I}_o$ and $p\in\{o\}\cup\mathscr{E}_o$.

Our goal is to prove $\hat{h}_o\geq 0$ for any electron number when the parameters satisfy the conditions described in Lemma 11.22. We first study this problem focusing only on ferromagnetic states, i.e., states with $S_{\mathrm{tot}}=n/2$, where n denotes the electron number. Then, as we saw in the proof of Proposition 11.2 (p. 373), we only need to study the system in which all electrons have up-spin. The problem reduces to that of a non-interacting fermion system, which can be readily solved if one knows the single-electron spectrum corresponding to $\hat{h}_o$. Since $\hat{h}_o$ contains an extra constant, it is better to look at the shifted operator $\hat{h}'_o = \hat{h}_o - (1 + 2d\nu^2)s$, whose non-interacting part is $-s \sum \hat{a}^\dagger_{o,\sigma}\hat{a}_{o,\sigma} + \{(t - \lambda)/2\} \sum \hat{b}^\dagger_{u,\sigma}\hat{b}_{u,\sigma}$. One finds that there is a "band" of single-electron eigenstates generated by the $\hat{a}^\dagger$ operators. Among them the state corresponding to $\hat{a}^\dagger_o$ has the lowest eigenvalue $-(1 + 2d\nu^2)s$, and the remaining $2d$ states have eigenvalue zero. There is another "band" consisting of $2d$ states generated by the $\hat{b}^\dagger$ operators, whose eigenvalues are proportional to $t - \lambda > 0$. Going back to (non-interacting) many-electron problem, we see that the lowest eigenvalue of $\hat{h}'_o$ is $-(1 + 2d\nu^2)s$, and the second largest eigenvalue is 0. This means that, in the ferromagnetic sector, the lowest eigenvalue of the local Hamiltonian $\hat{h}_o$ is 0, and there is an energy gap $(1 + 2d\nu^2)s$ above it. Let us call these eigenstates (of $\hat{h}_o$) with zero eigenvalue the ferromagnetic zero-energy states.

It is easy to explicitly write down the ferromagnetic zero-energy states. Take an arbitrary subset A (including the empty set) of $\mathscr{E}_o$, and let

$$|\Phi_A\rangle := \hat{a}^\dagger_{o,\uparrow}\Big(\prod_{p\in A} \hat{a}^\dagger_{p,\uparrow}\Big)|\Phi_{\text{vac}}\rangle. \tag{11.4.58}$$

By using the anticommutation relations (11.4.55), (11.4.56), and (11.4.57), and the relation $(\hat{a}^\dagger_{o,\uparrow})^2 = 0$ (and recalling that the Coulomb interaction vanishes), we verify that $\hat{h}_o|\Phi_A\rangle = 0$. Since $\hat{h}_o$ is SU(2) invariant, $(\hat{S}^-_{\text{tot}})^m|\Phi_A\rangle$ with $m = 1, \ldots, |A| + 1$ is also a ferromagnetic zero-energy state.

We now take the limit $t, U \uparrow \infty$ in the Hamiltonian $\hat{h}_o$, with all the other parameters fixed. Clearly the ferromagnetic zero-energy states are not affected by the limits, and remain to be zero-energy eigenstates. We then prove the following.

Lemma 11.23 *Suppose that $\nu > 0$ satisfies the condition stated in Lemma 11.22. Then, in the limit $t, U \uparrow \infty$, the ferromagnetic zero-energy states are the only zero-energy eigenstates of $\hat{h}_o$, and any other eigenstates have strictly positive eigenvalues.*

Recall that $\hat{h}_o$ is nothing more than a finite dimensional matrix independent of the system size L. Then, we see from Lemma 11.23 and continuity that all the nonzero eigenvalues of $\hat{h}_o$ are strictly positive when t/s and U/s are sufficiently large. (Recall that we take λ and κ to be proportional to s.) This proves the desired Lemma 11.22.

Proof of Lemma 11.23 The statement of the lemma has already been verified for ferromagnetic states. We shall prove that if $|\Phi\rangle$ is any normalized state (not necessarily an energy eigenstate) with $S_{\text{tot}} < n/2$, then one has

$$\lim_{t,U\uparrow\infty} \langle\Phi|\hat{h}_o|\Phi\rangle > 0, \tag{11.4.59}$$

where the limiting value may be infinite, provided that ν satisfies the condition in Lemma 11.22. Note that, in the left-hand side, only $\hat{h}_o$ depends on t or U. Then the desired Lemma 11.23 follows from the variational principle.

Clearly we only need to examine states such that $\lim_{t,U\uparrow\infty}\langle\Phi|\hat{h}_o|\Phi\rangle < \infty$ (where again $|\Phi\rangle$ does not depend on t or U), which we shall call finite energy states. Since $\hat{b}^\dagger_{u,\sigma}\hat{b}_{u,\sigma}$ and $\hat{n}_{x,\uparrow}\hat{n}_{x,\downarrow}$ are nonnegative, a necessary and sufficient condition that $|\Phi\rangle$ is a finite energy state is $\hat{b}^\dagger_{u,\sigma}\hat{b}_{u,\sigma}|\Phi\rangle = 0$ for any $u \in \mathscr{I}_o$ and $\sigma =\uparrow, \downarrow$, and $\hat{n}_{x,\uparrow}\hat{n}_{x,\downarrow}|\Phi\rangle = 0$ for any $x \in \Lambda_o$. Note that these finite-energy conditions are exactly the same as the zero-energy conditions that we encountered in the Proof of Theorem 11.11 (p. 392) for Tasaki's flat-band ferromagnetism. We can thus proceed in almost the same manner as in the Proof of Theorem 11.11. By using the conditions $\hat{b}_{u,\sigma}|\Phi\rangle = 0$ for all $u \in \mathscr{I}_o$ and $\sigma =\uparrow, \downarrow$, and $\hat{c}_{p,\downarrow}\hat{c}_{p,\uparrow}|\Phi\rangle = 0$ for all $p \in \{o\} \cup \mathscr{E}_o$, we find that any finite-energy state $|\Phi\rangle$ with $n_\uparrow$ up-spin electrons and $n_\downarrow$ down-spin electrons is written as

$$|\Phi\rangle = \sum_{\substack{A_\uparrow, A_\downarrow \subset \{o\}\cup\mathscr{E}_o \\ (A_\uparrow\cap A_\downarrow=\emptyset,\ |A_\uparrow|=n_\uparrow,\ |A_\downarrow|=n_\downarrow)}} \alpha(A_\uparrow, A_\downarrow)\Bigl(\prod_{p\in A_\uparrow}\hat{a}^\dagger_{p,\uparrow}\Bigr)\Bigl(\prod_{p\in A_\downarrow}\hat{a}^\dagger_{p,\downarrow}\Bigr)|\Phi_{\rm vac}\rangle, \tag{11.4.60}$$

where $\alpha(A_\uparrow, A_\downarrow) \in \mathbb{C}$ is a coefficient. The electron number $n = n_\uparrow + n_\downarrow$ can take any value in the range $\{1, \ldots, 2d + 1\}$. Recall that in the case of Tasaki's flat-band ferromagnetism, the same consideration led to the spin system representation (11.3.16). Here we do not get a spin system representation since the electron number is not necessarily equal to $2d + 1$. Also note that we still have not used the finite-energy condition $\hat{c}_{u,\downarrow}\hat{c}_{u,\uparrow}|\Phi\rangle = 0$ for $u \in \mathscr{I}_o$.

Let us now assume that the above state $|\Phi\rangle$ has $S_{\rm tot} < n/2$. We clearly must have $n > 1$. We also claim that, in the expansion (11.4.60), one must have $\alpha(A_\uparrow, A_\downarrow) = 0$ whenever $A_\uparrow$ or $A_\downarrow$ contains o. To see this, note that if both the states created by $\hat{a}^\dagger_o$ and $\hat{a}^\dagger_p$ for some $p \in \mathscr{E}_o$ are occupied, then the finite energy condition $\hat{c}_{u,\downarrow}\hat{c}_{u,\uparrow}|\Phi\rangle = 0$ for $u = p/2$ leads to ferromagnetic exchange interaction between the electrons at o and p, exactly as in the case of Tasaki's flat-band ferromagnetism. (See the part "Ferromagnetic exchange interaction" in the Proof of Theorem 11.11.) Thus, whenever the state $\hat{a}^\dagger_o$ is occupied, all the n electrons couple ferromagnetically, and hence the state has the maximum total spin $S_{\rm tot} = n/2$. The claim has been proved.

We have thus found that any finite-energy state with $n_\uparrow$ up-spin electrons, $n_\downarrow$ down-spin electrons (where we write $n = n_\uparrow + n_\downarrow$) and $S_{\rm tot} < n/2$ is written as[40]

$$|\Phi\rangle = \sum_{\substack{A_\uparrow, A_\downarrow \subset \mathscr{E}_o \\ (A_\uparrow\cap A_\downarrow=\emptyset,\ |A_\uparrow|=n_\uparrow,\ |A_\downarrow|=n_\downarrow)}} \alpha(A_\uparrow, A_\downarrow)\Bigl(\prod_{p\in A_\uparrow}\hat{a}^\dagger_{p,\uparrow}\Bigr)\Bigl(\prod_{p\in A_\downarrow}\hat{a}^\dagger_{p,\downarrow}\Bigr)|\Phi_{\rm vac}\rangle. \tag{11.4.61}$$

In the rest of the proof we show that

$$\langle\Phi|\hat{h}_o|\Phi\rangle \geq \frac{s}{1+\nu^2}\bigl\{1 + (2d+1)\nu^2 + 2(1-d)\nu^4\bigr\}, \tag{11.4.62}$$

[40] The converse is, of course, not true. There are states of the form (11.4.61) that have $S_{\rm tot} = n/2$.

for any normalized $|\Phi\rangle$ with $n \geq 2$ written in the form (11.4.61), and for any t and U. It is easily found that the right-hand side is strictly positive for any $\nu > 0$ if $d = 1$ and for any ν such that $0 < \nu < \nu_c(d)$ if $d \geq 2$. Since the right-hand side is independent of t or U, we get the desired (11.4.59) for any normalized $|\Phi\rangle$ with $S_{\text{tot}} < n/2$.

Since it holds for any finite energy state $|\Phi\rangle$ that

$$\langle\Phi|\hat{h}_o|\Phi\rangle = (1+2d\nu^2)s - s\langle\Phi|\sum_{\sigma=\uparrow,\downarrow}\hat{a}^\dagger_{o,\sigma}\hat{a}_{o,\sigma}|\Phi\rangle, \tag{11.4.63}$$

our task is to evaluate the expectation value $\langle\Phi|\sum_{\sigma=\uparrow,\downarrow}\hat{a}^\dagger_{o,\sigma}\hat{a}_{o,\sigma}|\Phi\rangle$. It is convenient to define the operator

$$\hat{d}_{o,\sigma} := \hat{a}_{o,\sigma} - \frac{\nu^2}{\sqrt{1+\nu^2}}\sum_{p\in\mathscr{E}_o}\hat{a}_{p,\sigma}, \tag{11.4.64}$$

for $\sigma = \uparrow, \downarrow$. From (11.4.56), we see that $\{\hat{a}^\dagger_{p,\sigma}, \hat{d}_{o,\tau}\} = 0$ for any $p \in \mathscr{E}_o$ and σ, $\tau = \uparrow, \downarrow$. Note that the anticommutation relation and $\hat{d}_{o,\sigma}|\Phi_{\text{vac}}\rangle = 0$ imply $\hat{d}_{o,\sigma}|\Phi\rangle = 0$ for any $|\Phi\rangle$ of the form (11.4.61). Then we see from (11.4.64) that

$$\hat{a}_{o,\sigma}|\Phi\rangle = \frac{\nu^2}{\sqrt{1+\nu^2}}\sum_{p\in\mathscr{E}_o}\hat{a}_{p,\sigma}|\Phi\rangle, \tag{11.4.65}$$

and hence

$$-s\langle\Phi|\sum_{\sigma=\uparrow,\downarrow}\hat{a}^\dagger_{o,\sigma}\hat{a}_{o,\sigma}|\Phi\rangle = -\frac{s\nu^4}{1+\nu^2}\langle\Phi|\sum_{\substack{p,q\in\mathscr{E}_o\\ \sigma=\uparrow,\downarrow}}\hat{a}^\dagger_{p,\sigma}\hat{a}_{q,\sigma}|\Phi\rangle, \tag{11.4.66}$$

again for any $|\Phi\rangle$ as in (11.4.61).

To evaluate the right-hand side of (11.4.66), it is convenient to map the problem to that of a simpler Hubbard model defined on $\mathscr{E}_o$. Denoting the fermion operators for the new model again as $\hat{c}_{p,\sigma}$ with $p \in \mathscr{E}_o$ and $\sigma = \uparrow, \downarrow$, we consider the effective Hamiltonian

$$\hat{h}_{\text{eff}} = -s'\sum_{\substack{p,q\in\mathscr{E}_o\\ \sigma=\uparrow,\downarrow}}\hat{c}^\dagger_{p,\sigma}\hat{c}_{q,\sigma}, \tag{11.4.67}$$

where $s' = s\nu^4/(1+\nu^2) > 0$. Then, from the anticommutation relation (11.4.56), we find an identity

$$-s\langle\Phi|\sum_{\sigma=\uparrow,\downarrow}\hat{a}^\dagger_{o,\sigma}\hat{a}_{o,\sigma}|\Phi\rangle = \langle\tilde{\Phi}|\hat{h}_{\text{eff}}|\tilde{\Phi}\rangle. \tag{11.4.68}$$

Here the state of the new model is defined by

$$|\tilde{\Phi}\rangle = \sum_{\substack{A_\uparrow, A_\downarrow \subset \mathscr{E}_o \\ (A_\uparrow \cap A_\downarrow = \emptyset,\ |A_\uparrow| = n_\uparrow,\ |A_\downarrow| = n_\downarrow)}} \alpha(A_\uparrow, A_\downarrow)\Big(\prod_{p\in A_\uparrow} \hat{c}^\dagger_{p,\uparrow}\Big)\Big(\prod_{p\in A_\downarrow} \hat{c}^\dagger_{p,\downarrow}\Big)|\Phi_{\text{vac}}\rangle, \tag{11.4.69}$$

where the coefficients $\alpha(A_\uparrow, A_\downarrow)$ are the same as in (11.4.61).

Thus the problem reduces to that of finding a lower bound of $\hat{h}_{\text{eff}}$ in the space with no doubly occupied sites. In other words we need to evaluate the ground state energy of the Hamiltonian $\hat{P}_0\hat{h}_{\text{eff}}\hat{P}_0$, where $\hat{P}_0 = \prod_{p\in\mathscr{E}_0}(1 - \hat{n}_{p,\uparrow}\hat{n}_{p,\downarrow})$ is the projection operator onto the space with no doubly occupancy, known as the Gutzwiller projection. See (11.1.10). The problem is trivial for $d = 1$, where $\mathscr{E}_o$ contains only two sites. We only need to treat basis states of the form $|\Gamma_{\sigma,\tau}\rangle = \hat{c}^\dagger_{1,\sigma}\hat{c}^\dagger_{2,\tau}|\Phi_{\text{vac}}\rangle$ with $\sigma, \tau = \uparrow, \downarrow$, where we wrote $\mathscr{E}_o = \{1, 2\}$. Noting that $\hat{P}_0\hat{h}_{\text{eff}}\hat{P}_0|\Gamma_{\sigma,\tau}\rangle = -2s'|\Gamma_{\sigma,\tau}\rangle$ for any σ, τ, we find $\hat{P}_0\hat{h}_{\text{eff}}\hat{P}_0 \geq -2s' = -2s\nu^4/(1+\nu^2)$. Recalling (11.4.63), we see that

$$\langle\Phi|\hat{h}_o|\Phi\rangle \geq (1 + 2\nu^2)s - \frac{2s\nu^4}{1+\nu^2}, \tag{11.4.70}$$

which is the same as the desired (11.4.62) with $d = 1$.

The case with $d > 1$ is not difficult, but we need a clever trick. As was pointed out in [78], this problem can be solved exactly by using the method found by Brandt and Giesekus [5]. By using easily verifiable relations[41]

$$-\hat{P}_0\,\hat{c}^\dagger_{p,\sigma}\hat{c}_{q,\sigma}\,\hat{P}_0 = \hat{c}_{q,\sigma}\,\hat{P}_0\,\hat{c}^\dagger_{p,\sigma}, \tag{11.4.71}$$

$$\hat{P}_0\Big(1 - \sum_{\tau=\uparrow,\downarrow}\hat{c}^\dagger_{p,\tau}\hat{c}_{p,\tau}\Big)\hat{P}_0 = \hat{c}_{p,\sigma}\,\hat{P}_0\,\hat{c}^\dagger_{p,\sigma}, \tag{11.4.72}$$

for any $p, q \in \mathscr{E}_o$ with $p \neq q$, and any $\sigma = \uparrow, \downarrow$, one can show that

$$\hat{P}_0\hat{h}_{\text{eff}}\hat{P}_0 = s'\sum_{\sigma=\uparrow,\downarrow}\Big(\sum_{p\in\mathscr{E}_o}\hat{c}_{p,\sigma}\Big)\hat{P}_0\Big(\sum_{p\in\mathscr{E}_o}\hat{c}^\dagger_{p,\sigma}\Big) - 2s'|\mathscr{E}_o|\,\hat{P}_0 + s'\hat{P}_0\Big(\sum_{\substack{p\in\mathscr{E}_o\\ \sigma=\uparrow,\downarrow}}\hat{n}_{p,\sigma}\Big)\hat{P}_0. \tag{11.4.73}$$

Noting that the first term in the right-hand side is nonnegative, and that we are working with states with more than one electrons, i.e., $\sum_{p,\sigma}\hat{n}_{p,\sigma} \geq 2$, we get[42]

$$\hat{P}_0\hat{h}_{\text{eff}}\hat{P}_0 \geq -2s'\,2d + 2s'. \tag{11.4.74}$$

Substituting this bound back into (11.4.63), we finally get

[41] One may check these relations by applying the left-hand and right-hand sides to relevant basis states and see that they yield the same results.

[42] In fact it is also easy to write down the ground states explicitly. See [78].

$$\langle \Phi | \hat{h}_o | \Phi \rangle \geq (1 + 2d\nu^2)s + \frac{s\nu^4}{1+\nu^2}(2 - 4d), \tag{11.4.75}$$

which is nothing but the desired (11.4.62). ∎

11.5 Toward Metallic Ferromagnetism

Metallic ferromagnetism is a fascinating phenomenon in which electrons exhibiting ferromagnetism also contribute to electric conduction. Needless to say iron is a typical substance which exhibits metallic ferromagnetism. Whether such a simple model as the Hubbard model can describe metallic ferromagnetism is an intriguing question.[43]

Nagaoka's ferromagnetism treated in Sect. 11.2 is certainly motivated by metallic ferromagnetism, and we believe that it sheds some light on possible mechanisms of metallic ferromagnetism. But, in the situation where the theorem is proved, the only dynamical freedom comes from the motion of a single hole. We never expect that the single hole can contribute to appreciable electric current in a bulk system. Nagaoka's ferromagnetism should better be interpreted as an insulator.

In Tasaki's flat-band model discussed in Sect. 11.3.1 and the related nearly-flat band models discussed in Sect. 11.4, the existence of ferromagnetism is proved for special electron numbers. These electron numbers correspond to the half-filling of the lowest bands, but since the ground states are ferromagnetic, the lowest bands become effectively fully filled. Then the systems should behave as Mott insulators.[44]

As we shall discuss below in Sect. 11.5.1, it is expected that the nearly-flat band models of Sect. 11.4 exhibit metallic ferromagnetism when the electron number N is strictly less than $L^d = |\mathscr{E}|$. Unfortunately we still do not have any rigorous results which confirm this conjecture. We believe that to rigorously justify the conjecture is one of challenging important problems in mathematical physics of quantum many-body systems.

The first rigorous example of metallic ferromagnetism in the Hubbard model was obtained in 1999 by Tanaka and Idogaki [61]. They followed Kubo's pioneering work in 1982 [22], and applied the Perron–Frobenius argument to a one-dimensional model with infinitely large band gap and Coulomb interaction. Their proof makes full use of the one-dimensional nature of the problem, and cannot be extended to models in higher dimensions. In 2007, Tanaka and Tasaki [63] proved that a (rather complicated) Hubbard model in any dimensions exhibits metallic ferromagnetism, again when the

---

[43]Recall that our aim here is not to build realistic models of existing materials, but to understand fundamental and universal mechanism of various physical phenomena including metallic ferromagnetism.

[44]In the ground state of Mielke's flat-band model (Sect. 11.3.2), the empty (dispersive) second lowest band touches the lowest flat-band, which is filled by ferromagnetically coupled electrons. It is likely that the model exhibits electric conduction, but the situation is slightly different from standard metals. In what follows we only focus on metals which have a partially filled band.

band gap and the Coulomb interaction are infinitely large. In Sect. 11.5.2 we prove, in the spirit of Kubo [22] and Tanaka and Idogaki [61], that a certain one-dimensional Hubbard model exhibits metallic ferromagnetism, and also discuss the theorem by Tanaka and Tasaki [63].

We admit that this final section of the whole book is a kind of anti-climax, in which we are able to present neither strong rigorous results nor promising nontrivial physical arguments. We nevertheless wish to have this section, in order to show some preliminary mathematical results provable within currently available techniques and to let the readers know of a widely open problem.

Given the fact that we are so familiar in our daily lives with metallic ferromagnetism stable at room temperatures, to prove the existence of metallic ferromagnetism (say, in a certain version of the Hubbard model) at low enough temperatures may appear as a modest goal. From theoretical and mathematical points of views, however, the problem looks formidably difficult.[45] It seems that not only mathematical techniques but fundamental understanding of physics of itinerant electron ferromagnetism is sill lacking.

11.5.1 Heuristic Arguments

Promising candidates of simple models exhibiting metallic ferromagnetism are the nearly-flat-band models studied in Sect. 11.4 with electron number N strictly less than $L^d = |\mathscr{E}|$, or, more precisely, with density N/L^d strictly less than one.[46] A rough argument is as follows. When N is less than L^d there appear $L^d - N$ "holes" in the $\hat{a}^\dagger$ state representation similar to (11.3.16) of the ground state. Even in such a situation the mechanism which generates the ferromagnetic exchange interaction between neighboring electrons should be still present. It is also important that electrons in the $\hat{a}^\dagger$ states are never at rest since the corresponding band is dispersive. Thus if N/L^d is close to 1, it is expected that all the electrons eventually interact with other electrons to generate a ferromagnetic state, i.e., a state with the maximum possible total spin. Then the model effectively reduces to that of non-interacting spinless fermions (as we have seen in the Proof of Proposition 11.2 in p. 373), and the ground states are written as Slater determinant states. Since the effective filling factor N/L^d of the lowest band is strictly less than one, the ground states correspond to a metallic ferromagnet.[47]

---

[45] Recall that we are not even able to prove the existence of ferromagnetic order in the three dimensional ferromagnetic Heisenberg model at low enough temperatures. See Sect. 4.4.4.

[46] The flat-band models of Sect. 11.3 with electron numbers less than the half-filling were also studied in [35, 40, 67], and it was proved that the models still exhibit ferromagnetism in two or higher dimensions when the electron fillings are sufficiently large. We however believe that the models (as they are) are not relevant to metallic ferromagnetism since the electrons in the lowest flat bands do not contribute to conduction.

[47] We are here relying on the standard criterion (which can be found in any textbook in condensed matter physics) that a non-interacting fermion system with a partially filled band describes a metallic

The above picture is partially supported by the Wannier state perturbation theory, as we shall discuss below. Penc, Shiba, Mila, and Tsukagoshi [45] made a detailed study of a related problem in one dimension, and found both theoretical and numerical evidences that the model exhibits metallic ferromagnetism for arbitrary filling factor $\nu = N/(2|\Lambda|)$ in the range $0 < \nu < \nu_0 = 1/4$ (where ν_0 corresponds to the half-filling of the lowest nearly-flat band). See [50] for a similar numerical results, and [79, 80] for numerical results for related models with filling factor larger than ν_0.

Wannier state perturbation theory Let us present here a heuristic perturbative argument which supports the above picture about metallic ferromagnetism in the nearly-flat band model with electron number less than L^d. We in particular show that the low-energy effective theory of the Hubbard model is a model known as the ferromagnetic t-J model.

We consider exactly the same one-dimensional model as in Sect. 11.4.1, but with electron number N strictly less than $L = |\mathscr{E}|$. Recall that in the unperturbed ground states (11.4.11) for $N = L$, every Wannier state centered at $p \in \mathscr{E}$ is occupied by a single electron, with either up or down spin. Now, as we have less electrons, there appear some unoccupied sites, which may be identified as holes. Unperturbed ground states in this case are thus given by

$$|\Omega_{A_\uparrow,A_\downarrow}\rangle = \Bigl(\prod_{p\in A_\uparrow} \hat{d}^\dagger_{p,\uparrow}\Bigr)\Bigl(\prod_{p\in A_\downarrow} \hat{d}^\dagger_{p,\downarrow}\Bigr)|\Phi_{\mathrm{vac}}\rangle, \tag{11.5.1}$$

with $A_\uparrow, A_\downarrow \subset \mathscr{E}$ such that $|A_\uparrow| + |A_\downarrow| = N$ and $A_\uparrow \cap A_\downarrow = \emptyset$. We denote by $\tilde{\mathscr{H}}_N$ the effective low energy Hilbert space spanned by the states (11.5.1). As in Sect. 11.4.1, we examine low energy properties of the model perturbatively within the space $\tilde{\mathscr{H}}_N$.

This is indeed not difficult since all necessary estimates have already been done in Sect. 11.4.1. Let us here make an additional assumption that the effective hopping amplitude τ, which appears in (11.4.6) and (11.4.7), is sufficiently small so that the "second order" exchange interaction $J_2 \simeq 4\tau^2/U$ in (11.4.20) is negligible. Therefore when two neighboring sites $p, p+1 \in \mathscr{E}$ are both occupied, we get a ferromagnetic exchange interaction exactly as in (11.4.17). When there is a vacant site adjacent to an occupied site, then the electron may hop to the vacant site with the effective hopping amplitude τ.[48]

Let $\hat{\tilde{P}}_0$ be the projection operator onto the space $\tilde{\mathscr{H}}_N$. We can write the the low-energy effective Hamiltonian, which is an operator on $\tilde{\mathscr{H}}_N$, as[49]

---

state. Although we are dealing with a system of strongly interacting electrons, it behaves as a non-interacting system within the space of states with $S_{\mathrm{tot}} = N/2$. There must be electric conduction, at least within the ferromagnetic sector.

[48] When there are vacant sites there appears new contribution from the "second order" perturbation which involves three sites. See, e.g., Sect. 5.1 of [10], Appendix 2.A of [9], and Sect. 3.2 of [2]. We can neglect this term since we assumed that τ^2/U is negligible.

[49] We have subtracted a trivial term $2\tau \sum_{p=1}^{L} \hat{\tilde{n}}_p$ from the effective Hamiltonian.

$$\hat{H}_{\text{eff}} = \tau\,\hat{\hat{P}}_0 \sum_{\substack{p\in\{1,\ldots,L\}\\ \sigma=\uparrow,\downarrow}} (\hat{d}^\dagger_{p,\sigma}\hat{d}_{p+1,\sigma} + \text{h.c.}) + J\sum_{p=1}^{L}\Big(\frac{\hat{\hat{n}}_p\hat{\hat{n}}_{p+1}}{4} - \hat{\hat{\mathbf{S}}}_p\cdot\hat{\hat{\mathbf{S}}}_{p+1}\Big), \tag{11.5.2}$$

where "h.c." stands for Hermitian conjugate, and $\hat{\hat{n}}_p = \sum_{\sigma=\uparrow,\downarrow}\hat{d}^\dagger_{p,\sigma}\hat{d}_{p,\sigma}$ is the number operator for the Wannier state $\boldsymbol{\omega}_p$. We here wrote $J \simeq J_1 > 0$.

We have thus obtained an effective model in which (i) double occupancy of a single site is inhibited (because of the strong Coulomb interaction), (ii) electrons may hop to neighboring sites, and (iii) the spin degrees of freedom of two electrons occupying neighboring sites interact ferromagnetically. This is in fact a standard model known as the (ferromagnetic) t-J model.

It should be clear that the above derivation works in much more general models in any dimensions. As in Sect. 11.3.1, let $\mathscr{E}$ denote the d-dimensional $L\times\cdots\times L$ hypercubic lattice with periodic boundary conditions, and consider a model obtained by slightly perturbing the hopping Hamiltonian of the flat-band Hubbard model on the decorated lattice $\Lambda = \mathscr{E}\cup\mathscr{I}$. Then by developing a similar Wannier state perturbation argument, we find that the low energy effective theory of the model with electron number N strictly less than L^d is the ferromagnetic t-J model with the Hamiltonian

$$\hat{H}_{\text{eff}} = \tau\,\hat{\hat{P}}_0 \sum_{\substack{p,q\in\mathscr{E}\\ (|p-q|=1)\\ \sigma=\uparrow,\downarrow}} \hat{d}^\dagger_{p,\sigma}\hat{d}_{q,\sigma} + \frac{J}{2}\sum_{\substack{p,q\in\mathscr{E}\\ (|p-q|=1)}}\Big(\frac{\hat{\hat{n}}_p\hat{\hat{n}}_q}{4} - \hat{\hat{\mathbf{S}}}_p\cdot\hat{\hat{\mathbf{S}}}_q\Big), \tag{11.5.3}$$

with $J>0$. Here $\hat{d}_{p,\sigma}$ is the fermion operator for the Wannier state (of the lowest band) centered around p, and $\hat{\hat{n}}_p$ and $\hat{\hat{\mathbf{S}}}_p$ are the corresponding number operator and the spin operator, respectively.

Since the ferromagnetic t-J model contains both electron hopping and ferromagnetic interaction, it is quite likely that the present model exhibits metallic ferromagnetism. But the story is not that simple, as we shall see in the next section. In short, ferromagnetism is always expected in one dimension, but, in higher dimensions, ferromagnetism may appear only when the electron density N/L^d is close to 1 and J/τ is large enough.

11.5.2 Rigorous Results

We shall discuss some rigorous results related to metallic ferromagnetism. As we have already stressed we still do not have any decisive theorems. We discuss results about the ferromagnetic t-J model and two different versions of the Hubbard model.

Ferromagnetism in the ferromagnetic t-J model Before dealing with the Hubbard model, we shall take a look at the ferromagnetic t-J model, which (at least apparently) looks easier. (But we will see that the essential problem may not be any easier.)

We have seen above that the ferromagnetic t-J model appears (approximately) as a low energy effective theory of the Hubbard model with a nearly-flat-band. Here we treat it as a given theoretical model.

Let us still denote by $\mathscr{E}$ the d-dimensional $L \times \cdots \times L$ hypercubic lattice with periodic boundary conditions, which is Λ_L in (3.1.2). We consider an electron model on $\mathscr{E}$ with electron number N strictly less than $L^d = |\mathscr{E}|$. Let $\mathscr{H}_N^{\rm hc}$ be the Hilbert space of N electron states with no doubly occupied sites. More precisely $\mathscr{H}_N^{\rm hc}$ is spanned by the basis states (9.2.35) which satisfy $x_i \neq x_j$ whenever $i \neq j$. We denote by $\hat{P}_{\rm hc}$ the projection operator onto $\mathscr{H}_N^{\rm hc}$. The Hamiltonian of the ferromagnetic t-J model, which is as an operator on $\mathscr{H}_N^{\rm hc}$, is given by

$$\hat{H}_{tJ} = -\tau\,\hat{P}_{\rm hc} \sum_{\substack{x,y\in\mathscr{E}\\(|x-y|=1)\\ \sigma=\uparrow,\downarrow}} \hat{c}^\dagger_{x,\sigma}\hat{c}_{y,\sigma} + J \sum_{\substack{x,y\in\mathscr{E}\\(|x-y|=1)}} \Bigl(\frac{\hat{n}_x\hat{n}_y}{4} - \hat{\boldsymbol{S}}_x\cdot\hat{\boldsymbol{S}}_y\Bigr), \tag{11.5.4}$$

where $\tau > 0$ is the hopping amplitude[50] and $J > 0$ is the ferromagnetic coupling constant.

Let us first focus on the ferromagnetic sector, i.e., the space of states which have maximum total spin $S_{\rm tot} = N/2$. It is easily seen that if $|\Phi_\uparrow\rangle$ is any state which consists only of up-spin electrons, then $(\hat{n}_p\hat{n}_q/4 - \hat{\boldsymbol{S}}_q\cdot\hat{\boldsymbol{S}}_p)|\Phi_\uparrow\rangle = 0$ for any $p \neq q$. Because both $\hat{n}_p\hat{n}_q$ and $\hat{\boldsymbol{S}}_q\cdot\hat{\boldsymbol{S}}_p$ are SU(2) invariant, we see that $(\hat{n}_p\hat{n}_q/4 - \hat{\boldsymbol{S}}_q\cdot\hat{\boldsymbol{S}}_p)|\Phi\rangle = 0$ for any $|\Phi\rangle$ in the ferromagnetic sector. Since there are no doubly occupied sites in a state in the ferromagnetic sector (see the Proof of Proposition 11.2 in p. 373), the problem reduces to that of a non-interacting spinless fermions with nearest neighbor hopping.

Let $\mathscr{K}_L$ be the k-space defined in (4.1.17). For each $k = (k_1, \ldots, k_d) \in \mathscr{K}_L$, we define the plane wave state by $\boldsymbol{\eta}^{(k)} = (L^{-d/2}\, e^{ik\cdot x})_{x\in\mathscr{E}}$, where we wrote $x = (x_1, \ldots, x_d) \in \mathscr{E}$ and $k \cdot x = \sum_{j=1}^d k_j x_j$. Take an arbitrary subset $\Gamma \subset \mathscr{K}_L$ with $|\Gamma| = N$ and an aribtrary $M = -N/2, \ldots, N/2$, and define

$$|\Phi_{\Gamma,M}\rangle := (\hat{S}^-_{\rm tot})^{(N/2)-M}\Bigl(\prod_{k\in\Gamma} \hat{C}^\dagger{}_\uparrow(\boldsymbol{\eta}^{(k)})\Bigr)|\Phi_{\rm vac}\rangle, \tag{11.5.5}$$

which is a state in the ferromagnetic sector with $\hat{S}^{(3)}_{\rm tot}|\Phi_{\Gamma,M}\rangle = M|\Phi_{\Gamma,M}\rangle$. It is then readily verified that

$$\hat{H}_{tJ}|\Phi_{\Gamma,M}\rangle = \Bigl(\sum_{k\in\Gamma}\varepsilon(k)\Bigr)|\Phi_{\Gamma,M}\rangle, \tag{11.5.6}$$

[50] For convenience, we here use a different sign convention for the hopping term.

where the single-electron energy eigenstate corresponding to $\eta^{(k)}$ is $\varepsilon(k) = -2\tau \sum_{j=1}^{d} \cos k_j$. Choosing the subset Γ that minimizes the energy $\sum_{k \in \Gamma} \varepsilon(k)$, we obtain the lowest energy states within the ferromagnetic sector. The essential question is whether and when theses lowest energy states also become the global ground states of the t-J model.

For $d = 1$ the situation is simple. We can show that, for any $J > 0$, the ground states are in the ferromagnetic sector when electron number N is odd. The proof is based on a simple application of the Perron–Frobenius theorem. It was Kubo who first applied the Perron–Frobenius theorem to the problem of metallic ferromagnetism in his pioneering work on the double exchange model [22].

Proposition 11.24 (Ferromagnetism in the $d = 1$ ferromagnetic t-J model) *Assume that the dimension is $d = 1$ and the electron number N satisfies $N < L = |\mathscr{E}|$ and is odd. Then the ground states of the t-J model with the Hamiltonian (11.5.4) have total spin $S_{\mathrm{tot}} = N/2$, and are non-degenerate apart from the trivial $2S_{\mathrm{tot}} + 1 = N + 1$-fold degeneracy.*

The reader might find it odd that the electron number should be odd in the above proposition. In fact if we use open boundary conditions instead of periodic boundary conditions, then the proof works for any electron number N which does not exceed L. Note that, for $N = L$, the model reduces to the ferromagnetic Heisenberg model, whose ground state is ferromagnetic both for even and odd $N = L$.

Proof of Proposition 11.24 The proof is based on an elementary observation, which may be called a rigorous "spin-charge separation" argument. (See [45] and references therein.) We consider a subspace of $\mathscr{H}_N^{\mathrm{hc}}$ which consists of states with $S_{\mathrm{tot}}^{(3)} = 1/2$. Note that Theorem A.17 in p. 473 guarantees that any eigenstate of $\hat{H}_{tJ}$ has its copy in this subspace. The subspace is spanned by the basis states (9.2.35) such that $\sum_{j=1}^{N} \sigma_j = 1/2$. (Here we identify $\uparrow$ and $\downarrow$ with $1/2$ and $-1/2$, respectively.) Without loss of generality, we can assume $1 \le x_1 < x_2 < \cdots < x_N \le L$, where we wrote $\mathscr{E} = \{1, 2, \ldots, L\}$.

We first claim that all the matrix elements of $\hat{H}_{tJ}$ in the present basis are nonpositive. For the hopping term, this is obvious since there is no exchange in electron ordering and one does not have to worry about fermion signs.[51] For the exchange term, the nonpositivity is easily verified by using the identity $\hat{\boldsymbol{S}}_x \cdot \hat{\boldsymbol{S}}_{x+1} = (\hat{S}_x^+ \hat{S}_{x+1}^- + \hat{S}_x^- \hat{S}_{x+1}^+)/2 + \hat{S}_x^{(3)} \hat{S}_{x+1}^{(3)}$. It is also verified that all the basis states are connected via nonvanishing negative matrix elements of $\hat{H}_{tJ}$. Note that the existence of the exchange term is essential for the connectivity. Therefore we can readily apply the Perron–Frobenius theorem (Theorem A.18 in p. 475) to conclude that the ground state $|\Phi_{\mathrm{GS}}\rangle$ is unique (in this sector), and it is a linear combination of all the basis states with positive coefficients. This means that the ground state has nonzero overlap

[51] A hop between sites 1 and L is exceptional, but it does not produce any sign if N is odd. When N is even, such a hop generates the "wrong" sign for the Perron–Frobenius theorem.

with the ferromagnetic state (2.4.11) (with a proper identification of sites), and hence has $S_{\rm tot} = N/2$. ∎

Unfortunately, Proposition 11.24 and its proof hardly shed light on the nature of the ground states of the higher dimensional t-J model. The one-dimensional problem is very special in the sense that no exchanges (in the spin configuration) are caused by hopping of electrons. This means that the ground states for $J = 0$ can have any values of $S_{\rm tot}$, from $1/2$ to $N/2$. This enormous degeneracy is lifted by infinitesimal $J > 0$, which produces ferromagnetism.

In higher dimensional lattices (or even in slightly more complicated one-dimensional systems, e.g., those with next-nearest neighbor hopping or on a ladder), the problem is essentially different since highly nontrivial exchange processes already take place when $J = 0$. There is no guarantee that the ground states are ferromagnetic when J is positive. In fact it is likely that, when the electron density is very low, the model in two or higher dimensions behaves essentially as an non-interacting system and the ground states exhibit Pauli paramagnetism (see Sect. 9.3.2). We believe that it is not very difficult to extend Theorem 11.4 in p. 376 to the t-J model to prove that the ground states never exhibit saturated ferromagnetism for any $J > 0$ provided that $d > 2$ and the density is sufficiently low.[52]

We of course believe that the ferromagnetic t-J model in higher dimensions also exhibits ferromagnetism when the ferromagnetic coupling $J/|t|$ is sufficiently large and the density N/L^d is sufficiently close to one. But nothing has been rigorously shown as far as we know, and we have no idea at all about how such a result can be proved, even in the singular limit $J \uparrow \infty$. Although "establishing ferromagnetism in the ferromagnetic t-J models" might sound like a tautology at first glance, it is indeed a very difficult problem, whose solution should shed light on various aspects of strongly interacting itinerant electron systems.

A version of the Hubbard model related to the t-J model Let us now turn to the Hubbard model. It would be desirable if we could treat, for example, the nearly-flat-band model studied in Sect. 11.4.3 with electron number N strictly less than $L^d = |\mathcal{E}|$. The model is of course artificial, but is probably sufficiently simple as a theoretical toy model. Unfortunately this problem is very difficult, probably much more difficult than one might expect. For the moment we cannot prove any nontrivial statements about the model.

We here study a more complicated version of the Hubbard model proposed by Tanaka and Tasaki [64], for which an exact equivalence to the t-J model can be proved in a certain limit. By using the equivalence we can establish the emergence of ferromagnetism in the ground state, provided that we restrict ourselves to the one-dimensional model.

Let us introduce the model in any dimension. Exactly as in Sect. 11.3.1, we denote by $\mathcal{E}$ the d-dimensional $L \times L$ hypercubic lattice, and by $\mathcal{I}$ the set of sites taken at the middle of bonds in $\mathcal{E}$. See Fig. 11.10. Here we follow Sekizawa [51], and consider

[52] Instead of the Gutzwiller projection, we use the projection operator on states in which an up-spin electron and a down-spin electron never come to neighboring sites.

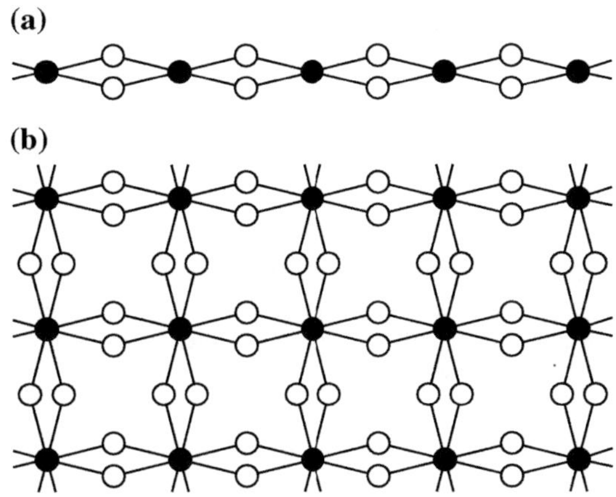

Fig. 11.24 The decorated hypercubic lattice used to define the model studied in the present section. Compared with the decorated hypercubic lattice in Fig. 11.10, the internal sites (i.e., white dots) are duplicated. The present lattice was studied by Sekizawa, who considered a version of flat-band ferromagnetism [51] (© Hal Tasaki 2020. All Rights Reserved)

a decorated hypercubic lattice $\Lambda = \mathscr{E} \cup \mathscr{I} \times \{1, 2\}$. Compared with the decorated hypercubic lattice in Sect. 11.3.1, which is $\mathscr{E} \cup \mathscr{I}$, the sites in $\mathscr{I}$ are duplicated. See Fig. 11.24. We denote sites in $\mathscr{E}$ again as $p, q, \ldots$, and sites in $\mathscr{I} \times \{1, 2\}$ as $(u, 1)$ and $(u, 2)$ with $u \in \mathscr{I}$.

We shall again define localized states $\boldsymbol{\alpha}_p$ and $\boldsymbol{\beta}_{(u,\zeta)}$ on the present decorated lattice. For $j = 1, \ldots, d$, we denote by $\boldsymbol{b}_j = (0, \ldots, 0, \frac{1}{2}, 0, \ldots, 0)$ the half of the unit vector in the jth direction. For $p \in \mathscr{E}$, we define $\boldsymbol{\alpha}_p = (\alpha_p(x))_{x\in\Lambda}$ by

$$\alpha_p(x) = \begin{cases} 1 & \text{if } x = p, \\ \nu & \text{if } x = (p + \boldsymbol{b}_j, 1), \\ -\nu & \text{if } x = (p + \boldsymbol{b}_j, 2) \text{ or } (p - \boldsymbol{b}_j, \zeta), \\ 0 & \text{otherwise}, \end{cases} \tag{11.5.7}$$

where $j = 1, \ldots, d$ and $\zeta = 1, 2$. Note that $p \pm \boldsymbol{b}_j$ is a site in $\mathscr{I}$ for any $j = 1, \ldots, d$. For $u \in \mathscr{I}$ and $\zeta = 1, 2$, we define $\boldsymbol{\beta}_{(u,\zeta)} = (\beta_{(u,\zeta)}(x))_{x\in\Lambda}$ by

$$\beta_{(u,1)}(x) = \begin{cases} 1 & \text{if } x = u, \\ \nu & \text{if } x = u + \boldsymbol{b}_j, \\ -\nu & \text{if } x = u - \boldsymbol{b}_j, \\ 0 & \text{otherwise}, \end{cases} \tag{11.5.8}$$

and

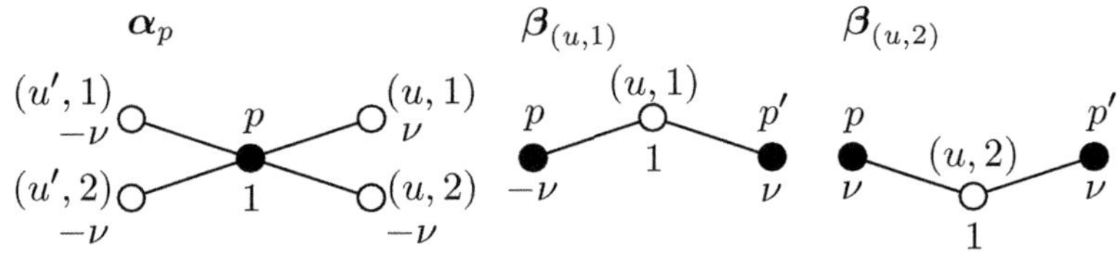

Fig. 11.25 Nonzero components of the localized states $\boldsymbol{\alpha}_p$, $\boldsymbol{\beta}_{(u,1)}$, and $\boldsymbol{\beta}_{(u,2)}$ for $d = 1$. The lattice sites are related by $u' = p - \frac{1}{2}$, $u = p + \frac{1}{2}$, and $p' = p + 1$. The particular choice of the signs of the components of $\boldsymbol{\alpha}_p$ ensures that these states with different p are orthogonal as in (11.5.11)

$$\beta_{(u,2)}(x) = \begin{cases} 1 & \text{if } x = u, \\ \nu & \text{if } x = u \pm \boldsymbol{b}_j, \\ 0 & \text{otherwise,} \end{cases} \tag{11.5.9}$$

where j is the unique index such that $u \pm \boldsymbol{b}_j \in \mathscr{E}$. Here $\nu > 0$ is a parameter of the model. See Fig. 11.25. It is not hard to show that these states are linearly independent and hence they altogether form a basis of the single-electron Hilbert space $\mathfrak{h}$.

We define fermion operators corresponding to these localized states by

$$\hat{a}_{p,\sigma} := \hat{C}_\sigma(\boldsymbol{\alpha}_p), \quad \hat{b}_{(u,\zeta),\sigma} := \hat{C}_\sigma(\boldsymbol{\beta}_{(u,\zeta)}), \tag{11.5.10}$$

for any $p \in \mathscr{E}$, $u \in \mathscr{I}$, $\zeta = 1, 2$, and $\sigma =\uparrow, \downarrow$. Then from the definition (11.5.7), one finds that the $\hat{a}$ operators satisfy

$$\{\hat{a}_{p,\sigma}, \hat{a}^\dagger_{q,\tau}\} = (1 + 4d\nu^2)\, \delta_{p,q}\, \delta_{\sigma,\tau}, \tag{11.5.11}$$

for any $p, q \in \mathscr{E}$ and $\sigma, \tau =\uparrow, \downarrow$, which, apart from the normalization, is the same as the basic anticommutation relation (9.2.28) for fermion operators. This fact makes the analysis of the present model a lot easier than that of the previous (similar) models. The main reason for the introduction of the duplicated internal sites was to guarantee the orthogonality. It is also easy to find, as in the previous models, that

$$\{\hat{b}_{(u,\zeta),\sigma}, \hat{a}^\dagger_{p,\tau}\} = 0, \tag{11.5.12}$$

for any $u \in \mathscr{I}$, $\zeta = 1, 2$, $p \in \mathscr{E}$, and $\sigma, \tau =\uparrow, \downarrow$.

We consider the Hubbard model on Λ with the Hamiltonian $\hat{H} = \hat{H}_{\text{hop}} + \hat{H}_{\text{int}}$, where the interaction Hamiltonian is

$$\hat{H}_{\text{int}} = U \sum_{x \in \Lambda} \hat{n}_{x,\uparrow} \hat{n}_{x,\downarrow}, \tag{11.5.13}$$

which is (9.3.29), and the hopping Hamiltonian is

$$\hat{H}_{\text{hop}} = t \sum_{\substack{u\in\mathscr{I},\,\zeta=1,2\\ \sigma=\uparrow,\downarrow}} \hat{b}^{\dagger}_{(u,\zeta),\sigma}\hat{b}_{(u,\zeta),\sigma} - s \sum_{\substack{p,q\in\mathscr{E}\\ (|p-q|=1)\\ \sigma=\uparrow,\downarrow}} \hat{a}^{\dagger}_{p,\sigma}\hat{a}_{q,\sigma}, \tag{11.5.14}$$

where $s \in \mathbb{R}\backslash\{0\}$ and $t > 0$. Note that this is similar to but is different from the hopping Hamiltonian (11.4.38) of the non-singular model studied in Sect. 11.4.3. Since $\hat{a}$ operators in this model satisfy the standard anticommutation relations (11.5.11), the diagonal term $\hat{a}^{\dagger}_{p,\sigma}\hat{a}_{p,\sigma}$ alone does not induce any electron hopping. We thus included the off-diagonal terms $\hat{a}^{\dagger}_{p,\sigma}\hat{a}_{q,\sigma}$ to make the band dispersive. We should note that the hopping Hamiltonian (11.5.14) becomes quite complicated when expressed in the standard form (9.3.17).

Exactly as in Tasaki's flat-band model (Sect. 11.3.1) and the corresponding perturbed models (Sect. 11.4), the single-electron energy spectrum of the present model consists of the lowest band spanned by the $\boldsymbol{\alpha}$-states and higher bands spanned by the $\boldsymbol{\beta}$-states. The lowest band thus contains L^d states. Recalling that the model with electron number $N = L^d$, which corresponds to the half-filling of the lowest band, should describe an insulator, we here concentrate on the case where N is strictly less than L^d (or, more precisely, the density N/L^d is strictly less than one). Unfortunately none of the techniques developed for the flat-band models or the nearly-flat band models apply to the present model with $N \neq L^d$. The only way (that we know of for the moment) to proceed is to consider the model in the limit where both the band gap and the Coulomb interaction become infinitely large, i.e., $t, U \uparrow \infty$.

Fix the electron number $N < L^d$, and let the finite-energy subspace $\mathscr{H}_0$ be a collection of states $|\Phi\rangle$ such that $\lim_{t,U\uparrow\infty}\langle\Phi|\hat{H}|\Phi\rangle < \infty$. (Note that the state $|\Phi\rangle$ is independent of t or U.) The Theorem A.12 (p. 470) implies that tha model in the limit $t, U \uparrow \infty$ is precisely described by the effective Hamiltonian $\hat{H}_{\text{eff}} = \hat{P}_0\hat{H}\hat{P}_0$, where $\hat{P}_0$ is the projection operator onto $\mathscr{H}_0$. We also find that a state $|\Phi\rangle$ belongs to $\mathscr{H}_0$ if and only if $\hat{b}^{\dagger}_{(u,\zeta),\sigma}\hat{b}_{(u,\zeta),\sigma}|\Phi\rangle = 0$ for any $u \in \mathscr{I}$, $\zeta = 1, 2$, and $\sigma =\uparrow, \downarrow$, and $\hat{n}_{x,\uparrow}\hat{n}_{x,\downarrow}|\Phi\rangle = 0$ for all $x \in \Lambda$. Lemma A.11 (p. 469) further implies that these conditions are equivalent to

$$\hat{b}_{(u,\zeta),\sigma}|\Phi\rangle = 0, \tag{11.5.15}$$

for any $u \in \mathscr{I}$, $\zeta = 1, 2$, and $\sigma =\uparrow, \downarrow$, and

$$\hat{c}_{x,\downarrow}\hat{c}_{x,\uparrow}|\Phi\rangle = 0, \tag{11.5.16}$$

for any $x \in \Lambda$. Note that these finite-energy conditions are exactly the same as the zero-energy conditions (11.3.11), (11.3.12), which played central roles in flat-band ferromagnetism. We shall therefore be sketchy, assuming that the reader has gone through the proof of Theorem 11.11. (In fact the first half of the following poof is almost identical to the proof of Lemma 11.23 in p. 434.)

Since $\boldsymbol{\alpha}$ and $\boldsymbol{\beta}$ states span the single-electron Hilbert space $\mathfrak{h}$, we can express any state by using $\hat{a}^{\dagger}$ and $\hat{b}^{\dagger}$ operators as in (11.3.13). Then from the condition (11.5.15) we see that any finite-energy state must be written only in terms of the $\hat{a}^{\dagger}$ operators, exactly as in (11.3.14). Next, by using the condition (11.5.16) for $x \in \mathscr{E}$, we find that there can be no double occupancies in $\hat{a}^\dagger$ states. Therefore any finite-energy state $|\Phi\rangle$ is expanded as

$$|\Phi\rangle = \sum_{\substack{A_\uparrow, A_\downarrow \subset \mathscr{E} \\ (A_\uparrow \cap A_\downarrow = \emptyset,\, |A_\uparrow|+|A_\downarrow|=N)}} c(A_\uparrow, A_\downarrow)\Big(\prod_{p\in A_\uparrow} \hat{a}^\dagger_{p,\uparrow}\Big)\Big(\prod_{p\in A_\downarrow} \hat{a}^\dagger_{p,\downarrow}\Big)|\Phi_{\text{vac}}\rangle, \tag{11.5.17}$$

with arbitrary coefficients $c(A_\uparrow, A_\downarrow) \in \mathbb{C}$. The model has thus reduced to an electron model on $\mathscr{E}$ with hard core repulsion. The effective Hamiltonian is of course given by

$$\hat{H}_{\text{eff}} = \hat{P}_0 \hat{H} \hat{P}_0 = -s \sum_{\substack{p,q\in\mathscr{E} \\ (|p-q|=1) \\ \sigma=\uparrow,\downarrow}} \hat{P}_0\, \hat{a}^\dagger_{p,\sigma} \hat{a}_{q,\sigma}\, \hat{P}_0, \tag{11.5.18}$$

which is the simple nearest neighbor hopping Hamiltonian on $\mathscr{E}$.

It only remains to take into account the condition (11.5.16) for x in $\mathscr{I} \times \{1, 2\}$. Exactly as in the proof of Tasaki's flat-band ferromagnetism (see the part "Ferromagnetic exchange interaction" in the proof of Theorem 11.11), we get ferromagnetic exchange interaction from these conditions. In this case we see that the coefficients $c(A_\uparrow, A_\downarrow)$ are such that the spins of any pair of electrons occupying $\hat{a}^\dagger_p$ and $\hat{a}^\dagger_q$ states with $|p-q| = 1$ are coupled ferromagnetically. This means that all the spins in each connected component of $A_\uparrow \cup A_\downarrow$ (which is the set of occupied sites) must be coupled to have the maximum total spin. (But note that this fact does not imply that $|\Phi\rangle$ exhibits ferromagnetism. In general $A_\uparrow \cup A_\downarrow$ consists of several connected components.)

Now inspection shows that one gets exactly the same finite-energy conditions in the $J \uparrow \infty$ limit of the t-J model (11.5.4) on the same lattice $\mathscr{E}$ with the same electron number N. This implies the desired equivalence of the Hubbard model and the t-J model.

Lemma 11.25 *The limit $t, U \uparrow \infty$ of the above Hubbard model with electron number $N \le L^d$ is equivalent to the $J \uparrow \infty$ limit of the t-J model (11.5.4) with the same electron number N and the hopping amplitude $\tau = (1 + 4d\nu^2)s$.*

The equivalence does not lead to any conclusions about the Hubbard model unless we know something about the t-J model. For the one-dimensional model, we get the following from Proposition 11.24 (p. 443) and Lemma 11.25.

Theorem 11.26 (Metallic ferromagnetism in the $d = 1$ Hubbard model)
Consider the above Hubbard model in one-dimension, and let the electron number N be odd and satisfy $N \le L$. Then in the limit $t, U \uparrow \infty$, the ground states of the model have total spin $S_{\text{tot}} = N/2$, and are non-degenerate apart from the trivial $2S_{\text{tot}} + 1 = N + 1$-fold degeneracy.

Fig. 11.26 A metallic ground state of the model by Tanaka and Tasaki. There are two dispersive bands, which we call the a-band and the b-band. One gets ferromagnetic ground states in which the a-band is fully filled and the b-band is partially filled (© Hal Tasaki 2020. All Rights Reserved)

When N/L^d is strictly less than one, the ground states describe metallic states since the lowest band is partially filled.[53] We have thus obtained a rigorous example of metallic ferromagnetism in the Hubbard model on the one-dimensional decorated lattice. We note that this result is a variant of earlier results by Kubo [22] and by Tanaka and Idogaki [61]. The proof is based on the Perron–Frobenius argument, which works only in simple one-dimensional systems.

The model and the theorem of Tanaka and Tasaki We finally discuss the work by Tanaka and Tasaki [63], which established that a certain version of (rather complicated) Hubbard model in any dimensions exhibits metallic ferromagnetism. At the time of writing, this is still the only rigorous example of Hubbard model which becomes a metallic ferromagnet in two or higher dimensions. As the reader can guess from Figs. 11.27 and 11.28, the model is artificial and complicated. We must note however that this is a fruit of our serious effort to construct the simplest possible Hubbard model in which we can rigorously control metallic ferromagnetism. It may be the case that we had gone too much into the mathematical game of designing artificial examples to prove some nontrivial phenomenon "for the first time".

Before introducing the model, we briefly discuss the nature of metallic ferromagnetism that we obtain in the present model. The model has a rather complicated band structure with two lowest dispersive bands, which we call the a-band and the b-band, and other higher bands. We take the limit in which both the Coulomb interaction U and the band gap separating the two lowest bands and the other higher bands become infinitely large. We consider the electron number which is larger than the half-filling of the lowest a-band but is smaller than the half-filling of both the a and b-bands. Then, in a suitable range of parameters, we get ferromagnetic ground states in which the a-band is fully filled and the b-band is partially filled as in Fig. 11.26. This certainly corresponds to a metallic state. (See footnote 47 above.) Very roughly speaking, in a ground state the present model, electrons filling the a-band first form a ferromagnetic state according to the mechanism familiar in the flat-band and nearly-flat band ferromagnetism, and then electrons in the b-band join the ferromagnetic state because of exchange interaction between the a-band and the b-bands.[54]

---

[53] Recall that the ground states can be expressed as Slater determinant states as in (11.5.5).

[54] This explanation should not be taken too literally since all electrons are exactly identical, and a ground state should be determined once and for all to minimize the total energy.

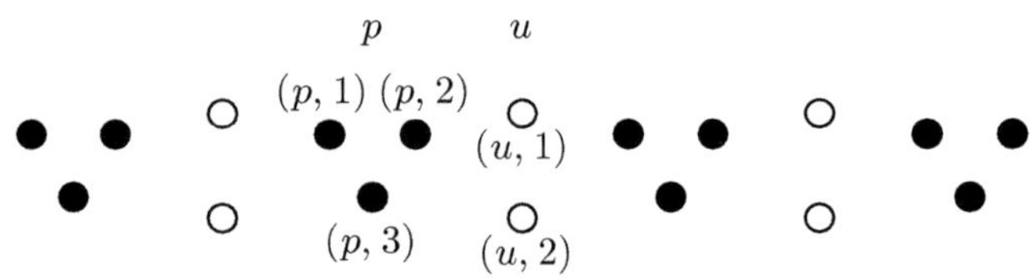

Fig. 11.27 Sites in the one-dimensional decorated lattice $\Lambda = \mathcal{E} \times \{1, 2, 3\} \cup \mathcal{I} \times \{1, 2\}$. Here $p \in \mathcal{E}$ is an integer and $u = p + \frac{1}{2} \in \mathcal{I}$ (© Hal Tasaki 2020. All Rights Reserved)

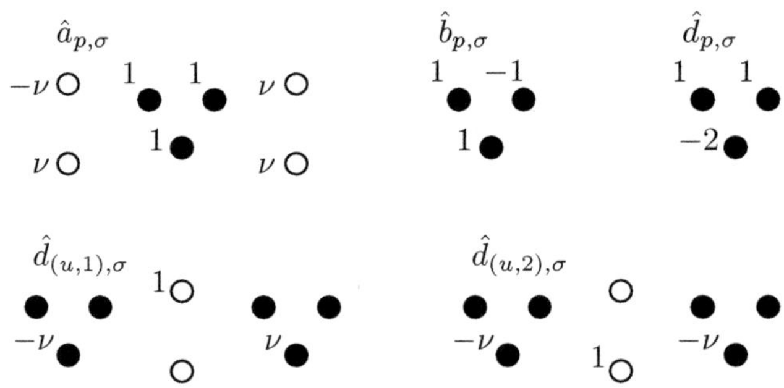

Fig. 11.28 Nonzero components of the fermion operators (11.5.19), (11.5.20), (11.5.21), and (11.5.22) for the one-dimensional model (© Hal Tasaki 2020. All Rights Reserved)

We again denote by $\mathcal{E}$ the d-dimensional $L \times L$ hypercubic lattice, and by $\mathcal{I}$ the corresponding lattice consisting of sites in the middle of bonds in $\mathcal{E}$. This time we consider a Hubbard model on a (considerably) decorated lattice $\Lambda = \mathcal{E} \times \{1, 2, 3\} \cup \mathcal{I} \times \{1, 2\}$. Now, sites in $\mathcal{I}$ are duplicated as before, and sites in $\mathcal{E}$ are triplicated. We denote lattice sites as $(p, 1)$, $(p, 2)$, $(p, 3)$, $(u, 1)$, and $(u, 2)$ with $p \in \mathcal{E}$ and $u \in \mathcal{I}$. See Fig. 11.27.

Let us define special fermion operators. We again let $\boldsymbol{b}_j = (0, \ldots, 0, \frac{1}{2}, 0, \ldots, 0)$ be the half of the unit vector in the jth direction, where $j = 1, \ldots, d$. For any $p \in \mathcal{E}$, $\zeta = 1, 2, 3$ and $\sigma = \uparrow, \downarrow$, we define

$$\hat{a}_{p,\sigma} = \frac{1}{\sqrt{3 + 4d\nu^2}} \Bigg[\sum_{\zeta=1}^{3} \hat{c}_{(p,\zeta),\sigma} + \nu \sum_{\substack{j=1,\ldots,d \\ \zeta=1,2}} \big\{ \hat{c}_{(p+\boldsymbol{b}_j,\zeta),\sigma} + (-1)^{\zeta}\, \hat{c}_{(p-\boldsymbol{b}_j,\zeta),\sigma} \big\} \Bigg], \tag{11.5.19}$$

$$\hat{b}_{p,\sigma} = \frac{1}{\sqrt{2}} \big\{ \hat{c}_{(p,1),\sigma} - \hat{c}_{(p,2),\sigma} \big\}, \tag{11.5.20}$$

$$\hat{d}_{p,\sigma} = \hat{c}_{(p,1),\sigma} + \hat{c}_{(p,2),\sigma} - 2\hat{c}_{(p,3),\sigma}, \tag{11.5.21}$$

and for any $u \in \mathcal{I}$, $\zeta = 1, 2$, and $\sigma = \uparrow, \downarrow$, we define

$$\hat{d}_{(u,\zeta),\sigma} = \hat{c}_{(u,\zeta),\sigma} - \nu\big\{\hat{c}_{(u-\boldsymbol{b}_j,3),\sigma} + (-1)^{\zeta}\,\hat{c}_{(u+\boldsymbol{b}_j,3),\sigma}\big\}, \tag{11.5.22}$$

where j is the unique index such that $u \pm \boldsymbol{b}_j \in \mathscr{E}$. Again $\nu > 0$ is a model parameter. See Fig. 11.28. As in the previous models, one can represent any state of the system by using the $\hat{a}^\dagger$, $\hat{b}^\dagger$, and $\hat{d}^\dagger$ operators. From (11.5.19) and (11.5.20), we see that the $\hat{a}$ and $\hat{b}$ operators satisfy the standard anticommutation relations

$$\{\hat{a}_{p,\sigma}, \hat{a}^\dagger_{q,\tau}\} = \delta_{p,q}\,\delta_{\sigma,\tau}, \quad \{\hat{b}_{p,\sigma}, \hat{b}^\dagger_{q,\tau}\} = \delta_{p,q}\,\delta_{\sigma,\tau}, \tag{11.5.23}$$

for any $p, q \in \mathscr{E}$ and $\sigma, \tau = \uparrow, \downarrow$. We also find that $\{\hat{b}, \hat{a}^\dagger\} = 0$, $\{\hat{d}, \hat{a}^\dagger\} = 0$, and $\{\hat{d}, \hat{b}^\dagger\} = 0$ for any combinations of indices.

We consider the Hubbard model on Λ with the Hamiltonian $\hat{H} = \hat{H}_{\text{hop}} + \hat{H}_{\text{int}}$, where the interaction Hamiltonian is the standard (11.5.13) or (9.3.29), and the hopping Hamiltonian is

$$\begin{aligned}\hat{H}_{\text{hop}} = &\sum_{\substack{p,q\in\mathscr{E}\\(|p-q|=1)\\ \sigma=\uparrow,\downarrow}} \big\{-s\,\hat{a}^\dagger_{p,\sigma}\hat{a}_{q,\sigma} - t\,\hat{b}^\dagger_{p,\sigma}\hat{b}_{q,\sigma}\big\} + u_1 \sum_{\substack{p\in\mathscr{E}\\ \sigma=\uparrow,\downarrow}} \hat{b}^\dagger_{p,\sigma}\hat{b}_{p,\sigma}\\ &+ u_2\Bigg\{\sum_{\substack{p\in\mathscr{E}\\ \sigma=\uparrow,\downarrow}} \hat{d}^\dagger_{p,\sigma}\hat{d}_{p,\sigma} + \sum_{\substack{u\in\mathscr{I},\,\zeta=1,2\\ \sigma=\uparrow,\downarrow}} \hat{d}^\dagger_{(u,\zeta),\sigma}\hat{d}_{(u,\zeta),\sigma}\Bigg\},\end{aligned} \tag{11.5.24}$$

where $s, t, u_1, u_2 \in \mathbb{R}$ are model parameters. From the above anticommutation relations, we see that $\hat{a}$-states and $\hat{b}$-states form separate bands with dispersion relations $\varepsilon_a(k) = -2s\sum_{j=1}^{d}\cos(k_j)$ and $\varepsilon_b(k) = u_1 - 2t\sum_{j=1}^{d}\cos(k_j)$. We take the limit $u_2 \uparrow \infty$, in which the energies of the other bands formed by $\hat{d}^\dagger$-states become infinite. Then any finite-energy state is written only in terms of $\hat{a}^\dagger$ and $\hat{b}^\dagger$ operators. From (11.5.19) and (11.5.20) (and also from Fig. 11.28) one finds that there should be a ferromagnetic exchange interaction between neighboring $\hat{a}^\dagger$ states and also between an $\hat{a}^\dagger$ state and a $\hat{b}^\dagger$ state sharing the same p. It is then likely that the ferromagnetic state as depicted in Fig. 11.26 emerges. In fact we have the following.

Theorem 11.27 *Let the dimension $d = 1, 2, 3, \ldots$ be arbitrary and suppose that the electron number N satisfies $L^d \le N \le 2L^d$, and $u_1 > 2d(|s| + 2|t|)$. Then, in the limit u_2, $U \uparrow \infty$, the ground states of the above model exhibit ferromagnetism in the sense that they have the maximum possible total spin $S_{\text{tot}} = N/2$.*

See the original paper [63] for the proof. The ground states are certainly metallic when N/L^d is strictly larger than 1 and strictly less than 2.

Given the simplicity of the picture of the ground states, one might expect that the theorem can be proved (with some effort) for finite but sufficiently large u_2 and U, as we have done for the insulating model in Sect. 11.4.3. But, as far as we understand, there is no hope of extending the proof to finite u_2 or U.[55] The treatment of models with a partially filled band seems to be essentially different from models with filled bands.[56] The proof of Theorem 11.27 in [63] is not long and does not use any advanced techniques than the variational principle. We nevertheless warn the reader that the proof is quite technical and difficult; we had to solve extremely complicated mathematical puzzles in order to justify some "physically plausible" facts. It seems that we still lack proper mathematical techniques as well as physical ideas for treating metallic systems.

References

1. R. Arita, Y. Suwa, K. Kuroki, H. Aoki, Gate-induced band ferromagnetism in an organic polymer. Phys. Rev. Lett. **88**, 127202 (2002), arxiv:cond-mat/0108413
2. A. Auerbach, *Interacting Electrons and Quantum Magnetism* (Springer, New York, 1994)
3. A. Barbieri, J.A. Riera, A.P. Young, Stability of the saturated ferromagnetic state in the one-band Hubbard model. Phys. Rev. B **41**, 11697 (1990)
4. E. Bobrow, K. Stubis, Y. Li, *Exact results on itinerant ferromagnetism and the 15-puzzle problem*. Phys. Rev. B **98**, 180101 (2018), arxiv:1804.02347
5. U. Brandt, A. Giesekus, Hubbard and anderson models on perovskitelike lattices: exactly solvable cases. Phys. Rev. Lett. **68**, 2648 (1992)
6. O. Derzhko, J. Richter, Dispersion-driven ferromagnetism in a flat-band Hubbard system. Phys. Rev. B **90**, 045152 (2014), arxiv:1404.2230v1
7. B. Douçot, X.-G. Wen, Instability of the Nagaoka state with more than one hole. Phys. Rev. B **40**, 2719 (1989)
8. F.J. Dyson, E.H. Lieb, B. Simon, Phase transitions in quantum spin systems with isotropic and nonisotropic interactions. J. Stat. Phys. **18**, 335–382 (1978)
9. F.H.L. Essler, H. Frahm, F. Göhmann, A. Klümper, V.E. Korepin, *The One-Dimensional Hubbard Model* (Cambridge University Press, Cambridge, 2010)
10. P. Fazekas, *Lecture Notes on Electron Correlation and Magnetism*. Series in Modern Condensed Matter Physics, vol. 5 (World Scientific, Singapore, 1999)
11. M. Garnica, D. Stradi, S. Barja, F. Calleja, C. Diaz, M. Alcami, N. Martin, A.L.V. de Parga, F. Martin, R. Miranda, Long-range magnetic order in a purely organic 2D layer adsorbed on epitaxial graphene. Nat. Phys. **9**, 368 (2013)
12. D. Ghosh, Nonexistence of magnetic ordering in the one- and two-dimensional Hubbard model. Phys. Rev. Lett. **27**, 1584 (1971)
13. Z. Gulácsi, A. Kampf, D. Vollhardt, Exact many-electron ground states on the diamond Hubbard chain. Phys. Rev. Lett. **99**, 026404 (2007), arxiv:cond-mat/0703818
14. T. Hanisch, E. Müller-Hartmann, Ferromagnetism in the Hubbard model: instability of the Nagaoka state on the square lattice. Ann. Phys. **2**, 381 (1993)
15. T. Hanisch, G.S. Uhrig, E. Müller-Hartmann, Lattice dependence of saturated ferromagnetism in the Hubbard model. Phys. Rev. B **56**, 13960 (1997), arxiv:cond-mat/9707286
16. W.J. Heisenberg, Zur theorie des ferromagnetismus. Z. Phys. **49**, 619–636 (1928)
17. M. Ichimura, K. Kusakabe, S. Watanabe, T. Onogi, Flat-band ferromagnetism in extended Δ-chain Hubbard models. Phys. Rev. B **58**, 9595 (1998)

[55]Of course, by continuity, one can show that the ground states are ferromagnetic for sufficiently large u_2 and U for each L. But we do not have any estimates that are independent of L.

[56]The biggest difference is that, in insulating systems, ferromagnetic ground states may be expressed in the k-space basis as in (11.5.5) or in the real space basis as in, e.g., (11.3.9). In a metallic system, one only has the former expression.

18. H. Katsura, I. Maruyama, How to construct tight-binding models with flat bands (in Japanese). Kotai Butsuri (Solid State Phys.) **50**, 257–270 (2015)
19. H. Katsura, I. Maruyama, A. Tanaka, H. Tasaki, *Ferromagnetism in the Hubbard model with topological/non-topological flat bands*. Europhys Lett. **91**, 57007 (2010), arxiv:0907.4564
20. H. Katsura, A. Tanaka, Nagaoka states in the SU(n) Hubbard model. Phys. Rev. A **87**, 013617 (2013), arxiv:1210.4774
21. T. Koma, H. Tasaki, Decay of superconducting and magnetic correlations in one- and two-dimensional Hubbard models. Phys. Rev. Lett. **68**, 3248–3251 (1992), arxiv:cond-mat/9709068
22. K. Kubo, Note on the ground states of systems with the strong hund-coupling. J. Phys. Soc. Jpn. **51**, 782–786 (1982)
23. K. Kusakabe, H. Aoki, Robustness of the ferromagnetism in flat bands. Phys. B **194–196**, 215–216 (1994)
24. K. Kusakabe, H. Aoki, Ferromagnetic spin-wave theory in the multiband Hubbard model having a flat band. Phys. Rev. Lett. **72**, 144 (1994)
25. K. Kusakabe, H. Aoki, Scaling properties of the ferromagnetic state in the Hubbard model. Phys. Rev. B **50**, 12991 (1994), arxiv:cond-mat/9407036
26. K. Kusakabe, M. Maruyama, *Magnetic nanographite*. Phys. Rev. B **67**, 092406 (2003), arxiv:cond-mat/0212391
27. S. Liang, H. Pang, Ferromagnetism in the infinite-U Hubbard model. Europhys. Lett. **32**, 173 (1995)
28. H.-H. Lin, T. Hikihara, H.-T. Jeng, B.-L. Huang, C.-Y. Mou, X. Hu, Ferromagnetism in armchair graphene nanoribbons. Phys. Rev. B **79**, 035405 (2009)
29. R.-J. Liu, W. Nie, W. Zhang, *Flat-band ferromagnetism of the SU(N) Hubbard model on the Tasaki lattice* (2019), arxiv:1901.07004
30. L. Lu, The stability of ferromagnetism in the Hubbard model on two-dimensional line graphs. J. Phys. A **42**, 265002 (2009)
31. M. Maksymenko, O. Derzhko, J. Richter, *Localized states on triangular traps and low-temperature properties of the antiferromagnetic Heisenberg and repulsive Hubbard models*. Eur. Phys. J. B **84**, 397–408 (2011), arxiv:1108.5453
32. M. Maksymenko, A. Honecker, R. Moessner, J. Richter, O. Derzhko, *Flat-band ferromagnetism as a Pauli-correlated percolation problem*. Phys. Rev. Lett. **109**, 096404 (2012), arxiv:1204.2110
33. A. Mielke, Ferromagnetic ground states for the Hubbard model on line graphs. J. Phys. A **24**, L73 (1991)
34. A. Mielke, Ferromagnetism in the Hubbard model on line graphs and further considerations. J. Phys. A **24**, 3311 (1991)
35. A. Mielke, Exact ground states for the Hubbard model on the Kagome lattice. J. Phys. A **25**, 4335 (1992)
36. A. Mielke, Ferromagnetism in the Hubbard model and Hund's rule. Phys. Lett. A **174**, 443 (1993)
37. A. Mielke, Ferromagnetism in single-band Hubbard models with a partially flat band. Phys. Rev. Lett. **82**, 4312 (1999), arxiv:cond-mat/9904258
38. A. Mielke, Stability of Ferromagnetism in Hubbard models with degenerate single-particle ground states. J. Phys. A: Mat. Gen. **32**, 8411 (1999), arxiv:cond-mat/9910385
39. A. Mielke, *Properties of Hubbard models with degenerate localised single-particle eigenstates*. Eur. Phys. J. B **85**, 184 (2012), arxiv:abs/1202.3158
40. A. Mielke, H. Tasaki, Ferromagnetism in the Hubbard model. Examples from models with degenerate single-electron ground states. Commun. Math. Phys. **158**, 341–371 (1993), https://projecteuclid.org/euclid.cmp/1104254245
41. T. Miyao, Stability of ferromagnetism in many-electron systems. J. Stat. Phys. **176**, 1211–1271 (2019), arxiv:1712.05529
42. Y. Nagaoka, Ferromagnetism in a narrow, almost half-filled s-band. Phys. Rev. **147**, 392 (1966)
43. Y. Nagaoka, The15 puzzle (in Japanese). Kotai Butsuri (Solid State Phys.) **22**, 756 (1987)
44. S. Nishino, M. Goda, Three-dimensional flat-band models. J. Phys. Soc. Jpn. **74**, 393 (2005)

45. K. Penc, H. Shiba, F. Mila, T. Tsukagoshi, Ferromagnetism in multi-band Hubbard models: from weak to strong Coulomb repulsion. Phys. Rev. B **54**, 4056 (1996), arxiv:cond-mat/9603042
46. P. Pieri, S. Daul, D. Baeriswyl, M. Dzierzawa, P. Fazekas, Low density ferromagnetism in the Hubbard model. Phys. Rev. B **45**, 9250 (1996), arxiv:cond-mat/9603163
47. W.O. Putikka, M.U. Luchini, M. Ogata, Ferromagnetism in the two-dimensional t-J model. Phys. Rev. Lett. **69**, 2288 (1992)
48. J.A. Riera, A.P. Young, Ferromagnetism in the one-band Hubbard model. Phys. Rev. B **40**, 5285 (1989)
49. L.M. Roth, Spin wave stability of the ferromagnetic state for a narrow s-band. J. Phys. Chem. Solids **28**, 1549 (1967)
50. H. Sakamoto, K. Kubo, Metallic ferromagnetism in a one-dimensional hubbard model; study using the density-matrix renormalization-group method. J. Phys. Soc. Jpn. **65**, 3732–3735 (1996)
51. T. Sekizawa, An extension of Tasaki's flat-band Hubbard model. J. Phys. A: Math. Gen. **42**, 10451 (2003), arxiv:cond-mat/0304295
52. B.S. Shastry, H.R. Krishnamurthy, P.W. Anderson, Instability of the Nagaoka ferromagnetic state of the $U = \infty$ Hubbard model. Phys. Rev. B **41**, 2375 (1990)
53. S.-Q. Shen, Ferromagnetic ground state in the Kondo lattice model. Eur. Phys. J. B **2**, 11–16 (1998)
54. A. Sütő, Absence of highest-spin ground states in the Hubbard model. Commun. Math Phys. **140**, 43–62 (1991), https://projecteuclid.org/euclid.cmp/1104247906
55. Y. Suwa, R. Arita, K. Kuroki, H. Aoki, Flat-band ferromagnetism in organic polymers designed by a computer simulation. Phys. Rev. B **68**, 174419 (2003), arxiv:cond-mat/0308154
56. M. Takahashi, $I = \infty$ Hubbard model on finite lattices. J. Phys. Soc. Jpn. **51**, 3475 (1982)
57. K. Tamura, H. Katsura, Ferromagnetism in the SU(n) Hubbard model with a nearly flat band. Phys. Rev. B **100**, 214423 (2019), arxiv:1908.06286
58. A. Tanaka, Ferromagnetism in the Hubbard model with a gapless nearly-flat band. J. Stat. Phys. **170**, 399–420 (2018), arxiv:1707.03944
59. A. Tanaka, *An Extension of the Cell-Construction Method for the Flat-Band Ferromagnetism*, preprint (2020), arxiv:2004.07516
60. A. Tanaka, T. Idogaki, Nonperturbative ferromagnetism in the Hubbard model for quarter-filling with one extra electron. J. Phys. Soc. Jpn. **67**, 401–404 (1998)
61. A. Tanaka, T. Idogaki, Metallic ferromagnetism in the Hubbard model on three coupled chains. J. Phys. A: Math. Gen. **32**, 4883–4895 (1999)
62. A. Tanaka, T. Idogaki, Ferromagnetism in multi-band Hubbard models. Phys. A **297**, 441–484 (2001)
63. A. Tanaka, H. Tasaki, *Metallic ferromagnetism in the Hubbard model: a rigorous example*. Phys. Rev. Lett. **98**, 116402 (2007), arxiv:cond-mat/0611318
64. A. Tanaka, H. Tasaki, Metallic ferromagnetism supported by a single band in a multi-band Hubbard model. J. Stat. Phys. **163**, 1049–1068 (2016), arxiv:1601.04265
65. A. Tanaka, H. Ueda, Stability of ferromagnetism in the Hubbard model on the Kagome lattice. Phys. Rev. Lett. **90**, 067204 (2003), arxiv:cond-mat/0209423
66. H. Tasaki, Extension of Nagaoka's theorem on the large-U Hubbard model. Phys. Rev. B **40**, 9192 (1989)
67. H. Tasaki, Ferromagnetism in the Hubbard models with degenerate single-electron ground states. Phys. Rev. Lett. **69**, 1608 (1992)
68. H. Tasaki, Exact resonating-valence-bond ground state and possibility of superconductivity in repulsive Hubbard models. Phys. Rev. Lett. **70**, 3303 (1993)
69. H. Tasaki, Uniqueness of the ground state in exactly solvable Hubbard, periodic Anderson, and Emery models. Phys. Rev. B **49**, 7763 (1994)
70. H. Tasaki, Stability of ferromagnetism in the Hubbard model. Phys. Rev. Lett. **73**, 1158 (1994)
71. H. Tasaki, Stability of ferromagnetism in Hubbard models with nearly flat bands. J. Stat. Phys. **84**, 535 (1996), arxiv:1901.02617

72. H. Tasaki, Ferromagnetism in Hubbard models. Phys. Rev. Lett. **75**, 4678 (1995), arxiv:cond-mat/9509063
73. H. Tasaki, From Nagaoka's ferromagnetism to flat-band ferromagnetism and beyond — An introduction to ferromagnetism in the Hubbard model. Prog. Theor. Phys. **99**, 489–548 (1998), arxiv:cond-mat/9712219
74. H. Tasaki, Ferromagnetism in the Hubbard model: a constructive approach. Commun. Math. Phys. **242**, 445–472 (2003), arxiv:cond-mat/9712219
75. D.J. Thouless, Exchange in solid ^{3}He and the Heisenberg Hamiltonian. Proc. Phys. Soc. Lond. **86**, 893 (1965)
76. B. Tóth, Failure of saturated ferromagnetism for the Hubbard model with two holes. Lett. Math. Phys. **22**, 321 (1991)
77. H. Ueda, A. Tanaka, T. Idogaki, Ferromagnetism in the Hubbard model on a lattice of three coupled chains. J. Magnetism Mag Mater. **272–276**, 950–951 (2004)
78. J.A. Vergés, F. Guinea, J. Galán, P.G.J. van Dongen, G. Chiappe, E. Louis, Ground state of the $U = \infty$ Hubbard model with infinite-range hopping. Phys. Rev. B **49**, 15400(R) (1994)
79. Y. Watanabe, S. Miyashita, New saturated ferromagnetism in the flat-band Hubbard model in high electron filling. J. Phys. Soc. Jpn. **66**, 2123 (1997)
80. Y. Watanabe, S. Miyashita, Properties of new saturated ferromagnetism in the 1D flat-band Hubbard model as the spiral state. J. Phys. Soc. Jpn. **66**, 3981 (1997)
81. R.M. Wilson, Graph puzzles, homotopy, and the alternating group. J. Combin. Theory, (B) **16**, 86–96 (1974), https://www.sciencedirect.com/science/article/pii/0095895674900987

Appendix A
Mathematical Appendices

A.1 Dirac Notation

Let us summarize the Dirac notation for quantum mechanics, which is standard in the physics literature, and used throughout the present book. Note that we do not intend here to provide a self-contained introduction to the formalism of quantum mechanics.

States We consider a finite dimensional Hilbert space $\mathscr{H}$, whose elements, i.e., quantum mechanical states, are denoted as $|\Phi\rangle, |\Psi\rangle, \ldots$ (and pronounced as "Phi-ket", "ket-Phi" etc.). For any $|\Phi\rangle, |\Psi\rangle \in \mathscr{H}$ and $\alpha, \beta \in \mathbb{C}$, one has $\alpha|\Phi\rangle + \beta|\Psi\rangle \in \mathscr{H}$. It is convenient to (fix an orthonormal basis and) identify $\mathscr{H}$ with $\mathbb{C}^D$, and its element with a column vector

$$|\Phi\rangle = \begin{pmatrix} \varphi_1 \\ \varphi_2 \\ \vdots \\ \varphi_D \end{pmatrix}, \tag{A.1.1}$$

where $\varphi_j \in \mathbb{C}$.

For any $|\Phi\rangle, |\Psi\rangle \in \mathscr{H}$, we denote the inner product of $|\Phi\rangle$ and $|\Psi\rangle$ as

$$\langle\Phi|\Psi\rangle := \sum_{j=1}^{D} (\varphi_j)^* \psi_j, \tag{A.1.2}$$

where we wrote $|\Psi\rangle = (\psi_1, \ldots, \psi_D)^{\mathrm{t}}$. The inner product satisfies

$$\langle\Phi|\Psi\rangle = \langle\Psi|\Phi\rangle^*, \tag{A.1.3}$$

and the linearity

H. Tasaki, *Physics and Mathematics of Quantum Many-Body Systems*,
Graduate Texts in Physics, https://doi.org/10.1007/978-3-030-41265-4

$$\langle \Phi | \big(\alpha | \Psi \rangle + \beta | \Xi \rangle \big) = \alpha \langle \Phi | \Psi \rangle + \beta \langle \Phi | \Xi \rangle, \tag{A.1.4}$$

for any $|\Phi\rangle, |\Psi\rangle, |\Xi\rangle \in \mathscr{H}$ and $\alpha, \beta \in \mathbb{C}$. The norm of a state $|\Phi\rangle \in \mathscr{H}$, which we write as $\|\Phi\|$ or $\||\Phi\rangle\|$, is defined by

$$\|\Phi\| := \sqrt{\langle \Phi | \Phi \rangle}. \tag{A.1.5}$$

We say that a state $|\Phi\rangle$ is normalized when $\|\Phi\| = 1$.

An orthonormal basis is a set of D states $\{|\Psi_j\rangle\}_{j=1,\ldots,D}$ with the property $\langle \Psi_j | \Psi_k \rangle = \delta_{k,j}$ for any $k, j = 1, \ldots, D$. An arbitrary state $|\Phi\rangle$ is expanded by using an orthonormal basis as $|\Phi\rangle = \sum_{j=1}^{D} \alpha_j |\Psi_j\rangle$, where the coefficients are given by $\alpha_j = \langle \Psi_j | \Phi \rangle$.

The symbol $\langle \Phi | \Psi \rangle$ may be simply understood as expressing the inner product of $|\Phi\rangle$ and $|\Psi\rangle$, but may also be interpreted as the product of the bra-state $\langle \Phi |$ and the ket-state $|\Psi\rangle$. A bra-state is an element of the dual space of $\mathscr{H}$, and is in one-to-one correspondence with a ket-state, an element of $\mathscr{H}$. More precisely the bra-state corresponding to the ket-state (A.1.1) is identified with a row vector

$$\langle \Phi | = (|\Phi\rangle)^\dagger = (\varphi_1^*, \varphi_2^*, \ldots, \varphi_D^*). \tag{A.1.6}$$

Here the conjugation $\dagger$ denotes the one-to-one mapping between the elements of $\mathscr{H}$ and its dual. Then (A.1.2) is consistent with the standard product of a row vector and a column vector. Note that the conjugate of a linear combination of ket-states are given by

$$\big(\alpha |\Phi\rangle + \beta |\Psi\rangle \big)^\dagger = \alpha^* \langle \Phi | + \beta^* \langle \Psi |. \tag{A.1.7}$$

Operators Operators are linear maps from $\mathscr{H}$ to itself. If we make the identification (A.1.1), the operators are nothing but $D \times D$ matrices with complex elements. We denote by $\hat{1}$ the identity operator. For any operators $\hat{A}$ and $\hat{B}$ and $\alpha, \beta \in \mathbb{C}$, the linear combination $\alpha \hat{A} + \beta \hat{B}$ is also an operator, and satisfies

$$(\alpha \hat{A} + \beta \hat{B}) |\Phi\rangle = \alpha \hat{A} |\Phi\rangle + \beta \hat{B} |\Phi\rangle, \tag{A.1.8}$$

for any $|\Phi\rangle \in \mathscr{H}$. For any operator $\hat{A}$, its adjoint (or Hermitian conjugate) $\hat{A}^\dagger$ is the unique operator that satisfies

$$\langle \Phi | \hat{A} | \Psi \rangle = \langle \Psi | \hat{A}^\dagger | \Phi \rangle^*, \tag{A.1.9}$$

for any $|\Phi\rangle, |\Psi\rangle \in \mathscr{H}$. Let $|\Xi\rangle = \hat{A}|\Psi\rangle$. Then (A.1.3) implies $\langle \Phi | \hat{A} | \Psi \rangle = \langle \Phi | \Xi \rangle = \langle \Xi | \Phi \rangle^*$. Since this should be identical to the right-hand side of (A.1.9) for any $|\Phi\rangle$, we may make an identification $\langle \Xi | = \langle \Psi | \hat{A}^\dagger$, which means

$$(\hat{A} |\Psi\rangle)^\dagger = \langle \Psi | \hat{A}^\dagger, \tag{A.1.10}$$

for any operator $\hat{A}$ and state $|\Psi\rangle \in \mathscr{H}$. It can be then shown that

$$(\hat{A}^\dagger)^\dagger = \hat{A}, \quad (\alpha\hat{A} + \beta\hat{B})^\dagger = \alpha^*\hat{A}^\dagger + \beta^*\hat{B}^\dagger, \quad (\hat{A}\hat{B})^\dagger = \hat{B}^\dagger\,\hat{A}^\dagger. \tag{A.1.11}$$

By using (A.1.10), one also finds that

$$\|\hat{A}|\Phi\rangle\|^2 = \langle\Phi|\hat{A}^\dagger\hat{A}|\Phi\rangle, \tag{A.1.12}$$

for any operator $\hat{A}$ and $|\Phi\rangle \in \mathscr{H}$.

For any $|\Phi\rangle$, $|\Psi\rangle \in \mathscr{H}$, we denote by $|\Phi\rangle\langle\Psi|$ the operator which maps an arbitrary state $|\Xi\rangle$ to $\langle\Psi|\Xi\rangle\,|\Phi\rangle$, which means the state $|\Phi\rangle$ multiplied by a complex number $\langle\Psi|\Xi\rangle$. This is quite natural from the interpretation (A.1.1) and (A.1.6) since the Kronecker product of a column vector and a row vector gives a matrix. The action of such an operator is (very conveniently) expressed as

$$\big(|\Phi\rangle\langle\Psi|\big)|\Xi\rangle = |\Phi\rangle\langle\Psi|\Xi\rangle. \tag{A.1.13}$$

In particular if $|\Phi\rangle$ is a normalized state then $|\Phi\rangle\langle\Phi|$ is the projection operator onto $|\Phi\rangle$. A necessary and sufficient condition that D states $|\Psi_1\rangle, \ldots, |\Psi_D\rangle$ form an orthonormal basis is written as

$$\sum_{j=1}^{D} |\Psi_j\rangle\langle\Psi_j| = \hat{1}. \tag{A.1.14}$$

An operator $\hat{A}$ that satisfies $\hat{A} = \hat{A}^\dagger$ is said to be self-adjoint (or Hermitian), and an operator $\hat{U}$ that satisfies $\hat{U}^\dagger\hat{U} = \hat{U}\hat{U}^\dagger = \hat{1}$ is said to be unitary. For any self-adjoint operator $\hat{A}$, there exists an orthonormal basis $\{|\Psi_j\rangle\}_{j=1,\ldots,D}$ such that $\hat{A}|\Psi_j\rangle = a_j|\Psi_j\rangle$ for all $j = 1, \ldots, D$, where $a_j \in \mathbb{R}$ is an eigenvalue of $\hat{A}$. Then the operator $\hat{A}$ admits the spectral decomposition $\hat{A} = \sum_{j=1}^{D} a_j|\Psi_j\rangle\langle\Psi_j|$. Likewise, for any unitary operator $\hat{U}$, there exists an orthonormal basis $\{|\Psi_j\rangle\}_{j=1,\ldots,D}$ such that $\hat{U}|\Psi_j\rangle = u_j|\Psi_j\rangle$ for all $j = 1, \ldots, D$, where $u_j \in \mathbb{C}$ with $|u_j| = 1$ is an eigenvalue of $\hat{U}$. We again have the spectral decomposition $\hat{U} = \sum_{j=1}^{D} u_j|\Psi_j\rangle\langle\Psi_j|$.

Let $\{|\Psi_j\rangle\}_{j=1,\ldots,D}$ be an arbitrary orthonormal basis. Then the trace of any operator $\hat{A}$ is defined by

$$\mathrm{Tr}[\hat{A}] = \sum_{j=1}^{D} \langle\Psi_j|\hat{A}|\Psi_j\rangle. \tag{A.1.15}$$

Recall that the trace has an important property $\mathrm{Tr}[\hat{A}\hat{B}] = \mathrm{Tr}[\hat{B}\hat{A}]$ for any operators $\hat{A}$ and $\hat{B}$.

Transformation of operators Let us make a brief comment on the conventions of unitary transformations of operators.

Let $\hat{U}$ be a unitary operator, and suppose that every state $|\Phi\rangle \in \mathscr{H}$ is transformed to $\hat{U}|\Phi\rangle$. Then the matrix element $\langle\Psi|\hat{A}|\Phi\rangle$ of an operator $\hat{A}$ is transformed to $\langle\Psi|\hat{U}^\dagger\hat{A}\hat{U}|\Phi\rangle$. One gets the same result if the states are unchanged, and every operator $\hat{A}$ is transformed as

$$\hat{A} \to \hat{U}^\dagger\hat{A}\hat{U}. \tag{A.1.16}$$

This is the most standard convention for the transformation of operators.

To see another convention, suppose that a self-adjoint operator $\hat{A}$ (such as a Hamiltonian) and a state $|\Phi\rangle$ satisfy the eigenvalue equation $\hat{A}|\Phi\rangle = a|\Phi\rangle$. Acted by an arbitrary unitary operator $\hat{U}$, this becomes $\hat{U}\hat{A}\hat{U}^\dagger\hat{U}|\Phi\rangle = a\hat{U}|\Phi\rangle$. Thus we have the same eigenvalue equation $\hat{A}'|\Phi'\rangle = a|\Phi'\rangle$ with $|\Phi'\rangle = \hat{U}|\Phi\rangle$ and $\hat{A}' = \hat{U}\hat{A}\hat{U}^\dagger$. We thus have the second convention

$$\hat{A} \to \hat{U}\hat{A}\hat{U}^\dagger. \tag{A.1.17}$$

Multiple systems Suppose that there are two quantum mechanical systems with Hilbert spaces $\mathscr{H}_1$ and $\mathscr{H}_2$, whose dimensions are D_1 and D_2, respectively. The Hilbert space of the joint system is given by the tensor product $\mathscr{H}_1 \otimes \mathscr{H}_2$.

The tensor product space $\mathscr{H}_1 \otimes \mathscr{H}_2$ consists of all the states of the form $|\Phi\rangle_1 \otimes |\Psi\rangle_2$ with $|\Phi\rangle_1 \in \mathscr{H}_1$ and $|\Psi\rangle_2 \in \mathscr{H}_2$ and their linear combinations, where we identify states according to linearity $(\alpha|\Phi\rangle_1 + \alpha'|\Phi'\rangle_1) \otimes (\beta|\Psi\rangle_2 + \beta'|\Psi'\rangle_2) = \alpha\beta|\Phi\rangle_1 \otimes |\Psi\rangle_2 + \alpha\beta'|\Phi\rangle_1 \otimes |\Psi'\rangle_2 + \alpha'\beta|\Phi'\rangle_1 \otimes |\Psi\rangle_2 + \alpha'\beta'|\Phi'\rangle_1 \otimes |\Psi'\rangle_2$ for any $|\Phi\rangle_1, |\Phi'\rangle_1 \in \mathscr{H}_1$, $|\Psi\rangle_2, |\Psi'\rangle_2 \in \mathscr{H}_2$ and $\alpha, \alpha', \beta, \beta' \in \mathbb{C}$.

The inner product of $|\Phi\rangle_1 \otimes |\Psi\rangle_2$ and $|\Phi'\rangle_1 \otimes |\Psi'\rangle_2$ is defined to be $\langle\Phi|\Phi'\rangle_1$ $\langle\Psi|\Psi'\rangle_2$. Then one can define inner products of any pair of states by linearity. Therefore, if $\{|\Phi_j\rangle_1\}_{j=1,\ldots,D_1}$ and $\{|\Psi_k\rangle_2\}_{k=1,\ldots,D_2}$ are orthonormal bases for $\mathscr{H}_1$ and $\mathscr{H}_2$, respectively, then an orthonormal basis of $\mathscr{H}_1 \otimes \mathscr{H}_2$ is given by $\{|\Phi_j\rangle_1 \otimes |\Psi_k\rangle_2\}_{j=1,\ldots,D_1,\, k=1,\ldots,D_2}$. Thus the dimension of $\mathscr{H}_1 \otimes \mathscr{H}_2$ is $D_1 D_2$.

Let $\hat{A}$ and $\hat{B}$ be arbitrary operators on $\mathscr{H}_1$ and $\mathscr{H}_2$, respectively. Their tensor product $\hat{A} \otimes \hat{B}$ is an operator on $\mathscr{H}_1 \otimes \mathscr{H}_2$ whose action is determined by

$$(\hat{A} \otimes \hat{B})|\Phi\rangle_1 \otimes |\Psi\rangle_2 = (\hat{A}|\Phi\rangle_1) \otimes (\hat{B}|\Psi\rangle_2), \tag{A.1.18}$$

and linearity.

We shall freely change the order of tensor products, keeping track of the space to which each state belongs. We can thus write either $|\Phi\rangle_1 \otimes |\Psi\rangle_2$ or $|\Psi\rangle_2 \otimes |\Phi\rangle_1$ to indicate the same state. We sometimes (following physicists' convention) omit the tensor product symbols in states, and simply write $|\Phi\rangle_1|\Psi\rangle_2$ instead of $|\Phi\rangle_1 \otimes |\Psi\rangle_2$.

It should be clear that the above discussion automatically extends to a joint system of more than two systems.

Entanglement and Schmidt decomposition Consider, as above, a quantum system that consists of two parts, and denote its Hilbert space as $\mathscr{H} = \mathscr{H}_1 \otimes \mathscr{H}_2$. A state $|\Xi\rangle \in \mathscr{H}$ is said to be separable if it can be written in a tensor product form as

$|\Xi\rangle = |\Phi\rangle_1 \otimes |\Psi\rangle_2$ with some states $|\Phi\rangle_1 \in \mathscr{H}_1$ and $|\Psi\rangle_2 \in \mathscr{H}_2$. A state $|\Xi\rangle \in \mathscr{H}$ that is not separable is said to be entangled.

A quantitative measure of entanglement is given by entanglement entropy S_{12}. Let $|\Xi\rangle \in \mathscr{H}$ be a normalized state, and define the corresponding reduced density matrices as

$$\hat{\rho}_1 := \mathrm{Tr}_2\big[|\Xi\rangle\langle\Xi|\big], \quad \hat{\rho}_2 := \mathrm{Tr}_1\big[|\Xi\rangle\langle\Xi|\big], \tag{A.1.19}$$

where $\mathrm{Tr}_\nu[\cdots]$ denote trace over $\mathscr{H}_\nu$. Then the entanglement entropy is defined as

$$S_{12} := -\,\mathrm{Tr}_1[\hat{\rho}_1 \log \hat{\rho}_1] = -\,\mathrm{Tr}_2[\hat{\rho}_2 \log \hat{\rho}_2]. \tag{A.1.20}$$

One easily finds that $S_{12} = 0$ if $|\Xi\rangle$ is separable.

It is known that an arbitrary normalized state $|\Xi\rangle \in \mathscr{H}$ is written in the Schmidt decomposition form

$$|\Xi\rangle = \sum_{j=1}^{d} q_j |\widetilde{\Phi}_j\rangle_1 \otimes |\widetilde{\Psi}_j\rangle_2, \tag{A.1.21}$$

with d not exceeding $\min\{D_1, D_2\}$, where D_1 and D_2 are the dimensions of $\mathscr{H}_1$ and $\mathscr{H}_2$. Here $\{|\widetilde{\Phi}_1\rangle_1, \ldots, |\widetilde{\Phi}_d\rangle_1\}$ and $\{|\widetilde{\Psi}_1\rangle_2, \ldots, |\widetilde{\Psi}_d\rangle_2\}$ are orthonormal sets of states in $\mathscr{H}_1$ and $\mathscr{H}_2$, respectively, and q_j are constants such that $q_j > 0$ and $\sum_{j=1}^{d}(q_j)^2 = 1$.[1]

Schmidt decomposition is useful for examining entanglement properties of a given state. In particular, we have

$$\hat{\rho}_1 = \sum_{j=1}^{d} (q_j)^2 |\widetilde{\Phi}_j\rangle_1\,{}_1\langle\widetilde{\Phi}_j|, \quad \hat{\rho}_2 = \sum_{j=1}^{d} (q_j)^2 |\widetilde{\Psi}_j\rangle_2\,{}_2\langle\widetilde{\Psi}_j|, \tag{A.1.22}$$

and hence

$$S_{12} = -\sum_{j=1}^{d} (q_j)^2 \log(q_j)^2. \tag{A.1.23}$$

Note that (A.1.23) is nothing but the Shannon entropy of a classical probability distribution $(p_1, \ldots, p_d)$ with $p_j = (q_j)^2$.

Suppose that $D_1 = D_2$. Then the entanglement entropy (A.1.23) takes the maximum possible value $\log D_1$ when $d = D_1$ and $q_j = 1/\sqrt{D_1}$ for all $j = 1, \ldots, D_1$. A state which has $S_{12} = \log D_1$ is said to be maximally entangled.

Proof of the Decomposition (A.1.21) Take arbitrary orthonormal bases $\{|\Phi_j\rangle_1\}_{j=1,\ldots,D_1}$ and $\{|\Psi_k\rangle_2\}_{k=1,\ldots,D_2}$ of $\mathscr{H}_1$ and $\mathscr{H}_2$, respectively, and expand $|\Xi\rangle$ as

$$|\Xi\rangle = \sum_{j=1}^{D_1}\sum_{k=1}^{D_2} A_{j,k} |\Phi_j\rangle_1 \otimes |\Psi_k\rangle_2, \tag{A.1.24}$$

[1] Sometimes one includes the terms with $q_j = 0$ in (A.1.21).

with coefficients $A_{j,k} \in \mathbb{C}$. We can assume $D_1 = D_2$ without loosing generality. This is because if $D_1 > D_2$, for example, we can add fictitious states $|\Psi_{D_2+1}\rangle_2, \ldots, |\Psi_{D_1}\rangle_2$ to the basis, and set $A_{j,k} = 0$ if $k > D_2$ to formally rewrite (A.1.24) into the same expression with D_2 replaced by D_1.

We then apply the singular value decomposition theorem (Theorem A.20 in p. 477) to the $D_1 \times D_1$ matrix $\mathsf{A} = (A_{j,k})_{j,k=1,\ldots,D_1}$ to find that there are unitary matrices $\mathsf{U} = (U_{j,\ell})_{j,\ell=1,\ldots,D_1}$ and $\mathsf{U} = (U_{k,\ell})_{k,\ell=1,\ldots,D_1}$ such that

$$\mathsf{A} = \mathsf{U} \begin{pmatrix} q_1 & 0 & \cdots & 0 \\ 0 & q_2 & \cdots & 0 \\ \vdots & \vdots & \ddots & \vdots \\ 0 & 0 & \cdots & q_{D_1} \end{pmatrix} \mathsf{V}^\dagger, \tag{A.1.25}$$

with $q_\ell \geq 0$ for $\ell = 1, \ldots, D_1$, or, equivalently,

$$A_{j,k} = \sum_{\ell=1}^{D_1} U_{j,\ell}\, q_\ell\, (V_{k,\ell})^*. \tag{A.1.26}$$

Substituting this into (A.1.24), we find

$$|\Xi\rangle = \sum_{\ell=1}^{D_1} q_\ell \Bigl(\sum_{j=1}^{D_1} U_{j,\ell} |\Phi_j\rangle_1\Bigr) \otimes \Bigl(\sum_{k=1}^{D_1} (V_{k,\ell})^* |\Psi_k\rangle_2\Bigr) = \sum_{\ell=1}^{D_1} q_\ell |\widetilde{\Phi}_\ell\rangle_1 \otimes |\widetilde{\Psi}_\ell\rangle_2, \tag{A.1.27}$$

with $|\widetilde{\Phi}_\ell\rangle_1 = \sum_{j=1}^{D_1} U_{j,\ell} |\Phi_j\rangle_1$ and $|\widetilde{\Psi}_\ell\rangle_2 = \sum_{k=1}^{D_1} (V_{k,\ell})^* |\Psi_k\rangle_2$. The orthonormality of $\{|\widetilde{\Phi}_\ell\rangle_1\}_{\ell=1,\ldots,D_1}$ and $\{|\widetilde{\Psi}_\ell\rangle_2\}_{\ell=1,\ldots,D_1}$ obviously follows from the unitarity. If necessary we throw away ℓ with $q_\ell = 0$ and relabel the basis states to get (A.1.21). (Then the added states $|\Psi_{D_2+1}\rangle_2, \ldots, |\Psi_{D_1}\rangle_2$ automatically disappear from the final expression.) ∎

A.2 Useful Properties of Operators

We discuss properties of operators used in the book. Some are quite standard and elementary, and some are slightly more advanced. As in Sect. A.1, $\hat{A}, \hat{B}, \ldots$ denote operators on the D dimensional Hilbert space $\mathscr{H}$ whose elements are denoted as $|\Phi\rangle, |\Psi\rangle, \ldots$.

A.2.1 Commutator and Operator Norm

The commutator of two operators $\hat{A}$ and $\hat{B}$ is defined as

$$[\hat{A}, \hat{B}] := \hat{A}\hat{B} - \hat{B}\hat{A}. \tag{A.2.1}$$

The commutator is clearly linear both in $\hat{A}$ and $\hat{B}$, i.e., $[\hat{A}, \beta\hat{B} + \gamma\hat{C}] = \beta[\hat{A}, \hat{B}] + \gamma[\hat{A}, \hat{C}]$, etc. Easily verifiable relation $[\hat{A}, \hat{B}\hat{C}] = [\hat{A}, \hat{B}]\hat{C} + \hat{B}[\hat{A}, \hat{C}]$, and its generalization

$$[\hat{A}, \hat{B}_1 \cdots \hat{B}_n] = \sum_{j=1}^{n} \hat{B}_1 \cdots \hat{B}_{j-1}[\hat{A}, \hat{B}_j]\hat{B}_{j+1} \cdots \hat{B}_n \tag{A.2.2}$$

are useful.

We define the norm (or the operator norm) of any operator $\hat{A}$ as[2]

$$\|\hat{A}\| := \max_{\substack{|\Phi\rangle\in\mathscr{H} \\ (|\Phi\rangle\neq 0)}} \frac{\|\hat{A}|\Phi\rangle\|}{\|\Phi\|} = \max_{\substack{|\Phi\rangle\in\mathscr{H} \\ (\|\Phi\|=1)}} \|\hat{A}|\Phi\rangle\|. \tag{A.2.3}$$

It readily follows from the definition that, for any $\hat{A}$ and for any unitary operators $\hat{U}$ and $\hat{V}$,

$$\|\hat{U}\hat{A}\hat{V}\| = \|\hat{A}\|. \tag{A.2.4}$$

It holds, for any operators $\hat{A}$ and $\hat{B}$, that

$$\|\hat{A} + \hat{B}\| \leq \|\hat{A}\| + \|\hat{B}\|, \tag{A.2.5}$$

and

$$\|\hat{A}\hat{B}\| \leq \|\hat{A}\|\,\|\hat{B}\|. \tag{A.2.6}$$

Proof By using the triangle inequality for state norm $\||\Phi\rangle + |\Psi\rangle\| \leq \|\Phi\| + \|\Psi\|$, we see

$$\begin{aligned} \|\hat{A} + \hat{B}\| &= \max_{\substack{|\Phi\rangle\in\mathscr{H} \\ (\|\Phi\|=1)}} \|(\hat{A} + \hat{B})|\Phi\rangle\| \leq \max_{\substack{|\Phi\rangle\in\mathscr{H} \\ (\|\Phi\|=1)}} \left\{\|\hat{A}|\Phi\rangle\| + \|\hat{B}|\Phi\rangle\|\right\} \\ &\leq \max_{\substack{|\Phi\rangle\in\mathscr{H} \\ (\|\Phi\|=1)}} \|\hat{A}|\Phi\rangle\| + \max_{\substack{|\Phi\rangle\in\mathscr{H} \\ (\|\Phi\|=1)}} \|\hat{B}|\Phi\rangle\| = \|\hat{A}\| + \|\hat{B}\|, \end{aligned} \tag{A.2.7}$$

which is (A.2.5). From the definition, we find

$$\frac{\|\hat{A}\hat{B}|\Phi\rangle\|}{\||\Phi\rangle\|} = \frac{\|\hat{A}\hat{B}|\Phi\rangle\|}{\|\hat{B}|\Phi\rangle\|}\,\frac{\|\hat{B}|\Phi\rangle\|}{\||\Phi\rangle\|} \leq \|\hat{A}\|\,\|\hat{B}\|. \tag{A.2.8}$$

We then get (A.2.6) by taking maximum over $|\Phi\rangle$. ∎

---

[2]It is standard to use sup (supremum) in the definition, but we can safely use max here since the maximum always exists in a finite dimensional Hilbert space (whose unit sphere is compact).

It is also useful to characterize the operator norm (A.2.3) as

$$\|\hat{A}\| = \max_{\substack{|\Phi\rangle, |\Psi\rangle \in \mathscr{H} \\ (\|\Phi\|=\|\Psi\|=1)}} \bigl|\langle\Psi|\hat{A}|\Phi\rangle\bigr|. \tag{A.2.9}$$

Proof Note for any state $|\Xi\rangle$ that $\|\Xi\| = \max_{|\Psi\rangle\,(\|\Psi\|=1)} \bigl|\langle\Psi|\Xi\rangle\bigr|$. Thus for any $|\Phi\rangle \in \mathscr{H}$ such that $\|\Phi\| = 1$, we have

$$\|\hat{A}|\Phi\rangle\| = \max_{\substack{|\Psi\rangle \in \mathscr{H} \\ (\|\Psi\|=1)}} \bigl|\langle\Psi|\hat{A}|\Phi\rangle\bigr|. \tag{A.2.10}$$

Then we get (A.2.9) from (A.2.3). ∎

Since $\langle\Psi|\hat{A}^\dagger|\Phi\rangle = \langle\Phi|\hat{A}|\Psi\rangle^*$, (A.2.9) implies

$$\|\hat{A}^\dagger\| = \|\hat{A}\|. \tag{A.2.11}$$

It also follows from (A.2.9) that[3]

$$\bigl|\langle\Phi|\hat{A}|\Phi\rangle\bigr| \le \|\hat{A}\|\ \langle\Phi|\Phi\rangle, \tag{A.2.12}$$

for any $|\Phi\rangle \in \mathscr{H}$. For any operators $\hat{A}$ and $\hat{X}$, it holds that

$$\bigl|\mathrm{Tr}[\hat{X}^\dagger\hat{A}\hat{X}]\bigr| \le \|\hat{A}\|\ \mathrm{Tr}[\hat{X}^\dagger\hat{X}]. \tag{A.2.13}$$

Proof We choose an arbitrary orthonormal basis $\{|\Psi_j\rangle\}_{j=1,\ldots,D}$, and observe that

$$\bigl|\mathrm{Tr}[\hat{X}^\dagger\hat{A}\hat{X}]\bigr| \le \sum_{j=1}^{D}\bigl|\langle\Psi_j|\hat{X}^\dagger\hat{A}\hat{X}|\Psi_j\rangle\bigr| \le \sum_{j=1}^{D}\|\hat{A}\|\langle\Psi_j|\hat{X}^\dagger\hat{X}|\Psi_j\rangle = \|\hat{A}\|\ \mathrm{Tr}[\hat{X}^\dagger\hat{X}], \tag{A.2.14}$$

where we used (A.2.12), and noted that $\langle\Psi_j|\hat{X}^\dagger\hat{X}|\Psi_j\rangle = \|\hat{X}|\Psi_j\rangle\|^2 \ge 0$. ∎

It is useful to note that for any self-adjoint operator $\hat{A}$ with eigenvalues $\lambda_1, \ldots, \lambda_D$, the definition (A.2.3) readily implies

$$\|\hat{A}\| = \max\{|\lambda_1|, \ldots, |\lambda_D|\}. \tag{A.2.15}$$

A.2.2 Exponential of an Operator

The exponential of any operator $\hat{A}$ is defined by the power series expansion as

---

[3] ***Proof*** Equation (A.2.9) implies $|\langle\Psi|\hat{A}|\Psi\rangle| \le \|\hat{A}\|$ for any normalized $|\Psi\rangle$. By setting $|\Psi\rangle = |\Phi\rangle/\|\Phi\|$, we get (A.2.12).

$$e^{\hat{A}} := \sum_{n=0}^{\infty} \frac{1}{n!} \hat{A}^n, \tag{A.2.16}$$

where the convergence is always guaranteed in a finite dimensional Hilbert space. Note that one has $[e^{\hat{A}}, \hat{B}] = 0$ if $[\hat{A}, \hat{B}] = 0$. If P is an invertible matrix, we have

$$\mathsf{P}\, e^{\hat{A}}\, \mathsf{P}^{-1} = \sum_{n=0}^{\infty} \frac{1}{n!} \mathsf{P} \hat{A}^n \mathsf{P}^{-1} = \sum_{n=0}^{\infty} \frac{1}{n!} (\mathsf{P} \hat{A} \mathsf{P}^{-1})^n = e^{\mathsf{P}\hat{A}\mathsf{P}^{-1}}, \tag{A.2.17}$$

which is used frequently. For a real variable t, it is readily verified that

$$\frac{d}{dt} e^{t\hat{A}} = \hat{A}\, e^{t\hat{A}} = e^{t\hat{A}}\, \hat{A}. \tag{A.2.18}$$

The relation

$$\frac{d}{dt} \left(e^{t\hat{A}}\, \hat{B} e^{-t\hat{A}} \right) = \hat{A}\, e^{t\hat{A}}\, \hat{B} e^{-t\hat{A}} - e^{t\hat{A}}\, \hat{B} e^{-t\hat{A}}\, \hat{A} = [\hat{A}, e^{t\hat{A}}\, \hat{B} e^{-t\hat{A}}] \tag{A.2.19}$$

is also useful.

Unlike the exponential function for numbers the identity $e^{\hat{A}+\hat{B}} = e^{\hat{A}} e^{\hat{B}}$ is not valid in general, but can be shown when $[\hat{A}, \hat{B}] = 0$.[4] The proof is essentially the same as that for the standard exponential function, and hence omitted. This in particular means

$$e^{(x+y)\hat{A}} = e^{x\hat{A}} e^{y\hat{A}}, \tag{A.2.20}$$

for any $x, y \in \mathbb{C}$. From this identity we find that $e^{i\hat{A}}$ is unitary if $\hat{A}$ is self-adjoint.

The following Lie product formula is useful in quantum physics.[5]

Theorem A.1 (Lie product formula) *For any operators $\hat{A}$ and $\hat{B}$, one has*

$$e^{\hat{A}+\hat{B}} = \lim_{N\uparrow\infty} \left(e^{\hat{A}/N} e^{\hat{B}/N} \right)^N. \tag{A.2.21}$$

---

[4] In general the Baker–Campbell–Hausdorff formula $e^{\hat{A}} e^{\hat{B}} = \exp[\hat{A} + \hat{B} + [\hat{A}, \hat{B}]/2 + [\hat{A}, [\hat{A}, \hat{B}]]/12 - [\hat{B}, [\hat{A}, \hat{B}]]/12 + \cdots]$ is valid.

[5] The relation (A.2.21) is sometimes referred to as the Trotter formula, but this is incorrect. The Trotter formula stands for the same relation for certain unbounded operators. See, e.g., Sect. VIII.8 of [8]. The relation for bounded operators (which we are dealing with) belongs to nineteenth century mathematics.

More generally for any operators $\hat{A}_1, \ldots, \hat{A}_n$*, one has*

$$e^{\hat{A}_1+\cdots+\hat{A}_n} = \lim_{N\uparrow\infty} \left(e^{\hat{A}_1/N} \ldots e^{\hat{A}_n/N}\right)^N. \tag{A.2.22}$$

Proof From (A.2.20), we find $e^{\hat{A}+\hat{B}} = (e^{(\hat{A}+\hat{B})/N})^N$, and from the definition (A.2.16), we observe

$$\begin{aligned} e^{(\hat{A}+\hat{B})/N} &= 1 + \frac{\hat{A}+\hat{B}}{N} + O\Big(\frac{1}{N^2}\Big) = \Big(1+\frac{\hat{A}}{N}\Big)\Big(1+\frac{\hat{B}}{N}\Big) + O\Big(\frac{1}{N^2}\Big) \\ &= e^{\hat{A}/N} e^{\hat{B}/N} + O\Big(\frac{1}{N^2}\Big). \end{aligned} \tag{A.2.23}$$

This means $e^{\hat{A}+\hat{B}} = (e^{\hat{A}/N}e^{\hat{B}/N})^N + O(1/N)$, which implies (A.2.21).[6] The proof of (A.2.22) is essentially the same. ∎

The following inequality, which was used in Sect. 4.4 (see (4.4.35)), is sometimes useful in quantum statistical mechanics. As far as we know the inequality was first proved in [7]. The following proof is due to [9]. See [3, 6] for another proof.

Lemma A.2 *For an arbitrary operator* $\hat{A}$ *and self-adjoint operators* $\hat{V}$ *and* $\hat{W}$*, one has*

$$\left|\mathrm{Tr}[\hat{A}\, e^{\hat{V}+i\hat{W}}]\right| \le \|\hat{A}\|\ \mathrm{Tr}[e^{\hat{V}}]. \tag{A.2.24}$$

Proof From the Schwarz inequality for trace,[7] one finds

$$\begin{aligned} \left|\mathrm{Tr}[\hat{A}\, e^{\hat{V}+i\hat{W}}]\right| &= \left|\mathrm{Tr}[\hat{A}\, e^{(\hat{V}+i\hat{W})/2}\, e^{(\hat{V}+i\hat{W})/2}]\right| \\ &\le \sqrt{\mathrm{Tr}[e^{(\hat{V}-i\hat{W})/2}\, \hat{A}^\dagger \hat{A}\, e^{(\hat{V}+i\hat{W})/2}]\ \mathrm{Tr}[e^{(\hat{V}-i\hat{W})/2}\, e^{(\hat{V}+i\hat{W})/2}]} \\ &\le \|\hat{A}\|\ \mathrm{Tr}[e^{(\hat{V}-i\hat{W})/2}\, e^{(\hat{V}+i\hat{W})/2}], \end{aligned} \tag{A.2.25}$$

where we used (A.2.13), (A.2.6), and (A.2.11). By using the Lie product formula (A.2.21), we can write

$$\mathrm{Tr}[e^{(\hat{V}-i\hat{W})/2}\, e^{(\hat{V}+i\hat{W})/2}] = \lim_{n\uparrow\infty} \mathrm{Tr}[(\hat{X}_n)^n (\hat{X}_n^\dagger)^n], \tag{A.2.26}$$

with $\hat{X}_n = e^{\hat{V}/(2n)} e^{-i\hat{W}/(2n)}$. Then from the bound (A.2.28) below, we see that

$$\mathrm{Tr}[(\hat{X}_n)^n (\hat{X}_n^\dagger)^n] \le \mathrm{Tr}[(\hat{X}_n \hat{X}_n^\dagger)^n] = \mathrm{Tr}[e^{\hat{V}}], \tag{A.2.27}$$

which proves the desired bound (A.2.24).

---

[6]The reader uncomfortable with the $O(1/N)$ notation may readily prove more formal estimates like $\|e^{\hat{A}+\hat{B}} - (e^{\hat{A}/N}e^{\hat{B}/N})^N\| \le (\text{constant})/N$.

[7]It is easily proved that $|\,\mathrm{Tr}[\hat{X}^\dagger \hat{Y}]|^2 \le \mathrm{Tr}[\hat{X}^\dagger \hat{X}]\,\mathrm{Tr}[\hat{Y}^\dagger \hat{Y}]$.

We finally prove

$$\mathrm{Tr}[\hat{X}^n(\hat{X}^\dagger)^n] \leq \mathrm{Tr}[(\hat{X}\hat{X}^\dagger)^n], \qquad \text{(A.2.28)}$$

for any operator $\hat{X}$.[8] Fix $\hat{X}$, and denote $\hat{X}$ and $\hat{X}^\dagger$ as $\hat{X}^+$ and $\hat{X}^-$, respectively. For a sequence $\boldsymbol{\sigma} = (\sigma_1, \sigma_2, \ldots, \sigma_{2n})$ such that $\sigma_j = \pm$ and $\sum_{j=1}^{2n} \sigma_j = 0$, let $f(\boldsymbol{\sigma}) = |\mathrm{Tr}[\hat{X}^{\sigma_1}\hat{X}^{\sigma_2} \cdots \hat{X}^{\sigma_{2n}}]|$ and $f_{\max} = \max_{\boldsymbol{\sigma}} f(\boldsymbol{\sigma})$. We also denote by $m(\boldsymbol{\sigma})$ the number of j such that $\sigma_j = \sigma_{j+1}$ in the sequence $\boldsymbol{\sigma}$, where we identify σ_{2n+1} with σ_1. Note that $m(\boldsymbol{\sigma}) = 0$ means $\boldsymbol{\sigma} = (+, -, +, -, \ldots, +, -)$ or $(-, +, -, +, \ldots, -, +)$. Our goal is to show that $f(\boldsymbol{\sigma}) = f_{\max}$ if $m(\boldsymbol{\sigma}) = 0$. This fact and $\mathrm{Tr}[(\hat{X}\hat{X}^\dagger)^n] \geq 0$[9] proves the desired (A.2.28).

Suppose that $f(\boldsymbol{\sigma}) = f_{\max}$ for some $\boldsymbol{\sigma}$ with $m(\boldsymbol{\sigma}) > 0$, i.e., there is a j such that $\sigma_j = \sigma_{j+1}$. By using the cyclicity of the trace, we can assume that $\sigma_1 = \sigma_{2n}$. From the Schwarz inequality for trace, we have $f(\boldsymbol{\sigma}) \leq \sqrt{f(\boldsymbol{\sigma}')\, f(\boldsymbol{\sigma}'')}$, where $\boldsymbol{\sigma}' = (\sigma_1, \sigma_2, \ldots, \sigma_n, -\sigma_n, \ldots, -\sigma_2, -\sigma_1)$ and $\boldsymbol{\sigma}'' = (-\sigma_{2n}, -\sigma_{2n-1}, \ldots, -\sigma_{n+1}, \sigma_{n+1}, \ldots, \sigma_{2n-1}, \sigma_{2n})$. The assumption $f(\boldsymbol{\sigma}) = f_{\max}$ implies $f(\boldsymbol{\sigma}') = f(\boldsymbol{\sigma}'') = f_{\max}$. An inspection shows that either $m(\sigma')$ or $m(\sigma'')$ is strictly less than $m(\boldsymbol{\sigma})$. We then repeat the above consideration for the new sequence σ' or σ'' with smaller m. By repeating the procedure, we finally see that $f(\boldsymbol{\sigma}) = f_{\max}$ if $m(\boldsymbol{\sigma}) = 0$. ∎

A.2.3 *Inequality Between Self-adjoint Operators and Nonnegative Operators*

We define the notion of inequality between self-adjoint operators (Hermitian matrices), and summarize basic facts.

Definition A.3 We say that an operator $\hat{A}$ is nonnegative (or positive semidefinite) and write $\hat{A} \geq 0$ if it is self-adjoint (Hermitian) and satisfies $\langle \Phi | \hat{A} | \Phi \rangle \geq 0$ for any $|\Phi\rangle \in \mathscr{H}$. For two self-adjoint operators $\hat{A}$ and $\hat{B}$, we write $\hat{A} \leq \hat{B}$ if $\hat{B} - \hat{A} \geq 0$.

We write $\hat{A} \leq b$ when we have $\hat{A} \leq b\hat{1}$ for a self-adjoint operator $\hat{A}$ and $b \in \mathbb{R}$. Note that (A.2.12) implies that $\hat{A} \leq \|\hat{A}\|$ for any self-adjoint $\hat{A}$.

Let us list some elementary properties of nonnegative operators which are used throughout the main body of the book.

Lemma A.4 *A self-adjoint operator $\hat{A}$ is nonnegative if and only if all the eigenvalues of $\hat{A}$ are nonnegative.*

Proof The proof is straightforward if one diagonalizes $\hat{A}$. ∎

---

[8]One might notice that the following proof is very similar to the proof of the chessboard estimate (Lemma 4.5 in p. 87).

[9]Note that $(\hat{X}\hat{X}^\dagger)^n = \hat{B}^\dagger\hat{B}$ with $\hat{B} = (\hat{X}\hat{X}^\dagger)^{n/2}$ if n is even, and $\hat{B} = \hat{X}^\dagger(\hat{X}\hat{X}^\dagger)^{(n-1)/2}$ if n is odd. Then the desired $\mathrm{Tr}[\hat{B}^\dagger\hat{B}] \geq 0$ follows from Lemma A.6 below.

Lemma A.5 *If $\hat{A} \geq 0$ and $\hat{B} \geq 0$, we have $\hat{A} + \hat{B} \geq 0$.*

Proof $\langle \Phi|(\hat{A} + \hat{B})|\Phi\rangle = \langle \Phi|\hat{A}|\Phi\rangle + \langle \Phi|\hat{B}|\Phi\rangle \geq 0$ for any $|\Phi\rangle$. ∎

Lemma A.6 *For an arbitrary operator $\hat{B}$, we have $\hat{B}^\dagger \hat{B} \geq 0$. Conversely, for any $\hat{A} \geq 0$, there is a unique $\hat{C} \geq 0$ such that $\hat{A} = \hat{C}^2$. One often writes $\hat{C} = \sqrt{\hat{A}}$.*

Proof It suffices to note that $0 \leq \|\hat{B}|\Phi\rangle\|^2 = \langle \Phi|\hat{B}^\dagger \hat{B}|\Phi\rangle$ for any $|\Phi\rangle$. To show the second part, we use the spectral decomposition $\hat{A} = \sum_{j=1}^{D} |\Psi_j\rangle a_j \langle \Psi_j|$ (where $\{|\Psi_j\rangle\}_{j=1,\ldots,D}$ is an orthonormal basis), and set $\hat{C} = \sum_{j=1}^{D} |\Psi_j\rangle \sqrt{a_j} \langle \Psi_j|$. The uniqueness is obvious if we diagonalize C. ∎

The following theorem is relatively unknown to physicists, but we believe it important and worth knowing.

Theorem A.7 *Let $\hat{A}$ and $\hat{B}$ be self-adjoint operators such that $\hat{A} \leq \hat{B}$. We denote by $a_1, \ldots, a_D$ and $b_1, \ldots, b_D$ the eigenvalues of $\hat{A}$ and $\hat{B}$, respectively, ordered so that $a_j \leq a_{j+1}$ and $b_j \leq b_{j+1}$. Then it holds that $a_j \leq b_j$ for any $j = 1, \ldots, D$.*

Proof For an arbitrary j-dimensional subspace $\mathscr{M}$ of $\mathscr{H}$, where $j \in \{1, \ldots, D\}$, let

$$\lambda_{\mathscr{M}}(\hat{A}) = \max_{\substack{|\Phi\rangle \in \mathscr{M} \\ (\|\Phi\|=1)}} \langle \Phi|\hat{A}|\Phi\rangle. \tag{A.2.29}$$

Let us minimize $\lambda_{\mathscr{M}}(\hat{A})$ by varying the j-dimensional subspace $\mathscr{M}$. Clearly the minimum is attainable and realized when $\mathscr{M}$ is the subspace spanned by the j eigenstates of $\hat{A}$ corresponding to the eigenvalues $a_1, \ldots, a_j$. We thus find an interesting expression

$$a_j = \min_{\substack{\mathscr{M} \\ (\dim \mathscr{M} = j)}} \lambda_{\mathscr{M}}(\hat{A}) = \min_{\substack{\mathscr{M} \\ (\dim \mathscr{M} = j)}} \max_{\substack{|\Phi\rangle \in \mathscr{M} \\ (\|\Phi\|=1)}} \langle \Phi|\hat{A}|\Phi\rangle, \tag{A.2.30}$$

which is known as the mini-max principle. Note that, when $j = 1$, it reduces to the familiar variational principle $a_1 = \min_{|\Phi\rangle\,(\|\Phi\|=1)} \langle \Phi|\hat{A}|\Phi\rangle$. Since $\langle \Phi|\hat{A}|\Phi\rangle \leq \langle \Phi|\hat{B}|\Phi\rangle$ for any $|\Phi\rangle$ by assumption, (A.2.30) implies $a_j \leq b_j$. ∎

The following corollary is useful in quantum statistical mechanics.

Corollary A.8 *For self-adjoint operators $\hat{A}$ and $\hat{B}$ such that $\hat{A} \leq \hat{B}$, we have*

$$\mathrm{Tr}[e^{\hat{A}}] \leq \mathrm{Tr}[e^{\hat{B}}], \tag{A.2.31}$$

or, more generally,

$$\mathrm{Tr}[f(\hat{A})] \leq \mathrm{Tr}[f(\hat{B})], \tag{A.2.32}$$

for any non-decreasing function[10] $f : \mathbb{R} \to \mathbb{R}$.

[10] On may define $f(\hat{A})$ from the series expansion of $f(x)$ as in (A.2.16). For a self-adjoint $\hat{A}$, let us use the spectral decomposition $\hat{A} = \sum_{j=1}^{D} |\Psi_j\rangle a_j \langle \Psi_j|$ and define $f(\hat{A}) := \sum_{j=1}^{D} |\Psi_j\rangle f(a_j) \langle \Psi_j|$.

Proof $a_j \le b_j$ implies $\sum_{j=1}^{D} f(a_j) \le \sum_{j=1}^{D} f(b_j)$, which is (A.2.32). ∎

The following lemma is used several times in the present book (and frequently in modern research of quantum many-body systems).

Lemma A.9 (Frustration-free Hamiltonian 1) *Consider an arbitrary Hamiltonian written as* $\hat{H} = \sum_{j=1}^{N} \hat{h}_j$ *with* $\hat{h}_j \ge \varepsilon_j$ *for all* $j = 1, \ldots, N$. *If there is a state* $|\Phi\rangle$ *such that* $\hat{h}_j|\Phi\rangle = \varepsilon_j|\Phi\rangle$ *for any* $j = 1, \ldots, N$, *then* $|\Phi\rangle$ *is a ground state of* $\hat{H}$, *and the ground state energy is* $E_{\mathrm{GS}} = \sum_{j=1}^{N} \varepsilon_j$.

We say that a Hamiltonian $\hat{H}$ is frustration free when it allows the decomposition as above and has a ground state with the above property. One should note that this notion is different from frustration in the context of antiferromagnetic spin systems.[11]

Proof By using Lemma A.5 repeatedly, one finds $\hat{H} \ge \sum_{j=1}^{N} \varepsilon_j$. Then $\hat{H}|\Phi\rangle = (\sum_{j=1}^{N} \varepsilon_j)|\Phi\rangle$ implies the $|\Phi\rangle$ is a ground state. ∎

The following lemma, although being almost trivial, also characterizes a frustration free system.

Lemma A.10 (Frustration-free Hamiltonian 2) *Consider an arbitrary Hamiltonian written as* $\hat{H} = \sum_{j=1}^{N} \hat{h}_j$ *with* $\hat{h}_j \ge \varepsilon_j$ *for all* $j = 1, \ldots, N$. *Suppose that the ground state energy is* $\sum_{j=1}^{N} \varepsilon_j$, *i.e., there is a state* $|\Phi\rangle$ *such that* $\hat{H}|\Phi\rangle = (\sum_{j=1}^{N} \varepsilon_j)|\Phi\rangle$. *Then the ground state* $|\Phi\rangle$ *satisfies* $\hat{h}_j|\Phi\rangle = \varepsilon_j|\Phi\rangle$ *for each* $j = 1, \ldots, N$.

Proof We can assume that $|\Phi\rangle$ is normalized. Then $\langle\Phi|\hat{H}|\Phi\rangle = \sum_{j=1}^{N}\langle\Phi|\hat{h}_j|\Phi\rangle = \sum_{j=1}^{N} \varepsilon_j$, along with $\langle\Phi|\hat{h}_j|\Phi\rangle \ge \varepsilon_j$, implies $\langle\Phi|\hat{h}_j|\Phi\rangle = \varepsilon_j$ for $j = 1, \ldots, N$. Since ε_j is the lowest eigenvalue, the variational principle implies $\hat{h}_j|\Phi\rangle = \varepsilon_j|\Phi\rangle$. ∎

The following lemma is sometimes powerful and useful.

Lemma A.11 *If* $\hat{A} \ge 0$, *the condition* $\langle\Phi|\hat{A}|\Phi\rangle = 0$ *implies* $\hat{A}|\Phi\rangle = 0$. *If* $\hat{A} = \hat{B}^\dagger\hat{B}$ *with any* $\hat{B}$, *the condition* $\langle\Phi|\hat{A}|\Phi\rangle = 0$ *(or* $\hat{A}|\Phi\rangle = 0$*) implies* $\hat{B}|\Phi\rangle = 0$.

Proof Because of Lemma A.6, we can always assume that $\hat{A} = \hat{B}^\dagger\hat{B}$ (because one can take $B = \sqrt{A}$). Then $\|\hat{B}|\Phi\rangle\|^2 = \langle\Phi|\hat{B}^\dagger\hat{B}|\Phi\rangle = \langle\Phi|\hat{A}|\Phi\rangle = 0$ implies $\hat{B}|\Phi\rangle = 0$. By operating $\hat{B}^\dagger$, we get $\hat{A}|\Phi\rangle = 0$. ∎

Here is a simple application to angular momentum operators (see Appendix A.3). Let $\hat{\boldsymbol{J}}^2 = (\hat{J}^{(1)})^2 + (\hat{J}^{(2)})^2 + (\hat{J}^{(3)})^2$ with self-adjoint $\hat{J}^{(1)}$, $\hat{J}^{(2)}$, and $\hat{J}^{(3)}$, and assume that $\hat{\boldsymbol{J}}^2|\Phi\rangle = 0$. Then $\langle\Phi|\{(\hat{J}^{(1)})^2 + (\hat{J}^{(2)})^2 + (\hat{J}^{(3)})^2\}|\Phi\rangle = 0$ obviously implies $\langle\Phi|(\hat{J}^{(\alpha)})^2|\Phi\rangle = 0$ and hence $\hat{J}^{(\alpha)}|\Phi\rangle = 0$ for $\alpha = 1, 2, 3$.

We finally present a useful theorem which allows us to construct low energy effective theory by making a parameter in a Hamiltonian infinitely large. The theorem

---

[11] See footnote 30 in p. 37.

is widely used in the physics literature. We also used it in Sects. 5.1 and 11.2. See (5.1.4) and (11.2.2).

Consider a general Hamiltonian written as $\hat{H}_\nu = \hat{H}_0 + \nu\hat{V}$, where $\hat{H}_0$ is an arbitrary self-adjoint operator, $\hat{V}$ is an arbitrary nonnegative operator, and $\nu \geq 0$ is a parameter. Let $\mathscr{H}_0$ be the subspace that consists of $|\Phi\rangle$ such that $\hat{V}|\Phi\rangle = 0$. We assume that $\mathscr{H}_0$ is not empty. Let us denote by $\hat{P}_0$ the orthogonal projection onto $\mathscr{H}_0$, and by D_0 the dimension of $\mathscr{H}_0$. Note that all the energy eigenvalues and eigenstates of $\hat{H}_\nu$ depend continuously on ν.

Eigenstates of $\hat{H}_\nu$ are classified into two classes, those with eigenvalues diverging as $\nu \uparrow \infty$ and those with eigenvalues converging to finite values as $\nu \uparrow \infty$. By continuity we see that the number of the latter is exactly D_0.

Theorem A.12 *The eigenstates and the eigenvalues in the second class (in the limit $\nu \uparrow \infty$) can be obtained from the effective Schrödinger equation $\hat{P}_0\hat{H}_0|\Phi\rangle = E|\Phi\rangle$ with $|\Phi\rangle \in \mathscr{H}_0$.*

Proof Take $|\Phi_\nu\rangle$ and E_ν that depend continuously on $\nu \geq 0$ and satisfy $\hat{H}_\nu|\Phi_\nu\rangle = E_\nu|\Phi_\nu\rangle$ and $\|\Phi_\nu\| = 1$ for all ν. We assume that they belong to the second class, i.e., $\lim_{\nu\uparrow\infty} E_\nu < \infty$. This means that $\lim_{\nu\uparrow\infty}\langle\Phi_\nu|\hat{V}|\Phi_\nu\rangle = 0$, and hence the limiting state[12] $|\Phi_\infty\rangle = \lim_{\nu\uparrow\infty}|\Phi_\nu\rangle$ satisfies $\langle\Phi_\infty|\hat{V}|\Phi_\infty\rangle = 0$. We then see from Lemma A.11 that $\hat{V}|\Phi_\infty\rangle = 0$, and hence $|\Phi_\infty\rangle \in \mathscr{H}_0$.

Since $\hat{P}_0\hat{V} = 0$, the Schrödinger equation $\hat{H}_\nu|\Phi_\nu\rangle = E_\nu|\Phi_\nu\rangle$ implies $\hat{P}_0\hat{H}_0|\Phi_\nu\rangle = E_\nu\hat{P}_0|\Phi_\nu\rangle$ for any $\nu \geq 0$. By letting $\nu \uparrow \infty$, one gets $\hat{P}_0\hat{H}_0|\Phi_\infty\rangle = E_\infty|\Phi_\infty\rangle$, which is the desired effective equation. Note that the effective Schrödinger equation has D_0 distinct eigenstates. They must coincide with the eigenstates in question. ∎

A.3 Quantum Mechanical Angular Momentum

Let us give a very brief (but hopefully self-contained) review of the theory of quantum mechanical angular momentum.

A.3.1 Definition and Basic Properties

Mathematically the theory of quantum mechanical angular momentum is equivalent to the representation theory of the Lie algebra $\mathfrak{su}(2)$. (See Appendix A.5.) Let $\hat{J}^{(1)}$, $\hat{J}^{(2)}$, $\hat{J}^{(3)}$ be self-adjoint operators (equivalently, Hermitian matrices) on a finite dimensional Hilbert space $\mathscr{H}$. They are assumed to satisfy the commutation relations

[12] The limit exists because the unit sphere of $\mathscr{H}$ is compact. Thus the proof relies essentially on the fact that the dimension of the Hilbert space is finite.

$$[\hat{J}^{(\alpha)}, \hat{J}^{(\beta)}] = i \sum_{\gamma=1}^{3} \varepsilon_{\alpha\beta\gamma}\, \hat{J}^{(\gamma)}, \tag{A.3.1}$$

for $\alpha, \beta = 1, 2, 3$. The Levi-Civita symbol $\varepsilon_{\alpha\beta\gamma}$ is defined as $\varepsilon_{123} = \varepsilon_{231} = \varepsilon_{312} = 1$, $\varepsilon_{321} = \varepsilon_{213} = \varepsilon_{132} = -1$, and $\varepsilon_{\alpha\beta\gamma} = 0$ for other components. We combine the three operators in a vector form as $\hat{\boldsymbol{J}} = (\hat{J}^{(1)}, \hat{J}^{(2)}, \hat{J}^{(3)})$, and write

$$\hat{\boldsymbol{J}}^2 = (\hat{J}^{(1)})^2 + (\hat{J}^{(2)})^2 + (\hat{J}^{(3)})^2. \tag{A.3.2}$$

Note that $\hat{\boldsymbol{J}}^2 \geq 0$. We also define

$$\hat{J}^{\pm} := \hat{J}^{(1)} \pm i \hat{J}^{(2)}, \tag{A.3.3}$$

which satisfy $(\hat{J}^{\pm})^{\dagger} = \hat{J}^{\mp}$. From the basic commutation relations (A.3.1), one can verify the following relations:

$$[\hat{\boldsymbol{J}}^2, \hat{J}^{(\alpha)}] = 0 \text{ for } \alpha = 1, 2, 3 \tag{A.3.4}$$

$$[\hat{\boldsymbol{J}}^2, \hat{J}^{\pm}] = 0 \tag{A.3.5}$$

$$[\hat{J}^{(3)}, \hat{J}^{\pm}] = \pm \hat{J}^{\pm} \tag{A.3.6}$$

$$\hat{J}^{-}\hat{J}^{+} = \hat{\boldsymbol{J}}^2 - \hat{J}^{(3)}(\hat{J}^{(3)} + 1) \tag{A.3.7}$$

$$\hat{J}^{+}\hat{J}^{-} = \hat{\boldsymbol{J}}^2 - \hat{J}^{(3)}(\hat{J}^{(3)} - 1) \tag{A.3.8}$$

Denote the eigenvalues of $\hat{\boldsymbol{J}}^2$ and $\hat{J}^{(3)}$ as $J(J+1)$ and M, respectively, where $J \geq 0$ and $M \in \mathbb{R}$. Since $\hat{\boldsymbol{J}}^2$ and $\hat{J}^{(3)}$ commute, one can decompose the whole Hilbert space according to the eigenvalues as $\mathscr{H} = \bigoplus_{J,M} \mathscr{H}_{J,M}$, where one has $\hat{\boldsymbol{J}}^2|\Phi\rangle = J(J+1)|\Phi\rangle$ and $\hat{J}^{(3)}|\Phi\rangle = M|\Phi\rangle$ for any $|\Phi\rangle \in \mathscr{H}_{J,M}$. The most basic result for quantum mechanical angular momentum is the following.

Theorem A.13 *The value of J is restricted to $n/2$, where n is a nonnegative integer. For a given J, the other eigenvalue M takes $2J + 1$ values $-J, -J + 1, \ldots, J - 1, J$.*

We shall prove the theorem below. For the moment we only assume that J and M may take any values with $J \geq 0$ and $M \in \mathbb{R}$.

Lemma A.14 *If there is a nonzero state $|\Phi\rangle \in \mathscr{H}_{J,M}$ with J and M such that $M < J$, then one has $\hat{J}^{+}|\Phi\rangle \neq 0$ and $\hat{J}^{+}|\Phi\rangle \in \mathscr{H}_{J,M+1}$. Similarly, if there is a nonzero state $|\Phi\rangle \in \mathscr{H}_{J,M}$ with J and M such that $M > -J$, then one has $\hat{J}^{-}|\Phi\rangle \neq 0$ and $\hat{J}^{-}|\Phi\rangle \in \mathscr{H}_{J,M-1}$.*

Proof Note that the conditions $M < J$ and $M > -J$ imply $M(M+1) < J(J+1)$ and $M(M-1) < J(J+1)$, respectively. From (A.3.7) and (A.3.8), one finds

$$\|\hat{J}^{\pm}|\Phi\rangle\|^2 = \langle\Phi|\hat{J}^{\mp}\hat{J}^{\pm}|\Phi\rangle = \langle\Phi|\{\hat{\boldsymbol{J}}^2 - \hat{J}^{(3)}(\hat{J}^{(3)} \pm 1)\}|\Phi\rangle$$
$$= \{J(J+1) - M(M \pm 1)\}\,\|\Phi\|^2, \tag{A.3.9}$$

which shows that $\hat{J}^{\pm}|\Phi\rangle \neq 0$ under assumed conditions. Then by using (A.3.5) and (A.3.6), one sees that

$$\hat{\boldsymbol{J}}^2\hat{J}^{\pm}|\Phi\rangle = \hat{J}^{\pm}\hat{\boldsymbol{J}}^2|\Phi\rangle = J(J+1)\,\hat{J}^{\pm}|\Phi\rangle, \tag{A.3.10}$$

$$\hat{J}^{(3)}\hat{J}^{\pm}|\Phi\rangle = \hat{J}^{\pm}\hat{J}^{(3)}|\Phi\rangle \pm \hat{J}^{\pm}|\Phi\rangle = (M \pm 1)\hat{J}^{\pm}|\Phi\rangle, \tag{A.3.11}$$

i.e., $\hat{J}^{\pm}|\Phi\rangle \in \mathscr{H}_{J,M\pm 1}$. ∎

Lemma A.15 *Both $J - M$ and $J + M$ must be nonnegative integers.*

Proof Let $M > J$, and suppose that there is a nonzero state $|\Phi\rangle \in \mathscr{H}_{J,M}$. Then (A.3.9) implies that $\|\hat{J}^{+}|\Phi\rangle\|^2 = \{J(J+1) - M(M+1)\}\,\|\Phi\|^2 < 0$, which is a contradiction. We thus see $M \le J$. Take J, M such that $J - M$ is not an integer, and suppose that there is nonzero $|\Phi\rangle \in \mathscr{H}_{J,M}$. We let n be a unique positive integer such that $M + n - 1 < J < M + n$. Then Lemma A.14 implies that $(\hat{J}^{+})^n|\Phi\rangle \neq 0$ and $(\hat{J}^{+})^n \in \mathscr{H}_{J,M+n}$, which is a contradiction. We have thus shown that $J - M$ is a nonnegative integer. The proof for $J + M$ is the same. ∎

Proof of the Theorem A.13 From Lemma A.15, we find that $2J$ is a nonnegative integer. The range of M then follows again from Lemma A.15. ∎

Let $J = n/2$ with a nonnegative integer n. A $2J + 1 = n + 1$ dimensional representation of the angular momentum operators is constructed by taking a Hilbert space spanned by normalized basis states $|\Psi_M\rangle$ with $M = -J, -J+1, \ldots, J$, and defining the action of $\hat{\boldsymbol{J}}$ as

$$\hat{J}^{(3)}|\Psi_M\rangle = M|\Psi_M\rangle, \tag{A.3.12}$$

$$\hat{J}^{\pm}|\Psi_M\rangle = \sqrt{J(J+1) - M(M \pm 1)}|\Psi_{M\pm 1}\rangle. \tag{A.3.13}$$

It is easily verified that (A.3.9) guarantees that (A.3.13) is consistent with normalization of $|\Psi_M\rangle$. Note that (A.3.12) and (A.3.13) are the same as (2.1.2) and (2.1.3).

A.3.2 *SU(2) Invariant Hamiltonian*

We shall prove useful theorems about the eigenvalues and eigenstates of an SU(2) invariant Hamiltonian. Let $\hat{\boldsymbol{J}} = (\hat{J}^{(1)}, \hat{J}^{(2)}, \hat{J}^{(3)})$ be a quantum mechanical angular momentum, and $\hat{H}$ be a Hamiltonian (i.e., a certain self-adjoint operator), acting on the same Hilbert space $\mathscr{H}$. We assume that they satisfy $[\hat{H}, \hat{J}^{(\alpha)}] = 0$ for any $\alpha = 1, 2, 3$. We again decompose the Hilbert space as $\mathscr{H} = \bigoplus_{J,M} \mathscr{H}_{J,M}$, according to the eigenvalues of $\hat{\boldsymbol{J}}^2$ and $\hat{J}^{(3)}$.

Theorem A.16 *Suppose that there is a nonzero state $|\Phi\rangle \in \mathscr{H}_{J,M_0}$ such that $\hat{H}|\Phi\rangle = E|\Phi\rangle$. Then for each $M = -J, -J+1, \ldots, J-1, J$, there is a nonzero state $|\Phi_M\rangle \in \mathscr{H}_{J,M}$ such that $\hat{H}|\Phi_M\rangle = E|\Phi_M\rangle$. Thus the energy eigenvalue E is at least $(2J+1)$-fold degenerate. Moreover theses eigenstates are "copy" of each other in the sense that*

$$|\Phi_M\rangle = c_M(\hat{J}^-)^{J-M}|\Phi_J\rangle, \tag{A.3.14}$$

for $M = -J, \ldots, J$, with $c_M \in \mathbb{C}$.

Proof By using Lemma A.14 repeatedly, we see that $|\Phi_J\rangle = (\hat{J}^+)^{J-M_0}|\Phi\rangle \in \mathscr{H}_{J,J}$ is nonzero. Because $[\hat{H}, \hat{J}^+] = 0$, we have $\hat{H}|\Phi_J\rangle = E|\Phi_J\rangle$. Again by Lemma A.14, the states (A.3.14) are all nonvanishing. That $\hat{H}|\Phi_M\rangle = E|\Phi_M\rangle$ again follows from $[\hat{H}, \hat{J}^-] = 0$. ∎

The following theorem is used repeatedly in the present book. It shows that one only needs to look at the sector with $J^{(3)} = 0$ or $1/2$ to find all the energy eigenvalues.

Theorem A.17 *Let E be an eigenvalue of $\hat{H}$. Then there exists a corresponding eigenstate $|\Phi\rangle$ that satisfies either $\hat{J}^{(3)}|\Phi\rangle = 0$ or $\hat{J}^{(3)}|\Phi\rangle = (1/2)|\Phi\rangle$.*

Proof Recall that 0 or 1/2 is the minimum possible value of J. Then the theorem is a straightforward consequence of Theorem A.16. ∎

A.3.3 Addition of Angular Momenta

We discuss the addition of angular momenta, which plays important roles throughout the book.

Addition of two spins with $S = 1/2$ Let us start by carefully discussing the simplest but an extremely important example of the addition of two spins with $S = 1/2$. The reader who is not familiar with the theory of quantum mechanical angular momentum is strongly encouraged to work out this example.

We consider a system of two spins with spin quantum number $S = 1/2$, which we call the spin 1 and spin 2. For $x = 1, 2$, we denote by $\mathfrak{h}_x \cong \mathbb{C}^2$ the Hilbert space and by $\{|\uparrow\rangle_x, |\downarrow\rangle_x\}$ the corresponding standard basis of the spin x. The Hilbert space of the whole system is $\mathscr{H} = \mathfrak{h}_1 \otimes \mathfrak{h}_2$, whose basis states are $|\sigma\rangle_1|\sigma'\rangle_2$ with $\sigma, \sigma' = \uparrow, \downarrow$. See (2.2.3). Let $\hat{\boldsymbol{S}}_x$ be the spin operator for the spin x as in (2.2.4) or (2.2.5), and define the total spin operator by $\hat{\boldsymbol{S}}_{\rm tot} = \hat{\boldsymbol{S}}_1 + \hat{\boldsymbol{S}}_2$. Then from the commutation relations (2.2.6), we see that $[\hat{S}^\alpha_{\rm tot}, \hat{S}^\beta_{\rm tot}] = i\sum_{\gamma=1,2,3}\varepsilon_{\alpha,\beta,\gamma}\hat{S}^\gamma_{\rm tot}$, and hence $\hat{\boldsymbol{S}}_{\rm tot}$ is also a quantum mechanical angular momentum. We denote the eigenvalues of $(\hat{\boldsymbol{S}}_{\rm tot})^2$ and $\hat{S}^{(3)}_{\rm tot}$ as $S_{\rm tot}(S_{\rm tot}+1)$ and $S^{(3)}_{\rm tot}$, respectively. Our goal is to find corresponding simultaneous eigenstates, i.e., $|\Phi_{S_{\rm tot}, S^{(3)}_{\rm tot}}\rangle \in \mathscr{H}$ such that

$$(\hat{\boldsymbol{S}}_{\text{tot}})^2|\Phi_{S_{\text{tot}},S_{\text{tot}}^{(3)}}\rangle = S_{\text{tot}}(S_{\text{tot}}+1)|\Phi_{S_{\text{tot}},S_{\text{tot}}^{(3)}}\rangle, \tag{A.3.15}$$

$$\hat{S}_{\text{tot}}^{(3)}|\Phi_{S_{\text{tot}},S_{\text{tot}}^{(3)}}\rangle = S_{\text{tot}}^{(3)}|\Phi_{S_{\text{tot}},S_{\text{tot}}^{(3)}}\rangle. \tag{A.3.16}$$

We first note that

$$(\hat{\boldsymbol{S}}_{\text{tot}})^2 = (\hat{\boldsymbol{S}}_1)^2 + (\hat{\boldsymbol{S}}_2)^2 + 2\hat{\boldsymbol{S}}_1\cdot\hat{\boldsymbol{S}}_2 = \hat{S}_1^+\hat{S}_2^- + \hat{S}_1^-\hat{S}_2^+ + 2\hat{S}_1^{(3)}\hat{S}_2^{(3)} + \frac{3}{2}, \tag{A.3.17}$$

where we used $(\hat{\boldsymbol{S}}_x)^2 = \frac{1}{2}(\frac{1}{2}+1) = 3/4$, and used (2.2.16). Then, from the basic actions (2.1.4) and (2.1.5) of spin operators onto the basis states, we find

$$(\hat{\boldsymbol{S}}_{\text{tot}})^2|\uparrow\rangle_1|\uparrow\rangle_2 = 2|\uparrow\rangle_1|\uparrow\rangle_2,\quad \hat{S}_{\text{tot}}^{(3)}|\uparrow\rangle_1|\uparrow\rangle_2 = |\uparrow\rangle_1|\uparrow\rangle_2, \tag{A.3.18}$$

$$(\hat{\boldsymbol{S}}_{\text{tot}})^2|\downarrow\rangle_1|\downarrow\rangle_2 = 2|\downarrow\rangle_1|\downarrow\rangle_2,\quad \hat{S}_{\text{tot}}^{(3)}|\downarrow\rangle_1|\downarrow\rangle_2 = -|\downarrow\rangle_1|\downarrow\rangle_2, \tag{A.3.19}$$

which shows these states are indeed the desired simultaneous eigenstates. For the remaining two basis states, we get

$$(\hat{\boldsymbol{S}}_{\text{tot}})^2|\uparrow\rangle_1|\downarrow\rangle_2 = |\uparrow\rangle_1|\downarrow\rangle_2 + |\downarrow\rangle_1|\uparrow\rangle_2,\quad \hat{S}_{\text{tot}}^{(3)}|\uparrow\rangle_1|\downarrow\rangle_2 = 0, \tag{A.3.20}$$

$$(\hat{\boldsymbol{S}}_{\text{tot}})^2|\downarrow\rangle_1|\uparrow\rangle_2 = |\uparrow\rangle_1|\downarrow\rangle_2 + |\downarrow\rangle_1|\uparrow\rangle_2,\quad \hat{S}_{\text{tot}}^{(3)}|\downarrow\rangle_1|\uparrow\rangle_2 = 0, \tag{A.3.21}$$

which shows that their linear combinations $(|\uparrow\rangle_1|\downarrow\rangle_2 \pm |\downarrow\rangle_1|\uparrow\rangle_2)/\sqrt{2}$ are the simultaneous eigenstates.

To summarize we have found that there are three simultaneous eigenstates

$$|\Phi_{1,1}\rangle = |\uparrow\rangle_1|\uparrow\rangle_2,\quad |\Phi_{1,0}\rangle = \frac{1}{\sqrt{2}}\big(|\uparrow\rangle_1|\downarrow\rangle_2 + |\downarrow\rangle_1|\uparrow\rangle_2\big),\quad |\Phi_{1,-1}\rangle = |\downarrow\rangle_1|\downarrow\rangle_2, \tag{A.3.22}$$

with $S_{\text{tot}} = 1$, which are called the triplet states (or the spin-triplet states), and one simultaneous eigenstate

$$|\Phi_{0,0}\rangle = \frac{1}{\sqrt{2}}\big(|\uparrow\rangle_1|\downarrow\rangle_2 - |\downarrow\rangle_1|\uparrow\rangle_2\big), \tag{A.3.23}$$

with $S_{\text{tot}} = 0$, which is called the singlet state (or the spin-singlet state). The spin-singlet is sometimes called the valence-bond. See p. 182. As we noted in p. 469 after Lemma A.11, the singlet state satisfies

$$\hat{S}_{\text{tot}}^{(\alpha)}|\Phi_{0,0}\rangle = 0, \tag{A.3.24}$$

for any $\alpha = 1, 2, 3$, which represents the SU(2) invariance of the singlet.

Addition of two general angular momenta The above example can be generalized. Let us briefly discuss the setting and the most important result.

Consider two quantum mechanical systems with Hilbert spaces $\mathscr{H}_1$ and $\mathscr{H}_2$. Suppose that there are angular momenta for these systems, which we denote as $\hat{\boldsymbol{J}}_1$ and $\hat{\boldsymbol{J}}_2$. For simplicity we assume that $\hat{\boldsymbol{J}}_1$ and $\hat{\boldsymbol{J}}_2$ have constant magnitudes, i.e., there are constants J_1 and J_2 such that $(\hat{\boldsymbol{J}}_1)^2|\Phi\rangle = J_1(J_1+1)|\Phi\rangle$ for any $|\Phi\rangle \in \mathscr{H}_1$, and $(\hat{\boldsymbol{J}}_2)^2|\Psi\rangle = J_2(J_2+1)|\Psi\rangle$ for any $|\Psi\rangle \in \mathscr{H}_2$. We then define the total angular momentum by $\hat{\boldsymbol{J}} = \hat{\boldsymbol{J}}_1 \otimes \hat{1}_2 + \hat{1}_1 \otimes \hat{\boldsymbol{J}}_2$, which of course satisfies the commutation relations (A.3.1). We denote the eigenvalue of $\hat{\boldsymbol{J}}^2$ as $J(J+1)$.

The problem of addition of angular momenta is to determine the range of the eigenvalues J, and to represent the corresponding simultaneous eigenstates of $\hat{\boldsymbol{J}}^2$ and $\hat{J}^{(3)}$ in terms of simultaneous eigenstates of $(\hat{\boldsymbol{J}}_j)^2$ and $\hat{J}_j^{(3)}$ for $j = 1, 2$. The solution and the derivation can be found in almost any advanced textbook in quantum mechanics. One fact that the reader must know (in order to read the present book) is that J takes the values $J_1 + J_2, J_1 + J_2 - 1, \ldots, |J_1 - J_2|$.

A.4 Some Linear Algebra

We discuss important results about matrices in Appendices A.4.1 and A.4.2, and summarize basic treatment of antilinear operators in Appendix A.4.3. Each subsection can be read independently.

A.4.1 Perron–Frobenius Theorem

The Perron–Frobenius theorem is an important and useful theorem for a class of matrices with real elements. In physics the theorem plays a fundamental role in the theory of Markov processes. Here, with applications to quantum physics in mind, we present and prove a simpler version of the theorem for a real symmetric matrix.

Theorem A.18 (Perron–Frobenius theorem for a real symmetric matrix) *Let* $\mathsf{M} = (m_{i,j})_{i,j=1,\ldots,N}$ *be an* $N \times N$ *real symmetric matrix (i.e.,* $m_{i,j} = m_{j,i} \in \mathbb{R}$ *) with the properties that (i)* $m_{i,j} \le 0$ *for any* $i \ne j$*, and (ii) All* $i \ne j$ *are connected via nonvanishing matrix elements of* M*, or, more precisely, for any* $i \ne j$*, we can take a sequence* $(i_1, \ldots, i_K)$ *such that* $i_1 = i$*,* $i_K = j$*, and* $m_{i_k,i_{k+1}} \ne 0$ *for all* $k = 1, 2, \ldots, K-1$*. Then the lowest eigenvalue of* M *is nondegenerate and the corresponding eigenvector* $\boldsymbol{v} = (v_i)_{i=1,\ldots,N}$ *can be taken to satisfy* $v_i > 0$ *for all* i*.*

Proof Let us present a standard elementary proof based on a variational argument. The essence of the argument is that a state without "nodes" has low energy, an idea standard in quantum mechanics. See (A.4.1). In what follows we assume (without loss of generality) that all components of vectors are real.

(1) We first prove that if an eigenvector $\boldsymbol{u} = (u_i)_{i=1,\ldots,N}$ of M satisfies $u_i \ge 0$ for all i, then it inevitably satisfies $u_i > 0$ for all i. For $\alpha > 0$ let $\tilde{\mathsf{M}} = \alpha\mathsf{I} - \mathsf{M}$, where

I denotes the identity matrix. By taking α large enough, we can assume that all the eigenvalues of $\tilde{\mathsf{M}}$ are positive, and $(\tilde{\mathsf{M}})_{i,i} > 0$ for any i. We also have $(\tilde{\mathsf{M}})_{i,j} \geq 0$ for any i, j from (i). By also using (ii) above, we see that there is n such that $(\tilde{\mathsf{M}}^n)_{i,j} > 0$ for any i, j. Note that $\boldsymbol{u}$ is an eigenvector of $\tilde{\mathsf{M}}^n$ with a positive eigenvalue, say, β. Then we have $\beta u_i = \sum_j (\tilde{\mathsf{M}}^n)_{i,j} u_j$, which implies $u_i > 0$ under the assumption that $u_j \geq 0$ and $\boldsymbol{u} \neq 0$.

(2) Let $\boldsymbol{v} = (v_i)_{i=1,\ldots,N}$ be a normalized eigenvector corresponding to the lowest eigenvalue μ_0. We shall prove that $\boldsymbol{v}$ satisfies either $v_i > 0$ for all i or $v_i < 0$ for all i. Define a vector $\boldsymbol{u}$ by $u_i = |v_i|$ for $i = 1, \ldots, N$. Because $\boldsymbol{v}$ is a normalized eigenvector, we have

$$\mu_0 = \boldsymbol{v}^{\mathrm{t}} \mathsf{M} \boldsymbol{v} = \sum_{i,j=1}^{N} v_i m_{i,j} v_j \geq \sum_{i,j-1}^{N} u_i m_{i,j} u_j = \boldsymbol{u}^{\mathrm{t}} \mathsf{M} \boldsymbol{u}, \tag{A.4.1}$$

where we used (i). Since $\boldsymbol{u}$ is normalized and μ_0 is the lowest eigenvalue, the variational principle implies that $\boldsymbol{u}$ is also an eigenvector with eigenvalue μ_0. Then (1) implies $u_i > 0$ (and hence $v_i \neq 0$) for all i. Since $\boldsymbol{u}^{\mathrm{t}} \mathsf{M} \boldsymbol{u} = \mu_0$, we have $\sum_{i,j} v_i m_{i,j} v_j = \sum_{i,j} u_i m_{i,j} u_j$. Since $(v_i)^2 = (u_i)^2$, this means $\sum_{i \neq j} v_i m_{i,j} v_j = \sum_{i \neq j} u_i m_{i,j} u_j$ (where the sum is over all i and j such that $i \neq j$). Since $u_i m_{i,j} u_j \leq 0$ for any $i \neq j$, we conclude that $v_i m_{i,j} v_j = u_i m_{i,j} u_j$ for any $i \neq j$. This means that $v_i v_j = u_i u_j > 0$ whenever $m_{i,j} \neq 0$. From (ii), we see that all v_i have the same sign.

(3) Finally suppose that the lowest eigenvalue of M is degenerate. Then we can find two mutually orthogonal eigenvectors $\boldsymbol{v}$ and $\boldsymbol{v}'$. But (2) implies that both have components with a fixed sign, which means $\boldsymbol{v} \cdot \boldsymbol{v}' \neq 0$. But this is a contradiction. ∎

A.4.2 Decomposition of Matrices

Here we state and prove two important decomposition theorems for general square matrices. Like any complex number z can be written in the polar form $z = r\, e^{i\theta}$ with $r \geq 0$ and $|e^{i\theta}| = 1$, any matrix A admits the polar decomposition as follows.

Theorem A.19 (Polar decomposition theorem) *Any $N \times N$ matrix A is written as*

$$\mathsf{A} = \mathsf{W}\mathsf{C}, \tag{A.4.2}$$

where W is a unitary matrix and C is a nonnegative (or, equivalently, positive semidefinite) matrix. When A is real (in the sense that all the entries are real) W can be chosen to be an orthogonal matrix.

The following singular value decomposition theorem, which is closely related to the above theorem, shows that any matrix can be written in a "nearly diagonalized" form. The decomposition is the basis of (or equivalent to) the Schmidt decomposition, which is a fundamental tool in many-body quantum physics. (See the end of

Sect. A.1.) Although we only discuss the theorem for square matrices, the extension to non-square matrices is not difficult, and is also useful.

Theorem A.20 (Singular value decomposition theorem) *Any $N \times N$ matrix A is written as*

$$\mathsf{A} = \mathsf{U}\mathsf{D}\mathsf{V}^\dagger, \tag{A.4.3}$$

where U and V are unitary matrices, and D is a diagonal matrix with nonnegative entries. When A is real (in the sense that all the entries are real) U and V can be chosen to be orthogonal matrices.

We see from the proof that the diagonal matrix D is written as

$$\mathsf{D} = \begin{pmatrix} \sqrt{\lambda_1} & 0 & \dots & 0 \\ 0 & \sqrt{\lambda_2} & \dots & 0 \\ \vdots & \vdots & \ddots & \vdots \\ 0 & 0 & \dots & \sqrt{\lambda_N} \end{pmatrix}, \tag{A.4.4}$$

where $\lambda_1, \dots, \lambda_N$ (with $\lambda_j \geq 0$ for $j = 1, \dots, N$) are the eigenvalues of the nonnegative matrix $\mathsf{A}^\dagger\,\mathsf{A}$, or, equivalently, of $\mathsf{A}\,\mathsf{A}^\dagger$. (We prove that these two matrices have the same eigenvalues.) The two unitary matrices are written as

$$\mathsf{U} = (\boldsymbol{u}_1, \boldsymbol{u}_2, \dots, \boldsymbol{u}_N), \quad \mathsf{V} = (\boldsymbol{v}_1, \boldsymbol{v}_2, \dots, \boldsymbol{v}_N), \tag{A.4.5}$$

where $\{\boldsymbol{u}_j\}_{j=1,\dots,N}$ and $\{\boldsymbol{v}_j\}_{j=1,\dots,N}$ are orthonormal basis such that

$$\mathsf{A}\,\mathsf{A}^\dagger\,\boldsymbol{u}_j = \lambda_j \boldsymbol{u}_j, \quad \mathsf{A}^\dagger\,\mathsf{A}\,\boldsymbol{v}_j = \lambda_j \boldsymbol{v}_j. \tag{A.4.6}$$

We prove Theorems A.19 and A.20 together. In our notation, $\boldsymbol{u} = (u_1, \dots, u_N)^{\mathrm{t}}$, $\boldsymbol{v} = (v_1, \dots, v_N)^{\mathrm{t}}$ are column vectors, and their Hermitian conjugate $\boldsymbol{u}^\dagger$, $\boldsymbol{v}^\dagger$ are row vectors. $\boldsymbol{u}^\dagger\boldsymbol{v} = \sum_{j=1}^{N}(u_j)^* v_j$ is a scalar, and $\boldsymbol{u}\boldsymbol{v}^\dagger$ is a matrix whose j,k entry is $u_j(v_k)^*$. We also recall that any orthonormal basis $\{\boldsymbol{w}_j\}_{j=1,\dots,N}$ satisfies the orthonormality $\boldsymbol{w}_j^\dagger\boldsymbol{w}_k = \delta_{j,k}$ and the completeness $\sum_{j=1}^{N} \boldsymbol{w}_j\boldsymbol{w}_j^\dagger = \mathsf{I}$.

Proof of the Theorem A.13 A.19 *and* A.20 Note that the matrix $\mathsf{A}^\dagger\,\mathsf{A}$ is nonnegative by Lemma A.6. Let $\{\boldsymbol{v}_j\}_{j=1,\dots,N}$ be the orthonormal basis formed by eigenvectors of $\mathsf{A}^\dagger\,\mathsf{A}$. See (A.4.6). We assume that the corresponding eigenvalues $\lambda_1, \dots, \lambda_N$ are ordered so that $\lambda_j > 0$ for $j = 1, \dots, M$, and $\lambda_j = 0$ for $j = M+1, \dots, N$. (Of course it is possible that $M = N$.)

For $j = 1, \dots, M$, we let $\boldsymbol{u}_j = (\lambda_j)^{-1/2}\mathsf{A}\boldsymbol{v}_j$. Note that $(\boldsymbol{u}_j)^\dagger\boldsymbol{u}_k = (\lambda_j\lambda_k)^{-1/2}$ $(\boldsymbol{v}_j)^\dagger\mathsf{A}^\dagger\,\mathsf{A}\boldsymbol{v}_k = \delta_{j,k}$. By choosing suitable unit vectors $\boldsymbol{u}_{M+1}, \dots, \boldsymbol{u}_N$, we can make $\{\boldsymbol{u}_j\}_{j=1,\dots,N}$ an orthonormal basis, i.e., $\boldsymbol{u}_j^\dagger\boldsymbol{u}_k = \delta_{j,k}$ for $j, k = 1, \dots, N$.

We define $\mathsf{W} = \sum_{j=1}^{N} \boldsymbol{u}_j\boldsymbol{v}_j^\dagger$, which is unitary, and $\mathsf{C} = \sum_{k=1}^{M} \sqrt{\lambda_k}\boldsymbol{v}_k\boldsymbol{v}_k^\dagger$, which is nonnegative. Then one confirms the desired polar decomposition (A.4.2) by observing that

$$\mathsf{WC} = \sum_{k=1}^{M} \boldsymbol{u}_k \sqrt{\lambda_k} \boldsymbol{v}_k^\dagger = \sum_{k=1}^{M} \mathsf{A}\boldsymbol{v}_k \boldsymbol{v}_k^\dagger = \sum_{j=1}^{N} \mathsf{A}\boldsymbol{v}_j \boldsymbol{v}_j^\dagger = \mathsf{A}. \tag{A.4.7}$$

Here we added terms with $j = M + 1, \ldots, N$ by noting that $\lambda_j = 0$ implies $\mathsf{A}^\dagger \mathsf{A}\boldsymbol{v}_j = 0$ and hence $\mathsf{A}\boldsymbol{v}_j = 0$ because of Lemma A.11.

Define D by (A.4.4) and V by the second relation in (A.4.5). Then C defined above is written as $\mathsf{C} = \mathsf{VDV}^\dagger$. From the polar decomposition (A.4.2) we find

$$\mathsf{A} = \mathsf{WC} = \mathsf{WVDV}^\dagger = \mathsf{UDV}^\dagger, \tag{A.4.8}$$

where we defined $\mathsf{U} = \mathsf{WV}$, which is unitary. This is the desired singular value decomposition (A.4.3). It is also easily verified that U is written as in the first relation in (A.4.5) because $\mathsf{W}\boldsymbol{v}_j = \boldsymbol{u}_j$.

Note that (A.4.3) implies

$$\mathsf{A}\,\mathsf{A}^\dagger \mathsf{U} = \mathsf{UDV}^\dagger\, \mathsf{VDU}^\dagger\, \mathsf{U} = \mathsf{UD}^2 = (\lambda_1 \boldsymbol{u}_1, \ldots, \lambda_N \boldsymbol{u}_N). \tag{A.4.9}$$

This is nothing but the first relation in (A.4.6). We have thus confirmed that $\mathsf{A}^\dagger \mathsf{A}$ and $\mathsf{A}\,\mathsf{A}^\dagger$ have the same eigenvalues.

To show the statements about real matrices, we only need to note that eigenvectors of a real symmetric matrices can be chosen to be real. ∎

A.4.3 Antilinear Operators

We summarize basic conventions of antilinear operators, which appear mainly in discussions involving time-reversal transformation. Let $V \cong \mathbb{C}^D$ be the space of D-dimensional complex vectors. We express a vector as $\boldsymbol{v} = (v_1, \ldots, v_D)^{\mathrm{t}} \in V$, and denote by $\langle \boldsymbol{u}, \boldsymbol{v} \rangle = \sum_{j=1}^{D} (u_j)^* v_j$ the inner product of $\boldsymbol{u}, \boldsymbol{v} \in V$.

A map $\mathsf{X} : V \to V$ is said to be an antilinear operator if

$$\mathsf{X}(\alpha \boldsymbol{u} + \beta \boldsymbol{v}) = \alpha^* \mathsf{X}\boldsymbol{u} + \beta^* \mathsf{X}\boldsymbol{v}, \tag{A.4.10}$$

for any $\boldsymbol{u}, \boldsymbol{v} \in V$ and $\alpha, \beta \in \mathbb{C}$. The complex conjugation map K defined by

$$\mathsf{K}\begin{pmatrix} v_1 \\ v_2 \\ \vdots \\ v_D \end{pmatrix} = \begin{pmatrix} (v_1)^* \\ (v_2)^* \\ \vdots \\ (v_D)^* \end{pmatrix} \tag{A.4.11}$$

is an important example. If A is a linear operator, and X and Y are antilinear operators, we see that AX and XA are antilinear, and XY is linear.

If X is an antilinear operator, its conjugate (or adjoint) $\mathsf{X}^\dagger$ is the unique antilinear operator such that

$$\langle \boldsymbol{u}, \mathsf{X}\boldsymbol{v}\rangle = \langle \mathsf{X}^\dagger \boldsymbol{u}, \boldsymbol{v}\rangle^*, \tag{A.4.12}$$

for any $\boldsymbol{u}, \boldsymbol{v} \in V$.[13] It is obvious that $(\mathsf{X}^\dagger)^\dagger = \mathsf{X}$. We also find that $\mathsf{K}^\dagger = \mathsf{K}$.

For any antilinear operators X and Y, we have

$$\langle \boldsymbol{u}, \mathsf{XY}\boldsymbol{v}\rangle = \langle \mathsf{X}^\dagger \boldsymbol{u}, \mathsf{Y}\boldsymbol{v}\rangle^* = \langle \mathsf{Y}^\dagger\mathsf{X}^\dagger \boldsymbol{u}, \boldsymbol{v}\rangle, \tag{A.4.13}$$

which implies

$$(\mathsf{XY})^\dagger = \mathsf{Y}^\dagger\mathsf{X}^\dagger. \tag{A.4.14}$$

Similarly for any antilinear X and linear A, we can show that

$$(\mathsf{AX})^\dagger = \mathsf{X}^\dagger\mathsf{A}^\dagger, \quad (\mathsf{XA})^\dagger = \mathsf{A}^\dagger\mathsf{X}^\dagger. \tag{A.4.15}$$

Note in particular that $(\alpha\mathsf{X})^\dagger = \mathsf{X}^\dagger\alpha^* = \alpha\mathsf{X}^\dagger$ for $\alpha \in \mathbb{C}$.

Let $\boldsymbol{u} = \mathsf{X}\boldsymbol{v}$ with some antilinear X. Then, for any linear operator A, we have

$$\langle \boldsymbol{v}, \mathsf{X}^\dagger\mathsf{AX}\boldsymbol{v}\rangle = \langle \boldsymbol{u}, \mathsf{A}\boldsymbol{u}\rangle^* = \langle \boldsymbol{u}, \mathsf{A}^\dagger\boldsymbol{u}\rangle. \tag{A.4.16}$$

An antilinear operator V is said to be antiunitary if

$$\langle \mathsf{V}\boldsymbol{u}, \mathsf{V}\boldsymbol{v}\rangle = \langle \boldsymbol{u}, \boldsymbol{v}\rangle^* \tag{A.4.17}$$

for any $\boldsymbol{u}, \boldsymbol{v} \in V$. Note that, if V is antiunitary, we have

$$\langle \mathsf{KV}\boldsymbol{u}, \mathsf{KV}\boldsymbol{v}\rangle = \langle \mathsf{V}\boldsymbol{u}, \mathsf{V}\boldsymbol{v}\rangle^* = \langle \boldsymbol{u}, \boldsymbol{v}\rangle \tag{A.4.18}$$

and

$$\langle \mathsf{VK}\boldsymbol{u}, \mathsf{VK}\boldsymbol{v}\rangle = \langle \mathsf{K}\boldsymbol{u}, \mathsf{K}\boldsymbol{v}\rangle^* = \langle \boldsymbol{u}, \boldsymbol{v}\rangle. \tag{A.4.19}$$

We thus find that both KV and VK are unitary. In other words, antiunitary operator V can always be written as $\mathsf{V} = \mathsf{KU}$ and $\mathsf{V} = \mathsf{U}'\mathsf{K}$ with some unitary operators U and U'. This in particular implies that $\mathsf{VV}^\dagger = \mathsf{V}^\dagger\mathsf{V} = \mathsf{I}$, which is also evident from (A.4.17).

A.5 Groups and Their Representations

The group theory provides a useful mathematical language for describing symmetry and transformations of physical systems. Although we do not make use of any non-

[13] Needless to say the conjugate of a linear operator satisfies $\langle \boldsymbol{u}, \mathsf{A}\boldsymbol{v}\rangle = \langle \mathsf{A}^\dagger\boldsymbol{u}, \boldsymbol{v}\rangle$.

trivial results from the group theory in the present book, we shall discuss the basics of the group theory for the reader's convenience.

Definition and basic examples Let G be a set, and suppose that there is a multiplication rule which assigns a unique "product" $ab \in G$ to any two elements $a, b \in G$. We say that G (along with the multiplication rule) is a group when (i) the multiplication is associative, i.e., $(ab)c = a(bc)$ for any $a, b, c \in G$, (ii) there is an element $e \in G$, which is called the identity, such that $ae = ea = a$ for any $a \in G$, and (iii) for every $a \in G$ there is an element $a^{-1} \in G$ such that $aa^{-1} = a^{-1}a = e$. It easily follows that a group has a unique identity, and for each $a \in G$ its inverse a^{-1} is unique.

The cyclic group of order 2, denoted as $\mathbb{Z}_2$, is the simplest nontrivial group. It consists of two elements as $\mathbb{Z}_2 = \{e, a\}$, and is specified by the multiplication rule $ea = ae = a$ and $a^2 = e$. We have defined slightly more complicated group $\mathbb{Z}_2 \times \mathbb{Z}_2$ in (2.1.27).

Sets of matrices with the standard matrix multiplication rule provide important and useful class of groups. Let n be a positive integer. The n-dimensional orthogonal group O(n) is the set of all $n \times n$ orthogonal matrices, and the n-dimensional special orthogonal group SO(n) is the set of all $n \times n$ orthogonal matrices with determinant 1. The group SO(n) is also known as the rotation group. SO(2) consists of the rotation matrix

$$\mathsf{O}_\theta = \begin{pmatrix} \cos\theta & -\sin\theta \\ \sin\theta & \cos\theta \end{pmatrix}, \tag{A.5.1}$$

with $\theta \in [0, 2\pi)$. The three dimensional rotation matrices (2.1.11) are elements of SO(3).

The unitary group of degree n, denoted as U(n), is the set of all $n \times n$ unitary matrices, and the special unitary group of degree n, denoted as SU(n), is the set of all $n \times n$ unitary matrices with determinant 1. The group U(1) is simply the set of complex numbers with absolute value 1, and its element is written as $e^{i\theta}$ with $\theta \in [0, 2\pi)$. From (A.5.1), we immediately see that U(1) and SO(2) are isomorphic.[14]

The group SU(2) and its generators Let us concentrate on the group SU(2), which plays a central role throughout this book.

Let U be an element of SU(2), i.e., a unitary matrix of degree 2 with determinant 1, and write it as

$$\mathsf{U} = \exp[-i\mathsf{A}], \tag{A.5.2}$$

where A is a Hermitian matrix. One easily finds by diagonalizing U that the rewriting (A.5.2) is always possible. Recalling that $\det \exp[-i\mathsf{A}] = \exp[-i\,\mathrm{Tr}[\mathsf{A}]]$ (which is verified by diagonalizing A) we see that U written as (A.5.2) is an element of SU(2) if and only if $\hat{A}$ is a traceless Hermitian matrix. Such A is in general written as $\mathsf{A} = \begin{pmatrix} a & b-ic \\ b+ic & -a \end{pmatrix}$ with arbitrary $a, b, c \in \mathbb{R}$. Equivalently, it is written as

[14] Two groups are isomorphic if and only if there is a one-to-one map between them which preserves the multiplication rules.

$$\mathsf{A} = \theta_1 \mathsf{S}^{(1)} + \theta_2 \mathsf{S}^{(2)} + \theta_3 \mathsf{S}^{(3)}, \tag{A.5.3}$$

with arbitrary $\theta_1, \theta_2, \theta_3 \in \mathbb{R}$, where

$$\mathsf{S}^{(1)} = \frac{1}{2}\begin{pmatrix} 0 & 1 \\ 1 & 0 \end{pmatrix}, \quad \mathsf{S}^{(2)} = \frac{1}{2}\begin{pmatrix} 0 & -i \\ i & 0 \end{pmatrix}, \quad \mathsf{S}^{(3)} = \frac{1}{2}\begin{pmatrix} 1 & 0 \\ 0 & -1 \end{pmatrix}. \tag{A.5.4}$$

are nothing but (the matrix representations of) the spin operators (2.1.7) for $S = 1/2$. The matrices $\mathsf{S}^{(1)}$, $\mathsf{S}^{(2)}$, and $\mathsf{S}^{(3)}$ are called the generators of the group SU(2).

The group SU(2) is an example of a Lie group. (This is also true for general O(n), SO(n), U(n), and SU(n).) The set of matrices (A.5.3) is the corresponding Lie algebra, which is written as $\mathfrak{su}(2)$.[15]

Representation of a group A representation,[16] or, more precisely, a linear representation, of a group G is a map ρ from G to the set of square matrices with a fixed degree that satisfies $\rho(e) = \mathsf{I}$ and $\rho(a)\rho(b) = \rho(ab)$ for any $a, b \in G$.

For example a two dimensional representation of $\mathbb{Z}_2 = \{e, a\}$ is given by

$$\rho(e) = \begin{pmatrix} 1 & 0 \\ 0 & 1 \end{pmatrix}, \quad \rho(a) = \begin{pmatrix} 0 & 1 \\ 1 & 0 \end{pmatrix}. \tag{A.5.5}$$

If $\hat{N}$ is any Hermitian matrix whose eigenvalues are integer, then $\hat{U}_\theta = \exp[-i\theta\hat{N}]$ gives a representation of U(1) or SO(2). This is the reason why we refer to the symmetry with respect to the change of the phase, which is an intrinsic symmetry in quantum theory, as the U(1) symmetry.

A quantum mechanical spin or a quantum spin system provide representations of SU(2). The spin operators $\hat{S}^{(1)}$, $\hat{S}^{(2)}$, and $\hat{S}^{(3)}$ with spin quantum number S (discussed in Sect. 2.1) give a $2S+1$ dimensional representation of the generators $\mathsf{S}^{(1)}$, $\mathsf{S}^{(2)}$, and $\mathsf{S}^{(3)}$ of SU(2). Likewise the total spin operators $\hat{S}^{(1)}_{\text{tot}}$, $\hat{S}^{(2)}_{\text{tot}}$, and $\hat{S}^{(3)}_{\text{tot}}$ in (2.2.7) give a $2|\Lambda|S+1$ dimensional representation of the same generators. Then the corresponding spin rotation operators $\hat{U}^{(\alpha)}_\theta$ give representations of elements of SU(2). It is common to make a slight abuse of terminology and refer to a representation of an element of G simply as an element of G. See, e.g., the remarks that follow (2.1.16) and (2.2.11).

We finally discuss a representation of SU(2) which sheds light on the relation between SU(2) and the rotation group SO(3). Let X be an arbitrary 2×2 traceless Hermitian matrix and write is as $\mathsf{X} = x\,\mathsf{S}^{(1)} + y\,\mathsf{S}^{(2)} + z\,\mathsf{S}^{(3)}$ with $x, y, z \in \mathbb{R}$. Take an arbitrary element U of SU(2), and consider a matrix $\mathsf{X}' = \mathsf{U}^\dagger\mathsf{X}\mathsf{U}$. It is clear that X' is also Hermitian. It is also traceless since $\text{Tr}[\mathsf{X}'] = \text{Tr}[\mathsf{U}^\dagger\mathsf{X}\mathsf{U}] =$

---

[15]In mathematics it is standard to regard $i\mathsf{A}$ as forming the Lie algebra.

[16]A representation is sometimes called a genuine representation, as opposed to a projective representation. A projective representation of a group G is a map ρ from G to the set of square matrices with a fixed degree that satisfies $\rho(e) = \mathsf{I}$ and $\rho(a)\rho(b) = e^{i\phi(a,b)}\rho(ab)$ for any $a, b \in G$, where $\phi(a, b) \in \mathbb{R}$ is a phase factor that satisfies the condition (8.3.26). See the remarks below (2.1.31) and (8.3.24) for more about the notion of projective representation..

$\mathrm{Tr}[\mathsf{X}\mathsf{U}\mathsf{U}^\dagger] = \mathrm{Tr}[\mathsf{X}] = 0$. We thus find that the new matrix is again written as $\mathsf{X}' = x'\mathsf{S}^{(1)} + y'\mathsf{S}^{(2)} + z'\mathsf{S}^{(3)}$ with $x', y', z' \in \mathbb{R}$. Since x', y', z' depend linearly on x, y, z, there exists a 3×3 matrix $\rho(\mathsf{U})$ with real components such that

$$\begin{pmatrix} x' \\ y' \\ z' \end{pmatrix} = \rho(\mathsf{U}) \begin{pmatrix} x \\ y \\ z \end{pmatrix}. \tag{A.5.6}$$

This defines a representation of SU(2), known as the adjoint representation. By using the properties of $\mathsf{S}^{(1)}$, $\mathsf{S}^{(2)}$, and $\mathsf{S}^{(3)}$ described just below (2.1.7) (or by a direct simple calculation) one confirms that $\mathsf{X}^2 = \{(x^2 + y^2 + z^2)/4\}\mathsf{I}$. It then follows that $(\mathsf{X}')^2 = \mathsf{U}^\dagger \mathsf{X}^2 \mathsf{U} = \{(x^2 + y^2 + z^2)/4\}\mathsf{I}$, and hence $(x')^2 + (y')^2 + (z')^2 = x^2 + y^2 + z^2$. This fact suggests that the matrix $\rho(\mathsf{U})$ describes a rotation in the three dimensional euclidean space. In fact it can be shown that $\rho(\mathsf{U})$ covers the whole SO(3) when U is taken from the whole SU(2). We have already made a key observation for this fact in (2.1.16). It should be noted however that SU(2) and SO(3) are not isomorphic. As can be seen from the fact that $\exp[-2\pi i \mathsf{S}^{(\alpha)}] = -\mathsf{I}$, the group SU(2) is in a sense "twice larger" than the group SO(3).

A.6 Wigner's Theorem

In 1931, Wigner proved a famous theorem (Theorem A.22) which asserts that any symmetry in a quantum system is represented by a unitary or antiunitary operator on the Hilbert space. (The theorem is proved in the appendix to Chap. 20 of his book [10, 11].) Our treatment of symmetry in quantum physics is fundamentally based on this theorem.

For us Wigner's theorem is not logically necessary since we always start with an explicit construction of a unitary or antiunitary transformation that describes the relevant symmetry. We nevertheless believe that the understanding of the theorem, especially its variant for automorphisms (Theorem A.21), gives a clearer perspective on symmetry transformations.

A.6.1 Statement and the Proof

Let $\mathscr{H}$ be a D dimensional Hilbert space, and denote by $M(\mathscr{H})$ the set of all (linear) operators on $\mathscr{H}$. We consider a map $\Gamma : M(\mathscr{H}) \to M(\mathscr{H})$ with the following properties.

(A1) Γ is one-to-one
(A2) For any $\hat{A}, \hat{B} \in M(\mathscr{H})$, we have $\Gamma(\hat{A}\hat{B}) = \Gamma(\hat{A})\Gamma(\hat{B})$
(A3) For any $\hat{A} \in M(\mathscr{H})$, we have $\Gamma(\hat{A}^\dagger) = \Gamma(\hat{A})^\dagger$

(A4) Γ is linear, i.e., $\Gamma(\alpha\hat{A}+\beta\hat{B})=\alpha\Gamma(\hat{A})+\beta\Gamma(\hat{B})$ for any $\hat{A},\hat{B}\in M(\mathscr{H})$ and $\alpha,\beta\in\mathbb{C}$

or

(A4') Γ is antilinear, i.e., $\Gamma(\alpha\hat{A}+\beta\hat{B})=\alpha^*\Gamma(\hat{A})+\beta^*\Gamma(\hat{B})$ for any $\hat{A},\hat{B}\in M(\mathscr{H})$ and $\alpha,\beta\in\mathbb{C}$.

When (A1), (A2), (A3), and (A4) hold, the map Γ is called a linear $*$-automorphism of $M(\mathscr{H})$, and when (A1), (A2), (A3), and (A4') hold, Γ is called an antilinear $*$-automorphism.[17,18]

Then we have the following theorem.[19]

Theorem A.21 (A variant of Wigner's theorem for automorphisms) *If Γ satisfies (A1), (A2), (A3), and (A4), then there exists a unitary operator $\hat{U}$ such that*

$$\Gamma(\hat{A})=\hat{U}^\dagger\hat{A}\hat{U}\quad \textit{for any } \hat{A}\in M(\mathscr{H}). \tag{A.6.1}$$

The unitary operator is unique up to a phase factor. If Γ satisfies (A1), (A2), (A3), and (A4'), then there exists an antiunitary operator $\hat{V}$ such that[20]

$$\Gamma(\hat{A})=\hat{V}^{-1}\hat{A}\hat{V}\quad \textit{for any } \hat{A}\in M(\mathscr{H}). \tag{A.6.2}$$

The antiunitary operator is unique up to a phase factor.

This theorem is closely related to but different from the original theorem proved by Wigner [10, 11], which treats a map between rays (or, equivalently, one-dimensional projections) and assumes the invariance of the inner product of two rays. See Theorem A.22 below. But Theorem A.21 is often called Wigner's theorem.

Proof We shall show that the properties (A1), (A2), and (A3), with (A4) or (A4') imply that there are two orthonormal bases $\{|\Phi_j\rangle\}_{j=1,\dots,D}$ and $\{|\Phi'_j\rangle\}_{j=1,\dots,D}$ such that

$$\Gamma(|\Phi_j\rangle\langle\Phi_k|)=|\Phi'_j\rangle\langle\Phi'_k|\quad \text{for any } j,k=1,\dots,D. \tag{A.6.3}$$

We then define $\hat{U}=\sum_{j=1}^{D}|\Phi_j\rangle\langle\Phi'_j|$, which is unitary and satisfies $\hat{U}^\dagger|\Phi_j\rangle\langle\Phi_k|\hat{U}=|\Phi'_j\rangle\langle\Phi'_k|$. When Γ satisfies (A4), we get the desired (A.6.1) from linearity. To see the uniqueness of $\hat{U}$, suppose that there is $\hat{U}'$ which also satisfies $\Gamma(\hat{A})=(\hat{U}')^\dagger\hat{A}\hat{U}'$ for any $\hat{A}\in M(\mathscr{H})$. But this and (A.6.1) imply $\hat{A}=\Gamma^{-1}(\Gamma(\hat{A}))=\hat{U}(\hat{U}')^\dagger\hat{A}\hat{U}'\hat{U}^\dagger$ for

[17] Note that (A1) and (A2) imply $\Gamma(\hat{1})=\hat{1}$.

[18] When (A3) is not required, Γ is called an automorphism. Here the symbol $*$ denotes the conjugate (in mathematicians' notation), which we denote as $\dagger$.

[19] The theorem is well-known, and can be found, e.g., in [1]. Although we only treat finite dimensional Hilbert spaces for simplicity, the theorem applies to infinite dimensional Hilbert spaces as well.

[20] See Appendix A.4.3 for antiunitary operators.

any $\hat{A} \in M(\mathscr{H})$, and hence $\hat{U}'\hat{U}^\dagger = e^{i\theta}\hat{1}$. When Γ satisfies (A4'), we set $\hat{V} = \hat{U}\hat{K}$ and $\hat{V}^{-1} = \hat{K}\hat{U}^\dagger$, where the complex-conjugation map $\hat{K}$ is defined by[21]

$$\hat{K}\Bigl(\sum_{j=1}^{D} \alpha_j|\Phi'_j\rangle\Bigr) = \sum_{j=1}^{D} (\alpha_j)^*|\Phi'_j\rangle. \tag{A.6.4}$$

The proof of the uniqueness is the same.

Let us prove (A.6.3). We note that $\Gamma(0) = 0$ because of (A4) or (A4'). Fix an arbitrary orthonormal basis $\{|\Phi_j\rangle\}_{j=1,\ldots,D}$, and let $\hat{P}_j = \Gamma(|\Phi_j\rangle\langle\Phi_j|)$. Since $\hat{P}_j \neq 0$ by (A1), $(\hat{P}_j)^2 = \Gamma(|\Phi_j\rangle\langle\Phi_j||\Phi_j\rangle\langle\Phi_j|) = \Gamma(|\Phi_j\rangle\langle\Phi_j|) = \hat{P}_j$ by (A2), and $\hat{P}_j^\dagger = \hat{P}_j$ by (A3), we find that $\hat{P}_j$ is an orthogonal projection operator. Moreover since $\hat{P}_j\hat{P}_k = \Gamma(|\Phi_j\rangle\langle\Phi_j||\Phi_k\rangle\langle\Phi_k|) = 0$ if $j \neq k$, we see that $\hat{P}_j$ with $j = 1, \ldots, D$ must be projections to mutually orthogonal one-dimensional subspaces. Thus there exists an orthonormal basis $\{|\widetilde{\Phi}_j\rangle\}_{j=1,\ldots,D}$ such that $\hat{P}_j = \Gamma(|\Phi_j\rangle\langle\Phi_j|) = |\widetilde{\Phi}_j\rangle\langle\widetilde{\Phi}_j|$. Our goal is to modify the phases of the basis states $|\widetilde{\Phi}_j\rangle$ to get the desired $\{|\Phi'_j\rangle\}_{j=1,\ldots,D}$. To begin with we set $|\Phi'_1\rangle = |\widetilde{\Phi}_1\rangle$. We thus have $\Gamma(|\Phi_1\rangle\langle\Phi_1|) = |\Phi'_1\rangle\langle\Phi'_1|$.

Let us note here that $\Gamma(|\Phi\rangle\langle\Phi|)$ is a one-dimensional projection for any $|\Phi\rangle$ such that $\|\Phi\| = 1$. To see this we only need to take an orthonormal basis which contains $|\Phi\rangle$, and repeat the above argument.

For $\ell = 2, \ldots, D$, let $|\Xi_\ell\rangle = (|\Phi_1\rangle + |\Phi_\ell\rangle)/\sqrt{2}$. Then there is $|\widetilde{\Xi}_\ell\rangle$ such that $\|\widetilde{\Xi}_\ell\| = 1$, and $\Gamma(|\Xi_\ell\rangle\langle\Xi_\ell|) = |\widetilde{\Xi}_\ell\rangle\langle\widetilde{\Xi}_\ell|$. From the definition, we have

$$|\Phi_1\rangle\langle\Phi_1|\Xi_\ell\rangle\langle\Xi_\ell|\Phi_1\rangle\langle\Phi_1| = \frac{1}{2}|\Phi_1\rangle\langle\Phi_1|. \tag{A.6.5}$$

Applying Γ to the left-hand side, we get

$$\begin{aligned}\Gamma(|\Phi_1\rangle\langle\Phi_1|\Xi_\ell\rangle\langle\Xi_\ell|\Phi_1\rangle\langle\Phi_1|) &= \Gamma(|\Phi_1\rangle\langle\Phi_1|)\Gamma(|\Xi_\ell\rangle\langle\Xi_\ell|)\Gamma(|\Phi_1\rangle\langle\Phi_1|)\\ &= |\Phi'_1\rangle\langle\Phi'_1|\widetilde{\Xi}_\ell\rangle\langle\widetilde{\Xi}_\ell|\Phi'_1\rangle\langle\Phi'_1|\\ &= |\langle\widetilde{\Xi}_\ell|\Phi'_1\rangle|^2\,|\Phi'_1\rangle\langle\Phi'_1|,\end{aligned} \tag{A.6.6}$$

where we used (A2). Since this is equal to $|\Phi'_1\rangle\langle\Phi'_1|/2$ because of (A4) or (A4'), we see that $|\langle\widetilde{\Xi}_\ell|\Phi'_1\rangle| = 1/\sqrt{2}$. Then by setting $|\Xi'_\ell\rangle = e^{i\theta_\ell}|\widetilde{\Xi}_\ell\rangle$ with suitable $\theta_\ell \in \mathbb{R}$, we have $\Gamma(|\Xi_\ell\rangle\langle\Xi_\ell|) = |\Xi'_\ell\rangle\langle\Xi'_\ell|$ and $\langle\Xi'_\ell|\Phi'_1\rangle = 1/\sqrt{2}$ for any $\ell = 2, \ldots, D$.

Repeating the same argument starting from another trivial identity

$$|\Phi_\ell\rangle\langle\Phi_\ell|\Xi_\ell\rangle\langle\Xi_\ell|\Phi_\ell\rangle\langle\Phi_\ell| = \frac{1}{2}|\Phi_\ell\rangle\langle\Phi_\ell|, \tag{A.6.7}$$

we find $|\langle\Xi'_\ell|\widetilde{\Phi}_\ell\rangle| = 1/\sqrt{2}$. Then setting $|\Phi'_\ell\rangle = e^{i\varphi_\ell}|\widetilde{\Phi}_\ell\rangle$ with suitable $\varphi_\ell \in \mathbb{R}$, we have $\Gamma(|\Phi_\ell\rangle\langle\Phi_\ell|) = |\Phi'_\ell\rangle\langle\Phi'_\ell|$ and $\langle\Xi'_\ell|\Phi'_\ell\rangle = 1/\sqrt{2}$ for any $\ell = 2, \ldots, D$.

[21] Note that the map $\hat{K}$ is defined with respect to the basis $\{|\Phi'_j\rangle\}_{j=1,\ldots,D}$.

Now we are ready to demonstrate (A.6.3). Noting that $\langle\Phi_1|\Xi_\ell\rangle\langle\Xi_\ell|\Phi_\ell\rangle = 1/2$ for $\ell = 2, \ldots, D$, we observe that

$$\begin{aligned}\Gamma(|\Phi_1\rangle\langle\Phi_\ell|) &= 2\,\Gamma(|\Phi_1\rangle\langle\Phi_1|\Xi_\ell\rangle\langle\Xi_\ell|\Phi_\ell\rangle\langle\Phi_\ell|)\\ &= 2\,|\Phi'_1\rangle\langle\Phi'_1|\Xi'_\ell\rangle\langle\Xi'_\ell|\Phi'_\ell\rangle\langle\Phi'_\ell|\\ &= |\Phi'_1\rangle\langle\Phi'_\ell|,\end{aligned} \tag{A.6.8}$$

and from (A3) that $\Gamma(|\Phi_\ell\rangle\langle\Phi_1|) = |\Phi'_\ell\rangle\langle\Phi'_1|$. We then have, for any $j, k = 1, \ldots, D$, that

$$\Gamma(|\Phi_j\rangle\langle\Phi_k|) = \Gamma(|\Phi_j\rangle\langle\Phi_1|\Phi_1\rangle\langle\Phi_k|) = |\Phi'_j\rangle\langle\Phi'_1|\Phi'_1\rangle\langle\Phi'_k| = |\Phi'_j\rangle\langle\Phi'_k|, \tag{A.6.9}$$

which is the desired (A.6.3). ∎

For the reader's reference, we discuss the original theorem of Wigner's. Consider a D dimensional Hilbert space $\mathscr{H}$ with $D \geq 2$, and let $\mathscr{P}_1$ be the set of all one-dimensional orthogonal projection operators on $\mathscr{H}$. Note that $|\Phi\rangle\langle\Phi| \in \mathscr{P}_1$ and $|\Psi\rangle\langle\Psi| \in \mathscr{P}_1$ are regarded as the same element if $|\Phi\rangle = e^{i\theta}|\Psi\rangle$. Also note that, for $\hat{P} = |\Phi\rangle\langle\Phi| \in \mathscr{P}_1$ and $\hat{P}' = |\Phi'\rangle\langle\Phi'| \in \mathscr{P}_1$, we have $\mathrm{Tr}[\hat{P}\,\hat{P}'] = |\langle\Phi|\Phi'\rangle|^2$.

Theorem A.22 (Wigner's theorem) *Suppose that there is a map* $\widetilde{\Gamma} : \mathscr{P}_1 \to \mathscr{P}_1$ *such that* $\mathrm{Tr}[\hat{P}\,\hat{P}'] = \mathrm{Tr}[\widetilde{\Gamma}(\hat{P})\,\widetilde{\Gamma}(\hat{P}')]$ *for any* $\hat{P}, \hat{P}' \in \mathscr{P}_1$. *Then there exists an operator* $\hat{V}$, *which is either unitary or antiunitary, such that*

$$\widetilde{\Gamma}(\hat{P}) = \hat{V}\,\hat{P}\,\hat{V}^{-1} \quad \textit{for any } \hat{P} \in \mathscr{P}_1. \tag{A.6.10}$$

The (anti)unitary operator $\hat{V}$ *is unique up to a phase factor.*

For details, we recommend Bargmann's paper [2], which contains a clear and readable proof as well as a careful discussion of the statement.

A.6.2 Applications

Theorem A.21 guarantees the existence of an (anti)unitary operator once we write down an (anti)linear $*$-automorphism of operators. This is certainly useful.

Spin systems Consider a general quantum spin system, as formulated in Sect. 2.2. We first recall that any operator of a quantum spin system is written as a polynomial of the spin operators $\hat{S}_x^{(\alpha)}$ with $x \in \Lambda$ and $\alpha = 1, 2, 3$. See Problem 2.1.a. Therefore to define a $*$-automorphism for a spin system, one only needs to specify the transformation of $\hat{S}_x^{(\alpha)}$ for each $x \in \Lambda$ and $\alpha = 1, 2, 3$. Here one also has to check whether the transformation is consistent with the commutation relations (2.2.6), such as $[\hat{S}_x^{(1)}, \hat{S}_x^{(2)}] = i\,\hat{S}_x^{(3)}$.

As a simple example, define the linear $*$-automorphism Γ_1 by $\Gamma_1(\hat{S}_x^{(1)}) = \hat{S}_x^{(1)}$, $\Gamma_1(\hat{S}_x^{(2)}) = -\hat{S}_x^{(2)}$, and , $\Gamma_1(\hat{S}_x^{(3)}) = -\hat{S}_x^{(3)}$. It is obvious from (2.1.21) that this automorphism is realized by the π rotation about the 1-axis, but the existence of a unitary transformation is guaranteed by Theorem A.21 once we check that the above transformation and linearity is consistent with the commutation relations. Similarly, we can be sure that the linear $*$-automorphism defined by $\Gamma_{231}(\hat{S}_x^{(1)}) = \hat{S}_x^{(2)}$, $\Gamma_{231}(\hat{S}_x^{(2)}) = \hat{S}_x^{(3)}$, and $\Gamma_{231}(\hat{S}_x^{(3)}) = \hat{S}_x^{(1)}$ can be realized by a unitary transformation, without examining how it can be done by combining $\pi/2$-rotations.

An important example is the time-reversal transformation discussed in detail in Sect. 2.3. In this case, corresponding to (2.3.23), we start from the basic transformation rules $\Gamma_{\text{tr}}(\hat{S}_x^{(1)}) = -\hat{S}_x^{(1)}$, $\Gamma_{\text{tr}}(\hat{S}_x^{(2)}) = -\hat{S}_x^{(2)}$, and $\Gamma_{\text{tr}}(\hat{S}_x^{(3)}) = -\hat{S}_x^{(3)}$. Now it is clear that this rule is consistent with the commutation relations only when we assume Γ_{tr} is an antilinear $*$-automorphism. Then the theorem shows that the time-reversal transformation is realized by an antiunitary operator, which we called $\hat{\Theta}$ in Sect. 2.3. This approach is illuminating since the definition of $\hat{\Theta}$ in Sect. 2.3 looks rather ad hoc; given only the definition, one hardly understands the basic principle which guarantees the existence of the operator. Similarly we see, without any examination, that the antilinear $*$-automorphism defined by $\Gamma_{321}(\hat{S}_x^{(1)}) = \hat{S}_x^{(3)}$, $\Gamma_{321}(\hat{S}_x^{(2)}) = \hat{S}_x^{(2)}$, and $\Gamma_{321}(\hat{S}_x^{(3)}) = \hat{S}_x^{(1)}$ can be realized by an antiunitary operator.

Electron systems Theorem A.21 also sheds light on the transformations for the Hubbard model discussed in Sect. 9.3.3. As a rather trivial example, the gauge transformation (9.3.37) is characterize by the liner $*$-automorphism Γ_θ specified by $\Gamma_\theta(\hat{c}_{x,\sigma}^\dagger) = e^{i\theta_x}\,\hat{c}_{x,\sigma}^\dagger$. Much more complicated Shiba transformation, for example, can also be fully characterized by only specifying the transformation rule (9.3.51) and checking its consistency with the anticommutation relations (9.2.27) and (9.2.28).

Let us finally formulate the time-reversal transformation for electron systems, although we do not make use of the transformation in the main text. It is convenient to first define an antilinear $*$-automorphism Γ_{tr} by $\Gamma_{\text{tr}}(\hat{c}_{x,\uparrow}) = \hat{c}_{x,\downarrow}$ and $\Gamma_{\text{tr}}(\hat{c}_{x,\downarrow}) = -\hat{c}_{x,\uparrow}$. The transformation is clearly consistent with the anticommutation relations (9.2.27) and (9.2.28). In this case the corresponding antiunitary operator $\hat{\Theta}$ is constructed explicitly, without invoking Wigner's theorem, by demanding $\hat{\Theta}|\Phi_{\text{vac}}\rangle = |\Phi_{\text{vac}}\rangle$ and $\hat{\Theta}(\hat{A}|\Phi_{\text{vac}}\rangle) = \Gamma_{\text{tr}}(\hat{A})|\Phi_{\text{vac}}\rangle$ for any operator $\hat{A}$. It is easily checked that the resulting $\hat{\Theta}$ is consistent with that defined in Sect. 2.3.

A.7 Operator Algebraic Formulation of Infinite Systems

In this appendix, we present a brief description of the operator algebraic formulation of quantum spin systems on infinite lattices. Although we have mostly avoided the use of this mathematical formulation in the book, it plays important roles in Sects. 4.3, 7.1.1, and 8.3.6. It is probably a good idea to discuss basic notions for the interested reader. We here describe basic definitions and theorems, without going into proofs. The motivated reader is invited to study Bratteli and Robinson's definitive textbook [4, 5] on the operator algebraic approach to quantum many-body systems.

Operator algebra We consider a quantum spin system with spin quantum number S on the infinite d-dimensional hypercubic lattice $\mathbb{Z}^d$. As in Sect. 2.2, we associate a local Hilbert space $\mathfrak{h}_x \cong \mathbb{C}^{2S+1}$ with each site $x \in \mathbb{Z}^d$. We do not however introduce the Hilbert space of the whole system. We again denote by $\hat{S}_x^{(\alpha)}$ for $x \in \mathbb{Z}^d$ and $\alpha = 1, 2, 3$ the α-component of the spin operator at site x. It acts as the operator $\hat{S}^{(\alpha)}$ determined by (2.1.2) and (2.1.3) on the space $\mathfrak{h}_x$ and as the identity operator on other $\mathfrak{h}_y$ with $y \in \mathbb{Z}^d \backslash \{x\}$. We can thus consider products and linear combinations of the operators $\hat{S}_x^{(\alpha)}$ for $x \in \mathbb{Z}^d$ and $\alpha = 1, 2, 3$. We can also take the conjugation (i.e., the mapping $\hat{A} \to \hat{A}^\dagger$) of all such operators. The set of operators formed in this manner plays a central role in this formulation.

Let $\mathfrak{A}_{\mathrm{loc}}$ be the algebra of local operators, i.e., the set of all polynomials (with complex coefficients) of the spin operators $\hat{S}_x^{(\alpha)}$ with $x \in \mathbb{Z}^d$ and $\alpha = 1, 2, 3$. By definition, each $\hat{A} \in \mathfrak{A}_{\mathrm{loc}}$ is an operator which depends only on a finite number of spins, but there is no limit for this finite number. For $\hat{A} \in \mathfrak{A}_{\mathrm{loc}}$ let $\operatorname{supp} \hat{A}$ be its support, i.e., the set of sites on which $\hat{A}$ acts nontrivially. For example, $\operatorname{supp} \hat{S}_x^{(\alpha)} = \{x\}$ and $\operatorname{supp}(a\, \hat{S}_x^{(\alpha)} + b\, \hat{S}_y^{(\beta)} \hat{S}_z^{(\gamma)}) = \{x, y, z\}$. (We define $\operatorname{supp} \hat{1} = \emptyset$.) Note that we can define the operator norm $\|\hat{A}\|$ for $\hat{A} \in \mathfrak{A}_{\mathrm{loc}}$ by identifying $\hat{A}$ with the corresponding operator on the finite dimensional Hilbert space $\bigotimes_{x \in \operatorname{supp} \hat{A}} \mathfrak{h}_x$ and by using the definition (A.2.3).

Physically speaking, $\mathfrak{A}_{\mathrm{loc}}$ is all we need to describe quantum spin systems on $\mathbb{Z}^d$. However it is mathematically convenient to slightly enlarge $\mathfrak{A}_{\mathrm{loc}}$ to make it into a C^*-algebra[22] as

$$\mathfrak{A} := \overline{\mathfrak{A}_{\mathrm{loc}}}, \tag{A.7.1}$$

where the bar denotes the completion with respect to the operator norm.

States Let us discuss the notion of states on the C^*-algebra $\mathfrak{A}$. Since we have not introduced Hilbert spaces, we are not able to talk about vector states such as $|\Phi\rangle$. Instead we shall characterize a state by specifying the corresponding expectation value of any operator in $\mathfrak{A}$.

Definition A.23 A state $\rho(\cdot)$ of a quantum spin system is a linear map from $\mathfrak{A}$ to $\mathbb{C}$ that satisfies $\rho(\hat{1}) = 1$ and $\rho(\hat{A}^\dagger \hat{A}) \geq 0$ for any $\hat{A} \in \mathfrak{A}$.

It follows from the definition that $|\rho(\hat{A})| \leq \|\hat{A}\|$ for any $\hat{A} \in \mathfrak{A}$.

The idea behind the definition is that $\rho(\hat{A})$ is the expectation value of the operator $\hat{A}$ in the state $\rho(\cdot)$. To give a simple (but important) class of examples, let $|\Phi\rangle$ be an arbitrary state in the Hilbert space $\bigotimes_{x \in \Lambda} \mathfrak{h}_x$, where Λ is a finite subset of $\mathbb{Z}^d$. Take an arbitrary monomial $\prod_{x \in S} (\hat{S}_x^{(\alpha_x)})^{n_x}$ of spin operators and define

$$\rho\Bigl(\prod_{x \in S} (\hat{S}_x^{(\alpha_x)})^{n_x}\Bigr) = \langle \Phi | \prod_{x \in S \cap \Lambda} (\hat{S}_x^{(\alpha_x)})^{n_x} | \Phi \rangle \prod_{x \in S \backslash \Lambda} \frac{\mathrm{Tr}_x\bigl[(\hat{S}_x^{(\alpha_x)})^{n_x}\bigr]}{\mathrm{Tr}_x[1]}, \tag{A.7.2}$$

[22] In general a C^*-algebra is a Banach $*$-algebra with the property $\|A^* A\| = \|A\|^2$ for any A.

where the trace is taken over the single spin Hilbert space $\mathfrak{h}_x$. Then we can define $\rho(\hat{A})$ for any $\hat{A} \in \mathfrak{A}_{\text{loc}}$ by linearity (and then extend $\rho(\cdot)$ to the whole $\mathfrak{A}$).

We have the following abstract but useful property of the set of all states.

Theorem A.24 (Banach–Alaoglu theorem) *The set of all states on $\mathfrak{A}$ is compact with respect to the weak-$*$ topology.*

The above statement probably does not make sense to most readers. Here is an elementary interpretation (in the case of quantum spin systems). Let $\rho_j(\cdot)$ with $j = 1, 2, \ldots$ be an arbitrary infinite sequence of states on $\mathfrak{A}$. Then there always exist a state ρ_∞ and a subsequence $j(i)$ with $i = 1, 2, \ldots$ and $j(i) < j(i+1)$ such that

$$\lim_{i\uparrow\infty} \rho_{j(i)}(\hat{A}) = \rho_\infty(\hat{A}), \tag{A.7.3}$$

for any $\hat{A} \in \mathfrak{A}$. This property is useful for us when considering infinite volume limits. See (4.3.7).

Ground states In order to define the notion of ground states, we need to introduce the Hamiltonian of the system. The central object is the local Hamiltonian $\hat{h}_x \in \mathfrak{A}_{\text{loc}}$ associated with each site $x \in \mathbb{Z}^d$. We assume, for any $x \in \mathbb{Z}^d$, that $\operatorname{supp} \hat{h}_x \subset \{y \in \mathbb{Z}^d \mid |y - x| \le R\}$ for a fixed constant R, and $\|\hat{h}_x\| \le h_0$ for another fixed constant h_0. Note that all the Hamiltonians for quantum spin systems studied in the book (except for that of the toy model (2.5.10)) can be written as a sum of such local Hamiltonians. We can formally write the total Hamiltonian as $\hat{H} = \sum_{x\in\mathbb{Z}^d} \hat{h}_x$, but note that $\hat{H}$ belongs neither to $\mathfrak{A}_{\text{loc}}$ nor $\mathfrak{A}$. More importantly we can define the commutator between the Hamiltonian and an arbitrary local operator $\hat{A} \in \mathfrak{A}_{\text{loc}}$ as

$$[\hat{H}, \hat{A}] := \Bigl[\sum_{x\in\Lambda_L} \hat{h}_x, \hat{A}\Bigr] \in \mathfrak{A}_{\text{loc}}, \tag{A.7.4}$$

for sufficiently large L, where Λ_L is the $L \times \cdots \times L$ hypercubic lattice defined in (3.1.2). We here noted that the commutator in the right-hand side does not depend on L when L is taken sufficiently large so that $[\hat{A}, \hat{h}_x] = 0$ for any x outside Λ_L. Then the following is the standard definition of a ground state.

Definition A.25 A state $\omega(\cdot)$ is a ground state of a quantum spin system on $\mathbb{Z}^d$ with the Hamiltonian $\hat{H}$ if and only if $\omega(\hat{A}^\dagger[\hat{H}, \hat{A}]) \ge 0$ for any $\hat{A} \in \mathfrak{A}_{\text{loc}}$.

The motivation for the definition is easily understood by noting that the finite volume version of the condition reads $\langle \Phi_{\text{GS}}|\hat{A}^\dagger[\hat{H}, \hat{A}]|\Phi_{\text{GS}}\rangle \ge 0$, which is rewritten as $\langle \Phi_{\text{GS}}|\hat{A}^\dagger \hat{H} \hat{A}|\Phi_{\text{GS}}\rangle \ge E_{\text{GS}}\langle \Phi_{\text{GS}}|\hat{A}^\dagger \hat{A}|\Phi_{\text{GS}}\rangle$. When $\hat{A}|\Phi_{\text{GS}}\rangle \ne 0$, this inequality becomes

$$\langle \Psi|\hat{H}|\Psi\rangle \ge E_{\text{GS}}, \tag{A.7.5}$$

with a normalized state $|\Psi\rangle = \hat{A}|\Phi_{\text{GS}}\rangle/\|\hat{A}|\Phi_{\text{GS}}\rangle\|$. Thus the condition in the definition simply says that one cannot lower the energy of the state ω by a local perturbation

$\hat{A}$. It is remarkable that Definition A.25 fully characterizes a state of the infinite system only in terms of the expectation values of local operators.

Let us give another characterization of the ground states in terms of the variational principle. For each $L = 1, 2, \ldots$, let

$$\hat{H}_L := \sum_{x \in \overline{\Lambda}_L} \hat{h}_x, \tag{A.7.6}$$

where $\overline{\Lambda}_L := \{x \in \mathbb{Z}^d \mid |x - y| \leq R \text{ for some } y \in \Lambda_L\}$. Note that we have enlarged the lattice so that to include in the sum all $\hat{h}_x$ whose support may overlap with Λ_L. We also define $\mathscr{C}_L^\omega$ to be the set of all states that coincide with $\omega(\cdot)$ outside Λ_L. More precisely, $\mathscr{C}_L^\omega$ consists of states $\omega'(\cdot)$ such that $\omega(\hat{A}) = \omega'(\hat{A})$ for any $\hat{A} \in \mathfrak{A}_{\text{loc}}$ such that $\operatorname{supp} \hat{A} \cap \Lambda_L = \emptyset$.

Theorem A.26 *A state $\omega(\cdot)$ is a ground state of a quantum spin system on $\mathbb{Z}^d$ with the Hamiltonian $\hat{H}$ if and only if*

$$\omega(\hat{H}_L) = \inf_{\omega' \in \mathscr{C}_L^\omega} \omega'(\hat{H}_L), \tag{A.7.7}$$

for any L.

The theorem states that a ground state is a state that minimizes the partial Hamiltonian $\hat{H}_L$ for any L. The reader with background in physics probably find this characterization plausible. For a proof, see the Proof of Theorem 6.2.52 in [5].

Let us finally state what we mean by a unique ground state with a nonzero gap.

Definition A.27 Let $\omega(\cdot)$ be a ground state of a quantum spin system on $\mathbb{Z}^d$ with the Hamiltonian $\hat{H}$, and further assume that $\omega(\cdot)$ is the unique ground state.[23] We say that $\omega(\cdot)$ is accompanied by a nonzero energy gap if and only if there is a constant $\gamma > 0$, and $\omega(\hat{A}^\dagger[\hat{H}, \hat{A}]) \geq \gamma\, \omega(\hat{A}^\dagger \hat{A})$ for any $\hat{A} \in \mathfrak{A}_{\text{loc}}$ such that $\omega(\hat{A}) = 0$.

The motivation of the definition becomes clear again by considering the corresponding relations in a finite system. By using the same $|\Psi\rangle$ as in (A.7.5), we see that the conditions read $\langle\Psi|\hat{H}|\Psi\rangle \geq E_{\text{GS}} + \gamma$ and $\langle\Phi_{\text{GS}}|\Psi\rangle = 0$. These are the conditions that $|\Phi_{\text{GS}}\rangle$ is a unique ground state with a nonzero gap.

Gelfand–Naimark–Segal (GNS) construction Let $\mathscr{H}$ be a Hilbert space, and denote by $B(\mathscr{H})$ the set of all bounded operators on $\mathscr{H}$. A pair $(\mathscr{H}, \pi)$, where π is a map from a C^*-algebra $\mathfrak{A}$ to $B(\mathscr{H})$, is said to be a representation of $\mathfrak{A}$ if

[23] Here it is only necessary in general to require that $\omega(\cdot)$ is the unique ground state in the corresponding GNS Hilbert space. This means that the energy gap associated with a symmetry breaking ground state of the quantum Ising model (see Sect. 3.3) can be characterized by this definition.

$$\pi(\alpha\hat{A} + \beta\hat{B}) = \alpha\,\pi(\hat{A}) + \beta\,\pi(\hat{B}), \tag{A.7.8}$$

$$\pi(\hat{A}\hat{B}) = \pi(\hat{A})\pi(\hat{B}), \tag{A.7.9}$$

$$\pi(\hat{A}^\dagger) = \pi(\hat{A})^\dagger, \tag{A.7.10}$$

for any $\alpha, \beta \in \mathbb{C}$ and $\hat{A}, \hat{B} \in \mathfrak{A}$.

When a C^*-algebra and a state are given, one can go through a standard procedure known as the Gelfand–Naimark–Segal (GNS) construction to define a representation of the C^*-algebra. In the present context, we get the following.

Theorem A.28 (GNS construction) *Let $\rho(\cdot)$ be a state on $\mathfrak{A}$. Then there exist a separable Hilbert space $\mathscr{H}_\rho$, a map $\pi_\rho : \mathfrak{A} \to B(\mathscr{H}_\rho)$, and a vector Ω_ρ such that $(\mathscr{H}_\rho, \pi_\rho)$ is a representation of $\mathfrak{A}$, and*

$$\rho(\hat{A}) = \langle \Omega_\rho, \pi_\rho(\hat{A})\,\Omega_\rho \rangle, \tag{A.7.11}$$

for any $\hat{A} \in \mathfrak{A}$. Here $\langle \cdot, \cdot \rangle$ denotes the inner product of $\mathscr{H}_\rho$. The set of vectors $\{\pi_\rho(\hat{A})\,\Omega_\rho \mid \hat{A} \in \mathfrak{A}\}$ is dense in the Hilbert space $\mathscr{H}_\rho$.

$(\mathscr{H}_\rho, \pi_\rho, \Omega_\rho)$ is called the GNS triple for the state $\rho(\cdot)$.

Remarkably (A.7.11) shows that an arbitrary state $\rho(\cdot)$ can be expressed as a vector state Ω_ρ in the Hilbert space $\mathscr{H}_\rho$. Physically speaking, $\mathscr{H}_\rho$ may be interpreted as a Hilbert space that contains the original state $\rho(\cdot)$ (now represented as Ω_ρ) and other states obtained by locally perturbing it. Such a "small" Hilbert space is indeed suitable for various applications in physics problems.

The Idea of the Proof of Theorem A.28 Let us discuss the basic idea of the GNS construction. We believe it is illuminating also to the readers who are not interested in abstract mathematical theories. See, e.g., Sect. 2.3.3 of [4] for a complete proof.

To construct the Hilbert space $\mathscr{H}_\rho$, we recall that the C^*-algebra $\mathfrak{A}$ itself is a linear space. We shall regard its elements $\hat{A}, \hat{B}, \ldots$ as "vectors", and define their inner products by using the state $\rho(\cdot)$ as

$$\langle \hat{A}, \hat{B} \rangle := \rho(\hat{A}^\dagger \hat{B}). \tag{A.7.12}$$

Of course $\mathfrak{A}$ is not yet a proper vector space with an inner product since there may be a nonzero $\hat{A}$ such that $\langle \hat{A}, \hat{A} \rangle = 0$. We thus define an equivalence relation $\sim$ by

$$\hat{A} \sim \hat{B} \iff \langle (\hat{A} - \hat{B}), (\hat{A} - \hat{B}) \rangle = 0. \tag{A.7.13}$$

The desired Hilbert space is then obtained as the completion of the quotient space as $\mathscr{H}_\rho := \overline{\mathfrak{A}/\!\sim}$.

By definition the quotient space $\mathfrak{A}/\!\sim$ consists of equivalence classes $\psi_{\hat{B}} = \{\hat{B}' \mid \hat{B}' \sim \hat{B}\}$, whose inner products are given by $\langle \psi_{\hat{B}}, \psi_{\hat{C}} \rangle = \langle \hat{B}, \hat{C} \rangle$. We can naturally define the representation π_ρ by $\pi_\rho(\hat{A})\psi_{\hat{B}} = \psi_{\hat{A}\hat{B}}$ for any $\hat{A}, \hat{B} \in \mathfrak{A}$. It is automatic to extend π_ρ to the whole Hilbert space $\mathscr{H}_\rho$. Finally we set $\Omega_\rho = \psi_{\hat{1}}$. Then for any $\hat{A} \in \mathfrak{A}$, we see that

$$\langle \Omega_\rho, \pi_\rho(\hat{A})\Omega_\rho \rangle = \langle \psi_{\hat{1}}, \psi_{\hat{A}} \rangle = \langle \hat{1}, \hat{A} \rangle = \rho(\hat{A}), \tag{A.7.14}$$

which is the desired (A.7.11). ∎

References

1. H. Araki, *Mathematics of Quantum Fields* (in Japanese). Iwanami Series in Modern Physics (Iwanami, 2001)
2. V. Bargmann, Note on Wigner's theorem on symmetry operations. J. Math. Phys. **5**, 862–868, (1964)
3. C. Benassi, J. Fröhlich, D. Ueltschi, Decay of correlations in 2D quantum systems with continuous symmetry. Ann. Henri Poincaré **18**, 2381–2487 (2017). https://link.springer.com/article/10.1007/s00023-017-0571-4
4. O. Bratteli, D.W. Robinson, *Operator Algebras and Quantum Statistical Mechanics I: C*-and W*-Algebras. Symmetry Groups. Decomposition of States* (Springer, Berlin, 1987)
5. O. Bratteli, D.W. Robinson, *Operator Algebras and Quantum Statistical Mechanics II: Equilibrium States Models in Quantum Statistical Mechanics* (Springer, Berlin, 1981)
6. J. Fröhlich, D. Ueltschi, *Some Properties of Correlations of Quantum Lattice Systems in Thermal Equilibrium*. J. Math. Phys. **56**, 053302 (2015). https://arxiv.org/abs/1412.2534
7. T. Koma, H. Tasaki, Decay of superconducting and magnetic correlations in one- and two-dimensional Hubbard models. Phys. Rev. Lett. **68**, 3248–3251 (1992). https://arxiv.org/abs/cond-mat/9709068
8. M. Reed, B. Simon, *Methods of Modern Mathematical Physics I. Functional Analysis* (Academic, Cambridge, 1980)
9. Y. Tada, T. Koma, Decay of superconducting correlations for gauged electrons in dimensions $D \leq 4$. J. Math. Phys. **59**, 031905-1–15 (2018). https://arxiv.org/abs/1612.00619
10. E.P. Wigner, *Gruppentheorie und ihre Anwendung auf die Quanten mechanik der Atomspektren*, (Braunschweig, Germany: Friedrich Vieweg und Sohn 1931)
11. E.P. Wigner, *Group Theory and its Application to the Quantum Mechanics of Atomic Spectra*, translation from German by J.J. Griffin (Academic, Cambridge, 1959)

Solutions

Problems of Chap. 2

2.1.a (p. 15) Note first that the projection operator onto the basis state $|\psi^\sigma\rangle$ is written as

$$|\psi^\sigma\rangle\langle\psi^\sigma| = \prod_{\substack{\tau=-S\\(\tau\neq\sigma)}}^{S} \left(\frac{\hat{S}^{(3)} - \tau\hat{1}}{\sigma - \tau}\right). \tag{S.1}$$

This is easily verified by applying the right-hand side to any basis state $|\psi^{\sigma'}\rangle$. Then, by suitably applying $\hat{S}^\pm$ and normalizing, one gets $|\psi^\tau\rangle\langle\psi^\sigma|$ for any τ and σ. Since any matrix can be written as a linear combination of $|\psi^\tau\rangle\langle\psi^\sigma|$, we see that the desired property holds.

2.1.b (p. 18) From (2.1.7) one finds $(2\hat{S}^{(\alpha)})^n = \hat{1}$ for even n and $(2\hat{S}^{(\alpha)})^n = 2\hat{S}^{(\alpha)}$ for odd n. By using the definition (A.2.16) of the exponential, we find

$$\begin{aligned} e^{-i\theta\hat{S}^{(\alpha)}} &= \sum_{n=0}^{\infty} \frac{1}{n!}\left(\frac{-i\theta}{2}\right)^n (2\hat{S}^{(\alpha)})^n = \sum_{n:\text{even}} \frac{1}{n!}\left(\frac{-i\theta}{2}\right)^n \hat{1} + \sum_{n:\text{odd}} \frac{1}{n!}\left(\frac{-i\theta}{2}\right)^n 2\hat{S}^{(\alpha)} \\ &= \left(\cos\frac{\theta}{2}\right)\hat{1} - 2i\left(\sin\frac{\theta}{2}\right)\hat{S}^{(\alpha)}. \end{aligned} \tag{S.2}$$

See the solution to Problem 2.1.c below for two other strategies.

2.1.c (p. 18) For $S = 1$, (2.1.9) implies $(\hat{S}^{(\alpha)})^3 = \hat{S}^{(\alpha)}$. One can then proceed as in the solution to Problem 2.1.b above, but let us show another derivation. The relation $(\hat{S}^{(\alpha)})^3 = \hat{S}^{(\alpha)}$ implies that one has $e^{-i\theta\hat{S}^{(\alpha)}} = a(\theta)\hat{1} + b(\theta)\hat{S}^{(\alpha)} + c(\theta)(\hat{S}^{(\alpha)})^2$ with coefficients $a(\theta)$, $b(\theta)$, and $c(\theta)$ whose initial conditions are $a(0) = 1$ and $b(0) = c(0) = 0$. By noting that $de^{-i\theta\hat{S}^{(\alpha)}}/d\theta = -i\hat{S}^{(\alpha)}e^{-i\theta\hat{S}^{(\alpha)}}$, we find that the coefficients satisfy the set of differential equations $a'(\theta) = 0$, $b'(\theta) = -i\{a(\theta) + c(\theta)\}$, and $c'(\theta) = -ib(\theta)$, whose unique solution is $a(\theta) = 1$, $b(\theta) = -i\sin\theta$, and $c(\theta) =$

H. Tasaki, *Physics and Mathematics of Quantum Many-Body Systems*, Graduate Texts in Physics, https://doi.org/10.1007/978-3-030-41265-4

$\cos\theta - 1$. We thus find

$$\hat{U}_\theta^{(\alpha)} = \hat{1} - i(\sin\theta)\hat{S}^{(\alpha)} + (\cos\theta - 1)(\hat{S}^{(\alpha)})^2. \tag{S.3}$$

Let us show the third derivation which is the simplest (but may look a bit tricky). From (2.1.9), one readily finds that $e^{-i\theta\hat{S}^{(3)}} = \begin{pmatrix} e^{-i\theta} & 0 & 0 \\ 0 & 1 & 0 \\ 0 & 0 & e^{i\theta} \end{pmatrix}$, from which one can read off the desired relation (S.3) for $\alpha = 3$. But this implies (S.3) for $\alpha = 1$ and 2 as well, since all the spin operators have the same algebraic properties.

2.1.d (p. 18) Substituting (2.1.7) into (2.1.26) we obtain

$$\hat{U}_\theta^{(1)} = \begin{pmatrix} \cos\frac{\theta}{2} & -i\sin\frac{\theta}{2} \\ -i\sin\frac{\theta}{2} & \cos\frac{\theta}{2} \end{pmatrix}, \quad \hat{U}_\theta^{(2)} = \begin{pmatrix} \cos\frac{\theta}{2} & -\sin\frac{\theta}{2} \\ \sin\frac{\theta}{2} & \cos\frac{\theta}{2} \end{pmatrix}, \quad \hat{U}_\theta^{(3)} = \begin{pmatrix} e^{-i\theta/2} & 0 \\ 0 & e^{i\theta/2} \end{pmatrix}. \tag{S.4}$$

By using this we find

$$|\psi_{\theta,\varphi}\rangle = \begin{pmatrix} e^{-i\theta/2} & 0 \\ 0 & e^{i\theta/2} \end{pmatrix} \begin{pmatrix} \cos\frac{\theta}{2} & -\sin\frac{\theta}{2} \\ \sin\frac{\theta}{2} & \cos\frac{\theta}{2} \end{pmatrix} \begin{pmatrix} 1 \\ 0 \end{pmatrix} = \begin{pmatrix} e^{-i\varphi/2}\cos\frac{\theta}{2} \\ e^{i\varphi/2}\sin\frac{\theta}{2} \end{pmatrix}. \tag{S.5}$$

We next see from (2.1.7) that

$$\hat{\boldsymbol{S}} \cdot \boldsymbol{n}(\theta, \varphi) = \frac{1}{2} \begin{pmatrix} \cos\theta & e^{-i\varphi}\sin\theta \\ e^{i\varphi}\sin\theta & -\cos\theta \end{pmatrix}. \tag{S.6}$$

Then the desired relation $\{\hat{\boldsymbol{S}} \cdot \boldsymbol{n}(\theta, \varphi)\}|\psi_{\theta,\varphi}\rangle = (1/2)|\psi_{\theta,\varphi}\rangle$ can be checked directly (with a little bit of calculation). Another (more illuminating) way to confirm the relation is to first note that (2.1.20) implies $\hat{U}\hat{S}^{(3)}\hat{U}^\dagger = \hat{\boldsymbol{S}} \cdot \boldsymbol{n}(\theta, \varphi)$ where $\hat{U} = \hat{U}_\varphi^{(3)}\hat{U}_\theta^{(2)}$. We then have

$$\{\hat{\boldsymbol{S}} \cdot \boldsymbol{n}(\theta, \varphi)\}|\psi_{\theta,\varphi}\rangle = \hat{U}\hat{S}^{(3)}\hat{U}^\dagger\hat{U}|\psi^\uparrow\rangle = \frac{1}{2}\hat{U}|\psi^\uparrow\rangle = \frac{1}{2}|\psi_{\theta,\varphi}\rangle. \tag{S.7}$$

2.1.e (p. 18) From (S.5), we have $|\psi_{(0,1,0)}\rangle = (e^{-i(\pi/4)}, e^{i(\pi/4)})^{\mathrm{t}}/\sqrt{2}$. By using (S.4), we find that

$$|\psi'_{(0,1,0)}\rangle = \hat{U}_{-\pi/2}^{(1)}|\psi^\uparrow\rangle = \frac{1}{\sqrt{2}} \begin{pmatrix} 1 & i \\ i & 1 \end{pmatrix} \begin{pmatrix} 1 \\ 0 \end{pmatrix} = \frac{1}{\sqrt{2}} \begin{pmatrix} 1 \\ i \end{pmatrix}. \tag{S.8}$$

We see that $|\psi'_{(0,1,0)}\rangle = e^{i(\pi/4)}|\psi_{(0,1,0)}\rangle$.

To work out the case with $S = 1$ explicitly, we use (S.3) and (2.1.9) to see that

$$\hat{U}^{(1)}_{-\pi/2} = \frac{1}{2}\begin{pmatrix} 1 & \sqrt{2}i & -1 \\ \sqrt{2}i & 0 & \sqrt{2}i \\ -1 & \sqrt{2}i & 1 \end{pmatrix}, \quad \hat{U}^{(2)}_{\pi/2} = \frac{1}{2}\begin{pmatrix} 1 & -\sqrt{2} & 1 \\ \sqrt{2} & 0 & -\sqrt{2} \\ 1 & \sqrt{2} & 1 \end{pmatrix},$$

$$\hat{U}^{(3)}_{\pi/2} = \frac{1}{\sqrt{2}}\begin{pmatrix} -i & 0 & 0 \\ 0 & 1 & 0 \\ 0 & 0 & i \end{pmatrix}. \tag{S.9}$$

Recalling that $|\psi^1\rangle = (1, 0, 0)^{\mathrm{t}}$, we find

$$|\psi_{(0,1,0)}\rangle = \hat{U}^{(3)}_{\pi/2}\,\hat{U}^{(2)}_{\pi/2}|\psi^1\rangle = \frac{1}{2}\begin{pmatrix} -i \\ \sqrt{2} \\ i \end{pmatrix}, \tag{S.10}$$

$$|\psi'_{(0,1,0)}\rangle = \hat{U}^{(1)}_{-\pi/2}|\psi^1\rangle = \frac{1}{2}\begin{pmatrix} 1 \\ \sqrt{2}i \\ -1 \end{pmatrix}, \tag{S.11}$$

which shows that $|\psi'_{(0,1,0)}\rangle = i\,|\psi_{(0,1,0)}\rangle$. In fact the final relation can be obtained from the result for $S = 1/2$ by noting that a spin with $S = 1$ can be represented by two spins with $S = 1/2$. See Appendix A.3.3 and the expression (2.4.11).

2.1.f (p. 19) A 180° rotation about the y-axis followed by a 180° rotation about the x-axis is equivalent to a 180° rotation about the z-axis. One my confirm this fact easily by, for example, turning a box around.

2.1.g (p. 20) The expression (2.1.34) for $\hat{u}_3$ is obvious. The basis state $|\psi^\sigma\rangle$ (with $\sigma = -S, \ldots, S$) of a spin S can be represented as

$$|\psi^\sigma\rangle = \binom{2S}{n_\uparrow}^{-1/2} \sum_{\substack{\sigma_1,\ldots,\sigma_{2S}=\pm 1/2 \\ (\sum \sigma_j = \sigma)}} |\psi_1^{\sigma_1}\rangle \otimes \cdots \otimes |\psi_{2S}^{\sigma_{2S}}\rangle, \tag{S.12}$$

where $|\psi_j^{\sigma_j}\rangle$ (with $\sigma_j = \pm 1/2$ and $j = 1, \ldots, 2S$) is the basis state of a spin 1/2. There are $n_\uparrow = S + \sigma$ spins with $\sigma_j = 1/2$, and $n_\downarrow = S - \sigma$ spins with $\sigma_j = -1/2$ in each term of (S.12). The expression (S.12) is nothing but (2.4.11). The original spin operator $\hat{S}^{(\alpha)}$ (which is the total spin operator in the present context) is then represented as $\hat{S}^{(\alpha)} = \sum_{j=1}^{2S} \hat{S}_j^{(\alpha)}$, where $\hat{S}_j^{(\alpha)}$ acts on $|\psi_j^\sigma\rangle$. Since $e^{-i\pi \hat{S}_j^{(\alpha)}} = -2i\,\hat{S}_j^{(\alpha)}$, one finds

$$e^{-i\pi\hat{S}^{(\alpha)}} = \prod_{j=1}^{2S}(-2i\,\hat{S}_j^{(\alpha)}) = \begin{cases} (-i)^{2S}\prod_{j=1}^{2S}(\hat{S}_j^+ + \hat{S}_j^-) & \text{if } \alpha = 1, \\ \prod_{j=1}^{2S}(-\hat{S}_j^+ + \hat{S}_j^-) & \text{if } \alpha = 2. \end{cases} \tag{S.13}$$

Recalling that $\hat{S}_j^\pm|\psi_j^{\mp 1/2}\rangle = |\psi_j^{\pm 1/2}\rangle$, one gets (2.1.34) by inspection.

2.2.a (p. 23) The first part is obvious from (2.1.25) for a single spin and the definition (2.2.11). To show the orthogonality, suppose that $\hat{U}_\pi^{(\alpha)}|\Phi\rangle = c|\Phi\rangle$. Note that we have $|c| = 1$. Then observe that $\hat{U}_\pi^{(\alpha)}(\hat{U}_\pi^{(\beta)}|\Phi\rangle) = -\hat{U}_\pi^{(\beta)}\hat{U}_\pi^{(\alpha)}|\Phi\rangle = -c\hat{U}_\pi^{(\beta)}|\Phi\rangle$. Thus $\hat{U}_\pi^{(\beta)}|\Phi\rangle$ is also an eigenstate of $\hat{U}_\pi^{(\alpha)}$ with eigenvalue $-c$. We thus see that $|\Phi\rangle$ and $\hat{U}_\pi^{(\beta)}|\Phi\rangle$ are orthogonal.

2.2.b (p. 23) From the solution to Problem 2.1.d (and a similar computation of $e^{-i\varphi\hat{S}^{(3)}}e^{-i\theta\hat{S}^{(2)}}|\psi^{\downarrow}\rangle$) we find

$$\begin{aligned}\hat{U}_\varphi^{(3)}\hat{U}_\theta^{(2)}\,|\uparrow\rangle_1|\downarrow\rangle_2 =&\big(e^{-i\varphi/2}\cos\tfrac{\theta}{2}|\uparrow\rangle_1 + e^{i\varphi/2}\sin\tfrac{\theta}{2}|\downarrow\rangle_1\big)\\ &\otimes\big(-e^{-i\varphi/2}\sin\tfrac{\theta}{2}|\uparrow\rangle_2 + e^{i\varphi/2}\cos\tfrac{\theta}{2}|\downarrow\rangle_2\big)\\ =&-e^{-i\varphi}\cos\tfrac{\theta}{2}\sin\tfrac{\theta}{2}|\uparrow\rangle_1|\uparrow\rangle_2 + e^{i\varphi}\cos\tfrac{\theta}{2}\sin\tfrac{\theta}{2}|\downarrow\rangle_1|\downarrow\rangle_2\\ &+(\cos\tfrac{\theta}{2})^2|\uparrow\rangle_1|\downarrow\rangle_2 - (\sin\tfrac{\theta}{2})^2|\downarrow\rangle_1|\uparrow\rangle_2.\end{aligned} \tag{S.14}$$

Then (2.2.14) follows by elementary integrations. Similarly we have

$$\begin{aligned}\hat{U}_\varphi^{(3)}\hat{U}_\theta^{(2)}\,|\uparrow\rangle_1|\uparrow\rangle_2 =&\big(e^{-i\varphi/2}\cos\tfrac{\theta}{2}|\uparrow\rangle_1 + e^{i\varphi/2}\sin\tfrac{\theta}{2}|\downarrow\rangle_1\big)\\ &\otimes\big(e^{-i\varphi/2}\cos\tfrac{\theta}{2}|\uparrow\rangle_2 + e^{i\varphi/2}\sin\tfrac{\theta}{2}|\downarrow\rangle_2\big)\\ =&-e^{-i\varphi}(\cos\tfrac{\theta}{2})^2|\uparrow\rangle_1|\uparrow\rangle_2 + e^{i\varphi}(\sin\tfrac{\theta}{2})^2|\downarrow\rangle_1|\downarrow\rangle_2\\ &+\cos\tfrac{\theta}{2}\sin\tfrac{\theta}{2}(|\uparrow\rangle_1|\downarrow\rangle_2 + |\downarrow\rangle_1|\uparrow\rangle_2),\end{aligned} \tag{S.15}$$

which gives (2.2.15) upon integration.

2.2.c (p. 23) The representation in the footnote and $\hat{U}|\Phi_{0,0}\rangle = |\Phi_{0,0}\rangle$ imply

$$\begin{aligned}\hat{U}|\uparrow\rangle_1|\downarrow\rangle_2 &= \sqrt{2}(\tfrac{1}{2} + \hat{U}\hat{S}_1^{(3)}\hat{U}^\dagger)(\tfrac{1}{2} - \hat{U}\hat{S}_2^{(3)}\hat{U}^\dagger)|\Phi_{0,0}\rangle\\ &= \sqrt{2}(\tfrac{1}{2} + \hat{\boldsymbol{S}}_1\cdot\boldsymbol{n})(\tfrac{1}{2} - \hat{\boldsymbol{S}}_2\cdot\boldsymbol{n})|\Phi_{0,0}\rangle,\end{aligned} \tag{S.16}$$

where the final expression clearly depends only on $\boldsymbol{n}$.

2.3.a (p. 31) Since V is antiunitary, we have $\langle \mathsf{V}\boldsymbol{u}, \mathsf{V}\boldsymbol{v}\rangle = \langle \boldsymbol{v}, \boldsymbol{u}\rangle$ for any $\boldsymbol{u}$ and $\boldsymbol{v}$. See (A.4.17). Setting $\boldsymbol{u} = \mathsf{V}\boldsymbol{v}$, we find $\langle \boldsymbol{v}, \mathsf{V}\boldsymbol{v}\rangle = \langle \mathsf{V}^2\boldsymbol{v}, \mathsf{V}\boldsymbol{v}\rangle = -\langle \boldsymbol{v}, \mathsf{V}\boldsymbol{v}\rangle$, which implies $\langle \boldsymbol{v}, \mathsf{V}\boldsymbol{v}\rangle = 0$.

2.4.a (p. 34) In each subspace $\mathscr{H}_M$ there is at least one ground state, i.e., $|\Phi_M\rangle$. It suffices to show the uniqueness of ground state in each $\mathscr{H}_M$. This is easily done by applying the Perron–Frobenius theorem to the matrix representation $\langle\Psi^\sigma|\hat{H}|\Psi^\sigma\rangle$ of the Hamiltonian.

2.4.b (p. 34) Such an expression is not unique. Here is a simple construction. Fix $\theta \in (0, \pi)$ and write $\hat{U}_\theta^{(2)}|\Phi^\uparrow\rangle = \sum_{M=-|\Lambda|S}^{|\Lambda|S} c_M|\Phi_M\rangle$. It is found that $c_M \neq 0$ for any M. Then we have

$$\int_0^{2\pi} d\varphi\, e^{iM\varphi} |\Xi_{\theta,\varphi}\rangle = \int_0^{2\pi} d\varphi \sum_{M'=-|\Lambda|S}^{|\Lambda|S} c_{M'}\, e^{i(M-M')\varphi} |\Phi_{M'}\rangle = 2\pi\, c_M |\Phi_M\rangle. \tag{S.17}$$

2.4.c (p. 34) Denoting the copy of $|\psi_{\theta,\varphi}\rangle$ (see Problem 2.1.d) on site $x \in \Lambda$ as $|\psi_{\theta,\varphi}\rangle_x$ we have

$$|\Xi_{\theta,\varphi}\rangle = \bigotimes_{x\in\Lambda} |\psi_{\theta,\varphi}\rangle_x = \bigotimes_{x\in\Lambda} \{e^{-i\varphi/2} \cos\tfrac{\theta}{2}\, |\psi_x^{\uparrow}\rangle + e^{i\varphi/2} \sin\tfrac{\theta}{2}\, |\psi_x^{\downarrow}\rangle\} \tag{S.18}$$

By expanding and using (2.4.11), we find

$$\begin{aligned} |\Xi_{\theta,\varphi}\rangle &= \sum_{M=-S_{\max}}^{S_{\max}} e^{-iM\varphi/2} (\cos\tfrac{\theta}{2})^{S_{\max}+M} (\sin\tfrac{\theta}{2})^{S_{\max}-M} \sum_{\substack{\sigma \\ (\overline{\sigma}=M)}} |\Psi^{\sigma}\rangle \\ &= \sum_{M=-S_{\max}}^{S_{\max}} \sqrt{\frac{(2S_{\max})!}{(S_{\max}+M)!\,(S_{\max}-M)!}}\, e^{-iM\varphi/2} (\cos\tfrac{\theta}{2})^{S_{\max}+M} (\sin\tfrac{\theta}{2})^{S_{\max}-M} |\Phi_M\rangle. \end{aligned} \tag{S.19}$$

2.4.d (p. 35) Expanding the right-hand side of (2.4.14), one finds

$$\begin{aligned} \text{(RHS)} &= \sum_{\{x,y\}\in\mathscr{B}} \{-|g_x|^2 - |g_y|^2 + g_x^* g_y + g_x g_y^*\} \\ &= -\sum_{x\in\Lambda} |\mathscr{N}(x)|\, |g_x|^2 + \sum_{\substack{x,y\in\Lambda \\ (\{x,y\}\in\mathscr{B})}} g_x^* g_y, \end{aligned} \tag{S.20}$$

which is identical to the left-hand side.

2.5.a (p. 38) This problem can be solved by using basic facts about addition of angular momenta. Note that $\hat{h}_0 = \hat{\boldsymbol{S}}_0 \cdot \hat{\boldsymbol{J}} = \{(\hat{\boldsymbol{S}}_0 + \hat{\boldsymbol{J}})^2 - (\hat{\boldsymbol{S}}_0)^2 - (\hat{\boldsymbol{J}})^2\}/2$ with $\hat{\boldsymbol{J}} = \sum_{j=1}^{z} \hat{\boldsymbol{S}}_j$. Consider a sector in which $\hat{\boldsymbol{J}}^2 = J(J+1)$. Then in order to minimize $(\hat{\boldsymbol{S}}_0 + \hat{\boldsymbol{J}})^2$ the whole system should have the minimum possible total angular momentum $|J - S|$. Then the eigenvalue of $\hat{h}_0$ is

$$\frac{1}{2}\Big\{|J-S|\big(|J-S|+1\big) - S(S+1) - J(J+1)\Big\} = -JS - \min\{J, S\}. \tag{S.21}$$

This is minimized when $J = zS$, and the minimum energy is $-S(1+zS)$.

2.5.b (p. 38) To get a lower bound for E_{GS}, note that $\hat{H} = \sum_{x\in A} \hat{h}_x$ with $\hat{h}_x = \sum_{y\in\mathscr{N}(x)} \hat{\boldsymbol{S}}_x \cdot \hat{\boldsymbol{S}}_y$. Here $\mathscr{N}(x) \subset B$ is the set of sites connected to x. Since the above estimate shows that the ground state energy of $\hat{h}_x$ is $-S(1 + |\mathscr{N}(x)|\, S)$, we find that $E_{\mathrm{GS}} \geq -\sum_{x\in\Lambda} S(1 + |\mathcal{N}(x)|\,S)$. In the d-dimensional hypercubic lattice with $\mathcal{N}(x) = 2d$ and $|\Lambda| = L^d$ (see p. 51), we get $E_{\mathrm{GS}} \geq -L^d S(1 + 2dS)/2$.

2.5.c (p. 39) The uniqueness of the ground state and the SU(2) invariance of the Hamiltonian implies $\langle\Phi_{\mathrm{GS}}|(\hat{S}_x^{(1)})^2|\Phi_{\mathrm{GS}}\rangle = \langle\Phi_{\mathrm{GS}}|(\hat{S}_x^{(2)})^2|\Phi_{\mathrm{GS}}\rangle = \langle\Phi_{\mathrm{GS}}|(\hat{S}_x^{(3)})^2|\Phi_{\mathrm{GS}}\rangle$ for each x. Since $(\hat{S}_x^{(1)})^2 + (\hat{S}_x^{(2)})^2 + (\hat{S}_x^{(3)})^2 = S(S+1)$, one readily gets (2.5.6).

2.5.d (p. 40) An essential observation is that the SU(2) invariance implies

$$\begin{aligned}\langle\Phi_{\mathrm{GS}}|\hat{\mathbf{S}}_x\cdot\hat{\mathbf{S}}_y|\Phi_{\mathrm{GS}}\rangle &= \frac{3}{2}\langle\Phi_{\mathrm{GS}}|(\hat{S}_x^{(1)}\hat{S}_y^{(1)} + \hat{S}_x^{(2)}\hat{S}_y^{(2)})|\Phi_{\mathrm{GS}}\rangle\\ &= \frac{3}{4}\langle\Phi_{\mathrm{GS}}|(\hat{S}_x^{+}\hat{S}_y^{-} + \hat{S}_x^{-}\hat{S}_y^{+})|\Phi_{\mathrm{GS}}\rangle.\end{aligned}\tag{S.22}$$

Therefore it suffices to show that $\langle\Phi_{\mathrm{GS}}|\hat{S}_x^{+}\hat{S}_y^{-}|\Phi_{\mathrm{GS}}\rangle$ is nonzero and has the sign specified in (2.5.7). It is easily found that $\hat{U}\hat{S}_x^{\pm}\hat{U} = (-1)^x\hat{S}_x^{\pm}$ for the unitary operator defined in the hint, where we set $(-1)^x = 1$ if $x \in A$ and $(-1)^x = -1$ if $x \in B$. Since $\hat{U}^2 = \hat{1}$, we have $\hat{U}|\Phi_{\mathrm{GS}}\rangle = \sum_{\sigma\,(\overline{\sigma}=0)} c_\sigma\,|\Psi^\sigma\rangle$. We then find

$$\begin{aligned}(-1)^x(-1)^y\langle\Phi_{\mathrm{GS}}|\hat{S}_x^{+}\hat{S}_y^{-}|\Phi_{\mathrm{GS}}\rangle &= \langle\Phi_{\mathrm{GS}}|\hat{U}\hat{S}_x^{+}\hat{S}_y^{-}\hat{U}|\Phi_{\mathrm{GS}}\rangle\\ &= \sum_{\substack{\sigma,\tau\\(\overline{\sigma}=\overline{\tau}=0)}} c_\sigma\,c_\tau\,\langle\Psi^\sigma|\hat{S}_x^{+}\hat{S}_y^{-}|\Psi^\tau\rangle,\end{aligned}\tag{S.23}$$

where the right-hand side is clearly positive. This is nothing but the desired result.

Problems of Chap. 3

3.3.a (p. 59) We start by examining the matrix elements of the Hamiltonian (3.3.1) with respect to these basis states. Diagonal elements are

$$\langle\Phi^{\uparrow}|\hat{H}|\Phi^{\uparrow}\rangle = \langle\Phi^{\downarrow}|\hat{H}|\Phi^{\downarrow}\rangle = E_{\mathrm{GS}}^{(0)} = -\frac{L-1}{4},\tag{S.24}$$

$$\langle\Phi_j^{\uparrow\downarrow}|\hat{H}|\Phi_j^{\uparrow\downarrow}\rangle = \langle\Phi_j^{\downarrow\uparrow}|\hat{H}|\Phi_j^{\downarrow\uparrow}\rangle = E_{\mathrm{GS}}^{(0)} + \frac{1}{2},\tag{S.25}$$

for $j = 1, \ldots, L-1$. Off-diagonal elements, which come from the spin flip by $\hat{S}_x^{(1)}$, are

$$\langle\Phi_j^{\uparrow\downarrow}|\hat{H}|\Phi_{j+1}^{\uparrow\downarrow}\rangle = \langle\Phi_j^{\downarrow\uparrow}|\hat{H}|\Phi_{j+1}^{\downarrow\uparrow}\rangle = -\frac{\lambda}{2},\tag{S.26}$$

for $j = 1, \ldots, L-2$, and

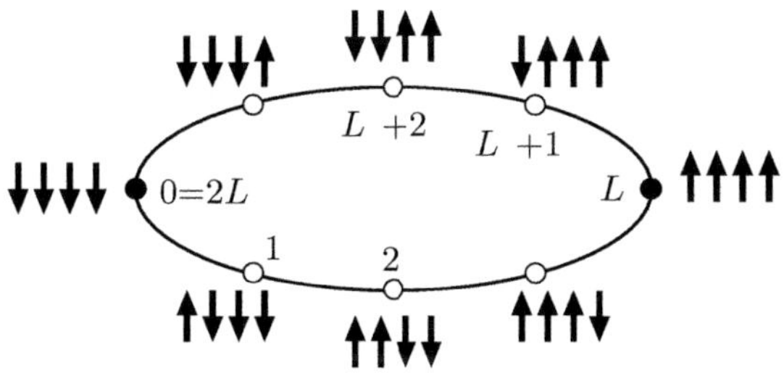

Fig. A.1 The tight-binding model (S.30) which describes low energy properties of the quantum Ising model (3.3.1). The black dots represent sites with potential energy 0, and white dots represent sites with potential energy 1/2. The spin configuration corresponding to each site is also shown (© Hal Tasaki 2020. All Rights Reserved)

$$\langle \Phi^{\uparrow}|\hat{H}|\Phi_1^{\downarrow\uparrow}\rangle = \langle \Phi^{\uparrow}|\hat{H}|\Phi_{L-1}^{\uparrow\downarrow}\rangle = \langle \Phi^{\downarrow}|\hat{H}|\Phi_1^{\uparrow\downarrow}\rangle = \langle \Phi^{\downarrow}|\hat{H}|\Phi_{L-1}^{\downarrow\uparrow}\rangle = -\frac{\lambda}{2}. \tag{S.27}$$

Matrix elements are symmetric, e.g., $\langle \Phi^{\uparrow}|\hat{H}|\Phi_1^{\downarrow\uparrow}\rangle = \langle \Phi_1^{\downarrow\uparrow}|\hat{H}|\Phi^{\uparrow}\rangle$. All other matrix elements are vanishing.

Let us expand low energy states (approximately) as

$$|\Phi\rangle = \varphi_0|\Phi^{\downarrow}\rangle + \sum_{j=1}^{L-1}\varphi_j|\Phi_j^{\uparrow\downarrow}\rangle + \varphi_L|\Phi^{\uparrow}\rangle + \sum_{j=1}^{L-1}\varphi_{j+L}|\Phi_j^{\downarrow\uparrow}\rangle, \tag{S.28}$$

and demand that it satisfies the Schrödinger equation

$$\hat{H}|\Phi\rangle = (E_{\mathrm{GS}}^{(0)} + \varepsilon)|\Phi\rangle. \tag{S.29}$$

By using the matrix elements obtained above, we can write down equations which determine the energy eigenvalues ε and the corresponding eigenstates $(\varphi_j)_{j=1,\ldots,2L-1}$, i.e.,

$$\varepsilon\,\varphi_j = -\frac{\lambda}{2}(\varphi_{j-1} + \varphi_{j+1}) + v_j\,\varphi_j, \tag{S.30}$$

for any $j = 1, \ldots, 2L-1$, where we use periodic boundary conditions $\varphi_{j+2L} = \varphi_j$. The potential is given by $v_0 = v_L = 0$ and $v_j = 1/2$ for $j \neq 0, L$.

Note that (S.30) is nothing but the standard tight-binding Schrödinger equation (see Sect. 9.3.1) on a periodic chain with two potential wells at $j = 0$ and L, which correspond to $|\Phi^{\uparrow}\rangle$ and $|\Phi^{\downarrow}\rangle$. See Fig. A.1. The Schrödinger equation (S.30) can be solved easily as similar exercises in elementary quantum mechanics.

The equation (S.30) is trivial for $j \neq 0, L$ and we see that $e^{\pm\kappa j}$ are independent solutions, where $\kappa > 0$ is a constant to be determined. The corresponding energy is

$$\varepsilon = -\frac{\lambda}{2}(e^{\kappa} + e^{-\kappa}) + \frac{1}{2}. \tag{S.31}$$

From the symmetry we see that there are two independent low energy eigenstates in which the "particle" is bound at $j = 0$ and L. They are the symmetric solution with $\varphi_0 = \varphi_L$, and the antisymmetric solution with $\varphi_0 = -\varphi_L$. We can thus write down the (unnormalized) solutions as

$$\varphi_j = \begin{cases} e^{-\kappa j} \pm e^{-\kappa(L-j)} & \text{for } j = 0, \ldots, L, \\ \pm e^{-\kappa(j-L)} + e^{-\kappa(2L-j)} & \text{for } j = L, \ldots, 2L, \end{cases} \tag{S.32}$$

where the signs $+$ and $-$ correspond to the symmetric and antisymmetric solutions, respectively. By substituting (S.32) into the Schrödinger equation (S.30) with $j = 0$ or L, we get another expression of the energy:

$$\varepsilon\,(1 \pm e^{-\kappa L}) = -\lambda\,(e^{-\kappa} \pm e^{-\kappa(L-1)}). \tag{S.33}$$

By substituting (S.31) into (S.33), we get an equation for κ alone, which, after some work, can be organized as

$$e^{\kappa} - e^{-\kappa} = \lambda^{-1}\,\frac{1 \pm e^{-\kappa L}}{1 \mp e^{-\kappa L}}. \tag{S.34}$$

If we let $L \uparrow \infty$, the right-hand side becomes λ^{-1}, and we get the same constant κ for the symmetric and antisymmetric eigenstates. This implies, as expected, that the two energy eigenvalues degenerate. We write the solution of (S.34) for $L \uparrow \infty$ as κ_∞, i.e.,

$$e^{\kappa_\infty} - e^{-\kappa_\infty} = \lambda^{-1}. \tag{S.35}$$

Because $0 < \lambda \ll 1$, we have $e^{\kappa_\infty} \simeq \lambda^{-1}$, or $\kappa_\infty \simeq -\log\lambda$.

We wish to find the lowest order correction to κ satisfying (S.34) for $0 < \lambda \ll 1$ and $L \gg 1$.[1] Writing $\kappa = \kappa_\infty + \delta$, we have

$$e^{\kappa_\infty+\delta} - e^{-\kappa_\infty-\delta} \simeq \lambda^{-1}\,(1 \pm 2\,e^{-\kappa_\infty L}). \tag{S.36}$$

Expanding the left-hand side in δ to the lowest order, we get

$$\delta \simeq \pm\lambda^{-1}\,\frac{2\,e^{-\kappa_\infty L}}{e^{\kappa_\infty} + e^{-\kappa_\infty}}. \tag{S.37}$$

Now, by substituting $\kappa = \kappa_\infty + \delta$ into (S.31), and expanding in δ, we get

$$\varepsilon = -\frac{\lambda}{2}(e^{\kappa_\infty+\delta} + e^{-\kappa_\infty-\delta}) + \frac{1}{2} \simeq \varepsilon_\infty - \frac{\lambda}{2}(e^{\kappa_\infty} - e^{-\kappa_\infty})\,\delta, \tag{S.38}$$

where

[1] More precisely we fix small λ, and then make L large so that $\lambda^{-1}e^{-\kappa_\infty L}$ is small.

$$\varepsilon_\infty = -\frac{\lambda}{2}(e^{\kappa_\infty} + e^{-\kappa_\infty}) + \frac{1}{2} = -\frac{\sqrt{1+4\lambda^2}}{2} + \frac{1}{2} \simeq -\lambda^2 \tag{S.39}$$

is the energy for $L \uparrow \infty$. Substituting (S.37), we find

$$\varepsilon_\pm \simeq \varepsilon_\infty \mp \frac{e^{\kappa_\infty} - e^{-\kappa_\infty}}{e^{\kappa_\infty} + e^{-\kappa_\infty}} e^{-\kappa_\infty L}. \tag{S.40}$$

We see that the symmetric solution has a lower energy, as it should be. The energy difference is given by

$$E_{1\text{st}} - E_{\text{GS}} = \varepsilon_- - \varepsilon_+ \simeq 2\frac{e^{\kappa_\infty} - e^{-\kappa_\infty}}{e^{\kappa_\infty} + e^{-\kappa_\infty}} e^{-\kappa_\infty L} \simeq 2\,\lambda^L. \tag{S.41}$$

We finally note that these eigenstates are concentrated at two sites $j = 0$ and L. Observe that $|\varphi_0| = |\varphi_L| \simeq 1$, while other components are smaller, e.g., $\varphi_1 \simeq e^{-\kappa} \simeq \lambda$, $\varphi_2 \simeq e^{-2\kappa} \simeq \lambda^2$. This confirms (3.3.8) and (3.3.9).

3.4.a (p. 67) Note that $[\hat{o}_z, [\hat{h}_x, \hat{o}_y]] \neq 0$ only when $|x - y| \le r$ and $|x - z| \le 2r$. For a fixed x, the numbers of such y and z do not exceed $(2r + 1)^2$ and $(4r + 1)^d$, respectively. Since there are L^d choices of x, we get the desired (3.4.13) by recalling that $\|[\hat{o}_z, [\hat{h}_x, \hat{o}_y]]\| \le 4h_0(o_0)^2$. We note that (3.4.13) is a very crude upper bound, and can be improved by estimating more efficiently the numbers of y and z with nonzero contributions.

3.4.b (p. 69) Observe, exactly as in (3.4.15), that

$$\begin{aligned}\langle \Xi_+|(\hat{\mathcal{O}}_L)^2|\Xi_+\rangle &= \frac{1}{2}\Biggl\{\Bigl(\langle\Phi_{\text{GS}}| + \frac{\langle\Phi_{\text{GS}}|\hat{\mathcal{O}}_L}{\|\hat{\mathcal{O}}_L|\Phi_{\text{GS}}\rangle\|}\Bigr)(\hat{\mathcal{O}}_L)^2\Bigl(|\Phi_{\text{GS}}\rangle + \frac{\hat{\mathcal{O}}_L|\Phi_{\text{GS}}\rangle}{\|\hat{\mathcal{O}}_L|\Phi_{\text{GS}}\rangle\|}\Bigr)\Biggr\}\\ &= \frac{1}{2}\Biggl\{\langle\Phi_{\text{GS}}|(\hat{\mathcal{O}}_L)^2|\Phi_{\text{GS}}\rangle + \frac{\langle\Phi_{\text{GS}}|(\hat{\mathcal{O}}_L)^4|\Phi_{\text{GS}}\rangle}{\langle\Phi_{\text{GS}}|(\hat{\mathcal{O}}_L)^2|\Phi_{\text{GS}}\rangle}\Biggr\},\end{aligned} \tag{S.42}$$

which, with (3.4.15), implies

$$\begin{aligned}&\langle \Xi_+|\Bigl(\frac{\hat{\mathcal{O}}_L}{L^d}\Bigr)^2|\Xi_+\rangle - \Bigl\{\langle \Xi_+|\frac{\hat{\mathcal{O}}_L}{L^d}|\Xi_+\rangle\Bigr\}^2\\ &= \frac{1}{2}\Bigl\{\langle\Phi_{\text{GS}}|\Bigl(\frac{\hat{\mathcal{O}}_L}{L^d}\Bigr)^2|\Phi_{\text{GS}}\rangle\Bigr\}^{-1}\Bigl[\langle\Phi_{\text{GS}}|\Bigl(\frac{\hat{\mathcal{O}}_L}{L^d}\Bigr)^4|\Phi_{\text{GS}}\rangle - \Bigl\{\langle\Phi_{\text{GS}}|\Bigl(\frac{\hat{\mathcal{O}}_L}{L^d}\Bigr)^2|\Phi_{\text{GS}}\rangle\Bigr\}^2\Bigr].\end{aligned} \tag{S.43}$$

Problems of Chap. 4

4.2.1.a (p. 104) The expression (2.5.11) of the toy Hamiltonian is still valid. We denote the eigenvalues of $(\hat{\boldsymbol{S}}_{\text{tot}})^2$, $(\hat{\boldsymbol{S}}_A)^2$, and $(\hat{\boldsymbol{S}}_B)^2$ as $S_{\text{tot}}(S_{\text{tot}}+1)$, $S^A_{\text{tot}}(S^A_{\text{tot}}+1)$, and $S^B_{\text{tot}}(S^B_{\text{tot}}+1)$, respectively. We then have $S^A_{\text{tot}} = |A|S - k$, $S^B_{\text{tot}} = |B|S - \ell$, and $S_{\text{tot}} = |S^A_{\text{tot}} - S^B_{\text{tot}}| + m$, where $k, \ell, m = 0, 1, 2, \ldots$. Thus the energy eigenvalues are given by

$$E_{k,\ell,m} = \frac{1}{2|\Lambda|}\Big\{\big(\big|(|A|-|B|)S - k + \ell\big| + m\big)\big(\big|(|A|-|B|)S - k + \ell\big| + m + 1\big) \\ - (|A|S-k)(|A|S-k+1) - (|B|S-\ell)(|B|S-\ell+1)\Big\}. \tag{S.44}$$

Clearly the ground state is attained with $k = \ell = m = 0$. Let us write $\Delta E_{k,\ell,m} = E_{k,\ell,m} - E_{0,0,0}$.

To confirm that there is a gap it suffices to check that $\Delta E_{1,0,0} = |B|S/|\Lambda|$, $\Delta E_{0,1,0} = |A|S/|\Lambda| + |\Lambda|^{-1}$, and $\Delta E_{0,0,1} = (|A|-|B|)S/|\Lambda| + |\Lambda|^{-1} = aS + |\Lambda|^{-1}$. It is useful to note that

$$\Delta E_{0,0,m} = aS\,m + \frac{m(m+1)}{2|\Lambda|}, \tag{S.45}$$

which, with $m = 1, 2, \ldots$, are the lowest excited energies when a is sufficiently small. We also recover the "tower" structure (4.2.8) when $a = 0$.

4.4.3.a (p. 130) First note that the basic inequality (4.4.36) is valid for any lattice geometry. In one-dimensional models with nearest neighbor interactions, it is convenient to chose the classical field configuration as

$$\varphi_x = \begin{cases} 2\mu\,(\ell - |x|) & \text{if } |x| \le \ell, \\ 0 & \text{if } |x| \ge \ell, \end{cases} \tag{S.46}$$

where $\mu > 0$ is a constant and ℓ is a positive integer. Note that $|\varphi_x - \varphi_y|$ is either 2μ or 0 for any $\{x, y\} \in \mathcal{B}_L$.

To show the decay of correlation, we set $h = 0$ and note that (4.4.37) is valid as it is. We set $\ell = |z|$, and observe that $\varphi_o = 2\mu\ell$ and $\varphi_z = 0$ to get

$$\big|\langle \hat{S}^+_o \hat{S}^-_z\rangle^L_{\beta,0}\big| \le 4S^2 \exp\big[-2\mu\ell + 2\beta S^2 \times 2\ell\{\cosh(2\mu) - 1\}\big]. \tag{S.47}$$

Let $\mu(\beta)$ be the positive solution of $\mu = 4\beta S^2\{\cosh(2\mu) - 1\}$, which behaves as $\mu(\beta) \simeq (8S^2\beta)^{-1}$ if $\beta \gg 1$. Then (S.47) yields

$$\big|\langle \hat{S}^+_o \hat{S}^-_z\rangle^L_{\beta,0}\big| \le 4S^2\, e^{-\mu(\beta)\,|z|}, \tag{S.48}$$

which establishes that the correlation function decays (at least) exponentially. In particular $1/\mu(\beta)$ is a rigorous upper bound of the correlation length.

To control the magnetization, we use (4.4.47) to see that

$$\bigl|\langle \hat{S}_o^+\rangle_{\beta,h}^L\bigr| \le 2S\,\exp\bigl[-\mu(\beta)\,\ell + h\,G(\beta,\ell)\bigr], \tag{S.49}$$

with

$$G(\beta,\ell) = 2\beta S \sum_{x=-\ell}^{\ell} \cosh\bigl[2\mu(\beta)\,(\ell - |x|)\bigr]. \tag{S.50}$$

We then get (4.4.22) exactly as in the two-dimensional case.

Problems of Chap. 5

5.2.a (p. 138) The single-particle Schrödinger equation corresponding to the Hamiltonian (5.1.3) is

$$\varepsilon\,\varphi_x = -\sum_{\substack{y\in\Lambda_L\\(\{x,y\}\in\mathcal{B}_L)}} \varphi_y. \tag{S.51}$$

For details, see Sects. 9.2, 9.3, and 9.4, where we discuss the basics of the Hubbard model. The ground state of (S.51) is given by $\varphi_x = |\Lambda_L|^{-1/2}$ for all $x \in \Lambda_L$. Let $\hat{b}^\dagger = |\Lambda_L|^{-1/2}\sum_{x\in\Lambda_L}\hat{a}_x^\dagger$ be the corresponding creation operator of the single-particle ground state. The ground state of the free N boson system is

$$|\Phi_{\rm GS}\rangle = \frac{1}{\sqrt{N!}}(\hat{b}^\dagger)^N|\Phi_{\rm vac}\rangle, \tag{S.52}$$

which is normalized because $\langle\Phi_{\rm vac}|\hat{b}^N(\hat{b}^\dagger)^N|\Phi_{\rm vac}\rangle = N!$. From the commutation relation (5.1.1), we find that $[\hat{a}_x,\hat{b}^\dagger] = |\Lambda_L|^{-1/2}$. We then see

$$\hat{a}_x|\Phi_{\rm GS}\rangle = \frac{1}{\sqrt{N!}}\hat{a}_x(\hat{b}^\dagger)^N|\Phi_{\rm vac}\rangle = \frac{1}{\sqrt{N!}}\frac{N}{\sqrt{|\Lambda_L|}}(\hat{b}^\dagger)^{N-1}|\Phi_{\rm vac}\rangle. \tag{S.53}$$

We thus find for any $x, y \in \Lambda_L$ that

$$\langle\Phi_{\rm GS}|\hat{a}_x^\dagger\hat{a}_y|\Phi_{\rm GS}\rangle = \frac{1}{N!}\frac{N^2}{|\Lambda_L|}\langle\Phi_{\rm vac}|\hat{b}^{N-1}(\hat{b}^\dagger)^{N-1}|\Phi_{\rm vac}\rangle = \frac{N}{|\Lambda_L|}, \tag{S.54}$$

which is the desired (5.2.1).

5.4.a (p. 143) Take any normalized state $|\mathscr{E}_M\rangle$ of the second (external) system which has $N_{\rm tot} - M$ particles. Then recalling (5.3.5), we define

$$|\Phi_{\text{tot}}\rangle = \frac{1}{\sqrt{2M_{\max}(L)+1}} \sum_{M=-M_{\max}(L)}^{M_{\max}(L)} |\Gamma_M\rangle \otimes |\mathscr{E}_M\rangle. \tag{S.55}$$

We then have

$$\langle\Phi_{\text{tot}}|(\hat{A}\otimes\hat{1})|\Phi_{\text{tot}}\rangle = \frac{1}{2M_{\max}(L)+1} \sum_{M=-M_{\max}(L)}^{M_{\max}(L)} \langle\Gamma_M|\hat{A}|\Gamma_M\rangle, \tag{S.56}$$

where the right-hand side is identical to (5.4.1).

Problems of Chap. 6

6.1.a (p. 157) Let $\hat{\mathscr{O}}'_L = \sum_{x\,(|x|\le BL/2)} \hat{o}_x$, and define a trial state by $|\Gamma\rangle = \hat{\mathscr{O}}'_L|\Phi_{\text{GS}}\rangle / \|\hat{\mathscr{O}}'_L|\Phi_{\text{GS}}\rangle\|$ as in (3.4.7). Note that the assumption (6.1.4) implies

$$\langle\Phi_{\text{GS}}|(\hat{\mathscr{O}}'_L)^2|\Phi_{\text{GS}}\rangle = \sum_{\substack{x,y\in\Lambda_L\\(|x|,|y|\le BL/2)}} \langle\Phi_{\text{GS}}|\hat{o}_x\hat{o}_y|\Phi_{\text{GS}}\rangle \ge C'L^{d+\kappa}, \tag{S.57}$$

and that the estimate as in Problem 3.4.a (p. 67) implies $\|[\hat{\mathscr{O}}'_L, [\hat{H}, \hat{\mathscr{O}}'_L]]\| \le C''L^d$. Then, exactly as in (3.4.8), (3.4.9), (3.4.10), and (3.4.11), we find

$$\langle\Gamma|\hat{H}|\Gamma\rangle - E_{\text{GS}} = \frac{\langle\Phi_{\text{GS}}|[\hat{\mathscr{O}}'_L, [\hat{H}, \hat{\mathscr{O}}'_L]]|\Phi_{\text{GS}}\rangle}{2\langle\Phi_{\text{GS}}|(\hat{\mathscr{O}}'_L)^2|\Phi_{\text{GS}}\rangle} \le \frac{C''L^d}{2C'L^{d+\kappa}} = CL^{-\kappa}, \tag{S.58}$$

which, with the variational principle, implies the desired bound.

Problems of Chap. 7

7.1.2.a (p. 184) We use the (standard) abbreviation $|\sigma\rangle_x$ for $|\psi_x^\sigma\rangle$. From the definition (7.1.11), we get

$$\begin{aligned}|\Phi_{\text{pre-VBS}}\rangle &= \frac{1}{2}\Big(|\!\uparrow\rangle_{1,\text{R}}|\!\downarrow\rangle_{2,\text{L}} - |\!\downarrow\rangle_{1,\text{R}}|\!\uparrow\rangle_{2,\text{L}}\Big)\Big(|\!\uparrow\rangle_{2,\text{R}}|\!\downarrow\rangle_{1,\text{L}} - |\!\downarrow\rangle_{2,\text{R}}|\!\uparrow\rangle_{1,\text{L}}\Big)\\
&= \frac{1}{2}\Big(|\!\downarrow\rangle_{1,\text{L}}|\!\uparrow\rangle_{1,\text{R}}|\!\downarrow\rangle_{2,\text{L}}|\!\uparrow\rangle_{2,\text{R}} - |\!\downarrow\rangle_{1,\text{L}}|\!\downarrow\rangle_{1,\text{R}}|\!\uparrow\rangle_{2,\text{L}}|\!\uparrow\rangle_{2,\text{R}}\\
&\qquad - |\!\uparrow\rangle_{1,\text{L}}|\!\uparrow\rangle_{1,\text{R}}|\!\downarrow\rangle_{2,\text{L}}|\!\downarrow\rangle_{2,\text{R}} + |\!\uparrow\rangle_{1,\text{L}}|\!\downarrow\rangle_{1,\text{R}}|\!\uparrow\rangle_{2,\text{L}}|\!\downarrow\rangle_{2,\text{R}}\Big).\end{aligned} \tag{S.59}$$

By using the definition (7.1.12) and the projection (7.1.9), one finds

$$|\Phi_{\text{VBS}}\rangle = \frac{1}{2}\Big(|0\rangle_1|0\rangle_2 - |+\rangle_1|-\rangle_2 - |-\rangle_1|+\rangle_2\Big). \tag{S.60}$$

Recalling (2.2.18), (2.2.16), (2.1.2), and (2.1.3), one easily checks that[2]

$$(\hat{\boldsymbol{S}}_1 + \hat{\boldsymbol{S}}_2)^2|\Phi_{\text{VBS}}\rangle = 0, \tag{S.61}$$

which means $|\Phi_{\text{VBS}}\rangle$ has $S_{\text{tot}} = 0$. This is of course a sufficient condition for $\hat{P}_2[\hat{\boldsymbol{S}}_1 + \hat{\boldsymbol{S}}_2]\,|\Phi_{\text{VBS}}\rangle = 0$.

7.1.2.b (p. 184) As in the above problem, we find from (7.1.11) that

$$\begin{aligned}
&2\sqrt{2}\,|\Phi_{\text{pre-VBS}}\rangle \\
=&\Big(|\uparrow\rangle_{1,\text{R}}|\downarrow\rangle_{2,\text{L}} - |\downarrow\rangle_{1,\text{R}}|\uparrow\rangle_{2,\text{L}}\Big)\Big(|\uparrow\rangle_{2,\text{R}}|\downarrow\rangle_{3,\text{L}} - |\downarrow\rangle_{2,\text{R}}|\uparrow\rangle_{3,\text{L}}\Big) \\
&\Big(|\uparrow\rangle_{3,\text{R}}|\downarrow\rangle_{1,\text{L}} - |\downarrow\rangle_{3,\text{R}}|\uparrow\rangle_{1,\text{L}}\Big) \\
=&|\downarrow\rangle_{1,\text{L}}|\uparrow\rangle_{1,\text{R}}|\downarrow\rangle_{2,\text{L}}|\uparrow\rangle_{2,\text{R}}|\downarrow\rangle_{3,\text{L}}|\uparrow\rangle_{3,\text{R}} - |\uparrow\rangle_{1,\text{L}}|\uparrow\rangle_{1,\text{R}}|\downarrow\rangle_{2,\text{L}}|\uparrow\rangle_{2,\text{R}}|\downarrow\rangle_{3,\text{L}}|\downarrow\rangle_{3,\text{R}} \\
&-|\downarrow\rangle_{1,\text{L}}|\uparrow\rangle_{1,\text{R}}|\downarrow\rangle_{2,\text{L}}|\downarrow\rangle_{2,\text{R}}|\uparrow\rangle_{3,\text{L}}|\uparrow\rangle_{3,\text{R}} + |\uparrow\rangle_{1,\text{L}}|\uparrow\rangle_{1,\text{R}}|\downarrow\rangle_{2,\text{L}}|\downarrow\rangle_{2,\text{R}}|\uparrow\rangle_{3,\text{L}}|\downarrow\rangle_{3,\text{R}} \\
&-|\downarrow\rangle_{1,\text{L}}|\downarrow\rangle_{1,\text{R}}|\uparrow\rangle_{2,\text{L}}|\uparrow\rangle_{2,\text{R}}|\downarrow\rangle_{3,\text{L}}|\uparrow\rangle_{3,\text{R}} + |\uparrow\rangle_{1,\text{L}}|\downarrow\rangle_{1,\text{R}}|\uparrow\rangle_{2,\text{L}}|\uparrow\rangle_{2,\text{R}}|\downarrow\rangle_{3,\text{L}}|\downarrow\rangle_{3,\text{R}} \\
&+|\downarrow\rangle_{1,\text{L}}|\downarrow\rangle_{1,\text{R}}|\uparrow\rangle_{2,\text{L}}|\downarrow\rangle_{2,\text{R}}|\uparrow\rangle_{3,\text{L}}|\uparrow\rangle_{3,\text{R}} - |\uparrow\rangle_{1,\text{L}}|\downarrow\rangle_{1,\text{R}}|\uparrow\rangle_{2,\text{L}}|\downarrow\rangle_{2,\text{R}}|\uparrow\rangle_{3,\text{L}}|\downarrow\rangle_{3,\text{R}}.
\end{aligned} \tag{S.62}$$

Then from (7.1.12) and (7.1.9), we get

$$\begin{aligned}
|\Phi_{\text{VBS}}\rangle =&\frac{1}{4}\Big(-|+\rangle_1|0\rangle_2|-\rangle_3 - |0\rangle_1|-\rangle_2|+\rangle_3 + |+\rangle_1|-\rangle_2|0\rangle_3 \\
&- |-\rangle_1|+\rangle_2|0\rangle_3 + |0\rangle_1|+\rangle_2|-\rangle_3 + |-\rangle_1|0\rangle_2|+\rangle_3\Big) \\
=&\frac{1}{4}\Big(|+\rangle_1|-\rangle_2|0\rangle_3 + |0\rangle_1|+\rangle_2|-\rangle_3 + |-\rangle_1|0\rangle_2|+\rangle_3 \\
&- |-\rangle_1|+\rangle_2|0\rangle_3 - |0\rangle_1|-\rangle_2|+\rangle_3 - |+\rangle_1|0\rangle_2|-\rangle_3\Big),
\end{aligned} \tag{S.63}$$

where we have rearranged the terms to see the structure. To determine the total spin on sites 1 and 2, we further rewrite (S.63) as

$$\begin{aligned}
|\Phi_{\text{VBS}}\rangle = \frac{1}{4}\Big(&\big\{|+\rangle_1|-\rangle_2 - |-\rangle_1|+\rangle_2\big\}|0\rangle_3 + \big\{|-\rangle_1|0\rangle_2 - |0\rangle_1|-\rangle_2\big\}|+\rangle_3 \\
&+ \big\{|0\rangle_1|+\rangle_2 - |+\rangle_1|0\rangle_2\big\}|-\rangle_3\Big).
\end{aligned} \tag{S.64}$$

[2]It is useful to note that (2.1.3) reads $\hat{S}^+|-\rangle = \sqrt{2}|0\rangle$, $\hat{S}^+|0\rangle = \sqrt{2}|+\rangle$, $\hat{S}^-|+\rangle = \sqrt{2}|0\rangle$, and $\hat{S}^-|0\rangle = \sqrt{2}|-\rangle$.

Recalling that the states with $S_{\text{tot}} = 2$ on sites 1 and 2 are

$$|+\rangle_1|+\rangle_2, \frac{|+\rangle_1|0\rangle_2 + |0\rangle_1|+\rangle_2}{\sqrt{2}}, \frac{|+\rangle_1|-\rangle_2 + 2|0\rangle_1|0\rangle_2 + |-\rangle_1|+\rangle_2}{\sqrt{6}},$$
$$\frac{|0\rangle_1|-\rangle_2 + |-\rangle_1|0\rangle_2}{\sqrt{2}}, |-\rangle_1|-\rangle_2, \tag{S.65}$$

one clearly sees that $|\Phi_{\text{VBS}}\rangle$ has no components with $S_{\text{tot}} = 2$, i.e., is orthogonal to (S.65).

7.1.2.c (p. 186) For $x = 1, \ldots, L-1$, we have

$$\hat{U}_{\text{LSM}}(|\psi_x^\uparrow\rangle|\psi_{x+1}^\downarrow\rangle - |\psi_x^\downarrow\rangle|\psi_{x+1}^\uparrow\rangle) = e^{i\Delta\theta/2}|\psi_x^\uparrow\rangle|\psi_{x+1}^\downarrow\rangle - e^{-i\Delta\theta/2}|\psi_x^\downarrow\rangle|\psi_{x+1}^\uparrow\rangle, \tag{S.66}$$

which means

$$|\Phi'\rangle = \bigotimes_{y=1}^{L/2} \frac{1}{\sqrt{2}}\Bigl(e^{i\Delta\theta/2}|\psi_{2y-1}^\uparrow\rangle|\psi_{2y}^\downarrow\rangle - e^{-i\Delta\theta/2}|\psi_{2y-1}^\downarrow\rangle|\psi_{2y}^\uparrow\rangle\Bigr). \tag{S.67}$$

From this one easily finds that

$$\langle\Phi_{\text{dimer}}^{\text{odd}}|\Phi'\rangle = \Bigl(\cos\frac{\Delta\theta}{2}\Bigr)^{L/2} \simeq \Bigl(1 - \frac{2\pi^2}{L^2}\Bigr)^{L/2} \to 1. \tag{S.68}$$

In short we see that $|\Phi'\rangle = \hat{U}_{\text{LSM}}|\Phi_{\text{dimer}}^{\text{odd}}\rangle \simeq |\Phi_{\text{dimer}}^{\text{odd}}\rangle$, i.e., the state essentially does not change after the twist. This is the same as what is expected for the $S = 1$ chain with Haldane gap. See (6.2.22). Lemma 6.2 does not apply here because $|\Phi_{\text{dimer}}^{\text{odd}}\rangle$ is not translation invariant.

Both Lemmas 6.1 and 6.2 (and hence Theorem 6.3) apply to the translation invariant ground state $|\Phi_{\text{dimer}}^{+}\rangle$. Thus we know that $|\Phi''\rangle$ is orthogonal to $|\Phi_{\text{dimer}}^{+}\rangle$ and has low energy. To see what $|\Phi''\rangle$ look like, we note that (S.66) should be modified (only) for $x = L$ as

$$\begin{aligned}\hat{U}_{\text{LSM}}(|\psi_L^\uparrow\rangle|\psi_1^\downarrow\rangle - |\psi_L^\downarrow\rangle|\psi_1^\uparrow\rangle) &= e^{-i(\theta_L-\theta_1)/2}|\psi_L^\uparrow\rangle|\psi_1^\downarrow\rangle - e^{i(\theta_L-\theta_1)/2}|\psi_L^\downarrow\rangle|\psi_1^\uparrow\rangle \\ &= -(e^{i\Delta\theta/2}|\psi_L^\uparrow\rangle|\psi_1^\downarrow\rangle - e^{-i\Delta\theta/2}|\psi_L^\downarrow\rangle|\psi_1^\uparrow\rangle),\end{aligned} \tag{S.69}$$

where we noted that $\theta_L - \theta_1 = 2\pi - \Delta\theta$. We thus find

$$\hat{U}_{\text{LSM}}|\Phi_{\text{dimer}}^{\text{even}}\rangle = -\bigotimes_{y=1}^{L/2} \frac{1}{\sqrt{2}}\Bigl(e^{i\Delta\theta/2}|\psi_{2y}^\uparrow\rangle|\psi_{2y+1}^\downarrow\rangle - e^{-i\Delta\theta/2}|\psi_{2y}^\downarrow\rangle|\psi_{2y+1}^\uparrow\rangle\Bigr) \simeq -|\Phi_{\text{dimer}}^{\text{even}}\rangle, \tag{S.70}$$

which implies $|\Phi''\rangle \simeq |\Phi_{\text{dimer}}^{\text{odd}}\rangle - |\Phi_{\text{dimer}}^{\text{even}}\rangle$.

7.2.2.a (p. 196) Since $\mathsf{A}^-(1,0)^t = (0,0)^t$, we may multiply $(1,0)^t$ either by A^+ or A^0. Multiplying n times by A^0, we get $(2^{-n},0)^t$. This generates n contiguous sequence of 0's. Multiplying by A^+, which corresponds to the insertion of a single $+$, the vector becomes a constant times $(0,1)^t$. Then we may multiply the vector either by A^0 or A^-. The process can be repeated to generate the whole configuration with complete hidden antiferromagnetic order.

7.2.2.b (p. 196) By using the standard formula $|\psi^\sigma\rangle = \sum_{\gamma=1,2,3} |p^{(\gamma)}\rangle\langle p^{(\gamma)}|\psi^\sigma\rangle$, the matrix product representation (7.2.14) is rewritten as

$$|\Phi_{\rm VBS}\rangle = \sum_{\gamma} \mathrm{Tr}[\mathsf{B}^{(\gamma_1)}\mathsf{B}^{(\gamma_2)}\cdots\mathsf{B}^{(\gamma_L)}]\,|\mathscr{P}^{\gamma}\rangle, \tag{S.71}$$

with $\mathsf{B}^{(\gamma)} = \sum_{\sigma=0,\pm}\langle p^{(\gamma)}|\psi^\sigma\rangle\,\mathsf{A}^\sigma$. Explicit calculations with (7.2.12) and (7.2.16) show, rather remarkably, that

$$\mathsf{B}^{(1)} = \frac{1}{2}\begin{pmatrix}0 & 1\\ 1 & 0\end{pmatrix}, \quad \mathsf{B}^{(2)} = \frac{1}{2}\begin{pmatrix}0 & -i\\ i & 0\end{pmatrix}, \quad \mathsf{B}^{(3)} = \frac{1}{2}\begin{pmatrix}1 & 0\\ 0 & -1\end{pmatrix}. \tag{S.72}$$

Thus the matrices $\mathsf{B}^{(\gamma)}$ in the matrix product representation (S.71) are exactly identical to the spin operators (2.1.7) for $S = 1/2$.

7.2.2.c (p. 200) Exactly as in (7.2.26), $\langle\Phi_{\rm VBS}|\hat{P}^0_{x_1}\cdots\hat{P}^0_{x_n}|\Phi_{\rm VBS}\rangle$ is written as the trace of the product of $L-n$ copies of $\tilde{\mathsf{A}}$ and n copies of $\tilde{\mathsf{C}}$, where

$$(\tilde{\mathsf{C}})_{\alpha,\beta;\alpha',\beta'} = \sum_{\sigma=-1,0,1}(1-\sigma^2)A^\sigma_{\alpha,\alpha'}\,A^\sigma_{\beta,\beta'}. \tag{S.73}$$

It is easily found that

$$\tilde{\mathsf{C}} = \begin{array}{c} \\ 1,1\\ 2,2\\ 1,2\\ 2,1\end{array}\!\!\begin{array}{c}\begin{array}{cccc}1,1 & 2,2 & 1,2 & 2,1\end{array}\\ \begin{pmatrix}\frac{1}{4} & 0 & 0 & 0\\ 0 & \frac{1}{4} & 0 & 0\\ 0 & 0 & -\frac{1}{4} & 0\\ 0 & 0 & 0 & -\frac{1}{4}\end{pmatrix}\end{array}. \tag{S.74}$$

This simple form makes the evaluation of the trace almost trivial (note that $\tilde{\mathsf{A}}$ and $\tilde{\mathsf{C}}$ commute). One finds

$$\begin{aligned}\langle\Phi_{\rm VBS}|\hat{P}^0_{x_1}\cdots\hat{P}^0_{x_n}|\Phi_{\rm VBS}\rangle &= \mathrm{Tr}[\tilde{\mathsf{C}}^n\,\tilde{\mathsf{A}}^{L-n}] = \left(\frac{1}{4}\right)^n \mathrm{Tr}\left[\begin{pmatrix}\frac{1}{4} & \frac{1}{2}\\ \frac{1}{2} & \frac{1}{4}\end{pmatrix}^{L-n}\right] + 2\left(-\frac{1}{4}\right)^L\\ &= \left(\frac{1}{4}\right)^n\left\{\left(\frac{3}{4}\right)^{L-n} + \left(-\frac{1}{4}\right)^{L-n}\right\} + 2\left(-\frac{1}{4}\right)^L,\end{aligned} \tag{S.75}$$

which implies

$$\lim_{L\uparrow\infty}\frac{\langle\Phi_{\mathrm{VBS}}|\hat{P}^0_{x_1}\hat{P}^0_{x_2}\cdots\hat{P}^0_{x_n}|\Phi_{\mathrm{VBS}}\rangle}{\langle\Phi_{\mathrm{VBS}}|\Phi_{\mathrm{VBS}}\rangle}=\left(\frac{1}{3}\right)^n. \tag{S.76}$$

This means that 0's appear independently with probability 1/3 in the VBS state.

7.2.2.d (p. 200) Fix $x<y$. Since 0's appear independently with probability 1/3 (in the $L\uparrow\infty$ limit), the probability that both x and y are occupied by $+$ or $-$ is $(2/3)^2=4/9$. Noting that there is complete hidden antiferromagnetic order, one finds $\lim_{L\uparrow\infty}\mathscr{S}^{(3)}_{x,y}(\Phi_{\mathrm{VBS}})=4/9$, which immediately implies (7.2.8).

7.2.2.e (p. 201) One can take, e.g., $\mathsf{A}^{\uparrow}=\begin{pmatrix}1&0\\0&0\end{pmatrix}$ and $\mathsf{A}^{\downarrow}=\begin{pmatrix}0&0\\0&1\end{pmatrix}$.

7.2.2.f (p. 201) One can take, e.g., $\mathsf{A}^{\uparrow}=\begin{pmatrix}0&1\\0&0\end{pmatrix}$ and $\mathsf{A}^{\downarrow}=\begin{pmatrix}0&0\\1&0\end{pmatrix}$.

7.2.3.a (p. 207) Since $\hat{S}^{(3)}_{\mathrm{tot}}|\Phi^{\sigma_{\mathrm{L}},\sigma_{\mathrm{R}}}_{\mathrm{VBS}}\rangle=(\sigma_{\mathrm{L}}+\sigma_{\mathrm{R}})|\Phi^{\sigma_{\mathrm{L}},\sigma_{\mathrm{R}}}_{\mathrm{VBS}}\rangle$, the independence of $|\Phi^{\uparrow,\uparrow}_{\mathrm{VBS}}\rangle$ and $|\Phi^{\downarrow,\downarrow}_{\mathrm{VBS}}\rangle$ is obvious. We consider the expansion (7.2.1) for the remaining two states. The state $|\Phi^{\uparrow,\downarrow}_{\mathrm{VBS}}\rangle$ leads to configurations where the left-most nonzero spin is $+$ (such as 00+0−+0−0) while the state $|\Phi^{\downarrow,\uparrow}_{\mathrm{VBS}}\rangle$ to those where the left-most nonzero spin is − (such as 0−0+0−+00). This proves that the two states are linearly independent. Note that the two states are not orthogonal since the configuration with all 0 appear in the both states.

7.2.3.b (p. 207) Let $|\Phi\rangle$ be a ground state. Then by the same logic as before, we find that the corresponding polynomial satisfies (7.1.24) for $x=1,\ldots,L-1$. Then the uniqueness of factorization implies

$$\Phi(\boldsymbol{u},\boldsymbol{v})=(c_1\,u_1u_L+c_2\,u_1v_L+c_3\,v_1u_L+c_4\,v_1v_L)\prod_{x=1}^{L}(u_xv_{x+1}-v_xu_{x+1}), \tag{S.77}$$

where $c_1,c_2,c_3,c_4\in\mathbb{C}$ are arbitrary. This means that the ground states are exactly four-fold degenerate.

7.2.3.c (p. 207) From the same consideration as $A^{\sigma}_{\alpha,\alpha'}$, the boundary vectors are obtained as $\ell^+_1=1$, $\ell^0_2=1/\sqrt{2}$, $r^+_2=-1/\sqrt{2}$, and $r^0_1=1/2$. Other components are vanishing. This gives a matrix product representation (7.2.45) of $|\Phi^{\uparrow,\uparrow}_{\mathrm{VBS}}\rangle$.

To compute normalization and correlation we need four component vectors $\tilde{\ell}^{\sigma}_{\alpha,\beta}=\sum_{\sigma}\ell^{\sigma}_{\alpha}\,\ell^{\sigma}_{\beta}$, $\tilde{m}^{\sigma}_{\alpha,\beta}=\sum_{\sigma}\sigma\,\ell^{\sigma}_{\alpha}\,\ell^{\sigma}_{\beta}$, and $\tilde{r}^{\sigma}_{\alpha,\beta}=\sum_{\sigma}r^{\sigma}_{\alpha}\,r^{\sigma}_{\beta}$. Using the representation as in (7.2.24), these vectors are written as

$$\tilde{\boldsymbol{\ell}}=(1,\tfrac{1}{2},0,0),\quad\tilde{\boldsymbol{m}}=(1,0,0,0),\quad\tilde{\boldsymbol{r}}=\begin{pmatrix}\frac{1}{4}\\\frac{1}{2}\\0\\0\end{pmatrix}. \tag{S.78}$$

Then the normalization factor is computed as

$$\langle\Phi_{\mathrm{VBS}}^{\uparrow,\uparrow}|\Phi_{\mathrm{VBS}}^{\uparrow,\uparrow}\rangle = \tilde{\boldsymbol{\ell}}\,\tilde{\mathsf{A}}^{L-2}\,\tilde{\boldsymbol{r}} = (1,\tfrac{1}{2})\begin{pmatrix}\frac{1}{4} & \frac{1}{2}\\ \frac{1}{2} & \frac{1}{4}\end{pmatrix}^{L-2}\begin{pmatrix}\frac{1}{4}\\ \frac{1}{2}\end{pmatrix}$$
$$= (1,\tfrac{1}{2})\,\mathsf{O}\begin{pmatrix}(\frac{3}{4})^{L-2} & 0\\ 0 & (-\frac{1}{4})^{L-2}\end{pmatrix}\mathsf{O}\begin{pmatrix}\frac{1}{4}\\ \frac{1}{2}\end{pmatrix} \simeq \frac{9}{16}\left(\frac{3}{4}\right)^{L-2}. \quad \text{(S.79)}$$

To evaluate the expectation value, we note that

$$\langle\Phi_{\mathrm{VBS}}^{\uparrow,\uparrow}|\hat{S}_1^{(3)}|\Phi_{\mathrm{VBS}}^{\uparrow,\uparrow}\rangle = \tilde{\boldsymbol{m}}\,\tilde{\mathsf{A}}^{L-2}\,\tilde{\boldsymbol{r}} = (1,0)\begin{pmatrix}\frac{1}{4} & \frac{1}{2}\\ \frac{1}{2} & \frac{1}{4}\end{pmatrix}^{L-2}\begin{pmatrix}\frac{1}{4}\\ \frac{1}{2}\end{pmatrix} \simeq \frac{3}{8}\left(\frac{3}{4}\right)^{L-2} \quad \text{(S.80)}$$

and, for $x \geq 2$, that

$$\langle\Phi_{\mathrm{VBS}}^{\uparrow,\uparrow}|\hat{S}_x^{(3)}|\Phi_{\mathrm{VBS}}^{\uparrow,\uparrow}\rangle = \tilde{\boldsymbol{\ell}}\,\tilde{\mathsf{A}}^{x-2}\,\tilde{\mathsf{B}}\,\tilde{\mathsf{A}}^{L-x-1}\,\tilde{\boldsymbol{r}}$$
$$= (1,\tfrac{1}{2})\begin{pmatrix}\frac{1}{4} & \frac{1}{2}\\ \frac{1}{2} & \frac{1}{4}\end{pmatrix}^{x-2}\begin{pmatrix}0 & -\frac{1}{2}\\ \frac{1}{2} & 0\end{pmatrix}\begin{pmatrix}\frac{1}{4} & \frac{1}{2}\\ \frac{1}{2} & \frac{1}{4}\end{pmatrix}^{L-x-1}\begin{pmatrix}\frac{1}{4}\\ \frac{1}{2}\end{pmatrix}$$
$$\simeq -\frac{3}{32}\left(-\frac{1}{4}\right)^{x-2}\left(\frac{3}{4}\right)^{L-x-1}, \quad \text{(S.81)}$$

which imply the desired (7.2.50).

7.3.1.a (p. 209) We write $\hat{\mathbf{S}}_x$, $\hat{\mathbf{S}}_{x+1}$ and $\hat{P}_J[\hat{\mathbf{S}}_x + \hat{\mathbf{S}}_{x+1}]$ as $\hat{\mathbf{S}}$, $\hat{\mathbf{S}}'$, and $\hat{P}_J$. We note, as in (7.1.6), that

$$\hat{\mathbf{S}}\cdot\hat{\mathbf{S}}' = \frac{1}{2}(\hat{\mathbf{S}}+\hat{\mathbf{S}}')^2 - 6 = \frac{1}{2}\{4\times 5\,\hat{P}_4 + 3\times 4\,\hat{P}_3 + 2\times 3\,\hat{P}_2 + 2\,\hat{P}_1\} - 6$$
$$= 4\,\hat{P}_4 - 3\,\hat{P}_2 - 5\,\hat{P}_1 - 6\,\hat{P}_0, \quad \text{(S.82)}$$

which means

$$(\hat{\mathbf{S}}\cdot\hat{\mathbf{S}}')^2 = 16\,\hat{P}_4 + 9\,\hat{P}_2 + 25\,\hat{P}_1 + 36\,\hat{P}_0, \quad \text{(S.83)}$$
$$(\hat{\mathbf{S}}\cdot\hat{\mathbf{S}}')^3 = 64\,\hat{P}_4 - 27\,\hat{P}_2 - 125\,\hat{P}_1 - 216\,\hat{P}_0. \quad \text{(S.84)}$$

It is then easy to check that (7.3.3) gives (7.3.2) with $a = 10$, $b = 10/7$, and $c = -10/7$.

7.3.2.a (p. 213) The representation is not unique, so yours may not be the same as ours. First we need a rule corresponding to (7.1.9). By setting $S_{\max} = 2$ in (2.4.11), we find

$$\mathscr{S}[|\uparrow\rangle|\uparrow\rangle|\uparrow\rangle|\uparrow\rangle]=|\psi^{2}\rangle,\quad \mathscr{S}[|\uparrow\rangle|\uparrow\rangle|\uparrow\rangle|\downarrow\rangle]=\frac{1}{2}|\psi^{1}\rangle,\quad \mathscr{S}[|\uparrow\rangle|\uparrow\rangle|\downarrow\rangle|\downarrow\rangle]=\frac{1}{\sqrt{6}}|\psi^{0}\rangle,$$
$$\mathscr{S}[|\uparrow\rangle|\downarrow\rangle|\downarrow\rangle|\downarrow\rangle]=\frac{1}{2}|\psi^{-1}\rangle,\quad \mathscr{S}[|\downarrow\rangle|\downarrow\rangle|\downarrow\rangle|\downarrow\rangle]=|\psi^{-2}\rangle. \tag{S.85}$$

As for the valence bonds $(1/\sqrt{2})\{|\psi_x^{\uparrow}\rangle|\psi_{x+e}^{\downarrow}\rangle-|\psi_x^{\downarrow}\rangle|\psi_{x+e}^{\uparrow}\rangle\}$, where $e=e_{\mathrm{x}}$ or e_{y}, we again set $\alpha_{x,x+e}=1$ for the first term and $\alpha_{x,x+e}=2$ for the second term. We assign the factor $\pm1/\sqrt{2}$ to A at the site x. Then by inspection one finds

$$\begin{gathered}
A^{2}_{1122}=\frac{1}{2},\\
A^{1}_{1112}=A^{1}_{1121}=\frac{1}{4},\quad A^{1}_{2122}=A^{1}_{1222}=-\frac{1}{4},\\
A^{0}_{2222}=A^{0}_{1111}=\frac{1}{2\sqrt{6}},\quad A^{0}_{1212}=A^{0}_{1221}=A^{0}_{2112}=A^{0}_{2121}=-\frac{1}{2\sqrt{6}},\\
A^{-1}_{2221}=A^{-1}_{2212}=\frac{1}{4},\quad A^{-1}_{1211}=A^{-1}_{2111}=-\frac{1}{4},\\
A^{-2}_{2211}=\frac{1}{2}.
\end{gathered} \tag{S.86}$$

7.3.3.a (p. 217) The identity (7.3.28) is easily confirmed by examining the action on the four basis states $|\psi_x^{\sigma}\rangle|\psi_y^{\sigma'}\rangle$ with $\sigma,\sigma'=\uparrow,\downarrow$. Recall that $|\Phi_{\mathrm{BR}}\rangle=(\prod_{y=1}^{L}\hat{C}_{y,y+1})\hat{\sigma}_x^{(1)}|\Phi_{\rightarrow}\rangle$. Clearly $\hat{\sigma}_x^{(1)}$ commutes with $\hat{C}_{y,y+1}$ unless $y=x-1$ or $y=x$. We then use (7.3.28) to see that $\hat{C}_{x-1,x}\,\hat{C}_{x,x+1}\,\hat{\sigma}_x^{(1)}=\hat{C}_{x-1,x}\,\hat{\sigma}_x^{(1)}\hat{\sigma}_{x+1}^{(3)}\,\hat{C}_{x,x+1}=\hat{\sigma}_x^{(1)}\,\hat{\sigma}_{x-1}^{(3)}\hat{C}_{x-1,x}\,\hat{\sigma}_{x+1}^{(3)}\,\hat{C}_{x,x+1}=\hat{\sigma}_{x-1}^{(3)}\hat{\sigma}_x^{(1)}\hat{\sigma}_{x+1}^{(3)}\hat{C}_{x-1,x}\,\hat{C}_{x,x+1}$, which implies the desired (7.3.18).

7.3.3.b (p. 219) Let $\hat{U}_{\hat{A}}=\prod_{c\in\mathscr{C}_{\hat{A}}}\hat{C}_c$ and $\hat{U}_{\hat{B}}=\prod_{c\in\mathscr{C}_{\hat{B}}}\hat{C}_c$. Noting that $\hat{C}_c$ commute with each other and $(\hat{C}_c)^2=\hat{1}$, we find

$$\begin{aligned}
\langle\Phi_{\mathscr{C}}|\hat{A}\hat{B}|\Phi_{\mathscr{C}}\rangle&=\langle\Phi_{\rightarrow}|(\hat{U}_{\hat{A}}\,\hat{A}\hat{U}_{\hat{A}})(\hat{U}_{\hat{B}}\,\hat{B}\hat{U}_{\hat{B}})|\Phi_{\rightarrow}\rangle\\
&=\langle\Phi_{\rightarrow}|\hat{U}_{\hat{A}}\,\hat{A}\hat{U}_{\hat{A}}|\Phi_{\rightarrow}\rangle\langle\Phi_{\rightarrow}|\hat{U}_{\hat{B}}\,\hat{B}\hat{U}_{\hat{B}}|\Phi_{\rightarrow}\rangle\\
&=\langle\Phi_{\mathscr{C}}|\hat{A}|\Phi_{\mathscr{C}}\rangle\langle\Phi_{\mathscr{C}}|\hat{B}|\Phi_{\mathscr{C}}\rangle,
\end{aligned} \tag{S.87}$$

where we noted that $|\Phi_{\rightarrow}\rangle$ is a trivial product state, and the supports of $\hat{U}_{\hat{A}}\,\hat{A}\hat{U}_{\hat{A}}$ and $\hat{U}_{\hat{B}}\,\hat{B}\hat{U}_{\hat{B}}$ (which are $\bar{\Lambda}_{\hat{A}}$ and $\bar{\Lambda}_{\hat{B}}$, respectively) do not overlap.

7.3.3.c (p. 220) Let us write the Hamiltonian (7.3.34) as $\hat{H}_{\mathscr{C}}=-\sum_{x\in\Lambda}\hat{h}_x$. The local Hamiltonian is $\hat{h}_x=\hat{\sigma}_x^{(1)}\hat{C}_{\{x_1,x_2\}}\hat{C}_{\{x_2,x_3\}}\hat{C}_{\{x_3,x_4\}}\hat{C}_{\{x_4,x_5\}}\hat{C}_{\{x_5,x_6\}}\hat{C}_{\{x_6,x_1\}}$, where $x_1,x_2,\ldots,x_6$ are the sites (ordered in a proper manner) of the hexagon centered at x.

Since $\hat{h}_x|\Phi_{\mathscr{C}}\rangle=|\Phi_{\mathscr{C}}\rangle$ for any x, any product of $\hat{h}_x$'s can be used to detect the hidden order. For example let B be a connected subset of one of the three sublattices. Then the operator $\prod_{x\in B}\hat{h}_x=(\prod_{x\in B}\hat{\sigma}_x^{(1)})(\prod_{j=1}^{M}\hat{C}_{\{x_j,x_{j+1}\}})$, where $x_1,x_2,\ldots,x_M$

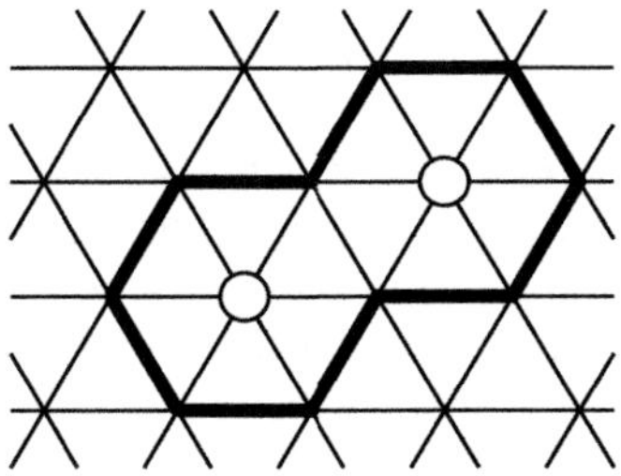

Fig. A.2 A simple example of the operator which measures the hidden order in the generalized cluster state (hyper graph state) on the hexagonal lattice. A white circle represents $\hat{\sigma}_x^{(1)}$ and a gray line represents $\hat{C}_{\{y,y'\}}$. This is operator is the product of two $\hat{h}_x$'s, where two x's correspond to white circles (© Hal Tasaki 2020. All Rights Reserved)

(with $x_{M+1} = x_1$) are sites in other sublattices surrounding the set B, has expectation value 1 in the generalized cluster state $|\Phi_{\mathscr{C}}\rangle$. See Fig. A.2.

Since it is obvious that $[\hat{h}_x, \hat{U}_\nu] = 0$ when $x \in S_\nu$, we only need to show that $[\hat{h}_x, \hat{U}_\nu] = 0$ when $x \notin S_\nu$. This is equivalent to showing that, for any configuration $\sigma_1, \ldots, \sigma_6 =\uparrow, \downarrow$, the sign factor $s(\sigma_1, \sigma_2)s(\sigma_2, \sigma_3) \ldots s(\sigma_5, \sigma_6)s(\sigma_6, \sigma_1)$ is invariant under the simultaneous spin flip $\sigma_2 \to -\sigma_2$, $\sigma_4 \to -\sigma_4$, and $\sigma_6 \to -\sigma_6$. By inspection one finds that, under the spin flip $\sigma' \to -\sigma'$, the sign factor $s(\sigma, \sigma')s(\sigma', \sigma'')$ is invariant if $\sigma = \sigma''$, and changes the sign if $\sigma \neq \sigma''$. But, in any configuration $\sigma_1, \ldots, \sigma_6$, the latter case does not take place or takes place exactly twice. This proves the desired invariance. It is interesting that one needs to treat the configuration of six spins on a closed loop (i.e., the hexagon) to see the invariance.

Problems of Chap. 8

8.3.3.a (p. 263) We assume that the ground state is $\mathbb{Z}_2 \times \mathbb{Z}_2$ is invariant and hence $\hat{U}_\pi^{(\alpha)} \hat{\rho}_\mathrm{R} (\hat{U}_\pi^{(\alpha)})^\dagger = \hat{\rho}_\mathrm{R}$ for $\alpha = 1, 2, 3$. This in particular means $[\hat{U}_\pi^{(3)}, \hat{\rho}_\mathrm{R}] = 0$, and hence we can assume that each $|\Psi_j\rangle_\mathrm{R}$ is an eigenstate of $\hat{U}_\pi^{(3)}$. We also see that the invariance (8.3.9) is valid with $|\Psi'_j\rangle_\mathrm{R} = \hat{U}_\pi^{(1)} |\Psi_j\rangle_\mathrm{R}$. Now, suppose that $|\Psi_j\rangle_\mathrm{R}$ can be regarded as states with half-odd-integer spins. Then from Problem 2.2.a (p. 23) we find that $|\Psi_j\rangle_\mathrm{R}$ and $|\Psi'_j\rangle_\mathrm{R}$ are orthogonal for each j. The rest is the same as the case with time-reversal symmetry.

8.3.4.a (p. 270) Let $\tilde{\mathsf{p}}(e) = \mathsf{p}(e) = \mathsf{I}$ and $\tilde{\mathsf{p}}(g) = e^{-i\phi(g,g)/2} \mathsf{p}(g)$ for $g \in \{a, b, c\}$. We see that $\tilde{\mathsf{p}}(\cdot)$ gives an equivalent projective representation with the property that $\tilde{\mathsf{p}}(g)\tilde{\mathsf{p}}(g) = \tilde{\mathsf{p}}(e)$ (and hence $\tilde{\phi}(g, g) = 0$) for all $g \in \mathbb{Z}_2 \times \mathbb{Z}_2$. Note also that $\tilde{\phi}(g, e) = \tilde{\phi}(e, g) = 0$ for $g \in \{a, b, c\}$ because $\tilde{\rho}(e) = \mathsf{I}$.[1]

[1] *Minor note:* The assumption $\mathsf{p}(e) = \mathsf{I}$ is indeed not necessary. If $\mathsf{p}(e) \neq \mathsf{I}$, we set $\tilde{\mathsf{p}}(e) = e^{-i\phi(e,e)} \mathsf{p}(e)$ and $\tilde{\mathsf{p}}(g) = e^{-i\{\phi(g,g)+\phi(e,e)\}/2} \mathsf{p}(g)$. We then have $\tilde{\mathsf{p}}(g)\tilde{\mathsf{p}}(g) = \tilde{\mathsf{p}}(e)$ (and hence $\tilde{\phi}(g, g) = 0$) for all $g \in \mathbb{Z}_2 \times \mathbb{Z}_2$. Next, multiplying the relation $\tilde{\mathsf{p}}(g)\tilde{\mathsf{p}}(e) = e^{i\tilde{\phi}(g,e)} \tilde{\mathsf{p}}(g)$ by $\tilde{\mathsf{p}}(e)$

The cocycle condition (8.3.26) for $\tilde{\rho}(\cdot)$ with $(g_1, g_2, g_3) = (a, b, c)$ then reads $\tilde{\phi}(a, b) = \tilde{\phi}(b, c)$. By setting $(g_1, g_2, g_3) = (b, c, a)$, we also see $\tilde{\phi}(b, c) = \tilde{\phi}(c, a)$, and hence that $\tilde{\phi}(a, b) = \tilde{\phi}(b, c) = \tilde{\phi}(c, a) =: \xi$ with some $\xi \in \mathbb{R}$. Similarly we have $\tilde{\phi}(a, c) = \tilde{\phi}(c, b) = \tilde{\phi}(b, a) =: \xi'$ with some $\xi' \in \mathbb{R}$. We next use (8.3.26) with $(g_1, g_2, g_3) = (a, b, b)$ to find $\tilde{\phi}(a, b) + \tilde{\phi}(c, b) = 0$, which implies $\xi + \xi' = 0$. We finally use (8.3.26) with $(g_1, g_2, g_3) = (a, b, a)$ to see that $\tilde{\phi}(a, b) + \tilde{\phi}(c, a) = \tilde{\phi}(b, a) + \tilde{\phi}(a, c) \pmod{2\pi}$, which, with the above results, shows that $4\xi = 0 \pmod{2\pi}$.

It is easily found that we only need to consider the cases with $\xi = 0$ and with $\xi = \pi/2$. (Other cases are equivalent to one of the two.) When $\xi = \xi' = 0$, we readily see that $\tilde{\rho}(\cdot)$ is a genuine representation, and hence the original projective representation $\rho(\cdot)$ is trivial. When $\xi = \pi/2$ and $\xi' = -\pi/2$, we have $\tilde{\rho}(a)\tilde{\rho}(b) = e^{i\pi/2}\tilde{\rho}(c)$ and $\tilde{\rho}(b)\tilde{\rho}(a) = e^{-i\pi/2}\tilde{\rho}(c)$, etc. By setting $\tilde{\tilde{\rho}}(e) = \tilde{\rho}(e) = \mathsf{I}$ and $\tilde{\tilde{\rho}}(g) = e^{-i\pi/2}\tilde{\rho}(g)$ for $g \in \{a, b, c\}$, we get an equivalent projective representation $\tilde{\tilde{\rho}}(\cdot)$, which satisfies $\tilde{\tilde{\rho}}(g)\tilde{\tilde{\rho}}(g) = -\tilde{\tilde{\rho}}(e)$ for $g \in \{a, b, c\}$, and $\tilde{\tilde{\rho}}(a)\tilde{\tilde{\rho}}(b) = -\tilde{\tilde{\rho}}(b)\tilde{\tilde{\rho}}(a) = \tilde{\tilde{\rho}}(c)$, etc. This is nothing but the nontrivial projective representation (2.1.31).

8.3.4.b (p. 271) Noting that $\hat{u}_2 = \exp[-i\pi\hat{S}^{(2)}] = \begin{pmatrix} 0 & 0 & 1 \\ 0 & -1 & 0 \\ 1 & 0 & 0 \end{pmatrix}$ as in (2.1.33), we find that the new matrices are given by

$$\tilde{\mathsf{A}}^+ = \mathsf{A}^-, \quad \tilde{\mathsf{A}}^0 = -\mathsf{A}^0, \quad \tilde{\mathsf{A}}^- = \mathsf{A}^+. \tag{S.88}$$

8.3.4.c (p. 272) The matrices $\tilde{\mathsf{A}}^\sigma$ above are written in the form (8.3.16) with $\mathsf{U}_2 = \begin{pmatrix} 0 & 1 \\ -1 & 0 \end{pmatrix}$, which is again $\exp[-i\pi\hat{S}^{(2)}]$ for $S = 1/2$. (Note that the choice of the unitary matrix is not unique.) This time we should set the phase factor as $e^{i\eta_L/L} = -1$.

8.3.4.d (p. 274) When no complex conjugation takes place as in the $\mathbb{Z}_2 \times \mathbb{Z}_2$ transformation, the matrices (8.3.39) transform in the same manner as (8.3.28) for the VBS state. This means that we can take the same unitary matrices U_1, U_2, and U_3, and hence the twisted VBS state is associated with the nontrivial projective representation of $\mathbb{Z}_2 \times \mathbb{Z}_2$.

There is a difference in time-reversal transformation (8.3.33) since it involves complex conjugation. One finds that (8.3.33) is recovered by (8.3.34) with $\zeta = \pi$ and $\mathsf{U} = \begin{pmatrix} 0 & 1 \\ 1 & 0 \end{pmatrix}$. We have $\mathsf{U}\mathsf{U}^* = \mathsf{U}^2 = \mathsf{I}$, which means that the associated phase is trivial, i.e, $\sigma_{\mathrm{tr}} = 1$.

Finally the VBS and the twisted VBS states can be connected smoothly via the θ-twisted-VBS state defined by the matrices

---

from left and right, one gets $\tilde{\rho}(e)\tilde{\rho}(g)\tilde{\rho}(e) = e^{i\tilde{\phi}(g,e)}\tilde{\rho}(e)\tilde{\rho}(g)\tilde{\rho}(e)$, which implies $\tilde{\phi}(g, e) = 0$. We similarly find $\tilde{\phi}(e, g) = 0$. This means that $\tilde{\rho}(g)\tilde{\rho}(e) = \tilde{\rho}(e)\tilde{\rho}(g) = \tilde{\rho}(g)$ for all $g \in \mathbb{Z}_2 \times \mathbb{Z}_2$. The rest of the proof is the same.

$$\mathsf{A}^{\pm}_{\theta\text{-VBS}} = \mathsf{A}^{\pm}_{\text{VBS}}, \quad \mathsf{A}^{0}_{\theta\text{-VBS}} = e^{i\theta}\,\mathsf{A}^{0}_{\text{VBS}}. \tag{S.89}$$

It can be checked easily that the state violates time-reversal symmetry except for $\theta = 0$ or π, but has $\mathbb{Z}_2 \times \mathbb{Z}_2$ symmetry for any θ.

8.4.a (p. 295) The conditions (8.4.2), (8.4.3) for ground states are written as $\hat{A}_v|\Phi_{\text{GS}}\rangle = |\Phi_{\text{GS}}\rangle$ and $\hat{B}_p|\Phi_{\text{GS}}\rangle = |\Phi_{\text{GS}}\rangle$ for any v and p. Since it is obvious that $\hat{A}_v|\Phi_{\uparrow}\rangle = |\Phi_{\uparrow}\rangle$, we find $\hat{A}_v|\Phi_0\rangle = |\Phi_0\rangle$ from $[\hat{A}_v, \hat{B}_p] = 0$. Note also that $(\hat{B}_p)^2 = \hat{1}$ means $\hat{B}_p(\hat{1} + \hat{B}_p) = \hat{1} + \hat{B}_p$, which implies $\hat{B}_p|\Phi_0\rangle = |\Phi_0\rangle$. Thus $|\Phi_0\rangle$ is a ground state. The state $|\Phi_{\uparrow}\rangle$ corresponds to the empty graph with $\nu = \nu' = 0$. Since the action of $\hat{B}_p$ precisely corresponds to the local modification (8.4.17), we see that $|\Phi_0\rangle = \text{const.}\,|\Phi_{0,0}\rangle$.

8.4.b (p. 295) We give an elementary constructive proof,[2] which may not be elegant.[3] We prove that any graph with $\nu = \nu' = 0$ can be modified into the empty graph. Other cases can be treated similarly (or can be reduced to this case).

Given a graph, concentrate on the configuration of vertical thick bonds which cross the horizontal line in Fig. 8.18. By using the rules (8.4.17) repeatedly, one can move the thick bonds to the left or right, and let them pairwise annihilate. Since there are even number of thick bonds to begin with, we can get a configuration with no thick vertical bonds passing through the horizontal line. By repeating the procedure for all horizontal lines, we end up with a graph with no thick vertical lines. Remaining horizontal thick bonds must form horizontal loops wrapping around the lattice. Note that the number of the loops must be even. By moving these loops and letting them pairwise annihilate, we get the empty graph.

8.4.c (p. 297) We shall evaluate the correlation

$$\langle \hat{Z}_{\text{hor}} \hat{Z}'_{\text{hor}} \rangle_{\beta,L} := \frac{\text{Tr}[\hat{Z}_{\text{hor}} \hat{Z}'_{\text{hor}}\, e^{-\beta \hat{H}_{\text{tc}}}]}{\text{Tr}[e^{-\beta \hat{H}_{\text{tc}}}]}. \tag{S.90}$$

Write the Hamiltonian (8.4.1) as $\hat{H}_{\text{tc}} = -\sum_{v\in\mathscr{V}} \hat{A}_v - \sum_{p\in\mathscr{P}} \hat{B}_p$ by using (8.4.18). Noting that all $\hat{A}_v$ and $\hat{B}_p$ commute, and using the relation in footnote 72 (p. 297), we find

$$\begin{aligned} e^{-\beta \hat{H}_{\text{tc}}} &= (\cosh\beta)^{2L^2} \prod_{v\in\mathscr{V}} \{\hat{1} + (\tanh\beta)\hat{A}_v\} \prod_{p\in\mathscr{P}} \{\hat{1} + (\tanh\beta)\hat{B}_p\} \\ &= (\cosh\beta)^{2L^2} \sum_{\substack{V\subset\mathscr{V}\\ P\subset\mathscr{P}}} (\tanh\beta)^{|V|+|P|} \Bigl(\prod_{v\in V} \hat{A}_v\Bigr)\Bigl(\prod_{p\in P} \hat{B}_p\Bigr), \end{aligned} \tag{S.91}$$

[2] Or a "destructive proof" since we shall eliminate loops.

[3] The reader looking for an elegant proof may try to show that graphs (of loops) with $\nu = \nu' = 0$ must be a boundary of a region in the square lattice.

where V and P are summed over all subsets (including the empty set) of $\mathscr{V}$ and $\mathscr{P}$, respectively. Note that $\left(\prod_{v\in V}\hat{A}_v\right)\left(\prod_{p\in P}\hat{B}_p\right)$ is a monomial in $\hat{\sigma}_x^{(1)}$ and $\hat{\sigma}_x^{(3)}$ with $x\in\mathscr{E}$. The trace of a monomial is easily calculated as

$$\mathrm{Tr}\Big[\prod_{x\in\mathscr{E}}\{(\hat{\sigma}_x^{(1)})^{n_x^{(1)}}(\hat{\sigma}_x^{(3)})^{n_x^{(3)}}\}\Big]=\begin{cases}2^{2L^d} & \text{when all } n_x^{(1)} \text{ and } n_x^{(3)} \text{ are even,}\\ 0 & \text{otherwise.}\end{cases}\tag{S.92}$$

By using (S.91) and (S.92), one can evaluate (S.90) easily. For the partition function $\mathrm{Tr}[e^{-\beta\hat{H}_{\mathrm{tc}}}]$ the only nonvanishing contribution from the expansion (S.91) come from terms with $(V,P)=(\emptyset,\emptyset)$, $(\mathscr{V},\emptyset)$, $(\emptyset,\mathscr{P})$, or $(\mathscr{V},\mathscr{P})$. We thus get

$$\mathrm{Tr}[e^{-\beta\hat{H}_{\mathrm{tc}}}]=(2\cosh\beta)^{2L^2}\{1+2(\tanh\beta)^{L^2}+(\tanh\beta)^{2L^2}\}.\tag{S.93}$$

For $\mathrm{Tr}[\hat{Z}_{\mathrm{hor}}\hat{Z}'_{\mathrm{hor}}\,e^{-\beta\hat{H}_{\mathrm{tc}}}]$, nonvanshing contribution come from $(V,P)=(V_1,\emptyset)$, $(V_2,\emptyset)$, $(V,P)=(V_1,\mathscr{P})$, or $(V_2,\mathscr{P})$ where V_1 and V_2 are the two regions in $\mathscr{V}$ which have the two horizontal lines as boundaries. We thus find

$$\mathrm{Tr}[\hat{Z}_{\mathrm{hor}}\hat{Z}'_{\mathrm{hor}}\,e^{-\beta\hat{H}_{\mathrm{tc}}}]=(2\cosh\beta)^{2L^2}\{(\tanh\beta)^{\ell L}+(\tanh\beta)^{(L-\ell)L}\}\{1+(\tanh\beta)^{L^2}\},\tag{S.94}$$

which implies

$$\langle\hat{Z}_{\mathrm{hor}}\hat{Z}'_{\mathrm{hor}}\rangle_{\beta,L}=\frac{(\tanh\beta)^{\ell L}+(\tanh\beta)^{(L-\ell)L}}{1+(\tanh\beta)^{L^2}}.\tag{S.95}$$

Problems of Chap. 9

9.2.3.a (p. 318) From the matrix elements listed above (9.2.38), one finds

$$\hat{S}_x^{(1)}=\frac{1}{2}(\hat{c}^\dagger_{x,\uparrow}\hat{c}_{x,\downarrow}+\hat{c}^\dagger_{x,\downarrow}\hat{c}_{x,\uparrow}),\quad \hat{S}_x^{(2)}=\frac{1}{2i}(\hat{c}^\dagger_{x,\uparrow}\hat{c}_{x,\downarrow}-\hat{c}^\dagger_{x,\downarrow}\hat{c}_{x,\uparrow}),$$
$$\hat{S}_x^{(3)}=\frac{1}{2}(\hat{c}^\dagger_{x,\uparrow}\hat{c}_{x,\uparrow}-\hat{c}^\dagger_{x,\downarrow}\hat{c}_{x,\downarrow}),\tag{S.96}$$

from which one gets (9.2.44). Then an explicit calculation shows that

$$(\hat{\boldsymbol{S}}_x)^2=\frac{3}{4}(\hat{n}_{x,\uparrow}+\hat{n}_{x,\downarrow}-2\hat{n}_{x,\uparrow}\hat{n}_{x,\downarrow})=\frac{3}{4}\hat{n}_x(2-\hat{n}_x),\tag{S.97}$$

where we repeatedly used $(\hat{n}_{x,\sigma})^2=\hat{n}_{x,\sigma}$. We thus see that if a state $|\Phi\rangle$ satisfies $\hat{n}_x|\Phi\rangle=|\Phi\rangle$ then there is a spin $S=1/2$ at x, while if it satisfies $\hat{n}_x|\Phi\rangle=0$ or $\hat{n}_x|\Phi\rangle=2|\Phi\rangle$ then there is no spin at x. In the latter case the two electrons at x are forming a spin-singlet. (See Appendix A.3.3.)

9.2.3.b (p. 319) We use the notation of Sect. 9.2.1, and write $u=(x,\sigma)$ etc. Thus $\hat{B}(\mathsf{A})=\sum_{u,v}\hat{c}^\dagger_u\,(\mathsf{A})_{u,v}\,\hat{c}_v$ and $\hat{C}^\dagger(\boldsymbol{\varphi})=\sum_u\varphi(u)\,\hat{c}^\dagger_u$. By using the anticommutation relations (9.2.27) and (9.2.28), one finds $\hat{c}^\dagger_u\hat{c}_v\hat{c}^\dagger_w=-\hat{c}^\dagger_u\hat{c}^\dagger_w\hat{c}_v+\hat{c}^\dagger_u\delta_{v,w}=$

$\hat{c}^\dagger_w \hat{c}^\dagger_u \hat{c}_v + \hat{c}^\dagger_u \delta_{v,w}$, which means $[\hat{c}^\dagger_u \hat{c}_v, \hat{c}^\dagger_w] = \delta_{v,w}\hat{c}^\dagger_u$. Then one gets $[\hat{B}(\mathsf{A}), \hat{C}^\dagger(\boldsymbol{\varphi})] = \sum_{u,v,w}(\mathsf{A})_{u,v}\varphi(w)\,[\hat{c}^\dagger_u \hat{c}_v, \hat{c}^\dagger_w] = \sum_{u,v}(\mathsf{A})_{u,v}\varphi(v)\hat{c}^\dagger_u = \hat{C}^\dagger(\mathsf{A}\boldsymbol{\varphi})$.

Similarly one finds that $[\hat{c}^\dagger_u \hat{c}_v, \hat{c}^\dagger_w \hat{c}_z] = \delta_{v,w}\hat{c}^\dagger_u \hat{c}_z - \delta_{u,z}\hat{c}^\dagger_w \hat{c}_v$. Then one gets

$$\begin{aligned}[\hat{B}(\mathsf{A}), \hat{B}(\mathsf{B})] &= \sum_{u,v,w,z}(\mathsf{A})_{u,v}(\mathsf{B})_{w,z}[\hat{c}^\dagger_u \hat{c}_v, \hat{c}^\dagger_w \hat{c}_z] \\ &= \sum_{u,z}(\mathsf{AB})_{u,z}\hat{c}^\dagger_u \hat{c}_z - \sum_{v,w}(\mathsf{BA})_{w,v}\hat{c}^\dagger_w \hat{c}_v = \hat{B}([\mathsf{A}, \mathsf{B}]). \end{aligned} \tag{S.98}$$

9.2.3.c (p. 323) From the commutation relations (9.2.67), one finds $\hat{S}^\pm_{\text{tot}}\hat{S}^\mp_{\text{tot}}\hat{c}^\dagger_{x,\uparrow}\hat{c}^\dagger_{y,\downarrow}|\Phi_{\text{vac}}\rangle = (\hat{c}^\dagger_{x,\uparrow}\hat{c}^\dagger_{y,\downarrow} + \hat{c}^\dagger_{x,\downarrow}\hat{c}^\dagger_{y,\uparrow})|\Phi_{\text{vac}}\rangle$ and $\hat{S}^{(3)}_{\text{tot}}\hat{c}^\dagger_{x,\uparrow}\hat{c}^\dagger_{y,\downarrow}|\Phi_{\text{vac}}\rangle = 0$. We thus get

$$\begin{aligned}(\hat{\mathbf{S}}_{\text{tot}})^2|\xi\rangle &= \left\{\frac{\hat{S}^+_{\text{tot}}\hat{S}^-_{\text{tot}} + \hat{S}^-_{\text{tot}}\hat{S}^+_{\text{tot}}}{2} + (\hat{S}^{(3)}_{\text{tot}})^2\right\}\sum_{x,y\in\Lambda}\xi(x,y)\,\hat{c}^\dagger_{x,\uparrow}\hat{c}^\dagger_{y,\downarrow}|\Phi_{\text{vac}}\rangle \\ &= \sum_{x,y\in\Lambda}\xi(x,y)\,\{\hat{c}^\dagger_{x,\uparrow}\hat{c}^\dagger_{y,\downarrow} + \hat{c}^\dagger_{x,\downarrow}\hat{c}^\dagger_{y,\uparrow}\}|\Phi_{\text{vac}}\rangle \\ &= \sum_{x,y\in\Lambda}\{\xi(x,y) - \xi(y,x)\}\,\hat{c}^\dagger_{x,\uparrow}\hat{c}^\dagger_{y,\downarrow}|\Phi_{\text{vac}}\rangle, \end{aligned} \tag{S.99}$$

which implies $(\hat{\mathbf{S}}_{\text{tot}})^2|\xi\rangle = 0$ if $\xi(\cdot,\cdot)$ is symmetric, and $(\hat{\mathbf{S}}_{\text{tot}})^2|\xi\rangle = 2|\xi\rangle$ if $\xi(\cdot,\cdot)$ is antisymmetric. The two cases correspond to spin-singlet with $S_{\text{tot}} = 0$, and spin-triplet with $S_{\text{tot}} = 1$, respectively. See Appendix A.3.3. This relation between the total spin S_{tot} and the symmetry of the spatial wave function $\xi(x,y)$ is essential for determining magnetic properties of many-electron systems.[1]

9.3.2.a (p. 329) The anticommutation relations (9.3.25) and (9.3.26) are straightforward consequences of (9.2.63) and (9.2.64). To show (9.3.27) we use the definition of $\hat{a}_{j,\sigma}$ and the Schrödinger equation (9.3.3) to observe

$$\begin{aligned}\sum_j \varepsilon_j\,\hat{n}_{j,\sigma} &= \sum_{j,x,y}\varepsilon_j\,\psi^{(j)}_x(\psi^{(j)}_y)^*\,\hat{c}^\dagger_{x,\sigma}\hat{c}_{y,\sigma} = \sum_{j,x,y,z}t_{x,z}\,\psi^{(j)}_z(\psi^{(j)}_y)^*\,\hat{c}^\dagger_{x,\sigma}\hat{c}_{y,\sigma} \\ &= \sum_{x,y,z}t_{x,z}\,\delta_{z,y}\,\hat{c}^\dagger_{x,\sigma}\hat{c}_{y,\sigma} = \sum_{x,y}t_{x,y}\,\hat{c}^\dagger_{x,\sigma}\hat{c}_{y,\sigma}. \end{aligned} \tag{S.100}$$

[1] The standard explanation in terms of wave function language is as follows: The total wave function of two electrons must be antisymmetric with respect to the change of the labels of the two electrons. If the spatial part of the wave function is symmetric (*resp.* antisymmetric), then the spin part must be antisymmetric (*resp.* symmetric), which means that the total spin is 0 (*resp.* 1). Note that the necessary symmetry is automatically satisfied in the Fock space representation.

From (9.3.25) and (9.3.26), we find that $[\hat{\tilde{n}}_{j,\sigma}, \hat{a}^\dagger_{k,\tau}] = \delta_{j,k}\delta_{\sigma,\tau}\,\hat{a}^\dagger_{j,\sigma}$ exactly as in (9.2.31). By applying $\hat{H}_{\text{hop}}$ represented as (9.3.27) onto (9.3.22), and using the above commutation relation repeatedly and noting $\hat{\tilde{n}}_{j,\sigma}|\Phi_{\text{vac}}\rangle = 0$, we recover (9.3.23).

9.3.3.a (p. 335) By rewriting $\hat{U}^{\text{ph},\theta}_{x,\sigma} = \hat{c}_{x,\sigma} - e^{-i\theta}\,\hat{c}^\dagger_{x,\sigma}$, one readily finds $(\hat{U}^{\text{ph},\theta}_{x,\sigma})^\dagger\,\hat{c}_{x,\sigma}\,\hat{U}^{\text{ph},\theta}_{x,\sigma} = -e^{-i\theta}\,\hat{c}^\dagger_{x,\sigma}$ and $(\hat{U}^{\text{ph},\theta}_{x,\sigma})^\dagger\,\hat{c}^\dagger_{x,\sigma}\,\hat{U}^{\text{ph},\theta}_{x,\sigma} = -e^{i\theta}\,\hat{c}_{x,\sigma}$.

9.3.3.b (p. 336) From (9.2.44) and (9.3.51), one readily finds $\hat{\eta}^+_x = (-1)^x\,\hat{c}_{x,\uparrow}\,\hat{c}_{x,\downarrow}$, $\hat{\eta}^-_x = (-1)^x\,\hat{c}^\dagger_{x,\downarrow}\,\hat{c}^\dagger_{x,\uparrow}$, and $\hat{\eta}^{(3)}_x = (1 - \hat{n}_x)/2$.

To show that $[\hat{\eta}^{(\alpha)}_{\text{tot}}, \hat{S}^{(\beta)}_{\text{tot}}] = 0$, it suffices to check that $\hat{\eta}^\pm_x$ and $\hat{\eta}^{(3)}_x$ are SU(2) invariant (with respect to the spin rotations). From (A.2.2), we see that

$$[\hat{c}_{x,\uparrow}\,\hat{c}_{x,\downarrow}, \hat{S}^+_{\text{tot}}] = [\hat{c}_{x,\uparrow}, \hat{S}^+_{\text{tot}}]\hat{c}_{x,\downarrow} + \hat{c}_{x,\uparrow}[\hat{c}_{x,\downarrow}, \hat{S}^+_{\text{tot}}] = (\hat{c}_{x,\downarrow})^2 = 0, \tag{S.101}$$

where we used the commutation relations (9.3.36). Similarly,

$$[\hat{c}_{x,\uparrow}\,\hat{c}_{x,\downarrow}, \hat{S}^-_{\text{tot}}] = [\hat{c}_{x,\uparrow}, \hat{S}^-_{\text{tot}}]\hat{c}_{x,\downarrow} + \hat{c}_{x,\uparrow}[\hat{c}_{x,\downarrow}, \hat{S}^-_{\text{tot}}] = (\hat{c}_{x,\uparrow})^2 = 0, \tag{S.102}$$

where we used $[\hat{c}_{x,\uparrow}, \hat{S}^-_{\text{tot}}] = 0$ and $[\hat{c}_{x,\downarrow}, \hat{S}^-_{\text{tot}}] = \hat{c}_{x,\uparrow}$, which follow by taking the conjugate of (9.3.36). We also see from (9.2.31) that

$$\begin{aligned}[\hat{c}_{x,\uparrow}\,\hat{c}_{x,\downarrow}, \hat{S}^{(3)}_{\text{tot}}] &= \frac{1}{2}\Big([\hat{c}_{x,\uparrow}\,\hat{c}_{x,\downarrow}, \hat{n}_{x,\uparrow}] - [\hat{c}_{x,\uparrow}\,\hat{c}_{x,\downarrow}, \hat{n}_{x,\downarrow}]\Big)\\ &= \frac{1}{2}(\hat{c}_{x,\uparrow}\,\hat{c}_{x,\downarrow} - \hat{c}_{x,\uparrow}\,\hat{c}_{x,\downarrow}) = 0.\end{aligned} \tag{S.103}$$

We have thus shown that $\hat{\eta}^+_x = (-1)^x\,\hat{c}_{x,\uparrow}\,\hat{c}_{x,\downarrow}$ is SU(2) invariant. Clearly $\hat{\eta}^-_x = (\hat{\eta}^+_x)^\dagger$ is also SU(2) invariant. Finally $\hat{\eta}^{(3)}_x = (1 - \hat{n}_x)/2$ is SU(2) invariant since we have shown that $\hat{n}_x$ is SU(2) invariant.

We have already shown that $[\hat{H}_{\text{hop}}, \hat{S}^{(\alpha)}_{\text{tot}}] = 0$ and $[\hat{H}'_{\text{int}}, \hat{S}^{(\alpha)}_{\text{tot}}] = 0$ for $\alpha = 1, 2, 3$. By applying the Shiba transformation, these relations immediately imply $[\hat{H}_{\text{hop}}, \hat{\eta}^{(\alpha)}_{\text{tot}}] = 0$ and $[-\hat{H}'_{\text{int}}, \hat{\eta}^{(\alpha)}_{\text{tot}}] = 0$.

Problems of Chap. 10

10.1.a (p. 342) From (10.1.3) and (10.1.4), we see that the Schrödinger equation $\hat{H}|\Phi\rangle = E|\Phi\rangle$ is

$$t(\beta + \gamma) = E\alpha, \quad 2t\alpha + U\beta = E\beta, \quad 2t\alpha + U\gamma = E\gamma. \tag{S.104}$$

This is an eigenvalue equation of a 3×3 matrix, but can be solved easily by using the symmetry between the two sites. Assume first that $\beta = \gamma$. Then (S.104) reduces to $2t\beta = E\alpha$, $2t\alpha + U\beta = E\beta$, which implies $E^2 - UE - 4t^2 = 0$. We thus find two energy eigenvalues $E = (U \pm \sqrt{U^2 + 16t^2})/2$, one of which is the ground state energy. The other energy eigenvalue behaves as $E \simeq U + 4t^2/U$ if $|t| \ll U$. Next

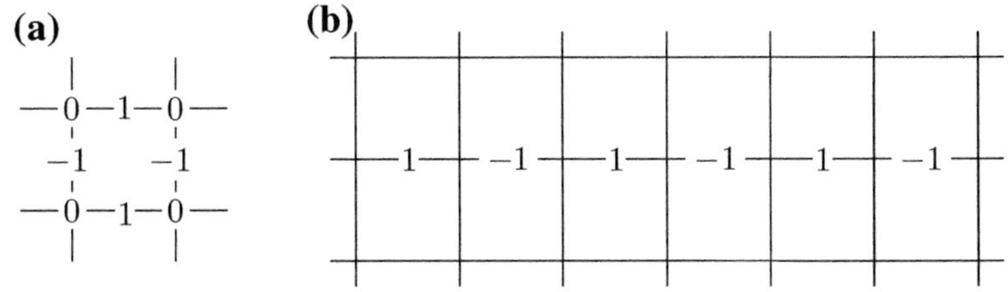

Fig. A.3 **a** A zero energy state that is localized on a single plaquette. **b** A nonlocal zero energy state that raps around the whole lattice ()

by assuming $\beta = -\gamma$ in (S.104), we get a trivial solution with $\alpha = 0$, in which case the energy eigenvalue is exactly U.

10.2.3.a (p. 358) For each plaquette (i.e., a unit square which has four black sites on the corners), consider the state depicted in Fig. A.3a, which is localized at the plaquette and has nonzero components only on the A sublattice. The state is clearly the energy eigenstate with zero energy. It seems like we are done since there are exactly L^2 plaquettes while Proposition 10.7 requires L^2 zero energy eigenstates.[1] However we find that these L^2 states are not linearly independent because we can take a linear combination of all these states to get zero. Thus we have found only $L^2 - 1$ zero energy states. One more is missing.

In fact there is a nonlocal zero energy eigenstate that lives on the horizontal line and has nonvanishing components only on the A sublattice. See Fig. A.3b. Such a state is well defined since we assume periodic boundary conditions. One may have noticed that the construction is somewhat similar to that of the ground states of Kitaev's toric code model discussed in Sect. 8.4. We have thus found one more eigenstate, but there clearly is one more independent state that raps around the lattice in the vertical direction. We now have $L^2 + 1$ zero energy eigenstates. Proposition 10.7 does not inhibit a model from having more than L^2 zero energy states, but the symmetry $-\varepsilon_j = \varepsilon_{|\Lambda|-j}$ requires the number of the zero-energy states to be even. So there must be at least $L^2 + 2$ zero energy states! What is one more state?[2]

Problems of Chap. 11

11.3.1.a (p. 399) The unit cell of this model consists of $d + 1$ sites. We can take one of them as the origin, and the rest as $\boldsymbol{e}_j/2$ with $j = 1, \ldots, d$, where $\boldsymbol{e}_j$ is the unit vector in the j-direction. We shall abbreviate them as 0 and j, respectively, and write $\mathscr{U} = \{0, 1, 2, \ldots, d\}$. See Fig. A.4.

---

[1]Probably this is sufficient when L is odd.

[2]Hint: How many zero energy eigenstates does the single-electron Schrödinger equation (10.2.18) have? The answer is L if L is odd, but is $L + 2$ if L is even.

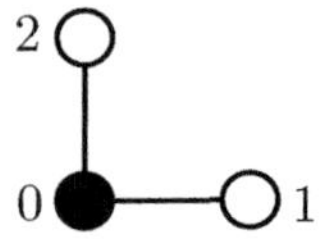

Fig. A.4 The unit cell of the flat-band model with $d = 2$ in Fig. 11.11 consists of three sites, which we call 0, 1, and 2 (© Hal Tasaki 2020. All Rights Reserved)

From (11.3.22) we see that the nonzero components of the effective hopping (9.3.12) are $S^{(k)}_{0,0} = 2\nu^2 t \sum_{j=1}^{d}(1+\cos k_j)$, $S^{(k)}_{j,j} = t$, and $S^{(k)}_{0,j} = (S^{(k)}_{j,0})^* = \nu t(1+e^{-ik_j})$, where $j = 1, 2, \ldots, d$. The effective hopping matrix thus looks like

$$\mathsf{S}^{(k)} = \begin{pmatrix} S^{(k)}_{0,0} & S^{(k)}_{0,1} & S^{(k)}_{0,2} & \cdots & S^{(k)}_{0,d} \\ S^{(k)}_{1,0} & t & 0 & \cdots & 0 \\ S^{(k)}_{2,0} & 0 & t & \cdots & 0 \\ \vdots & \vdots & \vdots & \ddots & \vdots \\ S^{(k)}_{d,0} & 0 & 0 & \cdots & t \end{pmatrix} \tag{S.105}$$

One then finds by inspection that

$$\begin{aligned} \det[\mathsf{S}^{(k)} - \varepsilon \mathsf{I}] &= (S^{(k)}_{0,0} - \varepsilon)(t-\varepsilon)^d - \sum_{j=1}^{d} |S^{(k)}_{0,j}|^2 (t-\varepsilon)^{d-1} \\ &= \varepsilon\{\varepsilon - (t + S^{(k)}_{0,0})\}(t-\varepsilon)^{d-1}, \end{aligned} \tag{S.106}$$

where we used the identity $\sum_{j=1}^{d} |S^{(k)}_{0,j}|^2 = t\, S^{(k)}_{0,0}$. The desired dispersion relations (11.3.23) readily follows from the eigenvalue equation $\det[\mathsf{S}^{(k)} - \varepsilon \mathsf{I}] = 0$.

11.3.2.a (p. 403) The d-dimensional checkerboard lattice Λ is identical to the set of internal sites $\mathscr{I}$ in the decorated hypercubic lattice considered in Sect. 11.3.1. See Fig. 11.10. Consider the single-electron Schrödinger equation for the Tasaki model

$$\hat{H}_{\rm hop} \sum_{u\in\mathscr{I}} \varphi(u)\, \hat{b}^\dagger_{u,\sigma} |\Phi_{\rm vac}\rangle = \varepsilon \sum_{u\in\mathscr{I}} \varphi(u)\, \hat{b}^\dagger_{u,\sigma} |\Phi_{\rm vac}\rangle, \tag{S.107}$$

which describes the bands other than the lowest flat-band. The definition (11.3.5) of $\hat{H}_{\rm hop}$ and the anticommutation relation

$$\{\hat{b}_{u,\sigma}, \hat{b}^\dagger_{v,\tau}\} = \begin{cases} (1+2\nu^2)\,\delta_{\sigma,\tau} & \text{if } u = v, \\ \nu^2\,\delta_{\sigma,\tau} & \text{if } \exists p \in \mathscr{E} \text{ s.t. } |p-u| = |p-v| = 1/2, \\ 0 & \text{otherwise} \end{cases} \tag{S.108}$$

imply that (S.107) becomes

$$(1+2\nu^2)t\,\varphi(u)+t\nu\sum_{\substack{v\in\mathcal{I}\\ \{u,v\}\in\mathcal{B}}}\varphi(v)=\tilde{\varepsilon}\varphi(u), \tag{S.109}$$

where $\mathcal{B}$ is the set of bonds for the generalized checkerboard lattice. The single-electron Schrödinger equation for the d-dimensional checkerboard lattice that we want to solve, on the other hand, reads

$$2t\,\varphi(u)+t\sum_{\substack{v\in\mathcal{I}\\ \{u,v\}\in\mathcal{B}}}\varphi(v)=\varepsilon\varphi(u). \tag{S.110}$$

Note that (S.109) reduces to (S.110) if we set $\nu=1$ and $\tilde{\varepsilon}=\varepsilon+t$. We thus get the desired (11.3.33) from (11.3.23).

11.3.2.b (p. 403) Nonzero components of the zero energy states are as in the following.

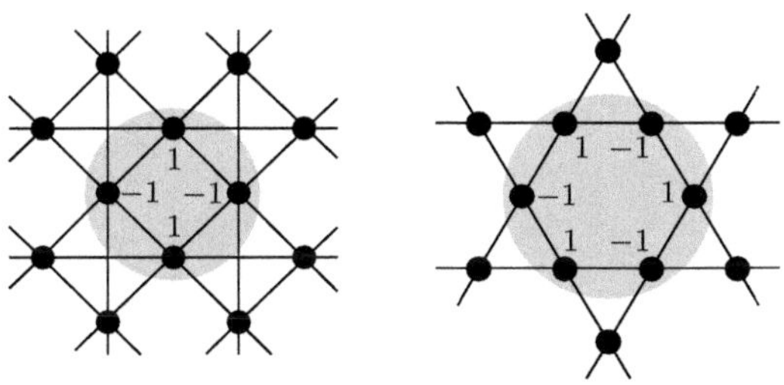

These localized states may or may not span the whole zero-energy subspace, depending on the boundary conditions. In fact the situation is quite similar to those for the zero-energy states in the Lieb lattice. See Problem 10.2.3.a.

11.3.2.c (p. 405) Nonzero components of the desired zero energy states are obtained by inspection as follows.

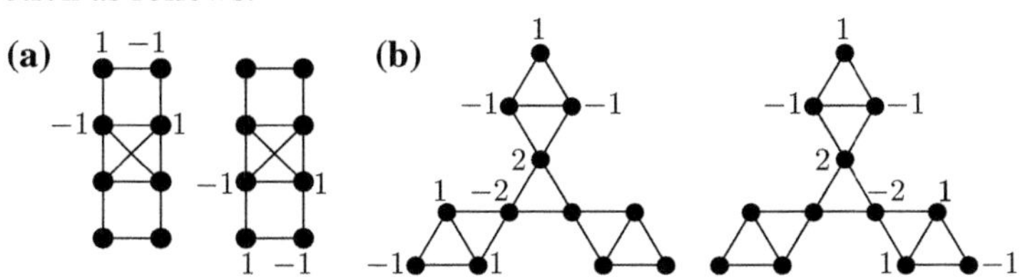

For each case, let $\hat{a}_{1,\sigma}$ and $\hat{a}_{2,\sigma}$ denote the fermion operators of the two states. For model (a), one finds that the state $\hat{a}^\dagger_{1,\sigma}\hat{a}^\dagger_{2,\tau}|\Phi_{\mathrm{GS}}\rangle$ with any $\sigma,\tau=\uparrow,\downarrow$ is a ground state for any $U>0$. Thus the ground states have both $S_{\mathrm{tot}}=0$ and 1. For model (b), one can repeat the proof of Tasaki's flat-band ferromagnetism in Sect. 11.3.1 by using the $\hat{a}^\dagger$ operators to prove that the model exhibits ferromagnetism.

11.4.1.a (p. 417) We can assume that $\omega_u(x)=O(\nu^3)$ if $|x-u|>1$. Then by writing $\omega_u(u)=A$ and $\omega_u(u\pm\frac{1}{2})=A\alpha$ (where α is expected to be of order ν), we find from the requirement $\langle\boldsymbol{\omega}_u,\boldsymbol{\omega}_{u+1}\rangle=0$ that $\omega_u(u\pm1)=-A\alpha^2/2+O(\nu^3)$. Now by demanding that $\langle\boldsymbol{\omega}_u,\boldsymbol{\omega}_{u+\frac{1}{2}}\rangle=0$, where $\boldsymbol{\omega}_{u+\frac{1}{2}}$ is given by (11.4.5), we find $\alpha=\nu/(1+\gamma)+O(\nu^3)$. We therefore get

$$\omega_u(x) = \Bigl(1 + \frac{2\nu^2}{(1+\gamma)^2}\Bigr)^{-1/2} \times \begin{cases} 1 + O(\nu^3) & \text{if } x = u, \\ \dfrac{\nu}{1+\gamma} + O(\nu^3) & \text{if } |x-u| = \frac{1}{2}, \\ -\dfrac{1}{2}\Bigl(\dfrac{\nu}{1+\gamma}\Bigr)^2 + O(\nu^3) & \text{if } |x-u| = 1, \\ O(\nu^3) & \text{otherwise.} \end{cases} \tag{S.111}$$

By operating the hopping matrix (11.4.1) explicitly, one finds

$$\mathsf{T}\omega_u = (1+\gamma)\, t\, \omega_u + \frac{\nu^2}{1+\gamma}\, t\, \{2\omega_u + \omega_{u-1} + \omega_{u+1}\} + O(\nu^3), \tag{S.112}$$

which is consistent with the dispersion relation (11.4.3).

Index

A
Addition of angular momenta, 473
Adjoint representation, 482
Affleck–Kennedy–Lieb–Tasaki (AKLT) model, 178, 281
AKLT state, *see* valence-bond solid state
Anderson's tower of states, 51, 95
Angular momentum, 470
Anisotropy, 43, 226, 239
Annihilation operator, 136, 310, 311, 314
Anticommutation relation, 312, 314
Antiferromagnetic Heisenberg model, 37, 49, 74, 118, 130, 154
  phase transition in the —, 130
Antiferromagnetic long-range order, 49
Antilinear operator, 26, 273, 277, **478**
Antiunitary operator, 26, 29, 273, 277, **479**, 483
Automorphism, 29, 284, **483**

B
Baker–Campbell–Hausdorff formula, 465
Band structure, 327, 357, 397
Baryon number conservation law, 137
BCS theory, 144
BEC, *see* Bose–Einstein condensation
Berezinskii–Kosterlitz–Thouless transition, 123, 157
Bethe ansatz, 154
Biconnectivity, 386, 403
Bilinear biquadratic interaction, 179
Bipartite lattice, 37
Bloch theorem, 327
Bloch theorem (for current), 165
Bogoliubov theory, 144
Bond, 32
  — in the d-dimensional hypercubic lattice, 52, 112
Bond dimension, 200
Bose–Einstein condensation(BEC), 135
Briegel–Raussendorf state, **214**, 256, 275, 291

C
C^*-algebra, 112, 283, **486**
CDW, 368
Checkerboard lattice, 401
Chern number, 229, 261
Chiral molecules, 60
Classical Ising model, 52
Classical spin, 25, 38, 52, 120
Cluster expansion, 59, 120, 180, 226, 241
Cluster-phase state, 218
Cluster state, **214**, 256, 275, 291
Coherent state, 18, 33, 143
Cold atoms, 50, 135, 230, 306, 337, 355
Commutator, 462
Complete graph, 405
Complex conjugation map, 26, 28, 273, 277, **478**
Connected lattice, 33
Connectivity condition, 385
Continuous group, 15
Creation operator, 136, 310, 314
Crystal field anisotropy, 43, 226, 239
$CsNiCl_3$, 157
Cuntz algebra, 205, 277
CuO lattice, 354
CZX model, 275

H. Tasaki, *Physics and Mathematics of Quantum Many-Body Systems*, Graduate Texts in Physics, https://doi.org/10.1007/978-3-030-41265-4

D

Decorated square lattice, 291
Den Nijs–Rommelse string order parameter, 193
Density of states, 380
Diamond chain, 357
Dimer state, 185
Dirac notation, 457
Direct exchange interaction, 396, 399, 419
Disordered flat phase, 229, 234
Dispersion-less band, 357, 397
Dispersion relation, 327
Dyson–Lieb–Simon theorem, 76, 130

E

Edge state, **206**, 216, **234**, 238, 246, 256
Effective Schrödinger equation, 470
Electromagnetism, 122
Electron-Spin-Resonance (ESR), 157
Energy gap in the operator algebraic formulation, 489
Entanglement, 147, 215, 254, 261, 288, **460**
Entanglement entropy, 461
Equilibrium state, 52
Ergodic state, 114
η-paring operator, 336
Exact but non-rigorous, 154
Exchange
— of electrons, 309

F

Fermi energy, 375
Fermion, 309
Fermion operator, 314
Ferrimagnetic long-range order, 78
Ferrimagnetism, 77, 103, 355
Ferromagnetic Heisenberg model, 32, 131
Ferromagnetism, 372
origin of —, 306
15 puzzle, 387
Filling factor, 165, **315**, 355, 359
Finitely correlated state, *see* matrix product state
Flat-band, 357, 397
Flat-band ferromagnetism, 389, 438
Fock space, 140, 316
Fock space representation, 136, 314
Fourier transformation, **80**, 368
Fractional spin, 206
Frustration, 37
Frustration-free Hamiltonian, 32, 181, 188, 216, 393, 425, **469**

G

Gap in the operator algebraic formulation, 489
Gauge transformation, 333
Gelfand–Naimark–Segal construction, 283, **489**
Genuine representation, 481
GNS construction, 283, **489**
GNS triple, 490
Gramm matrix, 320, 394
Graph state, 218
Graph theoretic distance, 42, 211
Graph theory, 32, 401
Ground state, 4
— in the operator algebraic formulation, 489
— of a frustration-free Hamiltonian, 469
—of an infinite system, 113
Ground state in the infinite volume, 207
Ground state phase transition, 61
Group, **480**
Group cohomology, 268
Gutzwiller projection, 376

H

Haldane gap, 155
— for AKLT model, 178, 188
— in anisotropic model, 227
Haldane phase, 228, 281
Haldane phenomena, 225, 239
Heisenberg model
antiferromagnetic —, 37, 49, 74, 118, 130, 154
ferromagnetic —, 32, 131
Hexagonal lattice, 210
Hidden antiferromagnetic order, 169, 191, 192
Hohenberg-Mermin-Wagner theorem, 33
Hopping Hamiltonian, 328
Hubbard model, 307, **332**
bosonic —, 135, 337
Hund's rule, 14, 181
Hypercubic lattice, 51
Hyper graph state, 218

I

Index theorem, 285
Infrared bound, 81
Injective matrix product states, 202
Inner dimension, 200
Interaction Hamiltonian, 330
Interference experiment, 148

Inversion symmetry, 257
Ising anisotropy, 43, 103, 239
Ising model
classical —, 52, 167
phase transition in the —, 53
quantum —, 55
— under transverse magnetic field, 55

K
Kagomé lattice, 401
Kennedy triplet, 234
Kennedy–Tasaki transformation, 241
Kinetic exchange interaction, 343
Kitaev model, 288
Kosterlitz–Thouless transition, *see* Berezinskii–Kosterlitz–Thouless transition
Kramers degeneracy, **30**, 257, 259, 262
Kronecker product, xviii

L
Large-D phase, 228, 272, 281
Lattice, 21
bipartite —, 37
connected —, 33
d-dimensional hypercubic —, 51
Lattice Laplacian, 35
Law of large numbers, 51, 54, 60, 66, 93
Levi-Civita symbol, 471
Levin-Gu model, 275
Lie algebra, 470, 481
Lieb lattice, 77, 291, 354
Lieb–Schultz–Mattis theorem, 162, 186
— without U(1) symmetry, 166, 276
Lieb's ferrimagnetism, 398
Lieb–Shen–Qiu superconductivity, 359
Lieb's theorem, 348, 349
Lie group, 481
Lie product formula, 170, 465
Line graph, 401
Local stability of ferromagnetism, 423
Long-Range Order (LRO), 49, 53, 57, 61, 65, 75, 78, 154, 162, 212, 237, 355
Long-range order parameter, 49, 53, 65, 75, 100, 101, 140
Low-lying energy eigenstate, 59, **67**, 95, 234
Low-lying state, **67**, 95, 140
LRO, *see* long-range order

M
Magnetic field, 54, 55, 118
staggered —, 70, 77, 102, 118
Magnon, 36
Majumdar–Ghosh model, 184, 209
Marshall-Lieb-Mattis theorem, 39
Martingale method, 179
Massless field theory, 154
Matrix Product States (MPS), 179, **194**, 264, 432
Mean field approximation, 379
Mean field theory, 115, 144
Mermin-Wagner theorem, *see* Hohenberg-Mermin-Wagner theorem
Metallic oxide, 390
Mielke model, 401
Mini-max principle, 468
Mott insulator, 438
MPS, *see* matrix product states

N
Nagaoka's ferromagnetism, **381**, 438
Nambu–Goldston mode, 97
Nearly-flat band, 416, 423
Néel order, 37, 50, 157, 193, 212, 228, 233, 239, 241
Néel state, **37**, 100, 167
Nematic order, 226
NENP, 157, 236
Neutron scattering, 50, 157
$Ni(C_2H_8N_2)_2NO_2(ClO_4)$, *see* NENP
NMR, 50
Nonlinear sigma model, 156
Nonnegative operator, 467
Norm
operator —, 463
Nuclear physics, 145
Numerical diagonalization, 96, 157

O
ODLRO, *see* off-diagonal long-range order
Off-diagonal long-range order (ODLRO), 138, 349, 359
Operator algebra, 112, 283, **486**
Operator norm, 463
Optical lattice, 135, 337
Order operator
— for ferrimagnetism, 103
— for general models, 65
— for many-boson system, 138
— for the classical Ising model, 53
— for the quantum Ising model, 56
—for the antiferromagnetic Heisenberg model, 74

Order parameter, 49, 53, 55, 65, 75, 97, 100, 102, 140, 141
Ornstein–Zernike form, 53, 156, 179, 204
Orthogonal group, 480

P
Paramagnetism, 306
  Pauli —, 329, 444
Particle-hole transformation, 334
Path integral, 170
Pauli exclusion principle, 311, 315
Pauli matrix, 15
Pauli paramagnetism, 329, 444
Percolation, 233
Perron–Frobenius theorem, 35, 39, 44, 57, 385, **475**, 496
Perturbation theory, 345
Phase, 63
Phase transition, 61
  — in the Ising model, 53
  — in the antiferromagnetic Heisenberg model, 130
Photon, 143
Physical "ground state", 60, 69, 99, **115**
Physical infinite volume ground state, 114
Polar decomposition theorem, 476
P-orbitals, 196
Positive semidefinite operator, 467
Power law correction, 53, 156, 179, 204
Primitive matrix product states, 202
Projective representation, 20, 268, 481
Protection by symmetry, 64, 254
Pseudo spin operator, 336

Q
Quantum computation, 296
  measurement based —, 183, 204, 212, 214
Quantum critical point, 154
Quantum field theory, 76, 154
Quantum fluctuation, 154
Quantum Hall effect, 160
  fractional —, 288
Quantum Ising model, **55**, 65, 115, 228, 246
Quantum Markov chain, 194
Quantum mechanical angular momentum, 470
Quantum spin liquid, 192, 230, 234
Qubit, 296

R
Reflection positivity, 76, 79, **90**, 131, 139, 212, **364**
Representation
  — of a C^*-algebra, 490
  — of a group, 481
Resonating Valence-Bond (RVB), 399
Ring exchange process, 378, 399
Roth state, 376

S
Saturated ferromagnetism, 372
Schmidt decomposition, 91, 261, **461**, 477
Schrödinger equation, 34, 324, 345, 499, 503
  time-dependent —, 25, 30
Schrödinger's cat, 58, 116, 144
Schwinger boson representation, 188
Second quantization, 136, 314
  — of an operator, 318
Shiba transformation, **335**, 349, 352
Singlet, 182, 185, 201, 235, 322, 324, 329, 349, **474**, 514, 515
Singular value decomposition theorem, 91, **476**
Slater determinant, 319
SO(3), 17, **480**
SO(4) symmetry, 337
Special orthogonal group, 480
Special unitary group, 480
Spin, 13
  classical —, 25, 38, 52, 120
  — configuration, 21
  — quantum number, 14
Spin coherent state, 18, 33, 103
Spin operator, 13
  — in electron systems, 316
  — in quantum spin systems, 21
Spin-orbit coupling, 325
Spin rotation operator, 15, 22
Spin-singlet, 182, 185, 201, 235, 322, 324, 329, 349, **474**, 514, 515
Spin-space reflection positivity, 364
Spin-triplet, 181, 234, 324, 342, 377, **474**, 515
Spin wave, 34, 81, 97, 423, 429
Split property, 205, 277
Spontaneous magnetization, 55
Spontaneous Symmetry Breaking (SSB), 50, 54, 55, 70, 102, 117, 148, 154
SPT, *see* symmetry protected topological phase
SPTO, *see* symmetry protected topological order

SSB, *see* spontaneous symmetry breaking
Staggered magnetic field, 70, 77, 102, 118
State in the operator algebraic formulation, 487
Stoner criterion, 379
String correlation function, 193, 245
String-net model, 289
String order operator, 236
String order parameter, 193
SU(2), 15, **480**
SU(2) invariance, **23**, 32, 39, 75, 178, 322, 332, 472
Superconductivity, 135, 359, 368
Super-exchange interaction, 343, 420
Surface code, 297
Symmetry breaking field, 54, 69, 102, 148
Symmetry breaking order parameter, 55, 97, 100, 102, 141
Symmetry Protected Topological (SPT) phase, 214, 216, 227, 254, **255**
— in higher dimensions, 275
Symmetry protected topological order, 255

T
Tasaki model, 390, 425
Tensor networks, 213, 432
Tensor product, 460
Tight-binding description, 308
Time crystal, 67
Time-reversal
— for electron systems, 486
— map, 26, 273
— of an operator, 28, 29
— symmetry, 257, 259, 273, 296
— transformation, 486
t-*J* model, 441, 442
Topological index, 229, 261, 264, 281
Topological insulator, 255
Topological order, 63, 135, 166, **288**
Topological phase transition, 228
Topological term, 156
Toric code model, 289
Tower of energy eigenstates, 51, 95
Transfer matrix, 197
Transformation of operators, 17, 23, 241, 333, 352, **459**
Translation invariant, 113, 160
Translation map, 113
Translation operator, 160, 423
Transverse magnetic field, 56
Triangle-phase state, 218, 220, 275
Triplet, 181, 234, 324, 342, 377, **474**, 515
Trotter formula, 465
Twist operator, 158
local —, 165

U
U(1), **480**
U(1) invariance, **23**, 43, 158
U(1) phase symmetry, 139
Undecidability of energy gap, 212
Uniaxial anisotropy, 43, 226, 239
Unitary group, 480

V
Valence bond, **182**, 474
Valence-Bond Solid (VBS) state, 178, **183**, 185, 208, 210, 248, 281
photonic —, 183
VBS state, *see* valence-bond solid state
von Neumann algebra, 205, 277, 284

W
Wannier state, 416
Wave-particle dualism, 332
Weyl representation, 187
Wigner's theorem, 29, 333, 482

X
XXZ model, 102, 137
XY model, 137

Z
$\mathbb{Z}_2$ symmetry, 56, 61, 69, 241
$\mathbb{Z}_2 \times \mathbb{Z}_2$ (as a group), 18, 270
$\mathbb{Z}_2 \times \mathbb{Z}_2$ symmetry, 238, 256, 270
Zero correlation length, 215, 218, 288
Zero-energy condition, 394, 412, 435, 447

Printed by Printforce, the Netherlands